Joep Paul Freddie
Girdwood, June 2001

Studies in Surface Science and Catalysis 136

NATURAL GAS CONVERSION VI

Studies in Surface Science and Catalysis 136

Advisory Editors: B. Delmon and J.T. Yates

Vol. 136

NATURAL GAS CONVERSION VI

Proceedings of the 6th Natural Gas Conversion Symposium,
June 17-22, 2001, Alaska, USA

Edited by:

E. Iglesia
Department of Chemical Engineering, University of California, Berkeley, CA, USA

J.J. Spivey
Department of Chemical Engineering, NC State University, Raleigh, NC, USA

T.H. Fleisch
BP, Upstream Technology Group, Houston, TX, USA

2001
ELSEVIER
Amsterdam – London – New York – Oxford – Paris – Shannon – Tokyo

ELSEVIER SCIENCE B.V.
Sara Burgerhartstraat 25
P.O. Box 211, 1000 AE Amsterdam, The Netherlands

First edition 2001

Library of Congress Cataloging in Publication Data
A catalog record from the Library of Congress has been applied for.

ISBN: 0-444-50544-X
ISSN: 0167-2991

The paper used in this publication meets the requirements of ANSI/NISO Z39.48-1992 (Permanence of Paper).
Printed in The Netherlands.

Preface

This volume contains peer-reviewed manuscripts describing the scientific and technological advances presented at the 6th Natural Gas Conversion Symposium held in Girdwood, Alaska on June 17-22, 2001. This symposium continues the tradition of excellence and the status as the premier technical meeting in this area established by previous meetings in Auckland, New Zealand (1987), Oslo, Norway (1990), Sydney, Australia (1993), Kruger National Park, South Africa (1995), and Taormina, Italy (1998).

The 6th Natural Gas Conversion Symposium is conducted under the overall direction of the Organizing Committee chaired by Theo Fleisch (BP) and Enrique Iglesia (University of California at Berkeley). The Program Committee was responsible for the review, selection, editing of most of the manuscripts included in this volume. A standing International Advisory Board has ensured the effective long-term planning and the continuity and technical excellence of these meetings. The members of each of these groups are listed below. Their essential individual contributions to the success of this symposium are acknowledged with thanks.

The Editors gratefully acknowledge the generous financial support given by the Sponsors, who are listed below.

The Editors acknowledge the talents and dedication of Ms. Sarah Key, who patiently handled the details of the review, acceptance, and submission processes for the manuscripts in this volume. The Editors also thank the authors of these manuscripts for their help and patience during the arduous task of assembling this volume.

The Editors

Enrique Iglesia, University of California at Berkeley

James J. Spivey, North Carolina State University

Theo H. Fleisch, BP

Program Committee

Enrique Iglesia, Program Chair, University of California at Berkeley
James J. Spivey, Program Co-Chair, North Carolina State University

Carlos Apesteguia, INCAPE (Argentina)
Manfred Baerns, Institut fuer Angewandte Chemie Berlin-Adler (Germany)
Calvin Bartholomew, Brigham Young University (USA)
Alexis Bell, University of California at Berkeley (USA)
Steven Chuang, University of Akron (USA)
Burtron Davis, University of Kentucky (USA)
Theo Fleisch, BP (USA)
Krijn de Jong, University of Utrecht (The Netherlands)
Daniel Driscoll, U.S. Department of Energy (USA)
Rocco Fiato, ExxonMobil (USA)
Joep Font Freide (USA)
Karl Gerdes, Chevron (USA)
Anders Holmen, Norwegian University of Science and Technology (Norway)
Graham Hutchings, University of Wales (United Kingdom)
Ben Jager, SASOL (South Africa)
Eiichi Kikuchi, Waseda University (Japan)
Jay Labinger, California Institute of Technology (USA)
Johannes Lercher, Technische Universitat Munchen (Germany)
Wenzhao Li, Dalian Institute of Chemical Physics (China)
Jack Lunsford, Texas A&M University (USA)
Mario Marchionna (Italy)
David Marler, ExxonMobil (USA)
Claude Mirodatos, CNRS-Villeurbane (France)
Julian Ross, University of Limerick (Ireland)
Jens Rostrup-Nielsen, Haldor Topsoe (Denmark)
Domenico Sanfilippo, Snamprogetti (Italy)
Lanny Schmidt, University of Minnesota (USA)
Stuart Soled, ExxonMobil (USA)
Samuel Tam (USA)
David Trimm, University of New South Wales (Australia)
Wayne Goodman, Texas A & M University (USA)
Charles Mims, University of Toronto (Canada)

Organizing Committee

International Advisory Board

Sponsors

Gold

Alaska Science & Technology Foundation
BP
ExxonMobil
Sasol-Chevron
Shell Global Solutions
United States Department of Energy

Silver

Conoco
Phillips Petroleum
Syntroleum
Texaco

Bronze

Air Liquide
ARCO
ITM Syngas Alliance
Natural Resources Canada
Nexant

TABLE OF CONTENTS

Studies in Surface Science and Catalysis
J.J. Spivey, E. Iglesia and T.H. Fleisch (Editors)

MODELING MILLISECOND REACTORS

Lanny D. Schmidt
Department of Chemical Engineering and Materials Science
University of Minnesota
Minneapolis MN 55455

ABSTRACT

Catalytic partial oxidation processes at very short contact times have great promise for new routes to chemical synthesis from alkanes because they are capable of producing highly nonequilibrium products with no carbon formation using reactors that are much smaller and simpler than with conventional technology. We summarize some of the considerations which may be important in modeling and in interpreting partial oxidation processes. The gradients in these monolith reactors are typically 10^6 K/sec and 10^5 K/cm, and reactions are fastest in the regions of highest gradients. Therefore a conventional one-dimensional model may be highly inaccurate to describe these processes, particularly when used to attempt to decide between different reaction mechanisms. We argue that detailed modeling which includes detailed descriptions of reactor geometry, gas and solid properties, and surface and homogeneous reaction kinetics will be necessary to develop reliable descriptions of these processes. Even with detailed modeling, it may be necessary to consider the validity of these parameters under extreme reaction conditions.

INTRODUCTION

Oxidation processes in monolithic catalysts exhibit features not observed in conventional packed bed reactors because they operate with gas flow velocities of ~1 m/sec with open channel catalyst structures for effective contact times of the gases on the catalyst of typically 1 millisecond and produce kilograms of product per day with less than a gram of catalyst. These processes are autothermal and nearly adiabatic because the exothermic oxidation reactions heat the gases and the catalyst from room temperature to typical operating temperatures of ~1000°C and the rate of heat generation is too large for effective wall cooling. After lightoff, the reactions usually run to completion of the limiting reactant, so conversion and selectivities are independent of flow rates over typically an order of magnitude of residence time.

A recently explored example is methane oxidation to synthesis gas

$$CH_4 + 1/2O_2 \rightarrow CO + 2H_2,$$

which occurs with 100% O_2 conversion, >90% CH_4 conversion, and >90% selectivity to CO and H_2 (based on C and H respectively) on Rh catalyst coated on α-alumina foam monolith[1]. Another example is alkane oxidation to olefins[2], for example,

$$C_2H_6 + 1/2O_2 \rightarrow C_2H_4 + H_2O,$$

which occurs with 100% O_2 conversion, >70% C_2H_6 conversion, and ~85% selectivity to ethylene on Pt- Sn catalyst coated on α-alumina foam monolith. As a final example, the total oxidation of alkanes to CO_2 and H_2O

$$CH_4 + 2O_2 \rightarrow CO_2 + 2H_2O$$

can be attained with >99% fuel conversion[3]. The oldest examples of monolith reactors with millisecond contact times are the Ostwald process to prepare nitric acid by ammonia oxidation

$$NH_3 + 5/4O_2 \rightarrow NO + 3/2H_2O,$$

and the Andrussow process to prepare HCN[4],

$$CH_4 + NH_3 + O_2 \rightarrow HCN + 2H_2O + H_2.$$

Both of these processes take place on multiple layers of woven Pt-10%Rh gauze catalysts operating at 800 and 1100°C.

All of these processes occur with approximately millisecond contact times with the exothermicity of the reactions providing the energy to heat the gases and catalyst from room temperature to operating temperatures from 800 to 1200°C in $<10^{-3}$ s.

We[3-6] and many others[7] have attempted to model these processes in detail to try to determine the mechanisms by which these product distributions are formed and to find conditions to optimize a particular product. We argue that these apparently simple processes are in fact far more complicated than the usual packed bed catalytic reactor assumptions used for typical modeling. First, the temperatures are sufficiently high that some homogeneous reaction may be expected to occur, even at very short reaction times. Second, the gradients in all properties are so large that all conventional assumptions may be inaccurate. It is the purpose of this manuscript to address these issues.

ONE-DIMENSIONAL MODELS

We first consider the geometry of millisecond reactors. These typically occur in open monolith catalyst structures which may consist of extruded, foam, or fiber ceramics or woven or sintered metal structures, as sketched in the left panel of figure 1. All of these structures can be approximated as a collection of tube wall reactors of length L and channel diameter d.

One Dimensional Models

Most simple models of these processes have assumed one-dimensional approximations[3] to the geometries of figure 1. Radial mass and heat transfer are included through effective mass and heat transfer coefficients to the walls. Either plug flow (no axial mass or heat transfer in the gas) or models including axial diffusion (a boundary value problem) are assumed, and the resultant model is easy to solve even for many equations.

The simplest approximation to the temperature is to assume a step change from the feed temperature (25°C) to the final catalyst and gas temperature (~1000°C) so that the energy balance can be ignored. Monolith temperatures typically vary by less than 100°.

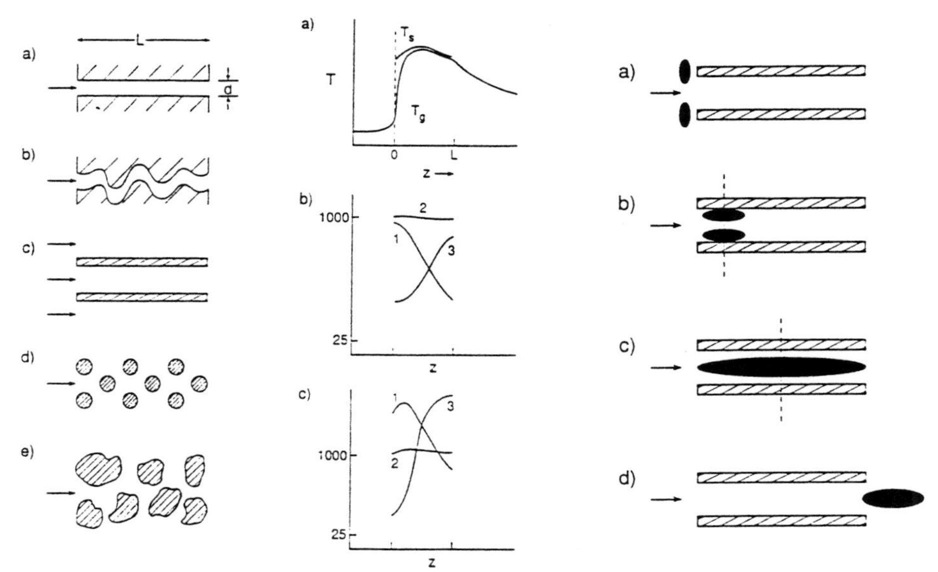

Figure 1. Left panels: Different monolith geometries approximated by a straight tube. Center panels: Possible temperature profiles down the reactor. Right panels: Possible locations of homogeneous reaction and flames.

from front to back, so the assumption of constant wall temperature is reasonable. Gases should attain the wall temperature within a few channel diameters, so the temperatures should be constant within less than 1 mm of the entrance. The expected temperature profiles for gas and catalyst are sketched in curve a of the center panel of figure 1.

Wall heat conduction is an important mechanism for backflow of heat which maintains the monolith and the gas isothermal, and the temperatures can be calculated by solving simultaneously for the gas and solid temperatures T_g and T_s, still in a one dimensional approximation.

Upon heating from 25°C to 1000°C, the kinematic viscosity increases by more than a factor of 10, as do the thermal diffusivity and mass diffusivity. Since reaction occurs very quickly upon entering the monolith, these variations in properties near the entrance must be included in any calculation of reactor performance.

Although Reynolds numbers are sufficiently small that laminar flow is a good approximation and heat transfer coefficients and solid thermal conductivities are sufficiently large that nearly isothermal gas and solid may be assumed, there may be serious problems in the one dimensional, plug flow assumption in approximating reactor behavior. First, most reaction appears to occur within a few tube diameters where the temperatures are varying strongly, so the large gradients in this region may be significant. Second, the gases strongly accelerate in the entrance region (typically by factors of 4 to 10), and the decoupling of the fluid flow from the reaction and temperature equations may lead to significant errors.

TWO-DIMENSIONAL SIMULATION

We have simulated quantitatively the temperature and velocity profiles for cold gases entering a hot tube for the reactions and conditions of the methane to syngas[6] and for ethane to ethylene[7]. These calculations were done using FLUENT to calculate fluid properties. All fluid properties are properly accounted for including temperature and mixture variations of diffusivities, thermal conductivity, and viscosity. We used the Hickman model[5,6] of syngas generation for the surface reaction mechanism.

The CH_4, temperature, and axial velocity profiles predicted by this model are shown in figure 2. The calculations shown are for three tube diameters: 0.025, 0.05, and 0.1 cm. These correspond to 80, 40, and 20 pores per linear inch which are typical foam ceramic sizes. The region shown is only the first millimeter near the entrance to the hot catalyst section. This section is preceded by an inert tube which produces a fully developed laminar flow profile before entering the catalytic section.

The predicted velocity profile is especially interesting. Even though $Re_d<30$ throughout the entrance region and the velocity profiles before and within the catalyst section are parabolic, the axial velocity is not parabolic in the entrance region, and the velocity has a minimum in the center. A very thin boundary layer is established near the entrance to the catalytic section as the temperture and all properties vary strongly in very short distances.

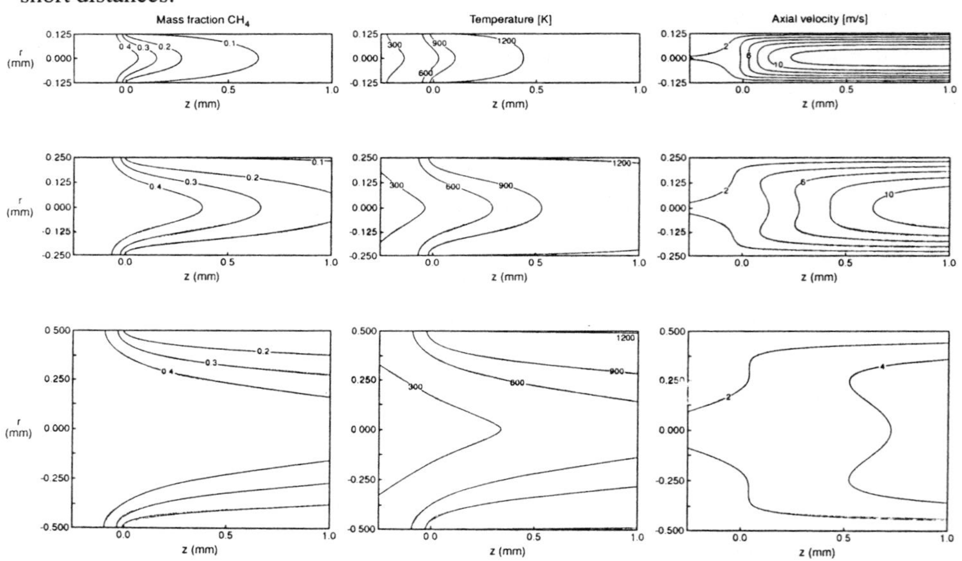

Figure 2. Calculated methane, temperature, and velocity profiles in a single cell of the reactor for syngas from methane.

The concentrations predicted by one dimensional calculations must therefore be very different from those calculated using these "exact" calculations. The gradients in temperature, velocity, and composition are so large in the entrance region that only a "complete" simulation should be expected to yield more than qualitative conclusions.

REACTIONS

Mechanisms and heat generation

Many modeling studies of these reaction systems have attempted to explain the reaction mechanism qualitatively by fitting proposed mechanisms to experimental data. In order to eliminate the large temperature gradients which exist in the millisecond reactor, these studies have frequently used either dilution with an inert such as He or N_2 or lower temperatures to slow down the rate of reaction so that intrinsic kinetics can be measured without interference from temperature or concentration gradients. It is our opinion that the changes caused by dilution or cooling can effectively disguise the kinetics to change the overall mechanism significantly.

Two zone model

Syngas formation has been suggested to proceed by either "direct" or "indirect" mechanisms. In the direct mechanism, reaction is initiated by CH_4 decomposition followed by H atom dimerization and C oxidation to produce syngas. The "indirect" mechanism assumes that the first reaction step is total oxidation to CO_2 and H_2O, and that CO and H_2 are formed by reaction of the remaining CH_4 with H_2O and CO_2 in methane reforming. These reactions would then occur in two distinct zones: an exothermic combustion zone where O_2 is present and an endothermic zone without O_2.

It is important to note that, by the indirect mechanism, the primary reaction is extremely exothermic, while CO_2 and H_2O reforming are extremely endothermic. These sequential processes would tend to first pull the temperature up strongly and then pull it down strongly, so one should expect a large variation in temperature from front to back in the catalyst, even larger than would be predicted by heat losses through conduction.

Olefins must be made by the initial formation of alkyl species, followed by H elimination to form the olefin,

$$C_2H_{6g} \rightarrow C_2H_5 \rightarrow C_2H_4 \rightarrow C_2H_4$$

for either a surface reaction or a gas phase (steam cracking) reaction.

Therefore, any indirect mechanism of partial oxidation suggests that there should be two different zones in the catalyst where different reactions dominate. These predictions are in strong contrast to a single mildly exothermic process, which is predicted by the direct processes.

Variable bed length experiments support direct processes for the majority of syngas formation, at least at temperatures above 900°C. At lower temperatures and lower flow rates, the reactions may become sequential, and, since heat removal cools the catalyst in the endothermic zone, there could be large temperature variations across the catalyst.

Surface reactions

One common approximation to surface reaction rate expressions is to use Langmuir Hinshelwood rate expressions. This involves assuming adsorption-desorption equilibrium in calculating the overall rate, so that the elementary adsorption and desorption steps are lumped into an effective rate before modeling the reactor. The s urface equilibrium assumption is clearly not appropriate for situations where the reaction probability upon adsorption is nearly unity, and it obviously cannot follow changes which occur as physical parameters such as T or composition are varied.

However, even an "elementary step" model of a surface reaction requires the lumping of many steps. For example methane adsorption $CH_4 \rightarrow C+4H$ almost certainly proceeds by the steps $CH_4 \rightarrow CH_3 \rightarrow CH_2 \rightarrow CH \rightarrow C$ or even more complicated steps in which adsorbed O assists in H abstraction. If these steps are fast and irreversible or if there is a surface equilibrium established between some species, then these lumping approximations are probably appropriate, but this can only be decided by fitting an expression to data and checking that it is consistent with surface science observations.

Reaction times

Next we note that at these high temperatures, surface coverages and reactions may be quite different than those calculated or measured at lower temperatures because times are extremely short. Consider a first order reaction process on a surface with a rate coefficient $k=k_o e^{E/RT}$ with k_o the preexponential factor for the process and E its activation energy. The characteristic time τ for such a process is given by

$$\tau = \frac{1}{k} = \frac{1}{k_o} e^{E/RT}.$$

Activation energies for surface processes are typically 10 to 25 kcal/mole or even smaller, and preexponential factors may be assumed to be vibrational frequencies, 10^{13} sec^{-1}

Adsorption lifetimes may be estimated from this expression, with E approximately the heat of adsorption. For H_2, CO, and H_2O, and CO_2, the measured heats of adsorption are 18, 30, 10, and 5 kcal/mole, so the estimated adsorption lifetimes should be 10^{-10}, 10^{-8}, $5x10^{-12}$, and 10^{-13} s, respectively.

Homogeneous reaction and flames

In contrast to surface reaction rates, the rates of homogeneous reactions of alkanes and O_2 in the absence of surfaces are well established. The 300 to 5000 elementary reaction step models of alkane combustion are sufficiently established that flame structures in premixed and diffusion limited combustion can be calculated quantitatively by solving for the 20 to 100 species in almost any geometry and flow situation.

However, most of these models were developed for high temperature flames in excess O_2, and they have not been fully tested in excess fuel and at low temperatures where catalytic oxidation processes occur. Pyrolysis models of alkane decomposition have also been developed for steam cracking of alkanes to form olefins[7], but much less has been done in excess fuel with O_2 present. The ignition and extinction characteristics of these systems[9] are especially difficult because the equations are especially stiff and the parameters have been adjusted to fit steady state behavior rather than ignition behavior where different reaction steps dominate.

The major characteristics of combustion and pyrolysis are that they occur by chain reactions consisting of initiation, propagation, and termination steps. While propagation steps make most product, initiation and termination steps control the concentrations of the free radical intermediates which in turn control the overall propagation rates. Therefore these processes are highly autocatalytic, exhibiting very large effective activation energies near ignition, but very fast mass transfer limited rates at high temperatures.

A major problem in describing homogeneous processes in catalytic partial oxidation is that all gases are always within 0.1 cm of a surface. Solids tend to thermostat the system, thus preventing thermal runaway which leads to flames. Surfaces also act as sinks and sources of free radicals which propagate homogeneous reaction steps. Thus the first effect of the proximity of a surface is that it acts as a "flame trap", quenching high temperature excursions and formation of free radicals which are necessary for stable flame propagation.

Another problem with homogeneous reactions is the possibility of a flame. In a free flame, this means the situation where a sharp concentration and temperature front is established. Published flammability limits are simply the range of fuel and O_2 where the flame velocity u_f of a spark initiated reaction in a static chamber will propagate because the flame velocity is greater than zero. In a flowing system a flame will be established if $u_f > u$ because otherwise the flame front will travel into the monolith which will presumably act as a flame trap. Since flame velocities depend strongly on composition, diluents, preheat temperature, pressure, and the presence of inhibitors and promoters, published flame limits have only qualitative significance in predicting flames in these flowing situations with large gradients.

The right panel of figure 1 shows a sketch of possible flame locations in a catalytic wall reactor. The first location (a) is at the upstream edge of the monolith where the cold reactant gases impinge on the hot monolith. This is a likely flame location because the gases have not yet reacted so that fuel and O_2 concentrations are highest. However, the temperature gradient is very large in this region, so that any flames must exist in the presence of large concentration and temperature gradients.

The region just inside the catalyst is also a possible location for flames, sketch (b), because the O_2 concentration is still large and gases have been heated to promote reaction. The gas temperature is largest near the walls, so flames may occur in the boundary layer near the entrance region of the monolith. Presumably, as soon as the O_2 is consumed by gas or surface reaction, these reactions would disappear.

The core of cells is another possible region of homogeneous reaction, sketch (c). This region has the highest O_2 concentration and radicals because it is farthest from the walls and the gas velocity is highest near the center. This is a likely region for homogeneous reaction in the monolith.

The fourth location of homogeneous reaction and flames is near the exit of the monolith, sketch (d). This will only occur if all O_2 is not consumed within the monolith, and we observe these flames only if the catalyst is sufficiently inactive that O_2 breakthrough occurs.

Homogeneous reaction should depend strongly on pressure[8] because surface reactions scale as the total pressure P (the flux of O_2 to the surface) while homogeneous reactions scales as P^2 (the collision rate for bimolecular reactions). We have shown that, for syngas, olefins, and HCN, successful operation without interference from flames can be attained above 5.5 atm if the flow rates are maintained sufficiently high. Another factor in homogeneous reaction and flames is the effect of pressure on mass and heat transfer between gas and walls. Since mass and thermal diffusivities are proportional to P^{-1}, high pressures should favor core flames, panel (c).

Boundary layer flames and coupling

In catalytic monoliths the homogeneous reaction rates cannot be considered independent of the surface. The presence of the surface will certainly thermostat the gases and prevent high temperature excursions which are a characteristic of combustion flames. However, the surface will rapidly heat the gases and thus provide an ignition source for homogeneous reaction.

However, the dominant effect of a catalyst is probably to alter the composition of the gases near the surface. Catalytic reaction removes both reactants, thus moving the reactant mixture further away from the stoichiometric composition. The concentration and temperature profiles to be expected in these situations are sketched in the center panel of figure 1 and in figure 3

. A hot inert surface should be much more effective in promoting homogeneous reaction than a catalytic surface because a catalytic surface will always remove the limiting reactant.

TEMPERATURE AND CONCENTRATION PROFILES

Axial temperature profiles

The center panel of figure 1 shows plots of possible axial temperature profiles expected in these reactors. As sketched at the top, the monolith temperature T_S is nearly constant in a typical millisecond reactor because of solid heat conduction and radiation in the monolith. At lower and higher flow rates the temperature profiles might be as sketched in panel (b). At low flow rates, marked 1, the heat generation rate is lower, while heat removal by conduction is high, so heat is generated early in the monolith and the temperature decreases down the monolith. At a higher flow rate, marked 2, the catalyst is nearly adiabatic and the temperature is nearly constant, the "normal" situation for millisecond reactors. As the flow rate is increased, solid conduction can not keep up with the increased gas load, so the front region of the monolith begins to cool off, marked 3. We observe these modes in syngas and olefins processes with inlet velocities between 0.1 and 10 m/sec producing rather flat profiles. Above 10 m/sec the front face cools down to nearly room temperature, and at higher velocities the reaction "blows out" and the reactor cools to room temperature.

For syngas from methane, the temperature profiles are probably even more complicated than those sketched in the center panels of figure 1. At shown in panel (c), at low flow rates the temperature at the leading edge of the catalyst is hotter than for moderate flow rates, while at very high flow rates the trailing edge of the monolith becomes much hotter than the uniform temperature of moderate flows. While these profiles are qualitatively as sketched in panel (b), the higher temperatures at low and high flow rates can be explained by the different reactions which occur under these conditions. At low flow rates there is considerable total oxidation early in the catalyst, and this generates much more heat than the syngas reaction which heats the front face very hot. At very high flow rates, the initial reaction occurs in the cooler sections of the catalyst, and at lower temperatures again the dominant reaction is total oxidation. This heats up the back face of the monolith much above that observed when syngas is the dominant product as sketched in curve 1 of panel (c).

Radial temperature and concentration profiles

The temperature and concentration profiles radially in the monolith cell structures may also be far from simple, especially with homogeneous reaction. Sketched in the right panel of figure 3 are some temperature and concentration profiles expected in various regions of the monolith for different conditions. Sketch (a) of the right panel shows the expected profile near the front edge of the monolith. The temperature profile is very steep as cold reactants impinge on the hot surface. The concentration of radical intermediates R· will attain a maximum within the boundary layer if homogeneous reaction is significant, and the temperature may in fact exhibit a maximum (dashed line) if heat generated by homogeneous reaction is significant.

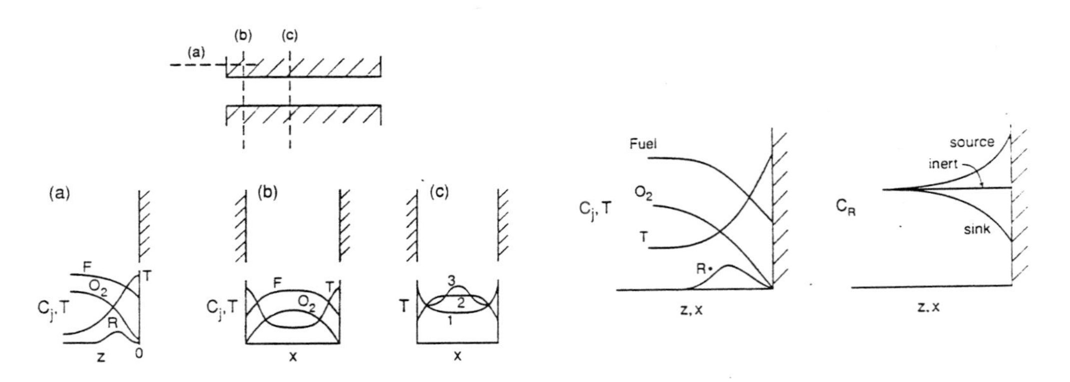

Figure 3. Left panels: Possible concentration and temperature profiles at three different locations in the monolith. Right: Possible radical profiles near a surface in a monolith.

Curve (b) of the right panel of figure 3 shows the expected T and C_j profiles expected within the cell for purely surface chemistry. Heat is generated on the surface and transferred back into the cooler gas, so the temperature profile is as shown. Since mass transfer and heat transfer are proportional, the O_2 and fuel concentrations will also decrease near the wall as sketched. Panel (c) shows possible radial temperature profiles farther in the monolith catalyst. For purely surface reaction, profile 1, the gas is somewhat cooler than the wall, but the gradient should be smaller because most reaction occurs early in the monolith. If homogeneous reaction is significant, the temperature may be higher near the center of the cell, profile 2. If surface reaction occurs at the monolith wall and homogeneous reaction only near the core of the cell far form the wall, then the temperature may exhibit both a minimum and a maximum, profile 3.

SUMMARY

Our intent has been to note how complex these apparently simple reaction processes actually are. The gradients are much larger than those normally encountered in chemical reactors, and the times involved are much shorter than those normally encountered in chemical reactions. It is therefore difficult to "show" any particular mechanism or to refute others by simple calculations, particularly from one dimensional, constant parameter calculations. To summarize our conclusions:

1. All properties of gases vary by a factor of more than 10 between feed and reactor conditions.
2. Most reaction occurs within one cell diameter where properties are varying.
3. Surface reaction events may be different at the 10^{-9} sec time scales predicted for these reactions than at low temperatures where rate parameters are measured.
4. Very large temperature and concentration gradients exist in these experiments.
5. Radial and axial concentration gradients cause quite unpredictable conditions for homogeneous reaction and flames.
6. Homogeneous reactions and flames are tightly coupled to the presence of surface reactions and to cold or hot surfaces which can alter free radical concentrations.
7. Multiple steady states may be expected in these systems for both homogeneous and heterogeneous processes.
8. Transitions between different steady states may be expected in these experiments which may lead to quite different behavior for nearly the same geometries or conditions.

REFERENCES

1. D. A. Hickman and L. D. Schmidt, "Syngas Formation by Direct Catalytic Oxidation of Methane", Science **259**, 343-346 (1993); J. Catalysis **138**, 267-282 (1992).

2. A. Bodke, D. Olschke, L. D. Schmidt, and E. Ranzi, "High Selectivities to Ethylene by Partial Oxidation of Ethane", Science **285**, 712, (1999); A Bodke, D. Henning, L. D. Schmidt, S. Bharadwaj, J. Siddall, and J. Maj, "Effect of H_2 Addition in Oxidative Dehydrogenation of Ethane", J. Catalysis, **191**, 62-74 (2000).

3. C. T. Goralski Jr. and L. D. Schmidt, "Lean Catalytic Combustion of Alkanes at Short Contact Times", Catalysis Letters **42**, 15-20 (1996); "Catalytic Incineration of Volatile Organic Compounds at Millisecond Contact Times", AIChE Journal **44**, 1880-1888 (1998).

4. L. D. Schmidt and D. A. Hickman, "Surface Chemistry and Engineering of HCN Synthesis", in Catalysis of Organic Reactions, Kosak and Johnson eds., pages 195-212 (1993); "Modeling Catalytic Gauze Reactors: Ammonia Oxidation", Industrial and Engineering Chemistry Research **30**, 50 (1991).

5. D. A. Hickman and L. D. Schmidt, D. A., "Elementary Steps in Methane Oxidation on Pt and Rh: Reactor Modeling at High Temperature", A.I.Ch.E Journal **39**, 1164-1177 (1993); "The Role of Boundary Layer Mass Transfer in Selectivity of Partial Oxidation, J. Catalysis **136**, 300-308 (1992).

6. O. Deutschman and L. D. Schmidt, "Partial Oxidation of Methane in a Short Contact Time Reactor: Two-Dimensional Modeling with Detailed Chemistry", AIChE Journal **44**, 2465-2477 (1998); "Two-Dimensional Modeling of Partial Oxidation of Methane in a Short Contact Time Reactor", Twenty-Seventh Symposium (International) on Combustion, The Combustion Institute, 2283-2291 (1998)

7. D. Zerkle, M. D. Allendorf, M. Wold, and O. Deutschmann, "Modeling the Partial Oxidation of Ethane to Ethylene", J. Catalysis **196**, 18-39 (2000).

8. A. Dietz III and L. D. Schmidt, " Effect of Pressure on Three Catalytic Oxidation Reactions", Catalysis Letters **33**, 15-29 (1995).

9. G. Veser and L. D. Schmidt, "Ignition and Extinction in Catalytic Oxidation of Hydrocarbons", AIChE Journal, **42**, 1077-1087 (1996); G. Veser, J. Frauhammer, L. D. Schmidt, and G. Eigenberger, "Catalytic Ignition during Methane Oxidation on Platinum: Experiments and Modeling", Studies in Surf. Sci. Catal. **109**, 273-284 (1997).

Studies in Surface Science and Catalysis
J.J. Spivey, E. Iglesia and T.H. Fleisch (Editors)

Molecular Design of Highly Active Methanol Synthesis Catalysts

Alexis T. Bell

Chemical Sciences Division, Lawrence Berkeley National Laboratory and Department of Chemical Engineering, University of California, Berkeley, CA 94720-1462, USA

Abstract

Active catalysts for the synthesis of methanol from H_2/CO_2 and H_2/CO can be prepared by dispersing Cu on ZrO_2. Mechanistic investigations demonstrate that these catalysts are bifunctional. ZrO_2 serves to adsorb CO_x and retain all carbon-containing intermediates, whereas Cu adsorbs H_2 dissociatively and provides H atoms to the surface of ZrO_2 by spillover. Methanol is formed by the stepwise hydrogenation of carbon-containing species on ZrO_2. The adsorption capacity of monoclinic ZrO_2 (m-ZrO_2) is significantly higher than that of tetragonal ZrO_2 (t-ZrO_2). Correspondingly, the methanol synthesis activity of Cu/m-ZrO_2 is higher than that of Cu/t-ZrO_2. However, the observed increase in the rate of methanol synthesis is less than that anticipated on the basis of the relative CO_x adsorption capacities of m-ZrO_2 and t-ZrO_2. A possible explanation for this is discussed.

1. Introduction

The effect of support composition on the activity of copper-based methanol synthesis catalysts has been the subject of considerable interest motivated by the desire to find more active catalysts. Such catalysts would make it possible to operate at lower reaction temperatures and thereby achieve higher equilibrium conversions. Although some authors have shown the role of the support in methanol synthesis is minimal [1, 2], other studies have shown that the support can significantly affect the activity for methanol synthesis [3-5]. Zirconia has emerged as a particularly interesting support material, as it enhances the activity of Cu for methanol synthesis from both H_2/CO and H_2/CO_2 [4, 6-19].

We have recently reported studies aimed at identifying the roles of zirconia and copper in ZrO_2-supported and ZrO_2-promoted Cu catalysts for methanol synthesis from both H_2/CO_2 [20] and H_2/CO [21]. *In situ* infrared spectroscopy was used to identify the nature of adsorbed species and the dynamics of their reaction. In the case of CO_2 hydrogenation, CO_2 was found to adsorb preferentially on ZrO_2 to form carbonate and bicarbonate species. Both species could undergo hydrogenation to methoxide species, but in the absence of Cu no evidence was found for methanol in the gas phase. When Cu was present, the rate at which carbonate species were converted to bicarbonate species and subsequently to methoxide species increased 10^3 times over that observed in the absence of Cu. The dynamics of gas phase methanol formation were identical to those for the formation of methoxide species on the surface of ZrO_2. The formation of methanol was established to be the hydrolysis of methoxide species rather than reductive elimination. The higher rate of hydrogenation of carbon-containing species in the presence of Cu was attributed to the dissociative adsorption of H_2 on Cu and the supply of H atoms to the surface of ZrO_2 by spillover.

The mechanism of methanol synthesis from CO and H_2 is similar to that for CO_2 and H_2. The principal difference is that CO adsorbs via reaction with hydroxyl groups on the surface of ZrO_2 to form formate species. These species then undergo hydrogenation to form methoxide species. However, in contrast to what happens in the case of methanol synthesis from CO_2, methanol is formed via reductive elimination of methoxide species. Here too, the dynamics of methoxide appearance and the dynamics of methanol appearance in the gas phase are identical.

The proposed mechanisms of methanol synthesis and in particular the role of hydrogen spillover from Cu to ZrO_2 are supported by H_2/D_2 isotopic tracer studies [22]. These studies confirm that H_2 dissociation does not occur readily on ZrO_2 and is enhanced 10^3 fold by the presence of Cu. A further finding is that water vapor suppresses the dissociative adsorption of H_2 on ZrO_2, but not on Cu, and, in fact, the rate of H/D exchange on Cu/ZrO_2 is enhanced by the presence of adsorbed water.

The mechanistic investigations summarized above suggest that increasing the surface concentration of adsorbed CO or CO_2 could result in an increase in the methanol synthesis activity of Cu/ZrO_2 catalysts. Since ZrO_2 has several crystalline phases, it is therefore of interest to ask whether the different phases exhibit different adsorption capacities. To this end, investigations were carried out to determine the effects of crystalline phase on the chemisorption capacity of ZrO_2 and the methanol synthesis activity of Cu/ZrO_2 catalysts.

Experimental

The preparation of the samples of tetragonal and monoclinic ZrO2 used for this work are described in detail in ref. [23]. Tetragonal ZrO2 was prepared by extended boiling of a zirconyl dichloride solution at a constant pH of 10. After drying, the material was calcined at either 973 K or 1323 K to obtain purely tetragonal ZrO_2 with BET surface areas of 187 m^2/g and 20 m^2/g, respectively. These samples are referred to as t-ZrO_2(187) and t-ZrO_2(20). Monoclinic ZrO_2 was produced by boiling zirconyl dichloride at a constant pH of 1.5. After drying, the precipitated material was calcined at either 573 K or 973 K to obtain purely monoclinic ZrO_2 with BET surface areas of 110 m^2/g and 19 m^2/g respectively. These samples are referred to as m-ZrO_2(110) and m-ZrO_2(20). Copper was dispersed onto the zirconia supports by incipient wetness impregnation using a solution of copper nitrate. After drying, the impregnated materials were reduced in H_2 at 573 K for 5 h.

Studies of CO_x adsorption and desorption were carried out by temperature- programmed desorption. Samples were placed in a small quartz reactor connected to gas handling manifold and a quadrupole mass spectrometer. The reactor could be heated at up to 1 K/s while He flowed through the sample. Methanol synthesis activities were measured in a separate quartz microreactor that could be operated at up to 0.65 MPa. Product analysis was done by a mass spectrometer. In situ infrared spectra of adsorbed CO and CO_2 were recorded using a specially designed low gas volume cell.

Results and Discussion

Table 1 lists the BET surface areas and CO_2 chemisorption capacities for the four samples of ZrO_2 investigated. It is evident that the CO_2 adsorption capacities of both the low and high surface area samples of m-ZrO_2 are more than order of magnitude higher than those of the low and high surface area samples of t-ZrO_2. It is also seen that independent of the adsorption temperature the adsorption capacities of the high surface area samples are consistently higher than the adsorption capacities of the low surface area samples. Infrared spectra for CO_2 adsorbed on m-ZrO_2 show bands characteristic of bicarbonate and carbonate species. Temperature-programmed infrared experiments reveal that the bicarbonate species decompose to release CO_2 in the temperature range of 300-400 K, whereas the carbonate species are stable up to much higher temperatures. In the case of t-ZrO_2, the bands for carbonate species are significantly more intense than those for bicarbonate species. Nevertheless, most of the CO_2 desorbed from t-ZrO_2 occurs at temperatures of less than 400 K.

Table 1 CO_2 Adsorption Capacity of t-ZrO_2 and m-ZrO_2

T_{ads} (K)	m-ZrO_2(110) ($\mu mol/m^2$)	m-ZrO_2(19) ($\mu mol/m^2$)	t-ZrO_2(187) ($\mu mol/m^2$)	t-ZrO_2(20) ($\mu mol/m^2$)
298	3.48	1.97	0.07	0.04
523	3.40	2.23	0.13	0.04

The CO adsorption capacities of t-ZrO_2 and m-ZrO_2 are shown in Table 2 for temperatures of 298 K and 523 K. The adsorption capacity of m-ZrO_2 is higher than that of t-ZrO_2. No data are shown for the t-ZrO_2(20), since the adsorption capacity of this sample was too low to be measured. Comparison of the data for the two samples of m-ZrO_2 reveals that the CO adsorption capacity of the high surface area sample is greater than that of the low surface area sample. It is also noted that in all cases the adsorption capacity of ZrO_2 for CO is less than that for CO_2. The infrared spectra of adsorbed CO are different at 298 K and 523 K. At the lower temperature the spectra resemble those for adsorbed CO_2, showing evidence for bicarbonate and carbonate species. When adsorption is carried out at 523 K, the only features seen are those for formate species. During temperature-programmed desorption, both CO and CO_2 are desorbed. The TPD spectra for the two samples of m-ZrO_2 are qualitatively similar. In both cases, the peak for CO desorption (650 K) occurs at a higher temperature than that for CO_2 desorption (600 K) and with increasing adsorption temperature, the fraction of the adsorbed CO desorbing as CO increases significantly (see Table 1). Only CO_2 was observed to desorb from t-ZrO_2 following adsorption of CO at either 298 K or 523 K. Differences are also evident between the two samples of m-ZrO_2. For the low surface area sample, CO desorbing as CO_2 is more weakly retained than CO desorbing as CO, but the reverse is observed for the high surface area sample.

Table 2 CO Adsorption Capacity of t-ZrO_2 and m-ZrO_2*

T_{ads} (K)	m-ZrO_2(110) (μmol/m^2)	m-ZrO_2(19) (μmol/m^2)	t-ZrO_2(187) (μmol/m^2)	t-ZrO_2(20) (μmol/m^2)
298	0.51/--	0.09/--	0.10/--	--/--
523	0.20/1.14	0.17/0.18	0.04/--	--/--

* The first entry designates adsorbed CO desorbing as CO_2 and the second entry designates adsorbed CO desorbing as CO

The effects of ZrO_2 phase and surface area and Cu loading on the methanol synthesis activity of ZrO_2-supporeted Cu was investigated at 548 K and a total pressure of 0.65 MPa. The feed in all cases was mixture containing a 3/1 ratio of H_2/CO or H_2/CO_2 flowing at 60 cm^3/min. The results are shown in Table 3.

Table 3 Effects of ZrO_2 phase and surface area and Cu loading on the methanol synthesis activity of Cu/ZrO_2

Catalyst	Rate from H_2/CO_2 (μmol/s/g-cat)	Rate from H_2/CO (μmol/s/g-cat)
5wt% Cu/t-ZrO_2(19)	0.014	0.004
5wt% Cu/m-ZrO_2(20)	0.06	0.03
5wt% Cu/m-ZrO_2(110)	0.11	0.13
15wt% Cu/m-ZrO_2(110)	0.34	0.16
25wt% Cu/m-ZrO_2(110)	0.28	0.13

Comparison of the first two entries of Table 3 shows that for identical BET surface areas and Cu loadings, the rate of methanol synthesis from H_2/CO_2 is a factor of four higher when m-ZrO_2 is used as the support rather than t-ZrO_2. For methanol synthesis from H_2/CO the rate of methanol synthesis is a factor of seven higher when the support is m-ZrO_2. Raising the support surface area from 20 m^2/g to 110 m^2/g, while keeping the loading of Cu at 5wt% results in a doubling in the methanol synthesis rate from H_2/CO_2 and a four fold increase in the rate of methanol synthesis from H_2/CO. Increasing the Cu loading on the 110 m^2/g m-ZrO_2 causes the methanol synthesis rate to first increase then decrease independent of whether H_2/CO_2 or H_2/CO is used as the feed. A maximum rate is obtained for 15 wt%.

The effect of ZrO_2 phase on the methanol synthesis rate is attributable to the higher adsorption capacity of m-ZrO_2 for both CO_2 and CO consistent with the data presented in Tables 1 and 2. However, the extent of increase is not nearly as large as the observed increase in CO_x adsorption capacity. Thus in the case of CO_2, one might have anticipated a fifty fold increase in methanol synthesis activity in going from t-ZrO_2 to m-ZrO_2 based on the ratio of the CO_2 adsorption capacities of the two phases. Likewise, one might have expected a nearly nine fold increase in the methanol synthesis rate in going from 5wt% Cu/m-ZrO2(19) to 5wt% Cu/m-ZrO_2(110) based on the higher surface area and somewhat higher CO_2 adsorption capacity of the high surface area monoclinic zirconia. Both of these observations suggest that the full benefit of increased CO_2 adsorption capacity is not achieved because other factors limit the rate of methanol synthesis. Since it has been established that the delivery of atomic hydrogen by spillover from dispersed Cu is important, it seems likely that the supply of atomic hydrogen limits the observed activity of both the 5wt% Cu/m-ZrO_2(19) and the 5wt% Cu/m-ZrO_2(110) catalysts. Consistent with this interpretation, it is seen that increasing the loading of Cu increases the methanol synthesis rate. Since metal dispersion often declines at high metal loadings, this might explain why the activity of the catalyst containing 25 wt% Cu is less than that of the one containing 15 wt% Cu for methanol synthesis from both H_2/CO_2 and H_2/CO. The effect of hydrogen supply by spillover appears to be less significant for methanol formation from H_2/CO than from H_2/CO_2. This can be attributed to the intrinsically slower rate of the former process. Table 3 shows that independent of BET surface area or Cu loading, the rate of methanol synthesis from H_2/CO is roughly two fold slower than that from H_2/CO_2 for all of the Cu/m-ZrO_2 catalysts. What these results suggest is that even higher rates of methanol synthesis could be achieved by increasing the surface area of Cu. This would best be achieved by increasing the Cu dispersion at given Cu loading. Experiments in this direction are currently in progress.

Conclusions

Methanol synthesis over zirconia supported Cu catalysts occurs via a bifunctional mechanism, whether the reactants are H_2/CO or H/CO_2. The zirconia adsorbs CO or CO_2 and these species then undergo hydrogenation to methoxide groups, which are then released from the zirconia surface via either hydrolysis or reductive elimination. Cu serves to adsorb H_2

dissociatively and provide atomic hydrogen to the zirconia via spillover. The strength of CO_x adsorption is a strong function of the zirconia phase. Monoclinic ZrO_2 is observed to have an adsorption capacity more than an order of magnitude higher than that of tetragonal zirconia. This effect is associated with the higher Lewis acidity of coordinatively unsaturated Zr cations exposed at the surface of m-ZrO_2, as well as the higher basicity and acidity of hydroxyl groups. The Lewis Acid centers contribute to the formation of carbonate species whereas the hydroxyl groups are involved in the formation of bicarbonate and formate species. Consistent with these findings, it is observed that the rate of methanol formation from both H_2/CO_2 and H_2/CO is significantly higher for Cu/m-ZrO_2 than for Cu/t-ZrO_2. The increase in activity achieved by the switch from t-ZrO_2 to m-ZrO_2 is less than that anticipated on the basis of the ratio of CO_x adsorption capacities of the two phases. This suggests that the supply of hydrogen by spillover from dispersed Cu is insufficient to enable all of the exposed ZrO_2 to be active for methanol synthesis.

Acknowledgment

This work was supported by the Office of Basic Energy Sciences, of the Division of Chemical Sciences, U. S. Department of Energy under Contract DE-AC03-SF7600098.

References

1. Chinchen, G.C., Waugh, K.C., and Whan, D.A., Appl. Catal., 25, 101 (1986).
2. Pan, W.X., Cao, R., Roberts, D.L., and Griffin, G.L., J. Catal., 114, 440 (1988).
3. Klier, K., Adv. Catal., 31, 243 (1982).
4. Bartley, G.J.J., and Burch, R., Appl. Catal., 43, 141 (1988).
5. Robinson, W.R.A.M., and Mol, J.C., Appl. Catal., 44, 165 (1988).
6. Denise, B., and Sneeden, R.P.A., Appl. Catal., 28, 235 (1986).
7. Chen, H.W., White, J.M., and Ekerdt, J.G., J. Catal., 99, 293 (1986).
8. Amenomiya, Y., Appl. Catal., 30, 57 (1987).
9. Denise, B., Sneeden, R.P.A., Beguin, B., and Cherifi, O., Appl. Catal., 30, 353 (1987).
10. Koeppel, R.A., Baiker, A., Schild, C., and Wokaun, A., in: Preparation of Catalysts V, Stud. Surf. Sci. Catal., Vol. 63, eds. G. Poncelet, P.A. Jacobs, P. Grange, and B. Delmon (Elsevier, Amsterdam, 1991) p.59.
11. Kanoun, N., Astier, M.P., and Pajonk, G.M., Catal. Lett., 15, 231 (1992).
12. Koeppel, R.A., Baiker, A., and Wokaun, A., Appl. Catal. A, 84, 77 (1992).
13. Sun, Y., and Sermon, P.A., J. Chem. Soc., Chem. Commun., 1242 (1993).
14. Nitta, Y., Suwata, O., Ikeda, Y., Okamoto, Y., and Imanaka, T., Catal. Lett., 26, 345 (1994).
15. Sun, Y., and Sermon, P.A., Catal. Lett., 29, 361 (1994).
16. Fisher, I.A., Woo, H.C., and Bell, A.T., Catal. Lett., 44, 11 (1997).

17. Schild, C., Wokaun, A., and Baiker, A., J. Mol. Catal., 63, 243 (1990).
18. Baiker, A., Kilo, M., Maciejewski, M., Menzi, S., Wokaun, A., in: New Frontiers in Catalysis, Proceedings of the 10th International Congress on Catalysis, eds. L. Guczi, et al., (Elsevier, Amsterdam, 1993) p.1257.
19. Weigel, J., Koeppel, R.A., Baiker, A., and Wokaun, A., Langmuir, 12, 5319 (1996).
20. Fisher, I. A., and Bell, A. T., J. Catal., 172, 222 (1997).
21. Fisher, I. A., and Bell, A. T., J. Catal., 178, 153 (1998).
22. Jung, K. D., and Bell, A. T., J. Catal., 193, 207 (2000).
23. Jung, K. T., and Bell, A. T., J. Mol. Catal., 163, 27 (2000).

Studies in Surface Science and Catalysis
J.J. Spivey, E. Iglesia and T.H. Fleisch (Editors)

Partial Oxidation of Methane to Syngas over NiO/γ-Al_2O_3 Catalysts Prepared by the Sol-Gel Method

Yuhong Zhang, Guoxing Xiong*, Weishen Yang, Shishan Sheng
State Key Laboratory of Catalysis, Dalian Institute of Chemical Physics,
Chinese Academy of Sciences, Dalian 116023, China

Partial oxidation of methane to syngas was investigated over NiO/γ-Al_2O_3 catalysts prepared by the impregnation method, the sol-gel method and the complexing agent-assisted sol-gel method. All catalysts had similar high conversions of methane and high selectivities to syngas. However, their resistance to carbon deposition differed greatly. The catalysts prepared by the sol-gel method showed high resistance to carbon deposition and sintering of nickel. Moreover, the catalyst prepared by the complexing agent-assisted sol-gel method had excellent thermal stability.

1. INTRODUCTION

In recent years, catalytic partial oxidation of methane to syngas (POM) as an alternative to the conventional steam reforming of methane has drawn much attention [1-8]. This process has advantages over the steam reforming of methane, because steam reforming is strongly endothermic and produces syngas having a H_2/CO ratio ≥ 3.0. On the other hand, the partial oxidation of methane is mildly exothermic and produces syngas with a more desirable stoichiometric ratio ($H_2/CO \approx 2$) which is suitable for the methanol and Fischer-Tropsch syntheses.

Supported nickel catalysts are very effective for the POM reaction because of their high conversion and low cost. However, the catalysts deactivate with time on stream by carbon deposition and sintering of the nickel [2,4]. Therefore, it is desirable to develop stable supported nickel catalysts with resistance to coke formation and to sintering of nickel. Al-Ubaid and Wolf [9] observed a much greater stability for Ni supported on the aluminate than on other supports, and Bhattacharyya and Chang [10] have proposed the use of a nickel aluminate spinel catalyst to reduce coke formation in the CO_2 reforming of methane. Ruckenstein [11] suggested that a stronger NiO-support interaction could improve the stability and inhibit carbon deposition. Therefore, one possible way to limit carbon deposition and to resist the sintering and loss of nickel is to strengthen the interaction between the metal and the support.

In our previous work, a stronger metal-support interaction was observed in NiO/γ-Al_2O_3 catalysts prepared by the sol-gel method [12]. The aim of this work is focused on developing stable NiO/γ-Al_2O_3 catalysts using the sol-gel method. The POM activity and the stability against carbon-deposition and sintering of nickel were investigated.

2. EXPERIMENTAL

2.1 Preparation of catalysts

Three types of NiO/γ-Al_2O_3 catalysts were used in this study. SG catalyst was prepared by the sol-gel method. The pure AlOOH sol was obtained by peptizing PURAL SB powder (Condea GmbH, Germany) with HNO_3 solution. A $Ni(NO_3)_2$ solution was added dropwise into the pure AlOOH sol with vigorous stirring to give a homogenous mixture. The mixture

* To whom correspondence should be addressed. Fax: +86 411 4694447; E-mail: gxxiong@dicp.ac.cn

was dried and calcined at 850°C in air to obtain the SG catalyst. SGB catalyst was prepared by the complexing agent-assisted sol-gel method, i.e, by adding $Ni(NO_3)_2$ solution dropwise to the complexed AlOOH sol, which was obtained by adding glycerol into the pure AlOOH sol in advance. The drying and calcining were the same as the SG catalyst. IT catalyst was prepared by impregnation. The content of nickel oxide was 10% for all catalysts.

2.2 Catalytic reaction

The catalytic reaction was carried out under atmosphere pressure and at 850°C in a quartz microreactor (I.D. 4mm) packed with 100mg catalysts (40-60 mesh size). The catalyst was not reduced before the tests. The catalytic activity and stability tests were conducted with a stoichiometric ratio of $CH_4/O_2 = 2$ and a space velocity (GHSV) of 1.8×10^5 $L\cdot kg^{-1}\cdot h^{-1}$. The reaction products were analyzed by a gas chromatography using a TCD detector [8].

2.3 Characterization of catalysts

The amount of carbon deposited on catalysts during POM reaction was measured by TG analysis (Perkin-Elmer 3600). The nickel contents of the catalysts were determined by atomic absorption spectroscopy with a Perkin-Elmer ICP-5000 spectrometer.

TPR of the catalysts was performed by heating the samples from 50°C to 1000°C at a rate of $10^oC\cdot min^{-1}$ in a 5% H_2/Ar gas flow. XRD was carried out on a Riguku D/Max-RB X-ray diffractometer by using a Cu Kα radiation source. The specific surface area and pore volume were obtained in a volumetric equipment Coulter OMNISORP-100CX by the BET method.

XPS analysis was undertaken on a VG ESCALAB MK II spectrometer, using a Al Kα X-ray source. The binding energies were calibrated with the Al 2p level (74.5ev) as the internal standard reference. The surface atomic ratios of the catalysts were calculated according to the spectra peak area of XPS [13]. EXAFS results were obtained by using the BL-2 facilities at BEPC (Beijing, China), with a position beam energy of 2.2 GeV and an average stored current of 150 mA. A procedure for EXAFS analysis has been given previously [14].

3. RESULTS AND DISCUSSION

3.1 Catalytic performance and carbon deposition

The catalytic activities of the IT, SG and SGB catalysts were measured with a stoichiometric feed gases ($CH_4/O_2=2$) at GHSV=1.8×10^5 $L\cdot kg^{-1}\cdot h^{-1}$ and 850°C. It can be seen from Fig. 1 that the three $NiO/\gamma\text{-}Al_2O_3$ catalysts all showed high activities and selectivities in the POM reaction. Both methane conversion and selectivity to syngas approached the thermodynamic equilibrium values. To further investigate their stabilities, life-tests as long as 80 h were carried out. The results showed that their catalytic activities remained stable during the 80 h operations (see Fig. 1), and no obvious catalytic deactivation was observed. Therefore, longer time-on-steam tests must be carried out to determine the stability of these catalysts.

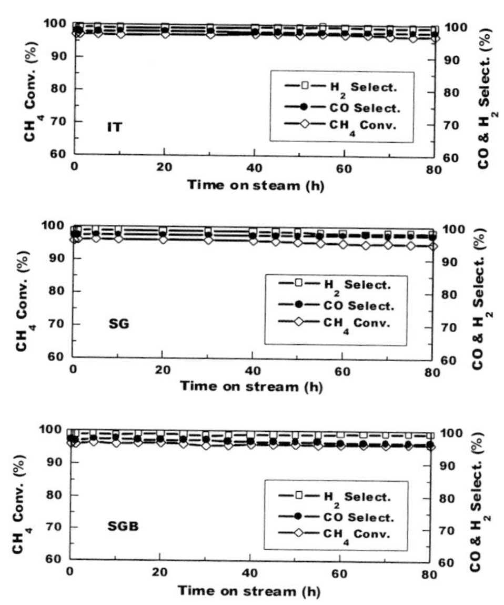

Fig. 1 CH_4 conversion and CO and H_2 selectivity as a function of time over the IT, SG and SGB catalysts

Although the IT, SG and SGB showed similar high catalytic activity and high selectivities for POM, their resistance to carbon deposition differed greatly. Carbon deposition on the catalysts after 80 h was measured by TG. As shown in Table 1, the amount of carbon deposited on the SG and SGB were much less than that on the IT. The weight of surface carbon on the IT was about 17.64% of its net weight, while the weight of surface carbon on the SG and SGB was nearly negligible. Fig. 2 shows the XPS spectra in the C 1s region for the IT, SG and SGB that had been on stream for 80 h. The peak at a binding energy (B.E.) of 284.6 eV is due to adventitious carbon, the peak at ~288 eV may be assigned to residual surface CO_3^{2-}, and the peak at ~283 eV is attributed to a graphitic or carbidic surface carbon species deposited on the catalysts during the reactions [2]. The graphitic or carbidic surface carbon will result in pore blocking as well as deactivation of the catalyst by covering the nickel surface. A prominent peak at the binding energy of 283 eV in the IT suggested that a large amount of surface carbon had deposited on the surface of the IT. Moreover, the surface C/Ni atomic ratio of the used IT increased remarkably to 1.24/0.012 from 0.16/0.038 for the fresh IT (see Table 1). However, for the SG and SGB, the C 1s XPS spectra and the surface C/Ni atomic ratio of the used catalysts are similar to those of the fresh ones, which indicated nearly no carbon deposition on the used SG and SGB. It can be concluded from these results that the catalyst prepared by the sol-gel method has excellent resistance to carbon deposition.

Table 1 Surface atomic ratios as well as carbon deposition on the surface of the catalysts

Sample	Surface relative atomic ratio								Carbon-deposition (%)
	Fresh catalyst				Used catalyst				
	C	O	Ni	Al	C	O	Ni	Al	
IT	0.16	1.55	0.038	1	1.24	1.51	0.012	1	17.64
SG	0.20	1.70	0.039	1	0.45	1.65	0.031	1	0.43
SGB	0.17	1.73	0.038	1	0.14	1.63	0.031	1	none

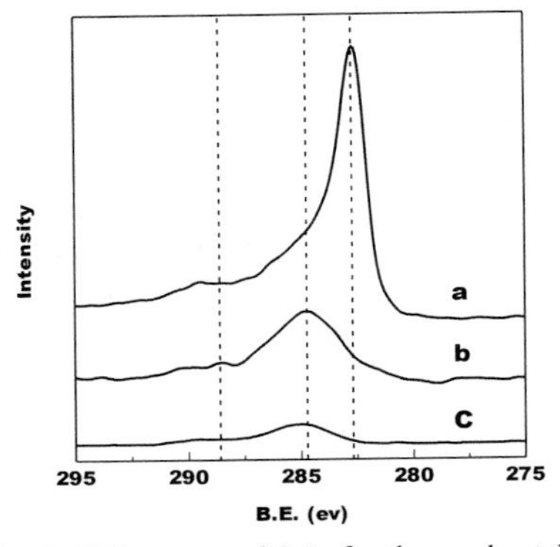

Fig. 2 XPS spectra of C 1s for the used catalysts:
(a) IT; (b) SG; (c) SGB

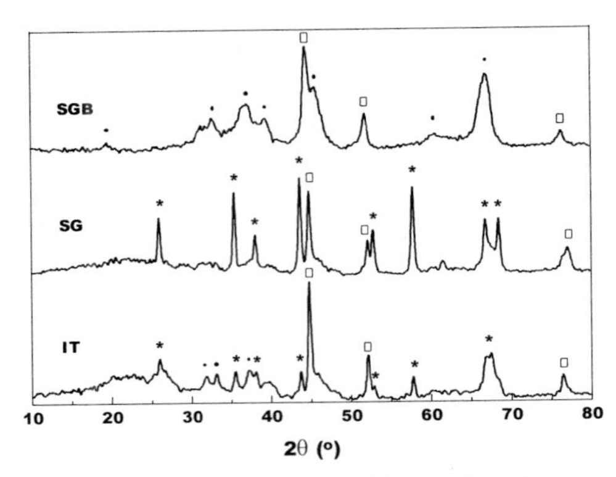

Fig. 3 XRD patterns of the used catalysts
(• γ-Al_2O_3, * α-Al_2O_3, □ Ni)

3.2 The sintering and loss of nickel

The Ni^0 species is active sites for the POM reaction [7]. However, the growth in size of the nickel particles as a function of time at higher temperatures due to sintering led to decrease the number of active sites. Fig. 3 shows the XRD patterns of these used catalysts. It was found that reduced nickel appeared on the catalysts after reaction. The XRD peak width of reduced nickel over the SG and SGB was boarder than that over the IT, which suggested that the Ni crystallite size is smaller on the SG and SGB than that on the IT. Fourier transforms of Ni K-edge EXAFS of the used IT and SG are shown in Fig. 4. The best fits to the Fourier transform and the Fourier-filtered EXAFS revealed that there were 12 nearest nickel neighbors for the IT and 9 nearest nickel neighbors for the SG. The peak at 2.49 Å confirmed the formation of zero-valent nickel atoms, which was in agreement with the XRD results. It is well known that the structure of metallic nickel is a face-centered cubic lattice and the coordination number of nearest nickel neighbors is 12. The number of low coordinated nickel atoms of the SG suggested that the nickel particles were smaller than that over the IT, while the nickel particles over the IT grew so large in size that they exhibit the same structure as bulk nickel. The larger nickel crystalline size will promote the loss of metal. The nickel content of those catalysts was measured by AAS before and after reaction (see Table 2). The results show that the loss of nickel on the IT is more than that on the SG and SGB. The serious decrease of Ni/Al atomic ratio over the IT also indicates that the loss and sintering of nickel on the IT is more serious than that of the SG and SGB. The results reveal that Ni tiny crystals can be stabilized on the catalysts prepared by the sol-gel method and such catalysts have high resistance to the loss and sintering of nickel.

The H_2-TPR results of the catalysts are given in Fig. 5. For the IT, a strong peak of reduction appears at about 840°C, which is close to that of bulk $NiAl_2O_4$, and two very small peaks also present at about 500°C and 600°C, respectively. The peak at 500°C is attributed to crystalline NiO reduction (by comparison with the reduction behavior of a bulk NiO), and the peak at 600°C is attributed to the reduction of dispersed NiO. For the SG and SGB, a strong peak appears at about 880°C. This indicates that it is more difficult to reduce NiO in the SG and SGB than that in the IT, and the metal-support interaction in the SG and SGB is stronger than in the IT. It is well known that the sol-gel method is a wet chemical preparation process. The nickel ions can be tightly bound to AlOOH colloidal particles in aqueous solution to form a mixed hydroxide surface compound, which is a precursor of nickel aluminate. During the drying and calcining, the interaction becomes stronger and a solid solution between $NiAl_2O_4$ and γ-Al_2O_3 is formed in these catalysts, which results in the nickel to be more uniformly dispersed in the support. Choudhary shows that NiO can be reduced in the POM atmosphere at high temperatures ($\geq 600^{\circ}C$). When reaction gases were used, only a portion of NiO can be reduced to Ni^0, which is strongly attached to the surface of the γ-Al_2O_3. Due to the strong interaction between Ni^0 and support, the aggregation of small Ni^0 particles is suppressed.

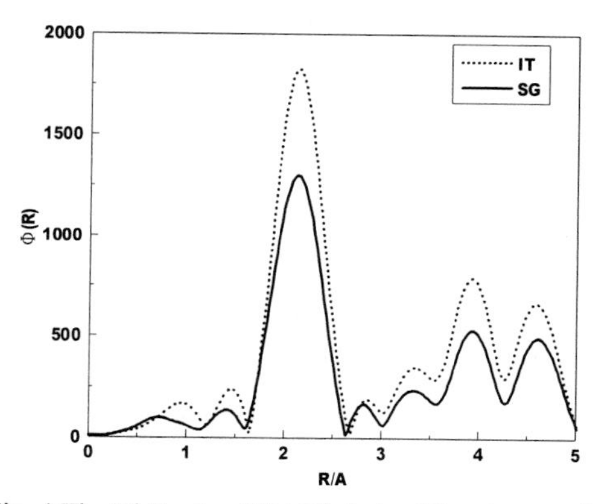

Fig. 4 The Ni K-edge EXAFS-derived Fourier transforms of the used SG (solid line) and IT (dashed line).

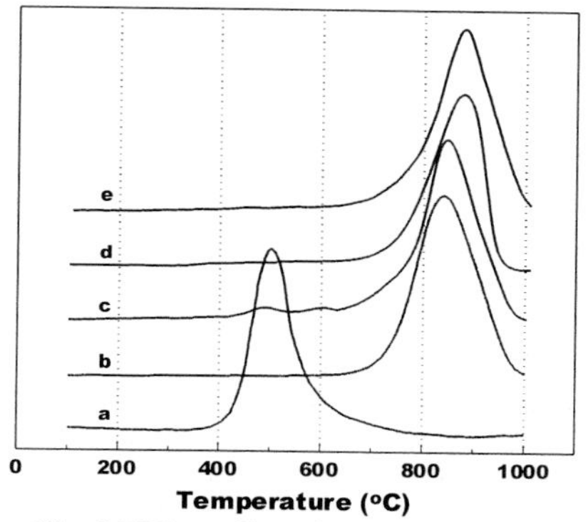

Fig. 5 TPR profiles of (a) NiO, (b) $NiAl_2O_4$, and the catalysts (c) IT, (d) SG and (e) SGB

Table 2 The comparison of nickel content on the fresh and used catalysts

Catalyst	Ni wt. (%)		Lost	S_{BET} ($m^2 \cdot g^{-1}$)		Decrease	V_p ($cm^3 \cdot g^{-1}$)		Decrease
	fresh	used	percentage	fresh	used	percentage	fresh	used	percentage
IT	7.2	6.9	41.7	123	54	56.1	0.28	0.15	46.4
SG	8.3	8.2	12.0	111	32	71.2	0.23	0.12	47.8
SGB	8.8	8.7	11.4	143	77	46.2	0.66	0.44	33.3

3.3 The thermal stability of support

XRD measurements were performed to determine the crystal phase of the catalysts. The slow transforming process of γ-Al_2O_3 into α-Al_2O_3 phase could slowly decrease the performance of the NiO/γ-Al_2O_3 catalyst [15]. For the fresh catalysts calcined at $1200^{\circ}C$ for 5h, it was found that the phase of the SGB was very stable and no phase transformation occurred, while the part of γ-Al_2O_3 supports on the IT and SG had transformed into α-Al_2O_3 phase (not shown). The XRD patterns and the surface area of the used catalysts after 80 h are shown in Fig. 4 and Table 2. It was found that a little α-Al_2O_3 appeared over the IT, and the surface area was 54 m^2/g, while γ-Al_2O_3 phase of support disappeared completely and was transformed into α-Al_2O_3 phase over the SG, and the surface area was only 32 m^2/g. As for the SGB, no α-Al_2O_3 phase appeared, and the surface area was 77 m^2/g. This indicates that the catalyst prepared by the complexing agent-assisted sol-gel method has better thermal stability.

4. CONCLUSION

$NiO/\gamma\text{-}Al_2O_3$ catalysts prepared by different methods in this study all have high catalytic activities and good selectivities for POM. However, the catalyst prepared by the sol-gel method exhibited excellent resistance to carbon deposition and sintering of nickel. Moreover, the catalysts prepared by the complexing agent-assisted sol-gel method showed better thermal stability. Stable $NiO/\gamma\text{-}Al_2O_3$ catalysts with high resistance to carbon deposition and sintering of nickel and excellent thermal stability can be obtained by the complexing agent-assisted sol-gel method.

ACKNOWLEDGEMENTS

The project was supported both by the National Sciences Foundation of China (Grant number 29392003) and by Chinese Academy of Sciences (Grant number KJ951-A1-508-2).

REFERENCE

1. A.T. Ashcroft, A.K. Cheetham, P.D.F. Vernon, Nature, 344 (1990) 319.
2. D. Dissanayake, M.P. Rosenek, K.C.C. Kharas, J.H. Lunsford, J. Catal., 132 (1991) 117.
3. P.M. Torniainen, X. Chu, L.D. Schmidt, J. Catal., 146 (1994) 1.
4. V.R. Choudhary, B.S. Uphade, A.S. Mamman, J. Catal., 172 (1997) 281.
5. S. Tang, J. Lin, K.L. Tan, Catal. Lett., 51 (1998) 169.
6. Y. Lu, Y. Liu, S. Shen, J. Catal., 177 (1998) 386.
7. Q. Miao, G. Xiong, S. Sheng, W. Cui and X. Guo, Stud. Surf. Sci. Catal., 101 (1996) 453.
8. S. Liu, G. Xiong, H. Dong, W. Yang, S. Sheng, W. Chu, Z.Yu, Stud. Surf. Sci. Catal. 130 (2000) 3567.
9. A. Al-Ubaid, E.E. Wolf, Appl. Catal., 40 (1988) 73.
10. A. Bhattacharyya, V.W. Chang, Stud. Surf. Sci. Catal., 88 (1994) 207.
11. E. Ruckenstein, Chemical Innovation, 30 (2000) 39.
12. Y. Zhang, G. Xiong, S. Sheng, S. Liu, W. Yang, Acta. Physico-chem. Sinica, 15 (1999) 735.
13. Y. Boudeville, F. Figueras, M. Forissier, J-L. Portefaix, J.C. Vedrine, J. Catal., 58 (1979) 52.
14. Y. Kou, Z. Suo, H. Wang, J. Catal., 149 (1994) 246.
15. Y. Zhang, G. Xiong, S. Sheng, W. Yang, Catal. Today, 63 (2000) 517.

Studies in Surface Science and Catalysis
J.J. Spivey, E. Iglesia and T.H. Fleisch (Editors)

Mo/MCM-22: a selective catalyst for the formation of benzene by methane dehydro-aromatization

Y.-Y. Shu, D. Ma, L.-L. Su, L.-Y. Xu, Y.-D. Xu*, X.-H. Bao*

State Key Laboratory of Catalysis, Dalian Institute of Chemical Physics, Chinese Academy of Sciences, 457 Zhongshan Road, P. O. Box 110, Dalian 116023, China.

The catalytic dehydro-aromatization of methane on Mo/MCM-22, compared to Mo/HZSM-5, is characterized by higher yields of benzene and lower yields of naphthalene, as well as higher tolerance to coke formation. The selectivity to light aromatics is ~ 80 % at a CH_4 conversion of ~10.0 % over a 6Mo/MCM-22 at 973 K. The effect of contact time shows that the reaction is severely inhibited by the products and/or intermediates. Experimental results of CH_4-TPSR, NH_3-TPD and C_6H_6- TPD reveal that the acidic function of the catalysts consists of strong and exchangeable Bronsted acid sites in 12 MR channels, while reduced Mo species are responsible for CH_4 activation.

1. INTRODUCTION

Recently, methane dehydro-aromatization (MDA) in the absence of oxygen over Mo/HZSM-5 catalysts in a continuous-flow mode has received considerable attention [1]. It is well accepted that Mo/HZSM-5 is a bi-functional catalyst. The Mo species is reduced and transformed into Mo_2C, or $Mo_2O_xC_y$, which may be responsible for activating CH_4, while Bronsted acid sites on the zeolite may play a key role in the aromatization of the C_2H_4 intermediates [2-4]. To improve the activity and selectivity to aromatics of the Mo modified zeolites, a series of zeolites have been screened as the support [5]. It was concluded that silica-alumina zeolites with a two-dimensional structure and a pore size close to the dynamic diameter of C_6H_6 are the best supports of Mo-based catalysts for the MDA.

MCM-22 is a new type of synthetic high-silica zeolite with both 10 and 12 MR pore structures [6]. It has been claimed that MCM-22 is a good support for light hydrocarbon conversion to aromatics [7]. Here we report on our finding that Mo/MCM-22 is also an active and selective catalyst for the formation of C_6H_6 from MDA. The selectivity to C_6H_6 has been found to be about 80% at a CH_4 conversion of about 10% at 973 K.

2. EXPERIMENTAL

MCM-22 was synthesized according to the procedure described in ref. [6]. The chemical composition of the as-synthesized MCM-22 was (in wt%) O 55.36%, Si 40.56%, Al 3.61%, Na 0.26%, Ca 0.068%, and traces of transition metal ions, as determined by XRF-1700 spectrometer. Mo/MCM-22 was prepared according to the same procedure as described previously in ref. [8], and the Mo content is measured in Mo atom wt%.

Catalytic evaluation was performed in a fixed bed down-flow reactor as described in ref. [9]. The catalyst (0.2 g) was first heated under a He stream (15 ml/min) to 973 K and maintained at this temperature for 30 min. Then, a 9.5% N_2/CH_4 gas mixture was introduced into the reactor at a space velocity of 1500 ml/gcat.h. N_2 was used as an internal standard so that coke formation can be determined on-line [3]. The effect of contact time was examined by changing the weight of the catalyst as well as the flow rate. NH_3-TPD measurements were performed in a conventional set-up as described in ref. [8]. The TPSR of CH_4 and C_6H_6-TPD experiments were carried out in a quartz tubular micro-flow reactor, using a Balzers QMS-200 mass spectrometer as the detector. The profiles of NH_3-TPD and C_6H_6-TPD were deconvoluted into Gaussian curves on a NEXT workstation. UV laser Raman spectra of carbonaceous deposits on the fresh and used catalysts were recorded on a spectrometer described in ref. [10].

3. RESULTS AND DISCUSSION

3.1. Methane dehydro-aromatization on Mo/MCM-22 catalysts

Table 1
Physical properties and Mo contents of Mo/MCM-22 catalysts analyzed by ICP

Sample	S_{BET}, m^2/g	Microp. vol. cc/g	Aver. diam. nm	Mo content, %
MCM-22(30)	458	0.16	1.97	0
2Mo/MCM-22	417	0.15	2.12	1.1
6Mo/MCM-22	398	0.14	2.30	3.6
10Mo/MCM-22	357	0.13	2.22	6.3

Table 2
The catalytic behavior of Mo/MCM-22 catalysts at 973 K[a]

Sample	CH_4 Conv., %	Selectivity, %				
		CO	C_2	C_6H_6+C_7H_8	$C_{10}H_8$	Coke
MCM-22	~0	~0	~0	~0	~0	~0
2Mo/MCM-22	5.7	0.9	4.9	67.8	7.9	18.5
6Mo/MCM-22	10.0	0.2	3.4	80.0	4.4	12.0
10Mo/MCM-22	5.8	0.8	4.5	61.4	3.6	29.7

a, Data were taken after running the reaction for 180 min.

The physical properties and the actual Mo contents of the Mo/MCM-22 catalysts are listed in Table 1. The effect of Mo loading on the catalytic performance of MDA shows that 6Mo/MCM-22 is the best among the tested catalysts, as shown in Table 2. Surprisingly, despite the fact that MCM-22 has a relatively high surface area, the CH_4 conversion decreased quickly with Mo loading from 6 to 10%. The change in selectivity to light aromatics (more than 90% was C_6H_6) follows the same trend as the CH_4 conversion. Selectivity to coke increased with Mo loading, indicating that in addition to the acid sites, the Mo species are also responsible for the coke formation.

By measuring directly the H_2 concentration in the tail gas, the H atom% in the coke was estimated, as listed in Table 3. It is interesting to notice that after running the reaction for 360 min, the H atom% in the coke was ca 2.3 % on the 6Mo/MCM-22, while it was ca 0.4% on the 6Mo/HZSM-5 catalyst. This implies that much more non-graphitic coke is formed on the 6Mo/MCM-22 than on the 6Mo/HZSM-5 during the reaction. The UV laser Raman spectra of carbonaceous deposits on the 6Mo/MCM-22 catalysts before and after running the reaction for 6 h are shown in Fig. 1. It is clear that there is a band centering at 1612 cm^{-1}, accompanied by a shoulder at ca 1422 cm^{-1}. The former can be attributed to polyolefinic and/or aromatic and polyaromatic species [10], indicating that the coke species are still rich in hydrogen even after running the reaction for 6 h at 973 K.

The profile of the TPSR by passing a CH_4 stream over the 6Mo/MCM-22 is similar to that over the 6Mo/HZSM-5 reported elsewhere [11]. There is an induction period in the early stage of the reaction. During the induction period the Mo species may be reduced by CH_4 into Mo_2C and/or $Mo_2O_xC_y$ [2-4].

Table 3
Hydrogen balance over the 6Mo/MCM-22 and 6Mo/HZSM-5 catalysts

Reaction time, min	6Mo/MCM-22			6Mo/HZSM-5		
	CH_4 Conv., %	H atom in tail gas, %	H atom in Coke, %	CH_4 Conv., %	H atom in tail gas, %	H atom in Coke, %
60	9.4	30.2	0.1	11.5	30.6	6.2
120	10.2	27.8	4.5	9.9	28.2	3.4
180	9.7	26.8	3.1	9.2	26.8	1.7
240	9.4	26.0	2.9	8.8	25.6	1.8
300	9.1	24.6	3.1	8.4	24.8	1.1
360	8.8	24.4	2.3	7.6	22.8	0.4

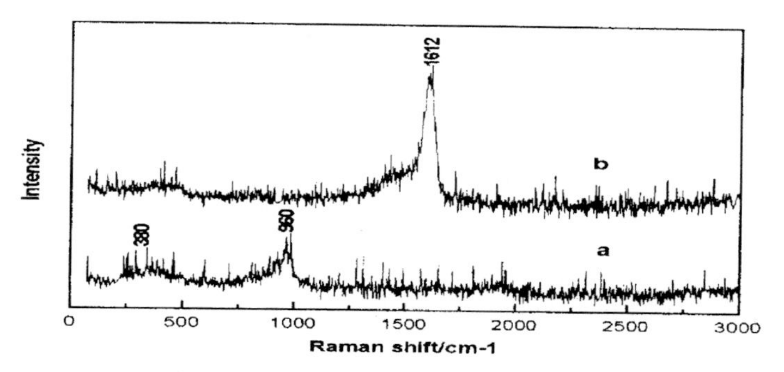

Fig. 1, UV-laser Raman spectra of 6Mo/MCM-22 before (a) and after (b) the reaction

3.2. Effect of time on stream and contact time

Figure 2 shows the changes in CH_4 conversion, X_{CH4}, and the yields of the products, such as Y_{Ben}, and Y_{Naph}, over the 6Mo/MCM-22 catalyst with time-on-stream. The corresponding data obtained on the 6Mo/HZSM-5 catalyst are also included in Fig. 2 for comparison. The catalytic behaviors of the 6Mo/MCM-22 and 6Mo/HZSM-5 are different, as shown in Fig. 2. It can be noted that during the early stage of the reaction (less than 3 h), the X_{CH4} decreased while the Y_{Ben} and Y_{Naph} increased with the time on stream on the 6Mo/MCM-22 catalyst. However, after a prolonged time-on-stream (more than 3 h), the X_{CH4} decreased, but the Y_{Ben} decreased only slightly. Interestingly, the Y_{Naph} was in a lower level on the 6Mo/MCM-22 catalyst than on the 6Mo/HZSM-5 catalyst, particularly in the early stage of the reaction.

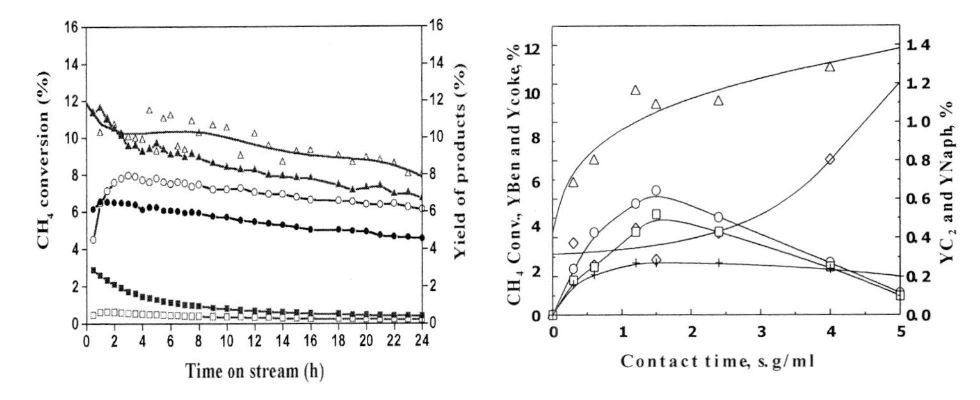

Δ, for CH_4 conv., O, □, +, and ◊, for the yields of C_6H_6, $C_{10}H_8$, C_2 & coke.

Fig. 2, Influence of time on stream on 6Mo/MCM-22 (open symbols) and 6Mo/HZSM-5 (solid symbols).

Fig. 3, Effect of contact time after running the reaction for 30 min over 6Mo/MCM-22.

The effect of contact time on methane conversion and the yields of C_6H_6, $C_{10}H_8$ and C_2 and coke after running the reaction for 30 min are shown in Fig. 3. The X_{CH4} increases sharply when the contact time is less than 0.5 s.g/ml, and then changes less rapidly with contact time. This means that MDA is severely inhibited by the products or intermediates due to their strong adsorption and slow desorption on the catalyst surface and slow diffusion in the channels. This behavior can be attributed to the competition between CH_4 and other hydrocarbon molecules or intermediates for the active sites. The Y_{C2}, Y_{Ben}, and Y_{Naph} pass through the origin and increase, whereas the Y_{Coke} almost remains constant with contact time in the range of ~0-0.5 s.g/ml. This implies that coke is an initial product. Part of the coke thus formed may be in the form of Mo_2C or $Mo_2O_xC_y$ in the early stage of the reaction and part of the coke may be present as the active carbon species, CH_x (x=2) [12, 13]. The products C_2, C_6H_6, and $C_{10}H_8$ are primary products formed in parallel and/or in succession with each other and/or are formed through the same reaction intermediate, i.e. the active carbon species, CH_x (x=2). With further increase of the contact time (longer than 1.0 but less than 1.5 s.g/ml), the Y_{Ben}, and Y_{Naph} still increase, but the Y_{C2} almost remains constant, indicating that the the C_2 species (mainly C_2H_4) may react subsequently to form C_6H_6. When the contact time is longer than 1.5 s.g/ml, the Y_{Ben} decreases very quickly. In accordance with this, the Y_{Coke} increases steeply with the contact time. This suggests that at a longer contact time the condensation reaction will result in the formation of coke. The effect of contact time on methane conversion and the yields of C_6H_6, $C_{10}H_8$ and C_2 and coke after running the reaction for longer than 30 min follows this same trend.

3.3. NH_3-TPD and C_6H_6-TPD

Similar to the NH_3-TPD profiles of HZSM-5 reported in the literature, two peaks, denoted as peak H and peak L, centering at about 532 and 681 K respectively, can be seen on the NH_3-TPD profiles of the MCM-22 zeolite. Table 4 lists the curve fitting results of NH_3-TPD profiles recorded from the Mo/MCM-22 and Mo/HZSM-5 with different Mo loadings. The distribution of acid strengths on MCM-22 is somewhat different from that of HZSM-5, particularly in the region of strong acid sites. The peak temperature of the peak H is about 739 K for HZSM-5 and about 681 K for MCM-22. However, the distribution of acid strengths on the 6Mo/MCM-22 is similar to that on the 6Mo/ HZSM-5, as listed in Table 4. For the Mo/MCM-22 catalysts, the areas of both peak L and peak H decrease with the introduction of the Mo component, but the area of peak H decreases more dramatically. With a Mo loading of 6%, the Bronsted acid sites decrease by 60%, indicating that the Mo species migrated into the channels and located closely to the Bronsted acid sites of the MCM-22 zeolite.

Table 4
Peak temperatures of NH_3-TPD and C_6H_6-TPD spectra.

Sample	Peak (L)		Peak (M)		Peak (H)	
	Temp., K	Area (a.u.)	Temp., K	Area (a.u.)	Temp., K	Area (a.u.)
NH_3-TPD						
MCM-22	532	448	--	--	681	640
2Mo/MCM-22	527	425	606	82	686	435
6Mo/MCM-22	529	362	604	226	680	261
10Mo/MCM-22	525	345	600	390	685	173
HZSM-5	551	322	--	--	739	457
6Mo/HZSM-5	541	305	606	104	680	195
C_6H_6-TPD						
MCM-22	365/400	56.5	450	187.2	515	44.8
2Mo/MCM-22	368/402	49.1	445	155.4	502	43.9
6Mo/MCM-22	368/400	45.4	435	96.5	480	16.1
10Mo/MCM-22	372/407	42.7	440	51.3	477	15.2

Since C_6H_6 is the main product of MDA, the results of the C_6H_6-TPD profiles on the Mo/HMCM-22 samples are particularly interesting. By deconvolution, two peaks at 365 and 400 K in the low temperature part, and other two peaks at 450 and 515 K in the high temperature part are resolved (see Table 4). The relative low desorption peak temperature in the C_6H_6-TPD, in comparison with the reaction temperature of MDA, also suggests that benzene desorption from the catalyst surface is not the rate determining step for the MDA reaction. The areas surrounded by each peak can be taken as a standard of comparison to the amount of desorbed C_6H_6. As shown in Table 4, the peak temperature in the low temperature part does not change significantly with the introduction of the Mo species. The amount of desorbed C_6H_6 at the low temperature shows a linear relationship to the BET surface area of the samples. Thus, it is reasonable to ascribe it to the physisorbed C_6H_6 on/in the samples. In contrast, the amount of desorbed C_6H_6 corresponding to the peaks centering at the 450 and 510 K respectively decreases with increasing Mo loading. By refering to the results reported by Otremba and Zajdel in the study of the HZSM-5 zeolite [14], these two peaks may be associated with Bronsted acid sites in two different channel systems (12 MR and 10 MR). Corma et al. in their molecular dynamic study claimed that the 10 MR sinusoidal channels of the HMCM-22 have a very slow diffusion rate for C_6H_6, whereas the channels between the 12 MR supercages are available for C_6H_6 diffusion, particularly if it is activated by a high temperature [15]. Therefore, it may be reasonable to assign the medium and high temperature peak of C_6H_6 to the C_6H_6 adspecies in the 12 MR located in different positions. The decrease in desorbed C_6H_6 in the medium and high temperature peaks with the increase of Mo loading are in good agreement with the fact that Mo species migrate into the channels and replace the protons of the Bronsted acid sites in the 12 MR. Therefore, we conclude that the Bronsted acid sites located in the 12 MR provide a key acid function for the Mo/MCM-22 catalyst.

It appears that the nature of the MDA on the Mo/MCM-22 catalyst is similar to that on the Mo/HZSM-5 catalyst. Both have an induction period that is necessary for the reduction of the Mo species into the active Mo species, and both display similar patterns recorded from NH_3-TPD, indicating that both have Bronsted acid sites. The optimum reaction temperature is about 973 K for both of the Mo/HZSM-5 and Mo/MCM-22 catalysts. The catalytic performance of the Mo/MCM-22, however, is characterized by a higher selectivity to C_6H_6 and a lower selectivity to $C_{10}H_8$ as well as a better tolerance to coke formation in comparison with the Mo/HZSM-5 catalyst under the same reaction condition. The reaction features of MDA on the Mo/MCM-22 catalyst may be attributed mainly to the unique pore systems and proper Bronsted acid strengths of the MCM-22 zeolite. Therefore, the Mo/MCM-22 catalysts also offer an opportunity to further explore and understand the nature of the MDA.

ACKNOWLEDGEMENTS

The authors are grateful to the support of the Ministry of Science & Technology of China.

REFERENCES

1. Y. Xu and L. Lin, Appl. Catal. A: 188 (1999) 53.
2. F. Solymosi, J. Cserenyi, A. Szoke, T. Bansagi and A. Oszko, J. Catal. 165 (1997) 150.
3. D. Wang, J.H. Lunsford and M.P. Rosynek, J. Catal. 169 (1997) 347.
4. R.W. Borry III, E.C. Lu, Y.-H. Kim, & E. Iglesia, Stud. Surf. Sci. Catal. **119** (1998) 403.
5. C. Zhang, S. Li, Y. Yuan, W. Zhang, T. Wu, and L. Lin, Catal. Lett. 56 (1998) 207.
6. S.L. Lawton, M.E. Leonowicz, R.D. Partridge, P. Chu, and M.K. Rubin, Micro. & Meso. Mater. 23 (1998) 109.
7. N. Kumer, L.E. Lindfors, Appl. Catal. A, 147 (1996) 175.
8. Y. Xu, Y. Shu, S. Liu, J. Huang, and X. Guo, Catal. Lett. 35 (1995) 233.
9. D. Ma, Y. Shu, X. Bao and Y. Xu, J. Catal. 189 (2000) 314.
10. S. Yuan, J. Li, Z. Hao, Z. Feng, Q. Xin, P. Ying and C. Li, Catal. Lett. 63 (1999) 73.
11. H. Jiang, L. Wang, W. Cui, and Y. Xu, Catal. Lett. 57 (1999) 95.
12. B.M. Weckhuysen, M.P. Rosynek and J.H. Lunsford, Catal. Lett. 52 (1998) 31.
13. Y. Xu, S. Liu, L. Wang, M. Xie, and X. Guo, Catal. Lett. 30 (1995) 135.
14. M. Otremba, W. Zajdel, React. Kinet. Catal. Lett. 51 (1993) 473.
15. G. Sastre, C.R.A. Catlow, A. Corma, J. Phys. Chem. B103 (1999) 5187.

Studies in Surface Science and Catalysis
J.J. Spivey, E. Iglesia and T.H. Fleisch (Editors)

The synthesis of dimethyl ether from syngas obtained by catalytic partial oxidation of methane and air

Hengyong XU, Qingjie GE, Wenzhao LI*, Shoufu HOU, Chunying YU and Meilin JIA

Dalian Institute of Chemical Physics, Chinese Academy of Sciences, Dalian, 116023, China

The synthesis of dimethyl ether (DME) from nitrogen-containing syngas produced by the catalytic partial oxidation (CPO) of methane and air has been investigated. In the first CPO step, a syngas with a 2/1 ratio of H_2/CO containing nearly 40% N_2 was obtained at 0.8MPa over Ni-based catalyst. This composition is close to thermodynamic equilibrium. In the DME synthesis step, 90% CO conversion and nearly 80% selectivity to dimethyl ether could be reached over Cu-based catalysts.

1. INTRODUCTION

At present, there are a number of obstacles to the direct conversion of methane. Thus, the indirect conversion of methane to chemicals and liquid fuels via syngas has been the subject of intense research. Because the cost of syngas production is generally about 50~60% of the total costs of converting syngas to chemicals, reducing the cost of syngas production is clearly important [1].

Steam reforming of natural gas to syngas is the conventional method for syngas production. Catalytic partial oxidation of methane (CPO) to syngas has attracted attention in recent years because the endothermic reforming reaction can be replaced by the mildly exothermic partial oxidation reaction, which has a higher reaction rate, shorter residence time, lower capital costs, and produces a H_2/CO ratio of 2/1, which is particularly suitable for F-T synthesis and methanol, dimethyl ether synthesis. However, the cost of oxygen separation, which is required to produce an undiluted syngas, can be prohibitive.

In this paper the use of air instead of pure oxygen for a syngas production by CPO is investigated. The key question is whether the syngas containing N_2 can be effectively used for the synthesis of downstream products. Because a great deal of the syngas feed needs to be recycled in methanol synthesis due to the low single-pass conversion, the presence of N_2 greatly increases the compression costs for this process, making a dilute syngas unfit for this process. According to the thermodynamic equilibrium for the syngas to DME reaction, about 90% equilibrium conversion of CO could be achieved at 4.0-5.0 MPa at 220-280°C. Experimental results have shown that a single-pass CO conversion up to 90% could be reached over many types of catalysts [2,3], meaning that the feedstock gases no longer need to be recycled. Therefore, the focus of this paper is an integration of the air CPO of methane with the synthesis of DME from syngas containing N_2 to determine whether this process offers a cheaper route for DME production.

The DME synthesis from methane consists of two major steps: First, syngas production by air CPO of methane over Ni-based catalysts; Then, DME synthesis over Cu-based catalysts from this syngas (Fig.1).

---

[1] *Corresponding author, E-mail: WZLi@ms.dicp.ac.cn, Fax: +86-411-4691570

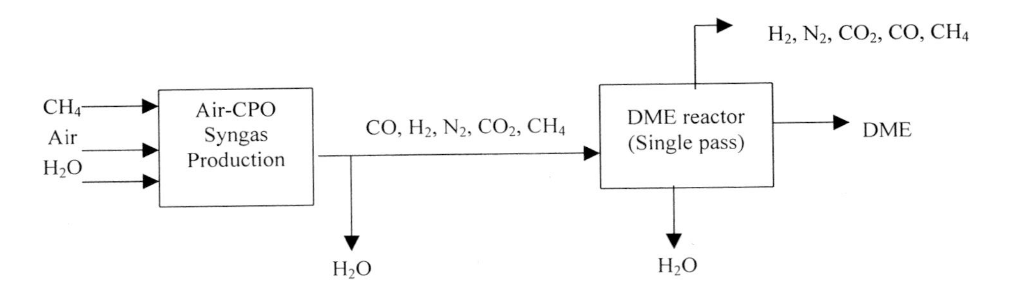

Fig. 1 The scheme of methane-air partial oxidation to DME via syngas

2. EXPERIMENTAL

The syngas production from methane, air and steam by CPO was carried out in a tubular fix-bed reactor with 5.0g catalyst. All gases must be firstly desulfurized before entering the reactor. The products were detected by an on-line GC-8A chromatography with 1.5m carbon molecular sieve column for H_2, O_2+N_2, CH_4, CO, CO_2 and 1.5m 5A molecular sieve column for H_2, O_2, N_2, CH_4, CO. The Ni-based catalysts used in activity testing are Ni-La-Mg/α-Al_2O_3. The preparation procedure is reported in reference [4].

The DME synthesis was carried out in a tubular fixed-bed reactor with 2.0g catalyst. The products were analyzed by an on-line GC-8A chromatography with TCD detector, 1m TDX-01 column for feed gases, and with FID detector, and a 2m GDX-401 column for CH_4, CH_3OH, DME and light hydrocarbons. The detailed preparation of Cu-based catalysts for activity tests is shown in reference [5].

3. RESULTS AND DISCUSSION

3.1 Syngas production

Syngas production from CH_4, air and H_2O (1: 2.4 : 0.8) over 5g Ni-based catalysts was carried out for 200 hours at 800°C, 0.8MPa, and 25000h^{-1} CH_4 space velocity. The results are listed in Table 1.

Table 1 The results of methane partial oxidation to syngas by air over Ni-based catalysts

Time hour	CH_4 conv. %	CO selec. %	Outlet dry-gas composition, %mol					R*
			CH_4	N_2	H_2	CO	CO_2	
15	91.1	71.3	1.82	38.6	41.0	13.3	5.34	1.91
50	90.2	71.6	2.00	39.0	40.5	13.3	5.26	1.90
100	90.5	71.5	1.94	38.9	40.6	13.3	5.29	1.91
150	90.2	71.2	2.01	38.8	40.6	13.2	5.34	1.90
200	90.6	70.2	1.92	38.7	40.9	13.0	5.52	1.91
T. V.**	90.5	73.9	1.98	38.9	40.3	13.9	4.90	1.88

* R= (H_2-CO_2)/(CO+CO_2) [6];

** T. V.: Theoretical calculation value.

It was shown that: (1) more than 90% CH_4 conversion and 70% CO selectivity could be obtained, which closely approaches the theoretical value of thermodynamic equilibrium; (2) no hot-spot temperature and coke formation are evident in reaction system during the 200-h test; (3) the R value was kept at 1.9, which is nearly the optimum value (2.0) for DME synthesis.

3.2 DME synthesis from N_2-containing syngas over Cu-based catalysts

A great deal of research on syngas to DME has been reported since the 1970s [7-10]. Most of the catalysts are Cu-based materials, over which 90% CO conversion and 80% DME selectivity can be obtained [11].

The syngas-to-DME reaction can be described as:

$$3CO + 3H_2 \rightarrow CH_3OCH_3 + CO_2,\ \Delta H_{600K} = -260.12\ kJ/mol \quad (a)$$

$$\text{or}\ 2CO + 4H_2 \rightarrow CH_3OCH_3 + H_2O,\ \Delta H_{600K} = -221.51\ kJ/mol \quad (b)$$

In the case of feed gases with a low H_2/CO ratio, the reaction is described by (a), the main byproduct of the reaction being CO_2. Reaction (b) shows the reaction of syngas to DME at high H_2/CO rfeed gas ratios; the main byproduct being H_2O. The choice of catalyst usually determines which reaction takes place.

3.2.1 DME synthesis over other catalysts

The feed gas for DME synthesis is as follows:H_2:CO:N_2:CO_2:CH_4=39.8:14.7:37.2:5.9:2.4, which is close to that of syngas produced by air-CPO in Table 1. The reaction conditions were 1500 h^{-1}, 8.0MPa.

Table 2. DME synthesis from syngas containing N_2 over different Cu-based catalysts

Catalysts	Temp.	Conv.	Selectivity , %					Selec.[1]	Yield
	^oC	CO%	DME	CO_2	CH_4	C_2	CH_3OH	DME%	DME%
CZZ[2]/HZSM-5	260	87.6	77.1	22.5	0.45	--	--	99.4	67.5
CZA[3]/HZSM-5	260	86.7	73.9	24.2	1.30	0.52	0.14	97.4	64.1
CZZ/HSY	270	84.1	75.1	23.6	0.93	0.30	0.11	98.3	63.2
CZZ/Al_2O_3-HZSM-5	280	86.9	75.5	23.0	1.12	0.34	0.13	97.9	65.6

Note: 1. DME selectivity in organic products; 2. CZZ: Cu-ZnO-ZrO_2; 3. CZA: Cu-ZnO-Al_2O_3.

It is seen from Table 2 that in the presence of nearly 40% N_2, DME synthesis takes place over all the catalysts studied here. Cu-ZnO-ZrO_2/HZSM-5 shows a higher catalytic activity: 87.6% CO conversion and 67.5% DME yield. DME selectivity to all organic products is over 99%. A slightly lower activity was observed over Cu-ZnO-ZrO_2/Al_2O_3-HZSM-5, but this catalyst has a higher mechanical strength.

3.2.2 Impact of N_2 content

It is clear that large concentrations of N_2 have a negative effect. The effects of N_2 content on the performance of DME synthesis have been investigated under 240°C, 6.0MPa, 1000h^{-1} over Cu-ZnO-ZrO_2/ HZSM-5 catalysts (Fig. 2). The ratio of $H_2/CO/CO_2/CH_4$ in the feed gas was kept at a constant value of 3.0/1.0/0.47/0.28, respectively, except that the N_2

content varied in the range of 10%-80%. The results show that CO conversion decreases slowly with increasing N_2 content. N_2 feed gas concentrations of more than 40% (i.e. the N_2 content in CPO products) led to a remarkable decrease of CO conversion , whereas the selectivity of DME was almost unchanged with increasing concentrations of N_2.

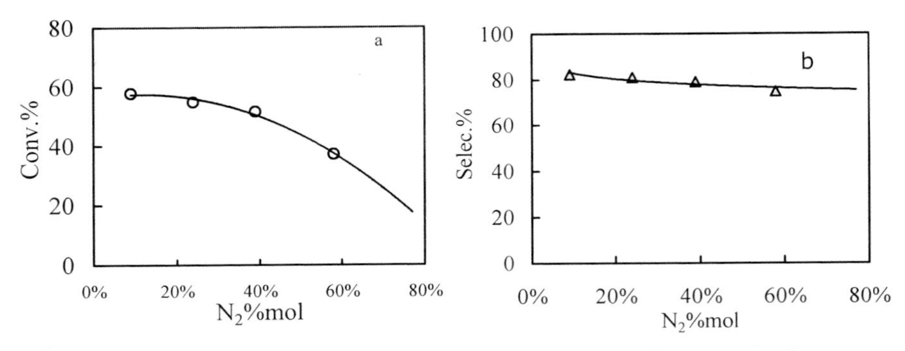

Fig.2 Influence of N_2 content in feed gases on the DME synthesis

3.2.3 Effect of reaction pressure

The presence of N_2 decreases the partial pressure of the reactants ($CO + H_2$). Generally, in order to maintain high enough effective partial pressure of ($CO+H_2$), the total reaction pressure must be increased. However, this leads to an increase in the compression costs. The effect of reaction pressure on DME synthesis was thus studied (Table 3). It was shown that at a total reaction pressure of 4.5MPa, the presence of N_2 decreases the partial pressure of ($CO+H_2$) from 3.9MPa to 2.4MPa, which decreases the CO conversion. If the pressure is increased from 4.5MPa to 6.0MPa to maintain the partial pressure of ($CO+H_2$) at 3.3MPa (i.e., the ($CO+H_2$)/N_2 partial pressure ratio is kept constant), the CO conversion is nearly the same as for syngas without N_2 at 4.5MPa. Preliminary results have shown that it would be possible to run the DME synthesis using N_2-containing syngas at conventional pressures (4.5-5.0MPa) with a high CO conversion if the catalyst would be improved further. Work toward this goal is underway.

Table 3 Effect of reaction pressure on DME synthesis

P_{Total} (MPa)	P_{CO+H2} (MPa)	P_{N2} (MPa)	P_{CO+H2}/P_{N2}	Conversion CO %	Selectivity / mol%				Yield DME%
					CO_2	C_2	CH_3OH	DME	
4.5	3.9	0	---	87.8	16.5	0.12	0.01	83.4	73.3
4.5	2.5	1.7	1.47	82.1	14.8	---	---	85.2	70.0
6.0	3.3	2.3	1.43	87.2	16.3	---	---	83.7	73.0
7.5	4.1	2.8	1.46	91.7	14.9	---	---	85.1	78.0

Reaction temperature: T= 260°C, GHSV=1000 h^{-1}; Catalyst: Cu/ZnO//ZrO_2/ZSM-5.
Feed gas composition (%mol): $41.0H_2/13.4CO/37.7N_2/5.6CO_2/2.3CH_4$.

3.2.4 Effect of space velocity

Effect of gas hour space velocity (GHSV) on the reaction is shown in Table 4. The increasing of GHSV does not increase the hydrogenation reaction, but would probably increase the DME space-time yield. It was shown that at $500h^{-1}$, 92.9% CO single-pass conversion, 85.8% DME selectivity, and nearly 80% carbon utilization could be obtained. To fully utilize the carbon resource in the feedstock, the GHSV used should be <1500 h^{-1}.

Table 4 Effect of GHSV on reaction properties over Cu-ZnO-ZrO_2 catalysts

GHSV (h^{-1})	Conversion CO %	Selectivity / mol%					Yield DME%
		CO_2	CH_4	C_2	CH_3OH	DME	
500	92.9	14.1	—	0.11	0.03	85.8	79.7
1500	87.3	16.6	—	—	0.02	83.4	72.7
2000	78.0	20.8	—	—	0.02	79.2	61.8
2500	66.5	22.6	0.60	—	0.03	76.8	51.1
3000	60.2	24.3	0.48	—	0.03	75.2	45.2

Reaction conditions: 260°C, 7.5MPa, Feed gas: $41.0H_2/13.4CO/37.7N_2/5.6CO_2/2.3CH_4$;

3.2.5 Effect of temperature

Influence of temperature on Cu-ZnO-ZrO_2/Al_2O_3-HZSM-5 is shown in Fig.3. The results show that in the range of 220°C~300°C, increasing, CO conversion first increases with reaction temperature, then decreases, with a maximum at 260°C. This maximum can be

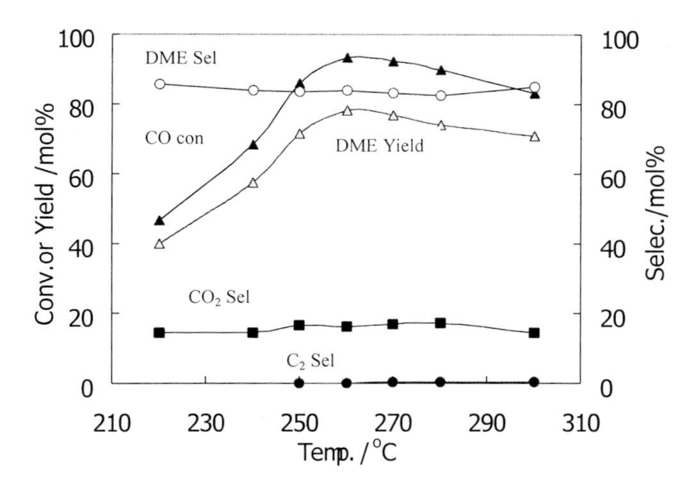

explained by

Fig. 3 Effect of temperature on the reaction of syngas containing N_2 to DME over Cu-ZnO-ZrO_2/HZSM-5 (7.5MPa, 1000 h^{-1}, Feed gas: $41.0H_2/13.4CO/37.7N_2/5.6CO_2/2.3CH_4$)

the relationship between the kinetics and thermodynamics. At lower temperatures, the reaction is controlled by kinetics and the reaction rate is strongly dependent on temperature. At higher temperatures, the reaction thermodynamics are not favorable, and the conversion is thermodynamically limited. In several tests between 260 and 290°C, the CO conversion increased reversibly when the reaction temperature decreased. This is consistent with a thermodynamic limitation. However, sintering of the active sites of the catalyst cannot be excluded as a possible reason for the loss of catalyst activity at high temperatures. No change in DME selectivity was found over this temperature range. Above 270°C, C_2 selectivity increased slightly due to the dehydration of DME.

4. CONCLUSION

For CPO of methane with air to syngas, more than 90% CH_4 conversion and a syngas with a 2/1 ratio of H_2/CO could be obtained at 0.8MPa over a Ni-based catalyst over a 200-h stability test.

The synthesis of DME using the N_2-containing syngas produced by the air-CPO process indicated that 90% CO conversion and nearly 80% dimethyl ether selectivity over Cu-ZnO-ZrO_2/HZSM-5 catalyst could be obtained.

The integration of air CPO with DME synthesis could serve as an option for DME production, although further work is needed.

ACKNOWLEDGEMENTS

The authors are grateful to Ministry of Chinese Science and Technology for its financial support.

REFERENCES

1. R. Lu, Petrochem. Ind. Trends (China), 7(6)(1999)53.
2. C. Sun, G. Cai, L. Yi, CN 1087033A (1994).
3. Y. Huang, Q. Ge, S. Li, CN 1153080A(1997).
4. Y. Chen, L. Cao, W. Li, CN 1154944A (1997).
5. Q. Ge, Y. Huang, S. Li, Chem. Lett., (1998) 209.
6. G. L. Farina, E. Supp, Hydro. Proc., 71(3)(1992)77.
7. G. Cai, Z. Liu, R. Shi, Appl. Catal. A., 125(1995)29.
8. Q. Ge, Y. Huang, F. Qiu, S. Li., Appl. Catal. 167(1998)23.
9. D. M. Brown, Catal Today., 8(1991)279.
10. G. Sam, Chem. Eng. 102(1995)17.
11. Q. Ge, Y. Huang, F. Qiu, S. Li., React. Kinet. & Catal. Lett., 63(1998)137.

Studies in Surface Science and Catalysis
J.J. Spivey, E. Iglesia and T.H. Fleisch (Editors)

Development of the High Pressure ITM Syngas Process

M.F. Carolan[a], P.N. Dyer[a,*], E. Minford[a], T.F. Barton[b], D.R. Peterson[b], A.F. Sammells[b], D.L. Butt[c,+], R.A. Cutler[c] and D.M. Taylor[c].

[a] Air Products and Chemicals Inc., 7201 Hamilton Boulevard, Allentown, PA 18195-1501
[b] Eltron Research Inc., 4600 Nautilus Court South, Boulder, CO 80301-3241
[c] Ceramatec Inc., 2425 South 900 West, Salt Lake City, UT 845119

The ITM Syngas Team led by Air Products and Chemicals and including Ceramatec, Chevron, Eltron Research, McDermott, Norsk Hydro and other partners, in collaboration with the U.S. Department of Energy (DOE), is developing ceramic Ion Transport Membrane (ITM) technology for the production of synthesis gas and hydrogen from natural gas. The ITM Syngas technology is in Phase 2 and the third year of a co-funded $90 MM, three phase, eight year development program.

Laboratory test reactors have been constructed at Air Products and Eltron Research to develop and test ITM Syngas membranes under conditions that simulate commercial processes. Test sample membranes and seals were fabricated and supplied by Ceramatec.

This paper describes the laboratory demonstration of the high pressure ITM Syngas process, and the significance of the results to the development and scale-up of the process.

1. THE ITM SYNGAS PROCESS

1.1 Introduction to ITM Syngas membranes

ITM membranes are fabricated from non-porous, multi-component metallic oxides that operate at high temperatures and have exceptionally high oxygen flux and selectivity. The ITM Syngas process uses ITM membranes to combine air separation and high-temperature syngas generation processes into a single compact ceramic membrane reactor, with the potential for significant cost reductions compared with conventional technology.

A conceptualization of the ITM Syngas process is illustrated in Figure 1. The membrane structure incorporates both the non-porous ITM and reduction and reforming catalyst layers. The mixed conducting membrane material must show long-term stability in reducing and oxidizing atmospheres, and long-term compatibility with the reduction and reforming catalysts. In addition, the ceramic membrane should exhibit a low chemical expansion effect to enhance operational lifetime. This effect, a vacancy concentration dependant expansion in turn dependant upon the oxygen partial pressure gradient, is pronounced in some types of ITM and can lead to significant stress gradients in the membrane.

* Corresponding author. Tel.: +1-610-481-5778, E-mail address: dyerpn@apci.com (P.N. Dyer)
+ Present address: Dept. of Materials Science, University of Florida, Gainesville, FL 32611-6400

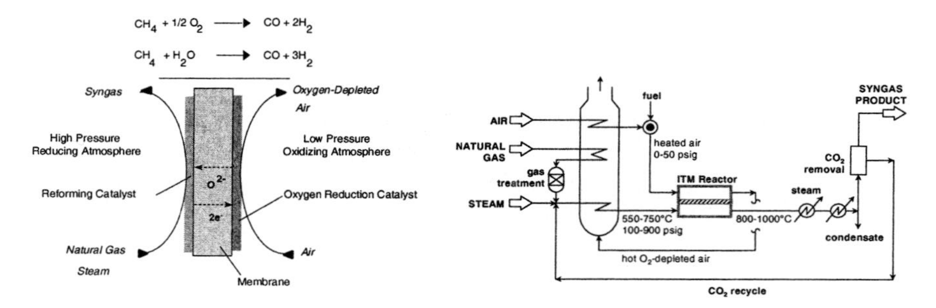

Fig. 1. The ITM Syngas process.

Fig. 2. Conceptual ITM Syngas process design

A major objective in Phase 1 was to identify suitable stable ITM materials and seals, and determine their performance under high-pressure process conditions.

1.2 Process engineering and economic evaluation

Eltron Research originally demonstrated the continuous partial oxidation of methane mixtures for over a year at atmospheric pressure [1,2]. The next step in materials development was to define the ITM Syngas process conditions, and set targets for the ITM material properties and performance. Air Products, Chevron, McDermott Technology and Norsk Hydro developed ITM Syngas processes to produce syngas with a H2/CO ratio of about 2, suitable for GTL applications [3-6]. Preliminary process designs and cost estimates were completed in Phase 1. For an offshore GTL plant processing 50 MMSCFD of associated gas and producing approximately 6000 BPD of syncrude, capital cost savings of greater than 33% were predicted for ITM Syngas when compared to conventional O2-blown ATR/ASU technology for syngas production. A conceptual process is illustrated in Figure 2.

The ITM Syngas process places severe requirements on the stability and performance of the ITM materials, with exposure to high partial pressures of CO_2 and H_2O and low partial pressures of oxygen. The typical process conditions and gas compositions experienced by the ITM membrane in the process illustrated in Figure 2 are listed in Table 1.

Table 1
Process conditions for ITM Syngas

Process gas composition	Equilibrium PO_2 (bar)	Temperature (°C)	Pressure (bar)
60 - 75 % H_2, CO, CH_4 40 - 25 % CO_2, H_2O	10^{-19} - 10^{-15}	750 - 950	18 - 31

1.3 Materials Development

Air Products, Ceramatec and Eltron Research carried out the development of stable ITM Syngas material compositions with input from Norsk Hydro. Thermodynamic predictions of stability under ITM Syngas process conditions were used to make initial selections of the compositions. These predictions were then confirmed by experimental determination of stability under atmospheres simulating the process environment. Selected compositions were synthesized in small batches at Ceramatec to produce samples for the measurement of ionic conductivity by oxygen permeation, syngas generation at atmospheric pressure, mechanical properties, and chemical and thermal expansion. Promising compositions were further scaled up at Ceramatec to produce samples for syngas generation at high pressure.

The development centered around two types of material composition, designated I2 and I4. Two important properties affecting membrane thickness and stability, ambipolar conductivity and isothermal chemical expansion, are shown in Table 2 for three compositions.

Table 2
Ambipolar conductivity and chemical expansion of I2 and I4 compositions

Composition	σ_{amb}(S/cm)	Final P_{O2} (atm)	Expansion (ppm)
I2 (X)	0.125	2.8 x10^{-13}	3918
I4 (A)	0.060	5.5 x10^{-21}	4060
I4 (C)	0.042	5.5 x10^{-21}	1820

The apparent ambipolar conductivity (σ_{amb}) was determined from the Wagner equation by measuring oxygen permeation through ~500 μm thick disc samples at 950 - 1000°C, using an atmospheric pressure air feed and a helium permeate sweep [7]. The isothermal chemical expansion of the materials was measured by first equilibrating samples in a dilatometer in 1 atm O_2 at 750°C, and secondly in a $H_2/H_2O/N_2$ mixture with the final oxygen partial pressure listed in the table. A chemical expansion of >3000 ppm results in a low predicted reliability in syngas service due to the mechanical stress caused in the membrane by the imposed oxygen partial pressure gradient. While I2 compositions have higher conductivity, enabling their use as membranes approximately 1 mm thick, they also show high chemical expansion effects. The less conductive I4 materials are subject to lower expansion stress and are more stable, but require a thinner membrane (< 500 μm) to achieve a useful flux. This thickness is within the range achievable using the process technology for fabricating supported thin membranes developed by Ceramatec. I4 (C) gave the best overall combination of properties consistent with the predicted requirements for use in syngas generation, and was therefore selected for further testing and scale-up.

2. LABORATORY TESTING AT HIGH PRESSURE

2.1 Laboratory-scale high pressure test reactors

Ceramatec fabricated closed-end tubular membranes from the I4 materials for testing at high pressures. The membranes were sealed into superalloy holders using double U-ring ceramic/metal seal assemblies developed by Air Products and Ceramatec [8]. High pressure, laboratory scale, test reactors were designed and installed at Eltron Research and Air Products. These are illustrated diagrammatically in Figure 3.

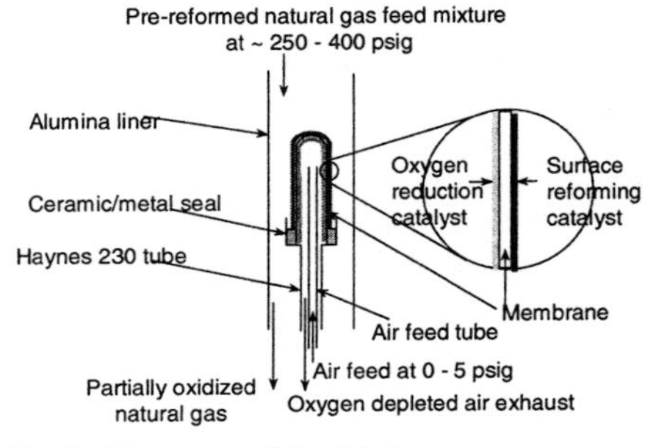

Fig. 3. Diagram of the high pressure ITM Syngas test apparatus.

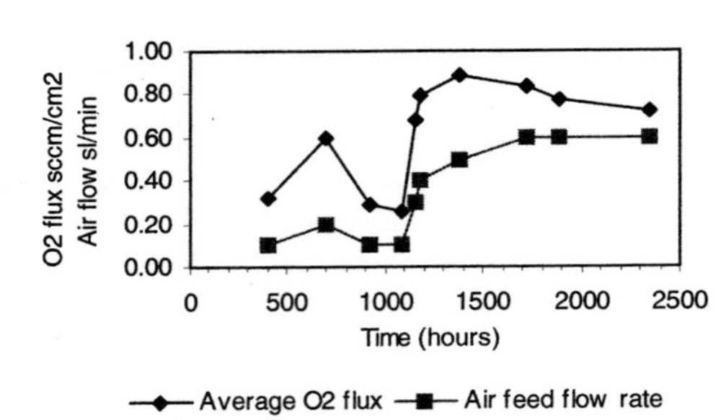

Fig. 4. 100-day high pressure test of an I4 (A) tubular membrane.

Air at atmospheric pressure was introduced to the interior of the tubular membrane, while a $CH_4/H_2O/CO/H_2/CO_2$ gas mixture representative of a section of the ITM Syngas reactor was contacted at pressure with the external surface. Oxygen was transported through the tubular membrane to partially oxidize the reducing gas mixture. The oxygen concentration of the exhaust air was measured by an oxygen analyzer and by GC. The exhaust, partially oxidized, high-pressure syngas mixture was analyzed by GC. Carbon, hydrogen and oxygen material balances were obtained for both the process gas and the air streams. Oxygen fluxes were calculated on both the air and process sides of the membrane, and compared for an overall oxygen material balance.

2.2 Eltron Research, 300 psig test reactors

Multiple tubular membrane/seal assemblies, fabricated by Ceramatec, were tested at Eltron Research at up to 300 psig of differential pressure. The longest high-pressure test under syngas was over 100 days, and the performance of this test is illustrated in Figure 4. The I4 (A) tubular membrane and seal operated with continuous oxidation of a reducing gas mixture for > 2400 hours. The test was conducted at 825°C, with a pressure differential of 250 psi applied across the ~ 1mm thick membrane. The U-ring ceramic/metal seal assembly was also exposed to these conditions over this time period, and no leakage was detected. As shown in Figure 4, changes in the air flow delivered at atmospheric pressure to the inside of the tube had a proportional effect on the oxygen flux, indicating the presence of limitations in this test due to mass transfer and mixing.

2.3 Air Products, 500 psig test reactors

Figure 5 illustrates the results of a 50 day test at Air Products of an I4 (C) tubular membrane/ seal assembly fabricated by Ceramatec. The oxygen flux measured by depletion on the air side of the membrane was constant at a given temperature and, within the limits of experimental error, was in agreement with the oxygen flux determined by GC analysis of the inlet and exhaust process gas streams. Carbon and hydrogen balances were within ± 10%.

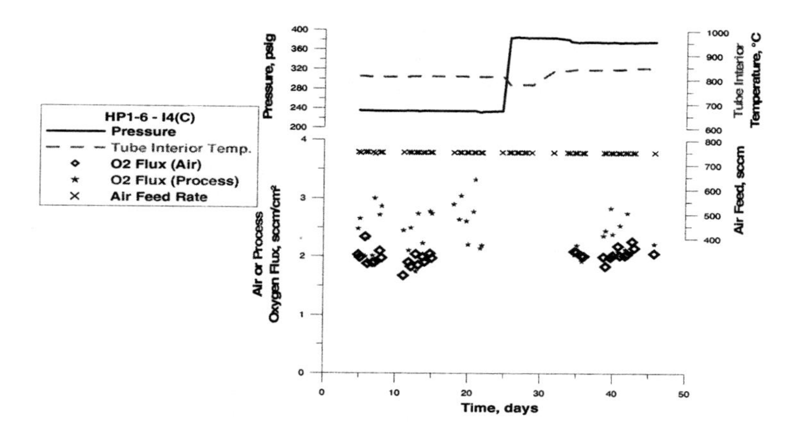

Fig. 5. High pressure test of an I4 (C) tubular membrane – air and process side oxygen flux

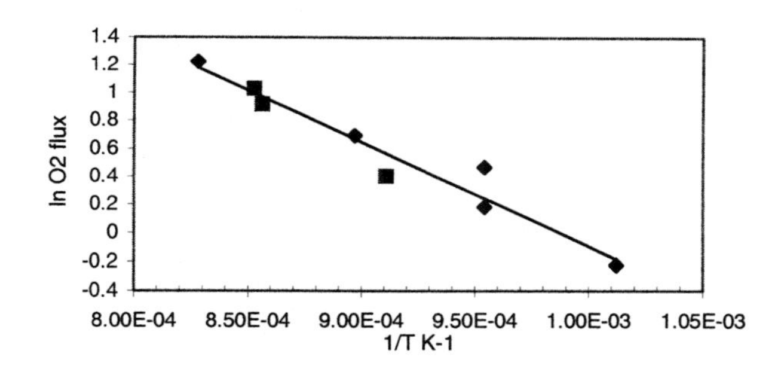

Process pressure: ■ 250 psig, ♦ 400 psig,

Fig. 6. Variation of oxygen flux of I4 (C) with temperature

The results of high-pressure tests of I4 (C) tube/seal assemblies tested at both Eltron Research at 250 psig, and at Air Products at 400 psig, are plotted as a function of temperature in Figure 6. The flux data (scm^3 O_2/cm^2/min) was obtained from long-term tests of up to 520 hours at each temperature using tubular membranes of I4 (C) under reducing gas mixtures. The activation energy is E_{act} = 62.5 kJ/mol. This series of experiments, carried out under the conditions predicted to exist in the scaled-up ITM Syngas reactor, confirmed the stability and performance of the I4 (C) composition, and the material was therefore selected for further scale-up. Ceramatec fabricated supported thin film I4 (C) membrane test samples, and initial tests of the performance of these membranes under low pressure syngas process conditions by Eltron Research have demonstrated oxygen flux values in the range of 8 -12 sccm/cm^2.

3. CONCLUSIONS

There are multiple technical hurdles to be overcome before the large potential economic and environmental benefits of ITM technology can be realized. The initial results of this development program, however, are very promising with the demonstration of the performance and stability of the selected ITM material and seal system under high-pressure laboratory process conditions.

Acknowledgments

The authors gratefully acknowledge the work of C.M Chen, J. Inga, M. Watson and other co-workers at Air Products, and the work of S.W. Rynders, M. Wilson and other co-workers at Ceramatec. Business Development and management support from D. L. Bennett and E. P. Foster are also acknowledged. The co-funding of the program by the U.S. DOE (DE-FC26-97FT96052) is also gratefully acknowledged.

References

[1] M. Schwartz, J.H. White, A. F. Sammells, "Solid state oxygen anion and electron mediating membrane and catalytic membrane reactors containing them", US Patent 6033632 (2000).
[2] A.F. Sammells, M. Schwartz, R.A. Mackay, T.F. Barton, D.R. Peterson, Catalysis Today, 56, 325 (2000).
[3] S. Nataraj, R.B. Moore, S.L. Russek, "Production of syngas by mixed conducting membranes", US Patent 6048472 (2000).
[4] S. Nataraj, S.L. Russek, "Synthesis gas production by ion transport membranes", US Patent 6077323 (2000).
[5] S. Nataraj, S.L. Russek, "Utilization of synthesis gas produced by mixed conducting membranes", US Patent 6110979 (2000).
[6] S. Nataraj, P.N. Dyer, S.L. Russek, "Syngas production by mixed conducting membranes with integrated conversion into liquid products", US Patent 6114400 (2000).
[7] H.J.M. Bouwmeester and A.J Burggraaf, "Fundamentals of Inorganic Membrane Science and Technology", Elsevier Science, chapter 10, page 451 (1996).
[8] E. Minford, S.W. Rynders, D.M. Taylor, R.E. Tressler, Europ. Patent 1067320 A2 (2001).

Studies in Surface Science and Catalysis
J.J. Spivey, E. Iglesia and T.H. Fleisch (Editors)

An Integrated ITM Syngas/Fischer-Tropsch Process for GTL Conversion

C.M. Chen[a,*], P.N. Dyer[a], K.F. Gerdes[b], C.M. Lowe[b], S.R. Akhave[c], D.R. Rowley[d], K.I. Åsen[e] and E.H. Eriksen[e]

[a] Air Products and Chemicals Inc., 7201 Hamilton Boulevard, Allentown, PA 18195 USA
[b] Chevron Research and Technology Co., 100 Chevron Way, Richmond, CA 94802 USA
[c] J. Ray McDermott Engineering, LLC, 801 North Eldridge Street, Houston, TX 77079 USA
[d] McDermott Technology Inc., 1562 Beeson Street, Alliance, OH 44601 USA
[e] Norsk Hydro ASA, P.O. Box 2560, N-3901 Porsgrunn, Norway

The ITM Syngas Team led by Air Products and Chemicals and including Ceramatec, Chevron, Eltron Research, McDermott, Norsk Hydro and other partners, in collaboration with the U.S. Department of Energy (DOE), is developing ceramic Ion Transport Membrane (ITM) technology for the production of synthesis gas and hydrogen from natural gas. The ITM Syngas technology is in the third year and Phase 2 of a co-funded $90 MM, eight year, three phase development program.

The ITM Syngas process combines air separation and high-temperature syngas generation processes into a single compact ceramic membrane reactor. The ITM Syngas process offers advantages in operation, efficiency, and integration with downstream Fischer-Tropsch processes. The ITM Syngas technology also offers significant capital cost savings compared to conventional technology. This paper describes the significant features of the ITM Syngas process and their advantages.

1. INTRODUCTION

1.1 Introduction to ITM Syngas membranes

ITM membranes are fabricated from non-porous, multi-component metallic oxides that operate at high temperatures and have exceptionally high oxygen flux and selectivity. A conceptualization of the ITM Syngas process is illustrated in Figure 1. The membrane structure incorporates both the non-porous ITM and reduction and reforming catalyst layers. The mixed conducting membrane material must show long-term stability in reducing and oxidizing atmospheres, and long-term compatibility with the reduction and reforming catalysts.

One of the conventional syngas generation technologies typically considered for Fischer-Tropsch (FT) GTL processes is an oxygen-blown Autothermal Reformer (ATR), which converts natural gas by partial oxidation to syngas with the appropriate 2:1 H_2:CO molar ratio. Oxygen-blown ATRs require a high-purity, high-pressure O_2 feed, which is usually supplied by a separate cryogenic Air Separation Unit (ASU).

* Corresponding author. Tel.: +1-610-481-3315, E-mail address: chencm@apci.com (C.M. Chen)

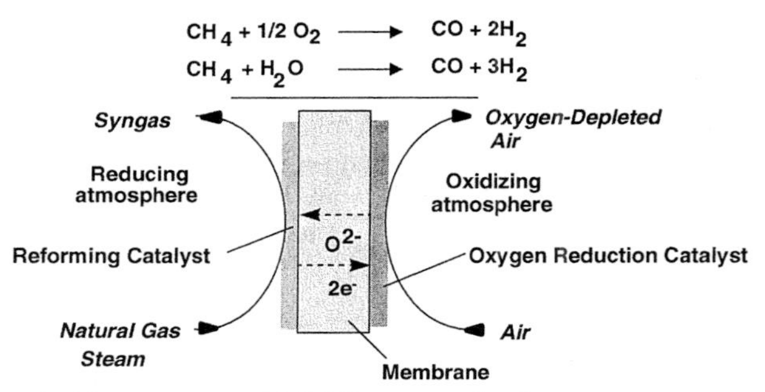

Figure 1. The conceptual ITM Syngas technology.

The ITM Syngas process combines the separate processes for air separation and high-temperature syngas generation into a single process where oxygen is separated and reacted with a hydrocarbon feed to produce syngas in a single unit operation. This combination of processes generates the potential for significant reductions in capital cost and in space requirement.

1.2 Development of a high pressure ITM Syngas process

Several authors have described previous laboratory demonstrations of the oxidative reforming of methane to syngas using mixed conducting membranes at atmospheric pressure and mainly isothermal conditions [1-5]. However, no work has been previously published on the design details of an industrial process based on the ITM technology to produce syngas from remote associated gas. A recent report [6] included information from the Alaska Dept. of Revenue on six different GTL process technologies, including a ceramic membrane reactor-based technology. However, no details on the ceramic membrane reactor or process were provided. To take full advantage of the potential economic benefits of the ITM Syngas technology, a commercial process must (a) operate without air compression other than that required to overcome pressure drop, and (b) maintain the syngas product at a pressure suitable for downstream processing. Additional advantages accrue from process integration. The result is a process in which the membrane is subjected to a large pressure differential, and is exposed to high-pressure oxidative reforming conditions. This paper describes work to design the high pressure ITM Syngas process, and define the pressure, temperature and gas composition to provide targets for the development of suitable ITM materials [7].

2. PROCESS DESIGN

2.1 ITM Syngas Process

The ITM Syngas process is shown in Figure 2 [8-11]. Hydrocarbon feedstocks are converted into syngas in a two-stage process comprising an initial steam reforming step followed by final conversion to syngas in an ITM Syngas reactor. As shown in Figure 2, natural gas is preheated in a heat recovery duct and treated in a catalytic hydrogenation and desulfurization reactor to hydrogenate olefins and remove sulfur compounds. The natural gas is typically provided at pressures of 200 to 500 psig. Steam is introduced into the treated feed. The steam-hydrocarbon feed is heated in a heat recovery duct to a temperature of about 700 to 1000 °F and the heated feed is introduced into an adiabatic reformer reactor (prereformer).

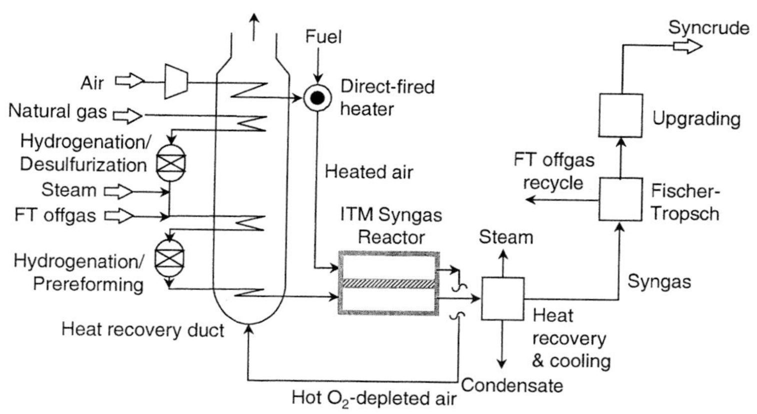

Figure 2. ITM Syngas process with adiabatic prereformer.

The prereforming step converts a portion of the methane and essentially all of the hydrocarbons heavier than methane into hydrogen, carbon monoxide, carbon dioxide, and methane. The partially reformed feed is further heated to 1100 to 1400 °F. The prereforming step minimizes processing problems, such as carbon formation, in the downstream heat exchanger and the ITM Syngas reactor.

Air is pressurized to about 10 psig in a blower, preheated in a heat recovery duct, and further heated, if necessary, by direct combustion to within ±200 °F of the temperature of the partially reformed feed gas. In the ITM reactor, oxygen selectively permeates from the oxidant side to the reactant side, and the prereformed feed gas undergoes reactions including partial and complete oxidation, steam reforming, carbon dioxide reforming, and water gas shift. The overall reactions are net exothermic and the reactant gas temperature increases.

Raw syngas product is withdrawn from the ITM Syngas reactor at a temperature of greater than about 1500 °F. The raw syngas product is rapidly cooled to a temperature below 800 °F against boiling water by indirect heat transfer and can be further cooled against other process streams. The final syngas product can be converted into liquid products such as liquid hydrocarbons by a Fischer-Tropsch reactor system. For FT conversion, compression of the product syngas is not normally needed since the ITM Syngas reactor operating pressure is matched to that needed in the FT reactor.

The non-permeate oxygen-depleted air is withdrawn from the ITM Syngas reactor at a temperature near the raw syngas product and contains about 1 to 5 vol% oxygen. The oxygen-depleted air heats the process feed stream in the heat recovery duct, which is similar to a conventional flue gas duct as used in steam-methane reformers.

2.2 Process Advantages

Elimination of the ASU and Operation with Low Pressure Air

Because the ITM Syngas process combines oxygen separation with oxidative reforming of natural gas, it eliminates the need for a separate ASU to supply high pressure oxygen. The ASU-based process has a considerably higher power requirement than the ITM Syngas process because power is required to compress the ASU air feed to approximately 100 psig or higher and to further compress the product O_2 to between 300 to 500 psig. In contrast, the air feed for the ITM Syngas process needs only to be compressed to about 10 psig, sufficient to overcome the pressure drop through the ITM Syngas reactor and associated piping and equipment.

Use of Prereforming

The prereforming step alleviates the problem of heavier hydrocarbon decomposition and coking at the operating temperatures of the ITM Syngas reactor. Other types of reformers can also be used for prereforming, such as Steam Methane Reformers (SMR) or Gas Heated Reformers (GHR). The prereforming and ITM Syngas reactors are heat-integrated for maximum operating efficiency. This two stage process allows flexibility in optimizing operation, cost, and syngas compositions suitable for a variety of final products, such as Fischer-Tropsch products, methanol, or hydrogen.

Non-isothermal Reactor Operation

One of the aspects of the ITM Syngas reactor's non-isothermal operation is that the syngas exit temperature can be within the preferred range of over 1500 °F, while the cooler prereformed feed inlet temperature brings advantages in operation and equipment design. Cooler ITM Syngas reactor inlet temperatures help to minimize carbon formation in the inlet piping and also permit a wider choice of piping materials, such as lower-cost alloys or non-refractory-lined piping.

2.3 Integration with Fischer-Tropsch GTL

The CO_2-containing offgas from a downstream Fischer-Tropsch (FT) GTL plant may be recycled to the feed of the ITM Syngas process to adjust the product syngas H_2/CO molar ratio to about 2, suitable for FT GTL. This avoids an acid gas removal system to recycle CO_2; and instead the FT offgas containing CO_2 is recovered at high pressure along with other useful components to be recycled, such as light hydrocarbons, hydrogen, and carbon oxides. This also serves to reduce emissions of CO_2 while increasing the efficiency of carbon conversion into useful products. Only a modest amount of compression is needed to overcome pressure drop through the ITM Syngas process and the FT process.

3. ECONOMIC ANALYSIS

An ITM Syngas process design and cost estimate were developed for an offshore GTL facility which processes 55 million SCFD of associated gas. The syngas is provided to the FT process at 335 psig without additional compression and with a H2 to CO molar ratio of about 2. The FT GTL process produces approximately 6,000 barrel per day (BPD) of syncrude.

Equipment costs were developed by McDermott for a GTL plant on a Floating Production, Storage and Offloading (FPSO) ship with a U.S. Gulf Coast location. The economic evaluation compared an ITM Syngas process to an oxygen-blown ATR process that included a cryogenic ASU. The ITM Syngas process used a steam-to-carbon (S/C) ratio of 1.5, and the ATR process was evaluated with a S/C ratio of 1.4. Electric power for the ITM and ATR processes was provided by steam turbines using high-pressure steam from the ITM process and medium-pressure steam from the FT process.

The benefits for ITM Syngas compared to O2-blown ATR are summarized in Table 1. The higher power requirement for the ATR/ASU process shifts the power demand balance from a balanced steam situation for ITM Syngas to a steam-deficient situation for O_2-blown ATR, requiring an additional boiler. The costs of the additional boiler and additional steam turbine generator capacity for the ATR process contribute to the ITM Syngas cost savings.

Table 1. Comparison of offshore ITM Syngas process and O2-blown ATR process with ASU (55 million SCFD natural gas and approximately 6000 BPD FT liquids)

	Benefit of ITM Syngas process
Capital cost of syngas process and power generation equipment	33 % lower for ITM Syngas
Overall GTL plant electric power required	up to 40 % lower for ITM Syngas
Deck space	up to 40 % lower for ITM Syngas

The additional power requirement for the ATR-based GTL plant reduces the overall fuel efficiency to 53 %, compared to 58 % for the ITM-based GTL plant.

Preliminary single-deck layouts were developed for the ITM Syngas process and the ATR/ASU process on an FPSO. The ITM Syngas process area, including steam turbines but excluding other utilities, required up to 40% less deck area than the ATR process with ASU. The reduced deck space required by the ITM Syngas process is expected to result in further cost savings.

Capital costs of the syngas generation plant, including equipment for power generation, were compared for the ITM Syngas process and the ATR process with ASU. The capital cost savings are shown in Figure 3 with sensitivity to ITM reactor cost. For the current estimated ITM Syngas reactor cost, a savings of 33% in capital cost is predicted. Overall cost savings are relatively insensitive to ITM reactor cost: a 20% change in ITM reactor installed costs translates to only about a 2% change in capital cost savings. The significant cost savings in the ITM Syngas process are due to combining the ATR and ASU into a single unit operation and the resulting reduction in the power generation equipment size. Approximately 60-70% of the cost savings for ITM are in the feed, reactor, and oxygen separation section, while 30-40% of the savings are in the heat recovery, steam, and power generation section.

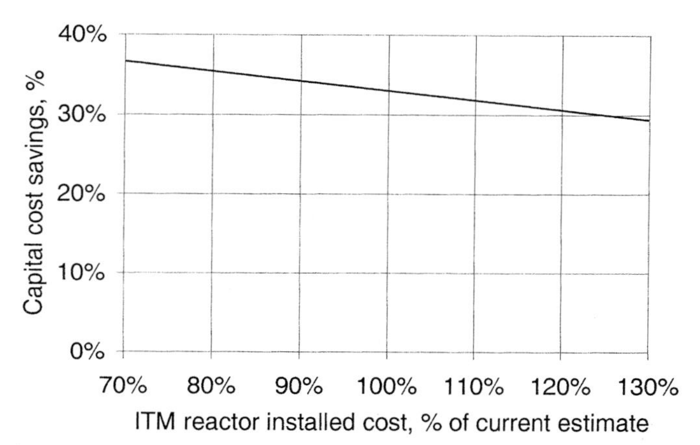

Figure 3. Capital cost savings for syngas process and power generation equipment: ITM Syngas compared to oxygen-blown ATR with ASU

The offshore ITM Syngas process design is being scaled up to a landbased ITM Syngas GTL plant that processes 500 MMSCFD associated gas and produces approximately 55,000 BPD of syncrude.

4. CONCLUSIONS

The ITM Syngas process offers numerous advantages in capital cost, efficiency, operation, and integration with downstream Fischer-Tropsch processes. The process is currently being scaled up to a Process Development Unit with an equivalent syngas throughput of 24,000 SCFD. Several GTL projects using current commercial technology are being established or evaluated. Technological improvements such as ITM Syngas will, in the future, broaden the applicability of GTL to additional opportunities for remote gas conversion.

REFERENCES

1. A.F. Sammells, M. Schwartz, R.A. Mackay, T.F. Barton, D.R. Peterson, Catalysis Today, 56, 325 (2000).
2. C.A. Udovich, Studies in Surface Science and Catalysis, 119, 417 (1998).
3. T.J. Mazanec, Solid State Ionics, 70/71, 11 (1994).
4. C.Y. Tsai, Y.H. Ma, W.R. Moser, A.G. Dixon, Chem. Eng. Comm., 143, 107 (1995).
5. U. Balachandran et al., Solid State Ionics, 108, 363 (1998).
6. E.P. Robertson, "Options for Gas-to-Liquids Technology in Alaska," INEEL/EXT-99-01023 (December 1999).
7. M. Carolan et al., "Development of the high pressure ITM Syngas process," Ibid (2001).
8. S.Nataraj, R.B.Moore, S.L.Russek, "Production of synthesis gas by mixed conducting membranes," U.S. Patent 6048472 (2000).
9. S.Nataraj, S.L.Russek, "Synthesis gas production by ion transport membranes," U.S. Patent 6077323 (2000).
10. S.Nataraj, S.L.Russek, "Utilization of synthesis gas produced by mixed conducting membranes," U.S. Patent 6110979 (2000).
11. S.Nataraj, P.N.Dyer, S.L.Russek, "Synthesis gas production by mixed conducting membranes with integrated conversion into liquid products," U.S.Patent 6114400 (2000).

ACKNOWLEDGEMENTS

The authors gratefully acknowledge the work of M.F. Carolan, J.R. Inga, E. Minford, S. Nataraj, S.L. Russek, T.R.Tsao, M.J. Watson, and other co-workers at Air Products, the work of M.J. Holmes and G.J. Montgomery at McDermott, and the work of N.H. Eldrup and K. Wilhelmsen at Norsk Hydro. Business development and management support from D.L. Bennett and E.P. Foster are also acknowledged. The co-funding of the program by the U.S. DOE (DE-FC26-97FT96052) is also gratefully acknowledged.

Studies in Surface Science and Catalysis
J.J. Spivey, E. Iglesia and T.H. Fleisch (Editors)

LiLaNiO/γ-Al_2O_3 catalyst for syngas obtainment by simultaneous catalytic reaction of alkanes with carbon dioxide and oxygen

Shenglin Liu [a*], Longya Xu [b], Sujuan Xie [a], Qingxia Wang [a] and Guoxing Xiong [b]

a. Laboratory of Natural Gas Utilization and Applied Catalysis, Dalian Institute of Chemical Physics, Chinese Academy of Sciences, P.O. Box 110, Dalian 116023, P. R. China
b. State Key Laboratory of Catalysis, Dalian Institute of Chemical Physics, CAS, P.R.China

The performance of a LiLaNiO/γ-Al_2O_3 catalyst for the simultaneous catalytic reaction of alkanes with CO_2 and O_2 to syngas was investigated in a fixed-bed flow microreactor using XRD and TG techniques. The catalyst exhibits high activity for the reactions of CH_4-C_2H_6-CO_2-O_2 and CH_4-C_2H_6-C_3H_8-C_4H_{10}-CO_2-O_2 to syngas. The life tests for the two reactions show that the catalyst is stable and has resistance to carbon deposition at high temperatures. At these high temperatures, the CH_4-C_2H_6-CO_2-O_2 mixture could be directly converted to syngas over the LiLaNiO/γ-Al_2O_3, and the presence of C_2H_6 does not lead to catalyst deactivation by carbon deposition. On the other hand, the CH_4-C_2H_6-C_3H_8-C_4H_{10}-CO_2-O_2 mixtures could not be directly converted to syngas due to a gas phase reaction which yielded coke on the reactor wall but not on the catalyst. Instead, syngas could be produced by following two steps: first, the C_3H_8 and C_4H_{10} in the mixture gases were converted to CH_4, C_2H_6, and CO_2 below 873K, then the CH_4-C_2H_6-CO_2-O_2 mixture was converted into syngas over the LiLaNiO/γ-Al_2O_3 catalyst at high temperatures (>873K).

1. INTRODUCTION

There are abundant supplies of gases containing CH_4, C_2H_6, C_3H_8 and C_4H_{10}, etc. from FCC (Fluidized Catalytic Cracking) tail gas, refinery gas, etc. Normally, these gas mixtures are combusted to carbon dioxide, since the complete separation of CH_4, C_2H_6, C_3H_8 and C_4H_{10} is not economical. If syngas could be produced from these mixtures over supported nickel catalysts with high selectivity and conversion, then syngas can be obtained directly from FCC tail gas, refinery gas, etc. This will lead to a better utilization of the light fractions of the FCC tail gas and refineries, etc. Both partial oxidation of alkanes (POA) and CO_2 reforming of alkanes are effective routes for syngas production. The POA reaction is an exothermic process, and the heat produced is strongly dependent on the selectivity, which makes the process very difficult to control and to operate safely. Meanwhile, CO_2 reforming of alkanes is a highly

* Corresponding author e-mail: slliu@ms.dicp.ac.cn

endothermic process which often leads to rapid carbon deposition, particularly on nickel catalysts. Coupling the endothermic CO_2 reforming with the exothermic POA reaction has the following advantages: (i) the coupled process can be made mildly endothermic, thermoneutral or mildly exothermic by manipulating the process conditions; (ii). the energy efficiency of the process can be optimized, and hot spots can be avoided, which eliminates runaway reaction conditions makes the process safe to operate; (iii). the process can be made environmentally beneficial.

Combined partial oxidation of methane and CO_2 reforming with methane has been reported by Vernon et al [1]. They reported that transition metals supported on inert oxides were active catalysts for combined partial oxidation and CO_2 reforming. They varied the $CH_4/CO_2/O_2$ ratio in order to achieve a thermoneutral reaction over a 1%Ir/Al_2O_3 catalyst. Ross et al [2] reported that hot spots in the catalyst bed were reduced significantly by combining the partial oxidation reaction with the CO_2 reforming reaction. Inui et al [3] carried out the reaction using a feed containing CH_4, C_2H_6, C_3H_8, CO_2, O_2 and N_2 over the Ni-Ce_2O_3-Pt-Rh catalyst, and found that an extraordinarily high space-time yield of syngas could be obtained using high flow rates at around 400 °C. Long et al [4] reported that the nickel supported catalysts showed were active for making syngas from a feed including a low molecular alkane, O_2, CO_2 and H_2O in a fluidized bed of catalyst.

Recent studies conducted in this laboratory [5,6,7,8] showed that high reaction rates and resistance to carbon deposition, combined with excellent stability could be achieved over the LiLaNiO/γ-Al_2O_3 catalyst for CH_4-O_2, C_2H_6-O_2, and CH_4-C_2H_6-O_2 reactions to syngas. In the present work, a fixed-bed flow microreactor, and XRD and TG techniques, were used to study the reactions of a series of (CH_4, C_2H_6, C_3H_8 and C_4H_{10}) with O_2 and CO_2 produces to syngas over a LiLaNiO/γ-Al_2O_3 catalyst. The obvious difference between the higher alkanes and methane is that C_2H_6, C_3H_8 and C_4H_{10} decompose rapidly at high temperatures, leading to catalyst deactivation by carbon deposition. Hence, the key to the simultaneous catalytic reactions is to optimize the supported nickel catalysts to reaction activity and selectivity, and to resist carbon deposition.

2. EXPERIMENTAL

2.1. Preparation of catalyst and test of reaction performance

Catalysts were investigated in a fixed-bed flow microreactor under atmospheric pressure. Reaction performance was tested using a microreactor with an internal diameter of 8 mm, using 1 mL of catalyst with an average particle size of 0.37-0.25mm. An EU-2 type thermocouple was placed in the quartz reactor, which was located in the middle of the catalyst bed to control the temperature in the electric furnace. This temperature was taken as the reaction temperature. The analysis methods for the reaction products and preparation of the LiLaNiO/γ-Al_2O_3 catalyst have been described earlier [5].

2.2. Characterization of catalyst

X-ray diffraction (XRD) characterization of the catalysts was performed with a Riguku D/Max-RB X-ray diffractometer using a copper target at 40KV x 100mA and a scanning speed of 8 degree/min. TG tests were recorded and treated by a Perkin-Elmer 3600 workstation at a programmed temperature rate of 10 K/min in air, with the flow rate 25 mL/min.

3. RESULTS AND DISCUSSION

3.1. Reaction performance of CH_4-C_2H_6 -CO_2-O_2 to syngas over the LiLaNiO/γ-Al_2O_3

The reactions of CH_4-CO_2-O_2 and C_2H_6-CO_2-O_2 mixtures to form syngas over the LiLaNiO/γ-Al_2O_3 were carried out. The results indicate that the catalyst possesses high activity for these two reactions. On this basis, the reaction of CH_4-C_2H_6-CO_2-O_2 to syngas was performed next. The effect of reaction temperature on the performance of the LiLaNiO/γ-Al_2O_3 is shown in Fig.1. When the reaction temperature is subsequently increased from 973 to 1123K, the CO_2 and CH_4 conversions increase, while the H_2 selectivity does not change significantly, and the H_2/CO ratio of the product remains 1.1/1. Ethane and oxygen are converted almost completely (not shown). Below 1053K and at a $CH_4/CO_2/C_2H_6/O_2/Ar$ ratio of 0.5/0.19/0.31/0.48/1.5, the influence of space velocity on the reaction over the LiLaNiO/γ-Al_2O_3 is studied. The results in Fig.2 indicate that the influence of space velocity is not appreciable. This means that the catalyst has sufficient active centers for the reaction over this wide range of space velocities.

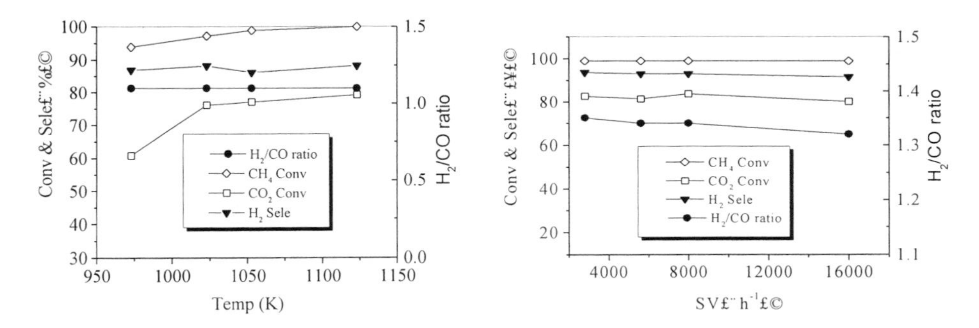

Fig.1. (Left) Effect of temperature on the performance of the LiLaNiO/γ-Al_2O_3 ($CH_4/CO_2/C_2H_6/O_2/Ar$ = 0.42/0.43/0.15/0.28/1.1; SV = $3.6x10^3h^{-1}$)

Fig.2. (Right) Performance of the LiLaNiO/γ-Al_2O_3 at different space velocities ($CH_4/CO_2/C_2H_6/O_2/Ar$ = 0.5/0.19/0.31/0.48/1.5; T = 1053K)

At a constant space velocity of 1.9×10^{3} h^{-1} (based on the combined flow rate of CH_4+CO_2+C_2H_6) and a temperature of 1053K, the flow rate of O_2 was changed to obtain different O_2/C_2H_6 ratios. As shown in Table1, as the O_2/C_2H_6 ratio changes from 1.35 to 2.19, the CO_2 conversion and the H_2 selectivity decrease, while the CH_4 conversion increases. The effect of diluting the reaction gases on reaction performance is investigated by addition of Ar, at an Ar/O_2 ratio of 4/1. As shown in Table 1, at the same O_2/C_2H_6 ratio, the CH_4 conversion, H_2 selectivity, and H_2/CO ratio without dilution of Ar are lower than those with dilution, and the CO_2 conversion of the former is higher than that of the latter. Since the methane reforming with carbon dioxide is a slow process compared to that of methane partial oxidation, adding Ar reduces the contact time of CO_2 and CH_4 with the catalyst, which decreases the CO_2 conversion. For methane partial oxidation, increasing the space velocity increases the rate of the direct partial oxidation (DPO) scheme[5], so that the H_2 selectivity increases. This work is under way. These results indicate that the addition of dilute Ar in the feed affects the reaction performance.

Table 1 Performance of the LiLaNiO/γ-Al_2O_3 at different O_2/ C_2H_6 ratios*
$CH_4/CO_2/C_2H_6$ = 0.5/0.19/0.31; SV $_{CH4-CO2-C2H6}$ = 1.9 x 10^3 h^{-1};T = 1053K

O_2/ C_2H_6 ratio	CH_4 Conv (%)	CO_2 Conv (%)	H_2 Sele(%)	H_2/CO ratio
1.35	76.2 (82.0)	98.3 (96.9)	94.7 (99.9)	1.4 (1.5)
1.58	82.2 (86.7)	98.0 (96.6)	88.9 (93.7)	1.3 (1.4)
1.84	93.0 (97.6)	88.9 (83.2)	83.8 (91.2)	1.3 (1.4)
2.19	95.8 (99.5)	84.0 (68.5)	71.8 (83.7)	1.1 (1.4)

*: 1. C_2H_6 and O_2 conversions are all 100%
2. The value in parentheses is for a $CH_4/CO_2/C_2H_6$ ratio of 0.5/0.19/0.31, Ar/O_2 ratio of 4/1, (CH_4+CO_2+C_2H_6) space velocity of 1.9 x 10^3 h^{-1} , and a temperature of 1053K.

The stability of the LiLaNiO/γ-Al_2O_3 was studied at 1053K, a $CH_4/CO_2/C_2H_6/O_2$ ratio of 0.42/0.43/0.15/x (where x represents the oxygen content of the mixture gases), a O_2/Ar ratio of 1/4, and a (CH_4+CO_2+C_2H_6) space velocity of 1.8x10^3 h^{-1} (Table 2). During the 50h life test, it is interesting that CH_4 and C_2H_6 are almost completely converted. Meanwhile, the CO and H_2 selectivities, and H_2/CO ratio remain ~100%, ~ 85% and ~1.1, respectively. The values approach to those of the thermodynamic equilibrium. The result of TG shows the carbon deposition of the used catalyst is only 0.22 wt%. These results indicate that the LiLaNiO/γ-Al_2O_3 is quite stable and resists carbon deposition during the high temperature reaction. A longer life test is under way.

Table 2 Reaction performance of the LiLaNiO/γ-Al_2O_3 as a function of time *
$CH_4/CO_2/C_2H_6/O_2$ = 0.42/0.43/0.15/x; Ar/O_2 = 4/1; SV$_{CH4-CO2-C2H6}$ = 1.8 x 10^3 h^{-1}; T = 1053K

Time (h)	x	CO_2 Conv (%)	CO Sele(%)	H_2 Sele (%)	H_2/CO ratio
0.2	0.45	54.6	100	77.7	1.1
2	0.37	71.9	100	83.6	1.1
7	0.34	77.4	100	85.5	1.1
14	0.32	75.2	100	88.6	1.1
27	0.36	67.9	100	85.8	1.1
29	0.37	69.4	100	84.3	1.1
39	0.35	70.3	100	86.3	1.1
43	0.31	79.2	100	87.8	1.1
49	0.33	74.2	100	87.9	1.1
50	0.34	71.7	100	87.2	1.1

*: CH_4, C_2H_6 and O_2 conversions are all 100%

3.2. Reaction performance of alkane-CO_2-O_2 to syngas over the LiLaNiO/γ-Al_2O_3

Because C_3H_8 and C_4H_{10} are present in FCC tail gas, refinery gas, etc., the simultaneous catalytic reaction of these alkanes (containing CH_4, C_2H_6, C_3H_8 and C_4H_{10}) with carbon dioxide and oxygen to syngas over the LiLaNiO/γ-Al_2O_3 catalyst was investigated. The alkane concentrations were: 47.2%(vol)CH_4, 40.0%(vol)C_2H_6, 9.6%(vol)C_3H_8 and 3.2%(vol)C_4H_{10}. Tests to measure the effect of temperature show that the conversions of CH_4 and CO_2 increase with increasing temperature, while H_2/CO ratio is constant (~ 0.90), at an O_2/ alkane ratio of 0.36, CO_2/ alkane ratio of 0.65 and alkane space velocity of $1.7x10^3$ h^{-1}. In addition, C_2H_6, C_3H_8, and C_4H_8 conversions, and CO selectivity all remain ~100% between 923K and 1073K. The CH_4 conversion increases, while CO_2 conversion and H_2 selectivity decrease with the increasing of O_2/ alkane ratio (from 0.19 to 0.61) at a temperature of1073K, CO_2/ alkane ratio of 0.65 and alkane space velocity of $1.7x10^3$ h^{-1}. The LiLaNiO/γ-Al_2O_3 catalyst is active over a wide range of space velocity at these conditions.

The life test experiment of the catalyst was performed using a microreactor with an internal diameter of 8mm and the volume of catalyst of 1 mL, while the stability of the LiLaNiO/γ-Al_2O_3 catalyst was studied at 1073K, under O_2/ alkane ratio of 0.36, CO_2/ alkane ratio of 0.65 and alkane space velocity of $1.7x10^3$ h^{-1} (Fig.3). During the 40-h life test, CH_4 conversion and H_2/CO ratio remain ~94%, ~ 0.9, respectively, CO_2 conversion and H_2 selectivity are >75%, >80% respectively. In addition, C_2H_6, C_3H_8, and C_4H_8 conversions, and CO selectivity all remain ~100%. These results indicate that the LiLaNiO/γ-Al_2O_3 is quite stable during the high temperature reaction.

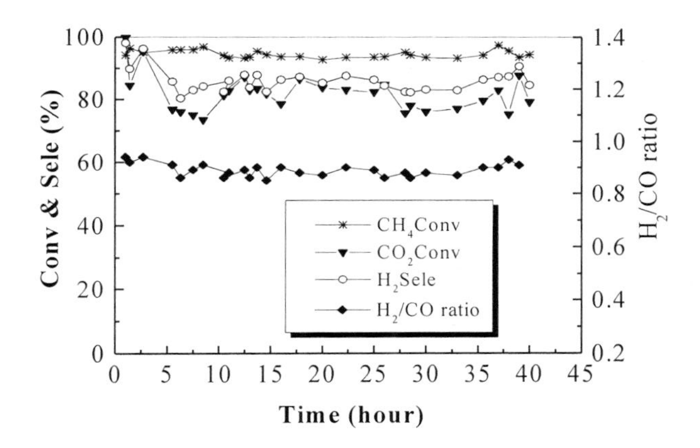

Fig.3 Reaction performance of the LiLaNiO/γ-Al_2O_3 as a function of time*
(O_2/ alkane = 0.36/1; CO_2/ alkane = 0.65/1; SV alkane =$1.7x10^3$ h^{-1}; T=1073K)
*: The alkane contains 47.2%CH_4, 40.0%C_2H_6, 9.6%C_3H_8 and 3.2%C_4H_{10}.

After the 40h life test, the reactor was cooled to room temperature in N_2 gas flow, then the catalyst was characterized by XRD and TG. The XRD measurements indicate that the support for the fresh and used catalysts are composed of γ-Al_2O_3. This demonstrates that the catalyst has a stable crystal phase during the high temperature reaction.

The TG results show that the carbon deposition on the used catalyst is only 0.1 wt%. This shows the LiLaNiO/γ-Al_2O_3 resists carbon deposition during the high temperature reaction. However, after the 40-h life test, there is some carbon on the reactor wall, resulting from the thermal decomposition of propane and butane. The coke will plug the reactor and cause the reaction rate to decrease over time. In other words, over the LiLaNiO/γ-Al_2O_3 catalyst, the mixture gases of CH_4-C_2H_6-C_3H_8-C_4H_{10}-CO_2-O_2 may not be directly converted to syngas due to coke formation on the reactor wall at high temperatures, but may be converted to syngas by two steps: first, below 873K there is little gas phase reaction, and the C_3H_8 and C_4H_{10} react to form CH_4, C_2H_6 and CO_2 [9,10]. Next, the CH_4+C_2H_6+CO_2+O_2 gas mixture is converted to syngas over the catalyst at high temperatures (>873K). This is being studied now. For the CH_4-C_2H_6-CO_2-O_2 reaction, no gas phase reaction is observed at these high temperatures, thus the CH_4-C_2H_6-CO_2-O_2 gas mixture can be converted directly to syngas over the LiLaNiO/γ-Al_2O_3 at high temperatures. Schmidt et al. [11] investigated partial oxidation of alkanes over noble metal coated monolith. They reported that the presence of C_2H_6 in the natural gas would not lead to catalyst deactivation by carbon deposition. C_3H_8 oxidation over Rh has led to carbon deposition, but only in severely fuel rich regimes. The amount of C_3H_8 in natural gas should not be sufficient for coking to be a concern.

4. CONCLUSION

At high temperatures(>873K), the CH_4-C_2H_6-CO_2-O_2 gas mixture can be directly converted to syngas over the LiLaNiO/γ-Al_2O_3 without any detrimental coke formation on the wall. While the CH_4-C_2H_6-C_3H_8-C_4H_{10}-CO_2-O_2 gas mixture may not be directly converted to syngas due to a gas phase reaction resulting in coke formation on the reactor wall (but not on the catalyst), this mixture may be converted to syngas in two steps.

REFERENCES

[1] P.D.F.Vernon, M.L.H.Green, et al., Catal. Today, 13 (1992) 417.
[2] A. M. O'Connor, J. R. H. Ross, Catal. Today, 46 (1998) 203.
[3] T. Inui, Sekiyu Gakkai Shi, 40 (1997) 243.
[4] Long, et al., Low hydrogen syngas using CO_2 and a nickel catalyst, US Patent No.5985178 (1999).
[5] Q. Miao, G. X. Xiong, S. S. Sheng, et al., React. Kinet. Catal. Lett. 66(1999)273.
[6] S. L. Liu, G. X. Xiong, S. S. Sheng, et al., Stud. Surf. Sci. Catal., 119 (1998) 747.
[7] S. L. Liu, G. X. Xiong, W. S. Yang, et al., Catal. Lett., 63 (1999)167.
[8] S. L. Liu, G. X. Xiong, S. S. Sheng, et al., Appl. Catal. A., 202 (2000) 141.
[9] S. L. Liu, G. X. Xiong, L.Y. Xu, et al., Appl. Catal. A., (In press).
[10] S. L. Liu, G. X. Xiong, L.Y. Xu, et al., Chin. Chem. Lett., (In press).
[11] M. Huff, P. M. Torniainen, L. D. Schmidt, Catal. Today, 21 (1994) 113.

Studies in Surface Science and Catalysis
J.J. Spivey, E. Iglesia and T.H. Fleisch (Editors)

Impact of Syngas Generation Technology Selection on a GTL FPSO

Cliff Lowe and Karl Gerdes – Chevron Research and Technology Company
Walt Stupin, Brian Hook and Paul Marriott, Fluor Daniel, Inc.

For the past two years Chevron Research and Technology Company has been working with Fluor Daniel, Inc. and Air Products and Chemicals, Inc. to evaluate a Fischer Tropsch based GTL (gas-to-liquids) FPSO (floating production storage offloading vessel). The objectives of the study were 1) to determine the technical feasibility of a ship-based GTL plant designed to handle associated gas; 2) to estimate the cost of such a facility and 3) to examine the impact on the overall facility of different syngas generation technologies. We intend to use these results to guide R&D activities.

Three different syngas generation technologies were evaluated: air-blown autothermal reformer (ATR), oxygen-blown ATR, and ITM Syngas (Ion Transport Membrane).

The study provided estimates for the overall facility for the following key areas: capital and operating requirements, deck space, topsides weights, hull dimensions, and liquid product rates.

1 Background

Worldwide, Chevron has several lease positions in deepwater or isolated offshore locations. In many of these fields, the associated gas cannot be easily marketed via a pipeline network, and alternative, low-cost means of disposition, gas injection or flaring, may not be viable or permitted. Chevron initiated this study to assess the feasibility of utilizing this gas by converting it into Fischer-Tropsch liquids on the ship-based production facility.

2 Study Objectives

There were three primary objectives to this study:

- Evaluate the technical feasibility of using GTL to utilize associated gas on an oil production FPSO
- Estimate capex/opex requirements to guide future technology development
- Evaluate the impact of different syngas generation technologies

3 Study Basis

The study was based on a field producing 50,000 barrels per day of crude, 100 MMSCFD of associated gas (feed to GTL plant), and up to 40,000 bpd of produced water. Gulf of Mexico metocean conditions were assumed.

4 Process Description

A block flow diagram of the overall process scheme is shown in Figure 1. In the syngas (synthesis gas) generation step, associated gas is converted into synthesis gas, containing primarily hydrogen and carbon monoxide. The Fischer-Tropsch reaction converts the synthesis gas to long-chain hydrocarbons and some byproducts. In Product Upgrading, only minimal processing is used to convert the FT products into a synthetic crude.

4.1 Syngas Generation

The process facilities were configured using three different syngas generation technologies: air-blown autothermal reforming (ATR), oxygen-blown autothermal reforming, and reforming using an Ion Transport Membrane (ITM Syngas) reactor. A brief description of the technologies, highlighting key differences, follows:

Figure 1 – Key Process Steps

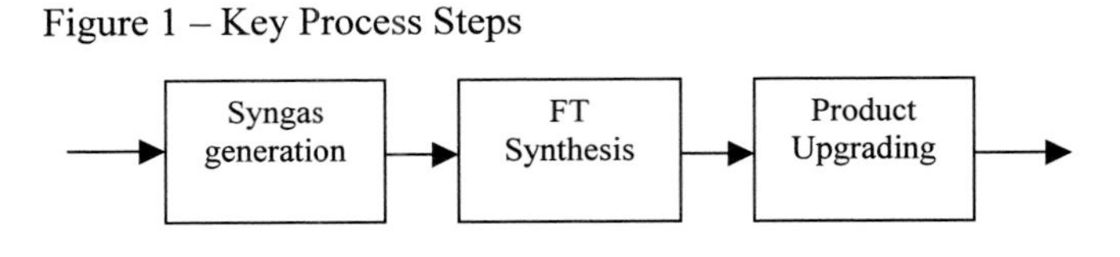

4.1.1 Air-blown autothermal reforming

Associated gas is first treated to remove trace sulfur compounds, heated, and then injected into the autothermal reactor along with steam and pressurized air. The refractory lined reactor is divided into two sections, a thermal zone where partial oxidation takes place, and a catalytic zone where steam reforming occurs. The synthesis gas from the reactor contains a significant percentage of nitrogen.

4.1.2 Oxygen-blown autothermal reforming

After sulfur removal, the associated gas is sent to a pre-reformer where the C_{2+} hydrocarbons are converted to H_2, CO and methane. The gas is then further heated and then injected into the ATR reactor along with steam and oxygen. The oxygen is provided by an air separation plant.

Figure 2 – Ion transport membrane

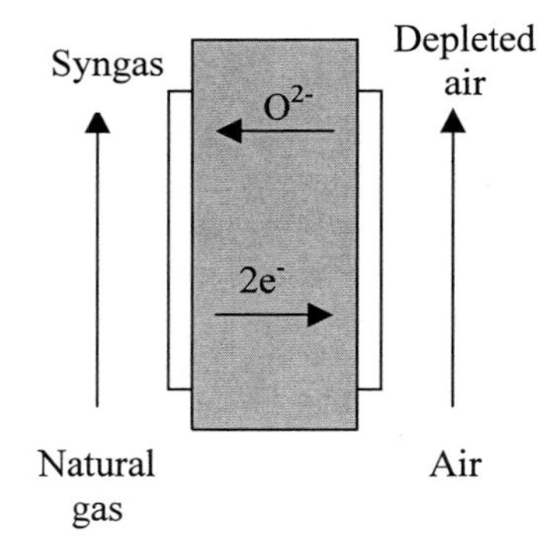

4.1.3 Ion Transport Membrane reforming

The U.S. Department of Energy is partially funding the development of Ion Transport Membrane technology for the production of synthesis gas and hydrogen. The development project, begun in 1998, is an eight year, 86$ million effort and is headed by Air Products and Chemicals, Inc. in partnership with Chevron Research and Technology Co., Ceramatec, Inc., Eltron Research, Inc., McDermott Technology, Inc., Norsk Hydro ASA, University of Alaska – Fairbanks, Penn State University, University of Pennsylvania, and Pacific Northwest National Laboratory.

After sulfur removal and pre-reforming, the associated gas is heated, mixed with steam and sent to the ITM Syngas reactor where partial oxidation and reforming take place.

Heated air is passed across one side of the Ion Transport Membrane. The steam/associated gas mixture is passed across the other side. Due to the unique properties of the membrane, oxygen ions pass through the membrane from the air side to the reforming side. The driving force for the ion migration is the difference in oxygen partial pressure on the two sides of the membrane. A sketch of the membrane system is shown in Figure 2[i,ii].

Due to the proprietary and developmental nature of the ITM Syngas technology, the detailed results of the ITM GTL work are not included in this paper. We have, however, included a qualitative analysis of ITM Syngas technology where appropriate.

4.2 Fischer-Tropsch Synthesis

The synthesis gas is sent to the FT reactors for the production of a mixture of hydrocarbons and some byproducts. The design of the reactors was based on published designs of FT fixed bed tubular reactors. A fixed bed design was chosen to minimize the effect of list and wave motion.

For the air-blown ATR case, the reactors were arranged in series because the high nitrogen content precluded recycling the unconverted gas. For the oxygen-blown and ITM Syngas cases, a single stage reactor system with off-gas recycle was chosen.

4.3 FT Product Upgrading

A mild hydrocracking unit is provided to convert the FT wax into a flowable syncrude product. Hydrogen is supplied from a pressure swing adsorption unit.

5 Evaluation Method

For each of the three syngas generation processes, an integrated conceptual design of the entire facility was generated. This included the production facilities, the syngas generation unit, the Fischer-Tropsch plant, the upgrading facilities, utilities (steam, power, water, etc.), and other facilities. Process simulation models were created for each of the key process steps as well as the primary utility systems. Complete heat and material balances were developed for the entire facility.

Figure 3 - GTL FPSO plan and side views

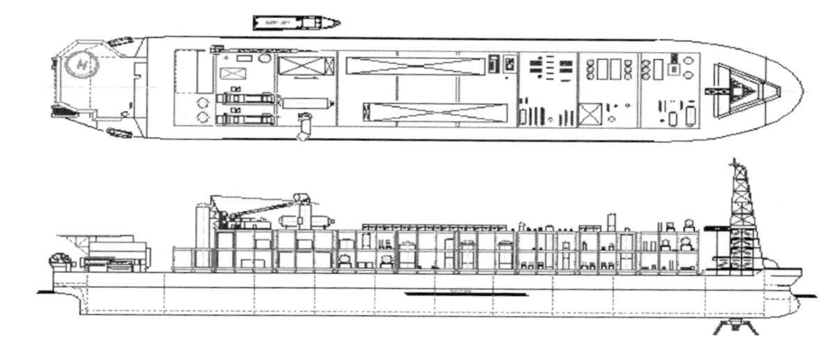

Major equipment was then sized and configured onto multi-level modules for installation on the vessel. The size of the vessel was adjusted to provide space for the process and utility modules, personnel accommodations, and a mooring system. In some cases, the hull size was also adjusted for stability considerations due to the weight of topsides.

It is important to note that this study was based on open literature information for the ATR and FT processes and, therefore, was not an evaluation of any specific proprietary process technology. It is also noted that ITM Syngas technology is in an early stage of development and results of this study are being used in guiding future development and process optimization.

In addition, since this is an offshore location with low feedstock values, an effort was made to minimize the size, number and costs of the process and utility equipment. This resulted in conceptual designs, which may have significant opportunities for optimization.

5.1 Fischer Tropsch synthesis yields

In the design of the FT system over fifty articles and papers were reviewed and a simplified model was developed for systems using cobalt based catalysts. A number of articles identify that the yield structure and reaction rates are dependent on both the temperature and pressure of the reactants. The Van Berge article[iii] clearly shows the impact of reactant pressures on yields for a specific FT reactor system. Specifically, this article shows that as the reactant partial pressure increases, the alpha, the tendency to produce longer chain molecules, of the reaction system increases. This results in a higher percentage of more valuable long chain hydrocarbons and a lower percentage of low value light gases.

Table 1 – FT yield distribution, wt%

	Air-blown	Oxygen-blown
C1/C2	18	12
C3/C4	11	7
C5-C11	34	29
C12-C18	19	23
C19+	18	29
Total	100	100

The effect of the reactant partial pressure was significant when the yield distribution was estimated for the three syngas generation cases. Of the three cases the reactant partial pressure in the air-blown case was the lowest because of the dilution effect of the nitrogen. The reactant partial pressure in the oxygen-blown case was higher, but was still lower than the ITM Syngas case because of the buildup of nitrogen and argon in the FT recycle loop. These inert gases accompany the 95% pure oxygen from the air separation unit. The reactant partial pressure in the ITM Syngas case was highest because of the absence of nitrogen and argon.

Table 1 shows how the yield distribution differs.

Table 2 – Deck and Hull requirements

	Air-blown	Oxygen-blown
Hull length, meters	275	300
Hull width, meters	50	53
Topsides weight, MT	31,100	39,700
No. of decks	3	3
No. of process modules	7	8.7
Total deck space, ft^2	184,000	238,000
FPSO size, DWT	156,000	215,000
Total storage cap., bbls	873,000	1,239,000

6 Results

6.1 Topsides Layout and Hull

The ship design was based on a new-built, non-powered, double hulled, vessel with passive turret mooring and stern thrusters.

A summary of the deck and hull requirements is provided in Table 2. Key points to note are:

- The deck space required for the process and utility equipment is very high (over 184,000 sq. ft.) in all the cases.
- The equipment in the air-blown case weigh less and require less space because of the absence of the air separation unit.
- The production and seawater injection require only 5% of the total deck space.
- The available storage volume for crude and syncrude was determined by hull size required to support the topsides equipment. In other words, the minimum storage requirements (500,000 bbls of crude storage) were easily met.

Figure 3 shows a side and plan view of the oxygen-blown GTL FPSO.

Table 3 – Operating cost summary, million$/yr

	Air-blown	Oxygen-blown
Operating manpower	6.6	7.4
Maintenance	21.1	24.9
Catalyst	5.9	5.3
Insurance	2.1	2.5
Taxes	4.2	5.0
Total	39.9	45.1

6.2 Operating Costs

Table 3 is a summary of the operating costs for the facility. The costs for the oxygen-blown case are slightly higher because of the requirement for an air separation unit.

6.3 Capital Costs

The capital cost estimate summarized in Table 4 has an estimated accuracy of plus 20% minus 40%. The estimate was based on major equipment costs and factors for bulks, subcontracts, and installation on an offshore facility. Key points to note are:

Table 4 – Capital expenditures summary, million$

	Air-blown	Oxygen-blown
Production/seawater injection.	16	16
Process and utilities	336	331
Hull/shipboard systems	158	165
Initial catalysts / internals	16	16
Engineering, fee, etc.	97	97
Contingency	125	125
Total	748	750
GTL portion of total capital	512	513

- The capital costs are based on fourth quarter 1999 dollars.
- A 20% contingency is provided to cover items not detailed and estimated as part of this study. There is no added contingency for technical uncertainty.
- The cost estimate excludes drilling, subsea facilities, risers, mooring (other than the

turret)

- The capital requirements for the air and oxygen blown cases were roughly the same. The cost of the oxygen plant in the oxygen blown case was balanced by lower costs for air compression and heat removal.
- The capital requirements for the ITM Syngas case were significantly lower than the ATR cases due to 1) the absence of an air separation unit and 2) significant savings in the utility portion of the facility.
- The cost of the GTL section of the plant is over 65% of the total facility costs.

6.4 Production rates

Table 5 summarizes the estimated production rates of crude and syncrude. The differences in the syncrude flowrates are primarily due to the effect of the reactant partial pressure in the FT reactor system (see section 5.1).

6.5 Cost of production

It is a common industry practice to analyze GTL projects on the basis of dollars of capital costs per barrel of product. The capital requirements that are attributable to the GTL section of the plant were estimated and are shown in Table 4. These values were then divided by the syncrude production rates. The results summarized in Table 5 show that the normalized costs are very high and well above industry targets of $20-25,000/bpd.

Table 5 – Production rates/Unit capital cost

	Air-blown	Oxygen-blown
Raw crude, bpd	50,000	50,000
Syncrude, bpd	8,400	10,800
Total, bpd	58,400	60,800
Capex $/bpd, syncrude	61,000	47,500

7 Conclusions

- Although expensive with today's technology of air-blown or oxygen-blown autothermal reforming, a GTL FPSO is a technically viable alternative for producing and processing oil and associated gas in an offshore location.
- Syngas generation technology can have a significant impact on capital requirements as well as syncrude production. ITM Syngas technology has the potential for significant capital cost savings.
- The conceptual designs in this study have room for optimization and efficiency improvement.

---

[i] P.N.Dyer and D.L.Bennett, "ITM Technology for GTL Processing," Gas-to-Liquids World Forum, Hart/IBC, London, November 1998.

[ii] P.N.Dyer and C.M.Chen, "Engineering Development of Ceramic Membrane Reactor Systems for Converting Natural Gas to H2 and Syngas for Liquid Transportation Fuels," US DOE H2 R&D Program Review, San Ramon, May 2000.

[iii] Van Berge PJ, Everson RC, "Cobalt as an alternative Fischer-Tropsch catalyst to iron for the production of middle distillates", Studies in Surface Science and Catalysis, Volume 107, 1997, pages 207-212

Studies in Surface Science and Catalysis
J.J. Spivey, E. Iglesia and T.H. Fleisch (Editors)

Developments in Fischer-Tropsch Technology and its Application

B. Jager, P. Van Berge and A. P. Steynberg

Sasol Technology Research and Development, P O Box 1, Sasolburg, 9570, Republic of South Africa.

1. ABSTRACT

The last couple of decades have seen important advances in the development of reactors used for Fischer-Tropsch conversions. This led to the introduction of commercial fluidised bed reactors for high temperature Fischer-Tropsch processes and commercial slurry bubble bed reactors for low temperature Fischer-Tropsch, both at SASOL. Important in these developments have been parallel developments in the catalysts used in these reactors. The present paper deals mainly with the development of the low temperature Fischer-Tropsch slurry bed reactor and catalyst.

2. INTRODUCTION

Fluidised beds have recently completely replaced the circulating fluidised bed reactors which were in operation since 1953 for high temperature Fischer-Tropsch (HTFT)[1]. Traditionally tubular fixed bed (TFB) reactors were used with iron based catalyst for low temperature Fischer-Tropsch (LTFT) synthesis. These reactors were further developed by SHELL to allow them to make use of the more demanding cobalt based catalyst. A major advance has been the development of the slurry phase reactor (SPR) which has partly replaced the TFB reactors for LTFT synthesis at SASOL.

There are several design issues with FT reactors:

- The FT reaction is highly exothermic and the effective removal of heat is a major design issue.
- Temperature and pressure affect both the conversion and selectivities obtained.
- The catalysts used commercially are either cobalt or iron based. They are differently affected by process conditions: temperature, pressure and H_2/CO ratio of the feed gas. Iron based catalyst is relatively more sensitive than cobalt based catalyst to temperature and pressure with respect to conversion and selectivities. Cobalt based catalyst is more sensitive towards the composition of the feed gas with respect to selectivities. Relatively low H_2/CO ratios are needed if a high Schultz-Flory distribution α-value (heavier products) is required.

For most effective performance of a system to produce a primary product which can be effectively worked up to final products, the interaction between the design of the reactor and the catalyst should be optimized.

For the continuing development of the FT processes, the development of the reactor concept and design has to interact with the development and design of the FT catalysts. This was done successfully for both LTFT and HTFT. In the case of LTFT, Sasol has developed a reactor-catalyst combination that makes commercial gas to liquid processing for monitising remote or stranded natural gas clearly viable on a commercial scale.

3. LTFT REACTOR DEVELOPMENT

Iron based catalyst was originally used and is still used by Sasol in TFB reactors in the so-called ARGE process. The ARGE process is costly and difficult to scale up from a mechanical point of view. Because of non iso-thermal operation and limits to the allowable operating temperature, the use of the catalyst is not very effective. The temperature profiles in the tubes imply that only that portion of the catalyst close to the temperature peak works optimally. Iron based catalyst has a life of months or weeks, which means considerable downtime and costs for periodic replacement of the catalyst. Cobalt based catalyst has a much longer life and downtime for catalyst replacement has no major effect on the availability of the reactor. Capital in this case is more effectively used.

At commercial operating conditions, iron based catalyst is inhibited by water, one of the products of the FT reaction. Cobalt based catalyst is not affected this way and has a higher activity at typical commercial per pass conversions. This higher activity however, makes it more likely to cause temperature run-aways if heat transfer is not sufficient. For Cobalt based catalyst smaller tubes have to be used in the TFB reactor, adding to its cost relative to TFB reactors using iron based catalyst.

Better use is made of the catalyst in the slurry bubble column reactor. This reactor is shown diagrammatically in Figure 1.

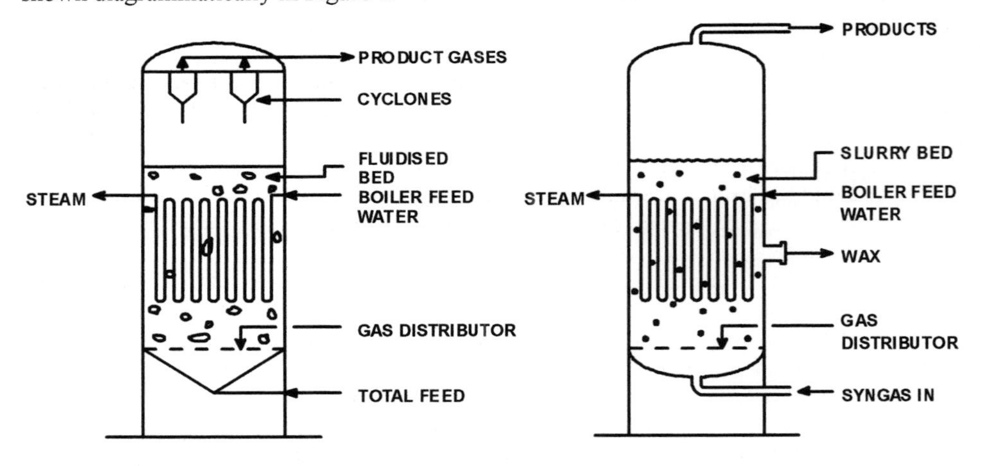

Fig. 1. Sasol Advanced Synthol (SAS) Reactor for HTFT

Fig. 2. Sasol Slurry Phase Distillate (SPD) Reactor for LTFT

The reactor was introduced commercially by SASOL in 1993 for LTFT operations. The SASOL slurry phase reactor (SPR) is still the only commercially operated slurry phase bubble column reactor.

The operation of the SPR is virtually iso-thermal, which means that all the catalyst operates at close to optimal temperature without danger of temperature runaways. Replacement of catalyst can be done simply on-line, resulting in much higher reactor availability and reduced operating costs where catalyst has to be replaced periodically. The design of the SPR is simple and lends it self to scale-up, which allows for considerable economy of scale. Capital cost for a cobalt based SPR is about 75% lower then that of iron based TFB system with similar capacities. The catalyst consumption is also reduced by about 75%.

The first, and up to now the only commercial SPR, has a capacity of 2 500 bpd. Its design is not optimal as a redundant fluidised reactor was modified for this purpose. This indicates robustness for the SASOL Slurry Phase Distillate (SPD) technology. Present designs of SPR's are for a capacity of nominally 15 000 bpd from a single reactor

In the SPR the syngas is bubbled through a slurry, consisting of process derived liquid product in which finely divided catalyst is suspended. The syngas diffuses from the bubbles through the liquid in the slurry to the catalyst surface where it reacts to form hydrocarbons. These hydrocarbons are partly liquid which report to the slurry phase and partly gaseous in which form they are removed from the bed through the gas phase bubbles. Heat is removed by cooling coils in which a low-pressure steam is generated.

The size of the catalyst particles has a major effect on the behavior of the catalyst but also on the solid-liquid separation process. Catalytic activity and, if olefins are required, even the selectivity distribution is improved by a reduction in particle size [2].

The smaller the catalyst particles, the easier the distribution of the catalyst in the bubble agitated slurry. Also small particles are required to keep the effectiveness factor of the catalyst close to unity and to ensure that too small catalyst surface area does not hinder or limit mass transfer.

Solid separation, however, becomes progressively more difficult as particle sizes decrease and separation of the liquid products from the slurry phase presented a critical step that had to be solved before the SASOL SPD process could become a commercial reality.

In a slurry bed a weak catalyst can produce smaller particles through attrition between particles and due to shear forces inherent in the turbulence in the bed. If the catalyst size distribution contains even a small proportion of micron or sub micron -sized particles, they will make solid separation very difficult.

Of the earlier work done in slurry phase reactors, was with a iron based catalyst obtained by precipitation and spray drying [3]. This catalyst was relatively weak and broke up easily, causing major solid separation problems. Work was done to develop catalyst, resistant to both break up and abrasion. Considerable experimentation was done on solid separation equipment and catalyst to find a combination giving an acceptable solid separation process. The present system reliably gives liquid products virtually free of solids.

Where cobalt based catalyst is much stronger than iron based catalyst, better control of the particle size distribution of the cobalt based catalyst is possible and the solid separation is considerably easier. For both catalysts, the solid separation is only a small fraction of the overall cost of the process.

4. LTFT CATALYST DEVELOPMENT

As mentioned, the first catalyst used commercially by SASOL for slurry phase Fischer-Tropsch synthesis catalyst was a precipitated and spray-dried iron based catalyst [3]

similar in composition to the well-known ARGE catalyst [4]. It produced a product with a product distribution similar [3] to that obtained from the TFB reactors used in the ARGE process. This simplified the integration of the SASOL SPD process into the existing operations.

Since 1975, it became clear that cobalt-based catalysts provide a viable alternative to iron based catalyst for the production of middle distillates by means of Fischer-Tropsch synthesis. Studies done at SASOL [5] showed that:

- The cobalt catalyst does not show any significant water-gas-shift activity and no water inhibition of the Fischer-Tropsch reaction rate was observed
- The cobalt catalyst is the preferred option if highs per pass conversions are required.
- Desired stabilized intrinsic Fischer-Tropsch synthesis activity levels can be obtained with cobalt catalysts, implying that extended slurry phase synthesis runs can be achieved. This is possible provided the feed gas is meticulously purified of catalyst poisons e.g. removal of sulphur containing compounds. A stabilized cobalt catalyst is understood to be a catalyst that has been fully conditioned with respect to deactivation mechanisms such as:
    - oxidation of metallic cobalt [6]
    - rejuvenatable poisoning [7,8], (Etc)
- Cobalt derived hydrocarbon product selectivities show greater sensitivity towards process conditions (e.g. reactor pressure) than that of iron. Iron, on the other hand, does show marked sensitivity towards chemical promotion [9]. The effectiveness of chemical promoters for cobalt can indeed be questioned [10], and SASOL has opted in favour of the geometric tailoring of pre-shaped support materials as an optimization tool for effecting increased wax selectivities, an approach that was also elegantly modeled at EXXON on the basis of a postulated structural parameter [11,2].

These observations contribute to the view that the combination of a highly active and stable cobalt catalyst with a high and gradientless concentration slurry phase reactor is the configuration of choice for the Fischer-Tropsch process step of a Gas-to-Liquids (GTL) plant.

As part of the studies a proprietary slurry phase Fischer-Tropsch catalyst was developed giving special attention to:

- Hydrocarbon product (i.e. syncrude) selectivities
- Specific stabilized intrinsic Fischer-Tropsch activities
- Mechanical integrity
- Production of cobalt "free" reactor wax

During the development of this catalyst it became clear that catalyst performances for slurry phase applications should be determined in slurry phase equipment at real process conditions. Testing in small fixed bed reactors or in slurry bed reactors at less severe conditions does not give an adequate indication of performance. Suitable adjustments must be made for the test procedures before comparing catalysts tested in different setups [12,13]. Adjusting for these differences, the Sasol developed catalyst compares very favourably with catalyst activities quoted in the literature. The activity of 0,66kg of total hydrocarbons per kg of catalyst per hour obtained for the Sasol catalyst is very competitive. Further developments have improved on this activity and are presently the subject of patent applications.

Impregnating pre-shaped support material with the catalytically active metal is best suited to secure the desired mechanical integrity of the final catalyst. For the SASOL proprietary catalyst, spraydried and calcined spherical alumina was selected and tailored taking into account:

- Particle size distribution
- Abrasion/attrition resistance
- Geometry

The mechanical strength of the final catalyst proves to be adequate and no catalyst break-up is observed by particle size distribution. This catalyst allows for the continuous production of high quality reactor wax during extended slurry bed synthesis runs. Using the procedures developed, several large batches of catalyst were made using commercially scaled equipment. The catalyst batches were all subjected to extended slurry phase Fischer-Tropsch synthesis trials in a nominally 1m slurry bed reactor. The catalyst is now at a stage of development were it can be used confidently on commercial scale.

5. HIGH TEMPERATURE FISCHER-TROPSCH DEVELOPMENTS

The developments described for LTFT have a parallel in the developments in HTFT. Early in 1999 the last of the circulating fluidised bed Synthol reactors were replaced by the SASOL Advanced Synthol (SAS) reactors which are basically conventional fluidised bed reactors, shown in Figure 2. The SAS reactors have a maximum capacity of 20 000 bpd as compared to the 6 500 bpd for the older circulating fluidised bed reactors. The newer reactors have produced large reductions in operating costs.

Although the catalyst originally used with the older type reactor was suitable for the new SAS reactors, catalyst development has taken place and further improvements are expected which will make better use of the opportunities created by the new generation of reactors. More than in the past the catalyst can be tailored to influence the product spectrum to suit market requirements. Considerable reductions in catalyst consumption were realised with the introduction of the SAS reactor. Further savings and even better control of the product spectra are expected from further HTFT catalyst developments. This will especially have significant implications for SASOLs drive towards greater production of chemicals.

6. FUTURE APPLICATIONS AND DEVELOPMENTS

The present LTFT designs are used for the Escravos GTL project, for which, Sasol, through its Global JV with Chevron, will support the establishment of a GTL business venture by Chevron Nigeria and the Nigerian National Petroleum Company. SASOL does this through the supply of technology, operating and marketing services and the supply of catalyst. These designs are also the basis for the gas to liquid plant for the venture between Sasol and the Qatar General Petroleum Corporation at Ras Laffan in Qatar. In both cases a total capacity of 30 000 bpd is obtained from two SPR's. Improved optimisation and product recovery however results in a total liquid production in the order of 33 000 bpd

The designs presently use the latest developments, which have as a base a calalyst with an activity and properties, which were less advanced than is now possible. For iron based catalyst the per pass conversion is limited by the water concentration in the reactor. In situe removal of this water would both increase the activity of the catalyst and would also allow a higher per pass conversion. A not so easy way of achieving this would be by the removal of water through membranes installed in the reactor. In the case of cobalt based catalyst this is less important.

. With higher activities and potentially higher per pass conversions new limits on the reactor design become applicable. Higher activities lead to higher conversion rates and to the need to remove more exothermic heat of reaction. In principle there are several options. Larger heat transfer area could be introduced in the reactor. Heat could be removed from the reactor in the vapour phase by flashing a light liquid hydrocarbon and removing the heat as latent heat. The temperature difference driving the heat removal can be increased preferably by developing a catalyst, which can operate at higher temperatures, or by accepting a lower quality steam. As in the case where liquid hydrocarbon is evaporated in the reactor, producing a lower quality steam will have an impact on the design of cooling equipment around the reactor and the total design needs to be optimised.

Plenty has been achieved over the last decade. More can still be done and is expected both in the development of the reactors and in development of the catalysts used for Fischer – Tropsch synthesis.

REFERENCES

1. B.Jager, Studies in Surface Science and Catalysis, Vol. 119 (1998) 25.
2. E.Iglesia, S.C. Reyes, S.L. Soled, Reaction-Transport Selectivity Models and the Design of Fischer-Tropsch catalysts (Chapter 7), Computer-Aided Design of Catalysts, edited: E.R. Becker, C.J. Perreira, Marcel Dekker, Inc., New York. Basel. Hong Kong (1993).
3. B. Jager, R.L. Espinoza, Catalysis Today, 23, (1995) 14.
4. C.D. Frohning, H. Kölbel, M. Ralek, W. Rottig, F. Schur, H. Schulz, Fischer-Tropsch-Synthese: Chemierohstoffe aus Kohle; editor: J. Falbe, Thieme-Verlag, Stuttgart (1977).
5. P.J. Van Berge, R.C. Everson, Studies in Surface Science and Catalysis, 107 (1997) 207.
6. P.J. Van Berge, J. van de Loosdrecht, S. Barradas, A.M. van der Kraan, Catalysis Today, 58 (2000) 321.
7. E Iglesia, S.L. Soled, R.A. Fiato, G.H.Via, Journal of Catalysis, 143 (1993) 345.
8. S.C. Leviness, C.J. Mart, W.C. Behrmann, S.J.Hsia, D.R. Neskora, WO 98/50487 (1998).
9. M.E. Dry, Brennstoff-Chemie, 50 (1969) 193.
10. H. Pichler, H. Schulz, D. Kühne, Brennstoff-Chemie, 49 (1968) 344.
11. E. Iglesia, S.L. Soled, R.A. Fiato, Journal of Catalysis, 137 (1992) 212.
12. R. Oukaci, A.H. Singleton, J.G. Goodwin, Applied catalysis A, General, 186 (1999) 129.
13. P.J. Van Berge, S. Barradas, E.A. Caricato, B.H. Sigwebela, J. van de Loosdrecht, 4th Worldwide Catalyst Industry Conference and Exhibition (CatCon 2000) Houston, Texas (June 2000).

Studies in Surface Science and Catalysis
J.J. Spivey, E. Iglesia and T.H. Fleisch (Editors)

An Innovative Approach for Ethylene Production from Natural Gas

ZHU Aimin**, TIAN Zhijian*, XU Zhusheng, XU Yunpeng, XU Longya, LIN Liwu

Dalian Institute of Chemical Physics, CAS, Dalian 116023, China

A natural gas conversion process (denoted here as the EPNG Process) in which a fixed-bed reactor is used to combine the exothermic oxidative coupling of methane (OCM) and the endothermic oxidative dehydrogenation of ethane with CO_2 (ODE), is described in this paper. The results at various operating conditions demonstrate that this process is an efficient way to product ethylene from natural gas. In a 130-hour-running of a 100mL-scale reactor (CH_4 1.7l/min, C_2H_6 0.5l/min, and O_2 0.63l/min), a conversion of 27% and 66% for methane and ethane was maintained, and a total carbon yield of 27% for ethylene was obtained.

1. INTRODUCTION

Among the direct routes that have been explored for methane conversion using the abundant worldwide natural gas resources, the oxidative coupling of methane (OCM) to ethane and ethylene appears to be most promising. However, the ethylene yield of less than 10% with an ethylene content of only about 5% in the product gas presents a challenge [1]. In order to improve the technical and economic feasibility of the OCM process, considerable research effort has been devoted in recent years to seeking new approaches to increase ethylene production. One approach is to enhance the ethylene yield during the OCM reaction by using a simulated countercurrent moving-bed chromatographic reactor (SCMCR)[2], a gas recycle reactor-separator [3], or a membrane contactor [4]. Another approach is the combination of exothermic OCM and endothermic conversion of ethane and naphtha to ethylene in the same reactor [5-9]. The IFP oxypyrolysis process [6] combines the OCM (a heterogeneously catalyzed reaction producing ethane as a primary product) with the steam cracking of ethane (a homogeneous gas-phase reaction producing ethylene and hydrogen). Methane, separated from natural gas, is mixed with oxygen or air, then added to the OCM reactor. After molecular oxygen has been nearly quantitatively consumed, the ethane separated from natural gas is added to the OCM effluent. The heat of the exothermic OCM reaction is then partially utilized in the endothermic conversion of ethane to ethylene and hydrogen. In the OXCO process [5,7], a single fluidized bed reactor is used to combine the OCM step with the pyrolysis of ethane and higher alkane components present in natural gas. Such a one-stage process could be attractive for its higher heat utilization efficiency. Moreover, an additional methanation process was proposed to hydrogenate the COx resulting from non-selective oxidation, for the recovery of valuable carbon [5-9].

** Present address: Laboratory of Plasma Physical chemistry, Box 288, Dalian University of Technology, Dalian 116024, P. R. China. E-mail: amzhu@dicp.ac.cn
* Corresponding author. E-mail: tianz@dicp.ac.cn

The oxidative dehydrogenation of ethane (ODE) with CO_2 over a catalyst, as an alternative route for producing ethylene, has been proposed [10-13]. It has been found that the addition of CO_2 promotes the ethylene yield and prevents carbon deposition [14-16]. Actually, the OCM reaction is always accompanied by the production of CO_x [17]. With most OCM catalysts, carbon dioxide is the major CO_x, and its selectivity is 4~6 times that of CO [5]. Therefore, the inevitable by-product CO_2 from the OCM step can act as one of the reactants required by the second step of ODE with CO_2. Moreover, the heat of the exothermic OCM reaction could be also utilized by the strongly endothermic reaction of ODE with CO_2:

$$C_2H_6+CO_2 \rightarrow C_2H_4+CO+H_2O(g) \quad (\Delta H_{298K}=+178.4 kJ/mol) \quad (1)$$

Thereby, an innovative approach to ethylene production from natural gas (denoted the EPNG Process), which combines the OCM process and the process of ODE with CO_2 in one reactor packed with the OCM catalyst and the ODE with CO_2 catalyst, has been proposed and explored by the present authors. In the EPNG Process, the inevitable by-product CO_2 from the OCM process is required by the process of ODE with CO_2 over a catalyst, and the heat generated by the exothermic OCM process is partially or even completely utilized by the endothermic reaction of ODE with CO_2. Also, ethane produced by the OCM process, together with added ethane, dehydrogenates to ethylene over the catalyst used in the ODE-CO_2 process, producing a high content of ethylene in the product gas.

2 EXPERIMENTAL

The $La_{0.2}Ba_{0.1}/CaO$ was selected as the catalyst for the OCM process for its high selectivity to CO_2 rather than CO [1]. The preparation of the $La_{0.2}Ba_{0.1}/CaO$ catalyst is described elsewhere [1]. The MnO/SiO_2 catalyst for the ODE with CO_2 was prepared by impregnating silica gel with an aqueous solution of $Mn(NO_3)_2$.

The flow diagram of the EPNG process and a tubular quartz reactor modeled on a jacket heat exchanger are shown schematically in Fig.1. The $La_{0.2}Ba_{0.1}/CaO$ catalyst was placed in the jacket of the reactor and the MnO/SiO_2 was placed in the inner tube. A 10ml-scale reactor packed 1.5ml $La_{0.2}Ba_{0.1}/CaO$ and 5 ml MnO/SiO_2 and a 100ml-scale reactor packed 30ml $La_{0.2}Ba_{0.1}/CaO$ diluted with 60ml quartz sand and 90 ml MnO/SiO_2 were electrically heated in flowing nitrogen. The catalyst zone of the 10ml-scale reactor was centred inside a tubular furnace (40mm I.D.). The catalyst zone of the 100ml-scale reactor was inserted into the bottom part of a tubular furnace (90mm I.D.), as shown in Fig.1. The feedstock mixtures in the jacket were heated indirectly by the effluent gas in the inner tube outside the furnace, before it contacted the heated catalyst. It is evident that this reactor cannot function as an adiabatic reactor.

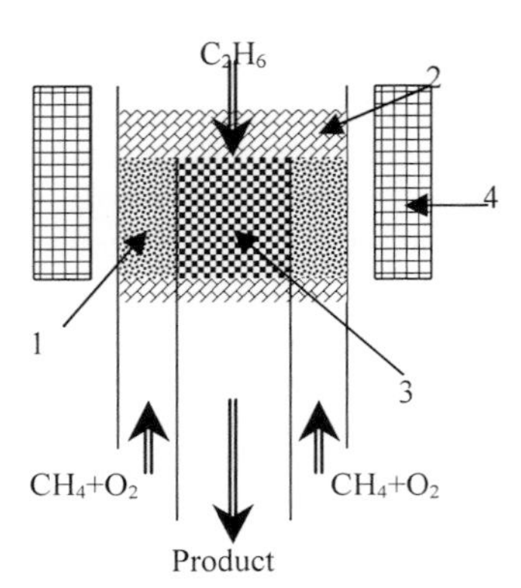

Fig.1 Schematics of the reactor
1 $La_{0.2}Ba_{0.1}/CaO$ 2 quartz sand
3 MnO/SiO_2 4 furnace tube

When the catalysts were at the required temperature the nitrogen flow was stopped and a methane and oxygen mixture was admitted to the reactor. By adjusting the furnace temperature or the composition of CH_4/O_2 mixture, a desired reaction temperature could be achieved at which the oxygen was totally consumed. The metered stream of ethane was then directly injected into the oxygen-free zone between the OCM catalyst and the ODE catalyst and mixed with the OCM product. Thus the OCM process and the reaction of ODE with CO_2 were combined in one fixed-bed reactor.

In the comparison experiments, the $La_{0.2}Ba_{0.1}/CaO$ catalyst or the MnO/SiO_2 catalyst were replaced by quartz sand, and the OCM and ODE with CO_2 were run separately in the 10ml-scale reactor without combining the reactions. For the ODE with CO_2 reaction, the nitrogen was introduced as a diluent with the CO_2 and ethane mixture to the MnO/SiO_2 catalyst in order to simulate the flow condition of the ODE stage of the EPNG process.

The composition of the reactant mixture was controlled by a set of mass flow controllers. The effluent gas from the reactor was analyzed by an on-line gas chromatograph equipped with a thermal conductivity detector after removing water by condensation at 273K. The overall results of the conversions and the ethylene yield were defined as follows:

CH_4 conversion (%)=100(moles of methane consumed/moles of methane introduced)

C_2H_6 conversion (%)=100(moles of ethane consumed/moles of ethane introduced)

CO_2 conversion (%)=100(moles of CO_2 consumed by ODE reaction/ moles of CO_2 produced by OCM reaction)

C_2H_4 yield (%)=200(moles of ethylene produced/total moles of carbon atom introduced)

3. RESULTS AND DISCUSSION

3.1 Coupled effect on the yield of ethylene

To examine the effect of ethylene on the yield, three comparative experiments shown in Fig.1were carried out in the 10-ml scale reactor: OCM process alone, ODE with CO_2 process alone and EPNG Process. The operating conditions are presented in Table 1.

Table 1. Operating conditions for the comparative experiments in Fig. 1

Parameters	OCM process alone[a]	ODE with CO_2 process alone[b]	EPNG Process[c]
Temperature of the OCM part (K)	1023	-	1023
Temperature of the ODE part (K)	-	1073	1073
Feed flow rate (ml/min):			
CH_4	80	0	80
C_2H_6	0	20	20

a CH_4/O_2=2, b CO_2/C_2H_6=1, c $CH_4/O_2/C_2H_6$=4/2/1

At 1023K, oxygen was fully consumed in the OCM experiment. The yield of C_2 hydrocarbons was 13.2% at 34.7% methane conversion and 38.0% C_2 hydrocarbons selectivity, giving an ethylene yield of 3.49 ml/min (Fig.2; the ethylene yield is presented at different flow rates for comparison). During the ODE-CO_2 process alone, nitrogen was added to the OCM catalyst bed and a stoichiometric gas mixture of CO_2 and C_2H_6 (CO_2/C_2H_6=1) was introduced to the catalyst bed. To make a corresponding comparison of these runs, the sum of the flow rate of CO_2 and N_2 was made equal to the flow rate of effluent gases in the OCM process alone. As a result, 3.87 ml/min of ethylene yield was formed in the ODE-CO_2 process (Fig.2).

In the EPNG process, a mixture of methane and oxygen ($CH_4/O_2=2$) was passed through the OCM catalyst bed first, then together with ethane for ODE with CO_2. As can be seen from Fig.2, the EPNG process gave 11.41mL/min of ethylene yield, which is much higher than the sum of the yields of ethylene produced in the OCM process alone (3.49 mL/min) and the ODE with CO_2 process alone (3.87 mL/min). As expected, the EPNG Process, composed of OCM and ODE with CO_2, greatly increases the yield of ethylene. This can be referred to as a coupled effect.

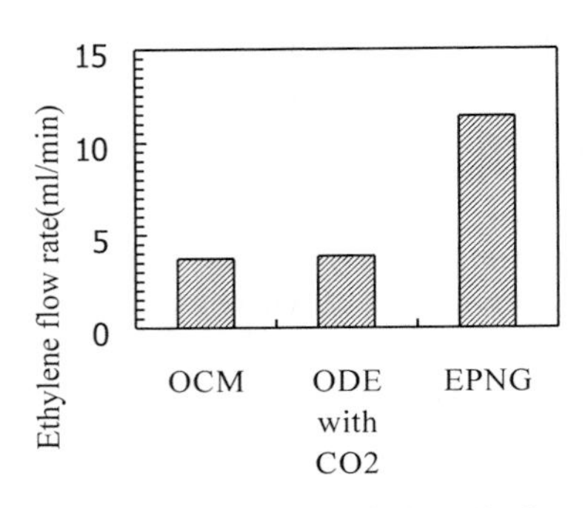

Fig.2 The yields of ethylene in the comparative experiment

3.2 The effect of C_2H_6/CH_4 molar ratio

Fig.3 indicates the conversion of methane, ethane, and carbon dioxide and the yield of ethylene as a function of C_2H_6/CH_4 molar ratio in the feed gases. The results in Fig.3 were obtained at the same conditions as the EPNG run in Table 1 except for the flow rate of ethane fed to the reactor. As the C_2H_6/CH_4 molar ratio in feed increased, the methane conversion gradually decreased, the ethane conversion did not change significantly, while the conversion of carbon dioxide as well as the yield of ethylene increased.

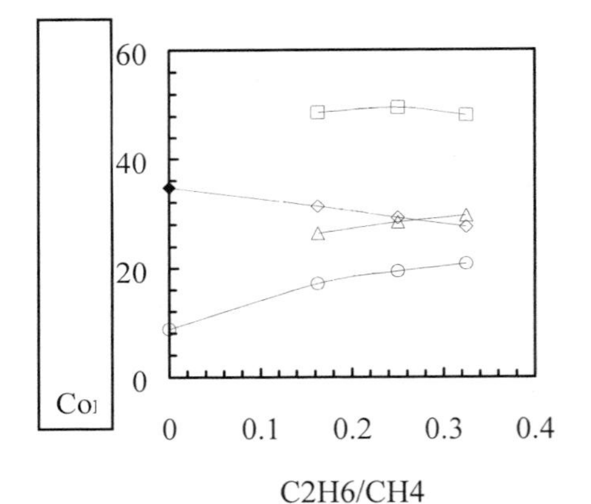

Fig. 3 The influence of C_2H_6/CH_4

According to the product distribution in the EPNG process at various C_2H_6/CH_4 molar ratios in feed, it was inferred that in the ODE with CO_2 zone, ethane is converted to ethylene via two parallel pathways: ethane dehydrogenation with CO_2 (reaction 1) and thermal pyrolysis of ethane. It is estimated that 40~50% of ethane converted ethylene is via reaction 1 and 50~60% is via pyrolysis. Moreover, the side reaction, reforming of ethane and CO_2 to syn-gas (reaction 2), is negligible when the C_2H_6/CH_4 molar ratio in feed less than 0.325.

$$C_2H_6+2CO_2 \rightarrow 4CO+3H_2 \qquad (2)$$

Another side reaction of hydrogenolysis of ethane (reaction 3) also occurred in the ODE with CO_2 zone.

$$C_2H_6+H_2 \rightarrow 2CH_4 \qquad (3)$$

Table 2 Variation of the relative C_2 increment with C_2H_6/CH_4 in the feedstocks*

C_2H_6/CH_4	C_2 in feedstocks (C mol%)	C_2 in products (C mol%)	relative C_2 increment (C_2 in products/C_2 in feedstocks)
0	0	13.2	
0.163	24.6	29.7	1.20
0.250	33.3	36.2	1.09
0.325	39.4	41.2	1.05

*C_2=C_2H_6 + C_2H_4

The amount of the C_2 in the products was always greater than that in the feed for C_2H_6/CH_4 feed molar ratios of 0 to 0.325 (Table 2). However, the higher the C_2H_6/CH_4 molar ratio, the greater the rate of reaction 3. The relative C_2 increment decreased from 1.20 to 1.05 (Table 2) and the overall methane conversion decreased from 34.7% to 27.6% (Fig.3), as the C_2H_6/CH_4 molar ratio in feed increased from 0 to 0.325.

3.3 A bench scale test for the EPNG Process

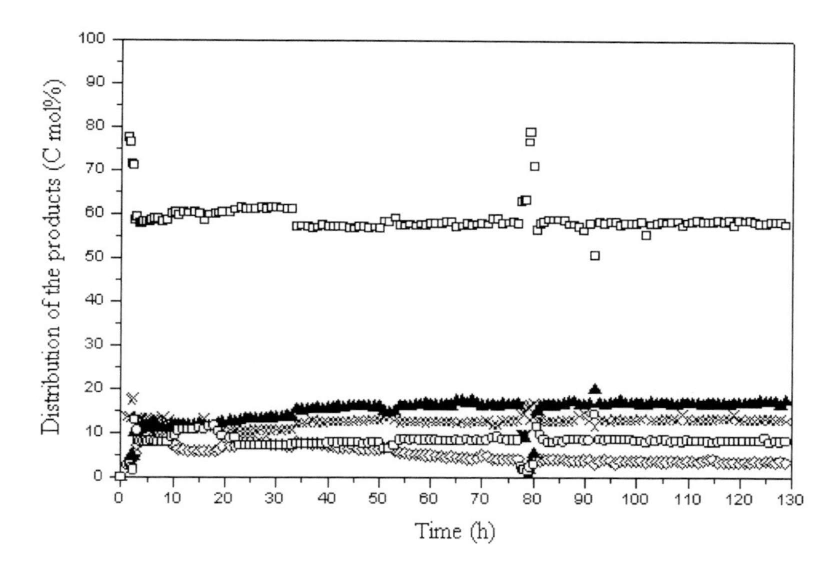

Fig.4 Product distribution of the EPNG Process over the $La_{0.2}Ba_{0.1}/CaO$ and MnO/SiO_2 catalysts at 1073 K (□, CH_4; ▲, C_2H_4; ×, CO_2; O, C_2H_6; ◊, CO). A(0~2.3h): T=1023K, F_{CH4}=1700ml/min, F_{C2H6}=0ml/min, F_{O2}=630ml/min; B(2.3~23.0h): T increased to 1073K, F_{CH4}=1700ml/min, F_{C2H6}=500ml/min, F_{O2}=630ml/min; C(23.0~77.0h): T=1073K, F_{CH4}=1700ml/min, F_{C2H6}=500ml/min, F_{O2}=630ml/min; D(77.0~80.0h): T=993K, F_{CH4}=1700ml/min, F_{C2H6}=0ml/min, F_{O2}=630ml/min; E(80.0~128.8h): T=1073K, F_{CH4}=1700ml/min, F_{C2H6}=500ml/min, F_{O2}=630ml/min.

A bench scale test for the EPNG was carried out with 30mL of the $La_{0.2}Ba_{0.1}/CaO$ catalyst (10~20 mesh) for OCM and 90ml of the MnO/SiO_2 catalyst (10~20 mesh) for ODE with CO_2. The distribution of the EPNG products is given in Fig.4. It should be noted that with an C_2H_6/CH_4 feed ratio of 0.29/1, the EPNG carbon-containing product contained ~17 mol% ethylene. This level of ethylene in the EPNG product gas is much higher than that in the product gas from the OCM alone (~5 mol%).

When the reaction temperature was kept at 1073K, and the flow rate of CH_4, C_2H_6 and O_2 were controlled at 1700 ml/min, 500ml/min and 630ml/min (STP), a constant methane conversion of 26%, ethane conversion of 67%, and a ethylene yield of 27% were achieved in a 130 hour test.

ACKNOWLEDGEMENTS
This work was supported by the Ministry of Science and Technology of China.

REFERENCES

[1] Zh. Sh. Xu, T. Zhang, L. F. Zang, T. Wang, J. R. Wu, L. Y. Chen and L.W. Lin, Chemical Engineering Science 51(1996)515
[2] A. L. Tonkovich, R. W. Carr and R. Aris, Science 262(1993)221
[3] Y. Jiang, I. V. Yentekakis and C. G. Vayenas, Science, 264(1994)1563
[4]E. M. Cordi, S. Pak, M. P. Rosynek, and J. H. Lunsford, Appl. Catal. A 155(1997)L1
[5] J. H. Edwards, K. T. Do and R. J. Tyler, in: Methane Conversion by Oxidative Processes: Fundamental and Engineering Aspects, ed. E. E. Wolf (Van Nostrand Reinhold, New York, 1992) p. 429
[6] H. Mimoun, A. Robine, S. Bonnaudet and C. J. Cameron, Appl. Catal. 58(1990)269
[7] J. H. Edwards, K. T. Do and R. J. Tyler, Fuel 71(1992)325
[8] M. Taniewski and D. Czechowicz, Ind. Eng. Chem. Res. 36 (1997) 4193
[9] M. Taniewski and D. Czechowicz, J. Chem. Technol. Biotechnol. 73 (1998) 304
[10] A. Kh. Mamedov, P. A. Shiryaev, D. P. Shashkin and O. V. Krylov, in: New Developments in Selective Oxidation, eds. G. Centi and F. Trifiro (Elsevier Science Publishers B. V., Amsterdam, 1990) p. 477
[11] S. R. Mirzabekova, A. Kh. Mamedov, V. S. Aliev and O. V. Krylov, React. Kinet. Catal. Lett. 47(1992)159
[12] O. V. Krylov, A. Kh. Mamedov and S. R. Mirzabekova, Catalysis Today 24(1995)371
[13] O. V. Krylov, A. Kh. Mamedov and S. R. Mirzabekova, Ind. Eng. Chem. Res. 34(1995)474
[14] F. Solymosi and R. Nemeth, Catal. Lett. 62(1999)197
[15] Sh. B. Wang, K. Murata, T. Hayakawa, S. Hamakawa and K. Suzuki, Catal. Lett. 63(1999)59
[16] L.Y. Xu, J. X. Liu, H. Yang, Y.Xu, Q. X. Wang and L. W. Lin, Catal. Lett. 62(1999)185
[17] S. Pak and J. H. Lunsford, Appl. Catal. A:General 168(1998)131

Studies in Surface Science and Catalysis
J.J. Spivey, E. Iglesia and T.H. Fleisch (Editors)

Methane conversion via microwave plasma initiated by a metal initiator**

XU Yunpeng, TIAN Zhijian*, XU Zhusheng, ZHU Aimin and LIN Liwu

Dalian Institute of Chemical Physics, the Chinese Academy of Sciences, Dalian 116023, China,

Usually, the conversion of methane in a continuous microwave plasma was conducted under low pressures. In this study, the atmospheric pressure microwave discharge plasma for methane conversion was investigated, and the main useful products were acetylene, ethylene, ethane and hydrogen. 88% acetylene selectivity and 6% ethylene selectivity at 76% methane conversion could be obtained with methane and hydrogen mixtures as the feed gas. A self-designed metal initiator was utilized to maintain the continuous microwave discharges under atmospheric pressure and ambient temperature. The effects of diluting gas were investigated. Hydrogen was found to be able to suppress the formation of coke, while the presence of argon favored the production of coke.

1. INTRODUCTION

Declining crude oil reserves and the great resources of natural gas lead to a large number of researches on methane processing for valuable chemical products [1]. Methane is the main component of natural gas, and it is thermodynamically one of the most stable hydrocarbons, so that selective conversion of methane to more useful organic chemicals is difficult [2]. A number of promising methods for natural gas conversion are under extensive development, one of them is the use of plasma technology for selective activation of methane [2-12]. For example, Kado et al. presented a process of non-catalytic direct conversion of methane to acetylene by using direct current pulse discharge under ambient temperature and pressure. The selectivity of acetylene from methane was >95% at methane conversions ranging from 16 to 52% [7]. Liu et al. converted methane to acetylene by using dc corona discharge with high selectivity and yield under atmospheric pressure in the temperature range of 343-773K, and catalysts were utilized in their study. A high yield of C_2 was obtained in a hydrogen-containing plasma at a flow rate of 10cm^3/min [8]. Other researchers have investigated microwave plasma methods for the conversion of methane to higher hydrocarbons at low pressure [4-6, 9-12], and the main products included acetylene, ethylene and ethane. Onoe et

** Supported by Youth Science Foundation of Laser Technology of China (No.98-11)
* To whom correspondence should be addressed

al. used a microwave plasma reaction to produce acetylene from methane with a selectivity of >90% at a low reaction pressure of 4.5KPa [4]. Wan et al. also reported that pulse high-power microwave radiation was used to convert methane to acetylene, ethylene and ethane over active carbon under a pressure of 100KPa, in their studies, pulse microwave discharge in catalyst bed could be observed [2].

In the present paper, we report on the conversion of methane via a continuous microwave discharge plasma under atmospheric pressure, which was carried out in a home-made reactor. During these experiments, a continuous microwave was used as the power source, and stable continuous microwave discharges could be observed. The main useful products of the experiments were acetylene, ethylene, ethane and hydrogen. The variations in selectivity and conversion with flow rates as well as gas dilutions were also explored.

2. EXPERIMENTAL

The methane used in our experiments was obtained directly from natural gas (98.5% methane, 1.3% ethane and 0.2% nitrogen). Hydrogen (purity of 99.99%) and argon (purity of 99.99%) were used as diluting gases for the experiments. Flow rates of the feed gases were controlled by bubble flow meters. Microwave radiation was supplied by a variable continuous microwave generator with work frequency of 2450 MHz and power of 10 KW. The reaction cavity was a modified section of the wave-guide where a quartz tubular reactor (30cm long and 2.5 cm in diameter) was inserted, and its longitudinal axis was perpendicular to the length-wide side of the rectangular wave-guide. A self-designed metal initiator of microwave discharge plasma was fixed in the center of the reaction zone. The feed gas flew into the reactor through the metal initiator. In the beginning of the experiment, the feed gas was fed into the reactor. After the air in the reactor was replaced by the feed gas completely, the reaction zone was subjected to continuous microwave irradiation with a power of 600 W. By varying the position of the plunger, the microwave cavity could be made to resonate at the working frequency, and the discharge was initiated. A spherical discharge plasma could be observed near the metal initiator, and the diameter of which was about 6 mm. Methane alone, mixtures of methane and argon or mixtures of methane and hydrogen were used as feed gases during the experiments.

The tail gases were analyzed by a GC-950 chromatograph with a flame ion detector (FID) and carbon molecular sieve columns. The molar percentages of methane (C_1), acetylene (C_2), ethylene (C_3) and ethane (C_4) in the tail gases were determined by the external standard method. Trace amounts of higher hydrocarbons except C2 hydrocarbons were found in our products. We could infer that the following four reactions were the main routes of methane conversion in our experiments:

$$a\,CH_4 \rightarrow a/2\,C_2H_6 + a/2\,H_2 \quad (1)$$
$$b\,CH_4 \rightarrow b/2\,C_2H_4 + b\,H_2 \quad (2)$$
$$c\,CH_4 \rightarrow c/2\,C_2H_2 + 3c/2\,H_2 \quad (3)$$

$$d\,CH_4 \rightarrow d\,C + 2d\,H_2 \quad (4)$$

a, b, c and d were respectively the molar percentages of methane conversion through the four reactions mentioned above. The molar percentage of methane in the feed gas was known to be C_1^0. Basing on mass balance, we could describe molar percents of methane, acetylene, ethylene and ethane in the tail gas (C_1, C_2, C_3 and C_4) by a, b, c, d and C_1^0 according to the following four equations:

$$(1\text{-a-b-c-d})C_1^0/((a + 3b/2 + 5c/2 + 2d)C_1^0 + 100 - C_1^0) = C_1/100 \quad (5)$$
$$(C_1^0\, a/2)/((a + 3b/2 + 5c/2 + 2d)C_1^0 + 100 - C_1^0) = C_2/100 \quad (6)$$
$$(C_1^0\, b/2)/((a + 3b/2 + 5c/2 + 2d)C_1^0 + 100 - C_1^0) = C_3/100 \quad (7)$$
$$(C_1^0\, c/2)/((a + 3b/2 + 5c/2 + 2d)C_1^0 + 100 - C_1^0) = C_4/100 \quad (8)$$

From these four equations we could calculate the values of a, b, c and d, and the conversion of methane as well as the selectivities to acetylene, ethylene, ethane and coke could be obtained.

3. RESULTS AND DISCUSSIONS

3.1. Plasma under atmospheric pressure

The data of Tables 1-3 show that methane can be effectively converted to higher hydrocarbons, such as acetylene, ethylene and ethane by the atmospheric pressure microwave discharge plasma. Usually, microwave plasma reactions of pure methane and its mixtures were conducted under low pressures [5] because stable continuous plasmas of those gas were difficult to maintain under atmospheric pressure. Accordingly, the vacuum technique had to be employed to control the plasma of pure methane or its mixtures in those studies.

In the present study, a metal microwave plasma initiator designed by us was utilized for the experiments, which could initiate easily continuous discharge plasmas under ambient temperature and pressure with continuous microwave irradiation. Thus, stable continuous microwave discharges could be produced when pure methane or its mixtures was used as the reagent gas, which usually were difficult to achieve under atmospheric pressure in a microwave field. Furthermore, a reactor designed by ourselves could make the plasma so produced to locate just at the center of the tubular reactor, while the wall of the reactor could not contact with the discharge plasma zone (the diameter of the discharge plasma was only 6mm, but the diameter of the reactor was about 24mm), so that the reactor would not be damaged by the high temperature produced by continuous microwave discharges. However, large amounts of coke produced in the experiment would stick on the wall of the tubular reactor and absorb the microwave to yield a high temperature, and this would exterminate the discharge plasma and damage the reactor. So suppressing the coke formation was important for obtaining stable continuous microwave discharge. A diluting gas could have certain effect on coke formation, and this effect will be discussed in Section 3.2. of the following text.

From Tables 1-3 we can see that the main product with or without diluting gas was acetylene, and the selectivity towards coke varied with different diluting gases fed into the reactor. The conversion of methane attained was between 27-79%, which increased with the total flow rate of the feed gas so long as the microwave power input was kept constant at 600 W during the experiments. The total selectivities towards acetylene, ethylene and ethane could reach as high as almost 100% under appropriate flow rate when certain amounts of hydrogen was used as the diluting gas.

Table 1.
Results of methane reactions without diluting gases via microwave discharge plasma

Total flow rate STP[a] (ml/min)	Conv. of CH_4 (%)	Molar Sele.(%)			
		Acetylene	Ethylene	Ethane	Coke
100	66	74.4	4.5	0.5	20.6
300	42	77.2	3.7	1.6	17.5
400	35	83.2	4.2	2.3	10.3

Reaction conditions: ambient temperature, atmospheric pressure, microwave power of 600W.
[a] Standard temperature and pressure

Table 2.
Results of the reaction of a methane and argon mixture via microwave discharge plasma

Total flow rate STP[a] (ml/min)	Molar ratio of CH_4/Ar	Conv. of CH_4 (%)	Molar Sele.(%)			
			Acetylene	Ethylene	Ethane	Coke
37	0.85	79	71.0	2.0	0	27.0
128	0.9	47	75.3	2.1	1.3	21.3
235	1.0	37	71.2	2.4	2.1	24.3

Reaction conditions: ambient temperature, atmospheric pressure, microwave power of 600W.
[a] Standard temperature and pressure

Table 3.
Results of the reaction of a mixture of methane and hydrogen via microwave discharge plasma

Total flow rate STP[a] (ml/min)	Molar ratio of CH_4/H_2	Conv. of CH_4 (%)	Molar Sele.(%)			
			Acetylene	Ethylene	Ethane	Coke
37	0.6	76	88.0	6.0	0	6.0
87	1.1	49	95.0	3.9	1.1	0
182	1.0	39	95.0	3.0	2.0	0
361	1.1	27	93.0	4.0	3.0	0

Reaction conditions: ambient temperature, atmospheric pressure, microwave power of 600W.
[a] Standard temperature and pressure

3.2. Effects of diluting gases

Table 2 shows that the presence of argon led to more coke formation as compared with the data in Table 1. However, from Table 3 we can see that the presence of hydrogen can suppress the formation of coke and the selectivity towards coke approached almost zero. Under such a condition, we did not observe any coke formation on the wall of the reactor or the metal initiator. This would stabilize the continuous microwave discharge during the experiments.

Previously, Suib et al. have reported that during microwave plasma reaction, H-atom abstraction is the major initial reaction of methane, while hydrocarbon radicals are produced at the same time [5].

$$CH_4 \rightarrow CH_3 + H \quad (9)$$
$$CH_3 \rightarrow CH_2 + H \quad (10)$$
$$CH_2 \rightarrow CH + H \quad (11)$$
$$CH \rightarrow C + H \quad (12)$$

Some steps of this process are reversible [2]:

$$CH_2 + H \rightarrow CH_3 \quad (13)$$
$$CH + H \rightarrow CH_2 \quad (14)$$
$$C + H \rightarrow CH \quad (15)$$

And recombination of the radicals is responsible for product formation [5]:

$$2\ CH_3 \rightarrow C_2H_6 \quad (16)$$
$$2\ CH_2 \rightarrow C_2H_4 \quad (17)$$
$$2\ CH \rightarrow C_2H_2 \quad (18)$$

When hydrogen is fed into the reactor, H atoms are produced in the discharge plasma [2]:

$$H_2 \rightarrow 2\ H \quad (19)$$

This reaction results in an increasing amount of H atoms, which are advantageous to reactions (13), (14) and (15). On the other hand, reaction (15) can eliminate the coke produced by a deep dehydrogenation of methane. Thus, the presence of hydrogen could reduce the formation of the coke in the plasma reaction of methane, enhance the selectivities of higher hydrocarbons and stabilize the continuous microwave discharge.

The effect of argon on methane conversion was different from that of hydrogen, because the former is a good "discharge conductor" [1] and can easily produce discharge under microwave irradiation. It was reported that argon could enhance the conversion of methane [1]. In our work, however, argon might have enhanced deep dehydrogenation of methane, thus resulted in more coke formation.

4. CONCLUSION

A self-designed metal initiator could initiate stable continuous microwave discharge plasma under ambient temperature and atmospheric pressure. Methane could be effectively converted to acetylene, ethylene, ethane and hydrogen by this approach. It was also found that the presence of hydrogen could prevent coke formation, while the presence of argon led to a converse result.

REFERENCES

1. Y.Y. Tanashev, V.I. Fedoseev, Y.I. Aristov, V.V. Pushkarev, L.B. Avdeeva, V.I. Zaikovskii and V.N. Parmon, *Catal. Today*, 42 (1998) 333-336.
2. M.S. Ioffe, S.D. Pollington, and J.K.S. Wan, *J.Catal.*, 151 (1995) 349-355.
3. G. Bond, R.B. Moyes and D.A. Whan, *Catal. Today,* 17 (1993) 427.
4. K. Onoe, A. Fujie, T. Yamaguchi and Y. Hatano, *Fuel*, 76 Iss 3 (1997) 281-282.
5. S. L. Suib and R. P. Zerger, *J.Catal.*, 139 (1993) 383-391.
6. J.H. Huang, and S.L. Suib, *Res. Chem. Intermed.*, 20 (1994) 133.
7. S. Kado, Y. Sekine and K. Fujimoto, *Chem. Commun.*, 24 (1999) 2485-2486.
8. C. Liu, R. Mallinson and L. Lobban, *J. Catal.*, 179 (1998) 326.
9. R.L.McCarthy, *J. Phys. Chem.*, 22 (1954) 1360.
10. Y. Kawahara, *J. Phys. Chem.*, 73 (1969) 1648.
11. R. Mach, H. Drost, J. Rutkowsky, and U. Timm, in "ISPC-7, Eindhoven, 1985,"p.531.
12. S.L. Suib, R.P. Zerger, and Z. Zhang, "Proceedings, Symposium on Natural Gas Upgrading II", p. 344. Div. Petroleum Chem., ACS, Washington DC, 1992.

Studies in Surface Science and Catalysis
J.J. Spivey, E. Iglesia and T.H. Fleisch (Editors)

Production of ethylbenzene by alkylation of benzene with dilute ethylene obtained from ethane oxidative dehydrogenation with CO_2

S-J Xie, L-Y Xu, Q-X Wang, J Bai and Y-D Xu

State Key Laboratory of Catalysis, Dalian Institute of Chemical Physics, Chinese Academy of Sciences, 457 Zhongshan Road, P.O. Box 110, Dalian 116023, China

The oxidative dehydrogenation of ethane with CO_2 over Fe-Mn/Silicalite-2 and K-Fe-Mn/Silicalite-2 catalysts was investigated in a continuous-flow fixed-bed reactor. The selectivity to C_2H_4 was much higher over the K-Fe-Mn/Silicalite-2 compared to the Fe-Mn/Silicalite-2 catalyst. The C_2H_4 thus produced can be used for the production of ethylbenzene via alkylation of C_6H_6 in the presence of a catalyst such as Al-ZSM-5, Al-ZSM-5-β or Al-β because the concentration of C_2H_4 in the reaction effluent-gas is close to the concentration of C_2H_4 in the tail gas of the FCC process. Among the catalysts tested, Al-ZSM-5 is the best for the vapor-phase alkylation of C_6H_6 with dilute C_2H_4 in a fixed-bed reactor. The total selectivity to ethylbenzene is 99%, and the conversion for dilute C_2H_4 is 99%. Meanwhile, the Al-β shows activity for C_6H_6 alkylation with dilute C_2H_4 at relatively low reaction temperatures in the liquid-phase by using either a fixed-bed or a slurry-bed reactor.

1. INTRODUCTION

Ethylbenzene (EB) is an important intermediate for the production of styrene. Conventionally, the production of EB is based on the alkylation of C_6H_6 with pure C_2H_4. A great variety of catalysts have been investigated for the production of C_2H_4 by the oxidative dehydrogenation of ethane (ODE) in the last decade [2, 3]. However, the strict purity requirement for the C_2H_4 prevented their further development and industrial application. Recently, a new process for the production of EB has been developed, which features the alkylation of C_6H_6 with dilute C_2H_4 present in the tail gas of the FCC process over an acidic zeolite catalyst. This process has been commercialized in China [1]. This novel process provides an alternative route for the utilization of dilute C_2H_4.

In the present paper, we report on a new route for the production of EB by combining the alkylation of C_6H_6 with the ODE using CO_2 as an oxidant. The ODE with CO_2 was reported previously in refs. [4-6]. The reaction can be carried out at a temperature lower than that required for the C_2H_6 pyrolysis. Over a newly developed catalyst, K-Fe-Mn/Silicalite-2 (hereafter, Silicalite-2 or Si-2), the ODE with CO_2 shows a high selectivity to C_2H_4 at a high conversion of C_2H_6. By combining the ODE with a CO_2 process and the alkylation of C_6H_6 in the vapor-phase or liquid-phase, EB can be directly produced in the presence of an alkylation catalyst such as ZSM-5, (ZSM-5 + β) or β-zeolite.

2. EXPERIMENTAL

2.1. Preparation of catalysts and their performance in the ODE with CO_2

The catalysts used in this work were Fe-Mn and K-Fe-Mn supported on Si-2 zeolite. Both the Fe-Mn and K-Fe-Mn catalysts were prepared by co-impregnation under a sub-atmospheric pressure. A solution with the desirable amount of $Fe(NO_3)_3$ and $Mn(NO_3)_2$ (and KNO_3 for the K-Fe-Mn) was impregnated onto the Si-2 zeolite. Then, the samples were dried at 400 K for 8 h and calcined at 800 K for 10 h. The samples were cooled, crushed and sorted to a suitable size for use.

The catalysts were investigated in a continuous-flow fixed-bed reactor made of stainless steel. The catalyst charge was 6 mL unless otherwise specified. Reaction conditions were as follows: reaction temperature of 1,073 K, C_2H_6/CO_2 of 1/1 (mole ratio), pressure of 0.1 MPa and GHSV of 1,000 h^{-1}. The products were analyzed on-line by a gas chromatograph with a squalane packed column and a TCD detector, using H_2 as the carrier gas.

2.2. Preparation of catalysts and the alkylation of C_6H_6 with dilute C_2H_4

ZSM-5 zeolite with a SiO_2/Al_2O_3 molar ratio of 50/1 was synthesized in our laboratory. Zeolite β was a product of the Catalyst Plant of the Fushun Petrochemical Corporation in China, having a SiO_2/Al_2O_3 molar ratio of 30/1. Both zeolites were highly crystalline, as determined by X-ray diffraction (XRD).

Table 1
Conditions for alkylation of C_6H_6 with dilute C_2H_4

Reaction condition	Fixed-bed reactor		Slurry-bed reactor
	Gas-phase	Liquid-phase	
Temperature, K	648	443	443
Pressure, MPa	0.6	1.0	1.0
C_6H_6/C_2H_4, molar ratio	5	5	5
WHSV of C_2H_4, h^{-1}	0.6	0.1	0.05-0.2

The as-synthesized zeolites, such as ZSM-5, β-zeolite, and the mixture of ZSM-5 and β-zeolite according to a definite weight ratio, were mixed at first with alumina as a binder. A proper amount of aqueous nitric acid solution was then added. After mixing and binding, it was extruded and calcined to remove some of the organic and inorganic components. The acidic forms of the zeolite catalysts (denoted as Al-ZSM-5, Al-β, Al-ZSM-5-β) were obtained by ion exchange in a 0.8 M NH_4NO_3 solution at 353K for three times and a subsequent calcination at 773 K for 3 h in air. It was then cooled, crushed and sorted for catalytic evaluation with a size of 16~32 mesh in a fixed-bed reactor or with a size of 100~130 mesh in a slurry-bed reactor, respectively.

Alkylation experiments were performed either in a continuous-flow fixed-bed apparatus or in a slurry-bed reactor, both made of stainless steel. The catalyst charge was 6 g for the fixed-bed reactor and 20 g for a slurry-bed reactor. The reaction conditions were listed in Table1.

The gaseous products were analyzed by GC with a squalane packed column and TCD detector, and using H_2 as the carrier gas. The liquid products were analyzed with a FFAP capillary column (30 m×0.25 mm×0.25 μm) and a FID detector, using N_2 as a carrier gas.

The conversion of C_2H_4 is defined as the weight of starting C_2H_4 fed to the reactor minus the recovered weight of C_2H_4 (products normalized to 100%) divided by the weight of the C_2H_4 feed. Selectivity of a product is defined as its moles divided by the moles of C_6H_6 consumed (products normalized to 100%).

2.3. NH_3-TPD measurements

NH_3-TPD measurements were performed in a conventional flow apparatus using a U-shaped microreactor (4mm i.d.) made of stainless steel, with He as the carrier gas. The NH_3–TPD process was monitored by a gas chromatograph with a TCD detector. A catalyst sample of 0.140 g, after pretreatment at 873 K for 0.5 h in a He stream with the flow rate of 25 mL/min, was cooled to 423 K. It was then saturated with a He stream containing NH_3 for 10 min. The sample was purged with a He stream for some time until a constant baseline was attained. NH_3–TPD was carried out in the range of 423-973 K at a heating rate of 18.8 K/min.

3. RESULTS AND DISCUSSION

3.1. Results obtained from the ODE with CO_2

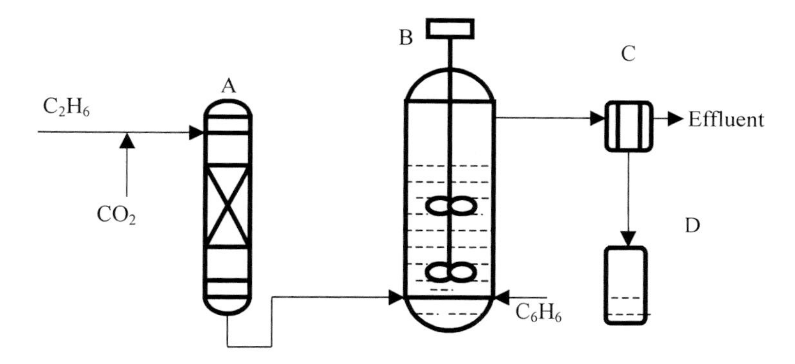

A. Fixed-bed reactor for the ODE
B. Slurry-bed reactor for the alkylation (or a fixed-bed reactor)
C. Cooling vessel for separation of liquid and gas
D. Collector for liquid products

Fig. 1. A sketched flow-chart of the combined process of the ODE with CO_2 and the alkylation of C_6H_6 with dilute C_2H_4.

For this new combined-process of the alkylation of C_6H_6 with dilute C_2H_4 produced from the ODE with CO_2, it is desirable to produce C_2H_4 with a concentration as high as possible from the reaction of ODE with CO_2. The catalytic behavior of the Fe-Mn/Si-2 and K-Fe-Mn/Si-2 catalysts (6 mL) for the ODE with CO_2 were examined. The results listed in Table 2 indicate that the selectivity to C_2H_4 over the K-Fe-Mn/Si-2 catalyst is much higher than that over the Fe-Mn/Si-2 catalyst, and the concentration of C_2H_4 in the effluent-gas over the K-Fe-Mn/Si-2 catalyst is about 25.8%. The reaction was further tested in a bench-scale reactor with K-Fe-Mn/Si-2 catalyst charge of 50 mL, and a C_2H_4 concentration of 23.6% in the effluent-gas could be attained, which is close to the concentration of C_2H_4 in the tail gas of the FCC process. Thus, the C_2H_4 concentration in the effluent-gas is high enough to directly alkylate C_6H_6 for the production of EB. A flow-chart of the combined process for the ODE with CO_2 and the alkylation of C_6H_6 with dilute C_2H_4 is shown in Fig. 1.

Table 2
Typical results for ODE with CO_2 over different catalysts

Catalyst	C_2H_6 conv., %	CO_2 conv., %	Select. to C_2H_4, %	C_2H_4 Content in tail gas, %
Fe-Mn/Si-2	66.2	21.2	89.2	23.0
K-Fe-Mn/Si-2	63.5	25.3	95.1	25.8
K-Fe-Mn/Si-2*	62.9	23.6	93.0	23.6

Reaction conditions: 1023 K, 0.6MPa, $1000h^{-1}$(GHSV), CO_2/C_2H_6 =1/1(mole ratio), and the catalyst charge: 6 mL.
*Catalyst charge: 50 mL.

3.2. Alkylation of C_6H_6 with dilute C_2H_4

3.2.1. Results obtained with the fixed-bed reactor

The effluent-gas from the ODE with CO_2 over the K-Fe-Mn/Si-2 catalyst (with a catalyst charge of 50 mL) is used directly to react with C_6H_6 in a fixed-bed reactor under the reaction conditions of 648 K, 0.6 MPa, C_6H_6/C_2H_4 molar ratio of 5/1 and C_2H_4 WHSV of 0.7 h^{-1} over the Al-ZSM-5, Al-ZSM-5-β and Al-β catalysts, respectively. From the results of alkylation over the three catalysts listed in Table 3, it is found that the Al-ZSM5 catalyst exhibits a very high catalytic activity, with about 99% of C_2H_4 conversion, whereas the catalytic activities of the Al-ZSM5-β and Al-β catalyst under vapor-phase reaction conditions are much poorer than that of the Al-ZSM5 catalyst. This means that the β zeolite is unsuitable for the alkylation of C_6H_6 with dilute C_2H_4 in the vapor-phase reaction. The surface acidities of the three catalysts were evaluated by NH_3-TPD. The peak temperatures of NH_3 desorption and the amounts of acid sites corresponding to the desorbed NH_3, calculated by a mathematical fitting program [7, 8], are summarized in Table 4. The total acidity of the Al-ZSM-5 catalyst is 848.2 μmol/g and the sum of medium and strong acid sites is 708.9 μmol/g, while they are 808.9 and 658.0 μmol/g, respectively, for the Al-ZSM-5-β catalyst, and 672.8 and 558.9 μmol/g, respectively, for the Al-β catalyst. Since the medium and strong acid sites are generally recognized as the main active sites for the alkylation of C_6H_6 [9, 10], the surface acidic properties of the three catalysts are directly related to their activities in the vapor-phase reaction, which are shown in the Table 3.

However, in the liquid-phase and at a low temperature, the Al-β catalyst is a better alkylation catalyst than the Al-ZSM-5 catalyst, which is listed in the bracket of Table 3. This may be attributed to the diffusion limitation, due to the 10-MR openings of the ZSM-5 and the12-MR openings of the β zeolite, besides the surface acidic properties of the catalysts. Smirniotis et al. [11] also reported that the alkylation mechanism is different between the ZSM-5 and the β zeolite catalysts. This demonstrates that the same catalyst behave differently in the vapor-phase and the liquid-phase because of different reaction conditions and/or transportation conditions for the reactants and products.

Table 3
Alkylation of C_6H_6 with dilute C_2H_4 in a fix-bed reactor*

Catalyst	Conv. of C_2H_4, %	Selectivity, %			C_2H_4 Content in Tail gas, %
		EB	DEB	By-products	
Al-ZSM-5	99.2 (11.4)	83.6(87.4)	11.7(12.3)	4.7 (0.3)	0.19 (20.9)
Al-ZSM-5-β	88.5	86.5	10.9	2.6	2.7
Al-β	52.3 (75.8)	88.6(89.4)	9.9(9.8)	1.5 (0.8)	11.2 (5.7)

1. C_2H_4 content in feed gas is 23.6%. ZSM5/β (weight ratio) is 1/1 in the Al-ZSM5-β. Ethylbenzene and diethylbenzene are denoted as EB, DEB, respectively.
2. The data in brackets are obtained under the reaction conditions: 443 K, 1.0 MPa, C_6H_6/C_2H_4 molar ratio of 5/1 and C_2H_4 WHSV of 0.1 h^{-1} (liquid-phase alkylation).

Table 4
Peak temperatures (T_{di}) and the amount of desorbed NH_3 from the NH_3-TPD profiles

Catalyst	Total acidity	Weak acid sites		Medium acid sites (A)		Strong acid sites (B)		μmol/g (A+B)
	μmol/g	μmol/g	T_{d1}, K	μmol/g	T_{d2}, K	μmol/g	T_{d3}, K	
Al-ZSM-5	848.2	139.3	537	194.2	612	514.7	794	708.9
Al-ZSM-5-β	808.9	150.9	531	200.5	598	457.6	741	658.0
Al-β	672.8	113.8	534	247.3	616	311.6	820	558.9

3.2 2. Results obtained on the slurry-bed reactor

Since the slurry-bed reactor possesses many advantages comparing to the fixed-bed reactor, such as a longer cycle of operation, high rates of mass-transfer between gas and liquid phases, and flexibility in loading-unloading of catalysts, the alkylation of C_6H_6 with dilute C_2H_4 was also evaluated in a slurry-bed reactor using the Al-β catalyst. The effect of C_2H_4 WHSV on the performance of the Al-β zeolite was investigated at 443 K, 1.0 MPa, C_6H_6/ C_2H_4 of 5/1(mole ratio), and 20 gram loading of catalyst having a particle size of 100-130 mesh. The results shown in Fig. 2 indicate that the variation of the C_2H_4 WHSV has obvious influence on the C_2H_4 conversion. With an increment of C_2H_4 WHSV from 0.05 to 0.20 h^{-1}, the C_2H_4 conversion decreased from 93.5 % to 45.0 %, while the total selectivity of EB and DEB remained stable (~ 99.5%). In addition, under the same reaction conditions (443 K, 1.0MPa, C_6H_6/C_2H_4 molar ratio of 5/1 and C_2H_4 WHSV of 0.1 h^{-1}), the Al-β catalyst exhibited higher alkylation activity in the slurry-bed reactor than in the fixed-bed reactor due to the better mass transfer in a slurry-bed. According to the above results, we can conclude that the slurry bed technique can enhances the production of EB from the alkylation of C_6H_6 with the dilute C_2H_4 from ODE with CO_2, as compared to the fixed-bed technique in the liquid-phase reaction.

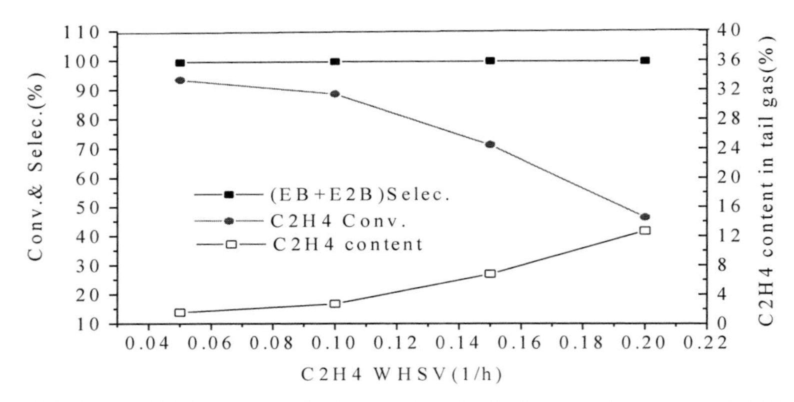

Fig. 2. Effect of C_2H_4 space velocity on the alkylation performance of Al-β catalyst

ACKNOWLEDGEMENT

The authors are grateful to the support of the National Natural Science Foundation of China.

REFFRENCES

1 . Q. Wang, S. Zhang, G. Cai, F. Li, L. Xu, Z. Huang and Y. Li, US Patent 5 869 021 (1999).
2. F. Cavani and F. Trifirò, Catal. Today 24 (1995) 307.
3. T. Blasco and J.M. Lopez-Nieto, Appl. Catal. A. 157 (1997) 117.
4. L. Xu, L. Lin, Q. Wang, L. Yang, D. Wang and W. Liu, Chin. J. Catal.18 (1997) 353.
5. L. Xu, L. Lin, Q. Wang, L. Yang, D. Wang and W. Liu, Stud. Surf. Sci. Catal.119 (1998) 605.
6. L. Xu, Q. Wang, D. Liang, X. Wang, L. Lin, W. Cui, Y. Xu, Appl. Catal. A.173 (1998) 19.
7. H.G. Karge, V. Dondur, J. Phys. Chem.94 (1990) 765.
8. H.G. Karge, V. Dondur and J. Weitkamp, J. Phys. Chem. 95 (1991) 283.
9. K.A. Becker, H.G. Karge and W.D. Streubel, J. Catal. 28 (1973) 403.
10. J.R. Anderson, T. Mole and V. Christov, J. Catal.61 (1980) 477.
11. P.G. Smirniotis, E. Ruckenstein, Ind. Eng. Chem. Res. 34 (1995) 1517.

Studies in Surface Science and Catalysis
J.J. Spivey, E. Iglesia and T.H. Fleisch (Editors)

Oxidative dehydrogenation of ethane with carbon dioxide to ethylene over Cr-loaded active carbon catalyst

YANG Hong, LIN Liwu, WANG Qingxia*, XU Longya, XIE Sujuan and LIU Shenglin

Dalian Institute of Chemical Physics, Chinese Academy of Sciences, P.O.Box 110, Dalian 116023, P. R. China

Oxidative dehydrogenation of ethane with carbon dioxide to ethylene was investigated in a fixed-bed flow micro-reactor under ambient pressure. Chromium oxide was found to be the best catalyst among activated carbon(AC) supported Fe-, Mn-, Mo-, W- and Cr-oxide catalysts. At 823K~923K, C_2H_4 selectivity of 69.6-87.5% at a C_2H_6 conversion of 8.5%~29.2% could be achieved over the Cr/AC catalyst. In this reaction, carbon dioxide facilitates the dehydrogenation of ethane and enhances ethane conversion and ethylene yield. Moreover, coke deposition on the Cr/AC catalyst was retarded greatly in the presence of carbon dioxide. Regeneration of the catalyst after dehydrogenation reactions with CO_2 was also attempted at 973K for 1h under an atmosphere of CO_2. The activity of the catalyst could be restored to a certain extent, but the original activity of the catalyst could not be recovered completely. Coke deposition and changes in the surface state of the catalyst are believed to be the reasons for catalyst deactivation.

1. INTRODUCTION

Up to now, production of C_2H_4 from C_2H_6 has been carried out by the conventional steam cracking process, in which a very high temperature (about 1103K) and a large amount of steam are needed. In the last decade, some researchers have become interested in the oxidative dehydrogenation of C_2H_6 in the presence of oxygen as an alternative to the conventional process[1]. However, due to the strong exothermic character of this reaction and the strong reactivity of oxygen, it is necessary to remove heat to avoid over-oxidation of ethylene to carbon oxides. Therefore, there is an incentive to develop more selective catalysts that can depress deep oxidation, and to explore less reactive oxidants to replace the oxygen in the conventional process.

Recently, a new process for the production of C_2H_4 from C_2H_6 was reported [2], based on the oxidative dehydrogenation of ethane (ODE) with CO_2. In this process, carbon dioxide, which is one of the major green house gases, is a mild oxidant that can replace oxygen. Catalytic hydrogenation of carbon dioxide into organic compounds has also been intensively studied[3,4]. Solymosi et al.[5] investigated oxidative dehydrogenation of ethane by carbon dioxide over a Mo_2C/SiO_2 catalyst, and obtained 90-95% ethylene selectivity and ethane conversion of 8-30% at 823-923K. Wang et al. [6] investigated the oxidative dehydrogenation of ethane by carbon dioxide over sulfate-modified Cr_2O_3/SiO_2 catalysts, and attained an

* Corresponding author.

ethylene yield of 55% at 67% ethane conversion at 923K.

In our laboratory, Xu et al. [7] have studied the performance of a K-Fe-Mn/Si-2 catalyst for the ODE with CO_2 to C_2H_4. Their results indicated that over this catalyst an ethane conversion of 66% with an ethylene selectivity of 93% at 1073K could be obtained. In the present paper, we report our results on a new catalyst of Cr_2O_3 supported on active carbon, which exhibits high activity and selectivity of ODE with CO_2 to C_2H_4 at 823K-923K.

2. EXPERIMENTAL

All chemicals were purchased from commercial sources, and were used without further purification. Granular active carbon (Bejing Guang-hua wood factory, 20-40 mesh, $1000m^2/g$) was used as a support.

Cr/AC, Fe/AC, Mn/AC, Mo/AC and W/AC were prepared by impregnating active carbon with an appropriate amount of $Cr(NO_3)_3$, $Fe(NO_3)_3$, $Mn(NO_3)_2$, $(NH_4)_2MoO_4$ and $(NH_4)_2WO_4$ aqueous solution, respectively, then dried in air at 393K for 6h, and calcined in argon at 823K for 2h.

The catalysts were investigated under ambient pressure in a fixed-bed flow micro-reactor with an internal diameter of 8mm, and 1 mL of catalyst with a particle size of 0.25-0.37mm was employed. An Eu-2 type thermocouple was placed in the center of the catalyst bed to control the reaction temperature. The reactant stream, consisting of 50% ethane and 50% carbon dioxide (or C_2H_6:Ar=1:1) was introduced into the reactor at the flow rate of 20 mL/min. Tail gases were taken for analysis after 20 min on-stream time at each temperature.

Reaction products were analyzed by an on-line gas chromatograph equipped with a thermal conductivity detector. H_2 was analyzed with a carbon molecular sieve column using argon as the carrier gas. Analyses of CO, CH_4, CO_2 and C_2 hydrocarbons were carried out by using the same column, but with hydrogen as the carrier gas. C_3 and C_4 hydrocarbon gases were analyzed with a squalane column, and hydrogen as the carrier gas. An internal standard method was employed to calculate the conversion of C_2H_6, the C_2H_4 yield, as well as the selectivities of C_2H_4, coking, CO and CH_4, all based on carbon balance of ethane in the system. The yield of H_2O was calculated based on the hydrogen balance of ethane in the system. Argon was used as the internal standard substance.

3. RESULTS AND DISCUSSION

3.1. Performance of catalysts with different components supported on AC

A series of M/AC(M denotes a metal oxide)catalysts were prepared with the same metal content (6%wt) and with the same preparation condition to compare the activities of different metals. The results are presented in Table 1. The AC support itself only showed 2.6% CO_2 conversion, 7.6% C_2H_6 conversion and 5.7% C_2H_4 yield at 923K. However, impregnation of Cr on AC brought about a striking change in the catalytic performance, which resulted in 28.9% C_2H_6 conversion, 23.5% CO_2 conversion and 20.3% C_2H_4 yield at 923K and a space velocity of $1200h^{-1}$. The results were different when Mn, Fe, Mo and W were used to replace the Cr component in the 6Cr/AC catalyst. For Fe/AC, Mn/AC and Mo/AC catalysts, the conversions of C_2H_6 and CO_2 higher than those with the AC support, but much lower than those of the Cr/AC catalyst. In addition, it was found that the C_2H_6 conversion and C_2H_4 yield were similar for the Fe/AC and Mn/AC catalysts and slightly higher than those of the Mo/AC catalyst. However, the W/AC catalyst showed the lowest activity, even lower than the AC support. This indicates that W had a retarding effect on the reaction of ethane dehydrogenation. Although the C_2H_4 selectivities of the AC-supported Fe, Mn, Mo, W catalysts were higher than that of the Cr/AC catalyst, based on the ethane conversions and ethylene yields, the catalytic activities of all catalysts follow the order of Cr/AC > Fe/AC ≈ Mn/AC > Mo/AC > AC > W/AC.

Table 1
Catalytic performance of different AC supported metal oxides for ODE with CO_2 to C_2H_4*

Catalyst	Conv.C_2H_6 (%)	Conv.CO_2 (%)	C_2H_4 Yield (%)	Selectivity (%)		
				C_2H_4	CH_4	CO
AC	7.6	2.6	5.7	75.3	16.0	8.7
6Cr/AC	28.9	23.5	20.3	70.5	14.8	14.6
6Fe/AC	9.9	13.8	7.5	76.0	11.5	12.5
6Mn/AC	10.0	11.8	7.5	75.2	13.2	11.6
6Mo/AC	8.8	5.6	6.4	72.5	13.4	14.1
6W/AC	7.5	0.74	5.5	73.2	21.3	5.5

Reaction conditions: 923K, 0.1MPa, $1200h^{-1}$,$C_2H_6/CO_2=1/1$
6Cr/AC means 6wt % Cr.
*Coke was not calculated.

3.2. Effect of temperature on the reaction performance of the 6Cr/AC catalyst

The effect of temperature on the performance of non-catalytic ODE with $CO_2(C_2H_6/CO_2=1:1)$ to C_2H_4 was first examined, and the results are shown in Table 2. Below 923K, the non-catalytic ODE with CO_2 to C_2H_4 was small. With an increase in the reaction temperature, both the C_2H_6 conversion and the C_2H_4 yield increased sharply, while the selectivity to ethylene decreased. During the reaction process, CH_4 and coke were by-products, and no CO could be detected. Therefore, for the non-catalytic ODE with CO_2 to C_2H_4, the main reaction is the dehydrogenation of ethane. From these results of non-catalytic reaction, we can infer that to avoid the effect of non-catalytic reaction during the catalytic reaction of ODE with CO_2 to C_2H_4, the reaction temperature should be lower than 923K. In addition, the reaction performance of active carbon with CO_2 over the 6Cr/AC catalyst was examined. The results indicated that active carbon does not react with CO_2 below 923K.

Table 3 shows the results of ODE with CO_2 to C_2H_4 over the 6Cr/AC catalyst below 923K. Even at a lower temperature of 823K, CO_2 conversion of 7.7% and C_2H_6 conversion of 8.5% could be obtained. By increasing the reaction temperature, both C_2H_6 conversion and CO_2 conversion increased, and C_2H_4 yield was improved from 7.5% at 823K to 20.3% at 923K. Unfortunately, C_2H_4 selectivity decreased from 87.5% at 823K to 69.6% at 923K. The rise in temperature also elevated the formation of CH_4, CO and coke. These results indicate that an increase in temperature favors the activation of CO_2 and C_2H_6, but does not increase C_2H_4 selectivity. Therefore, in order to obtain optimum C_2H_4 yield, we must choose an appropriate reaction temperature.

From the above results, we can also visualize that the process of ODE with CO_2 to C_2H_4 is comprised of the following reactions:

$$C_2H_6 \rightarrow C_2H_4 + H_2 \quad (1)$$
$$H_2 + CO_2 \rightarrow CO + H_2O \quad (2)$$
$$C_2H_6 + H_2 \rightarrow 2CH_4 \quad (3)$$
$$C_2H_6 \rightarrow 2C + 3H_2 \quad (4)$$
$$C + CO_2 \rightarrow 2CO \quad (5)$$

Table 2
Effect of temperature on the performance of noncatalytic ODE with CO_2 to C_2H_4

Temp(K)	C_2H_6 conv. (%)	C_2H_4 Yield (%)	Selectivity(%)		
			C_2H_4	CH_4	Coke
923	1.43	1.3	91.2	3.6	5.2
973	6.8	6.1	89.7	4.2	6.1
1023	22.3	19.5	87.5	6.9	6.5

Reaction conditions: 0.1MPa, $1200h^{-1}$, feed gas ratio: C_2H_6/CO_2=1/1

Table 3
Effect of temperature on the performance of ODE with CO_2 to C_2H_4 over the 6Cr/AC catalyst

Temp(K)	C_2H_6 conv. (%)	CO_2 conv. (%)	Yield(%)		H_2/CO	Selectivity(%)			
			C_2H_4	H_2O		C_2H_4	CH_4	CO	Coke
823	8.5	7.7	7.5	4.9	0.39	87.5	8.1	4.6	0.8
873	16.3	12.9	12.5	8.9	0.46	76.7	11.3	10.8	1.2
923	29.2	23.5	20.3	16.0	0.59	69.6	14.4	14.2	1.8

Reaction conditions: 0.1MPa, $1200h^{-1}$, feed gas ratio: C_2H_6/CO_2=1/1

3.3. The effects of CO_2 on the dehydrogenation of C_2H_6 to C_2H_4 over the 6Cr/AC catalyst

Takahara et al. [8] studied the effect of carbon dioxide in the dehydrogenation of C_3H_8 to C_3H_6, and found that CO_2 has a promoting effect on the yield of C_3H_6 and suppresses catalyst deactivation. Suzuki et al. [9] also investigated the dehydrogenation of isobutane to isobutene over an iron/activated carbon catalyst in the presence of carbon dioxide. They found that co-feeding with carbon dioxide promoted the dehydrogenation of isobutane. Moreover, the hydrogen produced by dehydrogenation was transformed into water in the presence of carbon dioxide, which in turn was reduced to carbon monoxide. So they concluded that carbon dioxide favored the redox cycle of Fe_3O_4 and metallic iron on activated carbon.

In order to understand the role of CO_2 in the dehydrogenation of C_2H_6 to C_2H_4, experiments were carried out in the present work with alternating feeds of C_2H_6/Ar and C_2H_6/CO_2 over the 6Cr/AC catalyst and the results are presented in Table 4. In the presence of CO_2, the C_2H_4 selectivity was lower than that when C_2H_6/Ar was used as the feed gas. This is because the addition of CO_2 promoted the production of CO from ethane. However, the conversion of C_2H_6 was much higher than in the case of C_2H_6/Ar. Consequently, the C_2H_4 yield was 60% higher than that when without CO_2. According to calculations based on thermodynamic equilibrium, the C_2H_6 conversion of $C_2H_6 \rightarrow C_2H_4 + H_2$ (1) and $CO_2 + C_2H_6 \rightarrow C_2H_4 + CO + H_2O$ (6) are 26% and 32%, respectively. Hence, theoretically the presence of CO_2 should increase the yield of C_2H_4, and our results have confirmed it. The reason for this enhancement is obvious. When CO_2 participates in the dehydrogenation of ethane, reaction (6) in fact involves two consecutive coupling reactions of reaction (1) and reaction (2). The reaction of

CO_2 with H_2 evolved from the dehydrogenation of C_2H_6 can continuously shift C_2H_6 to C_2H_4, thus enhancing the yield of C_2H_4. The mechanism for this reaction is still under investigation.

It is worth pointing out that CO_2 not only can promote the dehydrogenation of C_2H_6, but also can suppress the formation of coke. The coking selectivity when using C_2H_6/Ar as the feed was far greater than that of when using C_2H_6/CO_2 on the 6Cr/AC catalyst, i.e., coke deposition was retarded by CO_2.

Table 4
The results of C_2H_6 dehydrogenation to C_2H_4 in the presence and in the absence of CO_2 over the 6Cr/AC catalyst

Feed gas	C_2H_6 conv.(%)	CO_2 conv.(%)	Yield(%)		Selectivity (%)			
			C_2H_4	H_2O	C_2H_4	CH_4	CO	Coke
C_2H_6/CO_2	29.2	23.5	20.3	16.0	69.6	14.4	14.2	1.8
C_2H_6/Ar	15.7	----	12.6	---	80.2	10.4	---	9.4

Reaction conditions: 923K, 0.1MPa, C_2H_6/CO_2=1/1, C_2H_6/Ar=1/1, HSV=1200h^{-1}

3.4. Duration test of Cr/AC for the ODE with CO_2 to C_2H_4

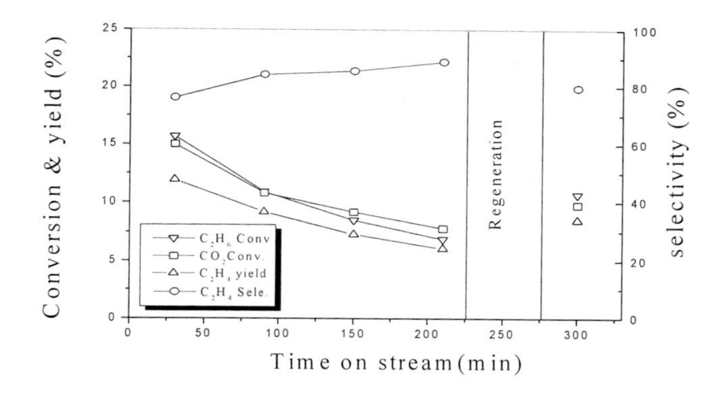

Fig.1. Regeneration of 6Cr/AC catalyst used for ODE with CO_2.
Reaction conditions: 873K, 0.1MPa, 1200h^{-1}, C_2H_6/CO_2=1/1;
Regeneration conditions: CO_2 flow rate: 30ml/min

The duration test of the 6Cr/AC catalyst for the ODE with CO_2 to C_2H_4 was carried out, and the result is shown in Fig.1. After running for 4h at 873K, 1200h^{-1} and C_2H_6/CO_2=1/1, the C_2H_6 conversion and the C_2H_4 yield decreased from 15.7% and 11.9% to 6.6% and 5.8%, respectively. Then, the reaction was interrupted. Under an argon stream, the temperature was raised to 973K, and CO_2 was introduced to regenerate the catalyst. After CO_2 treatment at 973K for 1h, the temperature was decreased to 873K in an atmosphere of Ar. The feed gas was introduced to the reactor instead of argon. Regeneration results show that C_2H_6 conversion and C_2H_4 yield increased to 10.7% and 8.5%, respectively, but the original activity of the catalyst could not be restored. Nevertheless, this implies that CO_2 is an effective regeneration agent for the 6Cr/AC catalyst.

Lunsford et al. [10] have reported that in the dehydrogenation of C_2H_6, the coke required to poison the Cr/SiO_2 catalyst was only ca. one carbon atom per chromium atom, and infrared evidence indicates that this coke is selectively deposited on the active sites. The authors suggest that the Cr/SiO_2 catalyst was poisoned by carbonaceous deposits during the dehydrogenation of ethane. Zhou et al. [11] reported that dehydrogenation of propane over CrO_X/SiO_2 catalyst under CO_2 atmosphere and found that the active centers of the catalyst for the reaction were Cr^{5+} species, which could be reduced to Cr^{3+} in the process of the dehydrogenation of propane. They suggested that the reduction of active sites was the major cause of catalyst deactivation for the dehydrogenation of propane with CO_2. The above observations show that coke deposition and the reduction of Cr oxides might be the main reasons for catalyst deactivation in the dehydrogenation of C_2H_6 with CO_2 to C_2H_4. During the regeneration process, CO_2 as a regeneration agent may have two effects: (1) to oxidize the coke which has selectively poisoned the active sites (2) to change the Cr surface state – from a low valance state to a high valance state.

REFERENCES

1. E. Morales and J.H. Lansford, J. Catal., 118(1989)255.
2. O.V. Krylov, A.Kh. Mamedov, S.R. Mirzabekova, Catal.Today, 24(1995)371.
3. S. Wang, G.Q. Lu and G.J. Millar, Energy Fuels, 10(1996)896.
4. T. Nishiyama and K. Aika, J.Catal., 122(1990)346.
5. F. Solymosi and R. Nemeth, Catal.Lett., 62(1999)197.
6. S. Wang, K. Murata, T. Hayakawa, S.Hamakawa and K. Suzuki, Catal. Lett.,63(1999)59.
7. L. Xu, J. Liu, H. Yang, Q. Wang and L. Lin, Catal. Lett., 62(1999)185.
8. I. Takahara and M. Saito, Chem.Lett., 11(1996)973.
9. H. Shimada, T. Akazawa, N. Ikenaga and T. Suzuki, Appl. Catal., A, 168(1998)243.
10. H.J. Lugo and J.H. Lunsford, J. Catal., 91(1985)155.
11. H. Zhou, X. Ge, M. Li, R. Shangguan and J. Shen, Chinese J. Inorg. Chem., 16(2000)775.

Studies in Surface Science and Catalysis
J.J. Spivey, E. Iglesia and T.H. Fleisch (Editors)

Catalytic performance of hydrothermally synthesized Mo-V-M-O (M=Sb and Te) oxides in the selective oxidation of light paraffins

Kenzo Oshihara, Tokio Hisano, Youhei Kayashima, and Wataru Ueda

Department of Materials Science and Engineering, Science University of Tokyo in Yamaguchi
1-1-1 Daigaku-dori, Onoda, Yamaguchi, 756-0884 JAPAN
E-mail: oshihara@ed.yama.sut.ac.jp, TEL/FAX: 81-836-88-4559

Hydrothermally synthesized $Mo_6V_2Sb_1O_x$ and $Mo_6V_3Te_1O_x$ mixed oxide catalysts were tested for methane, ethane and propane oxidation. These catalysts were very active for an oxidative dehydrogenation of ethane with 80% of the ethylene selectivity, which was almost independent of the reaction temperature from 300 to 400°C. The grinding-treatment made the catalyst more selective by suppressing the CO_x production. These catalysts were inactive for methane oxidation but they were highly active for propane oxidation to give acrylic acid. These catalysts were rod-shaped crystal and the cross-section was suggested to be active and selective face. The function of the active face of the catalysts for alkanes selective oxidations was discussed.

1. INTRODUCTION

Selective oxidation of alkanes with molecular oxygen in gas phase has been received much attention. Since alkanes are very stable molecules, a highly oxidative catalyst is required for their activation. However, since the partially oxidized products are generally more reactive than alkanes, highly oxidative catalyst cannot be used. In spite of the difficulty, some catalysts successfully achieved the selective partial oxidations.

V-P-O system is the most well known catalyst that is active for n-butane oxidation to maleic anhydride. It was reported that a (100) plane of hexagonal layered crystal of the catalyst is active and selective to maleic anhydride production.[1,2] This fact suggests that the structure of the active face is important to achieve the selective partial oxidation of alkane.

We have synthesized mono phasic Mo-V-M-O (M=Al, Cr, Fe) mixed oxide, having rod-shaped crystals, by hydrothermal method using Anderson-type heteropolymolybdates as precursors. The catalysts were active for ethane oxidation to acetic acid,[3] and we suggested that a cross-section of the rod-shaped crystal is an active face for the ethane oxidation.

Recently, Mo-V-Te-Nb-O mixed oxide was reported from Mitsubishi chemicals, which is active for propane ammoxidation.[4,5] We also synthesized $Mo_6V_3Te_1O_x$ mixed oxide, also having a rod-shaped crystal, by hydrothermal method using TeO_2 and $(NH_4)_6Mo_7O_{24}$, and $VOSO_4$. The catalyst exhibited high activity for propane oxidation to acrylic acid,[6] and the grind-treatment, increasing a cross-section of the rod-shaped crystal, further enhanced the activity and selectivity to acrylic acid. In this paper, we investigated methane, ethane and propane oxidations over $Mo_6V_3Te_1O_x$ and $Mo_6V_2Sb_1O_x$ catalysts. We will discuss the effect of pre-treatment and the functions of $Mo_6V_2Sb_1O_x$ and $Mo_6V_3Te_1O_x$ catalysts.

2. EXPERIMENTAL

2.1. Catalyst preparation

The hydrothermal synthesis of Mo-V-M-O was carried out as follows. Atomic content ratio of mother liquid was Mo: V : Te = 6 : 3 : 1 and Mo : V : Sb = 6 : 2 : 1, respectively. $(NH_4)_6Mo_7O_{24}$ was dissolved into 20 ml of water and heated at 80°C. $Sb_2(SO_4)_3$ or TeO_2 powder was added into the solution and stirred for 1 minute (Te) or 15 minutes (Sb). $VOSO_4$ aqueous solution was added to the solution (the liquid was changed into dark purple slurry). The slurry was moved to an autoclave with PTFE inner tube and was kept under hydrothermal condition at 175°C for 24 h ($Mo_6V_2Sb_1O_x$) or 72h ($Mo_6V_3Te_1O_x$), respectively. After hydrothermal synthesis, the autoclave was cooled to room temperature by water flow. Inside the tube, dark purple solid was produced on the wall and bottom, and the solid was filtered, and washed three times by 20ml of water, and it was dried in air at 80 °C for 6 hours. Before a catalytic test, three types of treatment were performed on the corresponding samples. The first was 'air calcination' and the second was 'N_2 calcination', and the third was 'grinding treatment'. Air calcination was carried out in a muffle furnace by heating at 280 °C for 2 hours before 'N_2 calcination'. N_2 calcination was performed on all the samples by heating at 600°C for 2h in 50 ml/min of N_2 flow. Grinding treatment was carried out by strongly grinding a sample in an agate mortar for 10 minutes after 'N_2 calcination'.

2.2. Characterization of Mo-V-M-O catalysts

All the samples of $Mo_6V_2Sb_1O_x$ and $Mo_6V_3Te_1O_x$ weakly ground and characterized by powder X-ray diffraction, RINT Ultima$^+$ (Rigaku) equipped with Cu Kα irradiation, after hydrothermal synthesis, after calcination, and after reaction.

SEM images of dried fresh samples were collected by JSM-T100 (JEOL). Samples were dispersed into acetone by ultra sonic. The suspension was then dropped on a holder and dried. After gold deposition, the samples were analyzed.

2.3. Ethane oxidation

A catalytic test was performed on a fixed bed flow reactor. The standard reaction conditions were as follows: 1.0 g of catalyst was loaded in Pyrex glass tube, and the sample was heated to 300 °C in N_2 flow. Reactant gas mixture composed with C_2H_6 : O_2 : N_2 : H_2O (vapor) = 15 : 5 : 20 : 10 (ml·min^{-1}) was introduced at 300°C. The reaction temperature was ranged from 300 to 400°C with an interval of 20°C. Inlet and outlet gases were analyzed with gas chromatographs GC-14B (Shimadzu) and GC-8A (Shimadzu) equipped with MS-13X and Porapak QS and Gasukuropack 54 columns.

3. RESULTS AND DISCUSSION

3.1. Characterization

XRD patterns of $Mo_6V_2Sb_1O_x$ and $Mo_6V_3Te_1O_x$ were displayed in Fig. 1. Both samples gave quite similar patterns. Intense and sharp peak at 2θ=22.1°, which was commonly observed for hydrothermally synthesized Mo-V based oxide, and characteristic three peaks at 2θ=6.5°, 7.8°, and 9.0° were observed. These XRD patterns indicated that the samples had orthorhombic unit cell with lattice parameters a=26.6Å, b=21.2Å, and c=4.01Å. After the

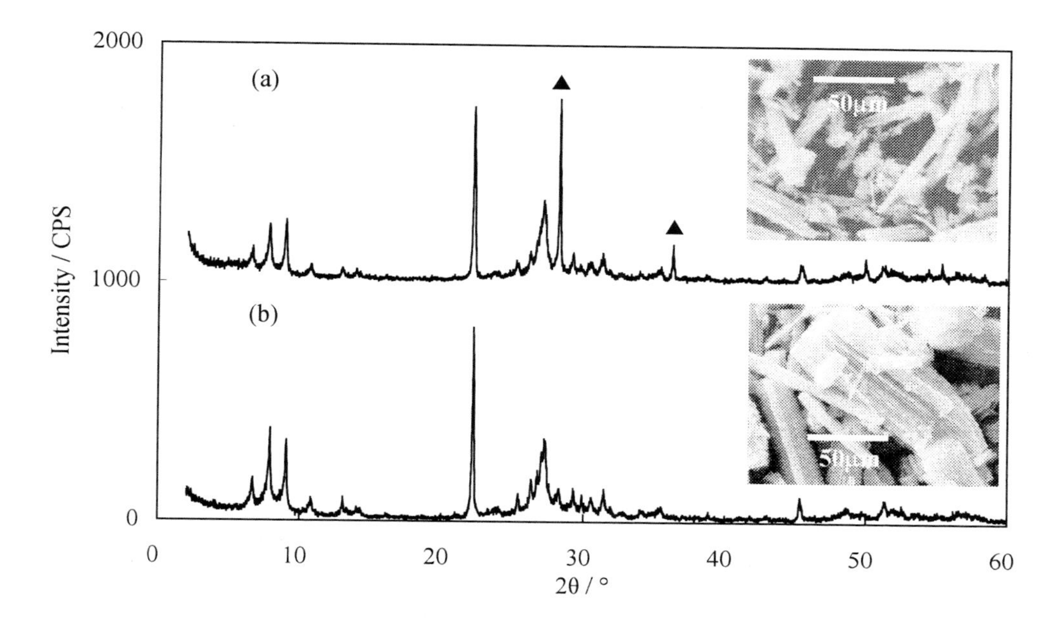

Fig. 2 XRD patterns and SEM images of hydrothermally synthesized $Mo_6V_2Sb_1O_x$ (a) and $Mo_6V_3Te_1O_x$ (b) catalysts. ▲: $Sb_4Mo_{10}O_{31}$

treatment in air at 280°C, the intensities of the peaks were scarcely changed. But after the treatment in N_2 at 600°C, the intensities were increased indicating the rearrangement of lattice structure by heating. The grinding treatment did not affect to the XRD patterns indicating that the catalyst structure was maintained after the treatment.

In the case of $Mo_6V_2Sb_1O_x$ sample, other peaks at 2θ=28.2°, 36.2°, and 49.9° were also observed. These peaks were assigned to a similar phase of $Sb_4Mo_{10}O_{31}$.[7] The intensity of the peaks of the phase was increased along with the stirring time of the suspension containing $(NH_4)_6Mo_7O_{24}$ and $Sb_2(SO_4)_3$ at the preparation step. In this research, the stirring time was fixed to 15 minutes because the catalyst was most active.

SEM images were also displayed in Fig. 1. In the both samples, rod-shaped crystals were observed. The characteristic rod-shaped crystal was also common to hydrothermally synthesized Mo-V based oxide.[8] The rod-shaped crystals were kept after heat-treatment at 280°C in air, and/or, at 600°C in N_2. However, after the grinding treatment, the rod-shaped crystals were disappeared, only rough particles were observed instead of them.

3.2. Effect of pretreatment

The results of ethane oxidation over variously treated catalysts were summarized in Table 1. The $Mo_6V_2Sb_1O_x$ and $Mo_6V_3Te_1O_x$ catalysts heat-treated in N_2 at 600°C showed high ethylene selectivity 84%. When the catalysts were ground, ethane conversion was increased and the CO_x production was suppressed simultaneously. The grinding treatment is found

Table 1
Ethane oxidation over hydrothermally synthesized Mo-V-M-O (M=Sb, Te) catalysts.

Catalyst	Treatment			BET area [c]	Conversion / % [d]		Selectivity / %			
	Air [a]	N_2 [b]	Grind	/ $m^2 \cdot g^{-1}$	C_2H_6	O_2	C_2H_4	AcOH	CO	CO_2
$Mo_6V_2Sb_1O_x$		Yes		2.9	4.0	8.8	84.2	1.0	9.8	5.0
		Yes	Yes	3.7	7.9	15.6	89.1	1.1	6.2	3.5
	Yes	Yes		3.1	6.2	10.6	82.6	1.5	10.8	5.1
	Yes	Yes	Yes	4.6	10.7	22.7	86.0	1.7	8.8	3.5
$Mo_6V_3Te_1O_x$		Yes		2.9	3.9	8.6	84.6	1.1	9.2	5.0
		Yes	Yes	3.7	9.4	19.3	88.4	1.5	6.4	3.7
	Yes	Yes		3.0	4.2	8.1	86.6	1.4	7.8	4.3
	Yes	Yes	Yes	5.8	9.0	16.7	89.6	1.3	5.9	3.2

[a] calcination in air at 280℃ for 2h; [b] calcination in N_2 flow at 600℃; [c] measured before reaction; [d] reaction condition; temperature 360℃, reactant gas mixture $N_2 : O_2 : C_2H_6 : H_2O = 20 : 5.0 : 15 : 10$ (ml/min), catalyst weight 1.0g .

effective to enhance the ethane conversion and the ethylene selectivity. However, enhancement of the ethane conversion was not due to the enlargement of surface area because the surface area was not so increased. This effect can be explained by increase of an active face as discussed in our previous report;[5] the cross-section of the rod-shaped crystal is active and selective for alkane partial oxidations and the flank is poorly active and selective to CO_x. If the catalyst keeps rod-shaped crystal, contribution of the cross-section to the reaction must be low because surface area of the cross-section is much smaller than that of the flank. However, the contribution of the cross-section to the reaction will increase after grind-treatment because the cross-section must be increased. As expected, CO_x production is suppressed and activity increased after grind-treatment, indicating that the cross-section is catalytically active and selective.

If the catalysts were calcined in air before N_2-treatment at 600°C, ethane conversions were increased on both the catalysts, however the CO_x selectivity was decreased in the case of $Mo_6V_3Te_1O_x$, it was increased in the case of $Mo_6V_2Sb_1O_x$. As mentioned above, $Mo_6V_2Sb_1O_x$ catalyst contained the impure phase similar to $Sb_4Mo_{10}O_{31}$ being less active and poorly selective. When the catalysts were heat-treated in air at 280°C, the main phase of $Mo_6V_2Sb_1O_x$ and $Mo_6V_3Te_1O_x$ became more active for ethane oxidation because ethane conversion was increased. However, $Sb_4Mo_{10}O_{31}$ phase of $Mo_6V_2Sb_1O_x$ catalyst was also became active by the treatment to produce CO_x. These contrary effects were canceled by each other, and resulted in the increase of CO_x selectivity in the case of $Mo_6V_2Sb_1O_x$.

The combination of the grinding-treatment and the heat-treatment in air at 280°C was very effective to increase ethane conversion and ethylene selectivity, indicating the effects of the treatments were synergistically enhance the catalytic activity.

We investigated the ethane oxidation at various reaction temperatures using the ground catalysts after calcined at 280°C in air and N_2-treated at 600°C. The results were displayed in Fig. 2. With increasing the reaction temperature to 400°C, the conversion of ethane was increased monotonously on both catalysts, and the selectivity to ethylene was remained over 80%. The reaction temperature seems to affect very weakly to the product distribution.

Strictly speaking, the ethylene selectivity was slightly decreased with increasing the reaction temperature for both catalysts due to the increase of CO_x production. The increase of the CO_x production on $Mo_6V_3Te_1O_x$ was explained due to the contribution of the flank of the rod-shaped crystal. However, at high temperature, the CO_x selectivity of $Mo_6V_2Sb_1O_x$ catalyst was clearly high in comparison with $Mo_6V_3Te_1O_x$ indicating another contribution of CO_x production. $Sb_4Mo_{10}O_{31}$ phase seemed to contribute to the CO_x production at high temperature though it was low at 360°C.

3.3. Comparison of methane, ethane and propane oxidation

Methane, ethane and propane oxidations were studied on $Mo_6V_2Sb_1O_x$ and $Mo_6V_3Te_1O_x$ catalysts. The results were summarized in Table 2. Both catalysts were inactive for methane oxidation even at 400°C. In contrast, both the catalysts were active enough for the ethane and propane oxidation at 360°C. The $Mo_6V_3Te_1O_x$ and $Mo_6V_2Sb_1O_x$ catalysts had enough activity to cleavage C-H bond of ethane and propane. This fact clearly indicated the distinction of methane molecule.

As a second step of the reaction after C-H bond cleavage, oxygen insertion may take place to alkene to give oxygenates. In the propane oxidation, these catalysts gave oxygenates like acrolein and acrylic acid, indicating that the catalysts were active enough to proceed allylic-oxidation relatively in high selectivity. On the other hand, only a little portion of acetic acid was observed in the case of ethane oxidation over $Mo_6V_3Te_1O_x$ and $Mo_6V_2Sb_1O_x$ catalysts. Since acetic acid seems to be produced by oxidative hydration-dehydrogenation from ethylene and the reaction needs acidic property, the low selectivity to acetic acid may be due to weak surface acidity of $Mo_6V_3Te_1O_x$ and $Mo_6V_2Sb_1O_x$ in comparison with $Mo_6V_2Al_1O_x$.

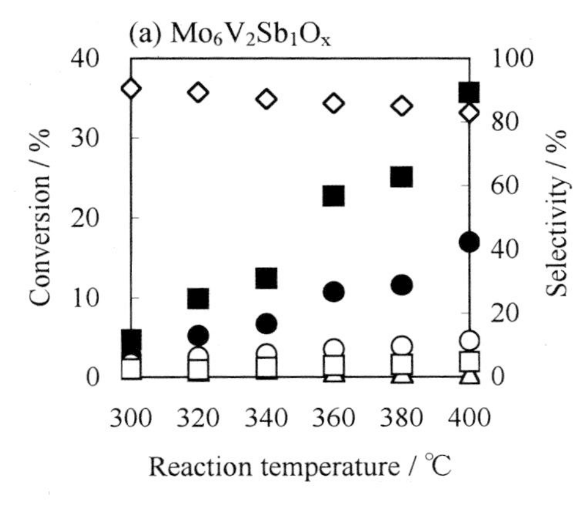

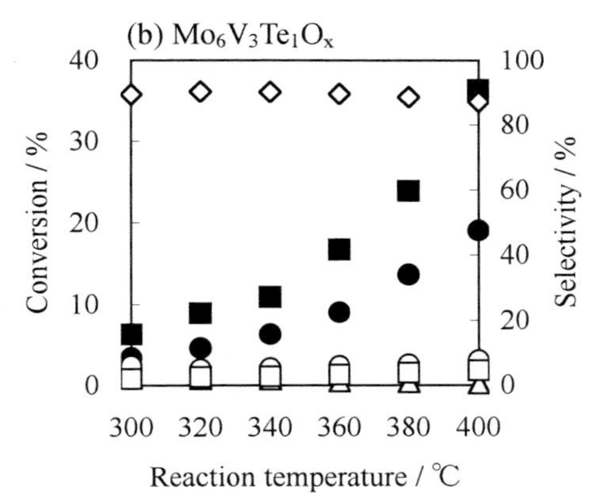

Fig. 2 Effect of reaction temperature on ethane oxidation over $Mo_6V_2Sb_1O_x$ and $Mo_6V_2Te_1O_x$ catalysts. Symbols indicate conversion of ethane (○) and oxygen (□), selectivity to ethylene (◆), acetic acid (▲), carbon monooxide (●), carbon dioxide (■). The catalyst was treated in air at 280°C, in N_2 at 600°C and ground before reaction. Reaction conditions; reactant gas, $C_2H_6:O_2:N_2:H_2O$=15:5:20:10 (ml/min); catalyst weight, 1.0g.

Table 2
Catalytic oxidation of light alkane over $Mo_6V_2Sb_1O_x$ and $Mo_6V_3Te_1O_x$ catalysts.

Catalyst	Reactant	Conversion / %		Selectivity / %				
		Alkane	O_2	Alkene	Aldehyde [a]	Acid [b]	CO	CO_2
$Mo_6V_3Te_1O_x$	C1	0.0	0.0	0.0	0.0	0.0	trace	trace
	C2	10.6	4.8	86.4	0.0	1.5	8.5	3.6
	C3	25.7	41.5	14.2	0.9	48.4	13.8	11.7
$Mo_6V_2Sb_1O_x$	C1	0.0	0.0	0.0	0.0	0.0	trace	trace
	C2	13.5	7.6	88.7	0.0	2.3	6.3	2.0
	C3	17.5	45.7	17.2	1.8	33.6	16.9	13.5

a total selectivity of aldehyde group with the same carbon number as alkane; b total selectivity of carbonate with thesame carbon umber as alkane.reaction temperature, 360 °C for ethane and propane, 400°C for methane oxidation; Reactant gas, alkane: O_2: N_2: H_2O =

The catalytic function for achieving high product selectivity obviously depends on the oxidation power of the catalyst as well as the stability of substances. As can be seen in Table 2, $Mo_6V_2Sb_1O_x$ and $Mo_6V_3Te_1O_x$ were quite suitable catalysts for ethane dehydrogenation and highly effective catalysts for propane selective oxidation to oxygenates. Both the $Mo_6V_3Te_1O_x$ and $Mo_6V_2Sb_1O_x$ catalysts showed enough oxidative power for oxidative dehydrogenation of ethane and propane although they were inactive to cleavage C-H bond of methane and they also have an ability to insert oxygen to olefinic intermediate to promote allylic oxidation.

4. CONCLUSION

Hydrothermally synthesized $Mo_6V_2Sb_1O_x$ and $Mo_6V_3Te_1O_x$ catalysts were tested for ethane oxidation and were found to be very active achieving over 80% selectivity to ethylene. The catalytic activity was enhanced by the air-calcination and the grinding-treatment. Especially, the grinding-treatment remarkably enhanced the ethane conversion and simultaneously suppressed the CO_x production. The effect was explained by an increase of the active and selective face that is the cross-section of the rod-shaped crystal of $Mo_6V_2Sb_1O_x$ or $Mo_6V_3Te_1O_x$ catalysts. The product distribution mainly depended on an exposure of the active face, and was almost independent on a reaction temperature. The $Mo_6V_2Sb_1O_x$ and $Mo_6V_3Te_1O_x$ catalysts have functions to cleavage C-H bonds of ethane and propane but not of methane, and have a sufficient ability of oxygen insertion in an allylic oxidation.

REFERENCES

1. K. Inumaru, T. Okuhara, M. Misono, Chem. Lett., (1992), 947.
2. G. Centi, Catal. Today, 49 (1993), 5.
3. W. Ueda, N. F. Chen, K. Oshihara, Chem. Lett., (1999), 517.
4. M. Vaarkamp, T. Ushikubo, Appl. Catal. A, 174 (1998), 99.
5. H. Watanabe, Y. Koyasu, Appl. Catal. A, 194-195 (2000), 479.
6. W. Ueda, T. Hisano, K. Oshihara, Topics in Catal., in press.
7. M. Parmentier et al, J. Solid State Chem., 31(1980), 305.
8. W. Ueda, K. Oshihara, Appl. Catal. A, 200 (2000), 135.

Studies in Surface Science and Catalysis
J.J. Spivey, E. Iglesia and T.H. Fleisch (Editors)

A novel two-stage reactor process for catalytic oxidation of methane to synthesis gas

Shikong Shen,* Zhiyong Pan, Chaoyang Dong, Qiying Jiang, Zhaobin Zhang, Changchun Yu

Key Laboratory of Catalysis CNPC, University of Petroleum-Beijing, Beijing 102200, China

A novel process for catalytic oxidation of methane to synthesis gas, which consists of two consecutive fixed-bed reactors with oxygen or air separately introduced into the reactors, was investigated. The first reactor, packed with a combustion catalyst such as a perovskite type oxide or a supported noble metal catalysts, is used for catalytic combustion of methane at low initial temperature (350~400 °C), and the second reactor, filled with a partial oxidation catalyst such as a Ni-based catalyst, is used for the partial oxidation of methane to syngas. A portion of methane (ca. 6~8 %) is oxidized to CO_2 and H_2O in the first reactor, in which the reactants are heated to the temperature required for methane partial oxidation (>700°C). The remaining oxygen (ca.75 %) is introduced between the exit of first reactor and the inlet of second reactor. In the second reactor, the exothermic partial oxidation of methane and the endothermic reforming reactions of CO_2 and H_2O provide thermal balance. Adiabatic operation can be achieved in this way. The ratio of methane to oxygen is not within the explosion limit in either reactor.

1. INTRODUCTION

Interest in the conversion of natural gas to liquid hydrocarbons (GTL) by Fischer-Tropsch synthesis has grown significantly over the last decade [1]. Significant improvement in this technology is still needed to strengthen its economic competitiveness. Most research and development work has been focused on the syngas production step, which accounts for about 60% of the total investment in conventional processes. Reducing the cost of syngas production would have great beneficial effects on the overall economics of GTL process. Catalytic partial oxidation of methane (CPOM) to syngas is a slightly exothermic, highly selective, and energy efficient process [2]. This process produces syngas directly with a near-ideal ratio of hydrogen to carbon monoxide for F-T synthesis. However, CPOM processes have not been practiced commercially. Two of the major engineering problems are the high temperature gradient and the risk of explosion with premixed CH_4/O_2 mixtures. In fluidized bed reactors the heat transfer is more rapid than in fixed beds because of the back mixing, which ensures a more uniform temperature and a safer operation. A technology for syngas production by contacting methane with limited amounts of steam and oxygen in a catalytic fluidized bed reactor has been developed by Exxon [3]. Alternative catalyst bed configurations, such as dual-bed or mixed-catalyst bed reactors have been examined by Ma

* Corresponding author. fax: +86-10-69744849, E-mail: shikongshen@163.net

and Trimm [4] for the purpose of determining the possible advantages of different configurations of the combustion (Pt) and reforming (Ni) catalysts. They found that the dual-bed system was inferior to the single bed system containing two mixed catalysts. Optimal performance, which is 60-65% conversion of methane with 80-85% selectivity, was obtained when both catalysts were on the same support.

In the present work, an alternative approach was investigated. This novel process consists of two consecutive fixed-bed reactors in which oxygen or air is introduced separately into the two reactors. The first reactor, which is packed with a combustion catalyst such as a supported noble metal or perovskite type oxide catalyst, is used for the catalytic combustion of methane at a low initial temperature (350~400 °C). The second reactor, which is filled with a partial oxidation catalyst such as a Ni-based catalyst, is used for the partial oxidation of methane to syngas. A portion of methane is oxidized to CO_2 and H_2O in the first reactor, in which the reactants are simultaneously heated to the temperature required for methane partial oxidation (>700°C). The remaining oxygen is introduced between the exit of first reactor and the inlet of second reactor. In the second reactor, the exothermic partial oxidation of methane and the endothermic reforming reactions of CO_2 and H_2O provide thermal balance. In this way, adiabatic operation can be achieved. The ratio of methane to oxygen is not within the explosion limit in either reactor. In this paper, the effects of reaction conditions on the catalytic oxidation of methane to syngas in the two-stage reactor are investigated. The experimental results show that catalytic oxidation of methane to syngas can be accomplished using two consecutive fixed-bed reactors and feeding O_2 or air separately.

2. EXPERIMENT

2.1. Catalysts preparation

2.1.1. Catalysts for combustion of methane

A supported noble metal /γ-Al_2O_3 catalyst (0.4 wt% Pd and 0.2 wt% Pt) was prepared by impregnation of γ-Al_2O_3 support (from Condea, γ-Al_2O_3 spheres of 2.5mm diameter, 210m^2/g) with a given amount of mixed aqueous solution of $PdCl_2$ and H_2PtCl_2. Then the catalyst was dried at 80 °C for 10 hr and calcined at 730 °C for 10 hr.

Lanthanum-based perovskite type oxide catalysts were prepared by the addition of citric acid to a given amount of mixed aqueous solution of the nitrates having a ratio of NO_3^- /$C_3H_4(OH)(COOH)_3$=1. The resulting solution was slowly evaporated until a vitreous material was obtained, subsequently decomposed at 300 °C for 2 hr, and then calcined at 800 °C for 5 hr. Finally the catalyst was ground, pressed into pellets, and crushed to 10~20 mesh particles.

2.1.2. Catalysts for catalytic oxidation of methane

A La_2O_3 promoted 7 wt% Ni/$MgAl_2O_4$-Al_2O_3 catalyst (atomic ratio of Ni/La/Mg =100/63/63.) was prepared by a two-step impregnation process according to [4]. First, the γ-Al_2O_3 support (from Condea, γ-Al_2O_3 spheres of 2.5mm diameter, 210m^2/g) was impregnated with an aqueous solution of magnesium nitrate, dried at 360 and calcined at 900 °C for 10 hr to form $MgAl_2O_4$ spinel compound on the surface of γ-Al_2O_3; second, the $MgAl_2O_4$-Al_2O_3 was impregnated with a mixed aqueous solution of nickel nitrate and lanthanum nitrate, then dried at 90 °C and calcined at 700 °C for 6 hr.

2.2. Reaction system

The schematic of two-stage reactor is shown in Fig.1. Both of the tubular reactors are made of HK-40 stainless steel with a 380 mm long ceramic lining of 20 mm inner diameter, in which a thermocouple is placed to measure the temperature in the center of reactor. The front combustion reactor is filled with 5g La-based perovskite oxide catalyst, while the rear partial oxidation reactor is filled with 5g La_2O_3-Ni/$MgAl_2O_4$-Al_2O_3 catalysts. The combustion catalyst bed is diluted with 1mm ceramic spheres (in double volumes) to improve thermal conductivity. The feed and products were analyzed on an on-line gas chromatograph equipped with a 1M Porapak N column and a 1.5M 5A Molecular Sieve column using TCD detector.

3. RESULTS AND DISCUSSION

3.1. Evaluation of catalysts for combustion of methane under lean oxygen condition

To keep the adiabatic temperature rise in the combustion reactor below 800°C, the evaluation of catalyst was carried out under lean oxygen condition (the ratios of CH_4 to O_2 = 8:1~6:1). Table 1 shows the results for various catalysts for the combustion of methane. The ignition temperature for the perovskite catalyst is higher than that for the Pd-Pt/γ-Al_2O_3 catalyst. Although the Pd-Pt/γ-Al_2O_3 catalyst exhibits higher activity, some partial oxidation products, such as CO and H_2, in addition to CO_2 and H_2O, were formed due a to lack of oxygen. To test the catalytic performance of the catalyst at high temperature and lean oxygen concentrations, the catalyst was also tested at 700 °C. The results in Table 1 show that a large amount of CO and H_2 is formed over the Pd-Pt/γ-Al_2O_3 catalyst at a reaction temperature of 700 °C, while only a small amount of CO is formed over perovskite catalysts under the same reaction conditions. It is well know that supported Pd, Pt catalysts have high activity for partial oxidation of methane, while perovskite type oxides catalysts have high activity for deep oxidation of methane. Therefore the $La_{0.7}Ca_{0.3}Fe_{0.3}Mn_{0.7}O_3$ catalyst was selected for combustion of methane. The experiment data in Fig.2 show that when the reaction pressure is raised to over 1.6 MPa, CO disappears in the products over $La_{0.7}Ca_{0.3}Fe_{0.3}Mn_{0.7}O_3$ catalyst, which implies that high pressure favors the deep oxidation of methane. The test of 100 hr

Table 1. Evaluation results of various catalysts for combustion of methane*.

Catalysts	Ignition Temp. °C	Conv. of CH_4, %	Selectivity, %			
			CO_2	H_2O	CO	H_2
Pd-Pt/γ-Al_2O_3	325 (477)**	6.68	82	98	18	2
Pd-Pt/γ-Al_2O_3	700 (750)	10.68	34	43	66	57
$La_{0.7}Ca_{0.3}MnO_3$	360 (522)	6.25	100	100	0	0
$La_{0.7}Ca_{0.3}MnO_3$	700 (817)	6.43	89	100	11	0
$La_{0.7}Ca_{0.3}Fe_{0.3}Mn_{0.7}O_3$	360 (521)	6.25	100	100	0	0
$La_{0.7}Ca_{0.3}Fe_{0.3}Mn_{0.7}O_3$	700(820)	6.39	93	100	7	0

*Reaction condition: CH_4/ O_2 =8, GHSV=30,000 hr^{-1} and 0.1 M Pa.
** In bracket shows highest temperature of the catalyst bed.

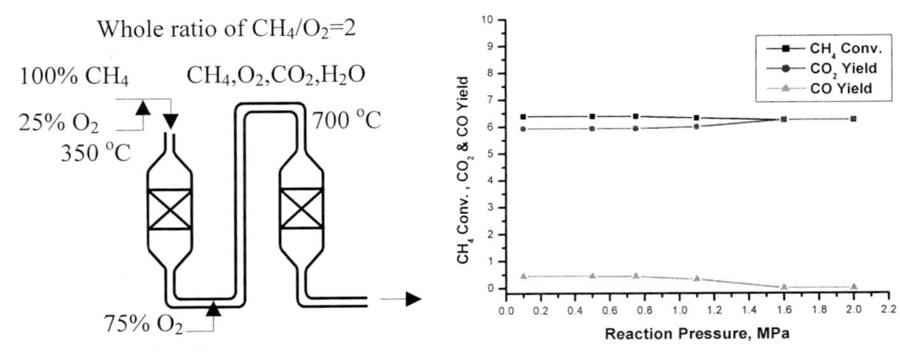

Fig.1. Schematic of the two-stage

Fig.2.Effect of reaction pressure on the combustion of CH_4 at 700 °C and GHSV=$3 \cdot 10^5$ h^{-1}

on stream at 700 °C, GHSV=30000/h, CH_4/O_2=8 and 0.1 MPa show that the activity of $La_{0.7}Ca_{0.3}Fe_{0.3}Mn_{0.7}O_3$ catalyst is stable. The carbon deposit on the used catalyst is 0.8 wt% after 100 hr run. These results show that the $La_{0.7}Ca_{0.3}Fe_{0.3}Mn_{0.7}O_3$ catalyst is more suitable for the complete oxidation of methane at high temperature and lean oxygen conditions than other catalysts.

3.2. Evaluation of catalysts for partial oxidation of methane to syngas

The La_2O_3 promoted 7 wt% Ni /$MgAl_2O_4$-Al_2O_3 catalyst prepared in our laboratory has excellent activity for the partial oxidation of methane to syngas [5]. The catalyst support, Al_2O_3, reacts with the MgO at high temperature forming $MgAl_2O_4$ spinel compound on the surface of Al_2O_3, which hinders the formation of $NiAl_2O_4$ at the POM reaction conditions and favors the dispersion of Ni on the catalyst surface. The $NiAl_2O_4$ spinel is known to be inactive for the partial oxidation of methane, and is very difficult to reduce into active metal Ni.

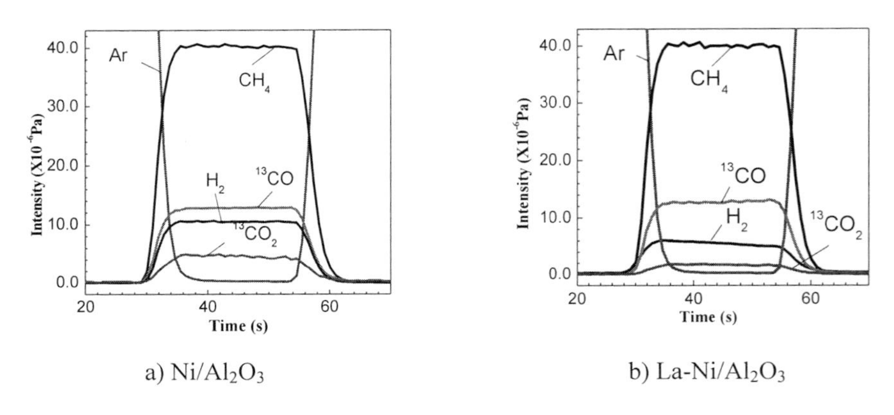

Fig.3. Response of switch Ar→ $CH_4/^{13}CO$($^{13}CO/CH_4$=1/3) →Ar at 600 °C

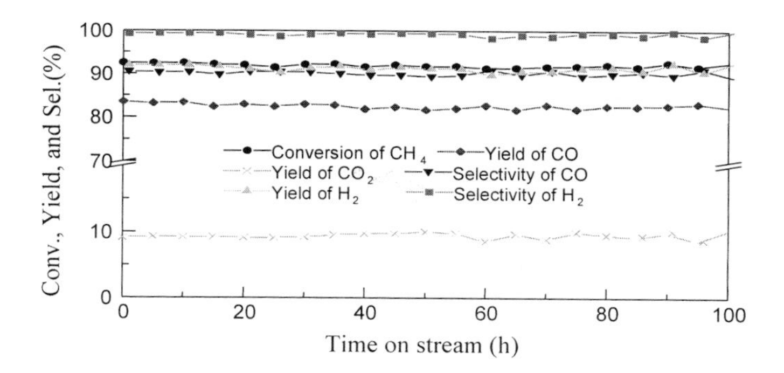

Fig. 4. The Stability of La_2O_3-Ni/$MgAl_2O_4$ for partial oxidation of methane

The effect of La promoter on carbon deposition has been investigated using an isotopic switch method. Fig. 3 shows that the addition of La decreases the activity of Ni-based catalysts for CH_4 decomposition and CO disproportionation. Carbon deposition is probably caused by the accumulation of adsorbed carbon atoms from these reactions, so La_2O_3-promoted Ni/Al_2O_3 exhibits resistance to carbon deposition. The stability test in Fig. 4 shows that the catalyst activity is stable during the 100 hr on stream at 700 °C, 0.1 MPa, GHSV=70000/h and CH_4/O_2=2. The carbon deposited on the used catalyst after 100 run is 1.1wt%. The results of BET and XRD show that the catalyst maintains structural stability. The result of X-ray fluorescence indicates that Ni is not lost in the reaction.

3.3. Effects of reaction conditions on two-stage oxidation of methane to syngas

5g $La_{0.7}Ca_{0.3}Fe_{0.3}Mn_{0.7}O_3$ perovskite catalyst and 5g Ni/La_2O_3/$MgAl_2O_4$-Al_2O_3 catalyst were loaded in the first and second reactor respectively. Because partial oxidation with air to produce synthesis gas for F-T process without a recycle loop reduces capital and

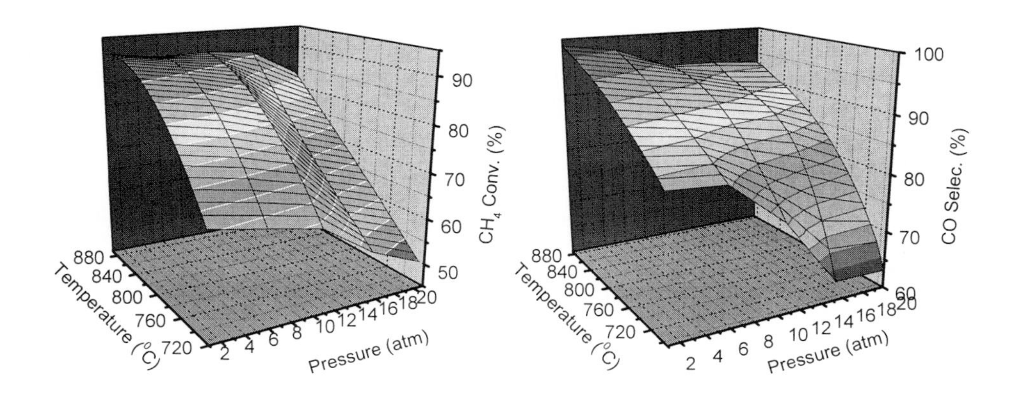

(a). Conversion of CH_4 vs. temperature and pressure. (b). Selectivity of CO vs. temperature and pressure.

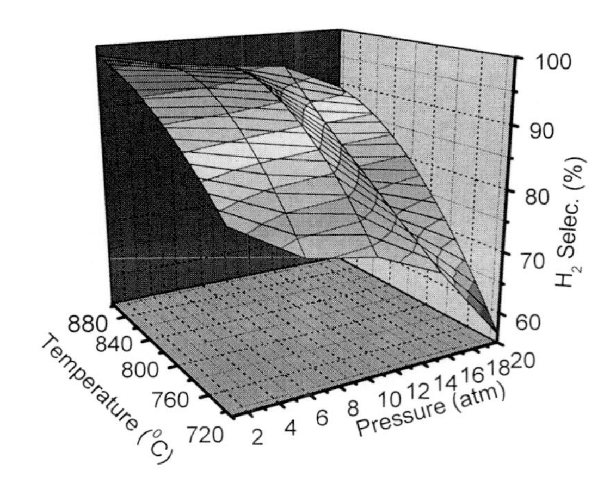

(c). Selectivity of H_2 vs. temperature and pressure.

Fig. 5. Effects of temperature and pressure on two-stage oxidation of methane to syngas

production costs of GTL [6], air was used as the oxygen resource in these experiments. The reaction conditions of the first reactor were fixed at air/CH_4 =5/8 and the furnace temperature was 400°C. The O_2 to CH_4 ratio in the feed gas reaches 0.5 by supplying 75 % air to the second reactor. Fig. 5 shows the effects of furnace temperature and pressure on the results of two-stage oxidation of methane to syngas at GHSV=10,000 h^{-1}. The experimental data are quite close to the thermodynamic equilibrium. The temperature difference between the furnace and the center of second reactor is much lower than that in the single fixed bed reactor. Methane conversion of 90% and CO and H_2 selectivity of 94% are obtained at 880 °C and 1.0 MPa, while the temperature difference between the furnace and center of second reactor is less than 30 °C. Further increase of GHSV will lead to a decrease in methane conversion and selectivity to CO and H_2. The probable reason is the slow reaction rate of steam and CO_2 with methane.

In summary, the two-stage reactor reduces the possibility of and magnitude of hot spots created during catalytic oxidation of methane, while allowing safer operation.

ACKNOWLEDGEMENTS

Financial support by the China National Petroleum Corporation is greatly acknowledged.

REFERENCES

1. M. J. Corke, Oil & Gas Journal, Sep. 21, 1998, 71.
2. S. S. Bharadwaj, L. D. Schmidt, Fuel Processing Technol., 42 (1995) 109.
3. B. Eisenberg, R. A. Fiato, T. G. Kaufman, R. F. Bauman, Chemtech, Oct.1999, 32.
4. L. Ma and D. L. Trimm, Appl. Catal., A 138 (1996) 265.
5. C. Zhang, C. Yu, S. Shen, CUIHUA XUEBAO (Chinese J. Catal.) , 21 (2000) 14.
6. A. K. Rhodes, Oil & Gas Journal, Dec. 30, 1996, 85.

Studies in Surface Science and Catalysis
J.J. Spivey, E. Iglesia and T.H. Fleisch (Editors)

Composite steam reforming catalysts prepared from Al_2O_3/Al matrix precursor

S. F. Tikhov, V. A. Sadykov, I. I. Bobrova, Yu. V. Potapova, V. B. Fenelonov, S. V. Tsybulya, S. N. Pavlova, A. S. Ivanova, V. Yu. Kruglyakov, N. N. Bobrov

Boreskov Institute of Catalysis SD RAS. pr. Lavrentieva, 5, Novosibirsk, 630090, Russia

Several composite catalysts based upon the Al_2O_3/Al matrix precursor have been prepared and tested in the methane steam reforming. The following sequence of the activity (cm^3 CH_4/gCat bar s) of ceramometals (cermets) was found: Ru-Ni-Cr-La-Al-O>Ni-Cr-La-Al-O=Ru-Sr-Zr-Al-O>Ni-La-Al-O>Co-La-Al-O>Rh-Sr-Zr-Al-O. The effect of the properties on the cermets' activity are discussed. The advantages of composite materials as tube coatings for the compact steam reformers design are demonstrated.

1. INTRODUCTION

The development of the compact steam reformers for small-scale Fischer-Tropsch synthesis and enhancement of effectiveness of gas turbines for power generation is one of the modern trends in the synthesis gas production [1-3]. The design of a steam reformer is limited by the heat transfer at the tube walls. Increasing the heat transfer surface area by decreasing the tube diameter results in a higher pressure drop [4]. The use of the reformer with a thin (< 100 μm) layer of an active catalytic material having a low pressure drop requires the high catalytic activity which can only be achieved using precious metals. Therefore, the thick catalytic coatings technology based upon Al_2O_3/Al porous ceramometals (cermets) has a considerable promise in overcoming the problems of the compact steam reformers design. Previously [5], we have shown that the porous cermets can be used to support thick (1-10 mm) catalytic coatings onto the inner or outer surface of the stainless steel tubes of a small diameter. In this work we report the physico-chemical properties and catalytic activity of a number of cermets and their catalytic activity in methane steam reforming.

2. EXPERIMENTAL

Several composite catalysts based upon Co-La-Al-O, Ni-La-Al-O, Ni-Cr-La-Al-O, Ru-Ni-La-Al-O, Ru-M-Zr-Al-O (M=Ca, Sr, Ba), Rh-Sr-Zr-Al-O were prepared and studied. The main step of the catalyst synthesis was the incapsulation of powdered components into the Al_2O_3/Al matrix during its formation via the hydrothermal aluminum oxidation, followed by calcination in air at 800^0C [2,5]. In this way, zirconia stabilized by alkaline earth metals (PSZ), lanthanum nickelates and cobaltites, and powdered Ni-Cr alloy were introduced into the cermet. The preparation procedure is described elsewhere [5-7]. These additives promote both the catalytic activity and thermal stability of cermets. Additionally, some amount of nickel was introduced via impregnation with the aqueous nitrate solution. The platinum group metals (0.5 wt.% of Ru or Rh) were introduced via the traditional incipient wetness impregnation method using the ethanol solutions of chloride salts ($Ru(OH)Cl_2$ and $RhCl_3$).

The specific surface area (SSA) of the cermets was determined by the BET method using Ar thermal desorption. The pore size distribution was estimated from the mercury porosimetry using an AutoPore 9200. The phase composition was analyzed with an URD-63 diffractometer using CuK_α radiation. The total pore volume was estimated from the measurements of the particle density using helium pycnometry and the apparent density of cermets.

The catalyst activity was determined using a flow-recycling isothermal reactor at atmospheric pressure and H_2O/CH_4 mole ratio of 2/1. The kinetic experiments and catalyst pretreatment are described in detail elsewhere [8]. The catalysts (0.5 g) diluted by quartz in a 1:10 ratio were typically tested as 0.5-1.0 mm particles. The catalytic activity per unit weight of catalyst or nickel was compared by calculating an effective first order rate constant at 750^{0}C. For the Ni-Cr-La-Al-O catalyst, the activity of a thick layer coated on the outer side of stainless steel tubes and loaded into a reactor without quartz chips was also estimated.

3. RESULTS AND DISCUSSION

3.1. Scale of catalysts activity

The specific activity of cermets (Table 1) varies by an order of magnitude. The most active is the Ru-containing Ni-Cr-La/Al_2O_3/Al composite. Supported precious metals without nickel are less active, Rh being less active than Ru. The catalyst with the active component based upon lanthanum cobaltite is less active than those containing nickel compounds.

Table 1
Some properties of composite catalysts

Catalysts		Specific surface area	Specific activity
Active component (wt.%)	Porous cermet matrix	m^2/g	k (cm^3 CH_4/g cat s bar)
Co (7)	La-Al- O	36	2
Ni (7)	La-Al- O	34	17
Ni (7)	La-Al- O (3-4 mm)	34	8
Ni (46)	Cr-La-Al- O	39	30
Rh (0.5)	Sr-Zr-Al- O	90	1
Ru (0.5)	Sr-Zr-Al- O	90	23
Ru-Ni (0.5Ru,46Ni)	Cr-La-Al- O	39	35

For the same active component, the specific catalytic activity was found to depend upon the cermet composition rather than the specific surface area (vide infra).

3.2. La-Ni/Al_2O_3/Al catalysts

After calcination at 800^0C, the XRD patterm show that the strongest peaks for the La-Ni/Al_2O_3/Al catalysts are those that are typical for the fcc lattice of powdered Al. This demonstrates incomplete oxidation of aluminum. XRD also show a phase of hexagonally distorted $LaNiO_3$ perovskite with small admixtures of the phase with a layered-type structure (La_2NiO_4) and NiO oxide.

As seen in the Fig.1, the specific activity (k) of La-Ni containing catalysts increases with the amount of capsulated La-Ni oxide. However, when normalized to the Ni loading, the activity (k*) of these catalysts varies non-monotonically, being the highest for sample with the La-Ni oxide content 40 wt.%. The total pore volume of composites except for the pure La-Ni ranges from 0.33 to 0.44 ml/g. The La-Ni oxide seems to promote the thermal stability of the Al_2O_3/Al matrix since there is a relationship between the content of the oxide and the surface area of composites (Fig.1). Possibly, a higher activity of La-Ni oxide in the cermet matrix is due to a more developed pore structure of the composite catalysts as compared with a pure La-Ni oxide. The role of La in stabilizing the specific surface ared of alumina is well-known [9].

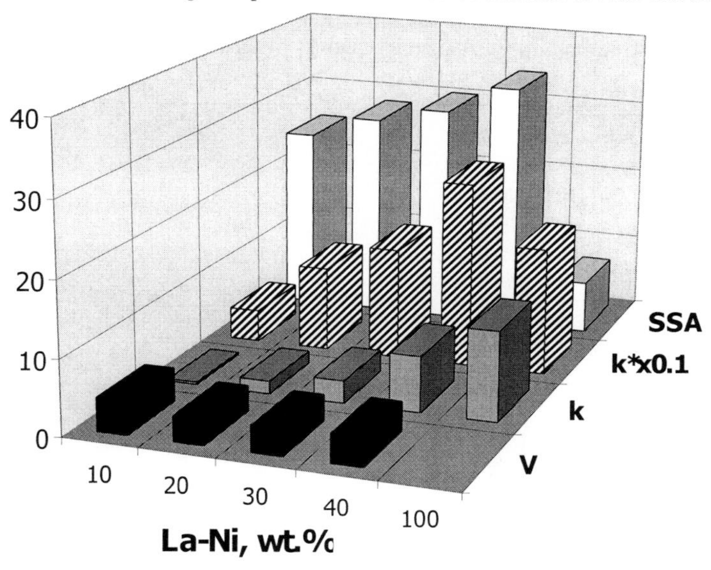

Fig.1. The effect of the La-Ni oxide content on the properties of Ni-La-Al-O catalysts: specific surface area (SSA, m^2/g), pore volume (V, ml/g) and specific activity related to the mass of catalyst (k, cm^3 CH_4/gCat s bar) or to the mass of Ni (k*, cm^3 CH_4/gNi s bar).

3.3. La-Ni-Cr/Al_2O_3/Al catalysts

According to XRD data, the initial catalysts are comprised of the mechanical mixtures of different phases: Al^0, alumina with the spinel structure, Ni-Cr (~18 wt.% Cr) solid solution with the fcc structure, the rhombohedral $LaNiO_{3-\delta}$ and layered La_2NiO_4 perovskites along with admixtures of simple oxides (NiO, La_2O_3). After catalyst testing, the rhombohedral and layered perovskites as well as simple oxides admixtures disappear and the cubic perovskite phase emerge. Taking into account that for the La-Co containing composite the same tendency is observed, one can conclude that the cubic perovskite phase is the most stable in the methane steam reforming conditions.

As is seen from Fig.2, the addition of large amounts of the Ni-Cr alloy results in a higher activity per unit of a total catalyst mass, which is important in practice. However, the effective rate constant per unit nickel content is much higher for pure La-Ni active component than with catalysts containing Ni-Cr alloy powder (Fig.2). This is probably because the La-containing perovskite matrix is more basic than chromia, which would favor more severe surface coking due to its inherent acidity, thus decreasing specific activity.

For 2 mm thick La-Ni-Cr oxide coating on tubes, the activity is more than one order of magnitude lower than for the same composites as particles without tubess (Fig.2).

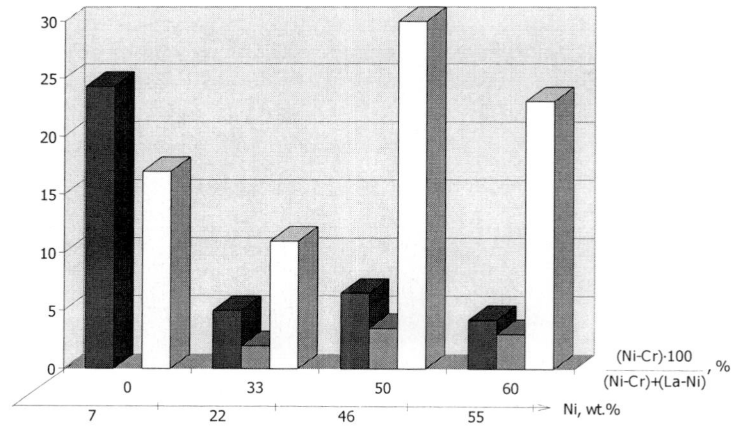

Fig.2. The effect of Ni- Cr content on the specific activity of Ni-Cr La-Al-O catalysts: ■ - k*x0.1 (cm^3CH_4/gNi s bar), fraction; □ - k (cm^3CH_4/gCat s bar), fraction; ▩ -k (cm^3CH_4/gCat s bar), a coating on tube.

3.4. Ru/M-Zr-O/Al_2O_3/Al catalysts

After HTT and air calcination at 800^0C, the specific surface area of the composites varies from ~20 m^2/g (alumina/aluminum cermet) to ~90 m^2/g (Sr-Zr-O/alumina/aluminum cermet). Detailed analysis revealed that for these composites, the specific surface area is higher than expected from a simple additive scheme. This means that partially stabilized zirconia hinders the alumina sintering during air calcination, due to the effect of zirconia and alkaline-earth metal oxides [9].

As is seen from Fig.3, for composites containing zirconia partially stabilized by alkaline-earth cations, the specific activity in the methane steam reforming reaction decreases in the order: Ca > Sr > Ba, which does not correlate with the catalysts specific surface area. A rather weak dependence of activity on the nature of alkaline-earth cation could be tentatively assigned to the levelling effect of the alumina matrix (Fig.3).

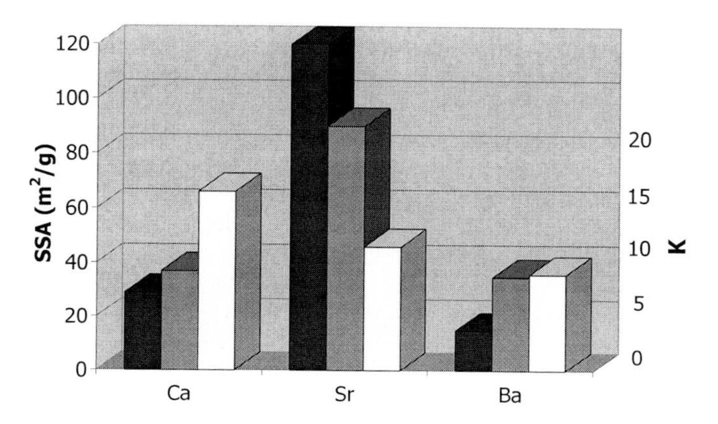

Fig.3. The effect of alkaline-earth metal (M=Ca, Sr, Ba)) on the properties of Ru/M-Zr-O/Al_2O_3/Al catalysts: ■ - SSA of encapsulated additive (M-Zr-O); ■ - SSA of catalyst (Ru/M-Zr-O/Al_2O_3/Al); □ - rate constant, k ($cm^3 CH_4$/gCat s bar).

3.5. Advantages of the thick catalytic coatings developed

The data of Bobrova et al. [8] show that methane steam reforming may proceed via a homogeneous- heterogeneous route. For such a route, the catalyst activity is greatly enhanced compared to a pure heterogeneous process. From this point of view, the foam-like materials are the most promising as steam reforming catalysts. Indeed, recently Ni-Cr foam-based catalysts (thickness 2-5 mm) containing 0.8 wt.% Ni/CeO_2 + δ-Al_2O_3/Ni-Cr have been proposed [10]. The first-order reaction rate constant estimated for this catalyst was found to be about 2.2 cm^3 CH_4/(ml Cat s bar) at 750^0C and GHSV of 7100 h^{-1}, corresponding to a conversion of ~50%. The first-order reaction rate constant estimated for La-Ni-Cr/Al_2O_3/Al cermet (thickness ~2 mm) containing 50 wt.% of Ni was found to be about 3.5 cm^3 CH_4/ (ml cat s bar) at 750^0C. The addition of Ni by impregnation doubles the rate constant, corresponding to a conversion of ~78% at these same conditions. Though the effectiveness of Ni is undoubtedly better for foam-like materials, the activity of composite catalysts per unit volume is certainly higher, which of a great practical importance.

For all cermet systems, the fraction of ultramacropores (>1 μm) was found to be very high (30-40%), the total pore volume varying from 0.23 to 0.40 ml/g. These cermets are thus materials that are intermediate between traditional porous ceramics and foams. Because of the well-developed transport pore network, the activity of these catalysts depends only slightly upon the size of granules or the coating thickness. It is probable that the cermets provide a higher dispersion of nickel particles than foam materials due to their higher SSA.

A rough estimate (without regard for the heat transfer) reveals that the capacity of a reformer equipped with tubes of small (10 mm inner and 18 mm outer) diameter with the inner surface coated by 2 mm thick layer and spaced at a 4 mm distance apart is 10-15 times higher than that of reformers with the same tubes packed by granulated catalysts like GIAP-16 (21 wt.% Ni). Moreover, the proposed technology allows to obtain tubes with both internal and external coatings. Hence, endothermic (methane steam reforming) and exothermic (methane combustion) processes can be efficiently integrated by heat transfer across the wall.

4. CONCLUSIONS

This work shows that mixed metal-oxide composites supported as a thick coatings on stainless steel tube walls could be used as efficient catalysts of methane steam reforming. A well-developed pore structure of cermets provides a higher activity per unit volume than foams and granulated catalysts. Hence, the challenge of developing compact steam reformers design can be met.

REFERENCES

1. E. A. Polman, J. M. Der Kinderen, F.M.A. Thuis, Catal. Today, 47 (1999) 347.
2. V. M. Batenin, D. N. Kagan, A. L. Lapidus, F. N. Pekhota, M. N. Radchenko, A. D. Sedykh, E. E. Shpilrain, «Natural Gas Conversion VI», Stud. Surf. Sci. Catal., (A Parmaliana et al., eds), Elsevier, Belgium, 119 (1998) 901.
3. J. R. Rostrup- Nielsen, K Aasberg- Petersen, P. S. Schoubuje, «Natural Gas Conversion IV», Stud. Surf. Sci. Catal., (M. de Pontes et al., eds), Elsevier, Belgium, 107 (1997) 473.
4. J. R. Rostrup- Nielsen, Catal. Today, 18 (1993) 305.
5. S. F. Tikhov, V. A. Sadykov, A. N. Salanov, Yu. V. Potapova, S. V. Tsybulya, G. S. Litvak, S. N. Pavlova, Mater. Res. Soc. Symp. Ser., MRS, USA, 497 (1998) 121.
6. V. A. Sadykov, A. S. Ivanova, V. P. Ivanov, G. M. Alikina, A. V. Kharlanov, E. V. Lunina, V. V. Lunin, V. A. Matyshak, N. A. Zubareva, A. Ya. Rozovskii, Mater. Res. Soc. Symp. Ser., MRS, USA, 457 (1997) 199.
7. S. F. Tikhov, V. A. Sadykov, Yu. A. Potapova, A. N. Salanov, G. N. Kustova, G. S. Litvak, V. I. Zaikovskii, S. V. Tsybulya, S. N. Pavlova, A. S. Ivanova, A. Ya. Rozovskii, G. I. Lin, V. V. Lunin, V. N. Ananyin, V. V. Belyaev, «Preparation of Catalysts VII», Stud. Surf. Sci. Catal., (B. Delmon et al.. eds), Elsevier, Belgium, 118 (1998) 795.
8. I. I. Bobrova, V. V. Chesnokov, N. N. Bobrov, V. I. Zaikovkii, V. N. Parmon, Kinet. Catal., 41 (2000) 25.
9. J. K. McCarty, M. Gusman, D. M. Lowe, D. L. Hilderbrand, K. N. Lau, Catal. Today, 47 (1999) 5.
10. Z. R. Ismagilov, O. Yu. Podyacheva, V. V. Pushkarev, N. A. Koryabkina, V. N. Antsiferov, Yu. V. Danchenko, O. P. Solonenko, H. Veringa, "12th International Congress on Catalysis" Stud. Surf. Sci. Catal. (A Corma et al., eds), Elsevier, Belgium, 130c (2000) 2759.

RELATION BETWEEN THE STRUCTURE AND ACTIVITY OF Ru-Co/NaY CATALYSTS STUDIED BY X-RAY ABSORPTION SPECTROSCOPY (XAS) AND CO HYDROGENATION

L. Guczi[1*], D. Bazin[2], I. Kovács[1], L. Borkó[1], and I. Kiricsi[3]

[1]Department of Surface Chemistry and Catalysis, Institute of Isotope and Surface Chemistry, CRC, HAS, P. O. Box 77, H-1525 Budapest, Hungary

[2]LURE, Université Paris XI, Bât 209D, 91405, Orsay, France

[3]Department of Applied Chemistry, Attila József University, Rerrich B. tér 1, H-6720, Szeged, Hungary

ABSTRACT

In order to follow the genesis of Ru and Co metallic particles from the "as prepared" to the reduced state, in situ EXAFS study was carried out over Ru-Co/NaY bimetallic samples. The results obtained earlier by means of X-ray photoelectron spectroscopy (*J. Catal.*, **167** (1997) 482) on the same sample the presence of Ru-Co bonds based upon the reversible oxidation/reduction behavior, was assumed. The heterometallic Ru-Co bonds were, however, not established by EXAFS. The cobalt particles were present in a somewhat larger cluster size than ruthenium. It is plausible to suggest that alloy does not exist only monometallic clusters in each other neighborhood. Metallic cobalt particles are produced from Co^{2+} ions using the spilt-over hydrogen atoms activated on the neighboring ruthenium particles. By comparison of the freshly reduced and air exposed samples it turned out that the size of cobalt slightly increased, while the size of Ru drastically decreased. The appearance of significant number of Ru-O bonds was explained by the easy access to Ru whereby cobalt was protected by ruthenium. The data obtained earlier from XPS and the CO hydrogenation appeared to be in good correlation with the model put forward from EXAFS measurements.

Keywords: In situ EXAFS study, genesis of Ru-Co/NaY samples, relation between structure and activity, CO hydrogenation.

1. INTRODUCTION

Cobalt and ruthenium are considered among the best catalysts for CO hydrogenation [1]. Catalytic activity and selectivity in the CO hydrogenation largely depend on the morphology of the catalyst, such as particle size, combination of the two metals, acid-base properties of the support, pretreatment conditions, etc. Earlier works on TiO_2 and SiO_2 supported cobalt catalysts promoted by small amount of ruthenium showed a large increase in C_{5+} selectivity and turnover frequency (TOF) which is indicative of synergism [2]. Similar effect was observed on the Ru-Co/NaY sample in CH_4 chemisorption [3]. Earlier studies in our laboratory [4] have shown that Co^{2+} ions can be reduced in the presence of ruthenium in

Ru-Co/NaY sample. Temperature programmed reduction (TPR) [3] and X-ray photoelectron spectroscopy (XPS) [4] indicated a reversible oxidation and reduction of cobalt upon heat treatment in oxygen and hydrogen, respectively. Transmission electron microscopy (TEM) on the reduced Ru-Co/NaY sample [4] showed nanoscale particles in agreement with other literature data [5]. Based on these results we may assume the formation of Ru-Co bimetallic particles, although binding energy shift of the Co 2p B.E. line, similar to that observed for Pt-Co/NaY [6], was not recorded for the Co species in Ru-Co/NaY sample [4].

In situ EXAFS technique has already been employed to receive information about the genesis of cobalt particles [7, 8]. The change in the Co L_{III}-edge revealed in an alteration in the oxidation state of the cobalt species from the "as prepared" to the final state. In order to obtain a deeper insight into the catalyst structure, *in situ* EXAFS technique was applied. Hence, significant relationships between structural characteristics and activity/selectivity of catalysts can be established [9].

In the present paper we use *in situ* EXAFS technique to measure the Ru and Co K-edges to find evidences whether Ru-Co bimetallic particles form an alloy phase or individual cobalt and ruthenium nanoparticles coexist in the zeolite supercage. The results of structural analysis are correlated to the XPS data and to the catalytic activity of samples in the CO hydrogenation.

2. EXPERIMENTAL

2. 1. Sample preparation

Co/NaY, Ru-Co/NaY and Ru/NaY samples were prepared by ion-exchange of NaY zeolite (Strem Chemicals, Lot No 031112104, atomic ratio of zeolite NaY was Na/Al/Si = 0.38/0.38/1.00) following the work of McMahon et al.[10]. The zeolite was first carefully suspended in deionized water (200 $cm^3/g_{zeolite}$) by stirring at RT for 1 h followed by dropwise addition of aqueous solution of $Ru(NH_3)_6Cl_3$ or $Co(NO_3)_2$ ($2x10^{-3}$ mol/cm^3) whilst stirring at 70°C. The pH of the slurry was set to 6.5. After 24 h ion exchange at the same temperature, the samples were filtered and thoroughly washed with deionized water, dried at RT overnight and at 370K. This specimens were stored in inert atmosphere and denotes as „as prepared" samples. For the preparation of bimetallic sample first the $[Ru(NH_3)_6]Cl_3$ solution then $Co(NO_3)_2$ one was added to the suspension of NaY. The composition was determined by X-ray flourescence (XRF) technique and the concentration of Ru and Co was found to be 3. 8 and 2. 8 wt %, respectively).. The exchange degrees for the various samples were in the range of 30-50 %.

2. 2. EXAFS measurements

In situ NEXAFS and EXAFS were used to follow the behaviour of the Ru-Co/NaY samples from the "as prepared" state through the reduced state (treatment in a stream of hydrogen at 673K for 2 h) to the oxidised state (the reduced sample was exposed to air at 573 K for 1 h). Experiments were performed at Lure using the DCI storage ring running at 1.85 GeV. XAFS IV-D44 beam line in transmission mode using a double-crystal arrangement: Si(111) for Co K-edge (7709 eV) and Ge(400) for Ru-K edge (22117 eV). At the Co K-edge the estimated energy resolution of the monochromator was approximately 3 eV. The catalyst sample was meshed and a fraction with 100-200 μm size was loaded on a sample holder introduced to a special furnace allowing in situ treatment. The furnace was equipped with entrance and exit windows made of boron nitride transparent to the X-ray beam

[11]. The EXAFS signals of cobalt and ruthenium metallic foils were used as a reference for monitoring Co-Co and Ru-Ru interactions. On-line data analysis was used in order to evaluate the changes in the chemical state of samples. A two-shell least-squares fitting procedure using the single scattering EXAFS formulation was used to extract the coordination number (N), distance (R) and Debye-Waller factor (σ). For all simulations the edge shift ΔE was maintained at a value less than 4 eV [7].

2. RESULTS AND DISCUSSION

In Fig. 1 the XANES part of the Co K-edge measured on reduced RuCo/NaY sample is presented (bottom curve). Comparing this spectrum to that of cobalt foil, an incomplete

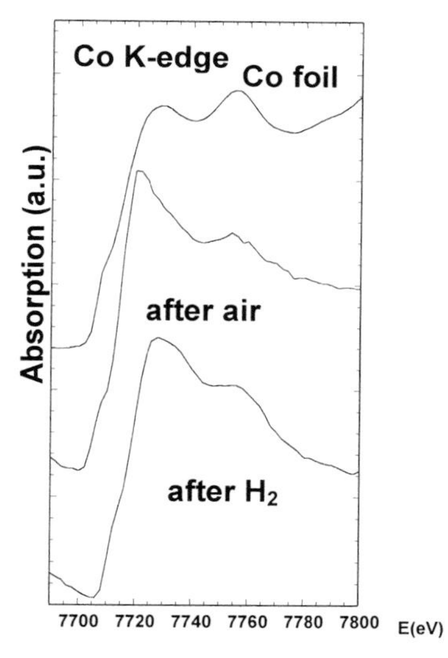

Fig. 1. XANES part at the Co K-edge for Ru-Co/NaY sample after reduction in H_2 at 673 K and after exposing it to air.

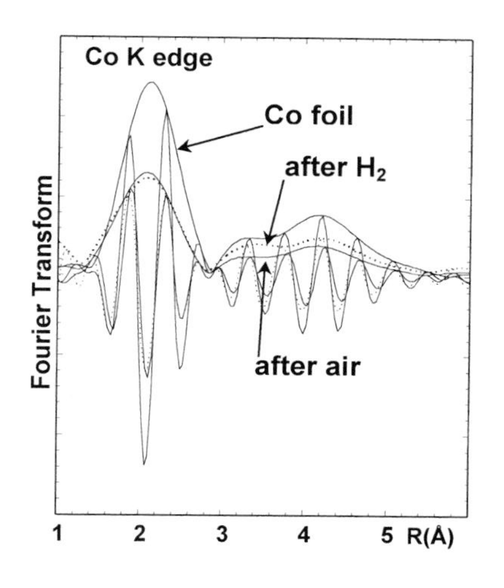

Fig. 2. Modules of the F.T. at the Co K edge for the catalyst after the reduction after replacing the hydrogen by air.

reduction of cobalt can be observed. This is in agreement with earlier TPR and XPS results [3, 4].

In Fig. 2. modulus of the Fourier transform associated with the metallic cobalt foil (solid line) and the reduced Ru-Co/NaY sample (dotted line). From these spectra the number of cobalt atoms in the first coordination sphere, $N_{(Co-Co)}$ = 7.0, their distance, $R_{(Co-Co)}$ = 0.2509 nm, and the Debye-Waller factor = 0.02 were computed. Results in Fig. 2 clearly show the presence of Co-Co metal bonds. Drastic decrease in the coordination number (from 12 for Co foil to 7.0 in zeolite) is due to the fact that in zeolite the cobalt atoms are in the form of metal nanocluster.

In Fig. 3 the absorption spectrum of XANES part of the Ru K-edge measured in the Ru-Co/NaY sample is shown. The shape of the curve due to the reduced sample is very similar to that measured on Ru metallic foil. The numerical simulation confirmed this qualitative estimation of the environment of ruthenium.

After reduction a lower coordination number ($N_{(Ru-Ru)} = 4.9$) was recorded for ruthenium than for cobalt. This may indicate that ruthenium is located near to the cobalt surface.

Surprisingly, no signal characteristic of Co-Ru heteroatomic bond seems to be present inside the material as confirmed by the numerical simulations (the Co environment refined from the EXAFS of the Ru-Co bimetallic catalyst).

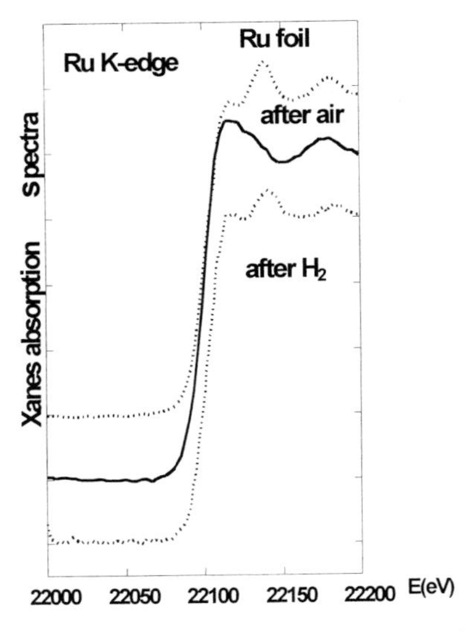

Fig. 3. XANES part measured at the Ru K-edge on Ru-Co/NaY after H_2 reduction at 673 K and after following air admission.

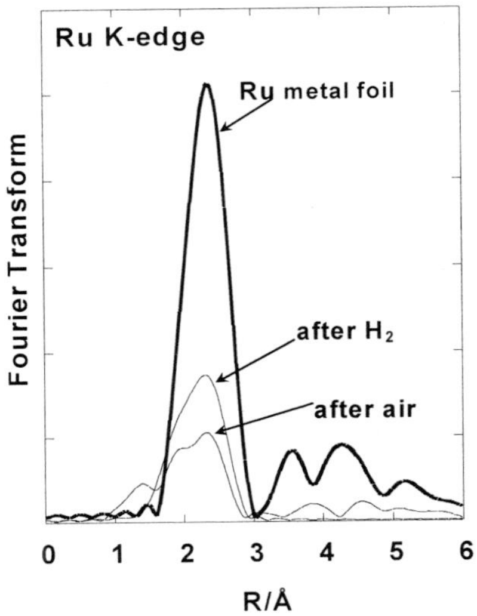

Fig. 4. Modules of Fourier transform of Ru K-edge on Ru-Co/NaY sample after reduction in H_2 and after following air admission

In order to study the stability of the reduced state of the Ru-Co/NaY sample, the reduced and helium flashed catalysts were exposed to air at 573 K for 1 h. The modulus and the Fourier transform (see Fig. 2) indicate that the state of cobalt does not change significantly. The cobalt keeps their metallic environment (the coordination number $N_{(Co-Co)}$ slightly decreases from 7 to 6) indicating that no major oxidation of cobalt species occurs.

On the other hand, a quantitative analysis shows that significant number of Ru-O bonds are formed during oxidation of the Ru-Co/NaY sample. The modulus of the Fourier Transform (Fig. 4) exhibits a change in the structure, which might be attributed to Ru-O bonds. The $N_{(Ru-Ru)}$ decreases from 4.9 to 2.3 and the $N_{(Ru-O)}$ is < 1 and Ru-O distances is 0.202 nm, i.e. the oxidized ruthenium cluster becomes small.

The results are in line with the XANES part of the X-ray absorption spectrum (Fig. 3) which exhibits a corresponding modification in its shape after exposure to air. This particular behavior is not surprising because nanosize metallic particles are highly reactive species.

A second point, which must be addressed, is related to the sensitivity of the absorption spectrum regarding the cluster size or the presence of heterometallic bonds. The EXAFS data collected at the Co K-edge give direct evidence on the formation of cobalt nanoparticles. The average size as measured through the coordination numbers is equal to 1.0-1.5 nm, which is comparable with the size of the zeolite supercage (1.3 nm). From the TEM data a particle size range of about 2-3 nm can be estimated [4]. Regarding the local environment of Ru, it seems that reduction under H_2 results also in the formation of nanometer scale metallic Ru clusters.

The XPS results on the Ru-Co/NaY samples indicated the reduction/oxidation cycles to be a reversible process [4]. That is, the role of ruthenium is to promote the reduction of Co^{2+} ions thereby increasing the number of metallic cobalt atoms in zeolite. From the XPS data it was not unambiguously proven whether this reversibility is attributed to alloy formation inside the NaY supercage. The lack of Ru-Co bonds found by EXAFS, suggests that either the concentration of the alloy species is under the analytical limit, or separated cobalt and ruthenium nanoparticles are present in close vicinity to each others. We may speculate whether or not during preparation of the bimetallic sample by treatment in hydrogen the Ru metal particles – after supplying hydrogen atoms for reduction of cobalt - decorate the cobalt nanoparticles. If the decoration does not result in alloy formation, but in a sort of physical interaction, the lack of Ru-Co bond may be understood. Nevertheless, Co^{2+} ions still exist in the sample due to the partial reduction. These ions may retard migration of the metal atoms in the interior of the zeolite or to its external surface.

Quite surprisingly after air exposure metallic Co-Co bonds still exist inside the zeolite despite the high reactivity of Co to oxidation. Regarding the second metal, we observe the formation of Ru-O bonds and the number of Ru-Ru decreases significantly. These experimental findings support the picture described above for the chemical nature of bimetallic phase in zeolite.

In order to refine the chemical character of metal clusters generated in zeolite, ab initio simulation was performed on part of the absorption spectrum at the Co K and the Ru K-edges [14]. The key issue of the study was that metallic cobalt and metallic ruthenium particles were generated in the zeolite cages.

The scenario put forward previously in the CO hydrogenation can also slightly be modified. The production of short chain hydrocarbons and low olefin/paraffin ratio [3, 12, 15] was explained by hydrogenolysis in the pores containing Ru-Co particles in the NaY sample. In the light of the present results we incline to attribute the short chain hydrocarbons observed to intensive hydrogenation on the ruthenium decorating the cobalt particles. Thus, whilst leaving the supercage, the olefins leading to higher hydrocarbons by re-adsorption, are competitively hydrogenated to produce paraffins resulting in low olefin/paraffin ratio. These catalytic results strongly support the features of the Ru-Co/NaY system obtained by physical methods. The structure is further supported by a comparison of the catalytic behavior of Co/NaY, Ru-Co/NaY and Ru/NaY. In the Co-containing samples no CO_2 is formed [12] and the main difference lies in the activity being the Ru-Co/NaY sample 10 fold more active than the pure Co/NaY.

The XAS results have certain, however, not unresolvable contradiction to the in situ XPS measurements, by which we suggested the presence of bimetallic Ru-Co particles inside the NaY zeolite [4]. EXAFS measurements highlights the fine structure, i.e. ruthenium is more accessible to the reacting substrates in the bimetallic particles than Co. Consequently, Ru may decorate the cobalt surface and/or being in its close vicinity. The catalytic measurements in the CO hydrogenation gives evidence of the above structure [6]. The Ru-Co/NaY samples have the same rate, temperature dependence of the Schulz-Flory distribution and the olefin/paraffin ratios regardless of the pretreatments.

3. CONCLUSION

In situ EXAFS proved to be a useful technique to characterize bimetallic samples, which consist of metal nanoclusters located inside the zeolite. It was theoretically established that the spectra give information on the size of the metal clusters in comparison with reference metal foils.

In the Ru-Co/NaY sample the presence of no Ru-Co bonds was established, however, the cobalt showed somewhat larger cluster size than ruthenium. It is plausible to suggest a model in which Co-nanoparticles are surrounded and/or decorated by ruthenium particles. Hydrogen is mediated by ruthenium to cobalt and its easy oxidation prevents the metalic cobalt particles.

ACKNOWLEDGEMENTS

The authors are indebted to the Hungarian Science and Research Fund (grant # T-22117 and 30343) for financial support and the COST D15 project (project # COST D15/0005/99).

REFERENCES

[1] C. H. Bartholomew, "New Trends in CO Activation" L. Guczi (ed), *Stud. Surf. Sci. Catal.*, **Vol. 64**, p. 158, Elsevier Sci. Publ. Co., Amsterdam, 1991.

[2] E. Iglesia, S. L. Soled, R. A. Fiato, *J. Catal.*, **137** (1992) 212; E. Iglesia, S. L. Soled, R. A. Fiato, G. H. Via, *ibid.*, **143** (1993) 345; A. Kogelbauer, J. G. Goodwin, R. Okachi, *ibid.*, **160** (1996) 125.

[3] L. Guczi, Zs. Koppány, K. V. Sarma, L. Borkó and I. Kiricsi, *Stud. Surf. Sci. Catal.* H. Chon, S. -K. Ihm and Y. S. Uh (eds.), **Vol. 105** (1997) 861.

[4] L. Guczi, R. Sundararajan, Zs. Koppány, Z. Zsoldos, Z. Schay, F. Mizukami, S. Niwa, *J. Catal.*, **167** (1997) 482.

[5] S. Wrabetz, U. Guntow, R. Schlögl and H. G. Karge, *Stud. Surf. Sci. Catal.*, H. Chon, S. -K. Ihm and Y. S. Uh (eds.), **Vol. 105** (1997) 583; J. Wellenbscher, M. Mueler, W. Madhi, U. Sauerlandt, J. Schutze, G. Ertl and R. Schlögl, *Catal. Lett.* **25** (1994) 61.

[6] Z. Zsoldos, G. Vass, G. Lu and L. Guczi, *Appl. Surf. Sci.*, **78** (1994) 467.

[7] D. Bazin, I. Kovács, L. Guczi, P. Parent, C. Laffon, F. De Groot, O. Ducreux and J. Lynch, *J. Catal.*, **189**, 456 (2000).

[8] D. Bazin, P. Parent, C. Laffon, O. Ducreux, J. Lynch, I. Kovács and L. Guczi, *J. Synchrotron Radiat. New.* **6** (1999) 430.

[9] L. Guczi and D. Bazin, *Appl. Catal. A.*, **188** (1999) 163.

[10] K. C. McMahon, S. L. Suib, B. G. Johnson, and C. H. Bartholomew, *J. Catal.* **106** (1987) 47.

[11] D. Bazin, H. Dexpert and J.Lynch, J in "X-ray absorption fine structure for catalysts and surfaces" (Ed.: Y. Iwasawa), World Scientific 1996.

[12] L. Guczi, G. Stefler, Zs. Koppány. V. Komppa and M. Reinikainen, *Science and Technology in Catalysis 1998*, H. Hattori and K. Otsuka (eds.), *Stud. Surf. Sci Catal.* **Vol. 121**, 209, Kodansha/Elsevier, Tokyo/Amsterdam, 1999.

[13] L. Guczi L., Z. Kónya, Zs. Koppány, G. Stefler and I. Kiricsi, *Catal. Lett.*, **44** (1997) 7.

[14] D. Bazin, D. Sayers, J. Rher, *J. Phys. Chem. B.* **101** (1997) 11040.

[15] G. Stefler, I. Kiricsi and L. Guczi, Porous Materials in Environmentally Friendly Processes, I. Kiricsi, G. Pál-Borbély, J. B. Nagy and H. Karge (eds.), *Stud. Surf. Sci. Catal,* **125**, 495, Elsevier Sci. Publ. Co., Amsterdam, 1999.

Studies in Surface Science and Catalysis
J.J. Spivey, E. Iglesia and T.H. Fleisch (Editors)

Development of a high efficiency GTL process based on CO_2/steam reforming of natural gas and slurry phase FT synthesis

T.Wakatsuki[1], Y.Morita[1], H.Okado[1], K.Inaba[1], H.Hirayama[2], M.Shimura[3], K.Kawazuishi[3], O.Iwamoto[4] and T.Suzuki[4]

[1] JAPEX Research Center, Japan Petroleum Exploration Co. Ltd., 1-2-1 Hamada Mihama-ku, Chiba 261-0025, Japan
[2] Technology Research Center, Japan National Oil Corporation, 1-2-2 Hamada Mihama-ku, Chiba 261-0025, Japan
[3] Chemical Technology Center, Chiyoda Corporation, 3-13 Moriya-cho, Kanagawa-ku, Yokohama 221-0022, Japan
[4] Research & Development Center, Cosmo Oil Co. Ltd., 1134-2 Gongendo Satte, Saitama 340-0193, Japan

Abstract

CO_2/steam reforming catalysts composed of noble or transient metal and MgO support and Ru-based slurry phase FT synthesis catalyst have been developed to improve the efficiency and lower the costs of GTL processes. Their high carbon and hydrogen conversion in the CO_2/steam reforming, low methane selectivity, high olefin/paraffin ratio of the light hydrocarbons used in the FT synthesis, together with their durability, have been verified through extensive bench scale tests. Based on the performance of these catalysts, an optimal commercial scale GTL plant configuration has been developed, along with a cash flow economic evaluation that confirms the promising economics of this process. The COE (Crude Oil Equivalent) price is around US$17/bbl-product oil.

Introduction

JAPEX, Chiyoda and Cosmo began a 6-yr development program for a high efficiency and low cost GTL process based on CO_2/steam reforming of natural gas and Ru catalyst-based slurry phase FT synthesis under the sponsorship of JNOC (Japan National Oil Corporation) in1998.

The program aims at manufacturing high grade and environmentally friendly liquid fuels from raw natural gas that have a high cetane number and are free of sulfur, nitrogen, and aromatics. It is also desirable to utilize natural gas reserves which have been discovered but left undeveloped commercially for the following reasons.

[1] The magnitude of the reserve is too small for an LNG project
[2] The location of the reserve is too remote from users for pipeline transportation
[3] The CO_2 content of the reserve is too high to be economically removed or disposed

The program is designed to overcome these disadvantages and pave the way for monetizing the stranded gas reserves. It consists of two phases, i.e. , basic study and development of catalysts for CO_2/steam reforming and slurry phase FT synthesis, followed by verification of their performance through extensive operation of a demonstration plant having a capacity of H_2+CO of 388 Nm^3/h and FT synthesis oil of 7 bbl/day.

This paper discusses the first phase; i.e., the results on the CO_2/steam reforming catalyst and slurry phase FT synthesis catalyst, and economic evaluation of the process.

Performance of CO_2/steam reforming catalyst

As illustrated in Fig.1, the maximum carbon and hydrogen conversion at an H_2/CO ratio of 2.0, most favorable for FT reaction, fall in the region of high coke formation in a commercial plant catalyst tube. This study focuses on developing a CO_2/steam reforming catalyst which avoids coke formation[1] even at a higher carbon activity[2] region in order to improve the efficiency and minimize the reformer size.

A hydrocarbon conversion of 75.1%, total carbon conversion of 68.4%, and total hydrogen conversion of 66.6% were obtained using transient or noble metal catalysts supported on MgO in a bench scale reactor having a catalyst volume of 30cc and GHSV of 6000 h^{-1}. This is equivalent to equilibrium conversion at a pressure of 19.8 kg/cm^2G,

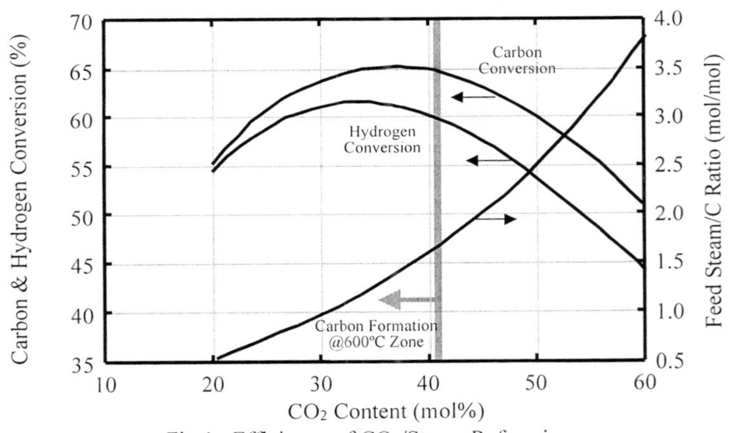

Fig.1 Efficiency of CO_2/Steam Reforming

Conditions ; Pressure : 25kg/cm^2G, Temperature : 900°C, H_2/CO = 2.0 in reformed gas

temperature (outlet of the catalyst bed), 900°C, feed gas composition of $CO_2/H_2O/CH_4$ = 0.43/1.05/1.00 and H_2/CO ratio of 1.95

A catalyst life test was also carried out under the following conditions; total pressure of 20kg/cm^2G, temperature of 850°C, feed gas composition of $CO_2/H_2O/CH_4$ = 0.4/0.85/1.0 and GHSV of 6000h^{-1}. As shown in Fig.2, the methane conversion remained at the equilibrium level during almost 8000 hours operation. No carbon accumulation on the catalyst was found after 8000 hours operation. Consequently the reforming catalyst life span is expected to be over 16000 hours for commercial plant operation.

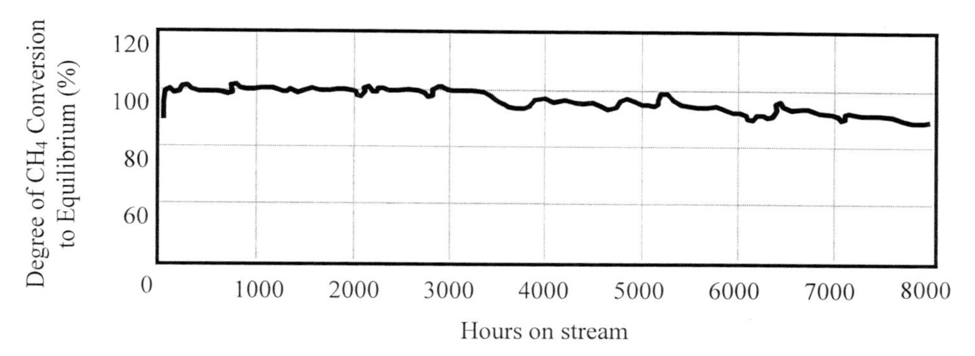

Fig.2 Laboratory Test Results for CO_2/Steam Reforming Catalyst

Conditions ;
Pressure : 20kg/cm^{2}G, Temperature : 850°C, Feed gas $CO_2/H_2O/CH_4$: 0.4/0.85/1.0, GHSV : 6000h^{-1}

The reason that this catalyst can avoid coke formation at conditions where carbon activity is greater than 1.0 is explained in Fig.3. In general coke is formed and accumulated on the active metallic atom from a coke precursor CH_X, which is produced through dehydrogenation of CH_4. However, if the precursor could react with an oxygen species (•O, O^{2-}, etc.) or hydroxy group species (•OH, OH^-, etc.) which migrate from the support surface, no coke formation or accumulation on the metallic atom occurs [2-4]. Our catalyst made this possible by means of high dispersion of active metallic atom and a specially designed surface of the support.

The reason that this catalyst can avoid coke formation at conditions where carbon activity is greater than 1.0 is explained in Fig.3. In general coke is formed and accumulated on the active metallic atom from a coke precursor CH_X, which is produced through dehydrogenation of CH_4. However, if the precursor could react with an oxygen species (•O, O^{2-}, etc.) or hydroxy group species (•OH, OH^-, etc.) which migrate from the support surface, no coke formation or accumulation on the metallic atom occurs [2-4]. Our catalyst made this possible by means of high dispersion of active metallic atom and a specially designed surface of the support.

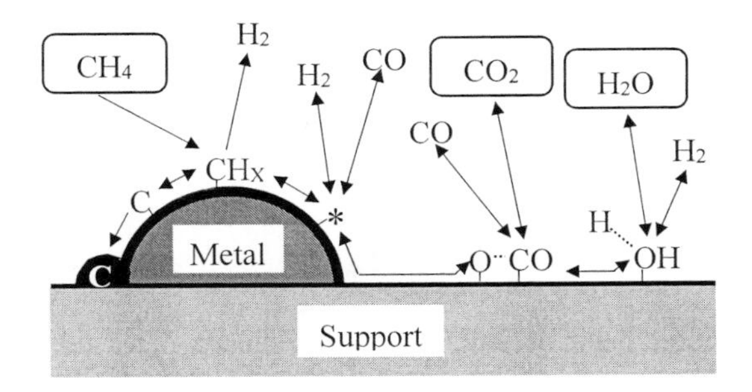

Fig.3 Reaction Model on Catalyst Surface

Performance of slurry phase FT synthesis catalyst

A supported ruthenium FT synthesis catalyst has been developed to yield higher liquid hydrocarbon fractions in the slurry phase system. Fig.4 shows the performance of this slurry phase FT synthesis catalyst. The CO conversion (once through) is 73%, α (chain growth probability) is 0.9, yield of C_5^+ fractions is 85.1wt%, yield of kerosene and diesel fractions is 71.0wt% at a total pressure of 20kg/cm²G, temperature of 270°C, and feed gas composition of H_2/CO = 2.0, in a 30-cc slurry phase reactor. This developed catalyst has a methane selectivity of 7wt%, lower than those of Fe-based or Co-based catalysts[5]. The olefin/paraffin molar ratio of light hydrocarbons in FT products is higher than Ru/Al_2O_3 or Co/SiO_2 catalyst, as shown in Fig.5. This higher olefin/ paraffin ratio, especially in the naphtha fraction, makes it possible to increase the yield of kerosene and diesel fractions through dimerization or oligomerization. An accelerated deactivation test shows that the catalyst life is expected to be over 3000 hours.

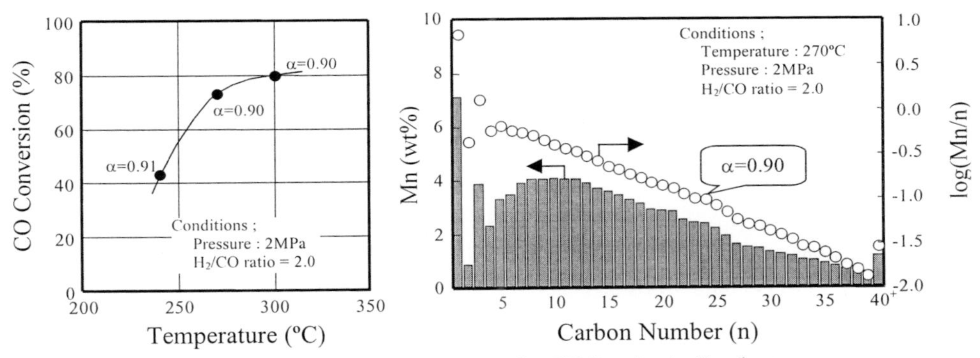

Fig.4 Laboratory Test Results for FT Synthesis Catalyst

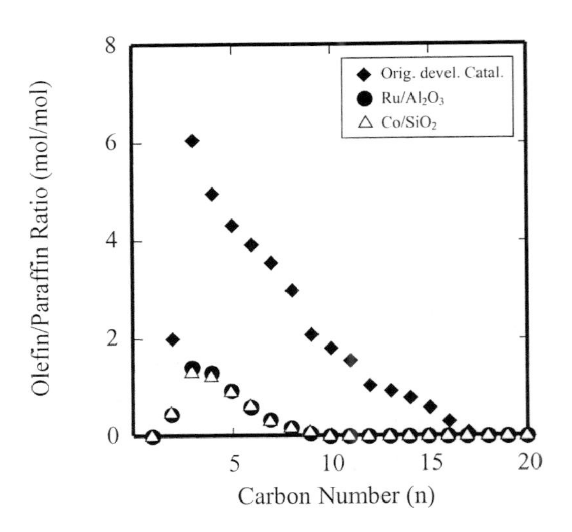

Fig.5 Olefin/Paraffin Ratio of FT Products

Economic evaluation of the process[6)]

Based upon the performance of the above catalysts for syngas production and for FT synthesis, the following economic evaluation of our GTL process has been conducted.

[1] The optimal process condition was studied for syngas production, FT synthesis, and upgrading.
[2] The process configuration for a commercial scale stand-alone and grass-root plant was established as shown in Fig.6.
[3] South-east Asian region was assumed as an actual commercial plant site; the CO_2 content of the raw natural gas was taken as 20vol%
[4] For a 15000 BPSD capacity plant, the material and heat balance was calculated and the energy consumption and CO_2 emission were estimated at 2066 Mcal/bbl-oil and 117 Nm^3-CO_2/bbl-oil. The basic specifications of equipment were also determined, and thus the investment cost was estimated at US$560 Million.
[5] Applying the standard economic parameters, the economic evaluation and sensitivity analysis were conducted by the cash flow method using the assumptions described in Table 1. The production cost of the liquid fuel was estimated at US$17.2/bbl-oil (COE price) under the basic conditions (raw natural gas price of US$0.5/MMBTU, IRR of 15%)

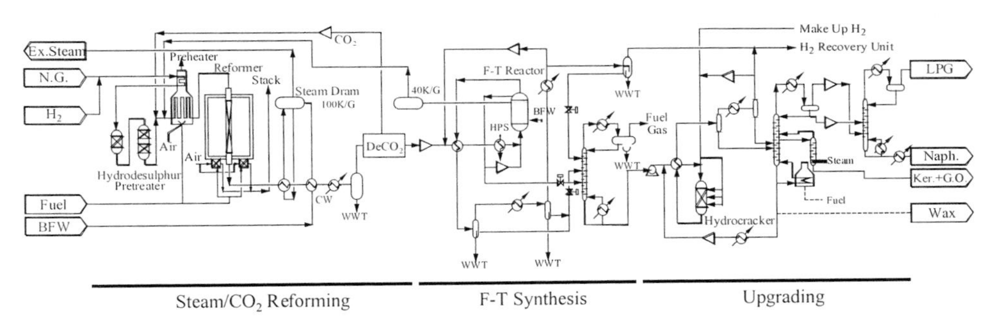

Fig.6 Process Configuration of the Originally Developed GTL Process for 15000 BPSD capacity.

Table 1 Assumptions of the Economic Evaluation and Sensitivity Analysis

Plant Capacity	5000 , 10000 , 15000 BPD
Operation Period Depreciation Liquid Product Value	25 years 10%/Y 143% Crude Oil Price
Unit Price of Natural Gas and Utilities as of 1994	
Natural Gas Utilities	US$0.5 , 1.0 /MMBTU Fuel Gas : Same as Natural Gas
Maintenance and Operation Labor Cost	3% of Plant Cost/Y

[6] Sensitivity analysis results for the smaller capacity at IRR 12% cases are illustrated in

Fig.7 at various natural gas prices. Should the natural gas price fall down to US$0.5/MMBTU, COE price can be improved by around 17% against US$1.0/MMBTU case. Fig.7 also teaches this GTL process has a COE price of US$20/bbl-prodution oil even for a rather small plant capacity of 5000 BPD which is equivalent to 0.63 TSCF gas reserve.

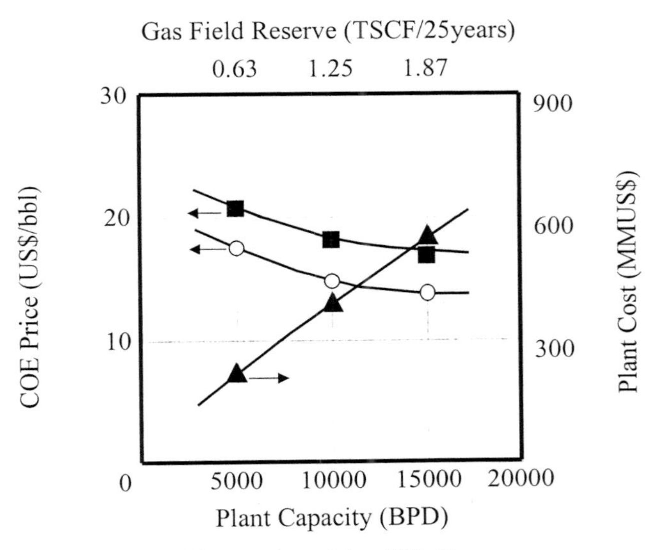

Fig.7 Economics of the GTL Process

IRR = 12%, Loan Interest = 3.51%/Y, Equity = 50%
Inflation : Crude Oil = 3.5%/Y, Natural Gas = 3.5%/Y
NG Price : 1.0 US$/MMBTU (■), 0.5 US$/MMBTU (○)
CO_2 Content = 20 vol%

This work is based upon a part of the activities of the project named "Indirect Conversion of Natural Gas to Liquid Fuels", a development program funded by Japan National Oil Corporation.

Reference

1 W. L. Holsten. J. Calal., 152, 42 (1995)
2 J. R. Rostrup-Nielsen, “Catalytic Steam Reforming” in Catalysis 5, 1 (1984)
3 Y. G. Chen, O. Yamazaki, K. Tomisige, K. Fujimoto, Catal. Lett., 39, 91 (1996)
4 T. Sodesawa, S. Sato, F. Nozaki, Stud. Surf. Sci. Catal., 77, 401 (1993)
5 H. Schulz, E. van Steen, M. Claeys, Stud. Surf. Sci. Catal., 81, 455 (1994)
6 G. N. Choi, S. J. Kramer, S. S. Tam, “Design/Economics of a Natural Gas based Fischer Tropsch Plant”, the AIChE 1997 Spring National Meeting, March 9-13 (1997)

Studies in Surface Science and Catalysis
J.J. Spivey, E. Iglesia and T.H. Fleisch (Editors)

Kinetic modeling of the slurry phase Fischer-Tropsch synthesis on iron catalysts

Lech Nowicki[a] and Dragomir B. Bukur[b]

[a] Technical University of Lodz, Faculty of Process & Environmental Engineering, 90-924 Lodz, Poland
[b] Texas A & M University, Department of Chemical Engineering, College Station, TX 77843-3122, USA

Two different approaches to modeling of hydrocarbon product distribution for Fischer-Tropsch synthesis have been tested using results from experiments conducted under industrially relevant conditions in a slurry reactor on iron catalysts. The first approach (two active site model) is based on the assumption that there are two types of active sites on the catalyst surface: type-1, where primary growth of hydrocarbon intermediates occurs, and type-2, where reversible carbon atom number dependent readsorption of 1-olefins takes place. The readsorbed 1-olefins form alkyl intermediates, C_nH_{2n+1}, on the surface, which in turn can participate in several reactions: chain growth propagation, hydrogenation to *n*-paraffins, and dehydrogenation to 2-olefins. The second approach (one active site model) accounts for secondary readsorption of 1-olefins on type-1 sites only. Model predictions provide information on hydrocarbon product distribution over the wide range (C_1 to C_{50}) of carbon numbers, and have been verified in experiments with several iron Fischer-Tropsch catalysts.

1. INTRODUCTION

The Fischer-Tropsch synthesis (FTS) is a viable route for the production of a feedstock for fuels and chemicals. In spite of the large number of product species typically formed during the FTS, only certain classes of products are present in significant quantities (*n*-paraffins, 1-olefins, and to smaller extent 2-olefins, branched paraffins and oxygenates). These products tend to follows regular trends with carbon number according to the so-called Anderson-Schulz-Flory (ASF) distribution. However, FTS products obtained on iron catalysts follow ASF distribution with some deviations. Methane is often reported to exceed the ASF value, while C_2 products tend to be less than expected. Also positive deviations from classical ASF distribution (single chain growth probability factor) have been frequently observed for higher molecular weight products. Kinetic models for prediction of product distribution in FTS are relatively scarce [1-6].

The objective of the present paper is to test the ability of two different models to predict experimentally observed product distributions obtained with representative iron Fischer-Tropsch catalysts under steady-state and industrially relevant process conditions in a stirred tank slurry reactor. The first model represents an extension of a model developed previously at Texas A&M University [1,2], and of a similar model proposed by Schulz and Claeys [4].

This model is based on the hypothesis that there are two types of active sites on the catalyst surface [7-9]. In our model it is assumed that chain growth, chain termination and 1-olefin readsorption all occur on type-1 sites, whereas secondary reactions of readsorbed 1-olefins occur on type-2 sites together with chain growth and termination reactions.

The second model assumes that carbon number dependent readsorption of 1-olefins occurs on the same catalytic sites, where the chain growth and termination (primary reactions) take place. The latter model was developed recently by van der Laan and Beenackers [5].

2. MODELING

The two site model is based on the reaction network shown in Fig.1A. According to this scheme, the hydrogenation of adsorbed methylene group (CH_2) initiates hydrocarbon chain growth on the catalyst surface. The same group (CH_2) is responsible for chain growth propagation. The CH_2 monomeric unit can be formed by hydrogenation of adsorbed carbon monoxide, as described elsewhere [1,2]. However, this step is not considered in the present model.

On primary FTS active sites (sites #1) the following reactions occur: chain growth propagation, and chain growth termination to form *n*-paraffins and 1-olefins. 1-olefins may readsorb on primary growing sites as well as on separate secondary reaction sites (sites #2). Hydrogenation of alkyl intermediates C_nH_{2n+1}#2 to form additional amounts of *n*-paraffin, and termination by dehydrogenation of alkyl species to form 1-olefin (reversible step) and/or 2-olefin (irreversible step) can occur on active sites #2.

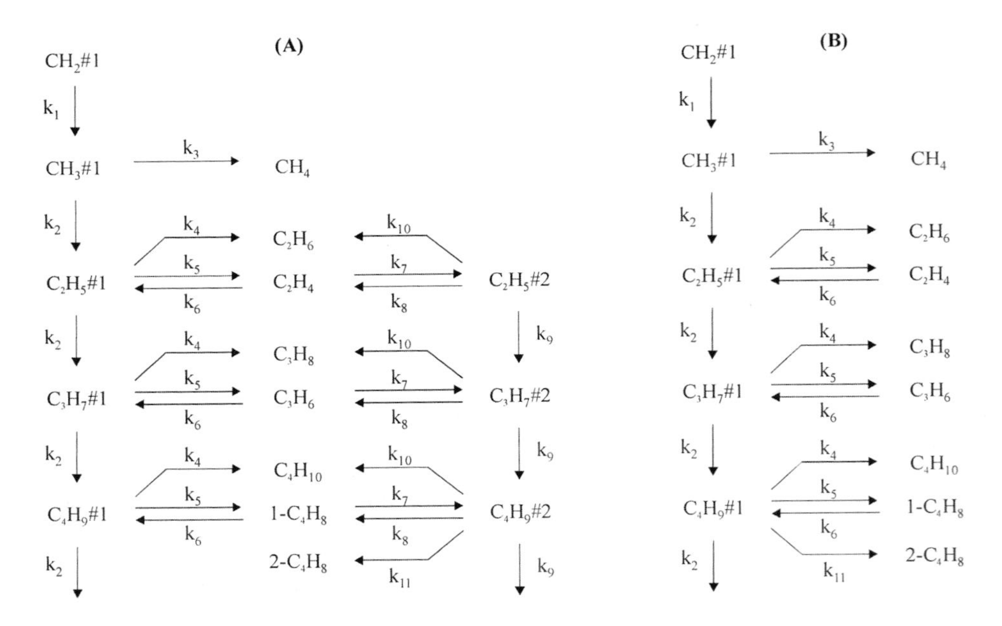

Fig. 1. Reaction network for primary and secondary FTS product formation in two active site (A) and one active site (B) models.

Table 1. Elemental reactions describing hydrocarbon products formation in FTS

Stoichiometric equations	Kinetic equations
$H_2C\#1 + H\#1 = H_3C\#1 + \#1$	$k_1 \cdot [H\#1] \cdot [H_2C\#1]$
$R_{n-1}\#1^a + H_2C\#1 = R_n\#1 + \#1$	$k_2 \cdot [H_2C\#1] \cdot [R_{n-1}\#1]$, n = 2,3,...
$H_3C\#1 + H\#1 = CH_4 + 2\#1$	$k_3 \cdot [H\#1] \cdot [H_3C\#1]$
$R_n\#1 + H\#1 = C_nH_{2n+2} + 2\#1$	$k_4 \cdot [H\#1] \cdot [R_n\#1]$, n = 2,3,...
$R_n\#1 + \#1 = 1\text{-}C_nH_{2n} + H\#1 + \#1$	$k_5 \cdot [\#1] \cdot [R_n\#1]$, n = 3,4,...
$1\text{-}C_nH_{2n} + H\#1 + \#1 = R_n\#1 + \#1$	$k_6 \cdot c_{o1,n} \cdot [H\#1] \cdot [\#1]$, n = 2,3,...
$1\text{-}C_nH_{2n} + H\#2 + \#2 = R_n\#2 + \#2$	$k_7 \cdot c_{o1,n} \cdot [H\#2] \cdot [\#2]$, n = 2,3,...
$R_n\#2 + \#2 = 1\text{-}C_nH_{2n} + H\#2 + \#2$	$k_8 \cdot [\#2] \cdot [R_n\#2]$, n = 2,3,...
$R_{n-1}\#2 + H_2C\#2 = R_n\#2 + \#2$	$k_9 \cdot [H_2C\#2] \cdot [R_{n-1}\#2]$, n = 2,3,...
$R_n\#2 + H\#2 = C_nH_{2n+2} + 2\#2$	$k_{10} \cdot [H\#2] \cdot [R_n\#2]$, n = 2,3,...
$R_n\#2 + \#2 = 2\text{-}C_nH_{2n} + H\#2 + \#2$	$k_{11} \cdot [\#2] \cdot [R_2\#2]$, n = 4,5,...

[a] $R_n = C_nH_{2n+1}$

The proposed reaction network neglects formation of oxygenates and branched hydrocarbons.

The kinetic model was derived from taking steady-state mass balances for each growing specie, assuming kinetic equations for all elementary reactions (chain growth and terminations reactions) given in Table 1. For example, the steady state mass balance for methyl group (chain initiator) adsorbed on active site #1 is

$$k_1^* = (k_2^* + k_3^*) \cdot [\mathrm{H_3C\#1}] \tag{1}$$

For alkyl species containing $n>3$ (for n=2,3 these equations have to be simplified according to the scheme in Fig. 1A) carbon atoms in a molecule adsorbed on active sites #1 and #2 the steady state mass balance equations are:

$$k_2^* \cdot [R_{n-1}\#1] + k_6^* \cdot c_{o1,n} = (k_2^* + k_4^* + k_5^*) \cdot [\mathrm{R_n\#1}] \tag{2}$$

$$k_7^* \cdot c_{o1,n} = (k_8^* + k_9^* + k_{10}^*) \cdot [\mathrm{R_n\#2}] \tag{3}$$

$R_n\#1$ and $R_n\#2$ denote alkyl groups adsorbed on sites #1 and #2 respectively, k^* is a pseudo first order reaction rate constant for the reaction path indicated by subscript. These rate constants incorporate true rate constants and surface concentrations of hydrogen atom, methylene group or vacant active sites depending on which of these species takes part in the elementary step. All rate constants are assumed to be carbon number independent, except for methane formation and ethylene readsorption to describe common deviations of these products from the ASF distribution.

The actual concentration of 1-olefin in the liquid phase has to be known to determine the surface concentrations of alkyl species from equation (2) and (3). To overcome this problem the following additional assumptions were made:

- The reaction is in steady state and all 1-olefins leave the reactor continuously via gas. This assumption is often used [5,10] because wax withdrawn from the reactor contains mainly *n*-paraffins.
- Gas and liquid phase are at thermodynamic equilibrium, which can be expressed by Henry's law.
- The values of Henry's constant are strongly carbon atom dependent. It can be described with the following equation:

$$He_{o1,n} = He_{o1,0} \cdot \exp(-a \cdot n) \tag{4}$$

where $He_{o1,0}$ and a are constants. Equation (4) reflects the well-known fact that solubility of 1-olefin increases exponentially with the chain length [11,12].
- The effects of intra and inter particle diffusional limitations of reactants and products have been neglected.

Taking into account the first two assumptions, it is possible to relate the concentration of 1-olefin in the liquid phase $c_{o1,n}$ to the rate of its formation $r_{o1,n}$ by the following mass balance equation (ideally mixed continuous reactor):

$$r_{o1,n} = \frac{SV(P,T) \cdot c_{o1,n} \cdot He_{o1,n}}{R_g \cdot T} \tag{5}$$

where T is temperature, P – pressure, R_g – gas constant and $SV(P,T)$ is space velocity at the reaction conditions.

From the other hand, formation rate of 1-olefin is equal to:

$$r_{o1,n} = k_5^* \cdot [\mathrm{R}_n\#1] + k_8^* \cdot [\mathrm{R}_n\#2] - (k_6^* + k_7^*) \cdot c_{o1,n} \tag{6}$$

Solving equations (2), (3), (5) and (6) yields the surface intermediate concentration and the formation rate of 1-olefins. Equations describing the formation rate of *n*-paraffins and 2-olefins must be also written using concentrations of alkyl intermediates.

One active site model takes into account secondary readsorption of 1-olefin on active sites where primary chain growth occurs (see Fig. 1B). Chain termination by dehydrogenation of adsorbed alkyl groups gives not only 1- but also 2-olefins. All other assumptions are identical to those made for the two-site model. Equations for one active site model can be easily obtained from those for the two-site model by putting reaction rate constants equal to zero for all elementary steps not existing in the one-site model.

3. RESULTS AND DISCUSSION

Experimental data in this study were obtained in a 1 dm^3 continuous stirred tank slurry reactor. A detailed description of our reactor system as well as experimental and analytical procedures was given elsewhere [13,14]. Experimental data from runs with two iron catalysts were used for testing of the validity of the two models. A precipitated iron catalyst, prepared by United Catalysts Inc. (UCI), with nominal composition 100 Fe/7.5 Cu/2.6 K/9.2 SiO_2 was tested in run SA-2052 at: 265°C, 2.1 MPa, 2.4 Ndm^3/g-Fe·h (where Ndm^3 denotes volume of gas at standard temperature and pressure) and H_2/CO feed ratio of about 0.7. During 300 h of testing at these conditions the CO conversion was 68-75%. Detailed information on the performance (activity and selectivity) of this and two other UCI catalysts can be found elsewhere [15]. Olefin hydrogenation and isomerization activities of this catalyst were rather low (i.e. total olefin content was high and 2-olefin content was low).

In contrast with the UCI catalyst used in run SA-2052, the silica supported catalyst, used in run SB-1509, had much lower total olefin and higher 2-olefin content in the whole range of carbon atom numbers, which reflects higher activity for secondary reactions. This catalyst was prepared by impregnation of commercial silica support (Davison 644) with aqueous solutions of iron and promoter salts. Nominal composition of the catalyst is 100 Fe/1 Cu/6 K/181 SiO_2 (on mass basis). The catalyst in run SB-1509 was tested at: 230-250°C, 3.1 MPa, 5.6-9.0 Ndm^3/g-Fe·h, and H_2/CO feed ratio of 1.7. During 220 h of testing at 230°C and 9 Nm^3/g-Fe·h, the CO conversion decreased from about 30 to 25 %.

The parameters appearing in model equations were calculated from experimental data by minimising the sum of absolute relative residuals for *n*-paraffins, 1- and 2-olefins.

Figure 2 shows predicted and experimental distributions of the various groups of hydrocarbon products for runs SB-1509 and SA-2052. The curved lines for the distribution of *n*-paraffins and olefins are fitted well with both two-site as well as one-site hydrocarbon selectivity models over the entire carbon number range considered. Significant differences in model predictions were observed for C_2 hydrocarbons only. This is shown more clearly in Figure 3, where total olefin [defined as 100%x(1-olefin+2-olefin)/(paraffin+1-olefin+2-olefin)] and 2-olefin [defined as 100%x(2-olefin)/(1-olefin+2-olefin)] content as a function of carbon number are presented.

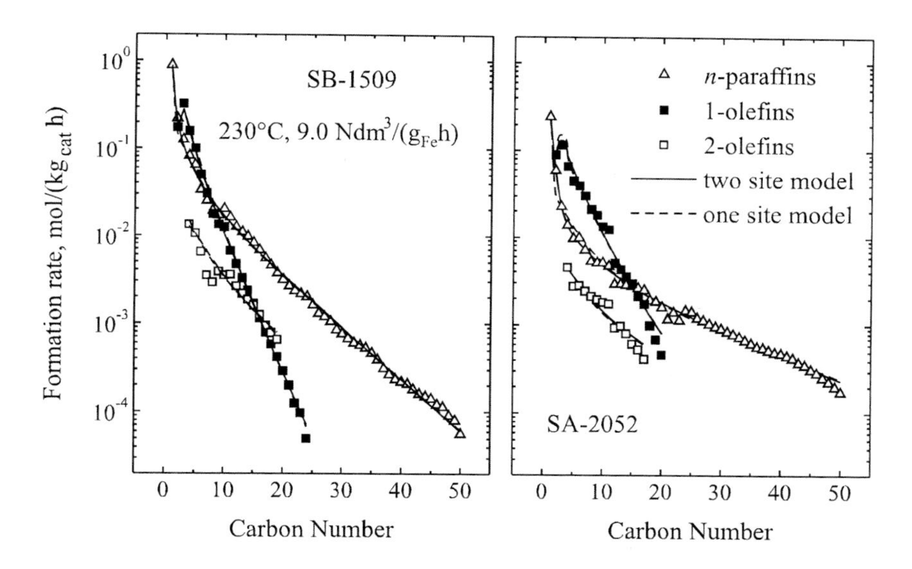

Fig.2. Product distribution for different groups of hydrocarbons in case of high (run SB-1509) and low (SA-2052) secondary reactions activity (symbols are used for experimental data and lines for models predictions).

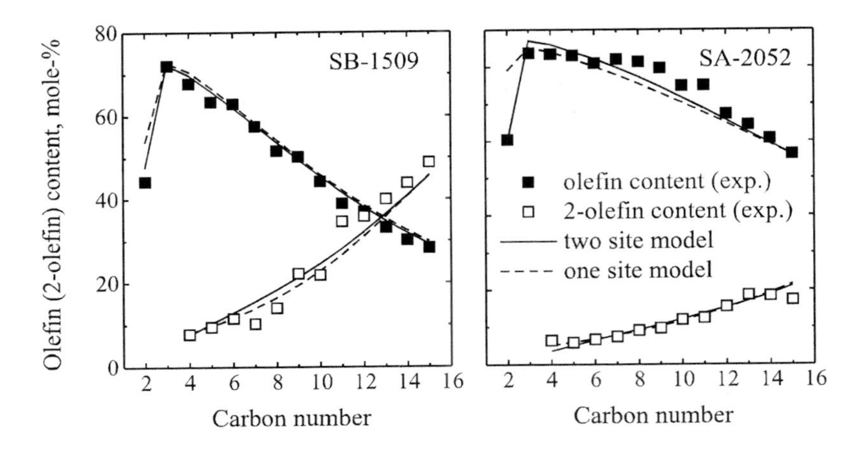

Fig. 3. Selectivity plots for data presented in Figure 2.

In summary, predictions from both models were found to be in very good agreement with experimental data over the entire range of carbon numbers considered, except for ethylene selectivity where the two-site model was found to be more accurate.

REFERENCES

1. W.H. Zimmerman, Kinetic modelling of the Fischer-Tropsch Synthesis, Ph.D. Thesis (1990), Texas A&M University, College Station, Texas, USA,
2. W.H. Zimmerman, D.B. Bukur and S. Ledakowicz, Chem. Eng. Sci., 47 (1992) 2707.
3. R.J. Madon, E. Iglesia, *J. Phys. Chem.*, 95 (1991) 7795.
4. H. Schulz and M. Claeys, Appl. Catal. A, 186 (1999) 91.
5. G.P. Van der Laan and A.A.C.M. Beenackers, Ind. Eng. Chem. Res., 38 (1999) 1277.
6. B. Wojciechowski, Catal. Rev.-Sci. Eng., 30 (1988) 629.
7. G.A. Huff and C.N. Satterfield, J. Catal., 85 (1984) 370.
8. L. Koenig, and J. Gaube, J, Chem-Ing.-Tech., 55 (1983) 14.
9. Schulz, H.; Rosch, S.; Gokcebay, H. in *Coal: Phoenix of '80s, Proc. 64th CIC. Coal Symp.* Vol. 2, Canadian Society for Chemical Engineering: Ottawa, Canada, 1982; pp. 486-493.
10. E. Iglesia, S.C. Reyes and S.L. Soled, Reaction-transport selectivity models and design of Fischer-Tropsch catalysts. Computer Aided Design of Catalysts and Reactors, Marcel Dekker, New York, 1993, p. 199.
11. M.C. Donohue, D.S. Shah, K.G. Connaly and V.R. Venkatachalam, Ind. Eng. Chem. Fundam., 24 (1985) 241.
12. B.B. Breman, A.A.C.M. Beenackers, E.W.J. Riettjens and R.J.H. Stege, J. Chem. Eng. Data, 39 (1994) 647.
13. D. B. Bukur, S. Patel, X. Lang, Appl. Catal., 61 (1990) 329.
14. D.B. Bukur, L. Nowicki, X. Lang, *Chem. Eng. Sci.*, 49 (1994) 4615.
15. D.B. Bukur, L. Nowicki, X. Lang, Natural Gas Conversion IV (M. de Pontes, R. L. Espinoza, C. P. Nicloaides, J. H. Sholz and M. S. Scurrell editors), Elsevier Science B. V., 1997, 163.

Studies in Surface Science and Catalysis
J.J. Spivey, E. Iglesia and T.H. Fleisch (Editors)

Mechanism of Carbon Deposit/Removal in Methane Dry Reforming on Supported Metal Catalysts

K. Nagaoka[a], K. Seshan[b], J. A. Lercher[c] and K. Aika[a]

[a]Department of Environmental Chemistry and Engineering, Interdisciplinary Graduate School of Science & Engineering, Tokyo Institute of Technology, 4259 Nagatsuta, Midori-ku, Yokohama 226-8502, Japan

[b]University of Twente, Faculty of Chemical Technology, P. O. Box 217, 7500, AE, Enschede, The Netherlands

[c]Technical University of Munich, Department of Physical Chemistry, D-85748 Garching, Germany

The greater resistance to coke deposition for Pt/ZrO_2 compared to Pt/Al_2O_3 in the CH_4/CO_2 reaction has been attributed to the higher reactivity of coke with CO_2 on Pt/ZrO_2 [1]. Hence, in this communication, the reaction of coke derived from methane (CH_x: which is an intermediate in the reforming reaction and also a source of coke deposition) with CO_2 was studied on Pt/Al_2O_3 and Pt/ZrO_2 at 1070 K. The reactivity of coke itself on Pt, as measured by its reaction with H_2, was higher on Pt/Al_2O_3 than on Pt/ZrO_2. However, the reactivity of coke toward CO_2 was lower. Hence, the difference between the two catalysts cannot be attributed to the difference in the reactivity of coke itself. Next, the ability of the active site to activate CO_2 (probably oxygen defect sites on the support), as shown by CO evolution measurement in CO_2 stream, was higher on Pt/ZrO_2 than on Pt/Al_2O_3. Therefore, the high reactivity of coke toward CO_2 on Pt/ZrO_2 is attributed not to the intrinsic reactivity of coke itself but to the high activity of CO_2 at oxygen defect sites of ZrO_2 that are in the vicinity of Pt particles.

1. INTRODUCTION

Pt/ZrO_2 has been found to be a stable catalyst for CH_4/CO_2 reforming reaction in a wide range of temperatures, while Pt/Al_2O_3 deactivates at these same conditions [2]. The cause of catalytic deactivation for Pt/Al_2O_3 is attributed to coke deposition on catalytic reactive site at temperatures lower than 1070 K [1,3]. In an earlier work, we have proposed the deactivation mechanism for Pt/Al_2O_3 at 1070 K [1]. A part of coke derived from methane (CH_x: $0 \leq x \leq 3$) remains on Pt/Al_2O_3 without being oxidized by CO_2 and gradually covers the Pt particles. After a while, only the Pt-Al_2O_3 interface remains as the active site. The coke deposition is brought about by an imbalance between coke (CH_x) formation on Pt particles and oxidation of the coke by activated CO_2. On the other hand, over Pt/ZrO_2, coke deposition is minimal. Therefore, the catalyst maintains stable activity for a longer time. A combination of three processes can be used to explain why coke deposited from methane is oxidized by CO_2 more rapidly on Pt/ZrO_2: (i) coke on Pt that is supported on ZrO_2 is more reactive toward CO_2 than on Pt supported on Al_2O_3; (ii) CH_4 decomposition is slower on Pt/ZrO_2 than on Pt/Al_2O_3; (iii) coke which may cover active sites is hardly formed on ZrO_2.

In this communication, we investigate (i); i.e., why coke on Pt supported on ZrO_2 is more reactive toward CO_2 than on Pt supported on Al_2O_3 at 1070 K. The reactants are coke derived from methane (CHx) and CO_2. Studying the reactivity of each species helps in the understanding of the difference in reactivity of coke toward CO_2. Temperature programmed hydrogenation (TPH) is performed to investigate the reactivity of coke itself on Pt/Al_2O_3 and Pt/ZrO_2 after exposure to CH_4/CO_2 (in the CH_4/CO_2 reforming reaction) or CH_4/He (which produces CHx). The results are compared with the results of temperature programmed reaction (TPRn) of coke with CO_2 (reproduced from [1]). Further, the reactivity of CO_2 on Pt/Al_2O_3 and Pt/ZrO_2 is compared by following the amount of CO evolved when CO_2 flows on over the catalysts.

2. EXPERIMENTAL

Pt/ZrO_2 and Pt/Al_2O_3 were prepared by the wet impregnation technique. For this purpose, a solution of H_2PtCl_6 $6H_2O$ in water, ZrO_2 (RC-100, Gimex, Japan), and Al_2O_3 (000-3AQ, AKZO, The Netherlands) were used. The catalysts were dried overnight at 395 K in static air and calcined for 15 h at 925 K (heating rate 3 K min^{-1}) in flowing air (30 ml min^{-1}).

TPH and TPRn with CO_2 were performed in an Altamira AMI-2000 apparatus. 0.1 g of catalyst was loaded into a tubular quartz reactor and a thermocouple was placed at the top of the catalyst bed to measure the temperature of the catalyst. After the catalyst was reduced *in situ* with H_2 for 1 h at 1120 K, the temperature was lowered to 1070 K in He stream and feed gas mixture (CH_4/He (1/1) or CH_4/CO_2 (1/1) with a total flow of 28 ml min^{-1}) was fed to the reactor for required time. Then, the catalyst was kept in He for 30 min at the reaction temperature and then cooled to 340 K. For the TPH experiment, the sample was then exposed to a flow of H_2/He (10/90 with a total flow of 30 ml min^{-1}) at 340 K and subsequently heated up to 1273 K (heating rate 10 K min^{-1}). CH_4 (m/e=16), H_2O (m/e=18), CO (m/e=28) and CO_2 (m/e=44) signals were measured with a mass spectrometer. For TPRn with CO_2, H_2/He in the above experiment was replaced with CO_2/He (10/90 with a total flow of 30 ml min^{-1}). The final dwell time at 1273 K was 10 min.

In order to perform CO evolution measurements during CO_2 flow, 0.2 g of a catalyst (Pt/ZrO_2 or Pt/Al_2O_3) was used. After reduction, the catalyst temperature was adjusted to 1070 K in He. The catalyst was purged by He for 1h to remove residue H_2 and CO_2/He (1/3 with a total flow of 40 ml min^{-1}) were fed to the reactor for 30 min (1st run). In subsequent experiments, the catalyst was again treated in H_2 followed by He at the reaction temperature of the 1st run (1070 K) and CO_2/He was again fed to the reactor for 30 min at the temperature (2nd run). The amount of CO evolved was analyzed by a gas chromatograph (Aera M200) equipped with a 2m-MS-5A column and a TCD.

3. RESULTS AND DISCUSSION

3.1. Temperature programmed hydrogenation (TPH)

TPH profiles (m/e=16, 18, 28, and 44) for Pt/Al_2O_3 and Pt/ZrO_2 after exposure to CH_4/He for 10 min are shown in Fig. 1. First, we will discuss the contribution of O (m/e=16) fragmentation from H_2O (m/e=18), CO (m/e=28), and CO_2 (m/e=44) to CH_4 (m/e=16). The peak intensities for H_2O on these catalysts decreased with temperature, and those for CO and CO_2 were small compared to those for m/e=16, indicating that the contribution was quite small.

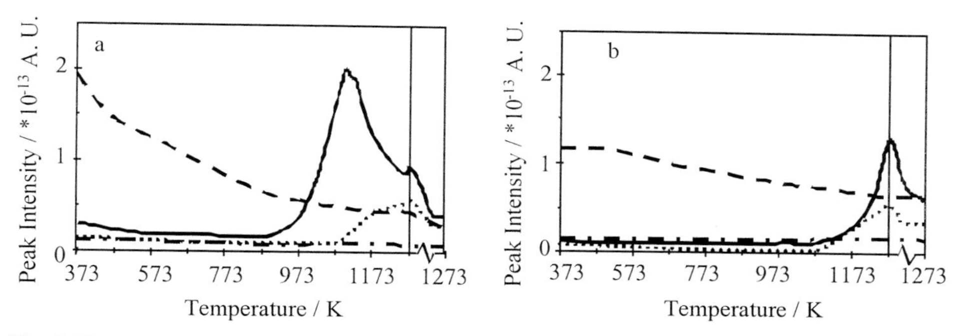

Fig. 1. Temperature programmed hydrogenation (TPH) profiles of m/e= (——) 16, (– –) 18, (·······) 28, and (– · –) 44 for (a) Pt/Al_2O_3 and (b) Pt/ZrO_2 after exposure to CH_4/He at 1070 K for 10 min.

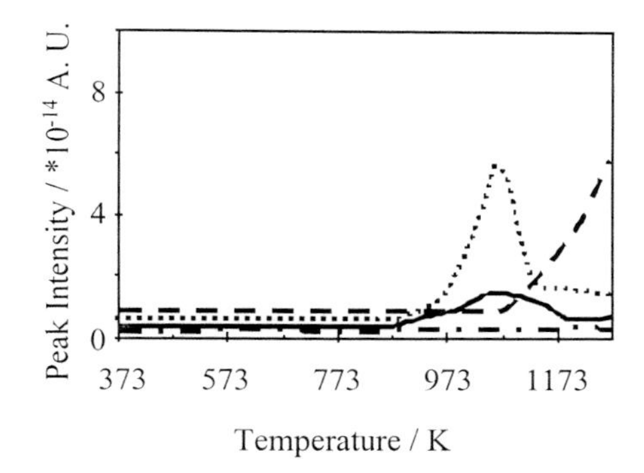

Fig. 2. Temperature programmed hydrogenation (TPH) profiles of m/e=16 for Pt/Al_2O_3, Pt/ZrO_2, and Al_2O_3 after exposure to CH_4/CO_2 at 1070 K for 1 or 12 h. (——); Pt/Al_2O_3 coked for 1 h, (·······); Pt/Al_2O_3 coked for 12 h, (– · –); Pt/ZrO_2 coked for 12 h, (– –); Al_2O_3 coked for 12 h.

Peaks for CH_4 were observed above 873 K for Pt/Al_2O_3, while a sole peak was visible above 1073 K for Pt/ZrO_2. The results indicate that some of the coke on Pt/Al_2O_3 is more reactive with H_2 than that on Pt/ZrO_2. For both catalysts, the peak intensities for CH_4 at 1273 K (after temperature was kept for 10 min) were higher than background level. Hence, some of coke seemed not to be removed from catalysts by the reaction with H_2 below 1273 K. On the other hand, the amount and peak temperature (above 1073 K) for CO formation were almost the same for both catalysts. Coke may have reacted with oxygen (lattice oxygen and/or hydroxyl) on the support to give CO above 1070 K.

It is necessary to confirm that coke deposited after catalyst was exposed to CH_4/He reflects the nature of the coke in the CH_4/CO_2 reaction. Thus, TPH is performed on Pt/Al_2O_3, Pt/ZrO_2, and Al_2O_3 after these catalysts are exposure to CH_4/CO_2 for 1 or 12 h. The CH_4 evolution profiles are shown in Fig. 2. No peak was visible on Pt/ZrO_2 even after reaction for 12 h, in accordance with the result of temperature programmed oxidation [1], which showed that no coke was deposited on Pt/ZrO_2 at the same reaction condition. On the other hand, for

Pt/Al_2O_3, the peak intensities at 1273 K were higher than background level, implying some of the coke was not activated by H_2 below 1273 K. Further, a single peak was observed on Pt/Al_2O_3 above 873 K with different exposure times (Fig. 2), and the peak temperature was identical with the temperature of the peak for the catalyst after exposure to CH_4/He for 10 min (Fig. 1a). The results indicate that the nature of the coke formed on Pt/Al_2O_3 at these conditions (CH_4/CO_2 and CH_4/He) is almost the same, and we assume this observation is applicable to the coke on Pt/ZrO_2 as well. Fig. 2 shows that CH_4 is produced above 1073 K for Al_2O_3 after exposure to CH_4/CO_2 for 12 h, and the temperature is higher than that of Pt/Al_2O_3 (873 K), indicating that the coke on Pt can be hydrogenated more easily than that on Al_2O_3. Therefore, most of the peak (CH_4) observed on Pt/Al_2O_3 after exposure to CH_4/He (Fig. 1) and CH_4/CO_2 (Fig. 2) would be attributed to coke on Pt. On the other hand, coke on Pt/ZrO_2 is attributed to coke on Pt, as confirmed in reference [1].

3.2. Temperature programmed reaction (TPRn) with CO_2

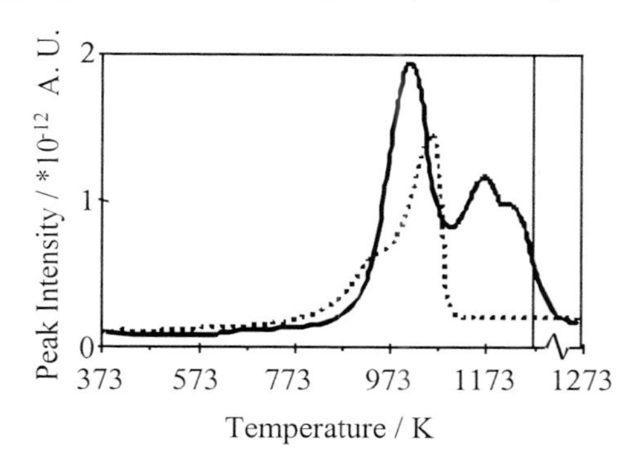

Fig. 3. Temperature programmed reaction (TPRn) with CO_2 profiles of m/e=28 for (——) Pt/Al_2O_3 and for (·······) Pt/ZrO_2 after exposure to CH_4/He at 1070 K for 10 min.

Figure 3 shows the TPRn profiles with CO_2 for Pt/Al_2O_3 and Pt/ZrO_2 after exposure to CH_4/He for 10 min. These results are reproduced from an earlier work [1]. During the experiments, coke was converted to CO (in all experiments the contribution of the CO_2 fragment to CO (m/e=28) has been subtracted). Our results indicate that CO is not evolved during experiments with the freshly reduced catalysts (not shown). This rules out the possibility of any significant contribution of the CO_2 dissociation ($CO_2 \rightarrow CO + O$) during TPRn with CO_2. For Pt/Al_2O_3, peaks were observed below and above reaction temperature (1070 K), and these were assigned to coke on Pt and on Al_2O_3, respectively. On the other hand, for Pt/ZrO_2, two peaks were observed below reaction temperature and these were assigned to coke on Pt. Varying the exposure time to CH_4/He on Pt/ZrO_2, we see that only the coke corresponding to lower temperature peak forms on Pt/ZrO_2 in the CH_4/CO_2 reaction. Therefore, we have concluded that coke on Pt is more reactive toward CO_2 on Pt/ZrO_2 than on Pt/Al_2O_3.

3.3. Comparison of the reactivity of coke (on Pt) with H_2 and CO_2 for Pt/Al_2O_3 and Pt/ZrO_2 catalysts

The results of TPH have been used to investigate the nature and reactivity of carbon species on the catalysts. The peak temperature is related to the composition of deposited coke [4-6]. Erdöhelyi *et al.* [6] distinguished three forms of carbon on supported Rh catalysts: (i) the highly reactive carbidic form, which can be hydrogenated even below 350-400 K, (ii) a less reactive amorphous layer, T_p=235-495 K, and (iii) the relatively inactive graphitic form, which reacts with hydrogen only above 650 K. In our experiments, the coke on Pt/Al_2O_3 and Pt/ZrO_2 were activated by H_2 (to produce CH_4) above 873 and 1073 K, respectively (Fig. 1). From above consideration, the forms of coke on Pt/Al_2O_3 and Pt/ZrO_2 would be graphitic. Further, coke reactivity with H_2 was higher on Pt/Al_2O_3 than on Pt/ZrO_2 (Fig. 1), indicating that the reactivity of coke itself would be better on Pt/Al_2O_3 than on Pt/ZrO_2. The difference in coke reactivity between Pt/Al_2O_3 and Pt/ZrO_2 could be attributed to the difference of coke crystallinity because in both cases graphitic coke was formed. The crystallinity of the coke is higher on Pt/ZrO_2 than on Pt/Al_2O_3.

We discuss here the factors that affect the peak temperature during TPH and TPRn with CO_2 in order to discuss the reactivity of coke toward CO_2. During TPH, coke activation would mainly depend on the reactivity of coke itself [6]. On the other hand, during TPRn with CO_2, it could depend on the nature of coke itself and/or the nature of activated CO_2. CO_2 activation ($CO_2 \rightarrow CO + O$) would be catalyzed on the active site which is associated with the oxide support, i.e., the presence of oxygen lattice defects on the support [10]. Our results (Fig. 1 and 3) show the reactivity of coke with H_2 was lower on Pt/ZrO_2 than on Pt/Al_2O_3. In contrast, the reactivity of coke on Pt/ZrO_2 (corresponding to lower temperature peak, which was related to catalyst stability) toward CO_2 (Fig. 3) was higher than on Pt/Al_2O_3. The reactivity of coke with H_2 would be mainly affected by the nature of coke itself, while the reactivity of coke toward CO_2 (for Pt/ZrO_2, lower temperature peak in Fig. 3) would be strongly affected by the reactivity of CO_2 on these catalysts. During the CH_4/CO_2 reaction, coke would be removed by the reaction with CO_2. This process would depend strongly on the ability to activate CO_2, i.e., on the nature of the support.

3.4. Activation of CO_2 by oxygen defects on the support

In earlier studies, CO_2 activation has been related to carbonate formation on supports [2,3,8,9]. We study here the CO_2 activation ($CO_2 \rightarrow CO + O$) on the oxygen defect sites. CO evolution is measured as a function of time in CO_2 flow tests on these catalysts at 1070 K, and the results are shown in Fig. 4. CO_2 flowing over reduced catalysts yielded only CO ($CO_2 \rightarrow CO + O$) in the gas phase, and the CO yield decreased with time on stream. These results are in accordance with those in earlier CO_2 pulsing tests at 875 K on Pt/ZrO_2 and IR spectroscopic measurements at 775 K [9]. The decrease of CO yield with time of exposure is explained by blocking of the sites that decompose CO_2 to adsorbed oxygen (O) and CO. However, the results of EXAFS for Pt/ZrO_2 have not given evidence for the presence of adsorbed oxygen on Pt or (surface) oxides of Pt-O, so it is not likely that oxygen is located on Pt. The results are also supported by XANES [9,10]. Thus, we have concluded that oxygen is consumed at the metal-support interface. In addition, earlier work has shown that CO_2 pretreatment increases CH_4 conversion (when CH_4 is pulsed on Pt/ZrO_2) compared with H_2 pretreatment [9]. It is likely that adsorbed oxygen left on the catalyst reacts with coke derived from methane under CH_4/CO_2 reaction conditions. The initial CO yield (rate of active oxygen production) and total amount of CO_2 converted (number of CO_2 activation sites) were higher

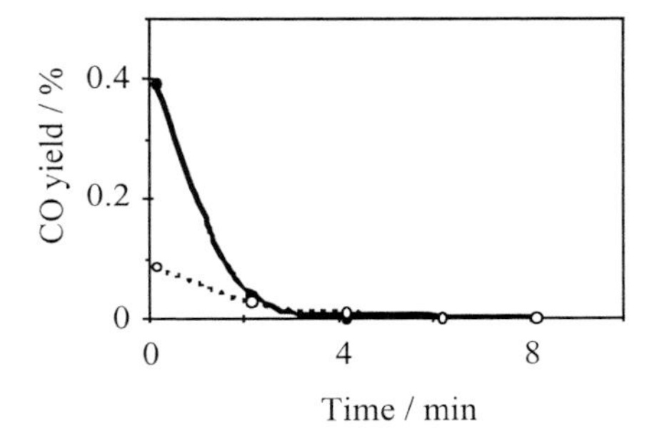

Fig. 4. CO evolution for (○) Pt/Al_2O_3 and (●) Pt/ZrO_2 (reduced at 1120 K) during CO_2 exposure at 1070 K (1st run).

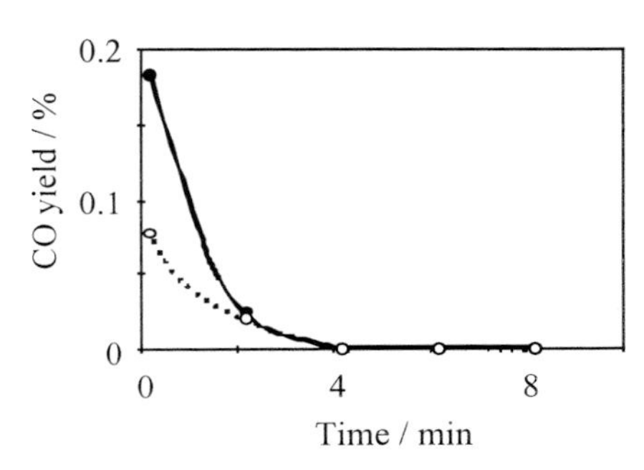

Fig. 5. CO evolution for (○) Pt/Al_2O_3 and (●) Pt/ZrO_2 (reduced at 1070 K) during CO_2 exposure at 1070 K after 1st run (2nd run).

on Pt/ZrO_2 than on Pt/Al_2O_3, indicating that the ability to produce active oxygen was higher on Pt/ZrO_2 than on Pt/Al_2O_3 at these conditions. Since the formation of oxygen defect sites may depend on reduction temperature, the catalysts were reduced with H_2 again after the 1st run, purged with He, and subsequently exposed to CO_2 (2nd run) at the same reaction temperature as in the 1st run (1070 K) (Fig. 5). At the same conditions as the 1st run (H_2 reduction at 1120 K, He purge, and subsequent CO_2 flow at 1070 K), we obtained similar CO yields after this 2nd run as for the 1st run. Thus the decrease in CO yield during our experiments for the 2nd run compared to the 1st run are due to a decrease in the sites accepting O from CO_2. Since the reduction temperature of the 2nd run is the same as that for the CH_4/CO_2 reaction, the CO yield during the 2nd run (Fig. 5) would be more related to a practical condition than that during the 1st run (Fig. 4). Pt/ZrO_2 showed higher initial CO yield and total number of CO_2 converted than Pt/Al_2O_3 during the 2nd run, indicating that the CO_2 activation was more rapid on Pt/ZrO_2 than on Pt/Al_2O_3 for the CH_4/CO_2 reaction. The difference between the catalysts can be attributed to the reducibility of Zr^{4+} (to Zr^{n+}) producing oxygen defect site. Under CH_4/CO_2 reaction conditions, coke would react with the active oxygen (produced on the oxygen defect site) to produce CO, and the oxygen defect site would be replenished by CO_2 (cycled).

REFERENCES

1. K. Nagaoka, K. Seshan, K. Aika and J. A. Lercher, J. Catal., 197(2001)34.
2. J. H. Bitter , Ph. D., thesis, University of Twente, 1997.
3. J. H. Bitter, K. Seshan and J. A. Lercher, J. Catal., 183(1999)336.
4. T. Koerts, M. J. A. G. Deelen and R. A. van Santen, J. Catal., 138(1992)101.
5. F. Solymosi, A. Erdöhelyi and J. Cserényi, Catal. Lett., 16(1992)399.
6. A. Erdöhelyi, J. Cserényi and F. Solymosi, J. Catal., 141(1993)287.
7. A. M. O'Connor, F. C. Meunier and J. R. H. Ross, Stud. Surf. Sci. Catal., 119(1998)819.
8. J. H. Bitter, K. Seshan and J. A. Lercher, J. Catal., 171(1997)279.
9. J. H. Bitter, K. Seshan and J. A. Lercher, J. Catal., 176(1998)93.
10. J. H. Bitter, K. Seshan and J. A. Lercher, Topics in Catalysis 10(2000)295.

Studies in Surface Science and Catalysis
J.J. Spivey, E. Iglesia and T.H. Fleisch (Editors)

New catalysts based on rutile-type Cr/Sb and Cr/V/Sb mixed oxides for the ammoxidation of propane to acrylonitrile

N. Ballarini[a], R. Catani[b], F. Cavani[a], U. Cornaro[b], D. Ghisletti[c], R. Millini[c], B. Stocchi[c], F. Trifirò[a]

[a]Dipartimento di Chimica Industriale e dei Materiali, Viale Risorgimento 4, 40136 Bologna
cavani@ms.fci.unibo.it
[b]Snamprogetti SpA, Via Maritano 26, 20097 S. Donato Milanese (MI), Italy
[c]EniTecnologie SpA, Via Maritano 26, 20097 S. Donato Milanese (MI), Italy

Cr/Sb and Cr/V/Sb mixed oxides with the rutile-type structure were prepared by coprecipitation from alcohol solutions, and calcined at 700°C. Over-stoichiometric-Sb rutile formed in Cr/Sb/O samples having Cr/Sb atomic ratios lower than 1/0.8. The reactivity in propane ammoxidation was affected by the Cr/Sb ratio. Cr/V/Sb mixed oxides with atomic ratios 1/x/2 formed rutile-type $Cr^{3+}{}_{1}V^{3+}{}_{x}Sb^{5+}{}_{1+x}O_{4+4x}$ solid solutions. With such materials, the catalytic performance in terms of activity was a function of the value of x, while the selectivity to acrylonitrile was less affected.

1. INTRODUCTION

The synthesis of acrylonitrile is carried out by ammoxidation of propylene, but interest exists in the development of a process which uses propane as the raw material, due to the lower cost of the paraffin as compared to the olefin. BP, Mitsubishi Chemical and Asahi have each announced the start-up of semicommercial units or of pilot plants for the ammoxidation of propane [1,2]. The catalysts claimed are based either on rutile-type metal antimonates (V/Sb/O, V/Sn/Sb/O, Fe/Sb/O) [1,3], or on mixed molybdates (Mo/V/Nb/Te/O for the Mitsubishi and Asahi catalysts) [2]. These catalysts can be considered as multifunctional systems, where one component is aimed at propane activation and oxidehydrogenation to adsorbed propylene, and the other component is aimed at allylic ammoxidation of this intermediate to acrylonitrile [4]. In rutile-type systems, the first role is played by the metal antimonate, or by the V/Sb-solid solution in SnO_2, while the active sites for the second role are provided by antimony oxide, dispersed on the surface of the rutile.

Rutile structures are extremely flexible, and a great number of mixed oxides and solid solutions are known to form [5]. In the present work, we characterized and checked the reactivity of Cr/Sb mixed oxides. Indeed, chromium antimonate might represent a good system for the reaction, since chromium oxide is known to be an efficient catalyst for the oxidehydrogenation of light paraffins [6]. Cr/V/Sb mixed oxides of composition 1/x/2 (atomic ratios) were also prepared and characterized.

2. EXPERIMENTAL

The catalysts were prepared with the coprecipitation method by addition of an alcohol solution containing $Cr(NO_3)_3.6H_2O$, $SbCl_5$, and $(VO)acac_2$, to an aqueous solution buffered at pH 7. After filtration and drying, the solids were calcined at 340°C for 1 h and at 700°C for 3 h, in air. The materials obtained were characterized by SEM/EDS, XRD (full profile fitting),

XPS, FT-IR spectroscopy, and Raman spectroscopy, and checked as catalysts for the ammoxidation of propane in a continuous flow, fixed-bed reactor loading 2 g of catalyst, shaped in particles 0.42-0.55 mm in size. The following reaction conditions were used: feed composition 25 mol.% propane, 10% ammonia, 20% oxygen, remainder helium; residence time 2.0 s.

3. RESULTS AND DISCUSSION

Cr/Sb/O catalysts

Cr/Sb mixed oxides were prepared with different atomic ratios Cr/Sb 1/x (Table 1). The X-ray diffraction patterns of the samples are reported in Figure 1, while the values for unit cell volume determined by full-profile-fitting methods (Rietveld analysis [7]) compared with that of reference stoichiometric $CrSbO_4$ [8] are reported in Table 1. All the samples were characterised by the rutile-type structure. Additional reflections were found only in the sample having the highest Sb content, corresponding to crystalline Sb_6O_{13}. The volume of the tetragonal unit cell for the Cr/Sb 1/0.8 system was close to that of stoichiometric $CrSbO_4$. The unit cell volume then increased with increasing Sb content, as a consequence of the expansion of the *a*-parameter, only partially compensated for by the slight contraction of the *c*-parameter.

Table 1. Composition, specific surface area, volume of the tetragonal unit cell and amount of secondary phases of the Cr/Sb/O samples

Catalyst, Cr/Sb 1/x at. ratio[a]	Surface area, m^2/g	Unit cell volume, $Å^3$	Secondary crystalline phases
1/0.8	37	64.43	-
1/1.6	67	65.29	-
1/2.8	54	65.52	14.6% Sb_6O_{13}
Ref. $CrSbO_4$		64.31	

[a]as determined by SEM-EDS

In the Raman spectra (Figure 2), absorption bands typical of rutile [9] were present for all samples, at 760, 670 and 560 cm^{-1}. Moreover, bands which can be attributed to Sb_2O_4 (non-detectable by X-ray diffraction and thus amorphous or microcrystalline) were present in the spectrum of sample Cr/Sb 1/1.6, at 140, 200, 260 and 400 cm^{-1}, while bands relative to Sb_6O_{13} (470 and 550 cm^{-1}) were present in the spectrum of sample Cr/Sb 1/2.8 [10], in agreement with the XRD pattern.

These data suggest that $CrSbO_4$ can host excess antimony in the structure thus forming over-stoichiometric solid solutions rich in antimony. The increase in cell volume (Table 1) can be attributed to a replacement of Cr^{3+} (0.69 Å) by the larger Sb^{3+} (0.76 Å) ions, thus with development of a solid solution of composition $Cr^{3+}{}_1Sb^{3+}{}_ySb^{5+}{}_{1+y}O_{4+4y}$ (expressed with reference to chromium). The maximum for y in each sample is equal to (x-1)/2, however in all cases it is actually lower than 0.3, since a fraction of the antimony forms extraframework oxide. Alternatively, reduction of Cr^{3+} to Cr^{2+} (necessary to compensate for the excess antimony ions) is possible, with progressive development of a trirutile-type cation ordering (analogous to what happens in the case of Fe/Sb/O systems [11]). However, the reduction of chromium under our conditions is unlikely.

The Cr/Sb surface atomic ratio, as determined by XPS, was equal to 0.21 for the Cr/Sb 1/0.8, and to 0.16 for the Cr/Sb 1/1.6 samples, thus considerably less than the corresponding bulk atomic ratio. This Sb surface enrichment, also observed by other authors in different antimonate compounds with rutile-type structures [12], can be attributed either to the dispersion of antimony oxides on the surface of the mixed oxide, or to the formation of metal concentration gradients in the rutile-type structure, with Sb-enrichment in the outermost atomic layers of the crystallites. The former hypothesis is less likely, since the Sb-enrichment is also observed for the essentially monophasic Cr/Sb 1/0.8 sample.

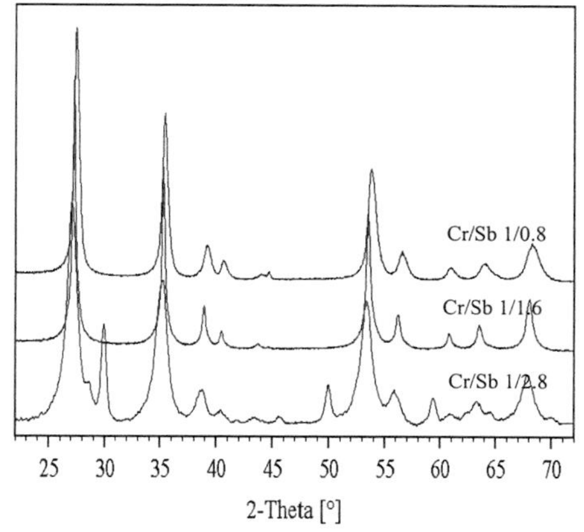

Figure 1. XRD patterns of Cr/Sb/O samples, with composition Cr/Sb 1/x (atomic ratios).

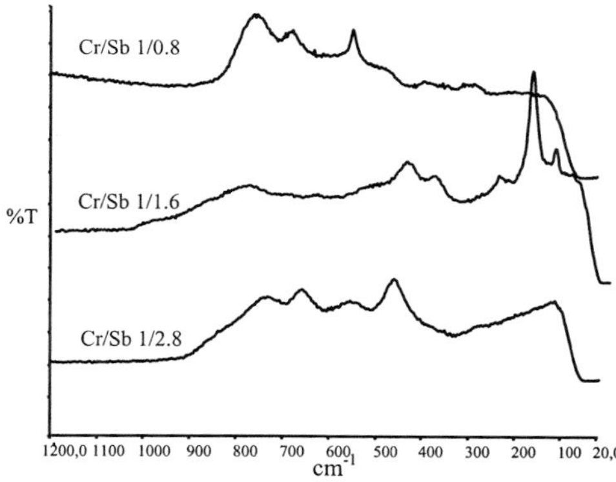

Figure 2. Raman spectra of Cr/Sb/O samples, with composition Cr/Sb 1/x (atomic ratios).

The effect of the reaction temperature on catalytic performance is reported in Figure 3 for the Cr/Sb 1/1.6 sample. One remarkable effect was the increase in selectivity to acrylonitrile with increasing reaction temperature, which occurred mainly at the expense of acetonitrile and hydrocyanic acid. These two latter products showed parallel trends as functions of the reaction temperature, and this would suggest a common precursor for acetonitrile and hydrocyanic acid. Therefore, the relative ratio between i) ammoxidation of this precursor to acrylonitrile and ii) its fragmentation and ammoxidation to yield acetonitrile + hydrocyanic acid, is a function of the reaction temperature. The data reported in Figure 3 also show that the selectivity to propylene is very low (less than 5%), even in correspondence with the lowest selectivity to acrylonitrile. This suggests that the catalyst is able to perform a quick transformation of intermediate olefinic species to the following products, without letting it desorb as propylene. This in practice corresponds to a direct, primary reaction for propane transformation to acrylonitrile.

The catalytic performances of Cr/Sb/O catalysts having Cr/Sb 1/x atomic ratios are compared in Figure 4. The following results are given: conversion of propane, conversion of ammonia and selectivity to acrylonitrile as a function of temperature, and distribution of the products at 500-505°C. The reactivity was modified as a consequence of the variation in Sb content. In particular, the activities of the Cr/Sb 1/0.8 and Cr/Sb 1/1.6 samples were similar (the surface area, however, was much higher for the latter sample, and therefore the Cr/Sb 1/0.8 catalyst was intrinsically more active), while the sample having the highest Sb content (Cr/Sb 1/2.8) was clearly the least active. The most active catalyst in ammonia conversion was the Cr/Sb 1/0.8 sample, which was also the most active in the combustion of ammonia to molecular nitrogen. The selectivities to acrylonitrile were similar for the Cr/Sb 1/1.6 and 1/2.8 catalysts, and higher than that of the Cr/Sb 1/0.8 catalyst. Moreover, for all catalysts the selectivity to propylene was very low. This constitutes one main difference with respect to the V/Sb/O catalyst. In the latter system, in fact, the selectivity to propylene becomes greater (and that to acrylonitrile correspondingly decreases), as the Sb content in the catalyst decreases, due to the lower amount of dispersed, extra-rutile antimony oxide in the catalyst [13]. Therefore, the selectivity to propylene and to acrylonitrile depends much more on the Me/Sb ratio in V/Sb/O catalysts than in Cr/Sb/O catalysts.

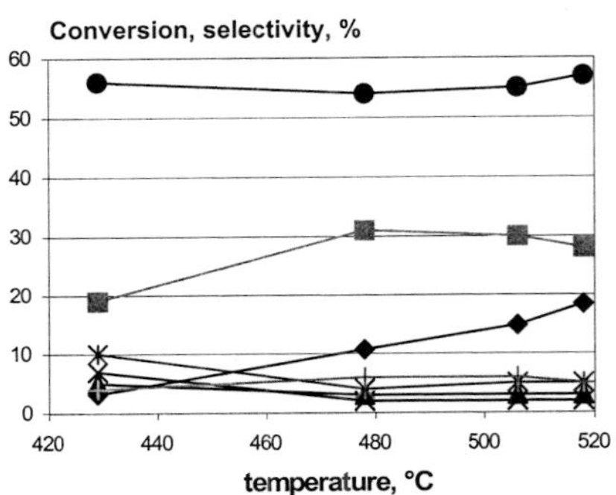

Figure 3. Conversion of propane (◆) and selectivity to propylene (▲), acrylonitrile (■), acetonitrile (✖), hydrocyanic acid (✳), CO (✚) and CO_2 (●), as functions of temperature. Catalyst Cr/Sb 1/1.6.

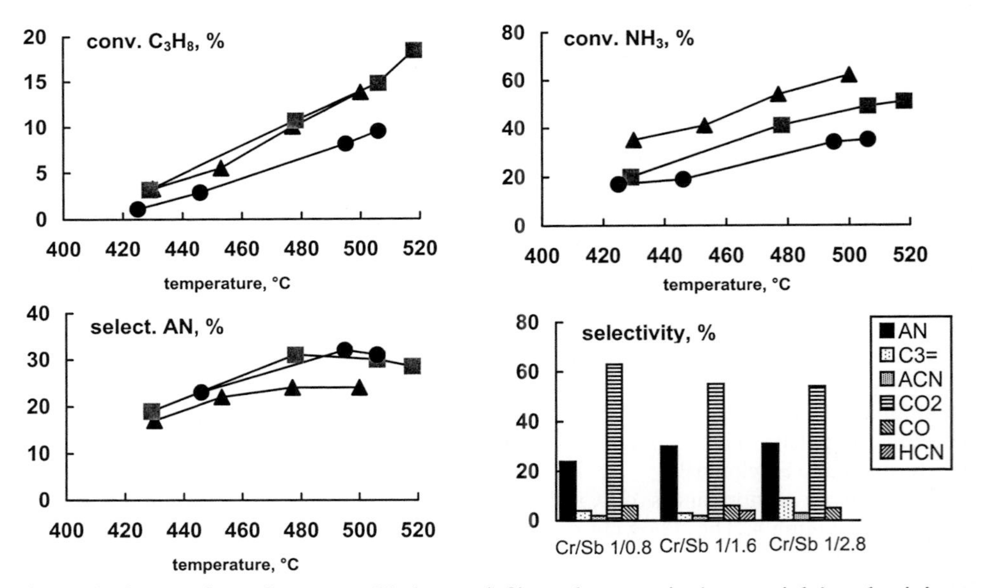

Figure 4. Conversion of propane C3 (upper left), and ammonia (upper right), selectivity to acrylonitrile AN (lower left) as functions of the reaction temperature, and distribution of products at 500-505 °C (lower right). Catalysts Cr/Sb/O, with Cr/Sb atomic ratios 1/0.8 (▲), 1/1.6 (■), 1/2.8 (●). ACN, acetonitrile; C3=, propylene.

It seems that the Cr/Sb/O catalyst works as a real bifunctional system, providing in the same structure active sites for both propane activation and allylic ammoxidation. This explains the low selectivity to propylene which is obtained under all conditions examined and for all Cr/Sb/O catalysts tested, regardless of the Cr/Sb ratio, and the relative independence of the distribution of products on the latter. In other words, the role which in the V/Sb/O system is played by dispersed antimony oxide, here is possibly played, at least in part, by intra-particle Sb concentration gradients, with enrichment in Sb in the outermost atomic layers of the rutile crystallites. The additional presence of antimony oxide, in Cr/Sb 1/1.6 and 1/2.8 samples, further improves the selectivity to acrylonitrile, with a decrease in CO_2 formation.

Cr/V/Sb/O catalysts

Cr/V/Sb/O catalysts were prepared having different atomic ratios between components: Cr/V/Sb 1/x/δ (with x between 0 and 1.3, and δ approximately equal to 2). The samples together with some of their characteristics are listed in Table 2.

Table 2. Composition, specific surface area, volume of the tetragonal unit cell and amount of secondary phases of the Cr/V/Sb/O samples.

Catalyst, Cr/V/Sb 1/x/δ, at. ratio*	Surface area, m^2/g	Unit cell volume, $Å^3$	Secondary crystalline phases
1/0/1.6	67	65.29	-
1/0.2/1.7	67	65.13	-
1/0.5/1.8	49	65.05	-
1/0.8/2.0	40	64.96	-
1/0.9/1.8	40	64.81	-
1/1.3/2.5	26	64.74	2.6% Sb_2O_4

*as determined by SEM-EDS

In the presence of vanadium, the following effects can be observed:

1) For values of δ lower than 2.5, the only crystalline compound was the rutile, and for increasing values of x the unit cell volume progressively decreased, as shown in Table 2. The contraction of the unit cell volume can be explained by taking into account the ionic radius of octahedral V^{3+} (0.64 Å), smaller than that of Cr^{3+} and of Sb^{3+}. Thus it is possible that starting from a Sb-enriched rutile (for Cr/V/Sb = 1/0/1.6), the addition of vanadium generates a solid solution of general composition $Cr^{3+}{}_1V^{3+}{}_xSb^{5+}{}_{1+x}O_{4+4x}$, which in practice corresponds to a mixed antimonate of V^{3+} and Cr^{3+}. Of course, the additional presence of Sb^{3+} can not be excluded, especially for low values of x. Excess Sb forms dispersed antimony oxide, in a relative amount which is equal to: (δ-1-x) SbO_z.
2) XPS characterization indicated that in all samples i) the surface Cr/V atomic ratio was similar to the bulk one, thus suggesting an homogeneous distribution of the two elements in the compound, ii) surface enrichment in Sb occurred, and iii) chromium was present exclusively as Cr^{3+} (BE $Cr_{2p3/2}$ 576.7-577.6) while vanadium was present mainly as V^{3+} (BE $V_{2p3/2}$ 516.1-516.2 eV) [9,14,15].

The results of the catalytic tests are summarized in Figure 5. An increase in the amount of vanadium led to a considerable increase in activity, both towards propane and ammonia. The selectivity to acrylonitrile was not greatly affected by the addition of vanadium. Only for catalysts with x = 0.9 and 1.3 did the maximum selectivity to acrylonitrile decrease, with a corresponding increase in CO_2. This is likely due to the fact that for increasing values of x the amount of antimony available for making dispersed antimony oxide decreased. The latter helps in providing a more efficient transformation of the intermediate compounds to acrylonitrile, rather than to CO_2.

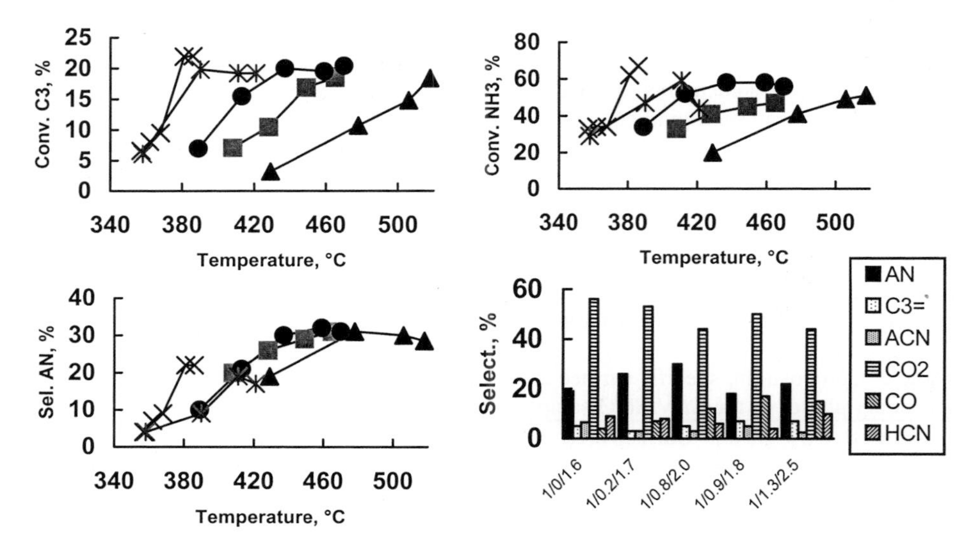

Figure 5. Conversion of propane C3 (upper left), conversion of ammonia (upper right), and selectivity to acrylonitrile AN (lower left) as functions of the temperature, and distribution of products at 420-430°C. Catalysts Cr/Sb/V/O, with Cr/V/Sb atomic ratios: 1/0/1.6 (▲), 1/0.2/1.7 (■), 1/0.8/2.0 (●), 1/0.9/1.8 (✱), 1/1.3/2.5 (✖).

Acknowledgements

Snamprogetti S.p.A. (S. Donato Milanese, Italy) is acknowledged for financial support.

References

1. A.T. Guttmann, R.K. Grasselli, J.F. Brazdil, US Patent 4,788,317 (1988), to BP Amoco.
2. T. Ushikubo, K. Oshima, T. Ihara, H. Amatsu, US Patent 5,534,650 (1996), to Mitsubishi Chem. Co.
3. G. Blanchard, P. Burattin, F. Cavani, S. Masetti, F. Trifirò, WO Patent 23,287 (1997), to Rhodia.
4. F. Cavani, F. Trifirò, Stud. Surf. Sci. Catal., 110 (1997) 19.
5. G. Centi, F. Trifirò, Catal. Rev.-Sci. Eng., 28 (1986) 165.
6. F. Cavani, E. Etienne, F. Trifirò, France Patent 2,748,021 (1997), to Elf Atochem.
7. R.A. Young, "The Rietveld Method", IUC, Oxford Univ. Press, 1993.
8. J. Amador, I. Rasines, J. Appl. Cryst., 14 (1981) 348.
9. R. Nilsson, T. Lindblad, A. Andersson, J. Catal., 148 (1994) 501.
10. C.A. Cody, L. DiCarlo, R.K. Darlington, Inorg. Chem., 18(6) (1979) 1572.
11. F.J. Berry, J.G. Holden, M.H. Loretto, J. Chem. Soc., Dalton Trans., (1987) 1727.
12. F.J. Berry, M.E. Brett, W.R. Patterson, J. Chem. Soc. Dalton Trans., (1983) 9 and 13.
13. G. Centi, P. Mazzoli, Catal. Today, 28 (1996) 351.
14. R. Nilsson, T. Lindblad, A. Andersson, Catal. Lett., 29(1994) 409.
15. F.J. Berry, M.E. Brett, R.A. Marbrow, W.R. Patterson, J. Chem. Soc., Dalton Trans., (1984) 985.

Studies in Surface Science and Catalysis
J.J. Spivey, E. Iglesia and T.H. Fleisch (Editors)

ALMAX catalyst for the selective oxidation of n-butane to maleic anhydride: a highly efficient V/P/O system for fluidized-bed reactors

S. Albonetti[a], F. Budi[a], F. Cavani[b], S. Ligi[b], G. Mazzoni[a], F. Pierelli[b] and F. Trifirò[b]

[a]Lonzagroup Intermediates and Additives, via E. Fermi 51, 24020 Scanzorosciate (BG), Italy

[b]Dipartimento di Chimica Industriale e dei Materiali, Viale Risorgimento 4, 40136 Bologna, Italy. cavani@ms.fci.unibo.it

The main features of the new ALMAX V/P/O-based catalyst, developed by Lonza, for the fluidized-bed selective oxidation of n-butane to maleic anhydride are presented. Preparation of the catalyst involves a fluidized-bed, superatmospheric thermal treatment in a steam/air flow, which leads to enhanced crystallinity features for the vanadyl pyrophosphate, the catalytically active compound. This new catalyst is compared with catalysts obtained by conventional activation procedures. The main features of the vanadyl pyrophosphate are discussed in relationship with the possible formation of different crystalline forms.

1. INTRODUCTION

The selective oxidation of n-butane currently represents the most important industrial process for the synthesis of maleic anhydride [1]. The catalyst of choice for this reaction is a V/P mixed oxide, essentially constituted of an active phase, vanadyl pyrophosphate $(VO)_2P_2O_7$, and of various dopants or additives, depending on the type of technology employed. In the 1980's Lonzagroup and Lummus developed a fluidized-bed technology, the ALMA process, and a V/P/O-based catalyst, ALMACAT, with excellent properties of mechanical resistance and performance [2].

The catalyst is usually prepared with the so-called organic procedure, where V_2O_5 is reduced by an alcohol (usually isobutanol) in the presence of phosphoric acid. A precipitate is obtained, $VOHPO_4 \cdot 0.5H_2O$, which when heated at temperatures higher than 350°C is transformed to a mixture of i) vanadyl pyrophosphate, ii) amorphous V/P/O compounds, and iii) crystalline V^{3+}/P/O or V^{4+}/P/O compounds. The procedure adopted for thermal treatment dramatically affects the ratio between the different components of the catalytic system. The transformation in the reaction environment of these less active and less selective compounds to vanadyl pyrophosphate may take hundreds of hours (to develop the so-called "equilibrated catalyst"), during which the yield to maleic anhydride is lower than the target one. It is thus important to develop catalysts which after the thermal treatment already have a large amount of the active compound, $(VO)_2P_2O_7$, in order to obtain the best performance from the very beginning of the catalyst lifetime.

In this work we compare the catalytic performance of different V/P/O catalysts prepared by various thermal treatments starting from the same precursor, and also compare the variation in catalytic performance with time-on-stream with the progressive modification of the catalyst features in the reaction environment. Lonzagroup has recently optimized the thermal treatment of the catalyst precursor, developing at an industrial level a new highly efficient V/P/O catalyst, referred to as ALMAX, which exhibits stable catalytic performance from the very beginning of the service life and gives superior yields to maleic anhydride [3].

2. EXPERIMENTAL

The precursor $VOHPO_4 \cdot 0.5H_2O$, prepared by the anhydrous procedure, was spray-dried in a pilot apparatus, and then calcined following two different procedures. One procedure (sample 1; the procedure formerly used for the ALMACAT catalyst) involves thermal treatment of the precursor in air at 350°C, followed by treatment in a nitrogen flow at 550°C. The second procedure (sample 2, currently employed for the ALMAX catalyst) consists of treatment in a mixture of air and steam, at 425°C in a fluidized-bed industrial calcinator [3]. The catalytic tests were carried out in a fluidized-bed bench scale plant, loading 1 liter of catalyst, with a feed composition of 4% n-butane in air.

3. RESULTS AND DISCUSSION

The fluidized-bed catalyst for the ALMA process of selective oxidation of n-butane to maleic anhydride, though giving excellent performance from both the maleic anhydride yield, fluodynamics and service life points of view, has been continuously studied in the Lonza labs, in order to further improve its characteristics. In particular, much effort has been dedicated to the modification of the calcination treatment. As an example, the main features of two V/P/O fresh samples, obtained by thermal treatment of the same precursor, following the procedures described in the Experimental section, are compared in Table 1. Reported in Figure 1 (left) are the X-ray diffraction patterns for the same samples. Fresh sample 1 (1a in Table 1) is in part amorphous; besides the reflections relative to $(VO)_2P_2O_7$, lines are present which can be attributed to $V(PO_3)_3$, a V^{3+} compound. Chemical analysis (Table 1) confirms the presence of a large fraction of V^{3+} (13%) in this sample, the remaining being exclusively present as V^{4+}. UV-Vis DRS spectra (Figure 1, right) also confirm the presence of reduced vanadium species. Absorption bands relative to CT absorptions and d-d transitions of V^{4+} are clearly visible at 300, 640 and 880 nm. Moreover, an intense absorption band at 500 nm in sample 1a can be attributed to V^{3+} species, as was confirmed by H_2-reduction tests made on an equilibrated vanadyl pyrophosphate [4]. The high energy of this transition is likely due to a different coordination geometry, arising from the loss of one O^{2-} ligand [5]. Fresh sample 2 (2a in Table 1), instead, is characterized by the presence of V^{5+}, formed as a consequence of a treatment made in an O_2-containing atmosphere. No V^{3+} was detected. X-ray diffraction data confirm the presence of a well crystallized vanadyl pyrophosphate. Small amounts of crystalline δ-$VOPO_4$ are also present. The electronic spectrum reported in Figure 1 (right) shows a band at around 420 nm, corresponding to a CT band of V^{5+} species.

The results of catalytic tests carried out in a pilot reactor are plotted in Figure 2. In particular, the yield to maleic anhydride is reported as a function of time-on-stream. Sample 1 exhibits first a rapid decrease in yield (mainly due to a corresponding decrease in n-butane conversion, as well as lower selectivity), and then a slower increase. On the contrary, sample 2 exhibits only minor variations in catalytic performance, thus it gives a stable catalytic performance from the very beginning of its service life.

Table 1. Main characteristics of V/P/O catalysts studied in the present work

Sample, ageing time, h	Surf. area, m^2/g	FWHM (200), degrees	FWHM (042) / FWHM (200)	V^{3+}/V^{5+}, at.%[a]
1a (0, fresh)	24±2	nd[b]	nd[b]	13 / 0
1b (70)	nd	0.72±0.05	0.50±0.08	nd
1c (800, equilibrated)	20±2	0.47±0.05	0.57±0.08	≤1 / 0
2a (0, fresh)	22±2	0.24±0.05	0.91±0.08	0 / 10
2c (800, equilibrated)	16±2	0.30±0.05	0.83±0.08	≤1 /< 0.5

[a] with respect to total V. [b] not-determined, due to the presence of reflections of $V(PO_3)_3$.

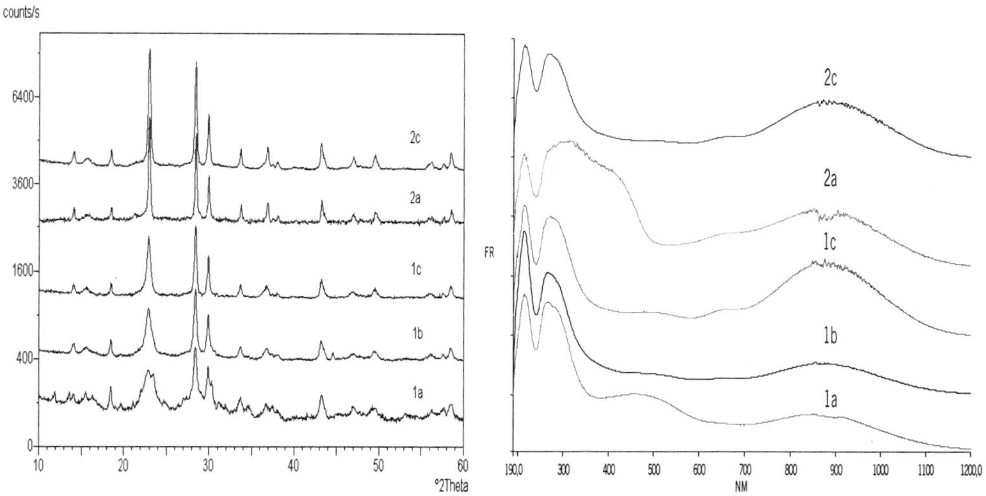

Figure 1. X-ray diffraction patterns (left) and UV-Vis-DR spectra (right) of V/P/O samples 1 and 2, for different ageing times: a (fresh catalyst, no ageing), b (after 70 h ageing), c (after more than 1 month ageing in a bench scale reactor).

It is known that V/P/O-based catalysts usually require a period of ageing, during which the fresh catalyst is progressively modified until it reaches an "equilibrated" state, after which the performance becomes stable [6]. Therefore, changes in performance can be related to changes occurring in the chemical-physical features of the catalyst. Sample 2, already immediately after calcination, exhibits well-crystallized vanadyl pyrophosphate, which does not undergo considerable changes from the chemical-physical features and reactivity points of view; the only modifications involve the oxidation state of vanadium (see Table 1). Sample 1, on the contrary, exhibits major changes with increasing time-on-stream, which correspond to variations in catalytic performance. The catalyst was unloaded in correspondence with the minimum in yield (sample 1b) (see Figure 2) and characterized. Figure 1 shows that during the initial 70 h of ageing the catalyst loses most of the initial V^{3+}, with disappearance of crystalline $V(PO_3)_3$ and of most of the amorphous component. The intensity of the absorption band at 500 nm considerably decreases. After this period (from sample 1b to 1c) further ageing corresponds to a slow increase in yield to maleic anhydride, and correspondingly to an increase in crystallinity, as shown in Table 1. Therefore the trend relative to catalytic performance of sample 1 in Figure 2 is due to the overlapping of the two different phenomena.

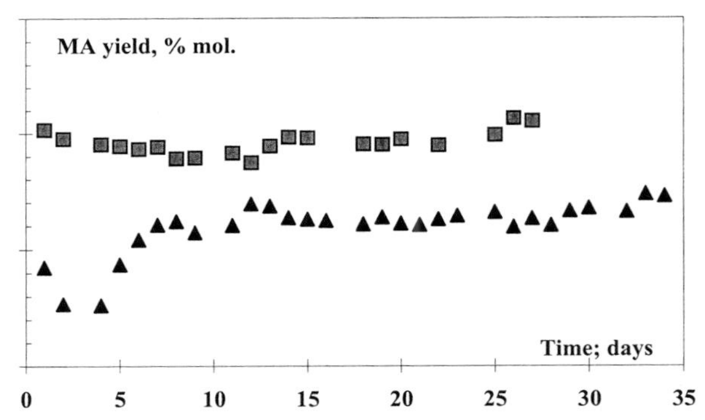

Figure 2. Effect of time-on-stream on the yield to maleic anhydride for sample 1-ALMACAT (▲) and sample 2-ALMAX (■) catalysts in a fluidized-bed bench scale reactor.

Another considerable difference concerns the maleic anhydride yield for the equilibrated samples, which is 8-10 points higher for sample 2 than for sample 1. Definitely, the new thermal treatment developed for ALMAX makes it possible to obtain a catalyst characterized by better and more stable performance. The novel procedure of thermal treatment involves a step which is realized in a steam-rich atmosphere at superatmospheric pressure. These hydrothermal-like conditions allow the development of larger crystals of vanadyl pyrophosphate to be achieved, and also avoid the formation of additional, undesired amorphous and crystalline V/P/O compounds.

Thermal treatments in a steam-containing atmosphere have also been claimed by other industries in the past [7]. However, greater advantages in terms of the vanadyl pyrophosphate crystallinity can be gained when the treatment is done with better control of the calcination temperature, that is in a fluidized-bed reactor configuration, and under superatmospheric pressure. Recently, it has been found that prolonged treatments of vanadyl pyrophosphate in wet air at temperatures higher than 450°C lead to an irreversible segregation of V_2O_5 and P_2O_5; this phenomenon is instead reversible for short treatment periods (a few hours) [8].

XRD patterns and FT-IR spectra of samples 1 and 2 unloaded from the pilot plant reactor after equilibration (referred to as samples 1c and 2c, respectively) are shown in Figure 1(left) and in Figure 3. Chemical analysis data are reported in Table 1. There are considerable differences in the samples, concerning both the relative intensity and width of some reflections in the XRD pattern, and the FT-IR spectra. In particular, as pointed out by Bordes et al. [9], it is as if some IR absorption bands of the vanadyl pyrophosphate were shifted and/or splitted. The main changes involve the absorption bands relative to stretching of V=O groups, while smaller changes involve PO_3 bending vibration and P-O-P stretching vibration bands.

One of the aspects of V/P/O catalysts which still has not received a definite answer concerns the existence of different crystallographic forms (the so-called polytypes, or polymorphs) of vanadyl pyrophosphate. Bordes and Courtine [10] were the first to talk of γ- and β-$(VO)_2P_2O_7$, obtained by utilizing different procedures, and characterized by different morphology (different exposure of the {100} platelets, leading to more layered-γ- or more bulky-β- particles), and by different catalytic performance, since the (200) plane is in general considered to contain the active sites, and thus more layered particles lead to more active catalysts. Later, Matsuura and Yamazaki [11] proposed the existence of three different crystalline forms, characterized by different XRD I_{042}/I_{200} intensity ratios, morphology, and IR spectra. The same differences in IR spectra were recently discussed again by Bordes et al. [9], and were correlated to XRD patterns. These authors postulated that they are associated with differences in morphology. The possibility of the existence of polytypes arising from the different orientation of O—V=O columns was taken into consideration by Ebner and Thompson [12]. A columnar reorientation of the vanadyl bonds could simulate the structural disorder associated with vanadium sites displaced with respect to the basal plane of the distorted octahedron. The two polytypes differed in the relative orientation of adjacent vanadyl dimer chains running perpendicular to the (100) plane. This led to differences in the relative intensity of X-ray reflections, as well as in the volumes of the orthorhombic unit cell. The authors also postulated that these phenomena might have consequences on the catalytic performance. Another possibility has been described by Sleight et al. [13], who proposed that indeed defectivity associated with the presence of V^{5+} may be responsible for the observed broadening effects in the XRD patterns. An increase in the V^{5+} defectivity led to a broadening of all reflections (rather than on the 200 reflection preferentially) in the pattern of a well-crystallized vanadyl pyrophosphate. In our case, indeed sample 2 initially contains V^{5+} (sample 2a), due to the oxidizing atmosphere used for calcination, but the permanence in the reaction atmosphere reduces it. Sample 2c (equilibrated) contains small amounts of V^{3+} and of V^{5+} (see Table 1), while the spent catalyst, i.e., unloaded after reaction under industrial conditions, contains V^{4+} and small amounts of V^{3+}, but no significant changes in the XRD pattern of the vanadyl pyrophosphate occur. Therefore in our case the observed phenomena are not to be ascribed to the presence of V^{5+} defectivity.

In all cases described in the literature, the polytypes for microcrystalline materials were synthesized starting from precursors prepared with different techniques (either organic or aqueous synthesis of the vanadyl orthophosphate hemihydrate, or reduction of vanadium phosphate). In the present case, instead, the different forms were obtained starting from the same precursor, calcining it under different conditions, and ageing the samples obtained under reaction conditions. However, it is clear that the differences observed in the equilibrated catalysts originate from differences in the fresh, calcined catalysts (samples 1a and 2a). Thus, it seems that the atmosphere of treatment (in particular, the presence of steam), has a considerable effect on the morphology of the samples.

The FWHM (042)/FWHM (200) ratio (often defined as the "aspect ratio") is plotted in Figure 4 as a function of the splitting of the V=O IR ν band, for several samples that were prepared either starting from different precursors, or using different thermal treatments, and only containing well-crystallized vanadyl pyrophosphate. It is clear that a relationship between the two features does exist. This relationship would seem to indicate that differences in crystal morphology, as suggested by variations in the "aspect ratio", dramatically affect the IR spectra. The same is known to occur in rutile-type mixed oxides [14], and has been attributed to the fact that the stretching vibrations are perturbed by the polarization charge induced by the coulombic fields developed at the particle surface. The superior performance of sample 2 as compared to sample 1, mainly due to the higher selectivity to maleic anhydride (as measured for the same level of conversion), might thus possibly be attributed to these morphological differences. As shown in Table 1, the FWHM(042)/FWHM(200) ratio (aspect ratio) is considerably higher for the equilibrated ALMAX catalyst (sample 2c) than for the equilibrated ALMACAT catalyst (sample 1c). This would mean a more plate-like morphology for sample 1, with preferential exposure of {100} crystal faces, and a more "bulky" morphology for sample 2. According to the literature, this would mean a lower activity but a higher selectivity for sample 2, in agreement with the results plotted in Figure 2; however, there is not much agreement in the literature about the effective advantage of having preferential exposure of the {100} crystal face.

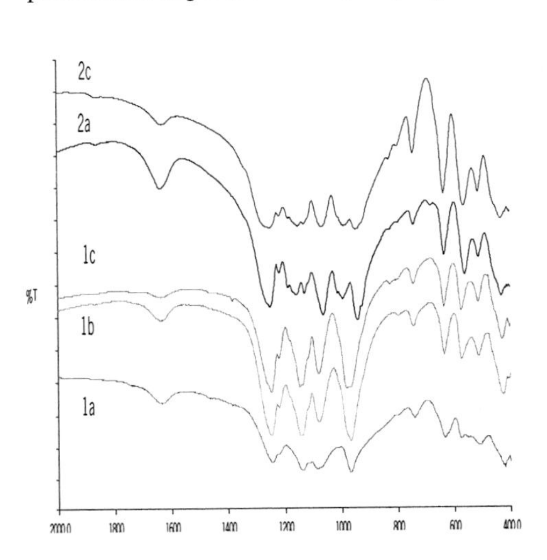

Figure 3. FT-IR spectra of samples 1 and 2. Captions as in Figure 1.

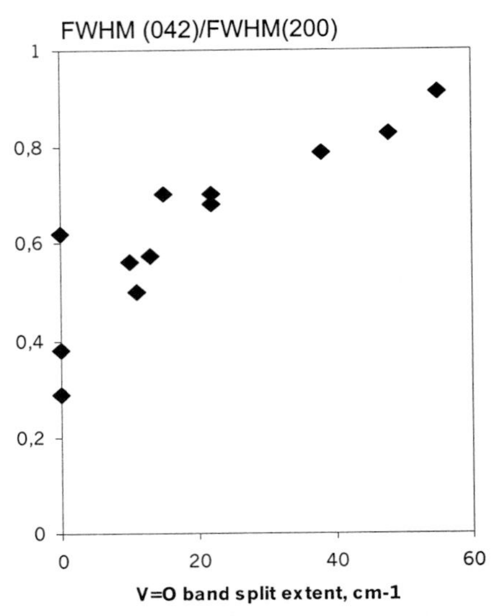

Figure 4. Relationship between FWHM (042) / FWHM (200) (aspect ratio) and splitting of the V=O IR ν band for several catalysts.

REFERENCES

1. F. Cavani, F. Trifirò, in "Catalysis, Vol. 11", (J.J. Spivey et al.Eds.), Royal Society of Chemistry, Cambridge, 1994, p.246.
2. S.C. Arnold, G.D. Suciu, L. Verde, A. Neri, Hydroc. Process., sept. 1985, 123
3. G. Mazzoni, G. Stefani, F. Cavani, Eur. Patent 804,963 (1997), to Lonzagroup.
4. F. Cavani, S. Ligi, T. Monti, F. Pierelli, F. Trifirò, A.Albonetti, G. Mazzoni, Catal. Today, 61 (2000) 203.
5. P.L. Gai, K. Kourtakis, Science, 267 (1995) 661.
6. C. Cabello, F. Cavani, S. Ligi, F. Trifirò, Stud. Surf. Sci. Catal., 119 (1998) 925.
7. J.R. Ebner, W.J. Andrews, US Patent 5,137,860 (1992), to Monsanto Co.
8. Z.-Y. Xue, G.L. Schrader, J. Phys. Chem. B, 103 (1999) 9549.
9. N. Duvauchelle, E. Bordes, Catal. Lett., 57 (1999) 81.
10. E. Bordes, P. Courtine, J. Catal., 57 (1979) 236.
11. I. Matsuura, M. Yamazaki, Stud. Surf. Sci. Catal., 55 (1990) 563.
12. J.R. Ebner, M.R. Thompson, Catal. Today, 16 (1993) 51.
13. P.T. Nguyen, A.W. Sleight, N. Roberts, W.W. Warren, J. Solid State Chem., 122 (1996) 259.
14. M. Ocana, V. Fornes, J.V. Garcia Ramos, C.J. Serna, J. Solid State Chem., 75 (1988) 364.

Studies in Surface Science and Catalysis
J.J. Spivey, E. Iglesia and T.H. Fleisch (Editors)

Oxygen Transport Membranes for Syngas Production

T. J. Mazanec [a]*, R. Prasad[b], R. Odegard[c], C. Steyn[d] and E. T. Robinson[e],

[a] BP Chemicals, 150 W. Warrenville Rd - MC H-5, Naperville, IL 60566 USA*

[b] Praxair, Inc., 175 East Park Drive, Tonawanda, NY 14151 USA

[c] Statoil, Arkitekt Ebbells veg 10, Rotvoll, 7005 Trondheim Norway

[d] Sasol Technology (Pty) Ltd, P.O. Box 5486, Johannesburg, 2000 South Africa

[e] Torix, Inc, 7165 Hart St., Mentor, OH 44060 USA

Ceramic membranes that transport oxygen hold great promise for integrating oxygen separation with natural gas conversion to produce syngas in a combined oxidation and steam reforming process. Commercialization of the Oxygen Transport Membrane (OTM) technology is being pursued by an Alliance among BP, Praxair, Sasol and Statoil. This Alliance has completed the first phase of its joint project and has begun process scaleup. High rates of oxygen separation and gas conversion have been achieved with stable materials. A scaleable fabrication technology is being developed that produces high yields of high quality reactor elements. A robust reactor design has been developed that minimizes stress on the ceramic elements, seals the elements into the reactor housing, and provides excellent heat management. Process economic calculations for the OTM syngas process show that the economic advantage can be greater than 35% of the syngas cost compared to the best conventional processes. Successful commercialization depends on solving many unique problems in the areas of ceramic materials and engineering, element fabrication, and reactor engineering. Selected highlights from the progress of the OTM Alliance are presented.

1. INTRODUCTION

Natural gas is a readily available, inexpensive feedstock that can be converted to clean burning liquid fuels and chemicals. This gas-to-liquids conversion, or GTL, proceeds via two stages, conversion to syngas, a mixture of hydrogen and carbon monoxide, followed by a syngas upgrading step such as Fischer-Tropsch (FT), methanol synthesis, ammonia synthesis, etc. The key to obtaining an economical process for fuels production rests in the syngas production step. Conventional technologies for syngas production include steam reforming, partial oxidation and autothermal reforming, all of which are very capital

* Author to whom correspondence should be addressed.

intensive. In an overall GTL scheme the syngas production stage consumes about 60% of the capital investment. Thus, reduction in the cost of syngas production could have a big impact on the cost of GTL fuels and chemicals.

Another incentive for natural gas upgrading is that GTL processes produce clean fuels. Natural gas derived fuels contain no sulfur or polynuclear aromatics (soot precursors), and produce less CO_2 than petroleum.

Dense oxygen transport membranes (OTM's) are now being applied to syngas generation [1-3]. The OTM syngas generation concept combines oxygen separation from air with methane oxidation and catalytic steam reforming into a single unit. Potential savings available to the OTM syngas process result from elimination of the oxygen plant and greater thermal efficiency due to close process integration. Estimates of savings range from 20 to 50% on installed capital.

BP has come together with three other global companies, Statoil, Sasol and Praxair, to form the OTM Alliance to commercialize this technology. The companies have world class skills in all of the critical fields needed to commercialize this technology, including air separation, gas processing, petrochemicals, syngas generation and utilization, Fischer-Tropsch, OTM materials, and ceramic powder and element fabrication. The OTM Alliance development program is divided into three phases:

1. Proof of technology. This phase was completed at the end of 1999 and focused mainly on the developments of materials and bench scale reactors.
2. Technical demonstration. This phase, which kicked off at the beginning of 2000, will focus on the scaleup of the technology and will culminate in the commissioning of a demonstration scale pilot plant.
3. Commercial demonstration. During the final phase, a commercial scale reactor will be built and operated.

The program was initiated in 1997 and is now in the second phase. This paper describes several key technical advances achieved by the OTM Alliance program in Phase 1.

2. CHALLENGES OF OTM DEVELOPMENT

The commercialization of ceramic OTM's for syngas production is an ambitious undertaking. Among the key challenges are 1) membrane development, 2) ceramic to metal seals, 3) stress on the ceramics, 4) reactor modeling, and 5) economics. Each of these areas has been advanced significantly by the OTM Alliance in Phase 1 of its program.

2.1. Membrane Development

A commercial OTM process requires robust materials that provide high oxygen separation rates, perhaps tens of cc/min/cm^2 [4]. Due to the severe process conditions stern demands are placed on the material. The membrane must withstand a highly oxidizing atmosphere on one side and a highly reducing atmosphere on the other at very high temperatures (> 900 C). Membrane elements must also be mechanically robust, resisting creep, corrosion, and poisoning.

Oxygen transport through the dense ceramic occurs in several steps:

- Oxygen molecules diffuse to the membrane surface.

- Oxygen dissociates to oxide ions at the membrane surface on the air side.
- Oxygen ions move through the membrane; electrons move in the opposite direction.
- Oxygen ions react with process gases on the process gas side in an oxidation reaction.
- Oxidation products diffuse away from the membrane surface.
- Process gas is catalytically reformed on the process gas side to form syngas.

Membrane performance depends on the membrane materials, architecture, and conditions extant on the two sides of the membrane. With ordinary OTM configurations the oxygen transport rate of the membrane rises rapidly with temperature, reaching the commercial flux targets only above about 950 C, as shown in Figures 1 and 2. This behavior provides high fluxes at high temperatures, but makes the membrane performance difficult to control. As the flux increases the temperature increases due to the exothermic oxidation reactions, which in turn drives flux still higher. An autothermal runaway can occur.

In order to provide for better process control the OTM Alliance has replaced these ordinary membranes with improved membranes. The improved membranes show very different behavior as temperatures are increased, as show in Figures 1 and 2. The improved membrane system consists of two components, one of which is a high flux membrane that permits high flux at low temperature, and a second component that is controlled by diffusion so that flux can increase only slowly after reaching a plateau value. The increase in flux with increasing temperature is much lower so that control is easier. As an added benefit the flux at lower temperatures is higher, so that lower temperatures can be used in the reactor. And cool and hot spots are less problematic since the membrane performance is not strongly dependent on the local temperature. Overall the OTM syngas process is easier to control since temperature variations due to local gas composition, membrane performance, catalyst activity and heat transfer are damped by the membrane performance.

2.2. Ceramic to Metal Seals

A unique reactor design challenge of the OTM syngas process is sealing the ceramic elements into the metal reactor housing under the severe conditions of temperature, gas composition and pressure present in the system. The thermal and chemical properties of the ceramic membrane and metal housing are quite different, and the ceramics are brittle, making this a formidable task. Ceramic to metal seals of this sort are not commercially available.

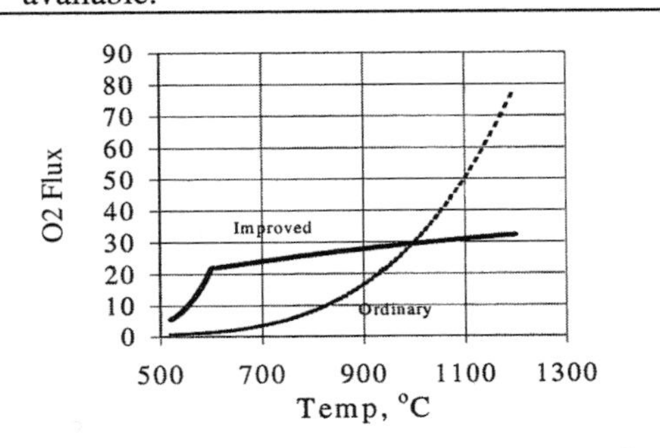

Fig. 1. Oxygen flux vs. temperature for ordinary and improved membranes.

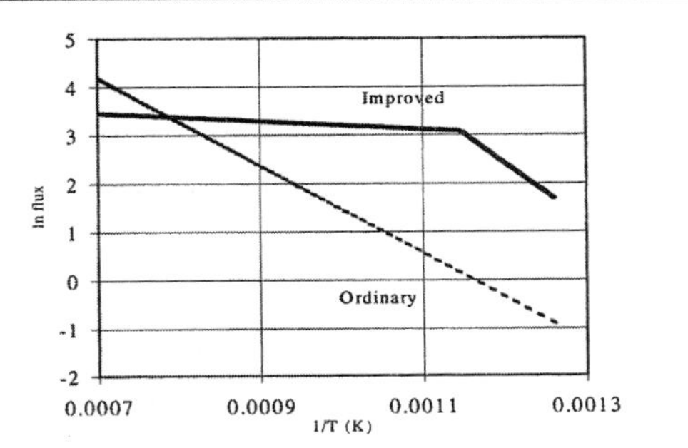

Fig. 2. Arrhenius plot of oxygen flux of ordinary and improved membranes.

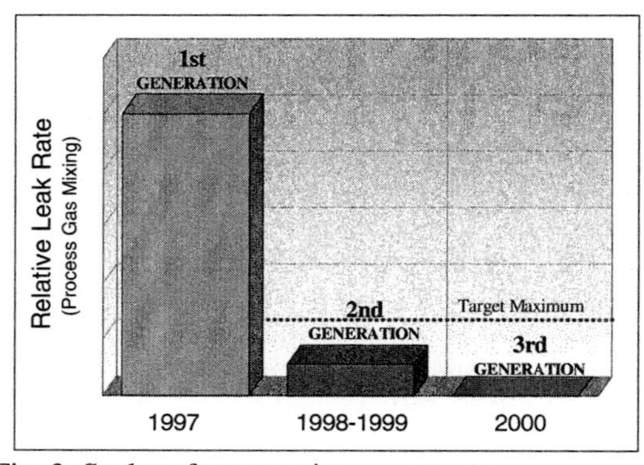

Fig. 3. Seal performance improvements

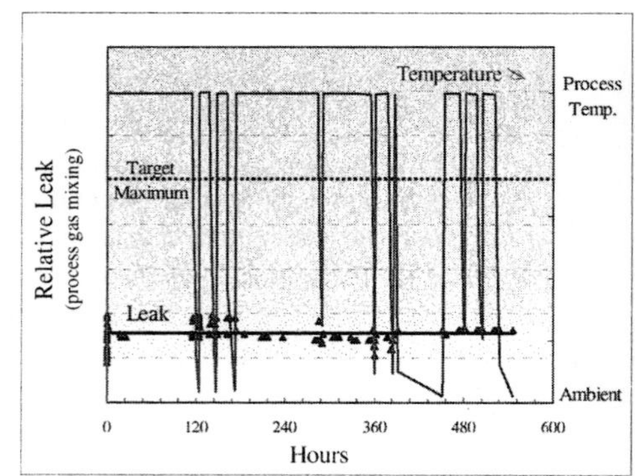

Fig 4. Seal performance during temperature cycling

A useful ceramic to metal seal must demonstrate low leakage rates, cyclability, and the ability to withstand pressure differentials. The OTM Alliance has developed proprietary seal technology that meets these criteria. Over a dozen seal designs were tested under severe conditions that represent the extremes of what could be encountered in the OTM syngas reactor. Figure 3 shows how the improved seal designs bring leak rates down to levels below the maximum tolerable in a commercial process. Figure 4 shows the response of the ceramic to metal joints subjected to temperature cycles from 1000 to 25 to 1000 °C over the course of 500 hours under a pressure differential of 100 psig. Leak rates remain far below the target level, demonstrating that the seals can be cycled without loss of performance.

Further development in seal technology is being focused on fabrication, maintenance and cost reduction so that the seals will be ready for commercial deployment.

2.3. Ceramic Stress Fields

OTM materials take up and lose oxygen with changes in environment. This reversible oxygen uptake is responsible for the membranes' ability to separate oxygen, but also results in some structural changes. As the oxygen content changes in a membrane, the lattice of the membrane expands or contracts. These chemically driven dilatations can result in stress buildup in the membrane.

Figure 5 shows the calculated stress response of a membrane to a rapid, small change in oxygen pressure based on dilatometry measurements. Initially there is a significant stress build up which is relieved slowly as oxygen exchanges between the

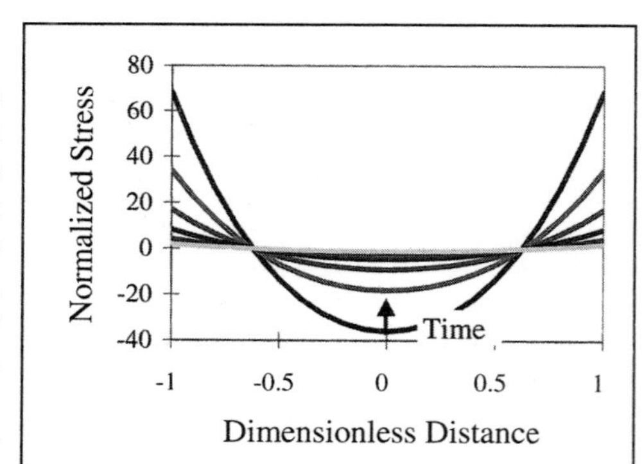

Fig. 5. Stress decay with time in a membrane exposed to an oxygen pressure change. Distance = 0 is the membrane centerline.

sample and the atmosphere. The rate of stress relief is fixed by the oxygen exchange rate.

Under dynamic conditions the membrane is exposed to a gradient of oxygen activity at steady state. From the activity gradient a stress gradient is produced within the OTM. In order to estimate the stress gradient, and evaluate its effect on membrane stability and reliability, dilatometry data on the membrane under reaction conditions has been collected. This data can be used to calculate the stress gradient. Figure 6 presents the stress gradient calculated from dilatometry results on a model membrane material at 1000 °C. The calculation shows that the stress is nonuniform through the membrane.

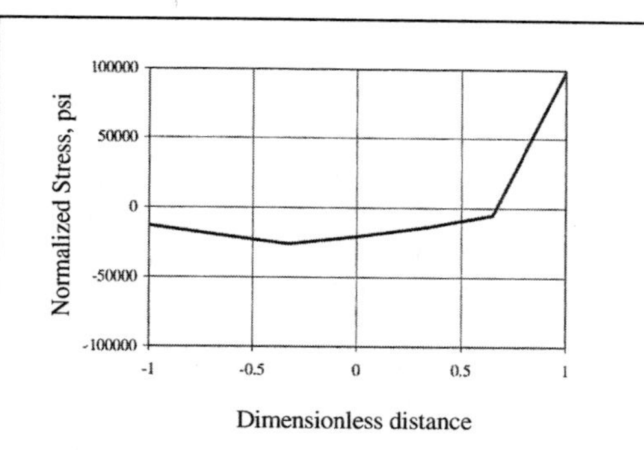

Fig. 6. Stress in a membrane under an oxygen gradient with air on one side and process gas on the other side.

2.4. Reactor Modeling

OTM syngas production combines oxygen separation with steam reforming and partial oxidation of methane. Heat integration of the very exothermic oxidation and endothermic reforming reactions is critical to achieve good process intensification and high yields of syngas. Computational models have been developed that permit various reactor designs to be evaluated. These models are based on fundamental equations for mass and heat transfer as well as kinetics of the chemical reactions.

Figure 7 presents the calculated temperature profile for a counter-current tubular OTM syngas process. In this model the air is fed from the bottom in the left hand segment and the process gas is fed from the top of the figure in the right hand segment. The white space represents the membrane.

3. CONCLUSIONS

OTM syngas production represents a step-out process for upgrading natural gas to synthesis gas. The OTM Alliance of BP, Praxair, Sasol and Statoil is conducting a three phase development program to bring the technology to commercial readiness. During the first phase of this program the Alliance has solved many significant problems associated with the unique features of ceramic membranes and oxygen transport materials.

Alliance scientists have developed new membranes that provide for better temperature control, higher flux and lower stress gradients than conventional membranes. Seals have been developed that have very low leak rates and survive temperature cycling. Stress in the membranes have been evaluated under static and dynamic conditions. Understanding the stress fields has led to more robust reactor designs and operating protocols. The thermal integration of the oxygen separation, steam reforming and oxidation processes has been studied to determine optimal syngas production conditions. This has led to advanced

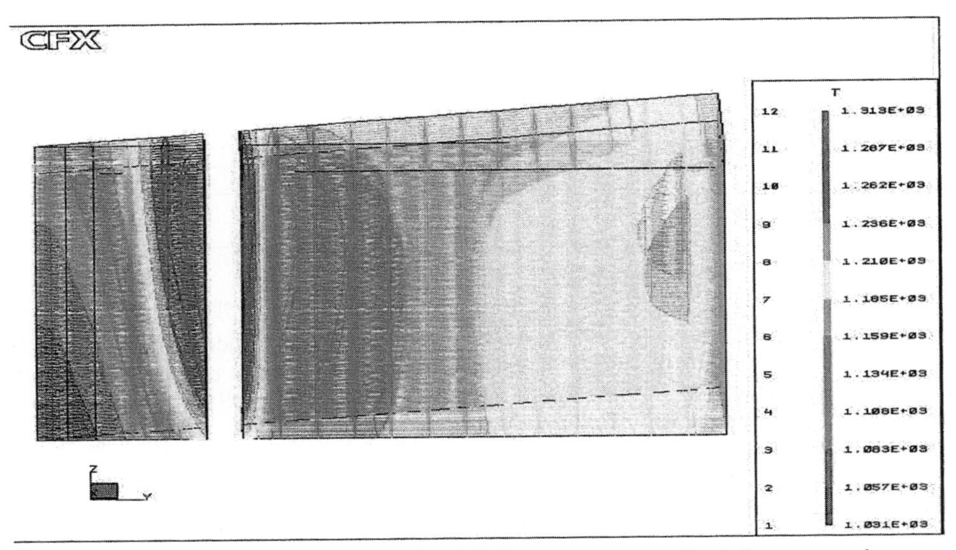

Fig. 7. Temperature profiles calculated for OTM syngas process. The left segment shows the air stream and the right segment shows the process gas. Temperature in Kelvin.

reactor designs that shrink the reactor footprint, minimize stress on the ceramic elements and maximize the syngas yield from the process.

The OTM Alliance has entered the second phase of its development program. In this phase the Alliance will focus on reactor scaleup in pilot plant reactors. At the end of this phase the Alliance intends to begin construction of a demonstration plant.

ACKNOWLEDGMENTS

The authors are grateful for the contributions of all of the OTM Alliance scientists and engineers, particularly S. J. Xu, J. C. Chen, R. Van Slooten, J. White, A. Swarts, A. Y Sane, and J. Collins.

REFERENCES

1. Cable, T. L., Mazanec, T. J., and Frye, J. G., Jr., European Patent Appl. 399833 (1990).
2. Kleefisch, M. F., Udovich, C. A., Bhattacharyya, A., Kobylinski, T. P.; US 5,980,840 (1999).
3. Mazanec, T. J., Cable, T. L., Frye, J. G. Jr., and Kliewer, W. R., US 5,306,411 (1994).
4. Bredesen, R.; Sogge, J.; "A Technical and Economic Assessment of Membrane Reactors for Hydrogen and Syngas Production" presented at Seminar on the Ecol. Applic. Of Innovative membrane Technology in the Chemical Industry", Cetraro, Calabria, Italy, 1-4 May 1996.

Studies in Surface Science and Catalysis
J.J. Spivey, E. Iglesia and T.H. Fleisch (Editors)

Deactivation of CrO_x/Al_2O_3 Catalysts in the Dehydrogenation of *i*-Butane

S.M.K. Airaksinen, J.M. Kanervo, A.O.I. Krause

Helsinki University of Technology, Department of Chemical Technology,
P.O.Box 6100, FIN-02015 HUT, Finland

The deactivation of two CrO_X/Al_2O_3 catalysts containing different amounts of Cr^{6+} was studied in the dehydrogenation of *i*-butane. The reduction-oxidation behaviour was also examined with H_2-TPR. The catalyst with the lower Cr^{6+} content was found to be more stable under the studied reaction conditions. The deactivation during a single dehydrogenation test run was caused by coke formation which was suggested to be related to the amount of redox Cr^{3+} on the catalyst. The deactivation in several dehydrogenation-regeneration cycles was attributed to the decline in the amount of Cr^{6+} and the clustering of the active species.

1. INTRODUCTION

Supported chromium oxide catalysts are widely used in various industrial processes. CrO_X/Al_2O_3 is effective in the dehydrogenation of light alkanes such as propane and *i*-butane, and the Phillips catalyst CrO_X/SiO_2 in ethene polymerisation. Because of their industrial importance, these catalysts have been extensively studied with different characterisation methods [1]. Especially CrO_X/Al_2O_3 has been the object of great interest, as recently reviewed by Weckhuysen and Schoonheydt [2]. However, there are still unanswered questions related to the structure of the catalyst under dehydrogenation conditions and thus to the nature of the catalytic site(s), the reaction mechanism and the catalyst deactivation. It is notable that the studies done on CrO_X/Al_2O_3 catalysts have often been performed on samples with a lower chromium content than what is used on the industrial dehydrogenation catalysts. These contain 12-14 wt-% chromium [3], which is also in the range of the chromium content at which the maximum activity in the dehydrogenation is reached [4,5]. Thus the studies on low chromium contents do not provide a proper description of the dehydrogenation catalyst.

An essential feature of the dehydrogenation process is the frequent regeneration of the catalyst. During the dehydrogenation the catalyst is fouled with coke. The short dehydrogenation-regeneration cycle enables the heat released in the burning of the coke to be used efficiently in the endothermic dehydrogenation reaction. Because of the demanding and rapidly changing conditions, the structure of the dehydrogenation catalyst should be stable. Therefore, information about the nature and the behaviour of the active sites is valuable in developing better and more durable catalysts.

A number of arguments can be given to support the idea that the Cr^{3+} sites formed in the reduction of Cr^{6+} (so called redox Cr^{3+} [4] or dispersed phase [2]) might be the active sites. However, it was recently shown [4,5] that the dehydrogenation activity of the CrO_X/Al_2O_3 catalysts reaches its maximum value at chromium loadings between 14 and 16 wt-% in the dehydrogenation of *i*-butane, even though the amount of redox Cr^{3+} levels off already at a chromium loading of about 8 wt-%. This means that the activity cannot be explained only by the amount of redox Cr^{3+}. It has been suggested [4,7] that in addition to the redox Cr^{3+}, also some non-redox Cr^{3+} could be responsible for the catalytic activity.

Irreversible alterations in the structure of the active phase have been proposed as potential causes for the long-term deactivation of the CrO_X catalysts. It has been observed [6] that the deactivation in repeated dehydrogenation-regeneration cycles is connected with a decline in the amount of Cr^{6+}. Two processes have been postulated [6] as possible explanations for this: 1) some of the chromium sites could migrate into the alumina lattice becoming inactive, or 2) deactivation could occur due to irreversible clustering of the active chromium species.

This study was overtaken in order to clarify the behaviour of the less-studied high chromium loading CrO_X/Al_2O_3 catalysts. Both the short-term and the long-term behaviour was of interest because the reasons for deactivation are still undetermined. With the aid of an on line FTIR equipment we can report the catalytic activity as a function of time on stream almost continuously during the dehydrogenation. Because the long-term deactivation is connected with the reduction-oxidation properties, it was examined — in addition to the consecutive dehydrogenation experiments — with temperature programmed reduction.

2. EXPERIMENTAL

The first catalyst was prepared by gas phase impregnation of $Cr(acac)_3$ and will be referred to as the ALD catalyst. The calcined support (AKZO alumina 000-1.5E, calcination temperature 600 °C) was treated in 12 successive cycles with $Cr(acac)_3$ at 200 °C and with air between the cycles at 520 °C. After the 12 cycles the ALD catalyst was calcined in air at 600 °C for 4 h. The second catalyst was a commercial CrO_X/Al_2O_3 catalyst without the promoter and developed for the fluidised bed operation (referred to as the FB catalyst). The total Cr content of the catalysts was 13-14 wt-%. The amount of Cr^{6+} was measured with UV-Vis spectrophotometry (UV-1201 Shimadzu) and was 2.9 wt-% for the ALD catalyst and 1.0 wt-% for the FB catalyst. XRD analysis (Siemens D500) showed no crystalline Cr_2O_3 phases. According to XPS measurements (SSX-100), the only chromium oxidation states present on the calcined samples were Cr^{3+} and Cr^{6+}.

The dehydrogenation experiments were carried out in a fixed-bed microreactor at 580 °C under atmospheric pressure. The products were analysed on-line with an FTIR gas analyser (Gasmet, Temet Instruments Ltd.) and with gas chromatography (HP 6890). The point at time on stream of 10 minutes was measured with GC. The amount of catalyst used was 0.1 g. One test series consisted of 12 (pre)reduction-dehydrogenation-regeneration cycles. First six cycles were done without prereducing the catalyst, the next five with prereduction and the last one without prereduction. The catalyst was heated to the required temperature under the flow of 5% O_2/N_2. Prereduction was accomplished at 590 °C with 10% H_2/N_2, using a reduction time of 15 minutes. The *i*-butane feed with weight hourly space velocity (WHSV) of 15 h^{-1} was diluted with N_2 at a mole ratio of 1:1. The catalyst was regenerated with 2-10% O_2/N_2 after each 15 minute dehydrogenation test run, until no carbon oxides were detected. The coke content of the catalysts was calculated based on the measured carbon oxides.

The H_2-TPR measurements were performed with Altamira Instruments AMI-100 catalyst characterisation system. The catalyst samples (30 mg) were stabilised prior to reduction by heating to 600 °C under the flow of 5% O_2/Ar. TPR was performed at heating rates of 6, 11 and 17 °C/min up to 600 °C under the flow of 11.2% H_2/Ar (50 cm^3/min). The hydrogen consumption was monitored using a thermal conductivity detector. The whole TPR procedure was also performed repeatedly five times in series for both ALD and FB samples to investigate their stability.

3. RESULTS AND DISCUSSION

3.1. Catalytic activity

The activity of the catalysts was studied in the dehydrogenation of *i*-butane. The behaviour of the ALD and FB catalysts was quite different during a single dehydrogenation test run. This can be seen from Figure 1, in which the conversion of *i*-butane and the selectivity to *i*-butene are shown as an example for the cycle no. 7 (the catalysts had been prereduced with H_2). The ALD catalyst showed high initial activity. However, the activity was not constant during the 15 minute experiment but declined continuously. Despite the high initial conversion, the yield of *i*-butene was hampered because of the generation of marked amounts of cracking products (mostly methane and propene). The selectivity to *i*-butene improved with increasing time on stream. The FB catalyst was less active but more stable. Also, the selectivity was better and did not change during the experiment. A slight deactivation could also be observed with the FB catalyst, but not to the extent as with the ALD. The coke content after the dehydrogenation was higher on the ALD catalyst (4.5 wt-%) than on the FB catalyst (0.9 wt-%). The similar behaviour as a function of time was observed during the other cycles.

The catalysts were subjected to 12 consecutive experiments in order to study their behaviour in several (pre)reduction-dehydrogenation-regeneration cycles. The conversion of *i*-butane in the series at time on stream of 10 minutes as measured with GC is shown in Figure 2.

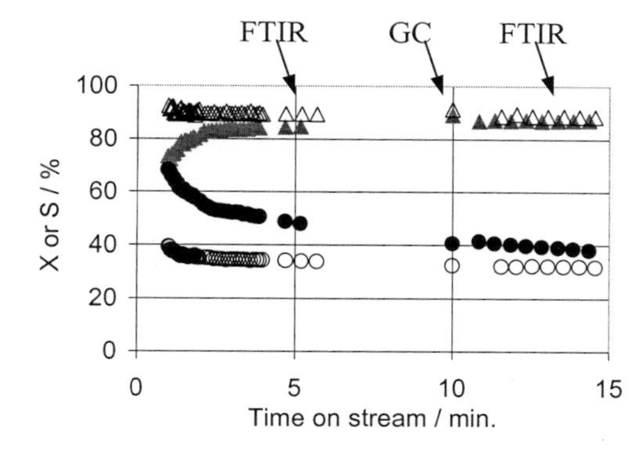

Fig. 1. Activities of the catalysts in a single dehydrogenation (data from test run 7).
• Conversion X (ALD)
o Conversion X (FB)
▲ Selectivity to *i*-butene S (ALD)
△ Selectivity to *i*-butene S (FB)

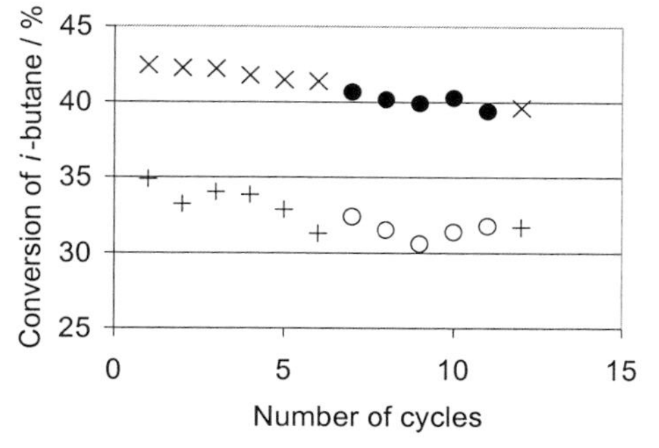

Fig. 2. The conversion of *i*-butane after 10 minutes during the test series.
× Conversion (non-prereduced ALD)
• Conversion (prereduced ALD)
+ Conversion (non-prereduced FB)
o Conversion (prereduced FB)

The activity decline in the series was fairly constant with both of the catalysts. The selectivity to *i*-butene did not change markedly as deactivation occurred. There was a clear decrease in the amount of coke deposited on the ALD catalyst. On the FB catalyst this process was not so obvious probably due to the low initial coke content. A decrease in the amount of reduction products was also noticed indicating a decline in the amount of redox Cr^{3+}. This is in accordance with the earlier results obtained by Hakuli *et al.* [6]. They attributed this decrease in the concentration of Cr^{6+} rather to the irreversible clustering of the chromium species than to the migration of the active sites into the alumina lattice. Interestingly, in our study it was noticed that the amount of reduction products was slightly increased with longer oxidation times compared to the previous cycle. This behaviour could support the clustering explanation since it seems unlikely that the chromium species would re-emerge from the alumina lattice if they had migrated into it. The increase in Cr^{6+} could indicate that the species clustered during reduction/dehydrogenation were better re-dispersed during longer oxidation treatments. However, the phenomenon was not correlated in the activities of the catalysts. This could support the idea that the redox Cr^{3+} is not the only active site on the catalyst.

The high initial activity of the ALD catalyst during a single dehydrogenation test run could most likely be related to its high redox Cr^{3+} content. However, the amount of side products was considerable. It has been proposed that the coke formation is related to the high alkene concentration in the gas phase [9]. Furthermore, it is possible that the redox Cr^{3+} sites generate more coke and side products than the non-redox sites. The poisoning of the redox sites could thus account for the increase in the selectivity after the initial period of the experiment. The more stable behaviour of the FB catalyst could thus be explained by its lower redox Cr^{3+} content which, however, decreases the overall activity in the beginning.

It was not possible to deduce clear differences between the stabilities of the catalysts based on the long-term deactivation experiments because of the coke formation on the ALD catalyst. So, it was necessary to study the stability in consecutive reduction-oxidation cycles with H_2-TPR. The data obtained in the reduction studies might give an indication about the number and the properties of the reducible sites.

3.2. Temperature programmed reduction

The reduction studies made for the ALD and FB catalysts indicated notable differences between the samples. The ALD samples reduced at substantially lower temperatures than the FB samples. The reduction rate maximum of the ALD catalyst was around 315 °C and that of the FB catalyst close to 340 °C under similar reduction conditions. The FB sites were evidently more difficult to reduce. The reduction data of the both samples exhibited clearly a single-peak behaviour. This was interpreted that only one clearly dominant reduction process was taking place on the both catalysts.

The Kissinger peak analysis [10] was performed for H_2 consumption data. It gave the apparent activation energy values of 99 and 73 kJ/mol for the ALD and the FB catalyst, respectively. Also the Friedman's method of constant conversion [11] was carried out. For the ALD sample data, this technique resulted in the activation energy values around 95 kJ/mol for the reduction degrees from 0.2 to 0.5 and then slightly higher values for the higher reduction degrees. Correspondingly, for the FB sample data, Friedman analysis gave the activation energy values around 75 kJ/mol for the reduction degrees below 0.6 and then again slightly higher values for the higher reduction degrees. The results suggest that the reduction behaviour of these two catalysts are fundamentally different.

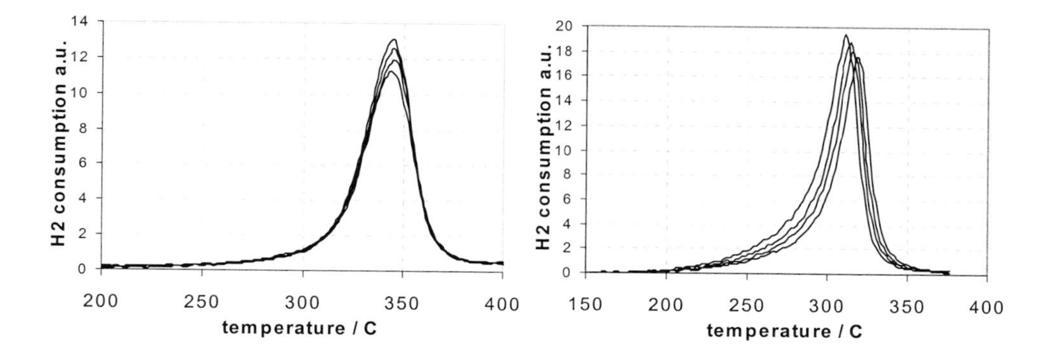

Fig. 3. Repeated TPR for the FB catalyst.

Fig. 4. Repeated TPR for the ALD catalyst.

The kinetic modelling of the TPR patterns was then performed with nonlinear regression analysis to test various reduction mechanisms [12]. The nuclei growth models (Avrami-Erofeev models) were adequate to describe the data, but the conventional models suggested that the dimension of the nuclei growth was different for the studied ALD and FB samples. The Avrami-Erofeev type model was therefore re-derived into a form that included both the nucleation and the nuclei growth with temperature dependent rate coefficients. This model with the first order nucleation and the 2-dimensional nuclei growth now best described the measured data for the both catalysts. The activation energies of the nuclei growth were 76 kJ/mol and 89 kJ/mol for the ALD and FB catalysts, resp. The nucleation parameters were not well-identified, but obtained such values that the nucleation for the FB catalyst was fast and for the ALD catalyst practically instantaneous. The modelling results imply differences in chromia structure on the two catalysts. The overall description of the kinetic model was better for the ALD catalyst. The structure of the reducible sites on the ALD catalyst might be closer to the structure assumed in the model derivation (i.e. large 2-dimensional islands) compared to the structure of the FB catalyst. The activation energies of nuclei growth deviate from the values received from the Kissinger and the Friedman analysis, which is likely due to the fact that these methods do not take into account any mechanistic assumptions. The two-dimensional nuclei growth model is adequate due to the assumed location of Cr^{6+} on the catalyst surface.

The repeated H_2-TPR procedures revealed qualitative differences between the studied catalyst samples. The reduction rate maxima of the ALD catalyst shifted to higher temperatures, whereas the reduction rate maxima of the FB catalyst remained within a temperature range of 3 °C for all the five TPR-runs, as seen in Figures 3 and 4. Furthermore, the five H_2-TPR curves, between which the oxidation at 590 °C was performed, showed that the decline in the amount of Cr^{6+} was different for the two catalysts: the decline was faster with the catalyst prepared by ALD. The ALD catalyst contained more reducible chromia than the fluidised bed catalyst with the equal chromium loading, and the reduction studies indicated a higher stability of the fluidised bed catalyst. The repeated reduction treatments were obviously able to change both the nature and the amount of reducible chromia species of the ALD catalysts.

4. CONCLUSIONS

An essential difference was found in the behaviour of the two catalysts during dehydrogenation. The activity of the ALD catalyst declined rapidly with an increase in the selectivity. The deactivation during one dehydrogenation test run was mainly caused by coke formation, which was more advanced on this catalyst because of the higher amount of redox Cr^{3+}.

The repeated TPR measurements showed a dissimilarity in the reduction behaviour of the catalysts. The ALD catalyst was easily reducible but the amount of Cr^{6+} decreased faster compared to the FB catalyst and the temperature of the maximum reduction rate shifted to higher temperatures. However, the observed decline in the amount of redox Cr^{3+} was faster than the decrease in the dehydrogenation activity on the two catalysts. In addition, the kinetic modelling indicated that the structure of the reducible sites on the ALD catalyst was closer to large 2-dimensional islands than the structure on the FB catalyst. From all these results we may conclude that on the FB catalyst the redox sites are more scattered and structurally stable. On the ALD catalyst the Cr^{6+} sites seem to be more mobile and thus susceptible to clustering, which could transform the reducible sites into more scattered non-redox species, approaching the structure of the FB catalyst. In that case the long-term deactivation of the catalysts could be explained by the decline in Cr^{6+}, in addition to the cluster formation on the ALD catalyst.

In a summary, it seems that the behaviour of the catalysts in the dehydrogenation cannot be explained by only one active site. This is in an agreement with the earlier observations that the redox and the non-redox Cr^{3+} sites are both active in dehydrogenation [4,7], but with different activities [7].

ACKNOWLEDGEMENTS

The Academy of Finland is gratefully acknowledged for supporting this study. We thank also Dr. Arla Kytökivi at Fortum Oil and Gas Oy for preparing the ALD catalyst.

REFERENCES

1. Weckhuysen, B.M., Wachs, I.E., Schoonheydt, R.A., Chem. Rev., 96 (1996) 3327-3349.
2. Weckhuysen, B.M., Schoonheydt, R.A., Catal. Today, 51 (1999) 223-232.
3. Buonomo, F., Sanfilippo, D., Trifirò, F., in Handbook of Heterogeneous Catalysis, vol 5, eds. Ertl, G., Knözinger, H., Weitkamp, J., Academic Press, Inc., New York, 1997, p. 2148.
4. Hakuli, A., Kytökivi, A., Krause, A.O.I., Appl. Catal. A,190 (2000) 219-232.
5. Kytökivi, A., Jacobs, J.P., Hakuli, A., Meriläinen, J., Brongersma, H.H., J. Catal., 162 (1996) 190-197.
6. Hakuli, A., Kytökivi, A., Krause, A.O.I., Suntola, T., J. Catal., 161 (1996) 393-400.
7. Cavani, F., Koutyrev, M., Trifirò, F., Bartolini, A., Ghisletti, D., Iezzi, R., Santucci, A., Del Piero, G., J. Catal., 158 (1996) 236-250.
8. Hakuli, A., Kytökivi, A., Lakomaa, E.-L., Krause, O., Anal. Chem., 67 (1995) 1881-1886.
9. Uchida, S., Osuda, S., Shindo, M., Can. J. Chem. Eng., 53 (1975) 666-672.
10. Kissinger, H.E., Anal. Chem., 29 (1957) 1702-1706.
11. Friedman, H.L., J. Polym. Sci. Part C, 6 (1965) 183.
12. Wimmers, O.J., Arnolds, P., Moulijn, J.A., J. Phys. Chem. 90 (1986) 1331-1337.

Studies in Surface Science and Catalysis
J.J. Spivey, E. Iglesia and T.H. Fleisch (Editors)

Selectivity of Fischer-Tropsch Synthesis: Spatial constraints and forbidden reactions

Hans Schulz and Zhiqin Nie

Engler-Bunte-Institut, Universität Karlsruhe, Kaiserstraße 12, 76128 Karlsruhe, Germany

Changes with time of reaction rate and selectivity have been determined with an iron, a cobalt and a nickel catalyst. It was observed that activity only developed slowly. FT-catalysts appear to be constructed under reaction conditions - by deposition of and reaction with carbon and by segregation of the metal phase by CO respectively. Distinct reactions are thereby selectively inhibited (forbidden) and the kinetic regime appears to be strongly ruled by spatial constraints.

1. INTRODUCTION

Selectivity of Fischer-Tropsch Synthesis refers to a product composition of numerous substances. Their individual amounts exhibit a high order in their distribution on compound classes and carbon number fractions, as used for kinetic modelling [1]. Examination of data for iron, cobalt and nickel catalysts reveals initial periods of self organization [2, 3, 4]. Linking kinetic insights to complementary work in the literature about restructuring of cobalt and other metals [5, 6, 7] towards a more dispersed surface with peaks and holes and about the stability of methyl species on hydrogenation metals [8] leads to the explanation of the distinct features of FT synthesis.

2. EXPERIMENTAL

A fixed bed reactor and methods for sampling and analysis have been designed for time resolution of rate and selectivity [9, 10, 11] at a high degree of accuracy. The low degree of data scattering (see Figs. 1-9) demonstrates the appropriateness of methods and procedures. An iron, a cobalt and a nickel catalyst have been used. The iron catalyst was prepared by precipitation and potassium addition to obtain the composition $100Fe:13Al_2O_3:10Cu:25K$ (weight ratios) [4]. Fine catalyst particles (d_p<0.1 mm) were mixed with coarse inert particles of fused silica (d_p=0.25-0.4 mm, weight ratio 1:10) to provide the catalyst bed (1 g of iron) of high isothermicity, no mass transfer restriction and uniform gas flow at negligible pressure drop. The catalyst was then dried/calcined, reduced at 673 K and used for synthesis at 523 K, 30 ml/min (NTP), H_2/CO=2.3 and P=1 MPa. The cobalt and nickel catalysts were prepared by quick precipitation, with suspended Aerosil (small SiO_2 particles of ca. 200 Å diameter) in the nitrate solution to obtain the compositions 97Co:12Zr:100Aerosil and $100Ni:12Zr:100SiO_2$ (by weight) [9]. After calcination (Argon flow 40 ml(NTP)/min, 1 bar, heating 2 K/min to 973 K, 2 h isothermal) and reduction (gas flow 40 ml(NTP, H_2:Ar=1:3) per minute, heating rate 2 K/min to 973 K, isothermal until no H_2O from reduction was to be detected in the exiting gas stream) of the diluted catalysts in the reactor (0.68 g of Co respectively Ni; volume of the diluted catalyst bed 6 ml), the synthesis on cobalt was performed at 463 K, 0.5 MPA, H_2/CO=2.05 and GHSV=300 h^{-1} and the synthesis on nickel at 453 K, 0.18 MPa (p_{CO+H2}=0.09 MPa, p_{Argon}=0.09 MPa), H_2/CO=2.0 and GHSV=450 h^{-1}.

Temperature programmed gas chromatography (-80 to 280 °C) of ampoule samples covered the compound range C_1 to C_{20}. From the data of product composition, probabilities of chain growth, chain branching and formation of paraffins and olefins were calculated.

3. RESULTS AND DISCUSSION

3.1 FT-Activity: Rate and selectivity depend on temperature, partial pressures and thus on reactor configuration in addition to their dependence on catalyst composition and structure. The FT base metals - iron, cobalt, nickel and ruthenium - each have their individual conditions for high performance, and each needs its specific mode of promotion, preparation and activation. Nevertheless, there are differences noticed which can be attributed to the particular base metal. Amazingly, with all the three catalysts, activity is low initially (Fig. 1) and only slowly increases with time. This slow rate of activity increase indicates a slow process of solid state catalyst transformation to generate "the real FT catalyst". With **iron**, carbon is deposited on the catalyst, specifically during the time span from 100 to 1000 minutes (see yields of volatile products Y_{VOC} and of retained carbon $Y_{\Delta C}$ in Fig. 1 left). Activity finally declines due to excessive carbon deposition.

With the cobalt catalyst (Fig. 1 middle above), the increase of activity is not accompanied by carbon deposition. The increase of $Y_{\Delta C}$ (retained carbon) from $t_{exp} \cong 1$ h onwards is related to hydrocarbons which are liquid under reaction conditions (see also in Fig. 2 the increase of chain growth probability with time). With **nickel** the changes with time are in principle similar to those with cobalt (Figs. 1 and 2, right).

Complementary observations with Ni, Co and Ru catalysts are now included to discuss the FT regime on these metals. Pichler proposed for cobalt (and likewise for Ni and Ru) the formation of "surface carbonyls", as the conditions of synthesis are close to those for thermodynamically allowing metal carbonyl formation [12]. For a Ni-FT catalyst Ponec [7] suggested a "gas induced segregation". Images of cobalt surfaces obtained with tunneling electron microscopy showed a surface rearrangement to a more dispersed structure [6] as caused by a CO/H_2-treatment. Now we propose that two differing sorts of metal sites (in holes or on peaks) of the surface are essential for the FT regime. These sites should exhibit different activity for individual steps of reaction and also different spatial constraints and they are thought to collaborate in the FT regime. (The mechanistic picture will be presented in detail in another publication.) This paper deals with the activity- and selectivity changes during the period of "FT-catalyst construction". The respective related elementary reactions will later be used to characterize/identify the active sites themselves.

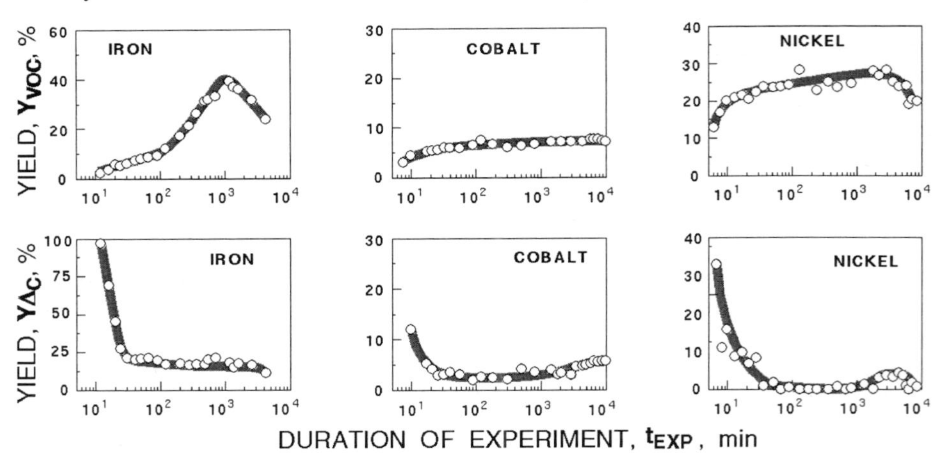

Fig. 1: Yield of volatile and non volatile products in dependence of reaction time. FT-synthesis with 3 catalysts. (Conditions in section Experimental).

3.2 Chain growth: Curves of chain growth probability in dependence of carbon number of the growing species are presented in Figs. 2, 3 and 4. Fig. 4 is only a principal drawing for explaining the curve shape. Comparing the curves for iron, cobalt and nickel at steady state (Fig. 3) one notices 1.) with cobalt and nickel - and not with iron - the formation of "extra methane", 2.) with cobalt and nickel and not with iron strong ethene readsorption for chain growth (and propagation of chains [13]) and 3.) increase of p_g with carbon number for all the three catalysts, indicating readsorption of higher olefins to occur on iron, cobalt and nickel [14]. This contributes to the average chain length, particularly in the higher carbon number range of products $N_C \geq 8$.

Regarding the influence of duration of the experiment (Fig. 2), one sees a different behaviour of iron as compared with cobalt and nickel. With the iron catalyst only a little change with time of chain growth probability is noticed. It is concluded, that as more and more FT-sites are generated, these sites are of the same FT-nature. Interestingly, a close look onto the curves in Fig. 2 left reveals increased olefin readsorption in the steady state, as consistent with the build up of a liquid phase and a related carbon number dependent increase of reactor residence time of the product molecules. At $N_C=3$ - where the secondary olefin reactions are the lowest - the value of p_g at steady state has become low, in contrast to cobalt and nickel where it has become high, indicating an opposite change with iron as compared with Co and Ni during construction of the FT-catalyst. With cobalt and nickel (Fig. 2 middle and right) the trends with time are 1.) strong increase of chain growth probability, indicating the polymerization nature of FT-synthesis – the combination reaction of surface species without product desorption only to develop with time. Methane formation (see p_g at $N_C=1$) declines. 2.) With nickel the contribution of olefin readsorption to chain growth appears particularly high.

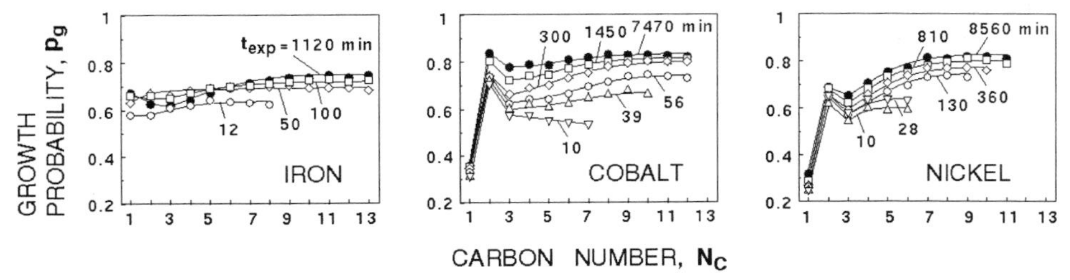

Fig. 2: Chain growth probability as a function of chain length (C_N) at different reaction times. FT synthesis with 3 catalysts (Conditions in section Experimental).

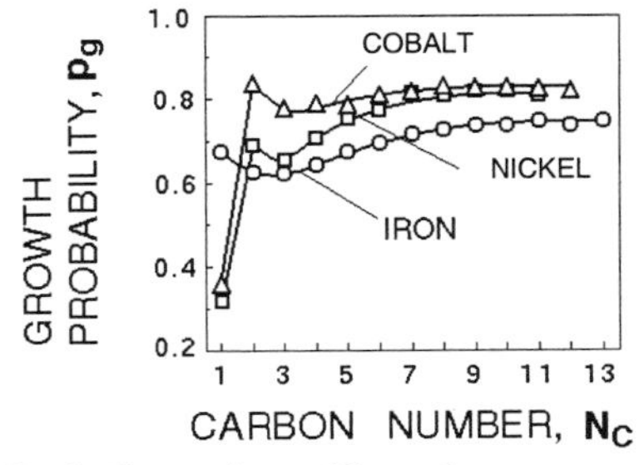

Fig. 3: Comparison of 3 steady state cases of FT-synthesis on the basis of chain growth probability (Conditions in section Experimental).

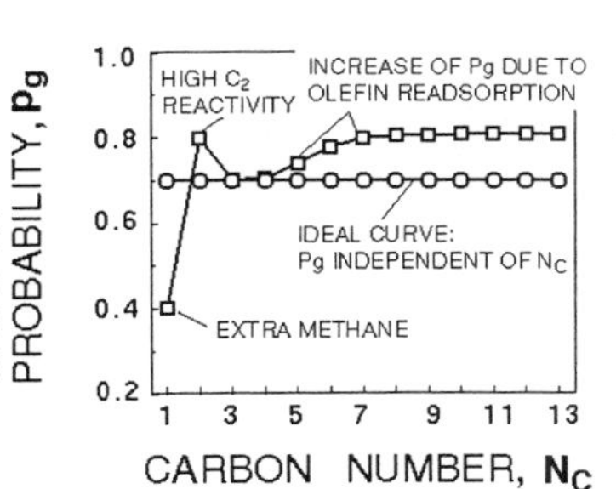

Fig. 4: Principal sketch of ideal and real chain length dependence of growth probability.

3.3 Chain branching: Branching probability as a function of carbon number of the growing species is shown in Figs. 5, 6, 7. The simpliest case, originally assumed by Anderson [15], would be chain length independent branching probability, the horizontal line in Fig. 7. In reality, for steady state FT-syntheses (see Fig. 6), branching probability declines with increasing carbon number from N_C=4 onwards. With size (carbon number) of the growing species - as spatial demands will increase - this effect reaches over 3 to 4 carbon atoms of chain length. Remarkably, the first value of $p_{g,br}$ is low. To explain this result one has to recall that in the model which is used for calculation, the desorption probability is assumed equal for branched as for unbranched species, however, desorption directly after branching might be very slow, as being demanding in space because a tertiary C-atom is involved in this step. At short times of the synthesis (see Fig. 5), particularly with cobalt, the shape of the curves is different and exhibits an increase with carbon number. The spatial constraints then will not be fully developed because "construction of the catalyst" is not completed and the selectivity controlling "alkyl grove" on the surface not ready. Then the readsorption of olefins will be possible not only at the terminal carbon atom but also e.g. at the second C-atom, giving rise to chain branching. The probability of this secondary reaction of olefin adsorption in the β-position would increase with carbon number of the olefin due to increasing reactor residence time. Comparing the influence of time on chain branching for the 3 catalysts (Fig. 5) reveals different trends with the iron - as compared with the cobalt- and the nickel catalyst. With the alkalized iron the branching probability is very low and this already at the beginning, consistent with the before derived conclusion that with time only the number of active sites increases, whereas, their nature remains the same. A weak increase of branching with time is observed with iron which is thus opposite to the pronounced decrease with cobalt and nickel at the beginning.

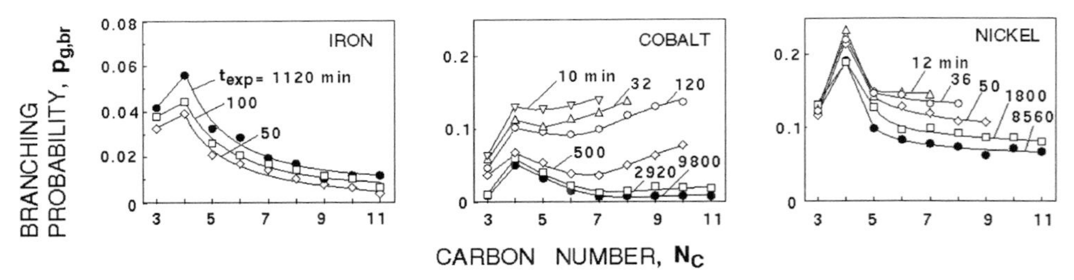

Fig. 5: Chain branching probability as a function of chain length with 3 catalysts. FT synthesis at different reaction times. (Conditions in section Experimental)

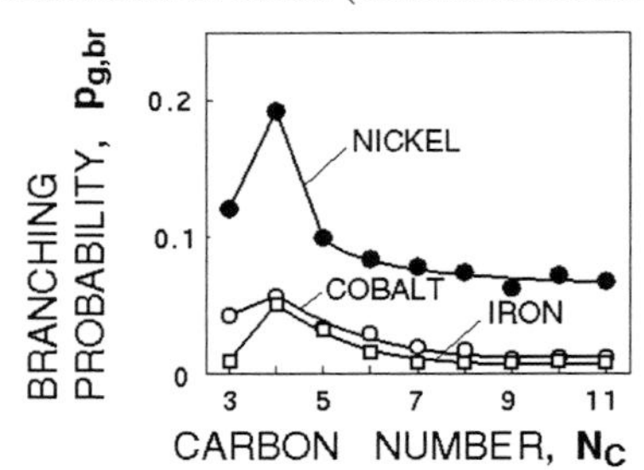

Fig. 6: Chain branching probability as a function of chain length for FT synthesis at steady state with 3 catalysts (Conditions in section Experimental).

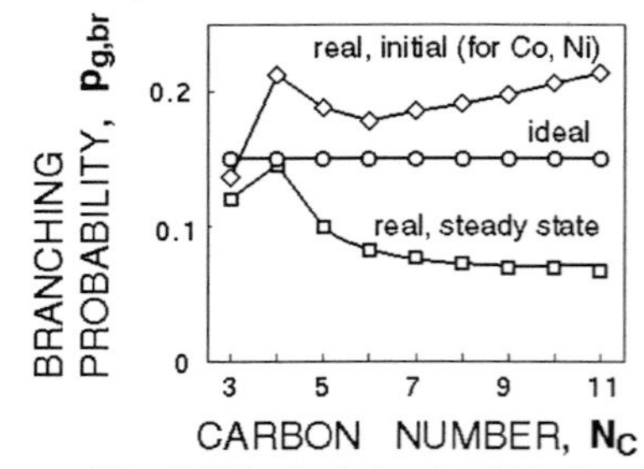

Fig. 7: Principal sketch of ideal and real chain length dependence of branching probability.

3.4 Olefinicity: Olefins-1 as products of FT-synthesis are commonly subject to secondary hydrogenation and double bond shift in addition to readsorption in alpha position for further chain growth [1, 2, 14, 16]. Olefin contents in hydrocarbon carbon number fractions for several times of reaction and the 3 catalysts are shown in Fig. 8. With the alkalized **iron** the curves represent approximately straight lines, as indicative for primary composition [1, 14, 16]. This primary selectivity increases with time, meaning that dissociative alkyl desorption (as olefin-1) is increasingly favoured against associative alkyl desorption (as paraffin). Reversible olefin adsorption for chain growth is consistent with a moderate decrease of olefin content with increasing carbon number (Fig. 8, left). As seen in Fig. 9, left, the olefins obtained with the iron catalyst are to a very high degree (90-97 %) olefins with terminal double bond, from the beginning, indicating suppression of olefin isomerization.

The shape of the curves of olefin content in carbon number fractions for **cobalt** (Fig. 8 middle) is characteristic of olefin secondary hydrogenation [14, 16]. The low value at N_C=2 reflects the high ethene reactivity. The decline of the curves in direction to higher carbon numbers indicates increasing readsorption of olefins on hydrogenation sites, as reactor residence time increases with increasing solubility of the olefins in the liquid phase, respectively in the "alkyl grove" attached to the catalyst surface. With cobalt, secondary olefin double bond shift (Fig. 9 middle) decreases until construction of the FT-catalyst is completed. With **nickel** the trends are the same as with cobalt. Amazingly, with nickel at high secondary olefin conversion the value of propene content in the C_3-fraction is specifically low (Fig. 9 right). It is thought that here, in concequence of spatial constraints at the hydrogenation sites, terminal double bonds are hydrogenated very preferrentially, and at C_3 there is only the one structure with terminal double bond possible.

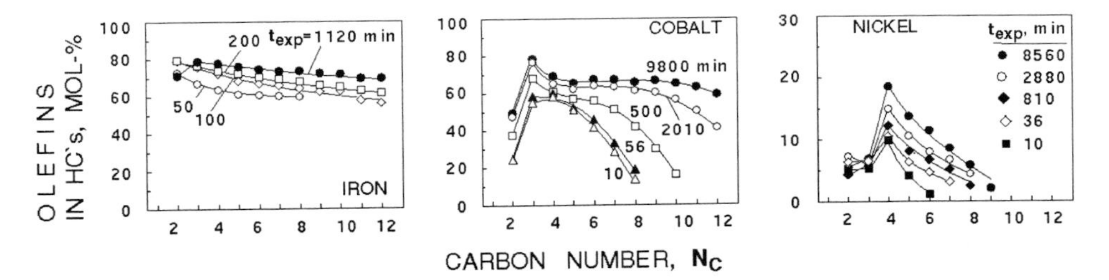

Fig. 8: Olefin content of hydrocarbons as a function of chain length at different times of syntheses with 3 catalysts (Conditions in section Experimental).

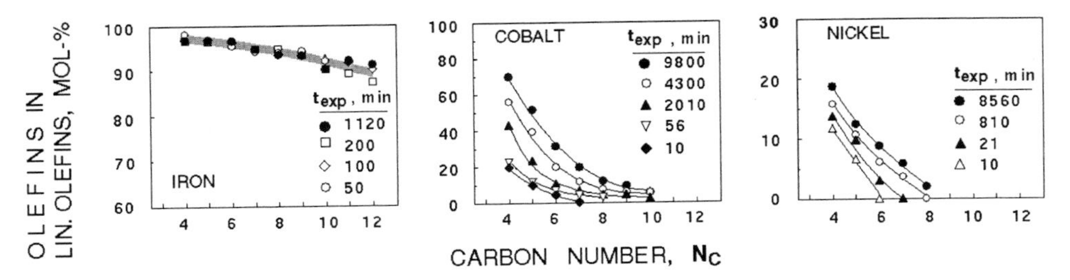

Fig. 9: α-Olefin content in linear olefins as a function of chain length at different times for FT synthesis with 3 catalysts. (Conditions in section Experimental).

4. CONCLUSIONS

Selectivity of FT synthesis concerns a product composition of many compounds. From their relative abundance conclusions about the probability of basic reactions can be drawn. It is noticed that the FT-active catalyst appears to be formed slowly under reaction conditions. Reaction with carbon in the case of iron and segregation of the metal by the action of carbon monoxide in the case of cobalt and nickel are visualized in this connection.

With the iron, from the beginning, the FT sites appeared to be the same, as selectivity merely changes, whereas the rate of synthesis increases drastically. This also means that the essential features - selective inhibition, forbidden reactions and spatial constraints - are present from the beginning. With cobalt and nickel the steady state activity is also only slowly developed. The fresh catalyst exhibits only poor FT-characteristics. "Construction" of the FT catalyst can again be assumed under reaction conditions.

The presented picture of Fischer-Tropsch catalysis relies also on our earlier work, on literature about stable "methyl species" on hydrogenation catalysts and on segregation of cobalt and nickel surfaces. In general terms the "mystery of FT synthesis" seems now to reveal its nature as a well ordered complex kinetic regime ruled by spatial constraints and forbidden reactions and only being established by self organization through in situ catalyst construction. A following publication will deal with the characterization/identification of FT-active sites on the basis of their capability to promote (or not) distinct basic reactions (among the theoretically many possible ones) and thereby to understand the FT-catalyst construction process.

5. LITERATURE

1. H. Schulz and M. Claeys, Appl. Cat. A, General 186 (1999)91
2. H. Schulz, E. van Steen, and M. Claeys, Stud. Surf. Sci. Catal. 81(1994)455
3. H. Schulz, Z. Nie and M. Claeys, Stud. Surf. Sci. Catal. 119(1998)191
4. H. Schulz, G. Schaub, M. Claeys and Th. Riedel, Appl. Cat. A, General 186 (1999)215
5. G. A. Somorjai, Rev. Phys. Chem. 45(1994)22
6. J. H. Wilson and C. P. M. de Groot, J. Phys. Chem. 99 (1995)7860
7. R. Ponec, W. L. van Dijk and J. A. Groenewegen, J. Catal. 45(1976)277
8. P.Albers, H.Angert, G.Prescher, H.Seibold and S.F.Parker, Chem.Commun. (1999)1619
9. Zh. Nie, PhD Thesis, Univ. of Karlsruhe, 1996
10. M. Claeys, PhD Thesis, Univ. of Karlsruhe, 1997
11. E. van Steen, PhD Thesis, Univ. of Karlsruhe, 1993
12. H. Pichler in "Adv. In Catal."IV, Eds. W.Frankenburg et al., Acad.PressInc.,NY,1952,271
13. H. Schulz and H. D. Achtsnit, Revista Portuguesa de Quimica, 19 (1977)317
14. H. Schulz and M. Claeys, Appl. Cat. A, General 186 (1999)71
15. R. B. Anderson, K. Friedel and H. Storch, J. Chem. Phys., 19 (1951)313
16. H. Schulz, K. Beck and E. Erich, Fuel Proc. Techn., 18 (1988)293

Studies in Surface Science and Catalysis
J.J. Spivey, E. Iglesia and T.H. Fleisch (Editors)

Effect of CaO promotion on the performance of a precipitated iron Fischer-Tropsch catalyst

Dragomir B. Bukur*, Xiaosu Lang[(1)] and Lech Nowicki[(2)]

Department of Chemical Engineering
Kinetics, Catalysis and Reaction Engineering Laboratory
Texas A & M University
College Station, TX 77843-3122, USA

Three catalysts containing CaO promoter with nominal compositions 100 Fe/3 Cu/4 K/x Ca/16 SiO_2, where x = 0, 2 or 6, were synthesized and their reduction behavior was characterized under both isothermal and temperature programmed conditions. In fixed bed reactor studies of Fischer-Tropsch synthesis it was found that the catalyst activity decreased, whereas the gaseous hydrocarbon selectivity increased, with increasing amount of CaO promoter. Decrease in activity is attributed to both lower surface area and lower reducibility with increasing amount of CaO promoter.

During testing in a stirred tank slurry reactor it was found that the 100 Fe/3 Cu/4 K/2 Ca/16 SiO_2 catalyst was less active (about 15%) than the baseline catalyst (x = 0) and its deactivation rate was higher. At reaction pressure of 1.48 MPa, methane selectivity on the CaO promoted catalyst was higher than that of the baseline catalyst. Selectivity of the CaO promoted catalyst improved during testing at a higher reaction pressure (2.17 MPa) whereas selectivity of the baseline catalyst was essentially independent of reaction pressure.

1. INTRODUCTION

Slurry phase Fischer-Tropsch (F-T) synthesis is potentially economically viable route for conversion of natural gas, coal and other carbonaceous raw materials into transportation fuels and chemicals. Slurry processing provides the ability to more readily remove the heat of reaction, minimizing temperature rise across the reactor and eliminating localized hot spots. As a result of the improved temperature control, yield losses to methane are reduced and catalyst deactivation due to coking is decreased [1]. Capital cost of a slurry bubble column reactor (SBCR) is 20-40% lower than that of a fixed bed reactor [2,3]. A semi-commercial scale SBCR (2500 barrels of liquid products/day) has been in operation at Sasol since 1993 [3,4]. Improvements in the catalyst performance (activity, selectivity and/or stability) are needed to accelerate commercialization of slurry phase F-T technology.

---

* The author to whom the correspondence should be addressed. [(1)] Present address: Dr. X. Lang, Department of Chemical Engineering, University of Saskatchewan, Saskatoon, S7N 0W0 Canada; [(2)] Dr. L. Nowicki, Faculty of Process and Environmental Engineering, Technical University of Lodz, 90-924 Lodz, Poland.
This work was supported in part by the U. S. DOE under contract DE-AC22-94PC93069.

Several iron F-T catalysts (Fe/Cu/K/SiO_2) synthesized and tested at Texas A&M University (TAMU) were found to be more active than iron catalysts employed in the two most successful slurry phase tests conducted by Mobil [5] and Rheinpreussen [6]. Also, TAMU's catalysts had high selectivity to liquid and wax hydrocarbons [7-9]. In some of the older German preparation procedures basic oxides such as: MnO, MgO and CaO were used as promoters [10]. In particular, calcium oxide was used as a standard promoter in catalysts synthesized by Ruhrchemie (e.g. 100 Fe/5 Cu/8 CaO/30 kieselguhr/3-3.5 KOH and 100 Fe/5 Cu/10 CaO/6.5 SiO_2 in parts by weight). Dolomite (a mixture of CaO and MgO) was used as a support in catalysts prepared by Rheinpreussen (e.g. 100 Fe/0.1 Cu/80 dolomite/3 K_2CO_3). According to Koelbel [Ref. 10, p. 133] "the calcium oxide in the dolomite increased activity and possibly increased the molecular weight of the product". However, there were no data presented to quantify these statements. These features (higher activity and increased selectivity to higher molecular weight products) are very desirable, particularly if they could be realized with TAMU's improved iron F-T catalyst. The present study was undertaken with the objective to investigate the effect of addition of CaO on performance of one of the best TAMU's catalysts (100 Fe/3 Cu/4 K/16 SiO_2). Catalysts with two levels of CaO promotion were synthesized, characterized by BET surface area (SA) and pore volume measurements, and their reduction behavior was studied under temperature programmed and isothermal conditions. Their catalytic performance was determined first in a fixed bed reactor, and the CaO promoted catalyst with x =2 was selected for further testing in a stirred tank slurry reactor (STSR).

2. EXPERIMENTAL

2.1 Catalyst Synthesis Procedure and Characterization Methods

Catalysts with nominal composition 100 Fe/3 Cu/4 K/x Ca/16 SiO_2 (where: x = 0, 2 or 6) were synthesized by a three step procedure: preparation of the iron-copper precursor, incorporation of silicon oxide binder, and finally impregnation of the corresponding Fe-Cu-SiO_2 precursor [11,12]. Dried precursors were impregnated by incipient wetness method first with calcium acetate monohydrate followed by potassium bicarbonate (x = 2) or first with potassium bicarbonate, followed by impregnation with calcium acetate monohydrate (x = 6). The final step was to dry the catalyst at 120°C for 16 hours in a vacuum oven. The dried catalyst was calcined in air at 300°C for 5 h, and then crushed and sieved to a desired particle size range.

The BET surface area and pore volume measurements were obtained by nitrogen physisorption at 77 K using Micromeritics Digisorb 2600 system. Catalyst samples were degassed prior to each measurement. Temperature-programmed reduction (TPR) studies were performed using 5% H_2/95% N_2 as reductant in Pulse Chemisorb 2705 unit (Micromeritics Inc.). In a typical TPR experiment about 10 to 20 mg of catalyst was packed in a quartz reactor and heated in a flow of 5% H_2/95% N_2 (flow rate of 40 cm^3/min) from room temperature to 800-900°C at a heating rate of 20°C/min. Hydrogen consumption was monitored by change in thermal conductivity of the effluent gas stream. Isothermal reduction in thermogravimetric analysis (TGA) experiments was conducted using approximately 20 mg catalyst samples in a simultaneous TGA/DTA apparatus (TA Instruments, Model SDT 2960).

2.2 Reactor System and Operating Procedures

Experiments were conducted in a conventional downflow fixed bed reactor (1 cm inside diameter, 27 cm^3 effective bed volume) embedded in an aluminum block with a two-zone heater. Slurry phase F-T experiments were conducted in a 1 dm^3 reactor (Autoclave Engineers). The feed gas (premixed synthesis gas, >99.7% purity) flow rate in both systems was adjusted with a mass flow controller and passed through a series of oxygen removal, alumina and activated charcoal traps to remove trace impurities. After leaving the reactor, the exit gas passed through a series of high and low (ambient) pressure traps to condense liquid products. The reactants and noncondensible products leaving the ice traps were analyzed on an on-line gas chromatograph. High molecular weight hydrocarbons (wax), collected in a high pressure trap (fixed bed reactor) or withdrawn from the slurry reactor through a porous cylindrical sintered metal filter, and liquid products, collected in a low pressure ice trap, were analyzed by capillary gas chromatography. Detailed description of our reactor systems, product analysis system and operating procedures can be found elsewhere [7,13,14].

3. RESULTS AND DISCUSSION

3.1 Catalyst Characterization Studies

Surface area of the baseline catalyst was 291 m^2/g and its pore volume was 0.43 cm^3/g. Catalysts containing 2 pbw and 6 pbw of Ca per 100 pbw of Fe had surface areas of 190 m^2/g and 105 m^2/g, respectively, whereas the corresponding pore volumes were 0.36 cm^3/g and 0.30 cm^3/g, respectively. The observed decrease in the BET surface area and the pore volume with increasing amount of CaO promoter is due to pore filling and/or partial blocking of small pores.

The effect of calcium oxide addition (x = 6) on reducibility of the baseline catalyst (x = 0) in temperature programmed mode is shown in Fig. 1. Both profiles show two distinct reduction peaks which are characteristic of the two-step reduction process of Fe_2O_3. Results show that the onset of reduction of iron oxides occurs earlier on the baseline catalyst (dominant peaks at 306°C and 530°C for x = 0 vs. 321°C and 609°C for x = 6), and that the reduction process is more facile (sharp narrow peaks for the baseline catalyst versus broad peaks for x = 6 catalyst). The second stage of the reduction process is much slower on both catalysts.

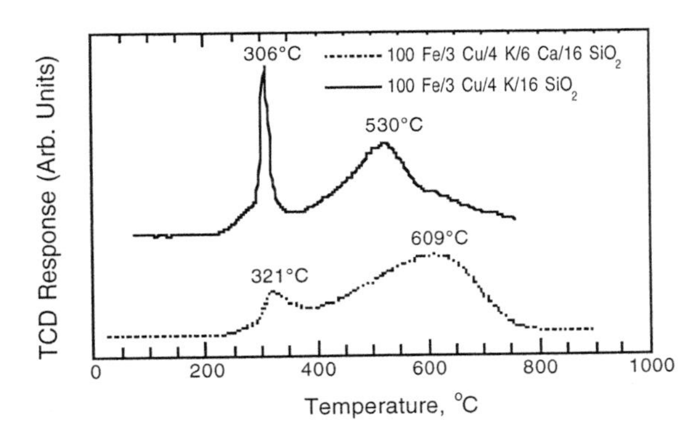

Fig. 1 Effect of CaO promotion on the reduction behavior

It has been postulated that the first peak corresponds to reduction of Fe_2O_3 to Fe_3O_4 (magnetite), whereas the second broad peak corresponds to subsequent reduction of Fe_3O_4 to metallic iron [15,16]. Results from hydrogen consumption measurements show that the degree of reduction after the first stage of the reduction process was 16% (for x = 6) and 23% (for x = 0). These values are higher than the theoretical amount corresponding to the reduction of Fe_2O_3 to Fe_3O_4 (11.1% degree of reduction). This is probably due to the fact that a small portion of iron oxide was reduced to metallic iron during the first stage of reduction (temperatures up to 400°C), whereas the bulk of iron oxides were reduced during the second stage of reduction. These results indicate that there are two types of iron sites with different reduction properties.

Isothermal reduction experiments were conducted in TGA unit with pure hydrogen as reductant at 280°C for 8 hours. The final degree of reduction of the baseline catalyst was significantly higher than that of the catalyst containing 6 pbw of Ca per 100 pbw of Fe (83% vs. 68%). This confirms that the addition of CaO inhibits reduction of iron in the baseline catalyst.

3.2 Fixed Bed Reactor Tests

The baseline catalyst and the two CaO promoted were tested in a fixed bed reactor to determine their activity and selectivity during Fischer-Tropsch synthesis. About 3 g of the catalyst (32 to 60 mesh) diluted 1:8 by volume with glass beads (of the same size as the catalyst) was used in fixed bed reactor tests. Prior to F-T synthesis the catalysts were reduced in-situ with hydrogen at atmospheric pressure, 7500 cm^3/min and 240°C for 2 h, and then gradually brought (conditioning period of 20-30 h) to desired process conditions (250°C, 1.48 MPa, 2 Nl/g-cat/h, $H_2/CO = 0.67$). Test duration was 120-140 h including the conditioning period.

Performance of CaO promoted catalysts (runs FA-1525 and FB-1515) is compared with that of the baseline catalyst (run FA-1605) in Fig. 2. Activity (syngas conversion) of the catalyst with a lower amount of CaO promoter was similar to that of the baseline catalyst, whereas the activity of the catalyst containing 6 pbw of Ca per 100 pbw of Fe was markedly lower (Fig. 2a). This is attributed to significantly lower surface area of the latter catalyst in comparison to the other catalysts (Table 1) and its lower degree of reduction. Hydrogen to CO usage ratios in all three tests were similar (about 0.6), indicating similar water-gas-shift activities. Methane (Fig. 2b) selectivity (expressed as carbon atom percent of hydrocarbons and oxygenates produced) was slightly higher on the CaO promoted catalysts, than on the baseline catalyst.

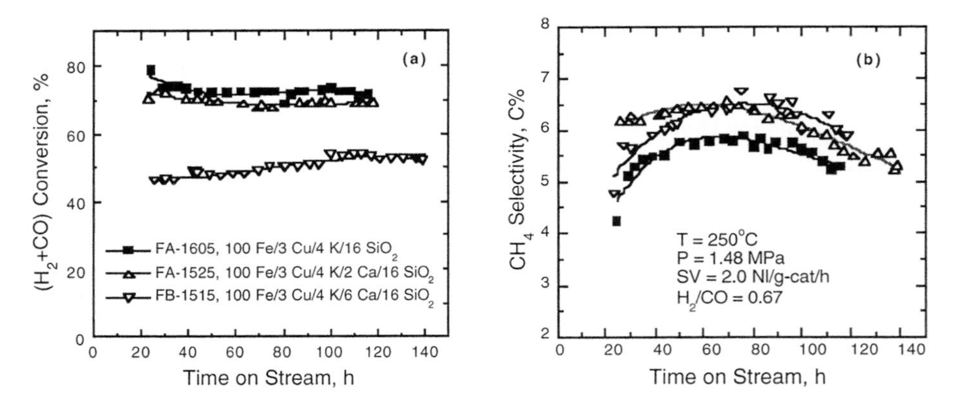

Fig. 2 Comparison of catalyst performance in fixed bed reactor tests.

3.3 Stirred Tank Slurry Reactor Tests

The baseline catalyst and the catalyst containing 2 pbw of Ca per 100 pbw of Fe were evaluated in a STSR to determine the impact of CaO promotion on the long term catalyst stability (deactivation), activity and selectivity. For these two tests about 10 g of the catalyst (smaller than 53 μm) was loaded into a slurry reactor filled with Durasyn 164 oil (a hydrogenated 1-decene homopolymer, ~C_{30}) to form 3.3 wt% slurry. Catalysts were reduced in-situ with hydrogen at 0.8 MPa, 7500 cm^3/min, and 240°C for 2 h. Initially, catalysts were tested at baseline process conditions of 260°C, 1.4 Nl/g-cat/h, 1.48 MPa and $H_2/CO = 0.67$.

Run SA-1665 with the baseline catalyst lasted about 500 h, but only the data from first 400 h are shown in Fig. 3. Figure 3a shows that the CaO promoted catalyst (SB-3115) had lower activity, i.e. lower syngas conversion, than the baseline catalyst. The CaO promoted catalyst started to deactivate around 220 h on stream, whereas the baseline catalyst was fairly stable during 400 h of testing. Usage ratios in both tests were between 0.57 and 0.61, indicative of a high water-gas-shift activity.

The CaO promoted catalyst had higher methane (Fig. 3b) selectivity during testing at 1.48 MPa. After the pressure was increased to 2.17 MPa (while proportionally increasing the gas space velocity to 2.0 Nl/g-cat/h) in run SB-3115, methane selectivity started to decrease (170 - 220 h), and then remained stable (220 - 350 h). Hydrocarbon selectivity of the CaO promoted catalyst depends on the reaction pressure, whereas gaseous hydrocarbon selectivity was independent of pressure in run SA-1665 with the baseline catalyst.

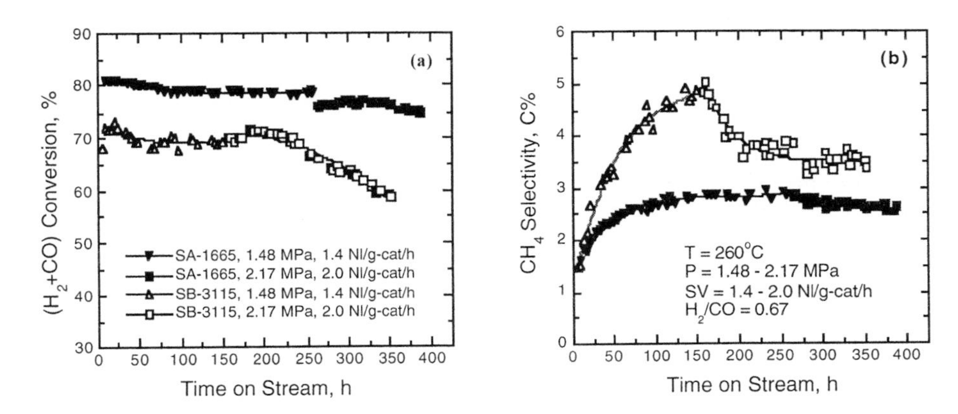

Fig. 3 Comparison of catalyst performance in stirred tank slurry reactor tests.

The addition of CaO promoter resulted in a decrease of the total olefin content and in increase of the 2-olefin content, at the reaction pressure of 1.48 MPa (Fig. 4). Change of pressure from 1.48 MPa to 2.17 MPa (at constant P/SV ratio) in run SB-3115, did not have an effect on the total olefin selectivity, whereas the 2-olefin selectivity decreased and became similar to that obtained in run SA-1665 with the baseline catalyst. The observed trends in olefin selectivity with increasing carbon number are typical for iron F-T catalysts, and have been attributed to either secondary hydrogenation and isomerization reactions of 1-olefins [17,18] or diffusion enhanced 1-olefin readsorption [19,20].

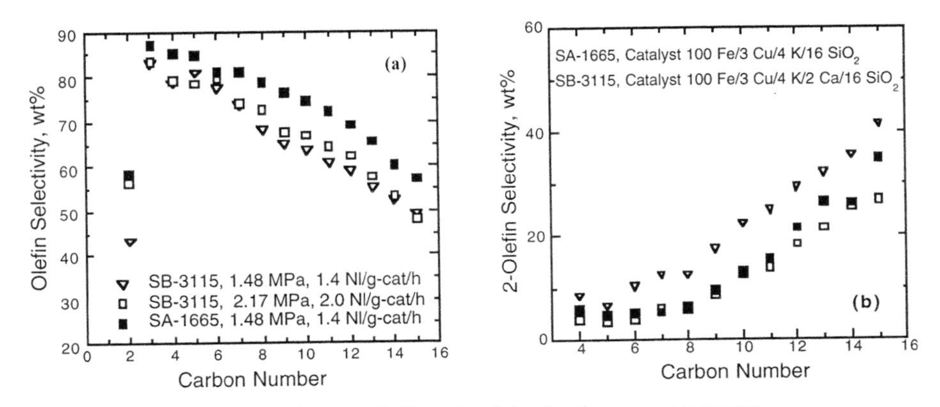

Fig. 4 Effect of CaO promotion on olefin selectivity in slurry reactor tests.

Carbon number distribution in run SB-3115 was fitted with a three parameter model of Huff and Satterfield [21]. The parameter values were: $\alpha_1 = 0.71$, $\alpha_2 = 0.90$ and $\beta = 0.90$ (α_1 and α_2 are chain growth probability factors on two types of sites, and β is a fraction of type 1 sites). The corresponding parameter values for the baseline catalyst were: $\alpha_1 = 0.69$, $\alpha_2 = 0.94$ and $\beta = 0.87$. Catalysts with $\alpha_2 > 0.92$ are referred to as "high-alpha" catalysts, and are suitable for production of high quality diesel fuels via hydrocracking of the F-T wax.

The present study shows that the addition of CaO did not improve performance of the baseline catalyst. Catalysts used in the earlier German work had substantially different catalyst composition than ours, and preparation and activation (pretreatment) methods were significantly different than those employed by us. Our results show that the CaO containing catalysts may be suitable for operation at higher reaction pressures, provided that their stability could be improved. The composition of CaO containing catalysts has not been optimized, and their performance (activity, selectivity and/or stability) may be improved with the use of different pretreatment procedures. The latter has been successfully demonstrated for the baseline catalyst [9,22] and several other precipitated iron F-T catalysts tested in our laboratory [11,14,23].

REFERENCES

1. H. Kölbel, H. and M. Ralek, Catal. Rev. - Sci. Eng., 21 (1980) 225.
2. G. N. Choi, S. J. Kramer, S. S. Tam and J. M. Fox. III, Proc. of the Coal Liquefaction and Gas Conversion Contractors Review Conference, U. S. Department of Energy, Pittsburgh, 1995; p. 269.
3. B. Jager and R. Espinoza, Catal. Today, 23 (1995) 17.
4. B. Jager, Stud. Surf. Sci. Catal., 107 (1997) 219.
5. J. C. W., Kuo, Final Report on US. DOE Contract No. DE-AC22-83PC60019, Mobil Res. Dev. Corp., Paulsboro, New Jersey, 1985.
6. H. Kölbel, P. Ackerman and F. Engelhardt, Proc. Fourth World Petroleum Congress, Section IV/C, Carlo Colombo Publishers, Rome, 1955, p. 227.
7. D. B. Bukur, L. Nowicki and X. Lang, Chem. Eng. Sci., 49 (1994) 4615.
8. D. B. Bukur and X. Lang, Stud. Surf. Sci. Catal., 119 (1998) 113.
9. D. B. Bukur and X. Lang, Ind. Eng. Chem. Res., 38 (1999) 3270.
10. R. B. Anderson, in Catalysis Vol. 4 (P. H. Emmett, Editor), Van Nostrand-Reinhold, New York, 1956, p. 29.
11. D. B. Bukur, X. Lang, J. A. Rossin, W. H. Zimmerman, M. P. Rosynek, E. B. Yeh and C. Li, Ind. Eng. Chem. Res. 28 (1989) 1130.
12. D. B. Bukur, X. Lang, D. Mukesh, W. H. Zimmerman, M. P. Rosynek, and C. Li, Ind. Eng. Chem. Res., 29 (1990) 1588.
13. D. B. Bukur, S. A. Patel and X. Lang, Appl. Catal., 61 (1990) 329.
14. D. B. Bukur, M. Koranne, X. Lang, K. R. P. M. Rao, G. P. Huffman, Appl. Catal., 126 (1995) 85.
15. R. Brown, M. E. Cooper, and D. A. Whan, Appl. Catal., 3 (1982) 177.
16. I. S. C. Hughes, J. O. H. Newman and G. C. Bond, Appl. Catal., 30 (1987) 303.
17. R. A. Dictor and A. T. Bell, J. Catal., 97 (1986) 121.
18. H. Schulz and H. Gokcebay, in Catalysis of Organic Reactions (J. R. Kosak, Editor), Marcel Dekker, New York, 1984, p. 153.
19. R. J. Madon, S. C. Reyes and E. Iglesia, J. Phys. Chem., 95 (1991) 7795.
20. R. J. Madon and E. Iglesia, J. Catal., 139 (1993) 576.
21. G. A. Huff, Jr. and C. N. Satterfield, J. Catal., 85 (1984) 370.
22. D. B. Bukur, X. Lang and Y. Ding, Appl. Catal., 186 (1999) 255.
23. D. B. Bukur, L. Nowicki, R. K. Manne, X. Lang, J. Catal., 155 (1995) 366.

Studies in Surface Science and Catalysis
J.J. Spivey, E. Iglesia and T.H. Fleisch (Editors)

PARTIAL OXIDATION OF METHANE AT HIGH TEMPERATURES OVER PLATINUM AND RHODIUM MONOLITH CATALYSTS.

F. Monnet, Y. Schuurman, F.J. Cadete Santos Aires, C. Mirodatos.

Institut de Recherches sur la Catalyse (CNRS-UPR 5401),
2 avenue A. Einstein, 69626 Villeurbanne Cedex, France.
Mirodato@catalyse.univ-lyon1.fr

ABSTRACT

Platinum and rhodium monoliths prepared by chemical impregnation were tested in the partial oxidation of methane at temperatures between 800 and 1100°C. The influence of contact time, nature of the metal and of the support was investigated. High conversions and selectivities to carbon monoxide and hydrogen can be obtained for contact times long enough to ensure total conversion of oxygen. Platinum is found intrinsically less active and selective than rhodium. A silicon nitride washcoat has a beneficial effect on metal sintering and volatilization. In the case of rhodium, a thin film of silicon oxide is formed around the metal, stabilizing it but at the expenses of surface accessibility.

INTRODUCTION

Syngas can be produced by partial oxidation of methane (POM) through a direct route at high temperatures and at short contact times (below millisecond) [1] or indirect route (including secondary reforming reactions) at longer contact times [2].

Whatever the applied mechanism, monoliths are interesting catalyst supports for POM due to their low flow resistance [2] and their thermal stability. Monoliths are generally used with a washcoat to increase the surface area and to improve mass and heat transfer [3]. Among the various washcoats tested in POM, ZrO_2 was found to give the best results in terms of syngas selectivity [3].

Noble metals (Pt, Rh, Ru, Pd) are generally used as the active phase with monolith supports [1-3]. However, the main drawback is their stability. Heitnes Hofstad et al. [4] reported on the loss of platinum by volatilization. The addition of a washcoat to a monolith may prevent metal loss by strengthening the metal-support interactions. In a recent study [5] we found that platinum supported on silicon nitride powder (Si_3N_4) demonstrated exceptional stability for the partial oxidation of methane.

In the present work the catalytic performance and stability of platinum and rhodium supported on cordierite monoliths are investigated, focussing at the effect of the silicon nitride washcoat.

EXPERIMENTAL

Catalyst preparation. A commercial cordierite from Corning (400 cells.in^{-2}) was used without further modification (referred to as cordierite) and the same cordierite was coated with α-Si_3N_4 (referred to as Si_3N_4-cord) by chemical impregnation.

These supports were then impregnated with an aqueous salt solution of H_2PtCl_6 or $Rh(NO_3)_3$, dried at 120°C overnight under vacuum conditions, then calcined in air at 500°C and reduced under hydrogen flow for 5 h at 500°C. Table 1 summarizes the characteristics of the catalysts prepared on cordierite and on silicon nitride.

Catalyst characterization. Before and after catalytic testing, the catalysts were characterized by Transmission Electron Microscopy (TEM) equipped with an Energy Dispersive X-ray

Spectroscopy (EDX) probe for microstructure study and by Inducted Coupled Plasma spectroscopy (ICP) for metal content analysis.

Table 1. Characterization data for Pt and Rh monolith catalysts (metal weight content, average diameter of metal particles determined from TEM and percent of metal loss after reaction).

catalyst	Me wt.%	d_{Me} before (nm)	d_{Me} after (nm)	% Me loss
Pt/cordierite	1.32	7.3	19.9	84
Pt/Si_3N_4-cord	1.27	7.5	7.8	60
Rh/cordierite	0.22	6.0	19	55
Rh/Si_3N_4-cord	0.25	8.0	8.2	12

Steady state testing experiments. Experiments were performed in a quartz tube reactor (11 mm internal diameter) in which the monolith (6 mm length) was held between two pieces of pure cordierite that acted as thermal screens. The feed consisted of a $CH_4/O_2/N_2$ (2/1/10) mixture. The product gas composition was analyzed by gas chromatography (HP 5890 series II equipped with two thermal conductivity detectors and four columns in a series / parallel arrangement).

Experiments were performed at atmospheric pressure in the temperature range 800 to 1100°C. The contact time was varied from 5 to 40 ms by changing both the total flow rate and the monolith length.

RESULTS

Under all experimental conditions, the products consisted of carbon monoxide, carbon dioxide, hydrogen and water. C_2 hydrocarbons were never detected and the deficiency in the mass balance was limited to 2 and 5% for carbon and oxygen, respectively.

Influence of contact time. The contact time is defined as the ratio between the volume of void in the catalytic zone and the volumetric total flow rate.
Figure 1 shows the effect of contact time on the catalytic performances of both 1.32 wt.% Pt/cordierite and the 0.22 wt.% Rh/cordierite catalysts at 900°C. Increasing the contact time results in increasing methane and oxygen conversions as well as in increasing syngas selectivity. Rhodium displays a better activity and selectivity than platinum, independent of the contact time.

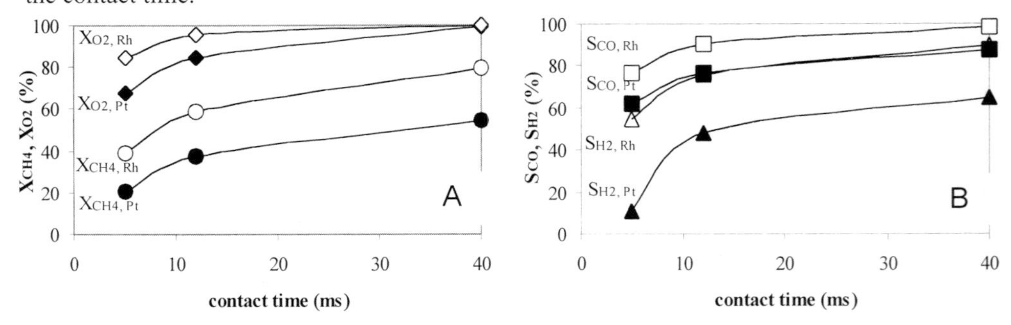

Figure 1 : Changes in methane and oxygen conversions (A) and syngas selectivity (B) as a function of contact time at 900°C for the 1.32 wt.% Pt/cordierite (full symbols) and the 0.22 wt.% Rh/cordierite (open symbols) catalysts.

Thus, for the 1.32 wt.% Pt/cordierite catalyst the maximum methane conversion obtained at a contact time of 40 ms is only 54.7% whereas it attains 79.1% for the 0.22 wt.% Rh/cordierite catalyst. The selectivities to carbon monoxide and hydrogen are 98.2 and 89.9% over rhodium compared to 87.5 and 65.3% over platinum, respectively, when oxygen conversion attains 100%. These trends for both metals simply derives from the fact that at complete oxygen conversion , the non selective conversion of methane is four times lower than the fully selective one. At a lower contact time of 5 ms, the conversion of oxygen is no longer complete and a much lower syngas selectivity is observed, especially for hydrogen over platinum (10.9%).

Influence of temperature. Figure 2 shows the performance of the 1.32 wt.% Pt/cordierite and the 0.22 wt.% Rh/cordierite catalysts as a function of temperature at a fixed contact time of 12 ms. Methane and oxygen conversions and syngas selectivity increase slowly with temperature for both catalysts. Over the whole temperature range activity and selectivity for rhodium are better than for platinum.

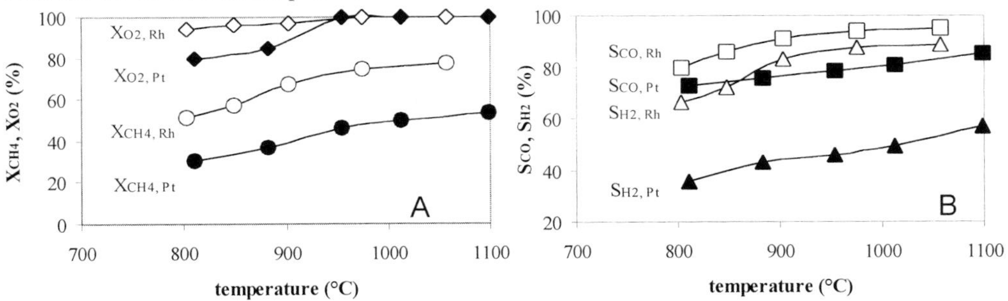

Figure 2 : Changes in methane and oxygen conversion (A) and selectivity in carbon monoxide and hydrogen (B) versus temperature for 1.32 wt.% Pt/cordierite (full symbols) and 0.22 wt.% Rh/cordierite (open symbols) catalysts, at τ=12 ms, $CH_4/O_2/N_2$=2/1/10.

Effect of a Si_3N_4 washcoat. Figure 3 shows the performances of the 1.27 wt.% Pt/Si_3N_4-cord and 0.25 wt.% Rh/Si_3N_4-cord catalyst as a function of temperature at a contact time of 12 ms. Like on cordierite, increasing temperature results in increasing syngas selectivity, especially when oxygen is fully consumed. However, while over platinum methane conversion increases from 25 to 60% by increasing temperature from 800 to 900°C, it decreases over rhodium from 80% to 75%. This trend reveals a deactivation of the rhodium monolith.

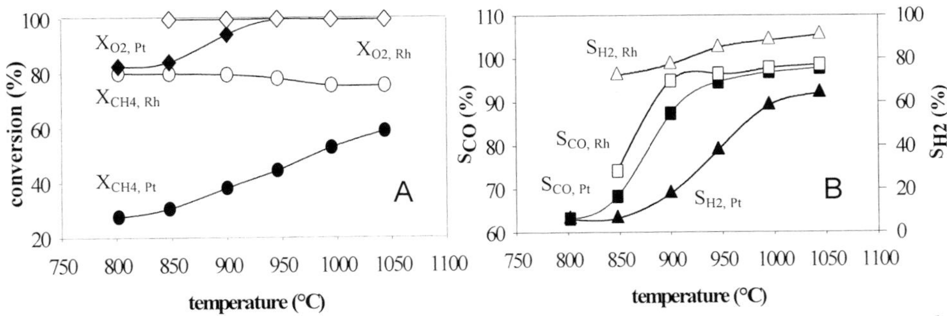

Figure 3 : Changes in methane and oxygen conversions (A) and of selectivities in carbon monoxide and hydrogen (B) over 1.27 wt.% Pt/Si_3N_4-cord (full symbols) and 0.25 wt.% Rh/Si_3N_4-cord (open symbols) catalysts versus temperature, at τ=12 ms, $CH_4/O_2/N_2$=2/1/10.

Catalysts characterization. TEM analysis before reaction shows that platinum and rhodium on cordierite exhibit a similar but rather heterogeneous particle size distribution with a 6 and 7.3 nm average particle diameter (see for example Figure 4A for the Rh/cordierite). With a wash-coat of Si_3N_4, a slightly higher average particle size is observed (8 nm) as shown in Figure 4C for the Rh/Si_3N_4 catalyst.

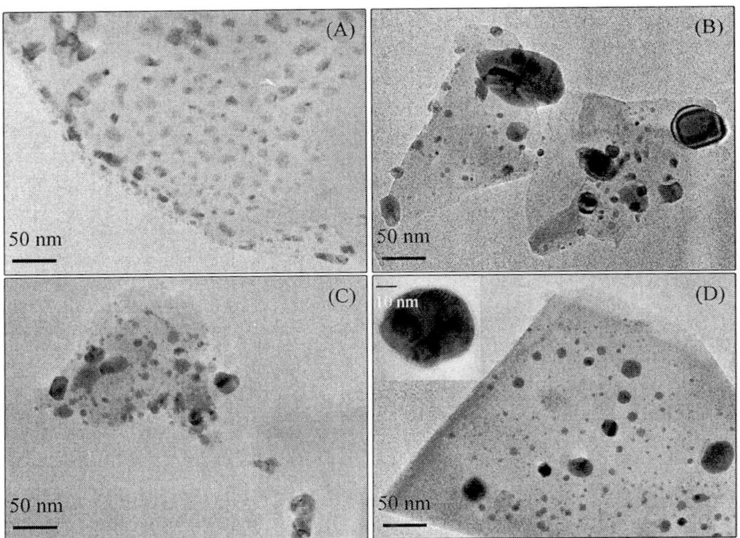

Figure 4 : TEM micrographs of rhodium catalysts before and after reaction. (A) 0.22% Rh/cordierite before reaction, (B) idem after reaction, (C) 0.25% Rh/Si_3N_4 before reaction, (D) idem after reaction. Insert in Figure 3d shows a rhodium particle covered with a nanometer-thick amorphous layer of $SiO_{1.5}$.

After reaction, the particle size has strongly increased and particles larger than 50 nm are observed on the cordierite samples (see Figure 4B for Rh/cordierite). In contrast, the average particle diameter remains quite constant on Si_3N_4-cord support and equal to 8.2 nm (compare Figures 4C and D for Rh/Si_3N_4). However, a thin film of amorphous $SiO_{1.5}$ over rhodium particles was detected by TEM/EDX (see the insert in Figure 4D). This layer was not detected over platinum.

Chemical analysis revealed that important metal losses occur during the reaction, but to an extent strongly influenced by the nature of the support. A platinum loss of 84% is observed for the Pt/cordierite catalyst, while this loss is limited to 60% when a wash-coat of Si_3N_4 was applied. For the rhodium catalyst, the loss of metal, much less important, varied from 55% for the cordierite to 12% for the silicon nitride.

DISCUSSION

Mechanistic aspect

Figures 2 and 3 show that increasing the temperature from 800 to 1100°C only results in a slight increase of the methane conversion or even decreasing it (for Rh/cordierite catalysts in Fig. 3A). This indicates that the reaction is governed i) by catalyst deactivation and possibly ii) by mass and/or heat transfer. Interpretation of the results is further complicated by complete or nearly complete oxygen conversion. For the former, the important metal loss reported in Table 1 may explain the unexpected conversion decrease with temperature. Inversely, in the case of Pt/Si_3N_4-cord which presents a limited loss of the active phase and a stable particle size, the increase of the methane conversion with temperature seems to correspond more to a kinetic regime as long as the oxygen conversion is not complete.

As presented in the introduction, the synthesis gas formation can be regarded as a one (direct) or a two step (indirect including total combustion followed by reforming) mechanism. As previously reported for Pt/Si_3N_4 powder in a shallow bed configuration [5,6] and Pt gauze [4,7-8], the syngas selectivity *decreases* with increasing contact time at very short contact times (below a millisecond). With a 2 mm monolith length, i.e., under conditions close to the gauze configuration, similar catalytic performances to those obtained on powder or gauze were found. This behavior is characteristic of the direct route. In this case oxygen conversion is not complete and primary CO and especially H_2 are rapidly oxidized along the catalytic zone at longer contact times.

At higher contact times achieved over 6 mm monoliths (12 to 40 ms as shown in Fig.1), oxygen conversion is complete. For this case, the secondary dry and steam reforming reactions take place and syngas selectivity strongly increases. This behavior was also reported under similar conditions (900°C, contact times between 5-40 ms) by Holmen et al [2].

Platinum versus rhodium

A comparison between intrinsic activity of platinum and rhodium catalysts can be attempted in the case of monoliths containing a wash-coat of Si_3N_4, by considering their rather stable metal dispersion and accounting for the metal losses measured after reaction (Table 1).

The ratio of the intrinsic activities can be evaluated by taking the ratio of the corresponding conversion corrected for the different metal contents and dispersions. The dispersion can be calculated from the average diameters reported in Table 1. Thus the concentration of platinum surface atoms was found about 2.2 times higher than for rhodium after reaction and by considering the methane conversions reported in Fig. 3A, the intrinsic activity of rhodium was found between 3 and 6 times higher than the one of platinum. This observation could hardly be compared to literature data under similar conditions since average particle sizes are not reported [1,3].

It is also worthy to note that the syngas selectivity is strongly affected by the nature of the metal. Thus the selectivity towards hydrogen is significantly higher over rhodium than over platinum. This trend already reported in the literature can be related to the faster reaction between surface oxygen and dissociated hydrogen to form surface hydroxyl groups (and therefore water) over platinum rather than over rhodium [1]. This could be due to the higher bond strength for Rh-O than for Pt-O, as deduced from their respective activation energies for oxygen desorption (resp. 70 and 52 kcal/mol [1]). This could also explain the significantly lower metal losses for rhodium as compared to platinum, as rhodium oxide is more stable and less volatile than platinum oxide.

Si_3N_4 support effect

As expected from the above discussion on mechanistic aspects stressing that the reaction was essentially governed by mass and/or heat transfer and eventually by catalyst deactivation, the nature of the support (with or without a silicon nitride wash-coat) was found not to improve the catalyst activity neither the syngas selectivity. In contrast, a strong support effect was evidenced on the catalyst stability. Thus, as shown in Table 1, the significant particle sintering and metal loss observed on cordierite monoliths are considerably reduced after addition of a Si_3N_4 wash-coat. However, in the case of Rh/Si_3N_4-cord, the slight deactivation observed in Figure 3 could also arise from the formation of an amorphous silicon oxide layer around the metal particle in addition to the metal loss. TEM analysis seems to indicate that the particles could either diffuse into the support or be decorated by the support, partly decomposing under reaction conditions. Such a strong metal/support interaction limiting metal losses was already revealed for Rh/Si_3N_4 powders by a high Rh $3d_{5/2}$ XPS binding energy indicating that the metal was in a slightly oxidized form [9]. This could arise from a reaction with the support to form a Me/Si interface as it has already been observed on silica supported Ni catalysts at high temperatures [10]. Note that this metal support interaction was not observed on platinum, probably due to the higher volatility of PtO_2 as compared to Rh_2O_3.

CONCLUSION

Both the nature of the noble metal deposited on cordierite monoliths and the addition of a wash-coat have been found to play key roles in the partial oxidation of methane at temperature ranging from 800 to 1100°C. Rhodium presents an intrinsic activity 3-6 times higher than platinum. Under most conditions prevailing in monoliths, i.e. involving relatively long contact times, the production of syngas is shown to occur via secondary reforming reactions.

Addition of a wash-coat of Si_3N_4 markedly limits metal losses and particle sintering under POM conditions. In the case of rhodium, a thin layer of silica arising from a partial decomposition of silicon nitride is formed around metal particles that contributes to stabilizing them by reinforcing metal/support interaction but at the expenses of surface accessibility. Further researches i) by combining platinum and rhodium properties and ii) by stabilizing the silicon nitride support by calcination at temperatures up to 1400°C before impregnation (the allotropic transformation between α and β phases occurs around 1200°C) are in progress to optimize the promising performance of these new materials.

ACKNOWLEDGEMENTS

This research was supported by the CNRS program "Catalyse et Catalyseurs pour l'Industrie et l'Environement". Dr. J.C. Bertolini is gratefully acknowledged for fruitful discussions as well as Mrs. S.Ramirez and Dr. E.Rogemond (SUNKISS) for providing Si_3N_4 materials.

REFERENCES

[1] D.A. Hickman and L.D. Schmidt, AIChE J., **39** (1993) 1164.
[2] K. Heitnes Hofstad, T. Sperle, O.A. Rokstad and A. Holmen, Catal. Lett., **45** (1997) 97.
[3] A.S Bodke, S.S. Bharadwaj and L.D. Schmidt, J. Catal., **179** (1998) 138.
[4] K. Heitnes Hofstad, O.A. Rokstad and A. Holmen, Catal. Lett., **36** (1996) 25.
[5] F. Monnet, Y. Schuurman, F.J. Cadete Santos-Aires, J.C. Bertolini, C. Mirodatos, Catal. Today **64** (2001) 51.
[6] F. Monnet, Y. Schuurman, F.J. Cadete Santos Aires, J.C. Bertolini and C. Mirodatos, C.R. Acad. Sci., **3** (2000) 577.
[7] M. Fathi, F. Monnet, Y. Schuurman, A. Holmen, C. Mirodatos, J. Catal., **190** (2000) 439.
[8] C.R.H. de Smet, M.H.J.M. de Croon, R.J. Berger, G.B. Marin and J.C. Schouten, Appl. Catal., **187** (1999) 33.
[9] F. Monnet, PhD thesis, University of Lyon (2000).
[10] H. Praliaud and G.A. Martin, J. Catal., 72 (1981) 394.

Studies in Surface Science and Catalysis
J.J. Spivey, E. Iglesia and T.H. Fleisch (Editors)

The Promoting Effect of Ru and Re Addition to Co/Nb_2O_5 catalyst in the Fischer-Tropsch Synthesis

F.Mendes[1], F.B.Noronha[2], R.R.Soares[3], C.A.C. Perez[1], G.Marcheti[4] and M. Schmal[1*]

[1] NUCAT-PEQ-COPPE, C.P.68502, CEP 21941 – Brazil; schmal@peq.coppe.ufrj.br
[2] Instituto Nacional de Tecnologia, CEP 20081-310 - Brazil
[3] Faculdade de Engenharia Química / UFU – Brazil
[4] Faculdad de Quimica, Universidad Litoral - Argentina

The promoting effect of Ru and Re addition to Co/Nb_2O_5 catalyst in the Fischer-Tropsch Synthesis was studied. The catalysts were characterized by TPR, XRD and H_2 chemisorption. The Re and Ru addition increased both the activity and the C_5^+ selectivity by different mechanisms. Re prevented the cobalt agglomeration whereas Ru increased the Co^o site density. The Ru addition to Co/Nb_2O_5 catalyst promoted diesel selectivity in the F.T.S. while the Re addition promoted the diesel as well as the gasoline one.

1. Introduction

Natural gas has becoming more attractive as an energy source due to the increasing prices of oil, the huge reserves of gas and environmental problems. This situation has driven many researches to develop more economic processes of upgrading the natural gas to higher hydrocarbon fuels. The Fischer-Tropsch (FT) synthesis is an alternative route to produce liquid fuels from synthesis gas.

Cobalt based catalysts have been widely used in the CO + H_2 reaction for the production of hydrocarbons (1). However, the main problem of the F-T synthesis is the wide range of product distribution by the conventional FT catalyst (2). In order to overcome the selectivity limitations and to enhance the catalyst efficiency in CO hydrogenation, several approaches have been studied, such as the use of reducible support (3) and the addition of a second metal (4).

Niobia (3,4) and titania-supported (5,6) cobalt catalysts showed a high selectivity toward long chain hydrocarbons in the FT synthesis after reduction at high temperature. These results might be explained by the formation of new sites during high temperature reduction.

The addition of noble metals to supported-cobalt catalyst has also affected the activity and selectivity of CO hydrogenation. Kappor et al. (7) reported that an increase in the C^{5+} hydrocarbon selectivity was observed on the CO hydrogenation, after addition of Ru, Pd or Pt to the Co/Al_2O_3 catalyst. Iglesia et al. (5,6) verified that Co-Ru bimetallic interactions were able to increase the reaction rate and selectivity towards C_5^+ hydrocarbons in the FT synthesis and to improve catalysts regeneration and reducibility. This synergetic effect was mainly related to an increase of the cobalt oxide reduction, providing a higher metallic site density. In the case of Co-Re bimetallic catalysts, the addition of rhenium prevent the cobalt sinterization (6). Recently, we observed that the presence of rhodium increased the C_5^+ selectivity and decreased the methane formation on Rh-Co/Nb_2O_5 catalysts (4). This promoting effect was attributed to an intimate contact between rhodium particles spread over a thin layer of cobalt ions.

The main goal of this work is to investigate both the effect of Ru and Re addition to Co/Nb_2O_5 catalyst and the role of niobium oxide support on the selectivity of CO hydrogenation.

2. Experimental

The Nb_2O_5 support was obtained by calcination of niobic acid (CBMM) in air at 823K, for 2h. The catalysts were prepared by incipient wetness impregnation or coimpregnation of the support with an aqueous solution of the precursor salts (cobalt nitrate, ruthenium and rhenium chloride. These catalysts contained 5 wt.% Co and 0.6 wt.% of Ru or Re. After impregnation, the samples were dried at 393K for 16h and calcined in air at 673K for 2h. TPR experiments were performed in a conventional apparatus. The samples were reduced with a mixture containing 5%H_2/N_2 at a heating rate of 10K/min from 298 to 1273K. X-Ray diffraction (XRD) measurements were carried out in a Rigaku Dmax 2200 PC diffractometer with an attached Anton Paar XRK 900 reactor . The data for the calcined and reduced catalysts were recorded using Cu K_α radiation (40 kV, 40 mA) and a secondary graphite monochromator. The X ray diffractogram was scanned with a step size of 0.05° (2θ) from 5° to 100° (2θ) and counting time of 1s per step. Crystallite sizes were estimated from the integral breadth of the Co(111) line using Scherrer equation. In order to calculate the crystallite size, the XRD pattern of the support was subtracted from the diffractogram of the reduced catalyst. Quantitative XRD phase analysis was carried out to determine the reduction degree by using the Fullprof software (8), following the procedure proposed by Hill and Howard (9). The weight fraction of each phase was calculated from the respective scale factor. The parameters refined for each phase were the scale factor, the lattice constants and the peak width parameter. Structural data for Co_3O_4 and metallic cobalt were taken from Wyckoff (10) and for t-Nb_2O_5 from Kato and Tamura (11). $CoNb_2O_6$ was used as an isomorphous phase to $FeTa_2O_6$ (PDF 83-0588) (10). The hydrogen chemisorption was measured in a volumetric adsorption system. The catalyst was reduced under H_2 at 773K, for 16h. Then, the samples were evacuated for 9h at reduction temperature and cooled to adsorption temperature under vacuum. Total adsorption isotherm was determined at 448K. The CO hydrogenation was carried out in a stainless steel microreactor (m_{cat} = 200mg) at 493K and 0.1 MPa using a mixture of H_2 / CO = 2.0 at differential conditions. Before the reaction, the samples were reduced under hydrogen flow at 773K for 16h. The products were analyzed by on line gas chromatography (Varian 3400, FID and TCD).

3. Results and discussion

3.1. Catalyst Characterization

The TPR profiles of niobia-supported catalysts are shown in Figure 1. The reduction profile of Co/Nb_2O_5 catalyst exhibited a shoulder at 643K and a broad peak at

754K. It has been reported the presence of Co_3O_4 particles and Co^{2+} species on the niobia-supported cobalt catalysts (4). According to the literature, the shoulder around 643K could be attributed to the reduction of Co_3O_4 particles, whereas the peak at 754K corresponds to the reduction of Co^{2+} either formed by reduction of Co_3O_4 and Co^{2+} species interacting with niobia. DRS and XPS analyses revealed that the Co^{2+} species were represented by cobalt niobate (4). Furthermore, TPR and magnetic measurements showed that the hydrogen uptake at high temperature was also related to a partial reduction of niobia. Therefore, TPR analyses can not be used to quantify the reduction degree of niobia supported catalysts. The TPR profile of Re/Nb_2O_5 catalyst exhibited a small peak around 580 K and a high hydrogen consumption above 1000K. The Ru/Nb_2O_5 catalyst presented two peaks at 450 and 480 K and a H_2 uptake above 1000 K, similar to that observed on Re/Nb_2O_5 catalyst.

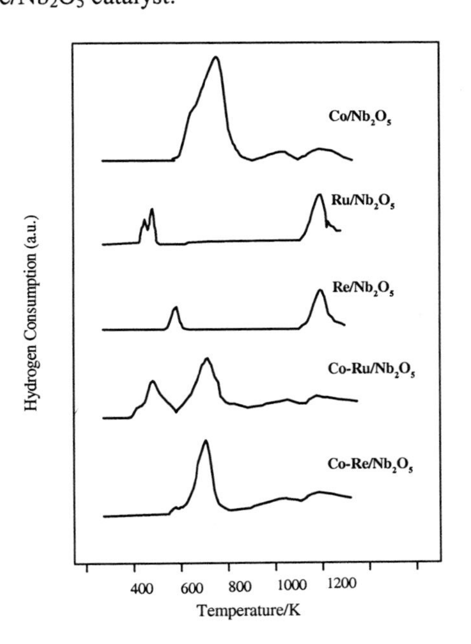

Figure 1- TPR profiles of the niobia-supported catalysts.

The ruthenium addition to Co/Nb_2O_5 catalyst promoted the reduction of Co_3O_4 particles since the hydrogen consumption at low temperature (400-600K) was higher than the one on the monometallic ruthenium catalyst. On the other hand, the hydrogen uptake attributed to cobalt oxide reduction was not shifted to lower temperature on the $Re\text{-}Co/Nb_2O_5$ catalyst. These results were in agreement with the literature. Hoff (12) also observed that rhenium did not promote the reduction of the cobalt on $Co\text{-}Re/Al_2O_3$. The addition of Ru to Co/Al_2O_3, Co/SiO_2 and Co/TiO_2 increased the reducibility of cobalt oxide (6).

Figures 2a and b showed the XRD patterns of the catalysts after reduction at 773 K. Nb_2O_5, $CoNb_2O_6$, and metallic Co phases were presented in all catalysts. However, the amount of $CoNb_2O_6$ phase calculated decreased in the following order: $Co/Nb_2O_5 \approx Re\text{-}Co/Nb_2O_5 > Ru\text{-}Co/Nb_2O_5$. After reduction, the Co_3O_4 lines disappeared while the lines corresponding to metallic Co were detected. On the other hand, the diffraction patterns of $CoNb_2O_6$ phase remained without any change on the Co/Nb_2O_5 and $Re\text{-}Co/Nb_2O_5$ catalysts. These results indicated that metallic cobalt is mainly formed from the reduction of Co_3O_4 phase, and the $CoNb_2O_6$ phase can not be reduced at temperatures below 773 K on these catalysts. But, the presence of ruthenium also promoted the reduction of the niobate phase. Rietvield method was used to calculate the cobalt reduction degree and the results were presented on Table 1. The rhenium addition did not change the cobalt reduction degree whereas the amount of metallic cobalt increased on the Ru-Co/Nb_2O_5 catalyst. These results agree well with the TPR analyses.

The position of the Co(111) line was practically the same on all catalysts (Table 1). Therefore, it was very difficult to confirm the formation of a Co-Ru alloy on the niobia-supported Co-Ru catalysts. Furthermore, the metal addition led to a decrease on the metallic cobalt crystallite size. The amount of adsorbed hydrogen was higher on the bimetallic catalysts. The dispersion and the particle size of the Co/Nb_2O_5 and Ru-Co/Nb_2O_5 catalysts were basically the same since the extension of reduction of the bimetallic catalyst was higher. The rhenium addition decreased the cobalt particle size. According to the literature, rhenium prevents the cobalt sinterization (6).

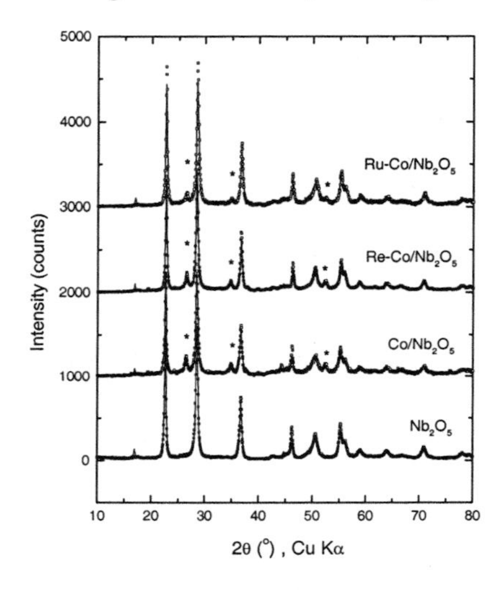

Figure 2 - XRD patterns of reduced catalysts and support.(a) $0 < 2\theta < 100$; (b) $40 < 2\theta < 60$. The symbol * mark angular positions of $CoNb_2O_6$ peaks.

3.2. Catalytic Activity and Selectivity

Table 2 showed CO conversion, reaction rate, turnover frequency (TOF) and selectivities on the CO hydrogenation over the niobia-supported catalysts. The metal addition increased the TOF, which followed the order: Co/Nb_2O_5 < Re-Co/Nb_2O_5 < Ru-Co/Nb_2O_5. The higher turnover rate of the Re-Co/Nb_2O_5 catalyst could be attributed to the presence of smaller Co particle size (6). However, Ru-Co/Nb_2O_5 catalyst exhibited approximately the same particle size than Co/Nb_2O_5 catalyst. As it was observed on the TPR analyses, ruthenium addition increased the cobalt oxide reducibility. Then, the presence of Ru enhanced the activity due to the increase of the Co site density. The same behavior was observed on Co/TiO_2 and Ru-Co/TiO_2 catalysts (5).

Co/Nb_2O_5 catalyst exhibited a very low methanation activity and a high selectivity towards long chain hydrocarbons. Silva et al. (3) compared the catalytic behavior of

Co/Al_2O_3 and Co/Nb_2O_5 catalysts reduced at 573 and 773K on the FT reaction. After reduction at 573K, Co/Al_2O_3 and Co/Nb_2O_5 catalysts presented both high methanation

Table 1- Reduction degree, angular position of Co(111) line and estimated crystallite sizes (dp) of the metallic cobalt particles obtained from XRD analyses and the amount of hydrogen chemisorbed.

Catalyst	Reduction degree (%)[a]	Co (111) (°)	d_p(nm)[b]	H_2 adsorbed (μmoles H_2/g_{cat})[c]	D (%)[d]	d_p(nm)[e]
Co/Nb_2O_5	48	44.27	25±2	16.2	7.9	12
$Ru-Co/Nb_2O_5$	68	44.33	14±2	20.7	7.2	13
$Re-Co/Nb_2O_5$	48	44.28	10±2	25.4	12.5	8

[a] Reduction degree of cobalt calculated using the Rietvield method
[b] Estimated from XRD
[c] Amount of hydrogen adsorbed mesured by H_2 chemisorption
[d] Cobalt metal dispersion calculated from hydrogen adsorption
[e] Cobalt particle size calculated by using the formula : D = 96/d (d = nm)

Table 2- CO conversion (X) and selectivity for methane (CH_4), light hydrocarbons (C_2-C_4), gasoline (C_5-C_{11}), diesel (C_{12}-C_{18}), hydrocarbons containing more than 19 carbons atoms (C_{19}^+) and alcohol, after reduction at 773K.

Catalyst	X(%)	Rate μmol/g.s	TOF s^{-1} $.10^3$	CH_4	C_2-C_4	Gasol	Diesel	C_{19}^+	Alcohol	C_5^+
Co/Nb_2O_5	6.7	0.050	1.54	10.3	29.0	33.8	18.8	2.3	5.8	55.0
Re/Nb_2O_5	0.9	0.011	--	4.3	5.2	8.3	59.0	22.8	0.3	90.1
Ru/Nb_2O_5	4.6	0.053	--	54.3	22.7	13.5	5.9	0.0	3.6	19.4
$Re-Co/Nb_2O_5$	7.4	0.130	2.56	9.4	8.7	49.4	26.0	5.6	0.8	81.0
$Ru-Co/Nb_2O_5$	6.8	0.160	3.86	3.5	32.4	28.7	29.3	1.8	4.2	59.8

activity and low C_5^+ formation. The reduction of the Co/Nb_2O_5 catalyst at high temperature led to a strong decrease in the CH_4 production and an increase of the C_5^+ selectivity. On the other hand, the Co/Al_2O_3 catalyst reduced at 773K showed the same product distribution as observed after reduction at 573K. These results were attributed to the decoration of cobalt particles by partially reduced niobium oxides species (NbO_x) which influences the cobalt adsorption properties. The rhenium addition to Co/Nb_2O_5 catalyst changed markedly the product distribution. The C_5^+ selectivity and, particularly, the selectivity to gasoline and diesel increased whereas the C_2-C_4 fraction decreased. These results could not be explained by the presence of isolated Re particles since the activity of $Re-Co/Nb_2O_5$ catalyst was much lower than the one of the Co/Nb_2O_5 catalyst. Furthermore, the selectivity towards gasoline on the $Re-Co/Nb_2O_5$ catalyst was higher than that observed on the monometallic catalyst. A similar increase in C_5^+ selectivity was verified increasing the cobalt dispersion on Co/TiO_2 catalysts (6). This behavior has already been observed by other authors and was related to the probability of increasing the olefin readsorption, leading to longer chain formation of saturated hydrocarbons (5,6). Although the increase of C_5^+ selectivity was less important on the Ru-Co/Nb_2O_5 catalyst, the Ru addition led to both strong reduction of methanation and an increase of diesel production. Kappor et al (7) studied the catalytic properties of Ru-Co/Al_2O_3 catalyst in the FT process. The addition of noble metals increased the yield of liquid hydrocarbons. Co-Ru catalysts were found highly selective towards C_5^+ hydrocarbons. According to TPR results, the addition of noble metals increased the reducibility of these catalysts. The authors suggested that the presence of a higher amount of metallic cobalt was responsible for the changes in selectivity. The same explanation was proposed by Iglesia (5,6) for alumina and silica supported Ru-Co bimetallic catalyst. Based on EXAFS analysis that revealed the presence of bimetallic Co-Ru particles (4), and on XRD results after reduction in situ, probably beside the bimetallic formation the observation of a higher density of metallic cobalt. TPR analysis showed that ruthenium addition promoted the cobalt oxide reduction, however any Ru-Co alloy was observed. It also shows that the presence of Ru and Re facilitated the interaction of niobium oxide species with cobalt that would explain the promoting effect of these metals to the formation of interfacial sites, which could explain the synergistic effects. Therefore, the increase of the C_5^+ selectivity on Ru-Co bimetallic catalysts could be explained by the promotion of the interfacial site of Co-NbOx sites. This behavior has already been observed by other authors and was related to the probability of increasing the olefin readsorption, leading to longer chain formation of saturated hydrocarbons (5,6,8,9). Undoubtedly, the nature of these sites at the interface of the Co- Nb_2O_x , as well as the formation of higher density of metallic cobalt, promoted by Re and Ru caused the markedly growth chain toward higher C_5^+ fractions. Further in situ measurements are needed to explain the nature of these interfacial sites.

It is important to stress that the high C_5^+ selectivity on the niobia-supported catalyst was obtained performing the reaction at atmospheric pressure and low CO conversion (<10%). Under these reaction conditions, the formation of long chain hydrocarbons should not be favored. These results revealed the important role of the niobia support on the selectivity of FT reaction.

4. Conclusion

The Re and Ru addition promoted the Co_3O_4 reduction and led to an interaction of Co with niobium species at the surface. The XRD and TPR results reinforce the formation of Co-NbOx interfacial sites. The niobate $Co_2Nb_5O_{14}$ bulk phase was observed on all catalysts and was not reducible at temperatures below to 773 K. The rhenium addition decreased the cobalt particle size. Ruthenium promoted the cobalt oxide reduction
The Re and Ru addition increased both the activity and the C_5^+ selectivity of Co/Nb_2O_5 catalyst. Particularly, the selectivity to diesel increased. The high selectivity to C_5^+ products emphasizes the role of the niobia support on the reaction..

References

1. Anderson, R.B., *The Fischer-Tropsch Synthesis*, Academic Press Inc., Orlando, 1984.
2. Schnke, D., in *New Trends of Methane Activation*, Elsevier, 1990.
3. Silva, R.R.C.M., Dalmon, J.A ., Frety, R. and Schmal, M., *J.Chem.Soc.Faraday Trans.* **89**, 3975-3980, 1993.

4. Noronha, F.B., Frydman, A., Aranda, D.A.G., Perez, C.A., Soares, R.R., Moraweck, B., Castner, D., Campbell, C.T., Frety, R. and Schmal, M., *Catalysis Today* **28**, 147-157, 1996.
5. E.Iglesia, S.L. Soled, A. Fiato and G.H. Via, J.Catal., 143 (1993) 345
6. Iglesia, E., *Applied Catal. A: General* **161**, 59-78, 1997.
7. Kapoor, M.P, Lapidus, A.L. and Krylova, A.Y., *Proceedings of the 10th Int. Cong. on Catal.* (Guczi,L., Solymosi, F. and Tetenyi, P., Eds.), part C, p2741-2744, Elsevier, Budapest, 1992.
8. J.Rodriguez-Carvajal, Abstracts of the Satellite Meeting on Powder Diffraction of the XV Congress of the IUCr, Toulouse, 1990, p 127.
9. R.J. Hill and C.J. Howard, *J.Appl.Cryst.* 1987, 20, 467.
10. R.W.G. Wyckoff, Crystal Structures, Wiley Interscience, New York, 1963.
11. S.Tamura, *Z. Anorg. Allg. Chem.* 1974, 410, 313.
12. A. Hoff, DSc. Thesis, NTH, Norway, 1993.

Studies in Surface Science and Catalysis
J.J. Spivey, E. Iglesia and T.H. Fleisch (Editors)

Comparative Studies of the Oxidative Dehydrogenation of Propane in Micro-Channels Reactor Module and Fixed-Bed Reactor

N. Steinfeldt, O.V. Buyevskaya, D. Wolf, M. Baerns

Institute for Applied Chemistry Berlin-Adlershof
Richard-Willstätter-Str.12, D-12489 Berlin, Germany

Catalytic VO_x/Al_2O_3 was deposited on the walls of the micro-channels within a micro-structured reactor module. The catalytic performance of this reactor type was compared for the oxidative dehydrogenation of propane with that of a fixed-bed reactor containing a diluted (1:9) granular catalyst. In spite of catalyst dilution temperature gradients occurred in the fixed bed reactor upon increasing reaction temperature and concentration of reactants in the feed gas. In the micro-channel reactor no gradients were observed under all conditions applied; its performance did not differ from that of fixed bed if isothermal conditions were ascertained in the latter one. It was shown that the micro-structured reactor offers a good temperature control and can be applied for studying reaction kinetics of a strongly exothermic reaction as the oxidative dehydrogenation of propane under isothermal conditions within a wide range of reaction conditions.

1. INTRODUCTION

For the oxidative dehydrogenation of propane being a strongly exothermic reaction, especially, redox-type catalysts such as supported vanadium oxides are very active resulting in significant temperature gradients. Catalytic micro-channels provide efficient heat transfer due to the high surface-to-volume ratio /1/. Therefore, it was of particular interest to prove whether the application of this type of reactor module containing a large number of channels allows to perform the oxidative dehydrogenation of propane isothermally in a wide range of conversions. The reactor operation was compared with that of a fixed-bed reactor. Different reaction parameters, e.g., temperature, partial pressure of reactants, flow rates were varied to find out differences in the performance of both reactors and to estimate the potential of the micro-structured reactor for kinetic studies of exothermal reactions.

2. EXPERIMENTAL

Catalyst preparation: V_2O_5 (Merck), γ-Al_2O_3 (Degussa) and boehmite (Disperal, Condea) were used for catalyst preparation. 0,3 g V_2O_5 was solved in oxalic acid at 75°C followed by adding 0.5 g of polyvinyl alcohol (PVA) under stirring until the solution became clear. 2 g of boehmite and 2 g of γ-Al_2O_3 were then added to the solution under continuous stirring. A part of the resulting suspension was deposited on the metallic walls of the micro-channels followed by drying in air at room temperature. The rest of the suspension was dried under vacuum conditions at 60°C. Both the deposited catalyst and the dried suspension were calcined at 600°C during 6 h. From the calcined powder the granular fraction (d_p = 255 – 355 μm) was obtained and was used in the fixed bed reactor for comparative studies.

Catalyst characterization: XRD studies of deposited and powder catalysts were performed using a transmission powder diffractometer (Stoe) with $CuK_{\alpha 1}$. BET surface areas of both catalysts were obtained by N_2 physisorption. TEM measurements were carried out with a CM 20 TWIN (Philips) apparatus. To study the deposited catalyst, it was removed from the metallic support by treating in alcohol in an ultrasonic bath. One drop of the resulting solution was deposited on a fine grid. The amount of vanadium was determined from the powder catalyst by ICP-OES (Perkin-Elmer: Optima 3000 XL).

Catalytic studies: Catalytic experiments were carried out in the micro-channels reactor module and in the fixed bed reactor ($Ø_{in}$ = 5 mm) at ambient pressure. In order to minimize temperature gradients in the quartz–made fixed-bed reactor the catalyst was diluted with quartz particles of the same size in a ratio of 1: 9. A thermocouple was placed in the middle of the fixed-bed reactor and could be shifted along the catalyst bed for measurements of axial temperature gradients. The total length of the bed amounted to 40 mm; the catalyst charge was 0.064 g. The micro-channels reactor module (Institut für Mikrotechnik Mainz) consisted of 10 structured stainless steel plates, each with 34 parallel channels (300 μm wide, 250 μm deep and 20 mm long). A single micro-structured plate was 0.5 mm thick. Its mass was ca. 5 g. The plates were sealed by graphite. The total amount of VO_x/Al_2O_3 deposited in the micro-channels reactor module was 0.064 g (as in the fixed bed).

Total flow rates were varied from 50 to 150 ml_{STP}/min applying a reaction mixture $C_3H_8/O_2/Ne$ = 0.3/0.15/0.55. At a constant flow rate of 120 ml_{STP}/min, C_3H_8 partial pressure was changed from 8.4 to 50.7 kPa at constant O_2 partial pressure (4.2 kPa). For the same flow rate, O_2 partial pressure was varied between 2.6 and 25.6 kPa at constant C_3H_8 partial pressure of 51 kPa. The measurements were carried out at temperatures between 460 and 525°C.

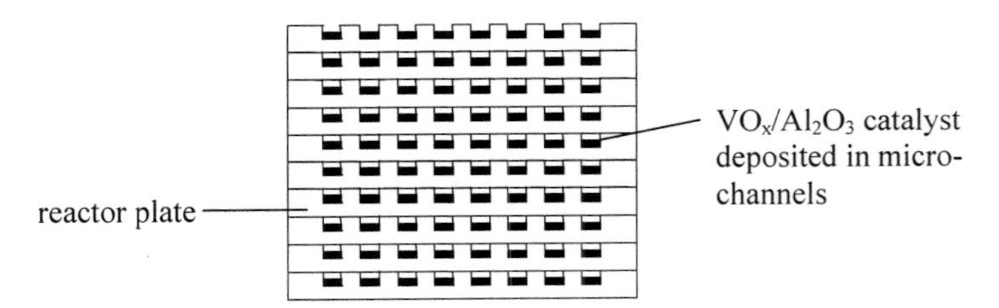

Figure 1: Cross section of the used micro-channel reactor.

3. RESULTS AND DISCUSSION

3.1 Catalyst Characterization

For both deposited and powder catalyst, no vanadium-containing phase could be detected by XRD. Al_2O_3 appeared in both cases as δ and γ phase. The amount of vanadium determined by ICP-EOS was 4.47 wt %. TEM patterns of the powder catalyst and of the deposited catalytic material did not reveal any differences. In both cases the vanadium oxide was distributed on the Al_2O_3 grains. The BET surface area of the granular catalyst amounted to 130 g/m^2 and of the deposited catalytic material to about 76 m^2/g. Thus, the only difference between the both catalysts which was found with the methods applied was the BET surface area.

3.2 Catalytic Performance of VO_x /Al_2O_3 in a Fixed-Bed and in a Micro-Channels Reactor Module

Selectivity and conversion data at different inlet temperatures: Dependencies of product selectivities on propane conversion obtained in both reactors are presented in Figure 2 for reaction temperatures of 460°C and 500°C, respectively. To obtain data at different degrees of propane conversion, total flow rates were varied from 50 to 150 ml_{STP}/min. The main products were C_3H_6, CO and CO_2 under all reaction conditions. Only minor amounts of C_2H_4 and C_2H_6 were detected.

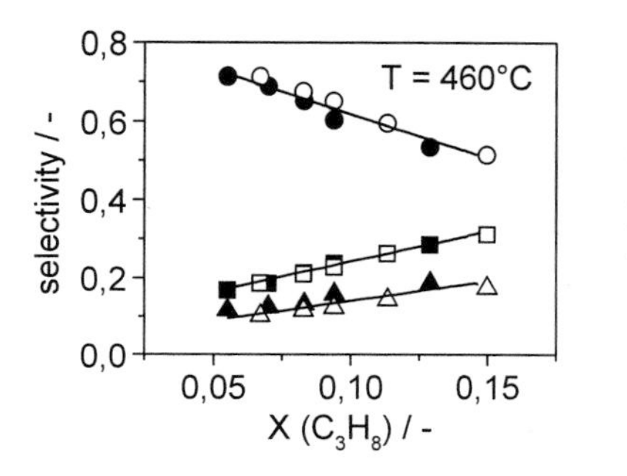

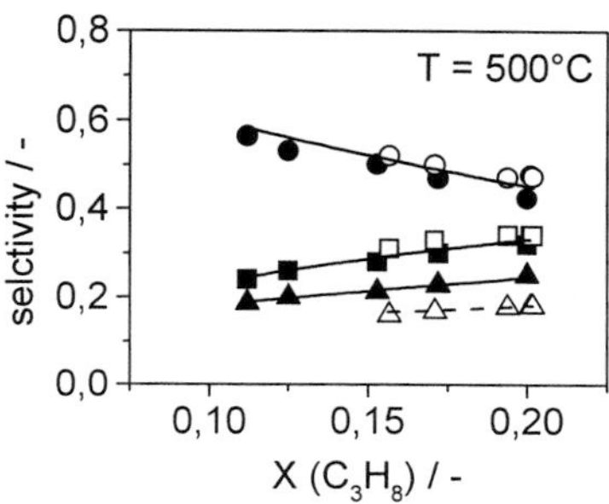

Figure 2: Dependencies of selectivities on propane conversion for the fixed bed (open symbols) and the micro-channel reactor module (closed symbols) at different inlet temperatures ($C_3H_8/O_2/Ne$ = 0.3/0.15/0.55): C_3H_6 (●,○), CO (■, □) and CO_2 (▲, △).

C_3H_6 selectivity decreased while CO and CO_2 selectivities increased with an increase of propane conversion. This agrees with data reported by different authors for the oxidative dehydrogenation of alkanes over VO_x/Al_2O_3 catalysts [2 - 4]. At 460°C, C_3H_6, CO and CO_2 selectivities at similar values of propane conversion were found to be similar in both reactors. At a reaction temperature of 500°C, no significant differences in product selectivities at similar conversions were observed, C_3H_6 and CO selectivities were slightly higher in the fixed-bed reactor while CO_2 selectivity was higher in the micro-channels reactor module. Thus, at similar degrees of propane conversion, nearly similar product distributions were achieved confirming that the catalysts in both reactors did not differ in their catalytic behavior. It should, however, be noted that for an identical inlet temperature, propane conversion was in general higher in the fixed-bed reactor than in the micro-channels reactor module. Figure 3 presents temperature dependencies of C_3H_8 and O_2 conversions for two reaction mixtures of different dilution. At 460°C, propane and oxygen conversions were nearly the same. Propane and oxygen conversion increased much strongly in the fixed-bed reactor compared to the micro-channel reactor module with increasing inlet temperature. In particular, the use of less diluted feed resulted in a pronounced difference in propane and oxygen conversion above 460°C. Complete oxygen conversion was achieved in a fixed-bed at lower inlet temperatures than in the micro-channel reactor module. To explain these findings hot spots temperatures and temperatures between reactor inlet and outlet, respectively, were measured in both reactors.

Temperature gradients: The measurements of temperature along the catalyst bed in the fixed-bed reactor revealed significant differences between inlet and maximum temperature. Depending on composition of the feed gas and on the flow rate the temperature gradients varied from ca. 3 K at 460°C to about 100 K at 500°C. For the micro-channels reactor module, maximum temperature gradients of 2 K were found between the inlet and the outlet of the reactor plates even when the inlet temperature was increased to 523°C.

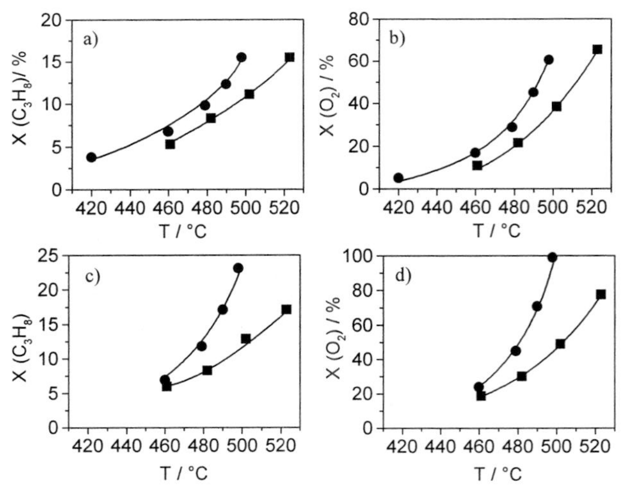

Figure 3: C_3H_8 and O_2 conversion depending on inlet temperature for fixed bed ● and micro-channel ■ reactor; (a,b) $C_3H_8/O_2/Ne$ = 0.3/0.15/0.55, F_{tot} = 150 ml/min , (c,d) – $C_3H_8/O_2/Ne$ = 0.5/0.25/ 0.25, F_{tot} = 120 ml/min.

Temperature differences between T_{max} und T_{inlet} for two reaction mixtures of different dilution are summarized in Figure 4. The data showed that even when using more diluted feed gas (left plot) the temperature gradients amounted to ca. 20 K at a reaction temperature of 500°C. An increase of the reactant concentration at constant propane-to-oxygen ratio led to a pronounced rise of the maximal temperature in the catalyst bed. Thus, neither the dilution of catalyst nor the diluted feed gas could ascertain isothermal operation of the fixed-bed reactor upon increasing the reaction temperature. On the contrary, the use of micro-channels reactor module allowed isothermal operation at all reaction conditions. Thus, results on C_3H_8 and O_2 conversions presented in Figure 3 can be explained in terms of different temperatures in the catalyst bed although the identical inlet temperatures were used for both reactors.

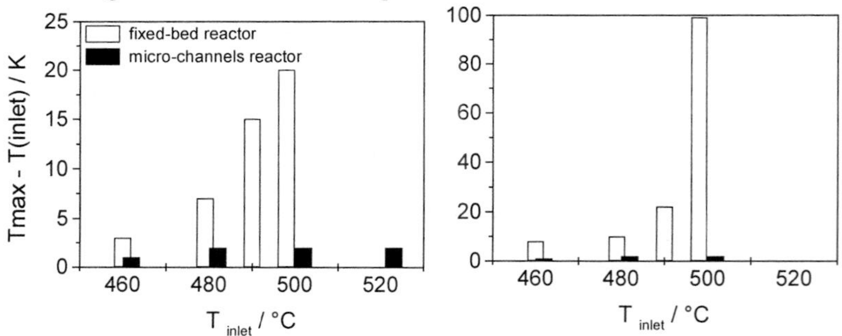

Figure 4: Differences between inlet and maximum temperature in micro-channel reactor and fixed-bed reactor depending on inlet temperature; (left) - $C_3H_8/O_2/Ne$ = 0.3/0.15/0.55, F_{tot} = 150 ml/min , (right) – 0.5/0.25/ 0.25 ($C_3H_8/O_2/Ne$), F_{tot} = 120 ml/min.

At 460°C, propane and oxygen conversion shows only small differences. At this temperature there existed only low axial temperature gradients (3 – 8) K in the fixed-bed reactor (cf. Figure 4). With an increase of inlet temperature propane and oxygen conversion increased stronger in fixed-bed compared to the micro-channel reactor due to higher temperature in the catalyst bed. As simulations of oxidative dehydrogenation of propane in one channel showed [8], there is an effective transfer of the produced heat from the deposited catalytic material to the metallic support and a high rate of heat removal due to high heat conducting of the stainless steel plates. This leads, intern, to the disappearance of both radial and axial temperature gradients.

Effect of C_3H_8 and O_2 partial pressure on product formation: The influence of the partial pressures of propane on the space-time-yield (STY) of C_3H_6, CO and CO_2 for both micro-channel reactor module and fixed bed reactor is presented in Figure 5a and b for 460°C and 490°C-500°C, respectively. The figure reveals that the STY of C_3H_6 increased approximately linear, its slope with increasing C_3H_8 partial pressure was considerably higher that the slopes of CO and CO_2 curves in both reactors. At 460°C, propane conversion was between 5 - 7% in fixed bed and 4 - 6% in the micro-channel reactor and oxygen conversion varied between 30 and 78% in both reactors. With increasing reaction temperature (cf. Figure 5b) some discrepancy in the results of the different reactors were observed. In particular, this concerned the formation of CO_2. The reason for higher CO_2 formation in the micro-structured reactor module compared to the fixed bed reactor should result either from different temperature gradients or the catalytic activity of the stainless-steel wall of the micro-channel reactor which can contribute to CO_2-formation at high temperature, since the catalytic material was deposited only on one site of the channels (cf. Figure 1).

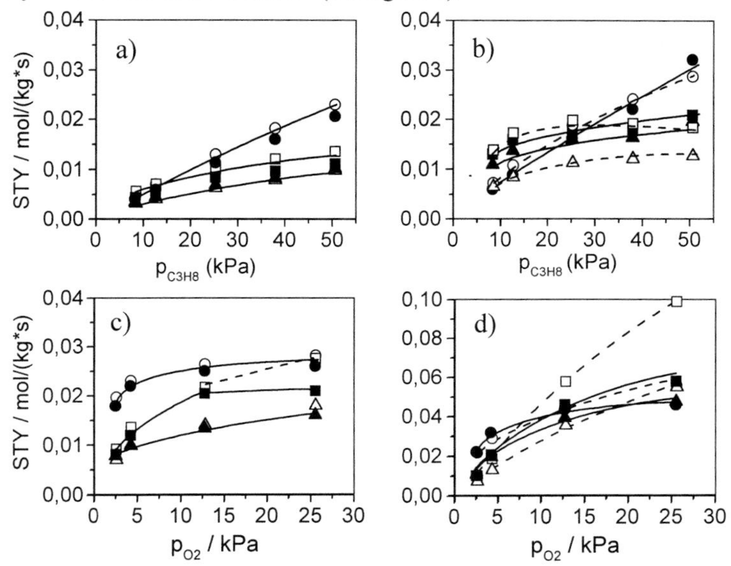

Figure 5a to d: STY of C_3H_6 •, o, CO ■, □ and CO_2 ▲, Δ for fixed-bed reactor (open symbols, doted lines) and micro-channel reactor (closed symbols, solid lines) in depending (a,b) on p_{C3H8} (p_{O2} = 4.16 kPa) and (c,d) on p_{O2} (p_{C3H8} = 51 kPa). T_{inlet} = 460°C (a, c) and T_{inlet} = 490°C in fixed-bed and 502°C in micro-channel reactor (b, d).

Blank experiments in the micro structured reactor module showed that the reactor material reveals only a small catalytic activity (X_{C3H8} max. = 1.5%) up to 500°C. However, the CO_2 selectivity was high (about 60%). The space time yields of the products depending on oxygen partial pressure are presented in Figure 5c and d. In both reactors STY of C_3H_6, CO and CO_2 were similar at 460°C. C_3H_8 conversion was between 4 and 7 % in fixed bed reactor and in micro-channel reactor module, respectively. At 490°C, STY of CO increased considerably with increasing oxygen partial pressure in fixed-bed reactor. Summarizing the above results it can be concluded that the reactor performance was similar at 460°C at all values of reactant partial pressures when isothermal operation was ascertained. STY of propene increased linearly with an increase of C_3H_8 partial pressure and did not depend on O_2 partial pressure. This was also observed in the fixed bed reactor at 420°C where conversion of propane (< 1,5%) and oxygen (< 16%) were still low. These results are in agreement with data from literature /5 - 7/ for redox-type catalysts where the rate of C_3H_6 formation was only weakly influenced by O_2 partial pressure. That means that gaseous oxygen is not directly responsible for the oxidative dehydrogenation, but lattice oxygen which is rapidly formed.

As expected from temperature gradients, STY data obtained in the fixed-bed reactor at 490°C differed significantly from those in the micro-channels reactor module. Thus, temperature gradients which cannot be avoided in the fixed bed reactor despite of catalyst dilution can result in uncertainties in the analysis of kinetic data.

4. CONCLUSIONS

Micro-channel reactor module and fixed bed reactor show approximately the same catalytic results under isothermal conditions. This is in good agreement with results of catalyst characterization, where no differences were found in the solid-phase and surface composition between the granular catalyst and the catalytic material on the wall of the micro-channel. With increasing of reaction temperature, however, axial temperature gradients appear in the fixed bed reactor which do not occur in the micro-channel reactor module. Consequently, at higher inlet temperatures, catalytic results differ for both reactor types. Since the micro-channels reactor module was operated isothermally, the respective experimental kinetic data can be used as basis of kinetic modeling.

Acknowledgement

This work was supported by the Federal Ministry for Science and Education (BMBF) in the frame of the project "Periodic Processes in Microreactors" (03C0282C). The authors thank Dr. M. Schneider for the XRD analysis and Dr. M.-M. Pohl for the TEM-measurements.

REFERENCES

1. K. Schubert, W. Bier, J. Brandner, M. Fichtner, C. Franz, G. Linder; 2rd International Conference on Microreaction Technology, New Orleans 1998.
2. .G.Eon, R. Olier, J.C.Volta, J.Catal. 145 (1994), 318 – 326.
3. J.M. Lopez Nieto, J. Soler, P. Concepcion, J. Herguido, M. Menendez, J. Santamaria, J.Catal.185 (1999) , 324 –332.
4. T. Blasco, A. Galli, J. M. Lopez Nieto, F. Trifiro, J.Catal. 169 (1997), 203 – 211.
5. M. Sautel, G. Thomas, A.Kaddouri, C. Mazzocchia, R. Anouchinsky, Appl. Catal. A 155 (1997), 217 – 228.
6. K. Chen, A. Khodakov, J. Yang, A.T. Bell, E.Iglesia, J.Catal. 186 (1999), 325 – 333.
7. J.M. Michaelis, D.L. Stern, R.K. Graselli, Catal. Lett. 42 (1996), 139 – 148.
8. N. Steinfeldt, N. Dropka, D. Wolf, "Oxidative Dehydrogenation of Propane in the Micro-Channels Reactor Module - Kinetic Data Analysis, Modelling and Simulation", in preparation

Studies in Surface Science and Catalysis
J.J. Spivey, E. Iglesia and T.H. Fleisch (Editors)

Catalyst-assisted oxidative dehydrogenation of light paraffins in short contact time reactors

Alessandra Beretta and Pio Forzatti

Dipartimento di Chimica Industriale e Ingegneria Chimica, Politecnico di Milano, piazza Leonardo da Vinci 32, 20133 Milano, Italy.
alessandra.beretta@polimi.it; pio.forzatti@polimi.it

In previous works, it was shown that the oxidative dehydrogenation of ethane and propane in the presence of a Pt/Al_2O_3 catalyst at high temperature and milliseconds contact times resulted in the production of olefins with over 50% yield. However, it was also found that the process was mainly governed by homogeneous reactions, while the Pt-catalyst seemed mainly involved in the initial reactor ignition. In this work, this hypothesis is further supported by the results of kinetic and mechanistic tests which suggest that both CO_2 and CO are *primary* and unique C-containing products of ethane decomposition over the Pt/Al_2O_3 catalyst. Analogous kinetic tests were performed in the presence of a Rh/Al_2O_3 catalyst, which showed a high activity not only in the deep and partial oxidations but also in steam and dry reforming routes. This activity could explain why, under autothermal conditions, the coproduction of syngas (which accompanies the formation of olefins) is more important in the presence of Rh than in the presence of Pt-catalysts.

1.INTRODUCTION

Oxidative dehydrogenation (ODH) of light paraffins represents one of the novel routes currently proposed for the conversion of natural gas into valuable chemicals; the synthesis of short chain olefins via the exothermal selective oxidation of ethane and propane is considered in fact a potential alternative to the traditional highly energy intensive pyrolysis of hydrocarbon feedstocks.

Schmidt and co-workers have first reported the obtainment of high selectivities to olefins at high degrees of paraffin conversion via selective oxidation of C_2-C_6 hydrocarbons over Pt- , Rh- and other noble metal coated foam monoliths [1-2]. Up to 70 % selectivity to ethylene was obtained from ethane/oxygen mixtures at 50 % conversion of ethane over the Pt-coated monoliths, while CO, H_2O and H_2 where the most important by-products observed [1]. This represented a considerable improvement with respect to the olefin yields so far reported over selective oxidation catalysts (e.g. V- or Mo-based catalysts), which usually range between 10 and 30%. Another innovative factor was represented by the non-conventional operating conditions proposed: the high olefin yields over the Pt-catalyst have been obtained by running the ODH in autothermal conditions, at extremely high temperatures (1000°C) and very short contact times (few milliseconds). Recently, Bodke et al. [3] have further improved the selectivity of ethane partial oxidation process by using H_2 enriched streams and a novel Pt-Sn catalyst. These results have been stimulating the discussion on the feasibility of novel routes for olefins production. Yet, a general consensus has not been reached on several mechanistic issues, and in particular on the role that homogeneous reactions play in the oxygen assisted activation of paraffins at high temperature and short contact times. While some authors [1, 2, 4] defended an essentially heterogeneous mechanism to explain the production of olefins, several pieces of evidence have been reported in the literature which would support an important role of gas-phase reactions [5, 6]. Also recent theoretical studies have shown that homogeneous reactions would be mostly responsible for olefin production [7, 8].

In previous works [6] the authors have addressed the study of the ODH of ethane and propane over a Pt/γ-Al_2O_3 catalyst. Homogeneous and heterogeneous reactions were studied separately by using of a non-adiabatic structured reactor; the bulk of results supported the hypothesis according to which the performance of the autothermal reactor was in reality governed by a homogeneous process of oxidative pyrolysis, while the present Pt-catalyst was mainly involved in the initial ignition of the reactor through the activation of oxidation reactions. The mechanism of alkane oxidation over the Pt-Al_2O_3 catalyst has been further studied in this work, in order to identify the mechanism through which CO_X are formed. The study was then extended to a Rh/Al_2O_3 catalyst; Rh-coated foam monoliths have also been studied in the literature for the partial oxidation of light alkanes and it was shown that under autothermal conditions the production of syngas prevailed over the production of olefins. An interpretation of the different behavior of Pt and Rh is herein searched through the means of the kinetic investigation.

2. ODH OF LIGHT ALKANES IN AUTOTHERMAL REACTORS

Pt-catalyst - As above mentioned, a Pt/Al_2O_3 catalyst was previously studied by the authors in the ODH of propane and ethane. Autothermal experiments were realized by depositing the catalyst powders onto a high void fraction metallic support, then loaded into the central portion of a quartz tube and insulated both in the axial and in the radial direction. Results of adiabatic tests have been reported in detail elsewhere [6, 7]; for the sake of completeness, it is briefly summarized that once the alkane/air mixture was fed into the reactor, light off occurred instantaneously. The final reactor temperature depended on pre-heat temperature and feed composition. In the case of ethane/air streams pre-heated at 450°C, final reactor temperature decreased from 1000 to 930°C at increasing C_2/O_2 ratio from 1.2 to 2.0. Ethane conversion decreased from 95 to over 60%, while ethylene C-selectivity increased progressively from 53 to 65% and CO C-selectivity decreased from 25 to 20%. Maximum ethylene yield was close to 50%. Through a theoretical analysis on the expected performance of a purely homogeneous reactor, it was found that these figures could be well explained by the main control of a gas-phase process. Conversely, kinetic tests in a structured non-adiabatic reactor indicated that the Pt-catalyst was active in the conversion of ethane (or propane) into CO_X, while no evidence could be found of a heterogeneous formation of olefins. On the basis of these and many other results, it was thus concluded that the present Pt-catalyst was simply an oxidation catalyst and that in an autothermal reactor this oxidation activity provided ignition to a homogeneous process, which in turn gave rise to the production of olefins. A major objection which has been made to such picture was that the isothermal tests on the single contribution of the catalyst phase might be not completely representative of the catalyst performance in the autothermal reactor (that is at very short contact time); in particular, it has been speculated that CO_X could possibly be terminal products of a kinetic scheme wherein olefins are formed and decomposed as intermediate species, being absent from the product mixture at longer contact times. In order to verify which kinetic routes are responsible for the production of CO_X over the Pt/Al_2O_3 catalyst, a focused investigation was addressed, and is presented in the following.

Rh-catalyst – Autothermal oxidative dehydrogenation of ethane over Rh-coated foam monoliths was reported to present a quite different behavior than Pt-coated monoliths [1, 2]. In particular, ethane/O_2 feed streams were mostly converted into CO and H_2; at varying ethane/O_2 feed ratio CO C-selectivity was always higher than 50%, while ethylene C-selectivity was below 40%. In order to gain insight of the different role of Pt and Rh in the high temperature oxidative dehydrogenation of light alkanes, the kinetic study of ethane and propane partial oxidation was herein extended to a $Rh-Al_2O_3$ catalyst.

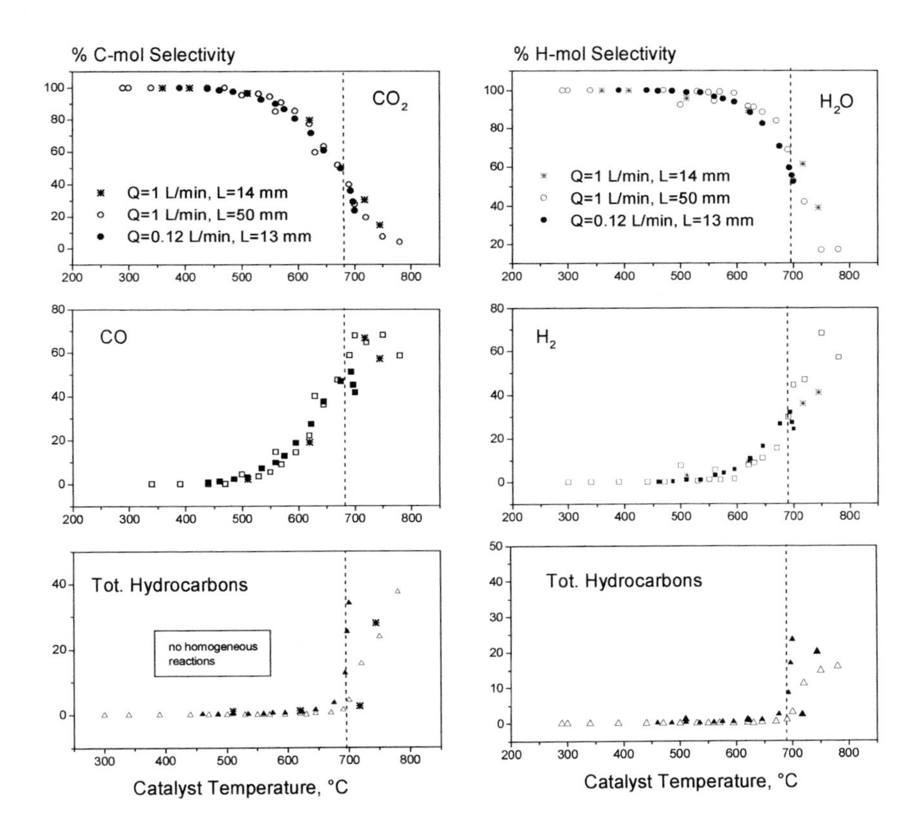

Figure 1 – Kinetic tests of ethane partial oxidation over a Pt/Al_2O_3 catalyst in annular reactor. Effect of contact time on product distribution at short contact time. GHSV= 0.72,1.2and5x106 L(NTP)/kgcat/h. Ethane/O_2 feed ratio=1.

3. KINETICS OF ALKANE PARTIAL OXIDATION OVER Pt/Al_2O_3 AND Rh/Al_2O_3 CATALYSTS AT HIGH SPACE VELOCITY.

Kinetic tests were performed in a non-adiabatic annular reactor, wherein the catalyst has the form of thin (50-100 μm) and short (0.5-5 cm) layers deposited onto a ceramic tube. The coated ceramic tube is inserted into a quartz tube (which is heated by an external furnace). The gas stream flows in between the tubes and contacts longitudinally the catalyst. As the flow regime is laminar, practically no pressure drop is established ever at high flow rates. Thus values of gas hourly space velocity (GHSV) as high as 10^5 – 10^7 Nl/kgcat/h can be easily realized, with extremely short contact times with respect to the catalyst phase. Other important advantages of this reactor are represented by the possibility to measure the catalyst temperature from inside the ceramic support (which is in fact exploited as a well for a sliding thermocouple) and to realize an efficient dispersion of heat via radiation [6].

Pt-catalyst. Experiments in the annular reactor [7] had shown that the heterogeneous reaction between C_2-C_3 alkanes and oxygen initiated at low heating temperature with the formation of CO_2 and water, but above about 450°C catalyst temperature, the production of CO_2 decreased, while a progressive formation of CO and H_2 initiated. At high temperature gas-phase reactions occurred and the product distribution was enriched by olefins and methane. In order to identify the origin of the production of CO_X, that is in order to discriminate between a direct and an indirect mechanism of CO_X formation, and verify the catalyst performance at extremely short contact times, new tests were performed to explore the effect of contact time in the range of high GHSV values.

Effect of contact time – A total flow rate of 1 LNTP/min (NTP = 273K, 1 atm) of a ethane/O_2/N_2 mixture with C_2/O_2=1 ratio was fed in the presence of a 5 cm long Pt/Al_2O_3 catalyst layer, corresponding to a weight load of about 50 mg. A GHSV value of $1.2x10^6$ L(NTP)/kgcat/h was thus realized. As expected, the conversion of ethane and O_2 started at catalyst temperature of 250°C; as shown in Figure 1, CO_2 and H_2O were the only products initially observed in the outlet stream. Above 450°C, traces of synthesis gas appeared; with increasing temperature CO and H_2 became then major products. Formation of ethylene and methane was observed only at high temperatures; however, the comparison with blank experiments in the absence of catalyst showed that in the same high T-range homogeneous reactions were active and could explain the production of ethylene. Thus, even at very short contact times only CO_2/H_2O and CO/H_2 were observed on the present catalyst.

Additional sets of experiments were then carried out at GHSV values of $7.2x10^5$ and $5x10^6$ LNTP/kgcat/h and the results concerning the observed product distributions are also reported in Figure 1. The superimposition of the three sets of data, which notably correspond to a wide variation of reactant conversions, formed an unique trend. For a given value of catalyst temperature, independently of the degree of ethane decomposition, the relative amount of CO to CO_2 and of H_2 to H_2O was constant. The absence of a contact time effect indicated that all the species observed were *primary* products of ethane oxidation; in other words, even at very short contact times, a parallel mechanism seemed to give rise to the formation of CO_2 and H_2O via deep oxidation and of CO and H_2 via partial oxidation, directly from ethane or a common intermediate. Additional tests at lower O_2 content and GHSV as high as $1.7x10^7$ LNTP/kg/h confirmed these findings.

Effect of CO_2 and H_2O addition – Further experiments were performed by co-feeding CO_2 and H_2O to the ethane/air or to the single ethane stream in order to verify the possible existence of secondary reactions of dry and stream reforming of ethane, which could contribute to the production of synthesis gas. Neither the enrichment of CO_2 or the addition of H_2O to an ethane/air mixture produced any enhancement in the formation of CO and H_2. In the absence of O_2, then, no appreciable conversion of ethane was observed with ethane/CO_2 and ethane/H_2O feed streams.

Pt vs. Rh-catalyst. The reactivity of a Rh/Al_2O_3 catalyst in ethane and propane partial oxidation was also studied in the annular reactor. Figure 2 reports the results obtained by feeding an ethane/air mixture with 1/1 ethane/O_2 feed ratio with total flow rate of 120 cm^3(NTP)/min and catalyst layer length of 1.3 cm (GHSV = 7.2×10^5 LNTP/kgcat/h). For comparison, the results of analogous experiments performed under the same operating conditions in the presence of the Pt/Al_2O_3 catalyst are reported too.

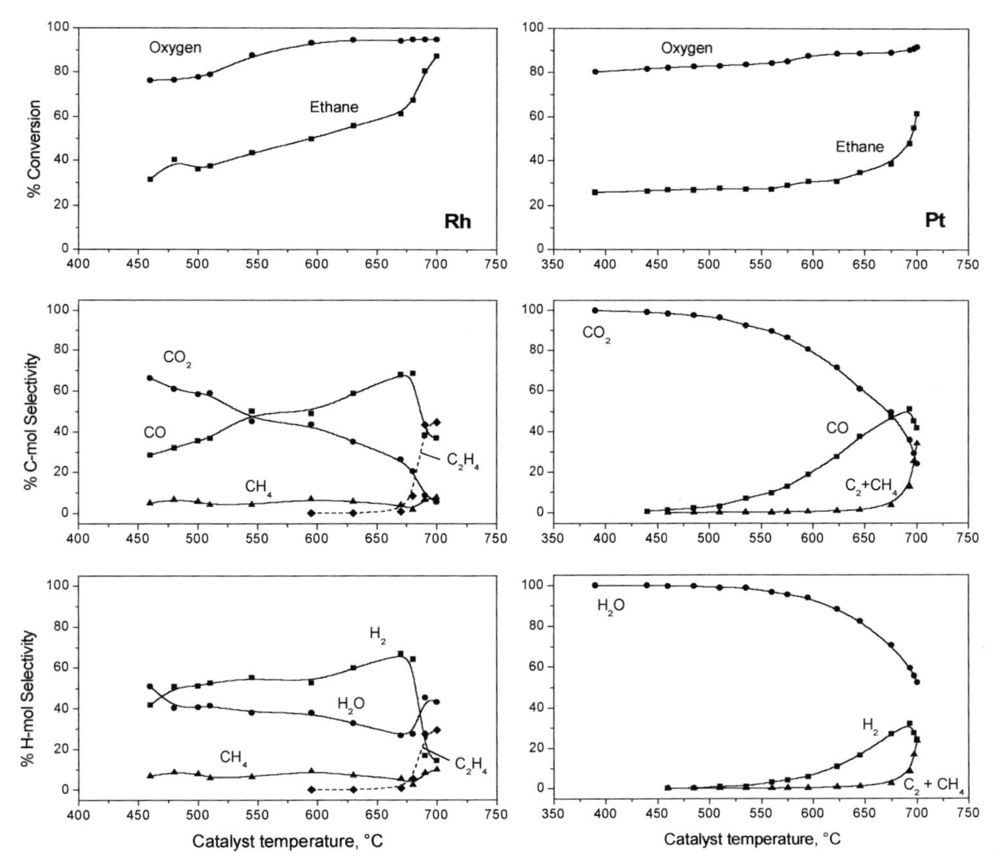

Figure 2 – Experiments of ethane partial oxidation over Rh/Al_2O_3 and Pt/Al_2O_3 catalysts (annular non adiabatic reactor). Feed: ethane/air with C_2/O_2 molar ratio=1. Flow rate = 120 cm^3NTP/kgcat/h, catalyst layer length = 1.3 cm.

Reaction of ethane and O_2 initiated over the Rh-catalyst at about 450°C, that is at higher values than over the Pt-catalyst. For almost equal values of O_2 conversions, ethane conversion was higher over the Rh catalyst. Data were not affected by homogeneous reactions up to 650-675°C catalyst temperature, but above this threshold gas-phase reactions contributed to the sharp increase of ethane conversion and the formation of ethylene and methane (as verified on basis of tests in the absence of any catalyst). Thus, limiting the analysis of data up to such threshold, it was apparent that the Rh catalyst was extremely active in the production of synthesis gas as soon as reaction started. At increasing temperature, CO and H_2 became by far the most abundant products. Also methane was formed over the Rh-catalyst in the whole range of temperature, while over the Pt-catalyst methane formation uniquely accompanied the onset gas-phase reactions. These results, together with additional mechanistic experiments, showed that the Rh/Al_2O_3 catalyst was extremely active in the partial oxidation of light alkanes to syngas, but also in secondary reforming reactions of ethane. This is in line with the known activity of the same catalyst in the partial oxidation of methane to CO/H_2 mixtures. The same evidence of methane formation herein observed can be explained by the existence of a reverse steam reforming reaction (methanation). The peculiar activity of the Rh-catalyst in converting light alkanes into CO and H_2 could explain why, under adiabatic conditions, these species are important co-products of ethane partial oxidation together with ethylene, whose formation could be still explained assuming a contribution from gas-phase reactions.

4. CONCLUSIONS

Kinetic tests at short contact times, non-adiabatic conditions and in the absence of gas-phase reactions showed that the present Pt/Al_2O_3 catalyst is an oxidation and partial oxidation catalyst which transforms light alkanes into CO_X, H_2O and H_2. CO_2 and water prevail at lower temperatures, but the production of CO and H_2 increases with increasing T. Still, the same catalyst, when tested in an autothermal reactor with sufficiently high void fraction, gives rise to the formation of high olefin yields. Apparently, the catalyst promotes ignition of a homogeneous process of oxidative pyrolysis of the alkane, which is very selective towards olefin production. The use of a Rh-catalyst in short contact time adiabatic reactors was proven to produce both olefins and syngas. This behavior could be reconciled with the specific high activity of Rh in converting light paraffin into CO/H_2 mixtures through both partial oxidation and reforming reactions, which are instead kinetically less important on Pt.

Financial support from Progetto Giovani Ricercatori – Politecnico di Milano is gratefully acknowledged

[1] Huff, Schmidt, *J. Phys Chem. 97*, 11815-11822 (1993).
[2] Liebmann, Schmidt, *Appl. Catal. A:General 179*, 93-106 (1999) and Ref. therein included.
[3]Bodke, Olschki, Schmidt, Ranzi, *Science 285* (July 1999), 712-715.
[4] Flick, Huff *Appl. Catal A: General*, 187, 13-24 (1999).
[5] Lodeng, Lindvag, Kvisle, Rier-Nielsen, Holmen, *Appl. Catal A: General, 187*, 25-31 (1999).
[6] Beretta, Piovesan, Forzatti *J. Catal.* 184, 455-468. Beretta, Ranzi, Forzatti, *J. Catal.* 184, 469-478 (1999). Beretta, Ranzi, Forzatti, *Chem. Eng. Sci.*, in press.
[7] Huff, Androulakis, Sinfelt, Reyes, *J. Catal.* 191, 46-54 (2000).
[8]Zerkle, Allendorf, Wolf, Deutschmann, *J. Catal.* 196, 18-39 (2000).

Catalytic Decomposition of Methane: Towards Production of CO-free Hydrogen for Fuel Cells

T. V. Choudhary, C. Sivadinarayana, A. Klinghoffer and D. W. Goodman

Department of Chemistry, Texas A&M University, P.O. Box 30012 College Station, Texas 77842-3012 USA
e-mail : **goodman@mail.chem.tamu.edu** , Fax No. 979-845-6822, Tel. No. 979-845-0214.

ABSTRACT

The catalytic decomposition of methane has been investigated as a method to obtain pure hydrogen for fuel cells. Since it is crucial to know the exact concentration of CO (for operation in fuel cells), in this work ppm levels of CO in the hydrogen stream have been monitored. Methane decomposition studies on a variety of Ni-based catalysts have revealed a strong dependence of stability of the catalyst, nature of carbon and CO content on the support. The CO content in the hydrogen stream was ca. 50 ppm, 100 ppm and 250 ppm for Ni/SiO_2, $Ni/SiO_2/Al_2O_3$ and Ni/HY, respectively, after the CO formation rates had stabilized. The rate of CO formation was found to increase with increasing reaction temperature and decrease with increasing space velocity.

1. INTRODUCTION

Fuel cells being non-polluting and highly efficient are touted as the energy sources of the future. Among the various fuels tested in fuel cells, hydrogen has been found to be the ideal primary fuel. Methane has the highest H/C ratio amongst the various hydrocarbons and thus is the most obvious source for hydrogen. The CO tolerance in the hydrogen stream for fuel cells is extremely low; less than 1.5% for phosphoric acid fuel cells and up to ppm levels for the current proton exchange membrane (PEM) fuel cells. Conventional methods such as steam reforming of methane, authothermal reforming, and partial oxidation produce significant amounts of CO along with hydrogen [1-3]. Removal of CO from hydrogen not only increases the complexity of the process but also has a detrimental effect on the process economics. Thus alternatives for the production of pure hydrogen are desirable.

Recently there has been a considerable interest in the production of hydrogen via catalytic methane decomposition [4-10]. The step-wise steam reforming of methane which was recently studied in our laboratory has shown the feasibility of the process for CO-free hydrogen production [7, 8]. Amiridis and co-workers performed similar experiments at higher temperature in a continuous flow system and showed that the Ni/SiO_2 catalyst could be successfully regenerated via steam gasification in a cyclic manner [9].

In this study we have investigated the decomposition of methane over several Ni-based catalysts. The low levels of CO formed due to the interaction of surface carbon (formed from methane decomposition) with the support have been quantitatively analyzed (ppm level) by methanation of the CO and subsequent analysis by flame ionization detector (FID). This study has employed a variety of supports, e.g. zeolites (HY and HZSM-5), silica-alumina, silica, and activated carbon and efforts have been directed towards investigating the role of the support in the decomposition process. A pulse mass analyzer (PMA) reactor was employed to follow the carbon growth and removal (via steam gasification) on $Ni/SiO_2/Al_2O_3$.

2. EXPERIMENTAL SECTION

2.1 Catalyst preparation

Ni-based catalysts supported on HZSM-5 (Degussa), HY (Degussa), activated carbon, silica (Cabot Corp.), were synthesized by the conventional wet impregnation method with a metal loading of 10 wt. %. The 65 wt. % $Ni/Al_2O_3/SiO_2$ catalyst was obtained from Aldrich chemicals. Prior to loading Ni on silica, the silica was calcined overnight at 823 K. After the impregnation procedure the catalysts were dried overnight at ca. 373 K and then calcined in air (in argon for Ni/C) at 823 K for 4 h. The powder catalysts were then pressed, crushed and sieved to a 20-40 mesh size.

2.2 Apparatus and analytical techniques

The continuous flow-reactor system used in the study is shown in Fig. 1. The gases to be analyzed first passed through the TCD and then were channeled into the FID via the methanizer. The methanizer was checked periodically for the CO to methane conversion efficiency. High purity gases : 20 % CH_4 in Ar(certified mix.), Ar(UHP), hydrogen and air (Zero grade) were utilized for this work. Experiments on the PMA reactor were carried out at Rupprecht & Patashnick Co., Inc. For these experiments similar conditions to those used in our laboratory were employed. In the steam gasification step the H_2O/Ar ratio was one and the total GHSV was 45000 $cm^3.g.h^{-1}$. The data presented has been corrected for some initial adsorption of water by catalyst/carbon filaments during the steam gasification step.

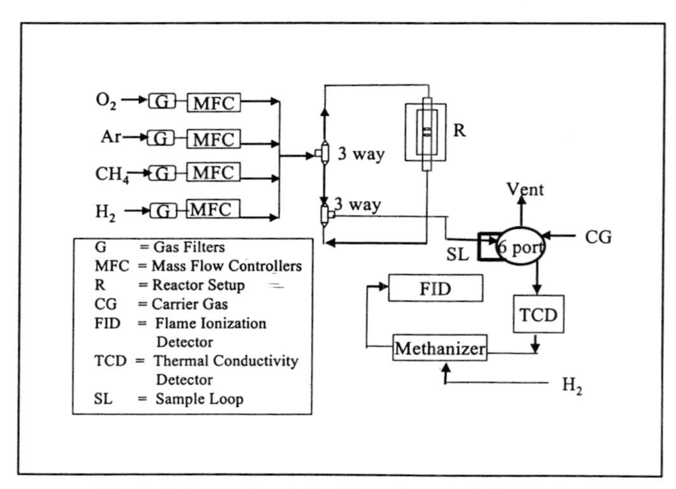

Fig. 1 Schematic of the Experimental Set-up

2.3 Catalyst Characterization

The dispersion of the Ni-based catalysts was measured by CO-pulse experiments. The dispersion for one of the samples, $Ni/SiO_2/Al_2O_3$, was measured by both CO-pulse and hydrogen chemisorption experiments to double-check. Transmission electron microscopy (TEM) characterization was carried out using a high resolution Jeol 2010 instrument (Electron Microscopy Center: Texas A&M University). Catalyst characterization via X-ray photoelectron spectroscopy (XPS) was carried out in an ion-pumped Perkin-Elmer PHI 560 system in our laboratory. The details regarding TEM and XPS characterization are reported in ref [10].

3. RESULTS AND DISCUSSION

The results obtained from the chemisorption experiments are shown in Table 1. Metal dispersions have been calculated from the CO and hydrogen uptake by assuming an adsorption stoichiometry of one CO (H) per catalyst site. 65 % $Ni/SiO_2/Al_2O_3$ was found to have the highest dispersion whereas 10 % Ni/HY had the least. The specific surface areas varied from ca. 0.5 m^2/g (Ni/HY) to ca. 68 m^2/g ($Ni/Al_2O_3/SiO_2$).

Table1. Summary of the Chemisorption experiments

Catalyst	Dispersion (%)
10 % Ni/HY	0.4[a]
10 % Ni/SiO_2	1.0[a]
10 % Ni/HZSM-5	3.1[a]
65 % $Ni/Al_2O_3/SiO_2$	8.5[a], 7.7[b]

[a] from CO chemisorption experiments
[b] from H_2 chemisorption experiments

The time on stream methane conversion activity for various catalysts is shown in Fig. 2. While the 65 % $Ni/Al_2O_3/SiO_2$, 10 % Ni/SiO_2, 10 % Ni/HY catalysts did not deactivate until

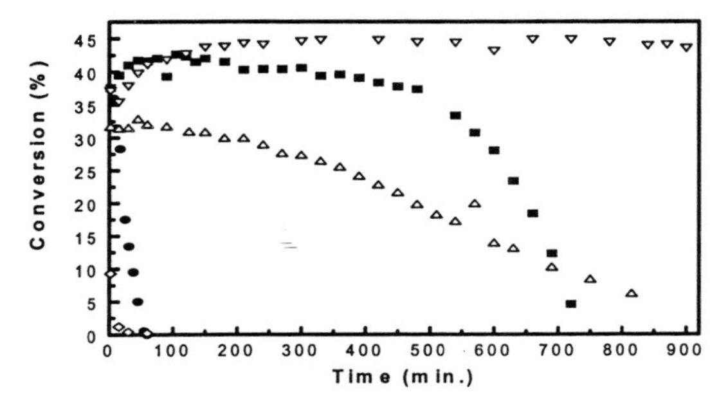

Fig. 2. Methane conversion activity for (∇) $Ni/Al_2O_3/SiO_2$, (■) Ni/SiO_2, (Δ) Ni/HY, (•) Ni/HZSM-5 and (◊) Ni/Carbon (20 % CH_4 in Ar, GHSV = 20000 $cm^3.g^{-1}.h^{-1}$) at 823 K.

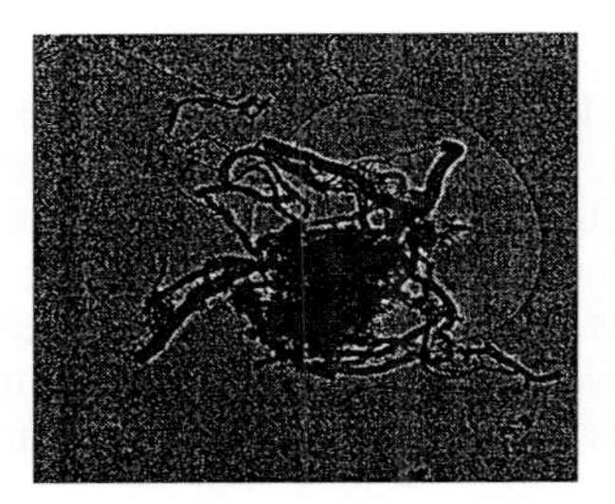

Fig. 3 TEM micrograph of carbon filaments on Ni/SiO_2.

several hours, a rapid deactivation was observed for the Ni/HZSM-5 and the Ni/C catalysts. Also the initial activity was found to be the lowest on the carbon supported catalyst. The $Ni/Al_2O_3/SiO_2$ catalyst showed a stable conversion for about 15 h. due to it large specific surface area. Though the dispersion of Ni/HZSM-5 was greater than that of Ni/SiO_2 and Ni/HY, it had a much smaller capacity for carbon. TEM micrographs revealed the presence of large amounts of filamentous carbon on 65 % $Ni/Al_2O_3/SiO_2$, 10 % Ni/SiO_2 and 10 % Ni/HY (representative TEM image shown in Fig. 3.), whereas an encapsulating type of carbon was observed in case of Ni/HZSM-5. In case of filamentous carbon, the Ni particle is present on the apex of the carbon filament resulting in a large lifetime of the catalyst. Encapsulation of the Ni particles, on the other hand, causes a rapid decline in catalytic activity. Each mole of methane decomposed to produce 2 moles of hydrogen for all catalysts.

The effect of the support on the CO content and rate of hydrogen production is illustrated in Fig. 4. The amount of CO in the hydrogen stream was large initially for all the catalysts, but rapidly decreased with time and finally stabilized. The CO content in the hydrogen stream was ca. 50 ppm, 100 ppm and 250 ppm for Ni/SiO_2, $Ni/SiO_2/Al_2O_3$ and Ni/HY respectively after the CO formation rate had stabilized. This support dependence on the CO content may be related to the amount/stability of hydroxyl groups present on the support.

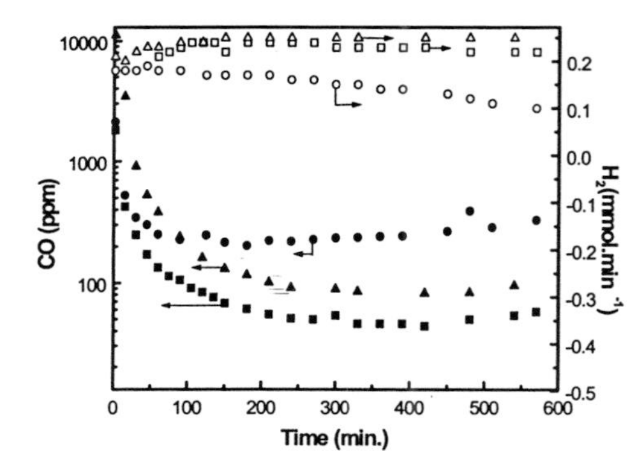

Fig. 4 CO content (solid symbols) and H_2 formation rates (open symbols) on (o,•) Ni/HY, (Δ,▲)$Ni/Al_2O_3/SiO_2$ and (□,■) Ni/SiO_2.

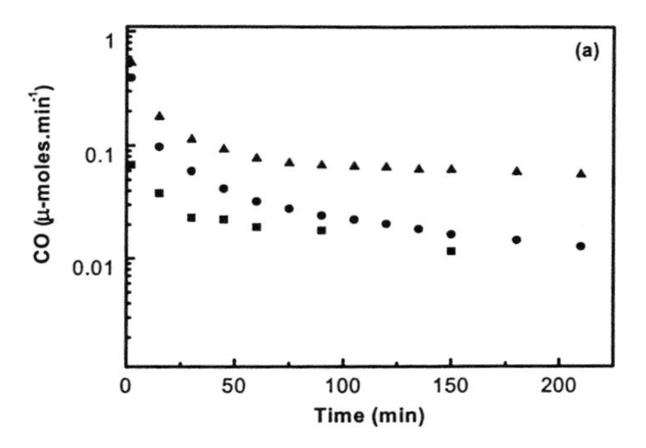

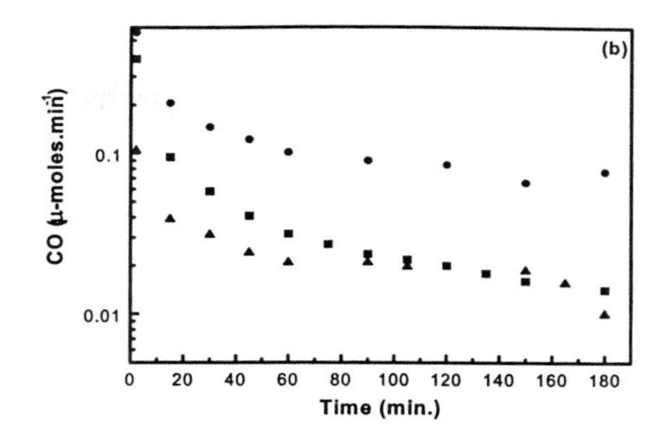

Fig. 5 CO formation rates at (a) (▲) 873 K, (•) 823 K (■) 723 K and (b) (•) 9000 $cm^3.g.h^{-1}$, (■) 20000 $cm^3.g.h^{-1}$ (▲) 60000 $cm^3.g.h^{-1}$ on Ni/SiO_2.

The effects of temperature (at GHSV = 20000 $cm^3.g.h^{-1}$) and space velocity (at temp. = 823 K) on the CO formation rates were investigated on the Ni/SiO_2 catalyst (Fig. 5a & b). Increase in decomposition temperature resulted in an increase in the rate of CO formation (consistent with fact that thermodynamics favors CO formation at higher temperatures). However since the hydrogen production rate is also dependent on temperature, the CO content in the hydrogen stream was the least at ca. 823 K.
The CO formation rate was observed to decrease with increasing space velocity. When the space velocity was increased from 20000 $cm^3.g.h^{-1}$ to 60000 $cm^3.g.h^{-1}$, the initial rate decreased from 0.4 to 0.1 μ-moles.min^{-1}, but the CO formation rates approached comparable values with time. The reason for the space-velocity effect on the CO formation rate is not yet clearly understood. The pre-treatment of $Ni/Al_2O_3/SiO_2$ (19.4 mg) and carbon uptake during

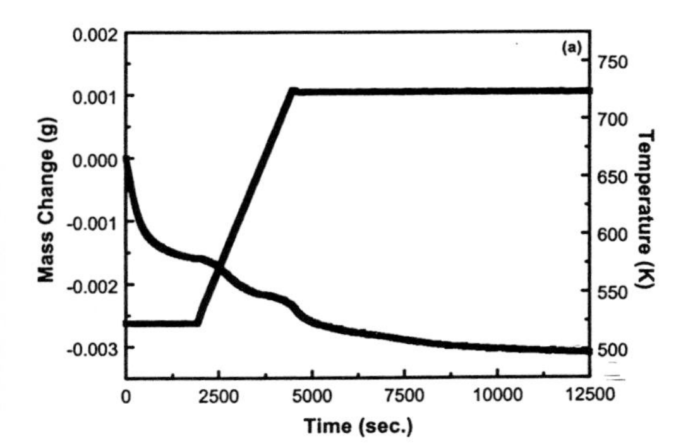

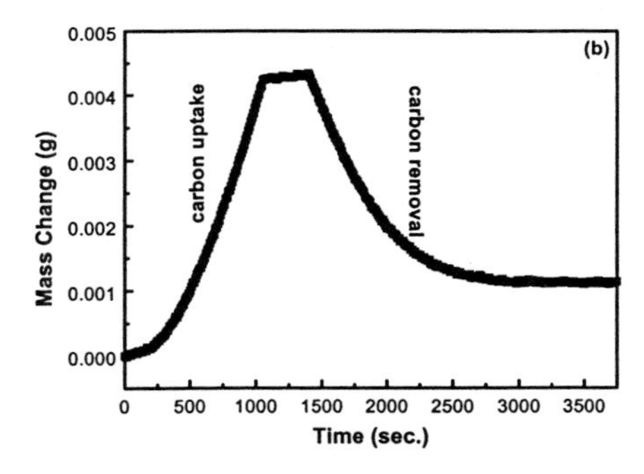

Fig. 6 (a) Catalyst mass change during pretreatment and (b) carbon uptake and removal during the decomposition and steam gasification at 823 K on $Ni/Al_2O_3/SiO_2$.

the methane decomposition step and subsequent carbon removal via steam gasification was investigated with a PMA reactor. Fig. 6a follows the decrease in the catalyst mass as a function of time during reduction (diluted H_2 stream) of the $Ni/Al_2O_3/SiO_2$ catalyst. Initially there is a rapid decrease followed by a gradual decline and finally the catalyst mass stabilizes. As can be seen from Fig. 6b the rate of carbon formation was higher than that of carbon removal during steam gasification at 823 K. The data suggests that ca. 74 % of the carbon deposited in the methane decomposition step can be removed in the steam gasification step at 823 K.

4. CONCLUSIONS

Methane decomposition on various Ni-based catalysts has highlighted the influence of the support on the catalyst conversion stability, nature of the surface carbon and CO content in the hydrogen stream. All the catalysts studied followed the same trend for CO formation; high initial rates followed by low stabilized rates. Increasing the space velocity resulted in a decrease in the initial CO formation rates. It has been shown that relative low CO contents (ppm levels) in the hydrogen stream can be obtained by the catalytic decomposition of methane on Ni/SiO_2 and $Ni/Al_2O_3/SiO_2$. These studies have revealed the potential of this process for obtaining pure hydrogen for applications in fuel cells. Studies are currently underway to explore other catalysts and gain further insights into the regeneration of the catalysts.

ACKNOWLEDGEMENTS

We acknowledge with pleasure the support of this work by the Department of Energy, Office of Basic Energy Sciences, Division of Chemical Sciences. TVC gratefully acknowledges the Link Foundation for the Link Energy Fellowship. We are grateful to Dr Cynthia Liu at R&P for performing the PMA experiments.

REFERENCES

1. J. R. Rostrup-Nielsen in J. R. Anderson and M. Boudart (eds.) Catalytic steam reforming, Catal. Sci. and Eng., Vol. 5, Springer, Berlin, 1984.
2. S. S. Bharadwaj and L. D. Schmidt, Fuel Proc. Tech., 42 (1995) 109.
3. V. R. Choudhary, B. S. Uphade and A. S. Mamman, J. Catal., 172 (1997) 281.
4. T. Zhang and M. D. Amiridis, Appl. Catal. A: Gen., 167 (1999) 161.
5. N. Z. Muradov, Energy and Fuels, 12 (1998) 41.
6. M. Steinberg, Int. J. Hydrogen Energy, 23 (1998) 419.
7. T. V. Choudhary and D. W. Goodman, Catal. Lett., 59 (1999) 93.
8. T. V. Choudhary and D. W. Goodman, J. Catal., 192 (1999) 316.
9. R. Aiello, J. E. Fiscus, H-C Z. Loye and M. D. Amiridis., 192 (2000) 227.
10. T. V. Choudhary, C. Sivadinarayana, C. Chusuei, A. Klinghoffer and D. W. Goodman, (submitted).

Studies in Surface Science and Catalysis
J.J. Spivey, E. Iglesia and T.H. Fleisch (Editors)

CO_2-CH_4 reforming with Pt-Re/γ-Al_2O_3 catalysts

James T Richardson, Jain-Kai Hung[a] and Jason Zhao[b]
Department of Chemical Engineering, University of Houston,
Houston, TX 77204-4792, U.S.A.

[a] Present address: ExxonMobil Chemical Company, P.O. Box 4900, Baytown, TX 77522-4900, U.S.A.
[b] Present address: Sud-Chemie, Inc., 1600 West Hill Street, Louisville, KY 40210, U.S.A.

In the search for carbon-resistant CO_2–CH_4 reforming catalysts, little attention has been given to the promotion of metals by other metals that limit carbon-forming reactions. Extensive experience in the petroleum refining industry with catalytic reforming has shown that addition of Re to Pt/γ-Al_2O_3 catalysts results in reduced carbon formation. This paper reports the application of this type of catalyst to dry reforming. Catalysts were prepared with 0.5 wt% total metal on γ-Al_2O_3 and Pt/Pt-Re ratios between 0.2 and 1. Dry reforming activity measurements showed that catalysts with low Re content deactivated very rapidly in the temperature range 600-800°C due to carbon formation. As the Re content increased, the catalysts became more stable and high conversions were maintained with a Pt/(Pt+Re) ratio of 0.2 for up to 750 hours at 800°C, with only minor amounts of carbon formation. However, the activity decreased rapidly at 700°C and 600°C. Thermogravimetric carbon deposition measurements showed this deactivation was not caused by carbon deposition. Furthermore, the catalyst was regenerated by increasing the temperature.

Possible explanations for improved stability include the ability of Re to dissociate CO_2, thereby releasing oxygen to remove the carbon, and inhibition of carbon-forming ensembles. The low temperature deactivation at high Re content is believed to come from poisoning of Re atoms by CO_2 molecules that strongly adsorb on the surface.

1. INTRODUCTION

Reforming of CH_4 with CO_2 via Reactions (1) and (2) has attracted much attention in

$$CH_4 + CO_2 = 2CO + 2H_2 \qquad (1)$$

$$H_2 + CO_2 = CO + H_2O \qquad (2)$$

recent years, and this has been reviewed by Bradford and Vannice [1]. The process is a source of CO-rich synthesis gas for hydroformulation, carbonylation and other applications, but of more interest is to combine Reaction (1) with Reaction (3) (steam reforming) to give

$$CH_4 + H_2O = CO + 3H_2 \qquad (3)$$

products containing H_2/CO ratios of about 2 that are more suitable for methanol and Fischer-Tropsch synthesis [2]. However, inlet and outlet compositions for dry and "mixed" reforming may favor carbon formation from Reactions (4) and (5) [3]. Previous research has

$$CH_4 = C + 2H_2 \quad (4)$$

$$2CO = C + CO_2 \quad (5)$$

focussed on developing catalysts that restrict the kinetic rates of carbon-forming reactions. Catalysts that are carbon-resistant for dry reforming are expected to perform as well or even better for mixed reforming.

Well-proven steam reforming catalysts based on supported nickel are active for dry reforming but quickly deactivate due to deposits of carbon. Attempts at controlling carbon formation have led to a number of promising results. These include (1) sulfur passivation of Ni surface sites [4], (2) inhibition of the carbon formation on the support by the addition of alkaline earth oxides [5], (3) enhancement of CO_2 adsorption on the support by using La_2O_3 and ZrO_2 [6, 7], (4) inducement of strong metal-support-interactions by promotion with reducible oxides [8], and (5) restriction of carbon-forming "ensembles" by decreasing the Ni crystallite size with difficult-to-reduce mixed Ni oxides [9, 10].

Certain noble metals (Rh, Ru, Pt, Pd) exhibit much higher intrinsic activities for dry reforming than Ni and are much less susceptible to carbon formation [1]. Rhodium is an excellent catalyst, with high activity and stability, but it may be impractical due to high cost and low availability [11]. Ruthenium is also effective but is more prone to form carbon [11]. Nevertheless, ruthenium is the only noble metal catalyst that has been used in a large-scale process [12]. Although Pt is active, it deactivates rapidly from excessive carbon formation. Some of the methods applied to Ni have also been successful with Pt [13, 14, 15, 16], but an alternate approach in which Pt is promoted with other metals that inhibit carbon formation is also a possibility. This was successful for catalytic reforming in the petroleum refining industry [16]. The process has similar, although not equivalent, carbon-forming problems to dry reforming. Rhenium was added to Pt/γ-Al_2O_3 to form Pt-Re clusters that not only greatly reduced hydrogenolysis but also inhibited sintering. Although some improvement for dry reforming with similar Pt-Sn combinations have been reported [18], to our knowledge there have been no reports published using Pt-Re. The purpose of this paper is to describe our experiments on the effectiveness and limitations of these Pt-Re catalysts for dry reforming.

2. EXPERIMENTAL

2.1 Catalysts

The catalysts used comprised 0.5 wt% Pt-Re/γ-Al_2O_3 with Pt/(Pt+Re) ratios from 0.2 to 1.0. Pellets of γ-Al_2O_3 (Johnson Matthey, 3.2-mm diameter, surface area 90-110 m^2g^{-1}, pore volume 1.28 cm^3g^{-1}) were crushed to 60-100 mesh size and heated at 120°C to remove pore moisture. Solutions of Johnson Matthey $H_2PtCl_6{\cdot}6H_2O$ and $ReCl_3$ or NH_4ReO_4 in doubly distilled water (containing HCl for $ReCl_3$) were mixed to give the required Pt/Re ratio in a sufficient volume to fill the pores of the catalyst at 0.5 wt% total metal loading. This was added drop by drop to the catalyst particles, which were soaked for one hour, dried in a preheated oven at 120°C, and calcined at 800°C. A commercial catalytic reforming catalyst (1.0 wt% Pt-Re/γ-Al_2O_3, Pt/(Pt+Re) = 0.20) was included for comparison.

2.2 Catalytic reactor.

The reactor was a quartz tube (1.27 cm i.d., 90 cm in length), with stainless steel fittings at the top and bottom for gas inlet and exit and for thermocouple access. The bottom half of the reactor was filled with 1-mm quartz particles for mixing and preheating. A 0.2-g bed of catalyst particles was located at mid section, followed by another 0.3 g of 1-mm quartz particles and a section of quartz wool. High-purity gases (CH_4, CO_2, H_2, He, and N_2) were mixed and passed through the reactor, which was maintained at the process temperature with an electric furnace. The effluent from the reactor was dried and analyzed with two on-line infrared gas analyzers (Anarad, Inc.). Water and H_2 concentrations were found from material balances. The catalyst was reduced in H_2 (200 sccm, 600-700°C for 2 hours) and the reactor flushed with He prior to each run. Thermogravimetric analyses (not reported here) indicate that no additional reduction occurred for longer times or higher temperatures. Normally, the feed was a 1:1 mixture of CH_4 and CO_2 flowing at GHSVs of 10 000 to 100 000 h^{-1}. Data were gathered with a central data acquisition system.

2.3 Rate of carbon deposition.

The rate of carbon deposition was measured with a modified thermogravimetric apparatus (TGS-2, Perkin-Elmer) equipped with a flow control system similar to the one described in the previous section. The sample holder and associated materials were made of quartz, and the tube was heated externally with a furnace. The sample comprised 50 mg of crushed catalyst (100 mesh), and the amount of carbon depositing during a run was determined from the recorded increase in weight.

2.4 Other analyses.

Other characterization tests included BET surface area (Quantasorb Surface Area Analyzer), total carbon deposition (WR-112 Leco Carbon Determinator), and metal surface area (H_2 chemisorption using a pulse technique at 25°C).

3. RESULTS

Figure 1 shows the performance of the commercial catalyst during a long-term test. Methane conversion was stable at 800°C for up to 150 hour, although it is difficult to be certain since it was close to equilibrium (0.907). However, only a small amount of carbon (0.056wt%) was detected at the end of the run, and tests at conversions as low as 0.15 for shorter periods (25 h) showed the same stability at 800°C. At 700°C, deactivation was very rapid and the initial conversion less than equilibrium (0.731). Deactivation at 600°C was even greater. Initial conversions of the experimental catalysts (Figure 2) declined with increasing Re content, passed through a minimum at about 0.5 Pt/(Pt+Re) and then increased. No activity was observed for 0.5 wt% Re. This is consistent with the catalytic reforming literature that states that Re_2O_7 is very difficult to reduce unless promoted with Pt [19, 20]. Dry reforming activity for Re has been reported but only following very high temperature reduction (900°C) [21]. Only a small amount of Pt is sufficient to ensure complete reduction at 600-700°C, although the Re so formed could be susceptible to re-oxidation.

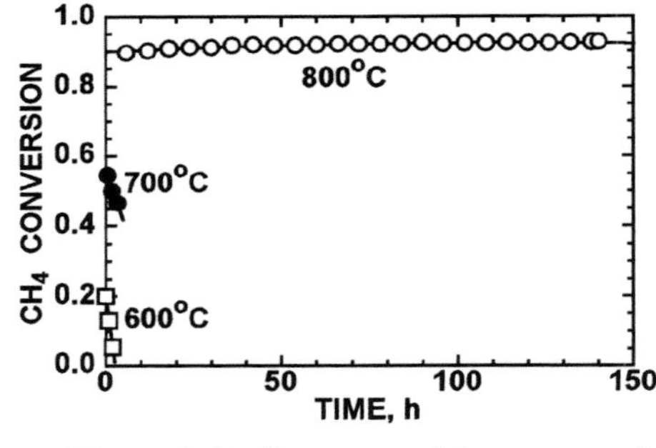

Figure 1. Performance of the commercial catalyst (GHSV = 30 000 h^{-1}).

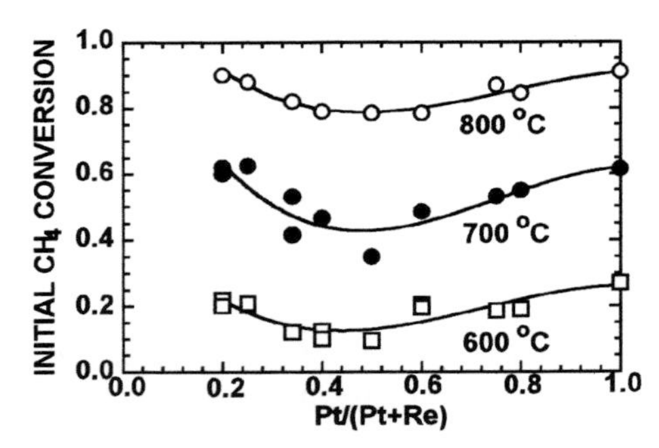

Figure 2. Initial conversion of experimental catalysts (GHSV=30 000h^{-1}).

Activity decline during the first three hours of operation at 800°C (Figure 3) indicated a significant decrease in deactivation as the Re content increased. The lowest deactivation rate was recorded for Pt/(Pt+Re) equal to 0.20. This catalyst gave a high and stable conversion (0.875) for up to 750 hours, but activity again declined rapidly at 700°C and 600°C.

Thermogravimetric measurements of carbon deposition on the Pt catalyst during CH_4 decomposition (Figure 4) showed lower rates of carbon formation as the temperature decreased from 800 to 600°C. Diluting Pt with Re greatly reduced carbon formation at 800°C, and there was no detectable carbon at 700°C and 600°C. Similar measurements (Figure 5) under dry reforming conditions (CO_2/CH_4 = 1) resulted in greatly reduced carbon formation rates, with a maximum at 700°C. Carbon formation was not detectable for dilute samples with Pt/(Pt+Re) < 0.4.

In summary: (1) carbon formation from CH_4 decomposition increases with increasing temperature but is reduced when Pt is diluted with Re, (2) carbon accumulation during dry reforming passes through a maximum at 700°C as carbon gasification by CO_2, H_2 or surface O compete with carbon deposition at higher temperatures, and (3) carbon is most likely the cause of deactivation at 800°C but not at 700°C and 600°C, since trends for the two processes are opposite. This last item was confirmed after deactivation at 800°C, when the original activity returned following treatment with CO_2 or H_2. At lower temperatures, increasing

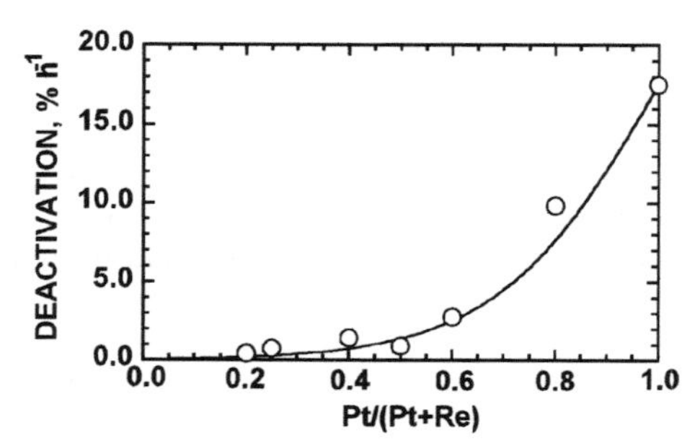

Figure 3. Deactivation rates at 800°C for the experimental catalysts.

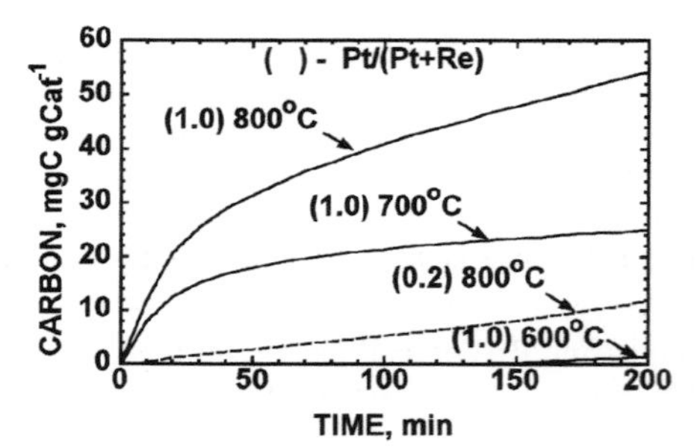

Figure 4. Carbon formation rates from CH_4 decomposition.

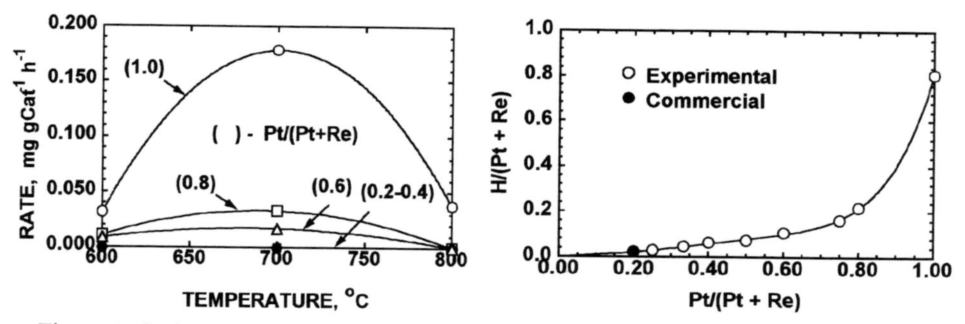

Figure 5. Carbon accumulation rates during dry reforming.

Figure 6. Hydrogen chemisorption data.

the temperature to 800°C (even in an inert) was sufficient to recover the activity, suggesting reversible poisoning.

Hydrogen chemisorption data (Figure 6) show a rapid initial decrease followed by a more gradual decline as Pt dilution increases. It is generally accepted that H_2 does not chemisorb on Re but only on surface Pt [22]. These data suggest that Re decreases the concentration of surface Pt non-linearly, i.e. Re enriches the surface. However, this conclusion may not apply at higher temperatures since the samples were reduced at 600°C. In some instances, we reduced the catalysts at 800°C and found similar trends to those in Figure 1. Nevertheless, there is no evidence that other effects, such as sintering, metal segregation, Re valence shifts or changes in Pt-Re-support interactions do not occur.

4. DISCUSSION

These results indicate that Re addition to Pt enhances resistance to carbon formation during dry reforming, without any significant change in initial conversion. Rhenium itself is not active, since it is difficult to reduce under these conditions, but small amounts of Pt provide enough activation of H_2 to stabilize the reduced state. The best results are obtained with Pt/(Pt+Re) ratios from 0.20 to 0.30. Long-term conversion close to equilibrium can be maintained for up to 150 hours at 800°C. However, there is rapid loss of activity at 700°C and 600°C, not due to carbon formation but recoverable by raising the temperature.

The key to understanding these phenomena is the role played by Re. In monometallic systems (e.g. Rh, Pt, etc.) surface metal sites dissociate CH_4 and possibly CO_2. There is competition between the two, but CH_4 dissociation is believed to be slower [1]. It is known that Re dissociates CO_2 rapidly and has a strong affinity for oxygen ions [23]. These ions are less strongly adsorbed on Pt-Re catalysts at high temperatures (e.g. 800°C) and could remove C species as CO before polymeric carbon deposits form. This last step may also be impeded by an ensemble effect, in which groupings of Pt sites needed for carbon polymerization are restricted. Electronic effects caused by alloying are also possible. Further research will be necessary to elucidate these features.

These events continue at lower temperatures (e.g. 700°C and 600°C), but the Re-O bond is stronger and more difficult to release. Rhenium sites are poisoned, and activity and carbon removal decline, although the benefits of the ensemble effect remain. Deactivation is reversed by increasing the temperature, possibly by releasing oxygen from the Re atoms.

Although these results demonstrate Pt-Re combinations are effective catalysts for dry reforming at 800°C, constraints imposed by low temperature deactivation need to be addressed.

ACKNOWLEDGMENTS

This research was supported by the State of Texas Higher Education Coordinating Board, Advanced Technology Program.

REFERENCES

1. M. C. J. Bradford, M. A. Vannice, Catal. Rev. Sci. Eng., 41 (1999) 1.
2. I. Dybkjaer, J. B. Haansen, Natural Gas Conversion IV, (Eds. M. dePontes, R. L. Espinoza, C. P. Nicolaides, J. H. Scholtz, M. S. Scurrell), Elsevier, Amsterdam (1997) p. 99.
3. J. R. Rostrup-Nielsen, J.-H. B. Hansen, J. Catal., 144 (1993) 38.
4. J. A. Stal, D. C. Hansen, J.-H. B. Bak Hansen, N. R. Udengaard. Oil & Gas J., 90(10) (1992) 62.
5. Z. Cheng, Q. Wu, J. Li, Q. Zhu, Catal. Today, 30 (1996) 147.
6. Z. Zhang, X. E. Verykios, Appl. Catal. A: Gen., 138 (1996) 109.
7. J-M Wei, B-Q X, J-L Li, Z-X Cheng, Q-M Zhu, Appl. Catal. A: Gen., 196 (2000) 167.
8. M. C. J. Bradford, M. A. Vannice, Appl. Catal. A:, 142 (1996) 73.
9. E. Ruckenstein, Y. H. Hu, Appl. Catal., 133 (1995) 149.
10. K. Tomishige, O. Yamazaki, Y. Chen, K. Yokoyama, X. Li, K. Fujimoto, Catal. Today, 45 (1998) 35.
11. J. R. Rostrup-Nielsen, Natural Gas Conversion II, (Eds. H. E. Curry-Hyde and R. F. Howe), Elsevier, Amsterdam (1994) p. 25.
12. M. Epstein and I. Spiewak, Proceedings of the 7th International Symposium on Solar Thermal Concentrating Technologies, Moscow, Russia, 1994, p 200.
13. J. H. Bitter, W. Halley, K. Seshan, J. G. V. Ommen, J. A. Lercher, Catal. Today, 30 (1996) 193.
14. K. Seshan, H. W. ten Barge, W. Hally, A. N. J. van Keulen, J. R. H. Ross, Natural Gas Conversion II, (Eds. H. E. Curry-Hyde and R. F. Howe), Elsevier, Amsterdam (1994) p. 285.
15. S. M. Stagg, D. E. Resasco, Natural Gas Conversion V, (Eds. A. Parmaliana, D. Sanfilippo, F. Frusteri, A. Vaccari, F. Arena), Elsevier, Amsterdam (1998) p.813.
16. M. C. J. Bradford, M. A. Vannice, Catal. Today, 50 (1999) 87.
17. D. M. Little, Catalytic Reforming, Penn Well Publishing Co., Tulsa (1985).
18. S. M. Stagg, D. E. Resasco, Stud. Surf. Sci. Catal., 111 (1997) 543.
19. M. F. L. Johnson, J. Catal., 39 (1975) 487.
20. N. W. Webb, J. Catal., 39 (1975) 485.
21. J. B. Claridge, M. L. H. Green, S. C. Tsang, Catal. Today, 21 (1994) 455.
22. H. C. Yao, M. Shelef, J. Catal., 44 (1976) 392.
23. F. Solymosi, J. Mol. Catal., 65 (1991) 337.

Studies in Surface Science and Catalysis
J.J. Spivey, E. Iglesia and T.H. Fleisch (Editors)

The Catalytic Properties of Alkaline Earth Metal Oxides in the Selective Oxidation of CH_4-O_2-NOx (x=1, 2)

T. Takemoto[a], K. Tabata[a,b*], Y. Teng[a], E. Suzuki[a,b] and Y. Yamaguchi[c]

[a]Research Institute of Innovative Technology for the Earth, 9-2, Kizugawadai, Kizu-cho, Soraku-gun, Kyoto 619-0292, Japan.

[b]Graduate School of Materials Science, Nara Institute of Science & Technology, 8916-5, Takayama-cho, Ikoma, Nara 630-0101, Japan

[c]Kansai Research Institute, Kyoto Research Park 17, Chudoji Minami-machi, Shimogyo-ku, Kyoto 600-8813, Japan

The catalytic properties of alkaline earth metal oxides in the selective oxidation of CH_4-O_2-NOx (x= 1, 2) were examined so as to enhance the selectivity of C_1 oxygenates (CH_3OH and CH_2O) in the products. The total yield of C_1 oxygenates in the products was enhanced in the presence of MgO and CaO in comparison to that in the absence of these metal oxides. We assumed the decomposition reaction between C_1 oxygenates and OH species was retarded in the presence of MgO and CaO. We could appropriately explain the decrease of decomposition reaction in the presence of MgO and CaO by using theoretical calculations.

1. INTRODUCTION

The direct selective oxidation of CH_4 with O_2 to C_1 oxygenates (CH_3OH and CH_2O) is a potentially important process for the effective use of natural gas resources. In recent years, many researchers have studied the selective oxidation of CH_4 with various types of catalysts; however, the only trace of CH_3OH and CH_2O in the products has been reported [1, 2]. The gas-phase selective oxidation of CH_4 seems to have the advantage of yielding C_1 oxygenates because the difficulty of desorption of the produced oxygenates from the catalyst surface could be liberated. Formation of CH_3OH and/or CH_2O in the direct gas-phase selective oxidation of CH_4 with O_2 has been reported [3-9]. The promotion effect of nitrogen oxides has been reported [10-15]. Tabata et al. reported that not only methane activation but the selectivity of C_1 oxygenates was enhanced by the addition of a small amount of NO_2 in the gas-phase selective oxidation of CH_4-O_2 [15]. They obtained ca. 7 % yield of C_1 oxygenates at SV= 7500 h^{-1}. They also suggested the decomposition route of the produced CH_3OH and CH_2O as follows:

$$CH_3OH + OH \rightarrow CH_2OH + H_2O \quad (1)$$
$$CH_2OH + O_2 \rightarrow CH_2O + HO_2 \quad (2)$$
$$CH_2O + OH \rightarrow CHO + H_2O \quad (3)$$

*Corresponding author. Tel.: +81-774-75-2305 ; Fax.: +81-774-2318. E-mail: kenjt@rite.or.jp

$CHO + O_2 \rightarrow CO + HO_2$ (4)

They also suggested the production route of OH species in the reactant gas,

$CH_4 + NO_2 \rightarrow$ *trans*-$HONO + CH_3$ (5)
$CH_3 + O_2 \rightarrow CH_3OO$ (6)
$CH_3OO + CH_4 \rightarrow CH_3OOH + CH_3$ (7)
$CH_3OOH \rightarrow CH_3O + OH$ (8)

The transition barrier of equation (1) was reported as 1.4 kcal/mol [3], and those of the others in eqs (2) - (4) were less than this value. If OH species are sufficient in a gas-phase, the reaction for the decomposition of CH_3OH and CH_2O will proceed easily. The enhancement of C_1 oxygenates selectivity in the direct selective oxidation of CH_4 was predicted making the assumption that eqs (1), (3) and the following eq (9) were omitted from the set of 288 elementary reactions, using the software packages of CHEMKIN III [16],

$CH_3OH + OH \rightarrow CH_3O + H_2O$ (9)

In this study, the catalytic properties of alkaline earth metal oxides in the direct selective oxidation of CH_4 are examined. It is expected that OH species in a reaction gas are trapped on the surface of alkaline earth metal oxide as the reactions between the produced C_1 oxygenates and the OH species are retarded.

2. EXPERIMENTAL SECTION

2.1. Activity test

All of the experiments were carried out using a single-pass flow reactor made of a quartz tube (i.d. 7.0 mm) at atmospheric pressure. A heated length of an electric furnace was 200 mm. Reaction temperature was controlled from the outside of the reactor with a thermocouple installed at the center position of the heated zone. The total flow rate was 120 cm^3min^{-1} (STP). The standard gas composition (CH_4: 55.5 mol%; O_2: 27.8 mol%; NO: 0.5 mol%; He: 16.2 mol%) was controlled with a mass flow controller. The ratio of CH_4 to O_2 was stabilized at 2.0. The products were analyzed with two on-line gas chromatographs serially connected. A thermal conductivity detector (activated carbon) and a flame ion detector (Gaskuropack 54), using He as a carrier gas, were used. The carbon balance before and after the reaction exceeded 95 %. Measurement of products was carried out after the reaction for 30 min at each experimental condition, and all experimental data were taken at least three times to assure the reproducibility. The selectivity variation in the products as a function of CH_4 conversion was calculated. CH_4 conversion was varied with raising the reaction temperature. The used alkaline earth metal oxides, MgO and CaO were obtained from the chemical reagents.

2.2. Calculation Procedure

All the calculations were carried out with the $DMol^3$ program package provided by Molecular Simulations Inc. (MSI) [17]. The two-layer $A_{16}O_{16}$ (100) cluster model (A= Mg, Ca) was used. The cluster geometry of MgO and CaO was fixed on an each crystal coordinate derived from the data in the Celius[2] software package [17]. The bond lengths of Mg-O and Ca-O were 2.106 and 2.405 Å, respectively. The single-point calculations of the energies were conducted by using the generalized gradient approximation (GGA) level. A Silicon Graphics Origin 2000R10000 workstation was used for calculations in this study.

3. RESULTS AND DISCUSSION

The variation of each selectivity of products at 10 % level of CH_4 conversion was examined as a function of installed position of MgO in the reactor. Feed gas composition was CH_4 (55.5 mol%) + O_2 (27.8 mol%) + He (16.2 mol%) + NO (0.5 mol%). The flow rate was 120 cm^3min^{-1} (STP). The weight of MgO was 0.05g. The distance of MgO was measured from the entrance of heated zone, and the distance of 100 mm was the center of heated zone. The observed variations of product selectivity at 10% level of CH_4 conversion at several positions are shown in Fig. 1. Every reaction temperature was included in the range of 560-575°C. The total selectivity of C_1 oxygenates at 10% CH_4 conversion slightly increased in the presence of MgO except for the case at 100 mm. The maximum value was obtained at 30 mm. Both CH_3OH and CH_2O selectivities were enhanced at there in comparison to those in the absence of MgO.

CH_4 conversions as functions of reaction temperatures in the presence of MgO were

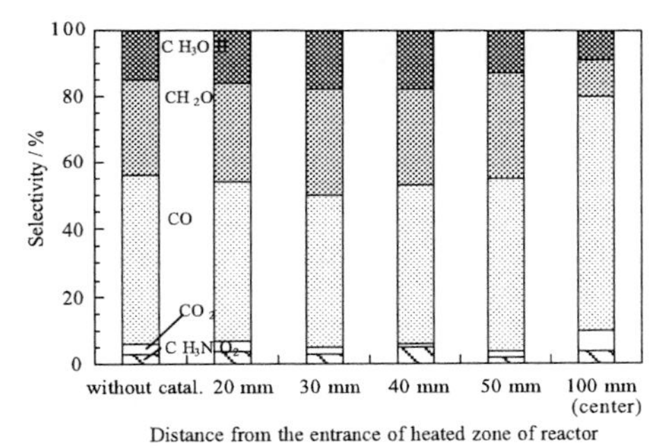

Fig.1. The effects of catalyst (MgO) position on the selectivities of products at 10 % level of CH_4 conversion in CH_4- O_2-NO

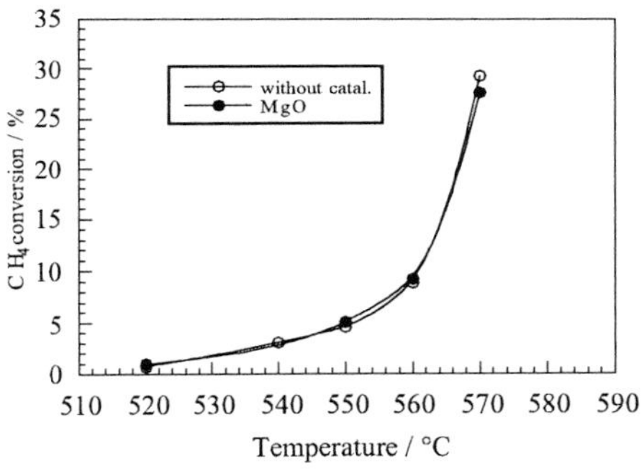

Fig.2. CH_4 conversion variation with and without MgO as a function of reaction temperature in CH_4(55.5 mol%) - O_2(27.8 mol%)-NO_2(0.5 mol%)

Table 1
The effects of catalyst weight on the selectivities of products at 10 % level of CH_4 conversion in CH_4-O_2-NO

Catal. weight (g)	Reaction temp. (°C)	Selectivity(%)				
		CH_3OH	CH_2O	CO	CO_2	CH_3NO_2
without catal.	572	15	29	50	3	3
0.05	564	18	32	45	2	3
0.1	564	13	26	54	4	3
0.5	570	13	26	52	6	3

Table 2
The effects of catalysts on the selectivities of products at 10 % level of CH_4 conversion in CH_4-O_2-NO

Catal.	Reaction temp. (°C)	Selectivity(%)				
		CH_3OH	CH_2O	CO	CO_2	CH_3NO_2
MgO	564	18	32	45	2	3
CaO	572	14	32	48	3	3
without catal.	572	15	29	50	3	3

examined with CH_4-O_2-NO_2 (Fig. 2). The conversions of CH_4 in the presence of MgO were almost the same as those in the absence of MgO. The presence of MgO did not affect the activation of CH_4 in the selective oxidation with CH_4-O_2-NO_2.

The effects of catalyst weight on the selectivity of C_1 oxygenates were also examined at 10% CH_4 conversion (Table 1) with the standard gas composition. MgO was installed at 30mm. The best selectivity of C_1 oxygenates was observed at 0.05 g, and further amount of MgO brought the decrease of C_1 oxygenates in the products.

The catalytic properties of CaO were examined with the standard gas composition of CH_4-O_2-NO. Table 2 shows the result of each selectivity for MgO and CaO. Both CH_3OH and CH_2O selectivities were enhanced in the presence of MgO. However, the selectivity of CH_2O was enhanced but that of CH_3OH was not enhanced in the presence of CaO.

Theoretical calculations with the density functional theory (DFT) were carried out so as to study the reason for the enhancement of C_1 oxygenates selectivity. The predicted adsorption energies of OH species on both MgO and CaO were 33.3 and 51.6 kcal/mol, respectively (Fig.3). Here the state of adsorbed OH species was taken as a standard. It was assumed from these large values that OH species in the reactant gas was trapped easily on the surface

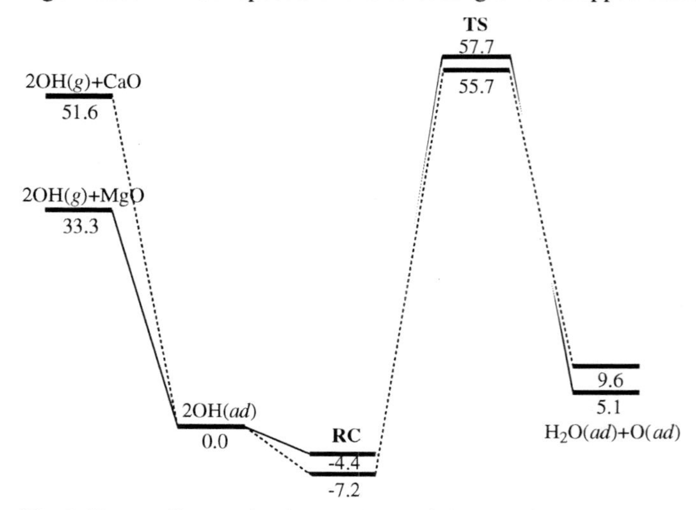

Fig. 3. Energy diagram for the coupling of OH on MgO and CaO catalysts. Relative energies are given in kcal/mol.

of MgO and CaO. Therefore the concentration of OH species in the reactant gas should be reduced in the presence of MgO or CaO in the selective oxidation of CH_4. The energy diagram of coupling reactions between adsorbed OH species over these catalysts were predicted (Fig.3). High transition barriers through a reactant complex (RC) were predicted. Consequently, the adsorbed OH species on MgO and CaO could be stabilized over the surface.

The predicted values of adsorption energy of CH_3OH and CH_2O on CaO were small as 3.4 and 2.1 kcal/mol. It was also predicted that both C_1 oxygenates did not adsorb on MgO. The calculated energy diagrams of decomposition reactions of CH_3OH and CH_2O in eqs (1) and (3) over MgO and CaO are shown in Figs. 4 and 5 respectively. Here the state of adsorbed OH species and the reactants in a gas-phase was taken as a standard. The predicted values of transition barrier of eq (1) over MgO and CaO were 29.3 and 20.9 kcal/mol respectively. These values were much higher than that of gas-phase reaction (1.4 kcal/mol).

The predicted values of transition barrier of eq (3) over MgO and CaO were 34.1 and 31.1 kcal/mol respectively. These values were much higher than that of gas-phase reaction (- 0.4 kcal/mol).

The OH species in the gas-phase should be trapped easily on the surface of both MgO and CaO (Fig. 3). The predicted values of transition barriers of decomposition reactions between C_1 oxygenates in the gas-phase and the trapped OH species over MgO and CaO were larger than those in the gas-phase decomposition reactions. Therefore the decomposition reaction between C_1 oxygenates and OH species should be retarded in the presence of MgO and CaO. The decrease of decomposition reaction of C_1 oxygenates brought about the enhancement of the total yield of C_1 oxygenates. The difference of enhancement of CH_3OH selectivity over between MgO and CaO could be explained by the difference of transition energies.

We assumed the concentration of OH species in the gas-phase increased gradually from the entrance of heated zone, and the temperature rose simultaneously. The position of catalyst at around 30 mm from the entrance of heated zone could be the most suitable position to retard the decomposition reactions.

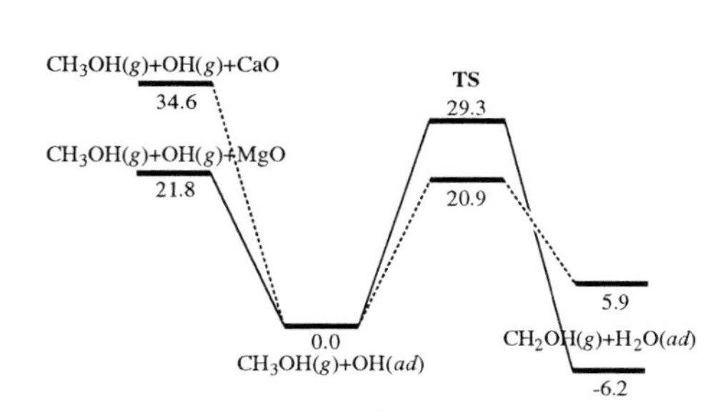

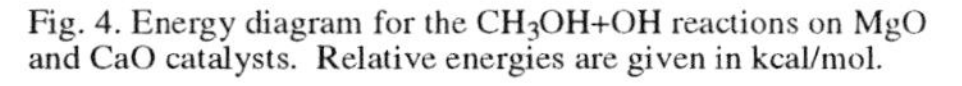
Fig. 4. Energy diagram for the CH_3OH+OH reactions on MgO and CaO catalysts. Relative energies are given in kcal/mol.

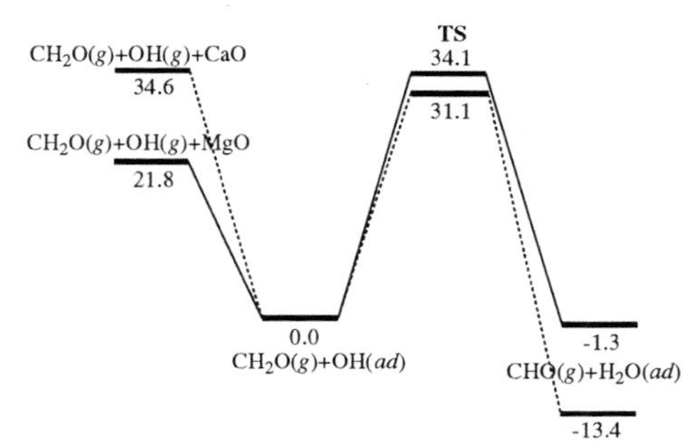

Fig. 5. Energy diagram for the CH_2O+OH reactions on MgO and CaO catalysts. Relative energies are given in kcal/mol.

4. CONCLUSIONS

The selectivity of C_1 oxygenates was enhanced by using MgO and CaO in the selective oxidation of CH_4-O_2-NOx (x=1, 2). MgO enhanced both CH_3OH and CH_2O selectivities in the products, but CaO enhanced only the selectivity of CH_2O. The total yield of C_1 oxygenates was enhanced from 4.4% to 5.0% at 10% conversion of CH_4 in the presence of MgO. The experimental results were appropriately explained by the theoretically calculated results.

ACKNOWLEDGEMENT

This study was supported financially by the New Energy and Industrial Technology Development Organization (NEDO, Japan). Y. Teng was supported by a Fellowship from the NEDO.

REFERENCES

1. S.H.Taylor, J. S. Hargreaves, G. J. Hutchings and W. J. Joyner, Methane and Alkane Conversion Chemistry, Plenum, New York, (1995) p339.
2. R. Pitchai and K. Klier, Catal., Rev. -Sci. Eng. 28 (1986) 13.
3. V. S. Arutyunov, V. Y. Basevich and V. I. Vedeneev, Russ. Chem. Rev., 65 (1996) 197.
4. J. C. Mackie, Catal. Rev. -Sci. Eng., 33 (1991) 169.
5. N. R. Foster, Appl. Catal., 19 (1985) 1.
6. H. D. Gesser, N. R. Hunter and C. B. Prakash, Chem. Rev., 85 (1985) 235.
7. O. V. Krylov, Catal. Today, 18 (1993) 209.
8. W. Feng, F. C. Knopf and K. M. Dooley, Energy Fuels, 8 (1994) 815.
9. J. H. Bromly, F. J. Barnes, X. Y. Muris and B. S. Haynes, Combust. Sci. Technol., 115 (1996) 259.
10. D. F. Smith and R. T. Milner, Ind. Eng. Chem., 23 (1931) 357.
11. S. Irusta, E. A. Lombardo and E. E. Miro, Catal. Lett., 29 (1994) 339.
12. K. Otsuka, R. Takahashi and I. Yamanaka, J. Catal., 185 (1999) 182.
13. M. A. Bañares, J. H. Cardoso, G. J. Hutchings, J. M. C. Bueno and V. L. G. Fierro, Catal. Lett., 56 (1998) 149.
14. Y. Teng, H. Sakurai, K. Tabata and E. Suzuki, Appl. Catal. A, 190 (2000) 283.
15. K. Tabata, Y. Teng, Y. Yamaguchi, H. Sakurai and E. Suzuki, J. Phys. Chem. A, 104 (2000) 2648.
16. T. Takemoto, K. Tabata, Y. Teng, S. Yao, A. Nakayama and E. Suzuki, Energy & Fuels in press.
17. B. Delly, J. Chem. Phys. 92 (1990) 503. $DMol^3$ ver. 3.9 of $Cerius^2$ program suite is available from Molecular Simulations Inc., San Diego, CA.

Studies in Surface Science and Catalysis
J.J. Spivey, E. Iglesia and T.H. Fleisch (Editors)

Production and storage of hydrogen from methane mediated by metal oxides

K. Otsuka*, A. Mito, S. Takenaka and I. Yamanaka

Department of Applied Chemistry, Graduate School of Science and Engineering,
Tokyo Institute of Technology, Ookayama, Meguro-ku, Tokyo 152-8552, Japan

A novel method for the storage and production of hydrogen from methane mediated by metal oxides has been proposed. The method combines the catalytic decomposition of methane, the redox of metal oxides and the utilization of the deposited carbon as a chemical feed stock for the production of CO or syngas. The hydrogen recovered through the redox of metal oxides does not contain a trace of CO, thus can be supplied directly to H_2-O_2 fuel cells.

1. INTRODUCTION

Hydrogen is a clean fuel that emits no CO_2 when it is burned or used in H_2-O_2 fuel cells. However, the current processes for the production of hydrogen from natural gas (methane) and water or from other fossil resources and water inevitably emit a huge quantity of CO_2 into the atmosphere. Moreover, the hydrogen produced by these processes inevitably contains CO as an impurity even after a thorough purification treatment. The CO strongly poisons anode electrocatalysts of the H_2-O_2 cells such as KOH-, H_3PO_4- and SPE-fuel cells. Nevertheless, we believe that one of the realistic resources for the production of hydrogen with less CO_2-emission is methane.

Another key technology to be developed for the H_2-O_2 fuel cells for automobiles and houses or factories is to explore a simple and safe method for the storage of hydrogen.

Under these circumstances, we here propose a new hydrogen production and storage method without CO_2-emission from methane, which is realized by the catalytic decomposition of methane into carbon and hydrogen over catalysts in the presence of metal oxides separated from the catalyst. The hydrogen produced by this method does not contain CO at all.

The concept of this method is made up of the following four steps.

I. Methane is completely decomposed into carbon and hydrogen (Eq. 1) over a solid catalyst in the presence of metal oxides (denoted as MO_x) which is reduced by the hydrogen (Eq. 2) produced in Eq. 1. The produced water in Eq. 2 should be condensed in a trap cooled at low temperatures (<200 K) in order to shift the equilibrium reaction of Eq. 2 to the right. The hydrogen is stored as a reduced solid metal oxide (MO_{x-1}).

$$CH_4 \rightleftarrows C + 2H_2 \quad (1)$$

$$H_2 + MO_x \rightleftarrows H_2O + MO_{x-1} \quad (2)$$

II. The carbon produced in Eq. 1 can be used as carbon black, graphite, fibers, plastics, composites, etc. [1-3]. The carbon could be used also as a chemical feed stock for the production of CO or synthesis gas with the reaction of CO_2 or steam.

III. The reduced metal oxide in Eq. 2 is transported, stored and installed for the systems of H_2-O_2 fuel cells at the local facilities, factories, individual houses, or on vehicles.

IV. Injection and contact of water vapor with the reduced metal oxides recover the pure hydrogen without carbon oxides (CO and CO_2). The hydrogen can be used directly as the fuel of H_2-O_2 fuel cells. The metal oxide (MO_x) regenerated here can be recycled to step I.

For the method described above, we have chosen Ni/Cab-O-Sil (Cab-O-Sil: Fumed silica from CABOT Co.) as a catalyst for the decomposition of methane in Eq. 1 and In_2O_3 and iron oxides (Fe_3O_4) as metal oxide mediators for the storage and production of hydrogen in Eq. 2. The Ni/Cab-O-Sil is one of the most active and stable catalysts we have tested [4]. In_2O_3 and iron oxides (Fe_3O_4) are the promising mediator oxides which can be quickly reduced and reoxidized at 673 K according to the following go and back reactions [5, 6].

$$3H_2 + In_2O_3 \rightleftarrows 2In + 3H_2O \quad (3)$$

$$4H_2 + Fe_3O_4 \rightleftarrows 3Fe + 4H_2O \quad (4)$$

In this work, firstly we demonstrate the repeated experiments of the reduction of the mediator oxides and the decomposition of water vapor by the reduced metal oxides. Secondly, since the complete decomposition of methane is thermodynamically prohibited at low temperatures < 973 K in the absence of the metal oxides, we show the 100% conversion of methane (Eq. 1) in the presence of mediator oxides and the 100% recovery of hydrogen through the decomposition of water by the reduced metal oxides.

2. EXPERIMENTAL

The Ni(10 wt%)/Cab-O-Sil catalyst used for the decomposition of methane was prepared by impregnation of an aqueous $Ni(NO_3)_2$ solution into Cab-O-Sil (Cabot Co.) at 363 K. The impregnated sample was dried at 373 K for 10 h and calcined at 873 K for 5 h in air. Prior to the methane decomposition, the catalyst was reduced by hydrogen at 823K. The Fe_2O_3 powder (starting iron oxide) was purchased from Wako Pure Chem. Co. The In_2O_3 (13.1 wt%) supported on MgO and ZrO_2 were prepared by calcination (at 823 K) of the powder supports immersed in aqueous solutions of $In(NO_3)_3$.

The decomposition of methane over the Ni/Cab-O-Sil and the reduction of the mediators were performed at the same time by circulating the gases in a gas-closed and gas-circulation system with two reactors and a cold trap (at dry-ice temp.) installed in the line in series. The regeneration of hydrogen from the reduced metal oxides (backward reactions of Eqs. 3 and 4) was performed by vaporizing the condensed water in the trap at 288 K and circulating the vapor with argon through the reactor containing the reduced metal oxide. The reactor containing the carbon and Ni/Cab-O-Sil was shut off by stopcocks during the decomposition of water. The analysis of hydrogen in the decomposition of methane and that during the decomposition of water were performed by G. C.

3. RESULTS AND DISCUSSION

For the reduction and reoxidation of indium oxides (Eq. 3), one of the authors reported that MgO and ZrO_2 enhanced the redox of In_2O_3 compared to the sample without these carriers [7]. Therefore, we examined the redox cycles of the In_2O_3/MgO and In_2O_3/ZrO_2. The results are shown in Figs.1 and 2, respectively. The reduction of the In_2O_3/MgO samples and In_2O_3/ZrO_2 by hydrogen and subsequent decomposition of water vapor by the reduced samples were repeated for 6 cycles at 623 K for both samples. The reduced extents of In_2O_3 evaluated from the amount of hydrogen consumed at the final points for each curve in Figs. 1(a) and 2(a) were ca. 50%. After degassing of the residual hydrogen in the experiments of reduction of In_2O_3, the vaporization of the condensed water at 287 K ($P(H_2O)$ = 1.60 kPa) and the circulation of the vapor through the reactor initiated a fast decomposition of water, forming hydrogen as can be seen in Figs 1(b) and 2(b).

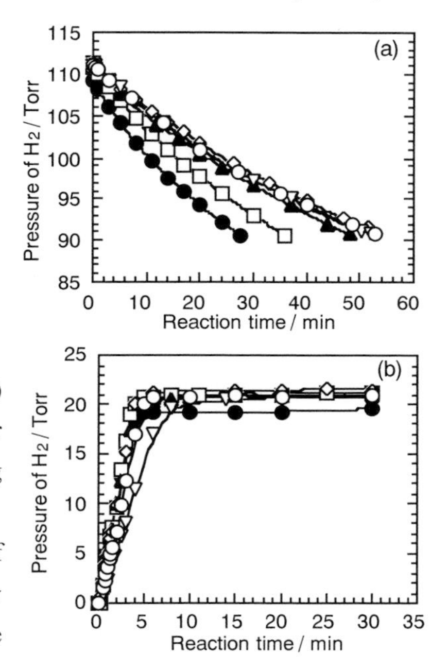

Fig. 1. Reduction and reoxidation of In_2O_3/MgO at 623 K. In_2O_3/MgO = 0.30 g. ●: 1st, □: 2nd, ▲: 3rd, ◇: 4th, ▽: 5th, ○: 6th.

In the case of In_2O_3/MgO in Fig.1(a) the rate of reduction of the sample decreased considerably with the cycles from the first to the third, but the decrease became very slight after the third cycle. On the other hand, the results in Fig.1(b) indicated that the rate of hydrogen formation did not change appreciably with the cycles. Each cycle recovered the hydrogen almost 100% of that consumed in the reduction of In_2O_3/MgO.

In the case of In_2O_3/ZrO_2, the rate of reduction of the sample increased noticeably in the second cycle compared to that observed in the first cycle (Fig.2(a)). However, the rate of reduction did not change appreciably after the second cycle. The results in Fig.2(b) indicated that the rate of water decomposition was very fast and did not change with the number of cycles. The reaction rate might be controlled by the supply rate of the water vapor. The consumed hydrogen in Fig.2(a) for each cycle was recovered by almost 100% in Fig.2(b).

The similar experiments without degassing of hydrogen were performed for iron oxide (starting oxide Fe_2O_3) at 623 K. The results are shown in Fig. 3. The reduction and the reoxidation of the oxides were repeated for three cycles. The condensation of water vapor at 0, 95 and 195 min brought about the reduction of the iron oxide. The first reduction of the iron oxide included the reduction of Fe_2O_3 to Fe_3O_4 and Fe. The formal degree of reduction of iron oxide at 45 min was 70% of Fe_2O_3. The vaporization of the condensed water at 50, 137 and 240 min initiated the decomposition of water according to the stoichiometric backward reaction of Eq. 4. The recovery of hydrogen was almost 100% for the second and third reduction-oxidation cycles in Fig. 3.

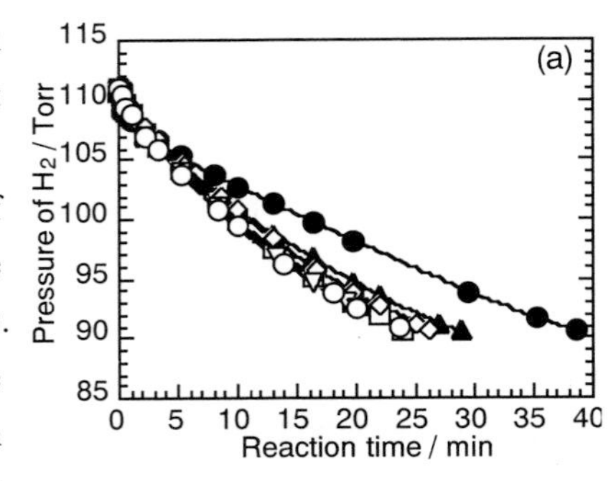

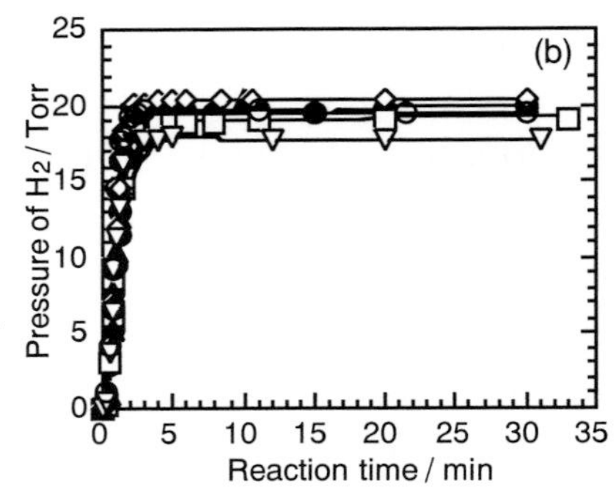

Fig. 2. Reduction and reoxidation of In_2O_3/ZrO_2 at 623 K. In_2O_3/ZrO_2 = 0.30 g. ●: 1st, □: 2nd, ▲: 3rd, ◇: 4th, ▽: 5th, ○: 6th.

Figure 4 indicates the decomposition of methane on Ni/Cab-O-Sil at 723 K in the absence and presence of In_2O_3/MgO in a separate reactor at 723 K by circulating methane and the produced hydrogen. The water vapor formed due to the oxidation of hydrogen with In_2O_3/MgO was condensed in a trap put at the exit of the reactor. It was confirmed that methane

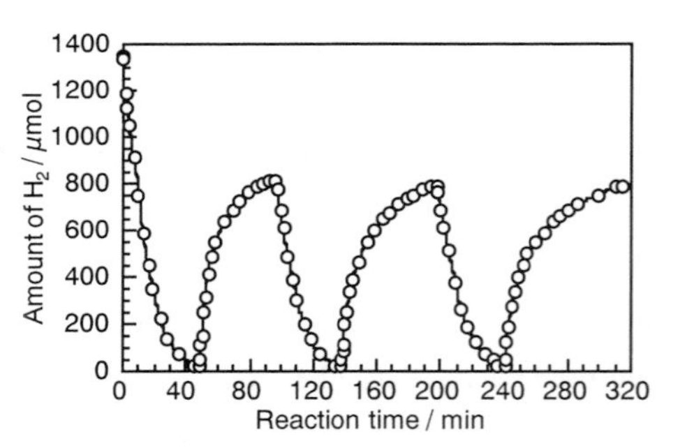

Fig. 3. Reduction and reoxidation cycles of Fe_2O_3 at 673 K. Fe_2O_3 = 0.10 g.

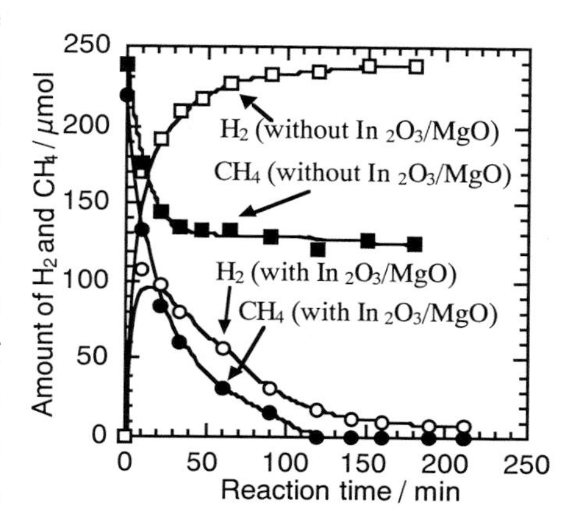

Fig. 4. Decomposition of methane over Ni/Cab-O-Sil in the presence or absence of In_2O_3/MgO. Reaction temperature = 723 K, Ni(10 wt%)/Cab-O-Sil = 0.10 g, In_2O_3/MgO = 1.00 g.

did not react with the In_2O_3/MgO at 723 K. The decomposition of methane on Ni/Cab-O-Sil catalyst in the absence of In_2O_3/MgO occurred at the early stage of the reaction, but the rate became very slow after 30 min probably due to the equilibrium of the reaction (Eq. 1). In contrast, the rate of decomposition of methane was remarkably increased in the presence of In_2O_3/MgO. The conversion of methane was completed at 120 min due to the consumption of hydrogen by the mediator oxide (In_2O_3/MgO). The 100% conversion of methane under the similar reaction conditions in Fig. 4 was confirmed also in the presence of In_2O_3/ZrO_2, In_2O_3 and Fe_2O_3 in the same gas-circulation reaction system.

The decomposition of methane on Ni/Cab-O-Sil and the subsequent decomposition of water were performed at 723 K for four cycles. The results are indicated in Fig. 5. After the complete decomposition of methane in the first cycle, the vaporization of the condensed water at 200 min caused the very fast formation of hydrogen due to the decomposition of water vapor by the reduced In_2O_3/MgO. CO was not detected within the analysis limit (< 100 ppm) during the reactions in Fig. 5. The recovery of hydrogen was almost 100%, i.e., all the hydrogen contained in methane was recovered as hydrogen through the decomposition of methane, the reduction of metal oxide with the formed hydrogen from methane, and the reduction of water by the reduced metal oxide. The

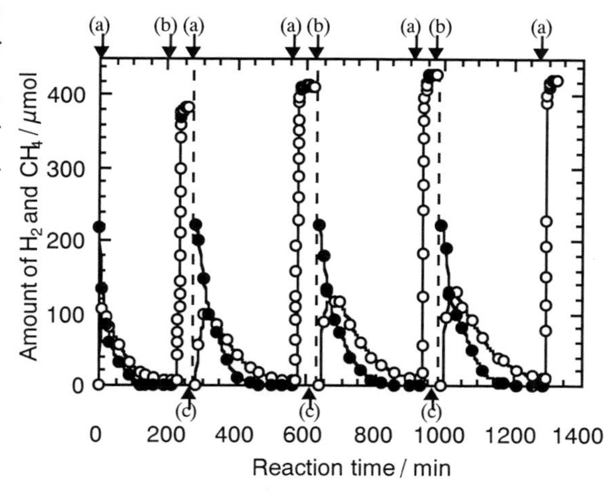

Fig. 5. Repeated decomposition of methane and the generation of hydrogen both at 723 K in the presence of In_2O_3/MgO. (a): addition of CH_4, (b): vaporization of H_2O, (c): evacuation of the gas-phase. Ni (10 wt%)/Cab-O-Sil = 0.10 g, In_2O_3/MgO = 1.00 g.

second cycle was continued after evacuating the formed hydrogen in the first cycle. The second to the fourth experiments gave the same results as can be seen in Fig. 5. The catalytic activity of the Ni/Cab-O-Sil did not decrease at least for 6 cycles. Very similar results to those of Fig. 5 were obtained for the experiments by using oxide mediators of In_2O_3/ZrO_2, In_2O_3 and Fe_2O_3.

The contact of the reduced indium oxides and reduced iron oxides with air under atmospheric pressure at room temperature did not decrease the reactivity of the oxides for the decomposition of water at > 673 K. In other words, the reduced states of these metal oxides (mixture of metal and oxide phases) are resistant under open-air at room temperature.

The gasification with CO_2 at 873 K for the carbon on the Ni/Cab-O-Sil after complete deactivation of the catalyst in the decomposition of methane produced CO selectively with conversions of the carbon greater than 92%. This treatment completely recovered the original catalytic activity of the Ni/Cab-O-Sil for the decomposition of methane.

4. CONCLUSION

The results described in this work suggest that In_2O_3-based oxides and iron oxides are promising mediator oxides for the storage of hydrogen generated in the decomposition of methane. The processes combining the catalytic decomposition of methane, the storage of hydrogen by using the redox of metal oxides and the gasification of the carbon with CO_2 may realized a zero CO_2-emission production of hydrogen which can be supplied directly to H_2-O_2 fuel cells.

ACKNOWLEDGEMENTS

A part of this work was financially supported by NEDO (International Join Research Grant) and Sankaken (Hiroshima Prefectural Institute of Industrial Science and Technology).

REFERENCES

1. M. Steinberg and H.C. Cheng, Int. J. Hydrogen Energy, 14 (1989) 797.
2. N.Z. Muradov, Int. J Hydrogen Energy, 18 (1993) 211.
3. B. Gaudernack and S. Lynum, Int. J Hydrogen Energy, 23 (1998) 1087.
4. K. Otsuka, T.Seino, S.Kobayashi and S. Takenaka, Chem. Lett., (1999) 1179.
5. K. Otsuka, T. Yasui and A. Morikawa, J. Catal. 72 (1981) 392; J. Chem. Soc., Faraday Trans I. 78 (1982) 3281.
6. S. Nakaguchi, Sekiyu Gakkaishi, 20 (1977) 69; 5th Synthetic Pipeline Gas Symposium (1973); B.S. Lee and P.B. Tarman, 6th Synthetic Pipeline Gas Symposium (1973).
7. K. Otsuka, S. Shibuya and A. Morikawa, J. Catal., 84(1983) 308.

Studies in Surface Science and Catalysis
J.J. Spivey, E. Iglesia and T.H. Fleisch (Editors)

Oxidative Dehydrogenation over Sol-Gel Mo/Si:Ti Catalysts: Effect of Mo Loading

Rick B. Watson and Umit S. Ozkan*

Department of Chemical Engineering, The Ohio State University,
Columbus, OH 43210, USA

Characterization has been performed over un-promoted Mo/Si:Ti catalysts with increasing Mo loadings to elucidate the effects of surface coverage. Normalized propylene formation rates (per Mo atom) diminish rapidly with Mo loading (2-20 wt.%) over Si:Ti 1:1. The most selective catalyst contained the least amount of supported molybdenum, 2wt.%Mo/Si:Ti 1:1. This behavior is explained in terms of the unique interaction of MoO_x structures with the support, Si:Ti, and the increasing formation of larger 2 and 3-dimensional MoO_3 structures. XPS results indicate, from a ~229eV Mo3d feature, that at low Mo loadings, isolated Mo-O-Si bonds are present on the catalyst surface. As Mo loading is increased, a steady shift of the Mo=O terminal Raman band was observed, from 970 to 995cm^{-1}. At higher loadings the isolated active Mo species may either be non-observable or preferentially replaced with less active MoO_3 crystallites.

1. INTRODUCTION

Strong demand for olefins has led to numerous studies on the use of a catalytic oxidative dehydrogenation (ODH) process to produce olefins from light alkanes. Alkali-promoted molybdate catalysts supported over the mixed oxides of silica-titania has led to promising results in this area [1-2]. Silica-titania mixed oxides represent a novel class of materials that have been studied extensively [3-8] for physical-chemical properties such as acidity, porosity, existence of Ti-O-Si chemical bonds, and TiO_2/SiO_2 phase separations. However, few studies have been done on their use as support material for transition metals [9-13]. Silica-titania mixed oxide supports, through sol-gel preparations, can provide stronger metal-support interactions, enhanced thermal stability, and smaller particle sizes that leads to better dispersion and higher surface area for a variety of catalyst systems. Ko et al. and Kumbhar [14-15] have shown that TiO_2/SiO_2 mixed oxides exert both direct and indirect support effects when used as supports for Ni catalysts. Furthermore, Baiker et al. [16] has proposed that by varying the TiO_x content in the mixed oxides, one can "tune" the interaction with VO_x species to form an optimal deNO_x catalyst.

---

*To whom correspondence should be addressed
Phone: (614) 292-6623
Fax: (614) 292-3769
E-mail: ozkan.1@osu.edu

An understanding of the structural characteristics of these mixed oxides and their relationship to physico-chemical properties is of great importance over a wide range of material applications [17]. Furthermore, these unique support effects on active metal oxides represent a relatively recent research focus.

In this work, a series of molybdena catalysts (2-20% wt. Loadings) have been studied in regard to their activity for the oxidative dehydrogenation (ODH) of propane. X-ray diffraction (XRD), Raman spectroscopy, and X-ray photoelectron spectroscopy (XPS) were used to determine the nature of the molybdena phases.

2. EXPERIMENTAL

2.1 Catalyst Preparation

Catalysts were prepared using a modified sol-gel/co-precipitation technique. Ammonium heptamolybdate (AHM) (Mallinkrodt) was used for the molybdenum precursor. For silica-titania mixed oxides, tetraethylorthosilicate (TEOS) (Aldrich) and titanium(IV)isopropoxide (TIPO) (Aldrich) were used. This method is referred to as a "one-pot" sol-gel/co-precipitation and is described in detail in [1]. Synthesized catalysts are listed in Table 1. Catalysts numbered 1 through 7 are a series of molybdate catalysts with increasing loading of Mo and a Si:Ti molar ratio of 1.

2.2 Characterization

BET surface area measurement and nitrogen adsorption-desorption isotherms were recorded using a Micrometrics AccuSorb 2100E instrument. X-ray diffraction patterns were obtained with a Scintag PAD-V diffractometer using Cu-Kα radiation. Raman spectra were recorded with a Dilor spectrometer using the 514.5nm line of an Innova 300 Ar Laser. Spectra were taken in the range 200-1800cm^{-1} in 180° back-scattering mode with a Spectrum One CCD detector. X-ray photoelectron spectroscopy (XPS) of catalysts was performed with an ESCALAB MKII ESCA/Auger Spectrometer operated at 14kV, 20mA, and using Mg-Kα radiation. Spectra were corrected using the C 1s signal, located at 284.6eV. Instrumental sensitivity factors were used to calculate relative elemental compositions.

2.2 Oxidative dehydrogenation of propane

Steady-state reaction experiments were carried out in a fixed-bed, quartz reactor, operated at ambient pressure. Catalyst samples, roughly 65m^2 of surface area in the reactor, were held in place by a quartz frit. To minimize effects from any homogeneous reaction or surface-initiated gas phase reaction and to provide a short residence time for propylene formed, the dead volume of the quartz microreactor was filled with quartz wool and/or ceramic beads.

Reaction temperature was 500°C. The feed consisted of propane (26%), oxygen (13%), and nitrogen(61%) at a flowrate of 25 cm^3/min. The product distributions maintained a carbon balance of 100% (+/- 5%).

Table 1
Catalyst Compositions

Composition	BET Surface Area (m^2/g)
Si:Ti 1:1	320
2%Mo/Si:Ti 1:1	343
5%Mo/Si:Ti 1:1	246
10%Mo/Si:Ti 1:1	210
15%Mo/Si:Ti 1:1	188
20%Mo/Si:Ti 1:1	116

3. RESULTS AND DISCUSSION

3.1 Effect of Mo loading on physical properties

Molybdate catalysts show a general decrease in surface area with increasing amounts of molybdenum added. X-ray diffraction of the Si:Ti 1:1 support yielded a pattern typical of a silica-titania sample [18]. One broad peak with center located at a d spacing of 3.59 Å was observed, which is the most intense diffraction line from anatase structure. A broad peak is indicative of a finely dispersed, small X-ray particulate anatase structure supported over amorphous silica. With the addition of molybdenum this band becomes narrower, indicating a change in the dispersion and segregation of titania in the Si:Ti matrix with the addition of Mo. Molybdena species are more finely dispersed on the mixed oxide supports than on silica or titania alone, showing no MoO_3 diffraction patterns up to 15% wt. loading. However, it may be possible for MoO_3 to exist as microcrystalline material below the approximately 40 Å XRD detection limit [19]. Although quite weak, the two most intense peaks from crystalline molybdenum oxide (110 and 021 planes) become detectable when the Mo loading level is increased to 20%.

Raman spectra of catalysts with different Mo loadings are presented in Fig. 1. An important feature of these spectra is that there is no evidence of crystalline MoO_3 up to 15% wt. loading. The broad bands associated with isolated terminal Mo=O stretching vibrations are observed to shift with Mo loading from 970 cm^{-1} at 2wt.% to 995 cm^{-1} at 20wt. %. These shifts are due to changes in the Mo=O bond length [20] and can be ascribed to a decreasing interaction with the support and the formation of 3-dimensional structures.

3.2 Equal surface area reaction experiments

Mo/Si:Ti catalysts were compared in the ODH reaction using equal surface area loading ($65m^2$) in the reactor and at a temperature of 500°C. The feed percentages for these experiments were $N_2/C_3/O_2$: 61/26/13. The formation rates of propylene and CO_x and the depletion rates of propane during the first hour of reaction are presented in Fig 2. The propylene formation rates, normalized to Mo atoms, show a decrease with Mo loading under these reaction conditions.

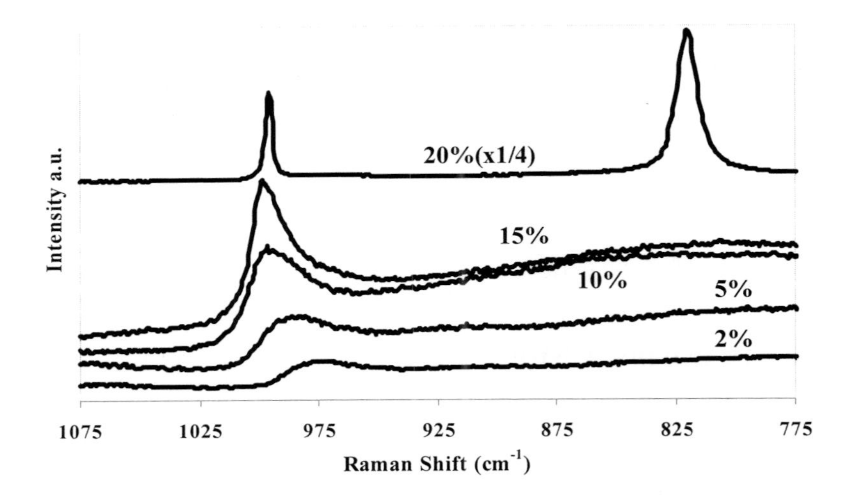

Fig. 1. Raman Spectra of Mo/Si:Ti 1:1 Catalysts

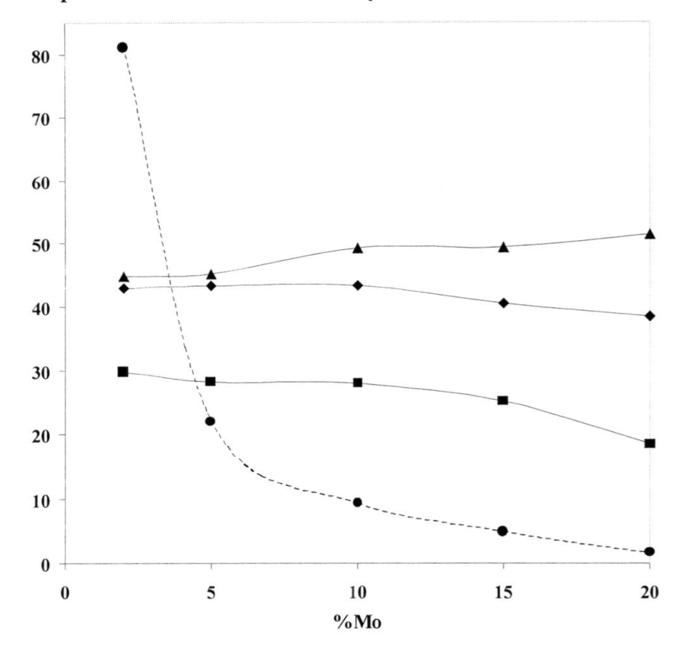

Fig. 2. Formation Rates (■propylene, ▲CO_x) and Depletion Rates (◆propane) for Equal Surface Area ($65m^2$) Reaction Experiments (μmol/min./m^2). ●Normalized Propylene Formation Rate (mol/min./Mo atom $x10^{12}$)

The observed decrease in propylene formation can be ascribed to the rising formation of 3-dimensional molybdate species and/or microcrystalline MoO_3. The formation of bulk MoO_3 with increasing Mo loading are not as favorable for the reaction as surface-supported MoO_x species over many support materials Observing the Raman spectra of these catalysts, the decrease in propane depletion and propylene formation is accompanied by a shift to higher frequency of the Mo=O terminal bonds. Similar Mo=O shifts have been related to an increase in the strength of the bond that lowers ODH activity [21].

3.3 XPS of Mo/Si:Ti Catalysts

Molybdenum 3d spectra of calcined catalysts with different Mo loadings are presented in Fig 3. Within the detection limits of XPS, molybdenum is seen to exist as Mo^{+6} on the surface of these catalysts. Comparing the location of the $3d_{5/2}$ peak, a positive shift of ~1eV is observed from 2 to 20wt.% Mo, approaching the $3d_{5/2}$ binding energy of crystalline MoO_3. At the lowest loading studied, 2wt.%, there is a low binding energy feature present around ~229eV. Since any effects from differential charging are expected to affect the high-B.E. side of the peaks, it appears correct to ascribe this feature to an interaction species (e.g., Mo-O-Si) forming at the catalyst surface at this lower loading of molybdenum.

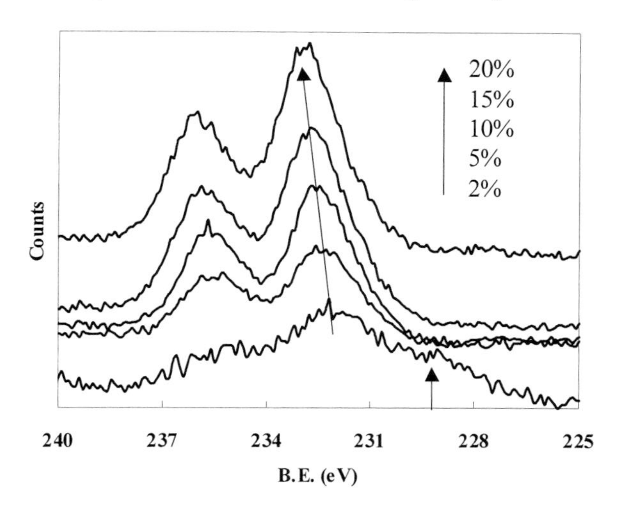

Fig. 3. XPS (Mo 3d region) of Mo/Si:Ti Catalysts

This is another indication of the structure becoming increasingly 3-dimensional with increasing Mo loading. The fact that this feature is observed only on the lowest loading level implies that, to facilitate Mo-Si interaction, it is necessary to have MoO_x as isolated species rather that 2 or 3-dimensional domains. Since the strength of these interaction species is often directly related to the dispersion of the catalytically active sites over supported molybdate catalysts [22], it is consistent that we observe a decrease in ODH activity with increasing Mo loading.

Furthermore, additional XPS characterization (Si 2p, Ti 2p, and O 1s) has shown that increasing Mo loading levels decrease Ti-Si interaction, leading to the formation of nanodomains of anatase. It appears that Mo loading affects not only the nature of the surface MoO_x species, but also influences the nature of TiO_x/MoO_x competition for the silica surface.

4. SUMMARY

Mo/Si:Ti catalysts are found to be active for the oxidative dehydrogenation of propane. Increasing molybdenum loadings are seen to provide a rather constant propylene formation up to 10wt.% followed by a decline with increasing Mo content. However, normalized propylene formation rates (per Mo atom) diminish rapidly with Mo loading (2-20 wt.%) over Si:Ti 1:1. The decrease in activity can be explained in terms of increasing crystalline MoO_3 or 3-dimenionality of the MoO_x surface units as indicated by XPS and Raman results. The accessibility to active sites decrease due to the formation of bulk phase MoO_3, therefore propylene formation rates decrease at higher Mo loadings. Molybdenum oxide typically interacts weakly with silica, which limits its dispersion [23]. However, on the Si:Ti mixed oxide support, the presence of Mo-support interactions that occur through surface hydroxyls or in Si:Ti surface interfaces provide a catalyst with better dispersion and higher activity than silica alone.

5. REFERENCES

1. Watson, R.B., Ozkan, U.S., J. Catal., 191 (2000) 12.
2. Watson, R.B., Ozkan, U.S., Stud. Surf. Sci, Catal., 130 (2000) 1883.
3. Stakheev, A.Y., Shipiro, E.S., Apijok, J., J. Phys. Chem., 97 (1993) 5668.
4. Lassaletta, G., Fernandez,A., Espinos, J.P., Gonzalez-Elipe, A.R., J. Phys. Chem., 99 (1995) 1484.
5. Walters, J.K., Rigden, J.S., Dirken, P.J., Smith, M.E., Howells, W.S., Newport, R.J., Chem. Phys. Letters, 264 (1997) 539.
6. Kumar, S.R., Suresh, C.,. Vasudevan, A.K, Suja, N.R., Mukundan, P., Warrier, K.G.K., Mat. Letters, 38 (1999) 161.
7. Klein, S., Thorimbert, S., Maier, W.F., J. Catal., 163 (1996) 476.
8. Liu, Z., Tabora, J., Davis, R.J., J. Catal., 149 (1994) 117.
9. Vogt, E.T.C., Boot, A., van Dillen, A.J., Guess, J.W., Janssen, F.J.J.G., van den Kerkhof, F.M.G, J. Catal., 114 (1988) 313.
10. Rieck, J.S., Bell, A.T., J. Catal., 99 (1986) 262.
11. Feng, Z., Liu, L., Anthony, R.G., J. Catal., 136 (1993) 423.
12. Cauqui, M.A., Calvino, J.J., Cifredo, G.C., Esguivias, L., Rodriguez-Izquierdo, J.M, J. Noncryst. Solids, 147 (1992) 758.
13. Udomsak, S., Anthony, R.G., Ind. Eng. Chem. Res., 35 (1996) 47.
14. Ko, E.I., Chen, J.-P., Weissman, J.G., J. Catal., 105 (1987) 511.
15. Kumbhar, P.S., Appl. Catal., 96 (1993) 241.
16. Baiker, A., Dollenmeier, P., Glinski, M., Appl. Catal. A, 35 (1987) 365.
17. Gao, X., Wachs, I.E., Catal. Today, 51 (1999) 233.
18. Izutsu, H., Nair, P.K.; Maeda, K.; Kiyozumi, Y.; Mizukami, F., Mat. Res. Bull., 32 (1997) 1303.
19. Datta, A. K., Ha, J.W., Regalbuto, J.R., J. Catal., 133 (1992) 55.
20. Wachs, I.E., Catal. Today, 27 (1996) 437.
21. Chen, K., Xie, S., Iglesia, E., Bell, A.T., J. Catal., 189 (2000) 421
22. Plyuto, Y.V., Babich, I.V., Plyuto, I.V., Van Langeveld, A.D., Moulijn, J.A., Appl. Surf. Sci., 119 (1997) 11.
23. Dhas, N.A., Gedanken, A., J. Phys. Chem. B, 101 (1997) 9495.

ACKNOWLEDGEMENTS

Financial support provided by the National Science Foundation (Grant# CTS-9412544) is gratefully acknowledged. The authors would also like to thank Dr. Gurkan Karakas for his technical assistance at the early stages of the project.

Studies in Surface Science and Catalysis
J.J. Spivey, E. Iglesia and T.H. Fleisch (Editors)

NO_x-Catalyzed Partial Oxidation of Methane and Ethane to Formaldehyde by Dioxygen

Ayusman Sen* and Minren Lin

Department of Chemistry, The Pennsylvania State University, University Park, Pennsylvania 16802. E-mail: asen@chem.psu.edu

At 600°C, NO_x was found to catalyze the partial oxidation of both methane and ethane by dioxygen to form formaldehyde as the predominant oxygenated product. The yield of oxygenates from methane was over 11%. The yield increased to over 16% when a trace of ethane (0.7%) was added to the gas mixture. As might be expected, the yield of oxygenates from ethane was higher: over 24%. A catalytic cycle involving NO_2 as the C-H activating species is proposed.

1. INTRODUCTION

Methane is the least reactive and the most abundant member of the hydrocarbon family. Ethane comes second in both categories. Together, their known reserves approach that of petroleum [1]. Thus, the selective oxidative functionalization of these alkanes to more useful chemical products is of great practical interest [2-7]. For example, one of the highest volume functionalized organics produced commercially is formaldehyde and the current technology for the conversion of alkanes to formaldehyde involves a *multi-step* process. Clearly, the *direct* conversion of the lower alkanes to formaldehyde would be far more attractive from an economic standpoint. Of particular interest would be the formation of the *same* end product from different starting alkanes, thus obviating the need to separate the alkanes. For example, natural gas is principally methane with 5-10% ethane. A system that converts both to the same C_1 product, would not require the prior separation of the alkanes. Herein, we report a NO_x-catalyzed process for the oxidation of methane and ethane to formaldehyde by dioxygen.

2. RESULTS AND DISCUSSION

Our procedure involves passing the gas mixture at 1 atmosphere pressure through a preheated empty quartz tube (length: 30 cm; diameter: 2 cm). The exiting mixture was passed through three gas bubblers each containing 4 ml of D_2O and finally analyzed by gas chromatography. The products trapped in water were analyzed by NMR spectroscopy.

Our results are summarized in Tables 1-3. In every case, increasing the flow rate resulted in decreased conversion and increased selectivity for the oxygenates. Several interesting observations were made. First, *both* methane and ethane gave formaldehyde as the predominant oxygenate (trapped either as aqueous solution or isolated as the solid polymer, paraformaldehyde). The total one-pass yield (= conversion x selectivity) of oxygenates from methane was over 11%. The yield increased to over 16% when a trace of ethane (0.7%) was added to the gas mixture; an interesting and apparently unprecedented synergistic effect. As might be expected, the yield of oxygenates from ethane was higher: over 24%. The major side product in all the partial oxidations was carbon monoxide; curiously, only a trace of carbon dioxide was formed in the oxidation of methane (vide infra). Finally, as shown in Tables 1-3, a good material balance was obtained. In the context of partial oxidation of methane and ethane to formaldehyde, we are unaware of any other system in the open literature that *simultaneously* exhibits such high one-pass conversion and selectivity (highest previously reported yield from methane: 6%). These literature reports include both metal-catalyzed oxidations [5-11], as well as oxidations initiated by nitrogen-oxo species [12-16].

A critical issue is whether the oxidations are indeed *catalytic* in NO_x, i. e. NO_x is not consumed in the course of the reaction (cf., Scheme 1). The two possible catalytic "deactivation" pathways are the formation of dinitrogen and organonitrogen compounds. No N_2 derived from NO_x was detected by gas chromatography in experiments where N_2 was replaced by He in the inlet gas mixture. When the partial oxidation of methane was carried out at a lower temperature (525°C) in an excess of dioxygen, up to 2% CH_3NO_2 was formed. Simply raising the reaction temperature from 525°C to around 600°C prevented CH_3NO_2 formation. Unfortunately, however, the elevated temperature also led to the increased formation of CO_x thereby decreasing the selectivity towards the useful oxygenates, *except* when the O_2:NO ratio was kept below 0.5 so that *no excess O_2 remained* after NO_2 was formed. Under these conditions, the oxidation proceeds *without any loss of NO_x*, either as dinitrogen or as organonitrogen compounds. Our work, in this respect, differ from those reported by others in which either a lower temperature was employed resulting in the formation of CH_3NO_2, or an excess of oxygen was employed resulting in lower selectivity to useful products [12-16].

Scheme 1

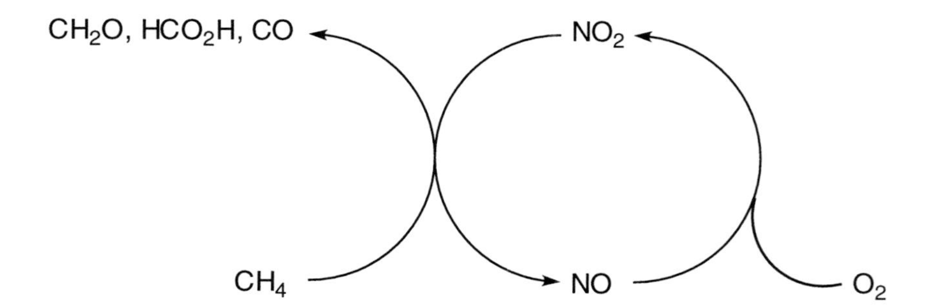

A key mechanistic question involves the identity of the radical initiator in the NO_x/O_2 system. Both NO and NO_2 are odd-electron species and are, in principle, capable of atom abstraction. However, no oxidation of methane was observed at either 600°C or 615°C using NO only. On the other hand, "normal" oxidation of methane was observed under identical conditions when NO was replaced by NO_2. These experiments suggest that the NO_x-mediated oxidations are initiated by attack of the substrate by NO_2. This conclusion is also supported by theoretical calculations which indicate a significantly lower barrier for H-atom abstraction from alkanes by NO_2. [17]. Scheme 1 outlines the mechanism of the NO_x-mediated oxidation of alkanes by dioxygen.

Our knowledge of the mechanism is incomplete; however, we believe that the principal role of NO_2 is to abstract a hydrogen-atom from the alkane, thereby generating a flux of alkyl radicals that is eventually converted to the products. Interestingly, some methane is invariably obtained from ethane (Table 3). This, together with the observation of formaldehyde as the principal product from ethane, appear to indicate that alkoxy radicals are the primary products which subsequently undergo the well-known β-bond cleavage reaction [18] to produce alkyl radicals together with formaldehyde. A control experiment indicated that methanol is converted to formaldehyde with a selectivity similar to that observed for methane (Table 4). The significantly higher yield of products obtained from methane in the presence of a trace of ethane may be ascribed to the easier formation of the ethyl radical which upon subsequent conversion to the corresponding alkoxy radical can continue the chain reaction by abstracting a hydrogen atom from methane.

ACKNOWLEDGEMENT

This research was funded by the CANMET Consortium on the Conversion of Natural Gas and in part by the National Science Foundation.

Table 1. Methane oxidation. Gas mixture: CH_4, (16.7%); NO, (11.1%); O_2, (5.6%); N_2, (16.7%); He, (50.1%).

Experiment No.	**1**	**2**
Temperature, °C.	600	600
Flow rate, mL/sec	4.52	5.65
Yield of CH_3OH*, %	---	0.24
Yield of CH_2O*, %	9.73	9.27
Yield of HCO_2H*, %	1.24	1.80
Yield of CO*, %	11.77	8.21
Total conversion*, %	**22.74**	**19.52**
Selectivity, %**	**48.24**	**58.00**
Yield*, %**	**10.97**	**11.31**
Total products, mmol.	4.35	4.57
CH_4 (inlet) - CH_4 (outlet), mmol.	4.20	4.40

*Based on CH_4. **Selectivity: $\Sigma(CH_2O + CH_3OH + HCO_2H)/\Sigma$(total products). ***Yield: Total conversion x Selectivity.

Table 2. Methane oxidation in presence of trace ethane. Gas mixture: CH_4, (16.6%); CH_3CH_3, (0.7%); NO, (11.0%); O_2, (5.5%); N_2, (16,6%); He, (49.7%).

Experiment No.	**1**	**2**
Temperature, °C.	600	600
Flow rate, mL/sec	2.26	5.65
Yield of CH_3OH*, %	---	1.00
Yield of CH_2O*, %	13.96	15.30
Yield of HCO_2H*, %	2.27	0.64
Yield of CO*, %	14.26	8.50
Total conversion*, %	**30.5**	**25.4**
Selectivity, %**	**53.2**	**66.5**
Yield*, %**	**16.2**	**16.9**
Total products, mmol.	3.08	6.21
[CH_4 (inlet) – CH_4 (outlet)] + CH_3CH_3 x 2, mmol	3.23	5.82

*Based on total carbon in CH_4 and CH_3CH_3. **Selectivity: $\Sigma(CH_2O + CH_3OH + HCO_2H)/\Sigma$(total products). ***Yield: Total conversion x Selectivity.

Table 3. Ethane Oxidation. Gas mixture: CH_3CH_3, (16.4%); NO, (11.0%); O_2, (6.9%); N_2, (16.4%); He, (49.4%).

Experiment No.	**1**	**2**	**3**
Temperature, °C.	600	600	600
Flow rate, mL/sec	2.26	6.78	9.04
Yield of CH_3OH*, %	0.93	0.44	1.60
Yield of CH_2O*, %	22.95	23.64	23.00
Yield of HCO_2H*, %	---	0.04	---
Yield of CO*, %	10.93	10.23	8.35
Yield of CO_2*, %	1.20	0.53	0.55
Yield of CH_4*, %	1.97	1.67	1.72
Yield of CH_2=CH_2*, %	2.40	2.62	2.60
Total conversion*, %	**40.38**	**39.20**	**37.82**
Selectivity, %**	**59.14**	**61.55**	**65.04**
Yield*, %**	**23.88**	**24.12**	**24.60**
Total products, mmol.	7.40	21.76	27.64
[CH_3CH_3(inlet) – CH_3CH_3(outlet)] x 2, mmol	7.40	21.52	27.40

*Based on total carbon in CH_3CH_3. **Selectivity: $\Sigma(CH_2O + CH_3OH + HCO_2H)/\Sigma$(total products). ***Yield: Total Conversion x Selectivity.

Table 4. Methanol Oxidation. Gas mixture: NO, (5.4%); O_2, (16.1%); N_2 (78.4%) was bubbled through methanol at 25°C.

Temperature, °C	525
Total CH_3OH, mmol.	84.0
Yield of CH_2O*, %	20.8
Yield of HCO_2H*, %	2.65
Yield of CO*, %	17.9
Yield of CO_2*, %	1.74
Unreacted CH_3OH, mmol.	44.8
Mass balance*, %	**96.4**

*Based on CH_3OH.

REFERENCES

1. M. G. Axelrod, A. M. Gaffney, R. Pitchai, and J. A. Sofranko, in Natural Gas Conversion II, H. E. Curry-Hyde and R. F. Howe (eds.), Elsevier, Amsterdam, 1994, p. 93.

2. A. Sen, Top. Organomet. Chem., 3 (1999) 81.

3. G. A. Olah, A. Molnár, Hydrocarbon Chemistry, Wiley, New York, 1995.

4. R. H. Crabtree, Chem. Rev., 95 (1995) 987.

5. T. J. Hall, J. S. J. Hargreaves, G. J. Huchings, R. W. Joyner, and S. H. Taylor, Fuel Process. Technol., 42 (1995) 151.

6. J. L. G. Fierro, Catalysis Lett., 22 (1993) 67.

7. M. J. Brown and N. D. Parkyns, Catalysis Today, 8 (1991) 305.

8. A. Parmaliana, F. Arena, F. Frusteri, and A. Mezzapica, Stud. Surf. Sci. Catal., 119 (1998) 551.

9. R. L. McCormick, G. O. Alptekin, A. M. Herring, T. R. Ohno, and S. F. Dec, J. Catal., 172 (1997) 160.

10. M. A. Banares, L. J. Alemany, M. López Granados, M. Faraldos, and J. L. G. Fierro, Catalysis Today, 33 (1997) 73.

11. T. Ono, H. Kudo, and J. Maruyama, Catal. Lett., 39 (1996) 73.

12. Y. Teng, H. Sakurai, K. Tabata, and E. Suzuki, Appl. Catal. A, 190 (2000) 283.

13. K. Otsuka, R. Takahashi, and I. Yamanaka, J. Catal., 185 (1999) 182.

14. K. Otsuka, R. Takahashi, K. Amakawa, and I. Yamanaka, Catalysis Today, 45, (1998) 23.

15. L.-B. Han, S. Tsubota, and M. Haruta, Chemistry Lett., (1995) 931.

16. S. Irusta, E. A. Lombardo, and E. E. Miró, Catalysis Lett., 29 (1994) 339.

17. Y. Yamaguchi, Y. Teng, S. Shimomura, K. Tabata, and E. Suzuki, J. Phys. Chem. A, 103 (1999) 8272.

18. J. K. Kochi, in Free Radicals, J. K. Kochi (ed.) Wiley, New York, 1973, Vol. II, p. 665.

Studies in Surface Science and Catalysis
J.J. Spivey, E. Iglesia and T.H. Fleisch (Editors)

Comparative study of partial oxidation of methane to synthesis gas over supported Rh and Ru catalysts using *in situ* time-resolved FTIR and *in situ* microprobe Raman spectroscopies

Wei Zheng Weng, Qian Gu Yan, Chun Rong Lou, Yuan Yan Liao, Ming Shu Chen and Hui Lin Wan*

Sate Key Laboratory for Physical Chemistry of Solid Surfaces, Department of Chemistry and Institute of Physical Chemistry, Xiamen University, Xiamen, 361005, CHINA

in situ time-resolved FTIR and *in situ* microprobe Raman spectra were used to study the partial oxidation of methane (POM) to synthesis gas over Rh/SiO_2, Ru/SiO_2 and $Ru/\gamma\text{-}Al_2O_3$ catalysts. It is found that direct oxidation of methane to synthesis gas is the main pathway of POM reaction over Rh/SiO_2 catalyst, while the combustion-reforming scheme is the dominant pathway of synthesis gas formation over Ru/SiO_2 and $Ru/\gamma\text{-}Al_2O_3$ catalysts. The significant difference in the mechanisms of POM reaction over supported Rh and Ru catalysts can be related to the difference in surface concentration of oxygen species (O^{2-}) over the two catalyst systems under the reaction condition.

1. INTRODUCTION

The partial oxidation of methane (POM) to synthesis gas (syngas) is a promising alternative to the conventional steam reforming process due to its mild exothermicity and the more favorable H_2/CO ratio in the syngas obtained [1]. A unified mechanism for partial oxidation of methane to syngas over metal catalysts has not been reached so far. Some people [2, 3] claim that direct oxidation of CH_4 to CO and H_2 is the major reaction pathway. While, others [4-6] believed that formation of syngas mainly proceeded via an initial exothermic oxidation of CH_4 to CO_2 and H_2O followed by the endothermic reforming of unconverted CH_4 with CO_2 and H_2O. A major problem encountered with the combustion-reforming scheme is the conceivably uneven axial temperature distribution within a fixed bed reactor due to large quantities of heat produced by the combustion of CH_4 at the entrance of the catalyst bed. If this mechanism predominantly contributes to the conversion of methane to syngas, the problems such as heat management, catalyst deactivation due to sintering and requirement of high temperature material for reactor construction have to be taken into account in the design of an industrial process. On the other hand, direct oxidation of methane to synthesis gas is of high industrial interest since the problem associated with severe uneven temperature distribution within the catalyst bed can be largely resolved.

---

* Corresponding author, e-mail: hlwan@xmu.edu.cn
This project is supported by the Ministry of Science and Technology (No. G1999022408), National Nature Science Foundation (No. 20023001) and the Doctoral Foundation from the Ministry of Education.

Obviously, for the direct oxidation scheme, CO is the primary product of the reaction, while for the combustion-reforming scheme, CO_2 should be formed before the formation of CO. Therefore, by monitoring the sequence of product (CO and CO_2) formation during the reaction, it is possible to distinguish the two different reaction mechanisms proposed for POM reaction. In the present study, *in situ* time-resolved FTIR (*in situ* TR-FTIR) spectroscopy with time resolution better than 0.30s is used to followed the reaction of a simulated POM feed ($CH_4/O_2/Ar$ = 2/1/45, molar ratio) over the H_2-reduced Rh/SiO_2, Ru/SiO_2 and $Ru/\gamma\text{-}Al_2O_3$ catalysts at 773~873K. The parallel experiments using *in situ* microprobe Raman spectroscopy to characterize the surface of the above three catalysts under O_2, H_2/Ar (5/95, molar ratio) and $CH_4/O_2/Ar$ (2/1/45, molar ratio) atmospheres were also performed at 773~873K. It is expected that the experiments can provide more information to the understanding of the correlation between the oxidation state of the metal in the catalyst under the reaction condition and the reaction schemes for partial oxidation of methane to syngas.

2. EXPERIMENTAL

The SiO_2 and $\gamma\text{-}Al_2O_3$ supported Rh and Ru catalysts were prepared by the method of incipient wetness impregnation, using metal chloride as the precursor compound for the metal. When the solvent evaporated, the solid material was dried at 383K and calcined at 773K for 2h.

The *in situ* TR-FTIR experiments were performed on a Perkin Elmer Spectrum 2000 FTIR spectrometer equipped with a liquid-nitrogen-cooled MCT detector and a home built high temperature *in situ* IR cell with quartz lining and CaF_2 windows. The catalyst was pressed to a self-supporting disk (~10mg for SiO_2 supported catalysts and ~20mg for $\gamma\text{-}Al_2O_3$ supported catalyst) and reduced with H_2 (99.999%) at 873K in the IR cell for 60min followed by evacuation at 773 or 873K for ca. 5min to remove the gas phase H_2. After evacuation, the IR cell was filled with 0.10MPa of Ar (99.999%) and then switched to a flow of simulated POM feed ($CH_4/O_2/Ar$ = 2/1/45, molar ratio) at 773 or 873K. The flow rate of the feed is higher than 68ml/min. The IR spectra of the surface and the gas phase species formed during the reaction were continuously recorded at the resolution of $16cm^{-1}$ with average of 2 scans. The time interval between two recorded spectra was less than 0.30s. The reference spectrum was that of the catalyst prior to the admission of reactant feed.

The experiments of *in situ* microprobe Raman characterization of the catalysts were performed using a home built high temperature *in situ* microprobe Raman cell, which allows us to record the Raman spectra of the catalyst from room temperature to 973K under different gas atmosphere. The Raman spectrum was recorded on a Dilor LabRam I confocal microprobe Raman system. The exciting wavelength was 632.8nm from Ar^+ laser with power of 10mW and a spot of ca. 3μm on the catalyst surface. The laser beam was focused on the top of the catalyst bed. In each experiment, the catalyst (ca. 2.5×10^{-3}ml) was first treated in a flow of O_2 (99.995%) at 773K for 30min. After O_2 pretreatment, the Raman spectra of the catalyst were recorded at each specified temperature point under a flow of O_2. The catalyst was then switched to a flow of H_2/Ar (5/95, molar ratio) at 773 or 873K to take the Raman spectra of H_2-reduced sample. Finally, the H_2-reduced sample was switched to a flow of simulated POM feed ($CH_4/O_2/Ar$ = 2/1/45, molar ratio) at 773 or 873K to take the Raman spectra of the catalyst under POM condition. In all Raman experiments, the flow rate of the gas (O_2, H_2/Ar and $CH_4/O_2/Ar$) is higher than 5ml/min.

3. RESULTS AND DISCUSSION

The TR-FTIR spectroscopic investigation on the reaction of $CH_4/O_2/Ar$ gas mixture over the H_2-reduced 1wt%Rh/SiO_2 catalyst was performed at 773K. The corresponding IR spectra were shown in Fig. 1. It was found that, when a H_2-reduced 1wt%Rh/SiO_2 was switched to a flow of $CH_4/O_2/Ar$ gas mixture (GHSV > 180,000h^{-1}) at 773K, the adsorbed CO species with IR band at 2017cm^{-1} was the first reaction product being observed (0.28s). The IR bands of CO_2 (2311, 2356cm^{-1}) and gas phase CO (2110, 2177cm^{-1}) were detected at 0.56 and 1.1s, respectively. Since no IR bands of CO_2 species were observed in the spectrum recorded at 0.28s, it is very unlikely that the adsorbed CO species detected at this stage is from the CO_2 reforming of CH_4 unless the rate of CH_4 reforming with CO_2 is higher than the rate of CH_4 combustion. The result of catalytic performance evaluation on the reaction of CH_4/CO_2 (1/1, molar ratio) gas mixture over a H_2-reduced 1wt%Rh/SiO_2 catalyst indicated that, at 773K with GHSV = 54,000h^{-1}, the conversions of CH_4 and CO_2 were less than 5%, which is much lower than the conversion of CH_4 in the partial oxidation reaction at the same temperature. At higher GHSV, the contribution of CH_4/CO_2 reforming to the CO formation will be even lower. It should also be realized that, the surface of Rh/SiO_2 catalyst is fully reduced under the reforming conditions, which will be favorable to the reaction of CH_4 with CO_2. While under the condition of methane partial oxidation, the surface of the catalyst is at least partially oxidized, on which the rate of reforming reaction will be reduced. Based on these results and analysis, it is reasonable to conclude that the adsorbed CO species detected at 0.28s in the IR spectrum shown in Fig. 1 is from the direct partial oxidation of methane.

In contrast to the situation over 1wt%Rh/SiO_2 catalyst, the reaction of $CH_4/O_2/Ar$ gas mixture (GHSV > 180,000h^{-1}) over a H_2-reduced 1wt%Ru/SiO_2 catalyst at 873K led to the formation of CO_2 (2311, 2356cm^{-1}) as the primary product (Fig. 2). IR bands of adsorbed CO (1976cm^{-1}) and those of gaseous CO (2113, 2181cm^{-1}) were detected at 2.0s. This result clearly indicates that CO_2 is the primary product of POM reaction over Ru/SiO_2 catalyst and the mechanism of partial oxidation of methane to syngas over the catalyst is most probably via combustion of CH_4 to CO_2 and H_2O followed by reforming of unconverted CH_4 with CO_2 and H_2O. When a H_2-reduced 2wt%Ru/γ-Al_2O_3 catalyst was switched to a flow of $CH_4/O_2/Ar$ gas mixture at 873K with GHSV > 160,000h^{-1}, CO_2 was the only product observed in the IR spectra. However, if the GHSV of the reactant feed was reduced to ~80,000h^{-1}, the IR bands of adsorbed CO species at 1950 and 2037cm^{-1} (2.2s) and those of gaseous CO at 2114 and 2177cm^{-1} (4.3s) was also found. The corresponding TR-FTIR spectra are shown in Fig. 3. This observation indicated that, for the POM reaction over Ru/γ-Al_2O_3 catalyst, the combustion-reforming mechanism is also the dominant pathway of syngas formation. These results are also consistent with the conclusion made by Guerrero-Ruiz et al. [7] and Verykios et al. [8] for the POM reaction over Ru/Al_2O_3 and Ru/SiO_2 catalysts based on results of isotopic tracing experiments.

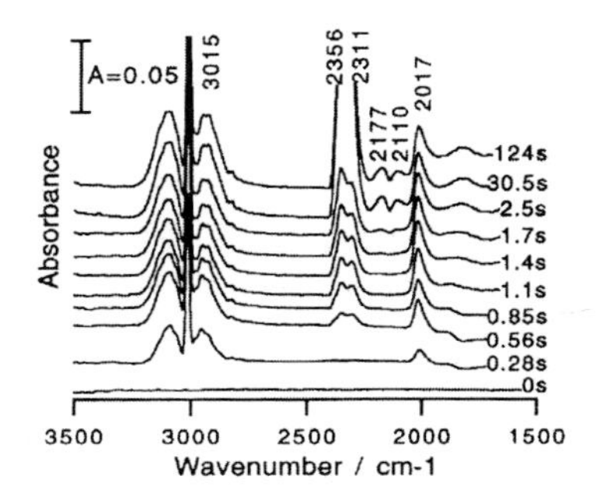

Fig. 1. *in situ* TR-FTIR spectra for a flow reaction of $CH_4/O_2/Ar$ gas mixture over H_2-reduced 1wt%Rh/SiO_2 at 773K.

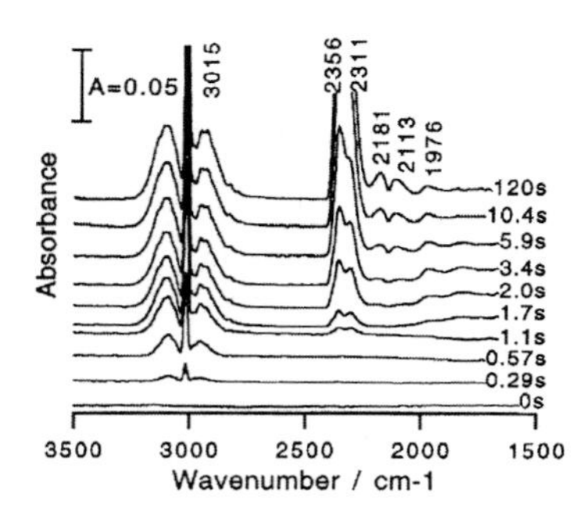

Fig. 2. *in situ* TR-FTIR spectra for a flow reaction of $CH_4/O_2/Ar$ gas mixture over H_2-reduced 1wt%Ru/SiO_2 at 873K.

The influence of gas hourly space velocity (GHSV) of the reactant ($CH_4/O_2/Ar$ = 2/1/45) on the catalytic performance of POM reaction over SiO_2 supported Rh and Ru catalysts were studied at 600°C. It was found that, when GHSV of the feed was increased from 300,000 to 1,000,000h^{-1}, the CO selectivity over 1wt%Rh/SiO_2 catalyst remained almost unchanged (~90%), while that over 4wt%Ru/SiO_2 catalyst gradually decreased from 88.1% to 81.7%, although the conversion of CH_4 over both catalysts decreased significantly. These results are also in support of the mechanisms proposed based on the above *in situ* TR-FTIR studies

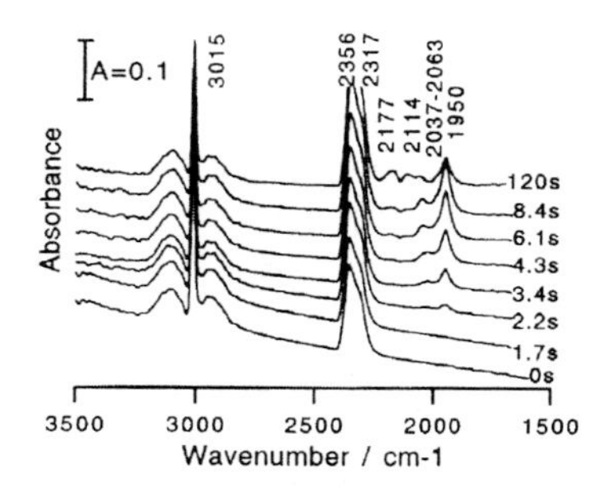

Fig. 3. *in situ* TR-FTIR spectra for a flow reaction of $CH_4/O_2/Ar$ gas mixture over H_2-reduced 2wt%Ru/γ-Al_2O_3 at 873K.

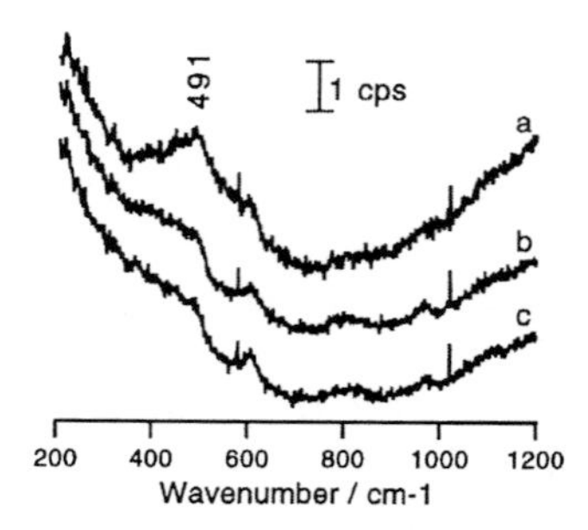

Fig. 4. Raman spectra of 4wt%Rh/SiO_2 at 773K under a flow of a) O_2, b) $CH_4/O_2/Ar$ and c) H_2/Ar.

In order to elucidate the relationship between the oxidation state of the metal in the above three catalysts under the POM reaction condition and the reaction schemes for partial oxidation of methane to syngas, the parallel experiments using *in situ* microprobe Raman spectroscopy to characterized the surface of 4wt%Rh/SiO_2, 4wt%Ru/SiO_2 and 2wt%Ru/γ-Al_2O_3 catalysts under O_2, H_2/Ar and CH_4/O_2/Ar atmospheres were performed at 773~873K. Fig. 4a shows the Raman spectrum of 4wt%Rh/SiO_2 catalyst recorded at 773K under a flow of O_2. A broad band with maximum at ca. 491cm^{-1} is observed. This band can be assigned to rhodium oxide. It has been reported by Beck et al. [9] that, under O_2 atmosphere, a complete oxidation of Rh to Rh_2O_3 took placed at temperature above 773K. When the O_2-pretreated 4wt%Rh/SiO_2 sample was switched to a flow of H_2/Ar at 773K, the band of Rh_2O_3 disappeared (Fig. 4c). No Raman band of Rh_2O_3 was observed when the H_2/Ar pretreated 4wt%Rh/SiO_2 sample was switched to a flow of CH_4/O_2/Ar gas mixture at the same temperature (Fig. 4b). Fig. 5a is the Raman spectrum of 4wt%Ru/SiO_2 catalyst recorded at 873K under a flow of O_2. Two bands attributable to RuO_2 were observed at 489 and 609cm^{-1}. These bands vanished when the O_2-pretreated 4wt%Ru/SiO_2 was switched to a flow of H_2/Ar at 873K (Fig. 5c), and reappeared when the H_2/Ar pretreated 4wt%Ru/SiO_2 catalyst was switched to a flow of CH_4/O_2/Ar gaseous mixture at the same temperature (Fig. 5b). The Raman spectra recorded over 2wt%Ru/γ-Al_2O_3 catalyst at 773K under a flow of O_2 (Fig. 6a), H_2/Ar (Fig. 6c) and CH_4/O_2/Ar (Fig. 6b) were very similar to those recorded over 4wt%Ru/SiO_2 catalyst under the same atmosphere. Obviously, the surface of Ru/SiO_2 and Ru/γ-Al_2O_3 catalysts contained a considerably larger amount of oxygen species (O^{2-}) than that of the Rh/SiO_2 catalyst under the POM reaction condition, indicating that the former two catalysts have a much higher tendency to be oxidized, as compared to the latter one due to the higher oxygen affinity of Ru. These results suggest that the significant difference in the reaction schemes of POM reaction over SiO_2 and γ-Al_2O_3 supported Rh and Ru catalysts can be related to the difference in the surface concentration of oxygen species (O^{2-}) over the catalysts under the POM conditions. A possible reason for this may be ascribed to the higher oxygen affinity of ruthenium compared with that of rhodium.

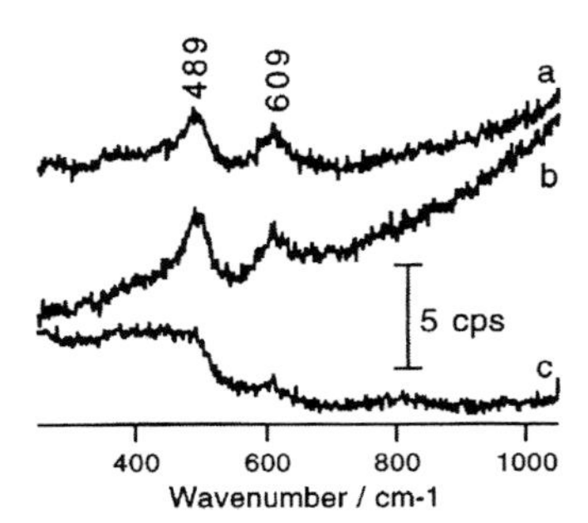

Fig. 5. Raman spectra of 4wt%Ru/SiO_2 at 873K under a flow of a) O_2, b) CH_4/O_2/Ar and c) H_2/Ar.

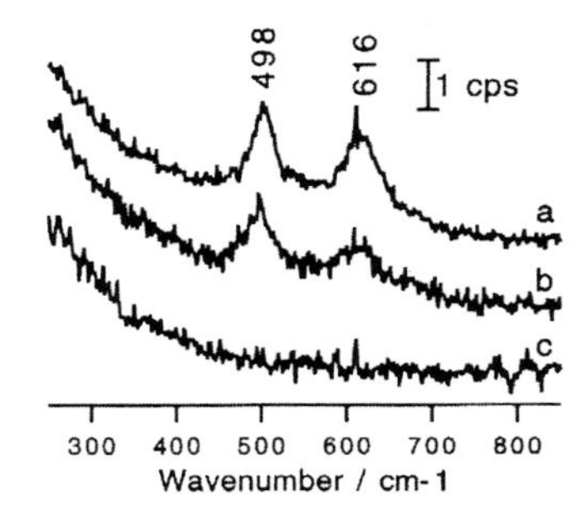

Fig. 6. Raman spectra of 2wt%Ru/γ-Al_2O_3 at 773K under a flow of a) O_2, b) CH_4/O_2/Ar and c) H_2/Ar.

On the Rh/SiO_2 catalyst, most of the Rh atoms in the catalyst bed are in the metallic state under the reaction condition. On the Rh^0 sites, CH_4 may be activated by dissociation to surface CH_X (x = 0~3) species, which was then converted to CO through a series of surface reactions, including the dissociation of H from CH_X and oxidation of surface C species by oxygen species on the catalyst surface, as proposed by Hickman and Schmidt [2, 3] based on the study of the catalytic performance of POM reaction on Pt and Rh impregnated monoliths and on Pt/10%Rh metal gauzes at temperature around 1300K and residence times between 10^{-2} and10^{-4}s.

On the Ru/SiO_2 and $Ru/\gamma\text{-}Al_2O_3$ catalysts, however, the Ru species near the top the catalyst bed is almost fully oxidized under the reaction condition. Over these catalysts, if the flow rate of the feed is high enough (i.e. the oxygen in the feed is not completely consumed), most of the Ru species in the catalyst bed will be in the oxidized state, which will predominantly catalyze complete oxidation of CH_4 to CO_2 and H_2O. This is the situation when a H_2-reduced 2wt%$Ru/\gamma\text{-}Al_2O_3$ was switched to a flow of $CH_4/O_2/Ar$ gas mixture with GHSV > 160,000h^{-1} at 873K. However, if the flow rate of the feed is not very high, oxygen in the $CH_4/O_2/Ar$ gas mixture will be completely consumed in a narrow zone near the front of the catalyst bed. The Ru species at the rear of the catalyst bed will remain in metallic state, which will catalyze the reforming reactions of the resulting $CH_4/CO_2/H_2O$ mixture to syngas.

5. CONCLUSIONS

Based on the above *in situ* TR-FTIR and *in situ* microprobe Raman spectroscopic characterization, it can be concluded that direct oxidation of CH_4 to syngas is the main pathway of POM reaction over Rh/SiO_2 catalyst, while the combustion-reforming mechanism is the dominant pathway of syngas formation over Ru/SiO_2 and $Ru/\gamma\text{-}Al_2O_3$ catalysts. The significant difference in the reaction schemes of POM reaction over SiO_2 and $\gamma\text{-}Al_2O_3$ supported Rh and Ru catalysts may be related to the difference in surface concentration of oxygen species (O^{2-}) over the catalysts under the reaction conditions mainly due to the difference in oxygen affinity of the two metals

REFERENCES

1. J.R. Rostrup-Nielsen, in J. R. Anderson and M. Boudart (eds.), Catalysis, Science and Technology, Vol. 5, p1, Springer, Berlin, 1984.
2. D.A. Hickman and L.D. Schmidt, J. Catal., 138 (1992) 267.
3. D.A. Hickman, E.A. Haupfear and L.D. Schmidt, Catal. Lett., 17 (1993) 223.
4. M. Prettre, C. Eichner and M. Perrin, Trans. Faraday Soc., 43 (1946) 335.
5. P.D.F. Vernon, M.L.H. Green, A.K. Cheetham and A.T. Ashcroft, Catal. Lett., 6 (1990) 181.
6. D. Dissanayake, M.P. Rosynek, K.C.C. Kharas and J.H. Lunsford, J. Catal., 132 (1991) 117.
7. A. Guerrero-Ruiz, P. Ferreira-Aparicio, M.B. Bachiller-Baeza and I. Rodriguez-Ramos, Catal. Today, 46 (1998) 99.
8. Y. Boucouvalas, Z.L. Zhang, A.M. Efstathiou and X.E. Verykios, Stud. Surf. Sci. Catal., 101 (1996) 443.
9. D.D. Beck, T.W. Capehart, C. Wong and D.N. Belton, J. Catal., 144 (1993) 311.

Studies in Surface Science and Catalysis
J.J. Spivey, E. Iglesia and T.H. Fleisch (Editors)

Optimisation of Fischer-Tropsch Reactor Design and Operation in GTL Plants

Dag Schanke(Corresponding author e-mail dsc@statoil.com), Petter Lian, Sigrid Eri, Erling Rytter, Bente Helgeland Sannæs and Keijo J. Kinnari
Statoil Research Centre, Trondheim, N-7005 Norway

Abstract
By using examples from fixed-bed and slurry bubble column reactors using supported cobalt Fischer-Tropsch catalysts, the influence of key process parameters on reactor performance is illustrated. It is shown that there is a strong coupling between reaction kinetics, intraparticle / interphase mass-transfer and reactor characteristics which, together with other process parameters, will determine the overall reactor performance.

1. Introduction

The interest in Gas-To-Liquids (GTL) processes as a way of utilizing associated and remote natural gas is growing. There is a strong focus on reducing capital costs, which is the main contributor to the overall production costs of gas based fuels. Technology improvements focus on new large scale synthesis gas production technology as well as new Fischer-Tropsch (F-T) catalyst and reactor technology. The latest F-T technology include improved cobalt based catalysts and three-phase slurry bubble columns (SBCR) with vastly increased train sizes (> 10 000 bbl/d). The present paper will discuss some key aspects of the use of cobalt catalysts in SBCR's.

2. Experimental

The experimental results were obtained either in fixed-bed laboratory reactors or in slurry bubble column reactors. In addition, a slurry CSTR was employed for kinetic studies. All the data shown are representative of steady state conditions (> 100 hours on stream). The slurry reactors were operated for extended periods (up to 2500 hours) with continuous liquid product (wax) separation. Al_2O_3-supported, rhenium-promoted cobalt catalysts with varying metal and promoter loading have been used in the present study.

3. Results and Discussion

3.1. Reaction kinetics for overall HC synthesis and selectivity prediction

Previous studies of F-T kinetics over Co catalysts have led to various rate expressions [1], none of which have proven to be satisfactory over a sufficient range of conditions. Our results show that F-T synthesis over cobalt catalysts is characterised by a low apparent "pressure order" at typical gas compositions, pressures and temperatures. However, there is a non-linear dependence of reaction rates on pressure, with a high (near 1st order) dependence at low pressures while at higher pressures the rates are almost insensitive to pressure (fig. 1). As a result, space-time vs. conversion plots typically give straight lines up to ca. 60% conversion (fig 2.).

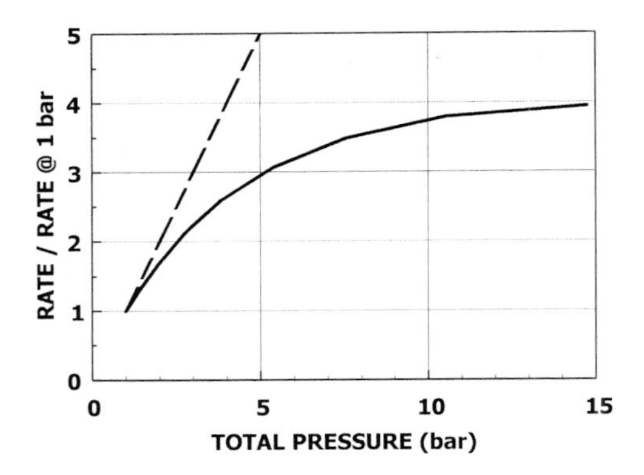

Fig. 1. Calculated hydrocarbon formation rate (solid curve) as a function of pressure at differential conditions (near zero conversion) at $H_2/CO = 2.1$. 1st order curve included for comparison (dashed line).

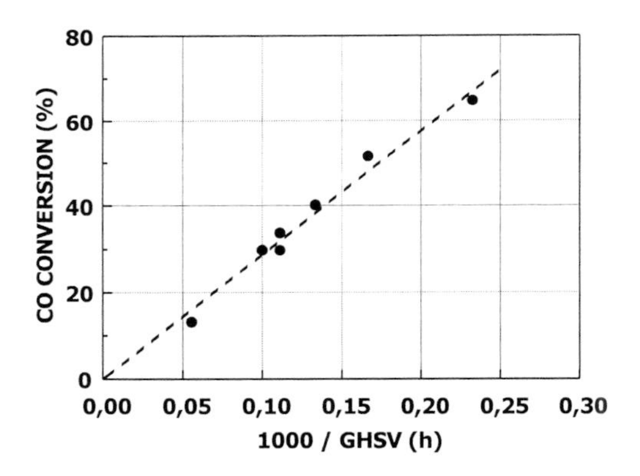

Fig. 2. Space-time vs. conversion plot in a plug-flow (fixed-bed) reactor. Cat: Co-Re/Al_2O_3. T = 210°C, P = 20 bar, H_2/CO (feed) = 2.1. GHSV = Ncm^3 total feed / g_{cat} / h

In order to predict overall CO consumption, total hydrocarbon formation and selectivities, separate rate expressions have been developed for the formation rates of the main product groups CH_4, C_2-C_4, C_5+ and CO_2. The values of the individual rate constants, adsorption constants and activation energies have been obtained from fitting the rate expressions to numerous experimental data obtained in continuous stirred tank reactors (slurry CSTR). A wide range of reaction conditions have been utilised in order to arrive at robust rate expressions that can be used in reactor simulation models and process optimisation studies.

The importance of a robust kinetic model which includes the product distribution is illustrated in fig. 3a. At typical reaction conditions for slurry phase F-T, the C_5+ selectivity increases with increasing conversion up to a certain level. The C_5+ selectivity then drops rather sharply and is accompanied by a significant increase in the CO_2 selectivity resulting from the water-gas-shift reaction. The behaviour shown in fig. 3a. is well described by the rate expressions developed, as shown by the calculated data in fig. 3b.

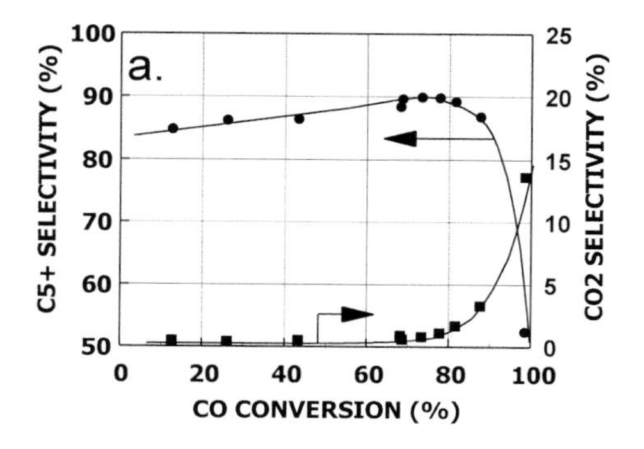

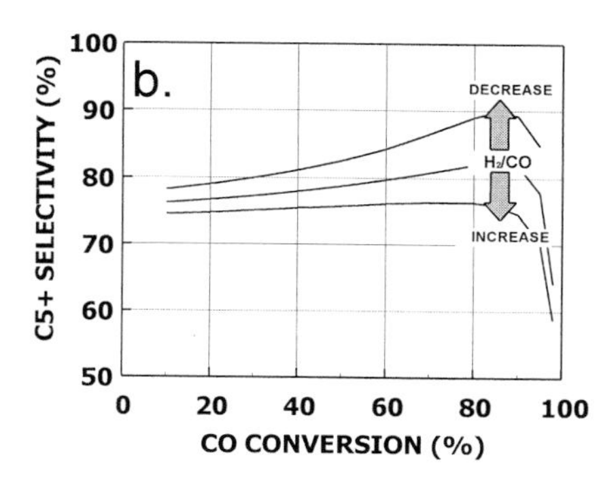

Fig. 3. C_5+ and CO_2 selectivity as a function of conversion.
a) Experimental data (slurry CSTR) obtained by varying space velocity at constant catalyst concentration. Catalyst: Co-Re / Al_2O_3, T =220°C, P = 20 bar, H_2/CO (feed) = 2.0.
b) C_5+ selectivity predicted by kinetic model at different feed gas compositions (CSTR)

3.2. Intraparticle mass transfer effects

One of the main advantages of a SBCR in F-T synthesis is the ability to use small catalyst particles. A catalyst effectiveness factor close to 1.0 can be expected as a result of the short diffusion path from the external catalyst surface to the catalytic site for catalyst particle sizes e.g. in the 10 - 150 micron range [2,3].

The effect of intraparticle diffusion effects on catalyst effectiveness and selectivity has been described in some detail by Iglesia et al. [2] and is obviously more relevant for fixed-bed reactors due to the larger particle sizes employed. However, some interesting additional conclusions can be derived from results using Al_2O_3 supported catalysts with different particle sizes in fixed-bed reactors (fig. 4). In contrast to the results shown by Iglesia et al., a "volcano plot" is not observed. Instead, the selectivity at 200°C / 20 bar is nearly constant up to χ= ca. $2000 \cdot 10^{16}$ m^{-1} (χ defined as in fig. 4 and ref. [2]) and then drops sharply as a result of diffusion limitations which change the intraparticle H_2/CO ratio. In a fixed-bed reactor, the most practical way of solving this problem is the use of so-called egg-shell or rim type catalysts where the active material is deposited in a thin layer at the outer surface of the catalyst pellets.

It is clear that more active catalysts and higher reaction temperatures will move the upper acceptable particle size towards lower values. The same is the case for lower pressures, which can be seen as equivalent to high conversions and/or significant gas-liquid mass transfer limitations in backmixed reactor systems. This is illustrated by the results obtained at 220°C and 6 bar syngas pressure, showing that under such conditions the maximum particle size will now correspond to $\chi <$ ca. $100 \cdot 10^{16}$ m^{-1}, giving particle sizes in the order of 0.1 mm or even less. This means that particle size effects may have to be taken into account even in a SBCR when highly active catalysts are used. It is also interesting to notice that trickle bed or ebullating bed reactors (both using large particles in the presence of a liquid-phase) which have been proposed for F-T operation will face severe constraints when simultaneously trying to maximise productivity and selectivity.

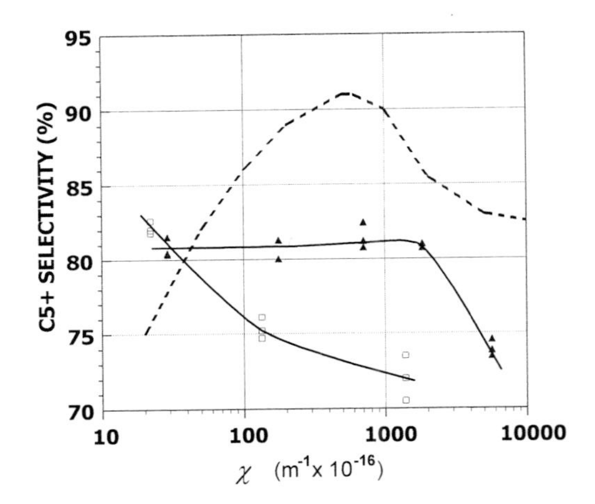

Fig. 4. Particle size effects on F-T selectivity in fixed-bed reactors . Catalyst: Co-Re / Al_2O_3 .
Solid triangles: T = 200°C, 20 bar inlet partial pressure of H_2+CO.
Open squares: T = 220°C, 6 bar inlet partial pressure of H_2+CO.
Dashed curve: Model curve redrawn from fig. 20 of Iglesia et al. [ref. 2]

$\chi = R_0^2 \_ \hbar / r_p$

R_0 = *Catalyst particle radius (m)*

φ = *Catalyst porosity*

θ = *Catalytic site density (sites/m²)*

r_p = *average pore radius (m)*

3.3. Gas-liquid mass transfer

Mass transfer between the bulk gas phase and the bulk liquid can be a determining factor for the overall volumetric productivity (space time yield, STY) that can be obtained in a SBCR. In addition and more important, mass transfer effects on product selectivity will become apparent before the overall hydrocarbon productivity is affected. The mechanism for selectivity loss at high volumetric production rates follows the same principle as described for particle diffusion, by increasing the liquid H_2/CO ratio relative to the bulk gas ratio.

An illustration of these effects is presented in fig. 5. In 1 - 2" I.D. SBCRs operating in the homogeneous (bubble flow) regime, indications of mass transfer limitations on the overall rate (expressed by the global effectiveness factor) can only be detected for catalyst concentrations corresponding to space time yields > 70 g_{HC} / l_{sl} /h with the catalysts and reaction conditions used here. A slight reduction in C_5+ selectivity can be observed at STY > 60 g_{HC}/ l_{sl} /h, but is becoming severe only above 80 g_{HC} / l_{sl} /h. The results shown in fig. 5 are obtained in small diameter pilot reactors which, due to wall effects, have to be operated at relatively low gas velocities and thus predominantly in the homogeneous bubble regime. It is demonstrated that high space time yields and high C_5+ and wax selectivities can be obtained without going to high gas velocities or larger diameter columns.

It has for a long time been known that higher mass transfer rates and therefore potentially even higher reactor productivities could be achieved by operating a SBCR at high gas velocities, i.e. in the so-called churn-turbulent or heterogeneous regime. Studies of gas holdup and mass-transfer at high solids concentration in pressurised non-aqueous systems indicate that the increase in the volumetric mass transfer coeffecient (k_La) at gas velocities above 10 cm/s in pressurised systems [4] is sufficient to allow space time yields of at least 100 g_{HC} / l_{sl} / h without severe loss of selectivity, as long as the catalyst activity and concentration is sufficient. However, no absolute upper limit can be given since it will depend on a number of factors such as system pressure, conversion level and feed composition in addition to the mentioned hydrodynamic parameters.

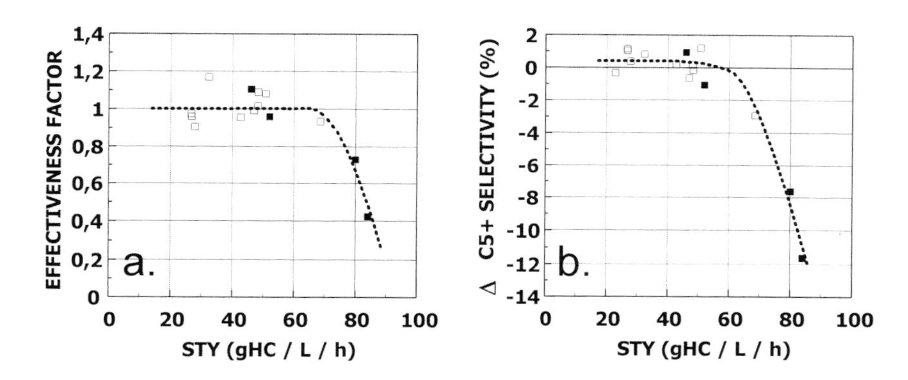

Fig 5. Global effectiveness factor (a) and selectivity effects (b) as a function of global reaction rate (STY) in 1" (filled symbols) and 2" (open symbols) SBCR.
(Cat.: Co-Re/Al_2O_3, P = 20 bar, H_2/CO = 2.2, 50% inerts in feed, $U_{g,0}$ = 3-4 cm/s).
The global effectiveness factor is the ratio between the observed catalytic activity and the activity measured at low catalyst concentration. ΔC_5+ means the difference in C_5+ selectivity between the actual (observed) value and the predicted value without gas-liquid mass-transfer resistance.

3.3. Reactor modeling

A successful simulation model for design and scale-up of a SBCR needs to describe quantitatively the reaction kinetics (including selectivity), the physical properties of the phases involved, phase holdup profiles (gas, liquid, solid), bubble dynamics, mass and heat transfer, fluid dynamics and phase dispersion.

For F-T systems, various types of models have been used to predict the SBCR performance [4,5], such as ideal reactors (plug-flow or perfectly mixed reactors), intermediate cases (e.g. gas phase in plug flow, liquid perfectly mixed) and axial dispersion models. Axial dispersion models, assuming no radial gradients in concentration or temperature and often using empirical correlations for phase holdup and phase dispersion, can be regarded as averaged models that can be applied with success within the range of conditions (e.g. gas velocities and column diameters) where the hydrodynamic correlations have been developed.

Fig. 6. illustrates the use of an axial dispersion model for simulation of a SBCR using a cobalt catalyst for F-T synthesis. The design and operation of the reactor becomes more complicated at high volumetric rates when there is a close coupling between reaction kinetics, mass transfer and phase mixing characteristics. Successful optimisation will require quantitative knowledge of all governing kinetic, hydrodynamic and physical parameters of the system and their interactions.

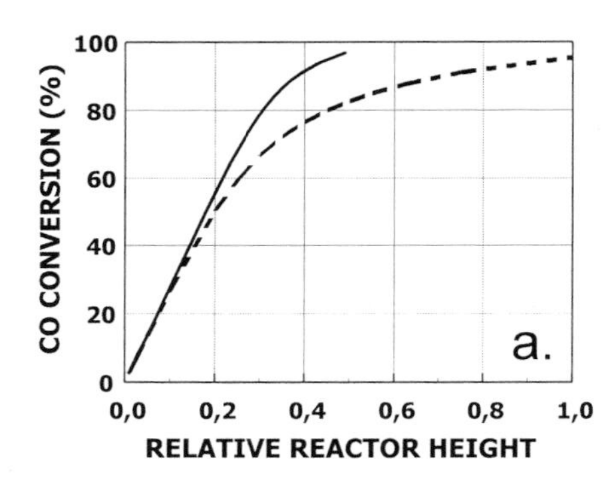

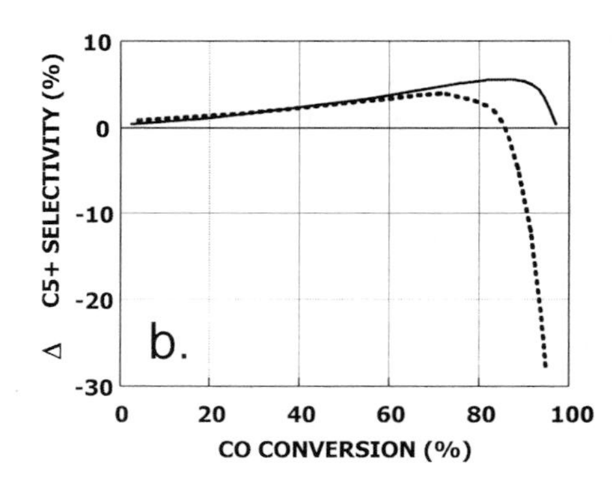

Fig. 6. Mass transfer effects on space-time (expressed as reactor height) vs. conversion (a) and conversion vs. selectivity (b) for a SBCR using an axial dispersion model. Solid line: System close to total kinetic rate control (low reaction rate). Dashed line: System influenced by gas-liquid mass transfer (high rate). (a) Reactor heights normalised to the same value at low conversion (b) ΔC_5+ = change in C_5+ selectivity relative to selectivity at low conversion.

The current development in multiphase reactor simulation is moving towards more advanced models based on hydrodynamic studies of more relevant model systems, more advanced diagnostic methods and improved computational capabilities [5]. Phenomenological models that include the known characteristics of a SBCR (e.g. the presence of two bubble classes and a radial gas holdup/velocity profile) represent a step forward [4,5]. Computational Fluid Dynamics (CFD) models have still not reached a level of maturity to become an universal tool for simulation and scale-up of SBCRs, but will play an increasing role in multiphase reactor engineering in the future.

4.Conclusions

Experimental studies and simulation of slurry bubble column reactors using highly active cobalt catalysts show that the ability to optimise the design and operation of such systems must rely on a complete and quantitative understanding of the coupling between catalyst properties, reaction kinetics, mass transfer effects and reactor hydrodynamics.

REFERENCES

1 G. P. van der Laan and A.A.C.M. Beenackers, Catal. Rev. - Sci. Eng. **41**, 255 (1999)
2 E. Iglesia, S.C. Reyes, R.J. Madon and S.L. Soled, Adv. Catal. **39**, 221 (1993)
3 M.F.M. Post, A.C. van't Hoog, J.K. Minderhoud and S.T. Sie, AIChE Journal **35**(7), 1107 (1989)
4 R. Krishna and S.T. Sie, Fuel Proc. Techn. **64**, 73 (2000)
5 M.P. Dudukovic, F. Larachi and P.L. Mills, Chem. Eng. Sci. **54**, 1975 (1999)

Studies in Surface Science and Catalysis
J.J. Spivey, E. Iglesia and T.H. Fleisch (Editors)

Catalytic partial oxidation of methane to syngas: staged and stratified reactors with steam addition

E. J. Klein, S. Tummala, and L. D. Schmidt

Department of Chemical Engineering and Materials Science, University of Minnesota, Minneapolis, MN 55455

The role of steam reforming reactions and water-gas shift reactions in the partial oxidation of methane to syngas process has been studied by both varying the catalytic lengths in a reactor and by adding steam to the partial oxidation system. It is seen through a series of stratified, staged, and steam addition experiments that both steam reforming and water-gas shift can occur in the presence of water in a millisecond contact time reactor.

1. Introduction

The production of synthesis gas or syngas, a mixture of carbon monoxide and hydrogen, from natural gas has been an extensively studied process in both industry and academia in the past decade. The driving forces for these studies have varied from a historical desire to inexpensively create the syngas feedstock associated with methanol production and Fischer-Tropsch synthesis of synthetic diesels, to recent needs to produce hydrogen rich streams for fuel cell operation and for hydrogen addition to gasoline in combustion engines.

Traditionally, the standard industrial process for syngas production has been endothermic steam reforming.

$$CH_4 + H_2O \rightarrow 3\ H_2 + CO,\ \ \Delta H_r = 205.9\ \text{kJ/mol} \qquad (1)$$

Because it is highly endothermic, this reaction is carried out on $Ni/\alpha\text{-}Al_2O_3$ catalysts in large tube furnaces at temperatures close to 900°C and pressures between 15 and 30 atm. [1]

An alternative syngas production process that has received extensive attention is the partial oxidation of methane to syngas.

$$CH_4 + \tfrac{1}{2}\ O_2 \rightarrow 2\ H_2 + CO,\ \ \Delta H_r = -35.6\ \text{kJ/mol} \qquad (2)$$

High yields of syngas have been recorded for this process under various reactor configurations, contact times, and for various metals. [2-6]

One important topic that has received attention in recent years has concerned the roles that side reactions play in the partial oxidation process; namely the steam reforming reaction and the water-gas shift reaction.

$$CO + H_2O \Leftrightarrow H_2 + CO_2,\ \ \Delta H_r = -\ 40.9\ \text{kJ/mol} \qquad (3)$$

The determination of the roles these two subsets of reactions play in the partial oxidation of methane is essential for enabling the engineering of reactor systems that can both increase production of syngas and can alter the relative concentrations of H_2 and CO in the syngas product stream. This paper discusses experiments conducted in which either reactive bed length, number of reactive stages, or inlet reactants (steam addition) are altered in order to determine the roles of steam reforming and water-gas shift in the partial oxidation of methane on rhodium at millisecond contact times.

2. Experimental

The general partial oxidation reactor used for these experiments has been described previously. [7] Inlet gas flow rates (CH_4, O_2, N_2) were controlled by Brooks mass flow controllers to an accuracy of ± .05 slpm. An appropriate catalytic monolith had blank monoliths placed both before and after it in order to prevent axial heat losses. This three-monolith system was wrapped in alumino-silicate cloth to prevent bypass of the reactant gases and placed within an 18 mm internal diameter quartz tube. Insulation was placed outside the quartz reactor tube to minimize radial heat losses.

Product gases were analyzed using an HP5890 gas chromatograph. Separation of the gases was achieved through the use of two packed columns, HaySep D followed by MoleSieve 5A, placed in series. The number of moles of each gas component was calculated based on the area of the chromatograph output using nitrogen, a system inert, as a standard. Carbon and hydrogen balances were calculated to determine the error associated with this procedure. All balances closed to within ±3%.

2.1. Stratified Reactor

The stratified reactor utilized 80 ppi α-Al_2O_3 monoliths of different bed lengths (1 mm, 3 mm, 5 mm, and 10 mm) as the appropriate catalytic supports. These monoliths were coated with 5 weight percent rhodium using an insipient wetness technique described previously. [8] Between a given rhodium monolith and the backside blank, a chromel-alumel thermocouple was placed in order to measure the reaction temperature associated with the given bed length.

2.2. Staged Reactor

The staged reactor utilized the standard partial oxidation reactor described above with a 5 weight percent rhodium, 10 mm bed length, 80 ppi α-Al_2O_3 foam monolith as the catalyst in the first stage. The second stage consisted of two rhodium loaded (5 weight percent) 80 ppi α-Al_2O_3 foam monoliths placed within a second 18 mm internal diameter quartz tube downstream of the first stage. This quartz tube was enclosed in a ceramic radiant heating tube furnace to enable external heating of this second reaction zone.

2.3. Steam Addition

The steam addition system utilized the standard partial oxidation reactor described above. However, in order to deliver steam as an additional reactant to the partial oxidation system, an ISCO LC5000 syringe pump was used in combination with a fluidized bed heater to deliver vaporized steam to the system. In order to insure no condensation of water before the partial oxidation zone, an external ceramic heater was placed upstream of the rhodium catalyst to preheat the inlet gases to 500°C.

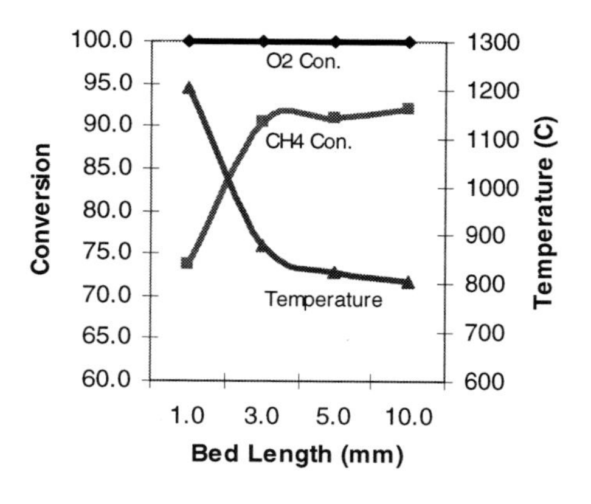

Fig. 1. Reactant conversions and reaction temperature versus bed length at 5 slpm ($CH_4/O_2 = 1.8$, 20% N_2 dilution)

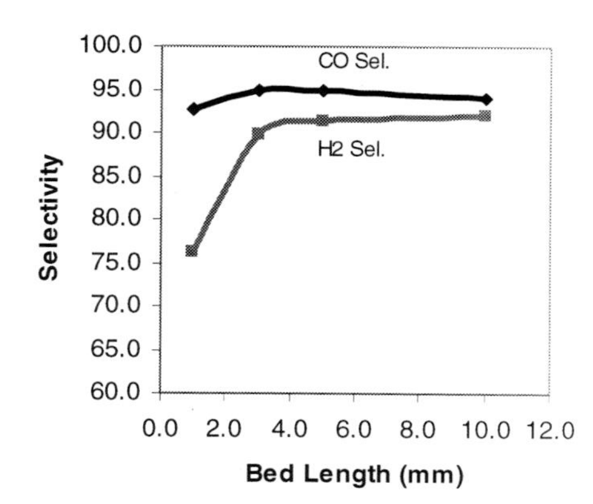

Fig. 2. Product selectivities versus bed length at 5 slpm ($CH_4/O_2 = 1.8$, 20% N_2 dilution)

3. Results

3.1. Stratified Reactor

The first set of experiments involved altering the catalytic bed length associated with partial oxidation to determine the extent of reaction at various positions in the rhodium bed. The idea that motivated this experiment was that the products and temperatures associated with a particular bed length mirrored those products and temperatures at that distance in the standard 10 mm reactive rhodium bed. Figure 1 depicts methane conversion, oxygen conversion, and reaction temperature versus reactive bed length while Figure 2 depicts both hydrogen and carbon monoxide selectivity versus reactive bed length.

The stratified reactor experiment showed that both oxygen conversion and CO selectivity remain relatively unchanged as bed length is altered while CH_4 conversion, H_2 selectivity, and reaction temperature vary greatly. Oxygen conversion was complete for all bed lengths while CO selectivity held near 95%. H_2 selectivity and CH_4 conversion, however, were low for the 1 mm bed length (76% and 74% respectively), but both increased to over 90% as the bed length was increased to 3 mm and beyond. This rise in CH_4 conversion and H_2 selectivity was accompanied by a measured drop in reaction temperature from 1200°C for a 1 mm bed to 800°C for longer beds. It is important to note that these measured temperatures all lie within 50°C of the calculated adiabatic temperatures for the given products.

3.2. Staged Reactor

The second set of experiments involved adding a second rhodium stage to the partial oxidation system. This second stage was heated externally with a ceramic furnace. The idea that motivated this research was to both test whether steam reforming could occur at the millisecond contact times associate with this reactor and to test whether conversion could be increased by addition of extra catalytic length beyond the standard 10 mm monolith. Figure

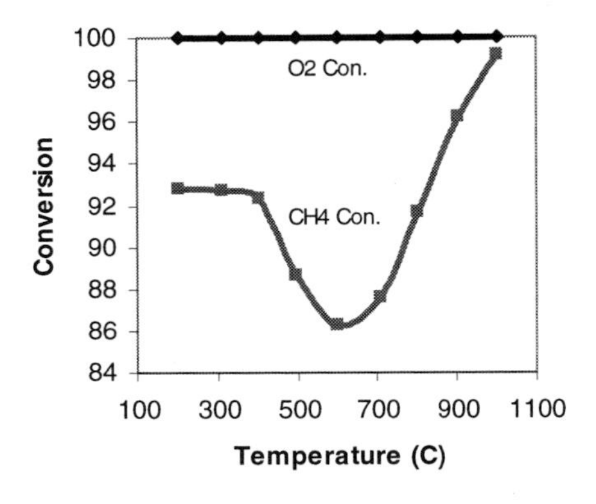

Fig. 3. Reactant conversions versus 2nd stage temperature at 5 slpm (CH_4/O_2 = 1.8, 20% N_2 dilution)

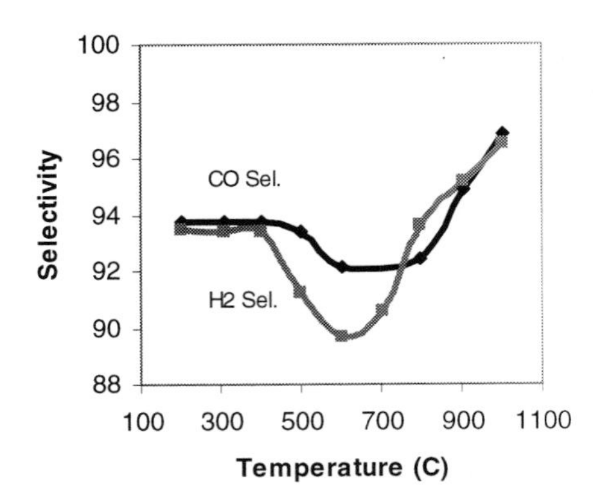

Fig. 4. Product selectivities versus 2nd stage temperature at 5 slpm (CH_4/O_2 = 1.8, 20% N_2 dilution)

3 shows both methane and oxygen conversion versus second stage (furnace) temperature while Figure 4 shows both CO selectivity and H_2 selectivity versus second stage temperature.

The staged reactor experiment showed that CH_4 conversion, H_2 selectivity, and CO selectivity can be increased with addition of extra catalytic length. Following an initial drop in CH_4 conversion and syngas selectivities due to methanation at low temperatures, CH_4 increased from 93% at 200°C to 99% at 950°C, while H_2 selectivity increased from 94% to 97% and CO selectivity increased from 94% to 97% over the same second stage temperature range. In addition, oxygen conversion remained complete as second stage temperature was altered.

3.3. Steam Addition

The final set of experiments involved adding steam in addition to the standard partial oxidation reactants in a one stage partial oxidation system (10 mm Rh-loaded monolith). The idea that motivated this research was to see whether steam addition to the partial oxidation system could promote both steam reforming and water-gas shift in the millisecond contact time reactor. Figure 5 shows both methane conversion and CO selectivity versus percent steam added (on a dry gas basis). Note that the H_2 selectivity was excluded in the graph because H_2 selectivity is defined with respect to water and excess water was being added. Figure 6 shows both H_2 production and CO production versus percent steam added. Further, although not graphically indicated, oxygen conversion was complete for all amounts of steam added.

The steam addition experiment showed that CH_4 conversion, CO selectivity, and H_2 production could be altered by addition of steam to a partial oxidation system. As steam addition was increased from 0 % to 80%, there was a slight increase in CH_4 conversion from 92% to 96%. This methane conversion increase was followed by a slight conversion decrease to 77% as steam addition was further increased to 400%. Similar to the methane conversion trend, there was an initial increase seen in hydrogen production for moderate

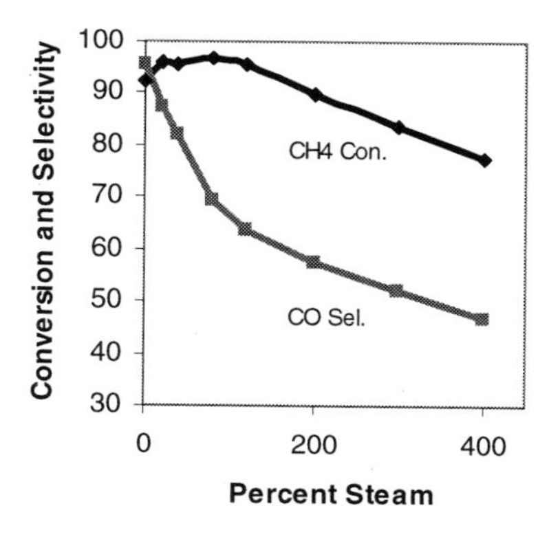

Fig. 5. CH_4 conversion and CO selectivity versus percent steam added for 2 slpm dry gas basis (CH_4/O_2 = 1.8, 20% N_2 dilution)

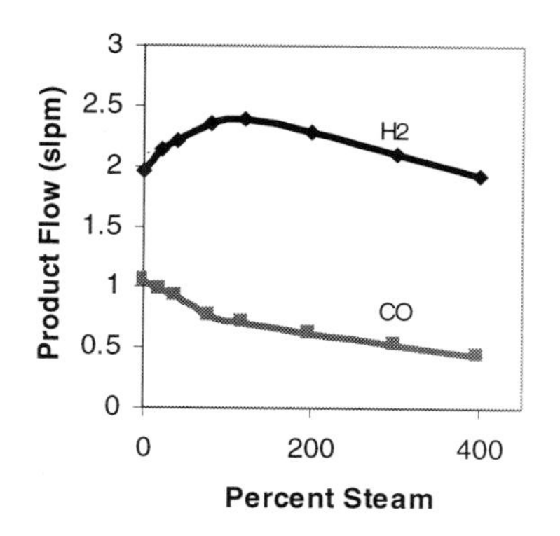

Fig. 6. H_2 and CO product flows versus percent steam added for 2 slpm dry gas basis (CH_4/O_2 = 1.8, 20% N_2 dilution)

amounts of steam addition (0 to 200%) followed by a H_2 production decrease at higher steam levels due to lower CH_4 conversions. Further, as steam addition was increased from 0% to 400%, there was seen a constant decrease in CO selectivity (95% to 47%) and CO production (1 to .5 slpm).

4. Discussion

The staged, stratified, and steam addition series of experiments provide evidence that both steam addition and water-gas shift can have measurable roles in partial oxidation of methane on rhodium at millisecond contact times. First, both the stratified and staged reactors indicate that methane conversion continues in later stages of the reactive bed without the presence of oxygen. In both systems, oxygen conversion is shown complete, while methane conversion continues to produce significant shifts in hydrogen selectivity. Furthermore, both systems show that energy is needed (whether by drop in reactive temperature or increase of the second stage temperature) for this continued conversion of methane. These trends suggest that endothermic steam reforming is occurring between the produced water and the unreacted methane in later stages of a rhodium-loaded monolith. However, even though both experiments indicate that steam reforming can occur at the millisecond contact times associated with this system, it is still unclear as to how dominant this reaction is in the system because the stratified reactor was not able to indicate the exact bed position where oxygen conversion becomes complete and steam reforming becomes the sole method for methane conversion. Therefore, direct partial oxidation could still be the method by which most syngas is produced on rhodium monoliths.

The experiment in which steam was added in addition to the standard partial oxidation reactants in a single stage rhodium monolith system strongly indicate that water-gas shift can be carried out to significant levels in a millisecond contact time reactor. The large drop seen in CO selectivity while CH_4 conversion remained relatively unchanged suggest that CO is converted into CO_2 (the only other product considered in CO selectivity calculations) in the presence of water. This idea of significant water-gas shift is further enhanced by the initial increases seen in H_2 production with steam addition (before low CH_4 conversions begin to hinder H_2 production). In addition, beyond the fact that the experiment show that water-gas shift reactions can occur at millisecond contact times, the results also indicate that the extent of this reaction can be adjusted by varying the amount of steam fed into the reaction system. This property could become extremely useful in creating syngas product streams of varying H_2/CO ratios such as low CO concentration streams for fuel cell systems.

The staged, stratified, and steam addition experiments all combine to show that, while direct oxidation to H_2 and CO dominate, reactions beyond direct partial oxidation can occur at millisecond contact times on rhodium. The evidence of the presence of these reactions has opened the door for new reactor and experiment designs that could enable greater conversions of methane to syngas streams that have variable amounts of CO.

References

1. S.S. Bharadwaj and L.D. Schmidt, *Catalytic Partial Oxidation of Natural Gas to Syngas*. Fuel Processing Technology, No. 42 (1995) 109.
2. D.A. Hickman and L.D. Schmidt, *Production of Syngas by Direct Catalytic Oxidation of Methane*. Science, No. 259 (1993) 343.
3. D. Dissenayake et. al., *Partial Oxidation of Methane to Carbon Monoxide and Hydrogen over a Ni/Al_2O_3 Catalyst*. Journal of Catalysis, No. 132 (1991) 117.
4. P.D.F. Vernon et. al., Catalysis Letters, No. 6 (1990) 181.
5. A.T. Ashcroft et. al., *Selective Oxidation of Methane to Synthesis Gas using Transition Metal Catalysts*. Nature, No. 344 (1990) 319.
6. V.R. Choudhary et. al., *Nonequilibrium Oxidative Conversion of Methane to CO and H_2 with High Selectivity and Productivity over Ni/Al_2O_3 at Low Temperatures*. Journal of Catalysis, No. 139 (1993) 326.
7. K.L. Hohn et. al., *Methane Coupling to Acetylene over Pt-coated Monoliths at Millisecond Contact Times*. Catalysis Letters, No. 54 (1998) 113.
8. P.M. Torniainen et. al., *Comparison of Monolith-Supported Metals for the Direct Oxidation of Methane to Syngas*. Journal of Catalysis, No. 146 (1994) 1.

Studies in Surface Science and Catalysis
J.J. Spivey, E. Iglesia and T.H. Fleisch (Editors)

Natural Gas Conversion in Monolithic Catalysts: Interaction of Chemical Reactions and Transport Phenomena

Olaf Deutschmann, Renate Schwiedernoch, Luba I. Maier, Daniel Chatterjee

Interdisciplinary Center for Scientific Computing (IWR), Heidelberg University
Im Neuenheimer Feld 368, D-69120 Heidelberg, Germany

Abstract

The interaction of transport and kinetics in catalytic monoliths used for natural gas conversion is studied experimentally and numerically. The paper focuses on a precise flow field agreement between experiment and model. Therefore, we use extruded monoliths with rectangular channel cross-section and a three-dimensional Navier-Stokes simulation including detailed reaction mechanisms and a heat balance. Latter also accounts for heat conducting channel walls and external heat loss. If a washcoat is used, a set of one-dimensional reaction-diffusion equations is additionally applied for modeling the transport and heterogeneous reactions in the washcoat. Partial oxidation of methane to synthesis gas on rhodium coated monoliths has been studied as example.

1. INTRODUCTION

Monolithic catalysts are often applied for natural gas conversion processes such as partial oxidation of light alkanes [1, 2] and catalytic combustion [3, 4]. In particular at short contact times, a complex interaction of transport and reaction kinetics can occur. Chemical reactions can not only take place on the catalytic surface but also in the gas phase as was shown for partial oxidation of methane on rhodium at elevated pressure [5] and oxy-dehydrogenation of ethane on platinum at atmospheric pressure [6]. In those studies, computational tools for the numerical simulation of heterogeneous reactive flows were developed and applied to a two-dimensional simulation of a single-channel in a foam monolith. In that work, the complex shape of the pores of the foam monolith was described by using the simplified model of a straight tube. A three-dimensional simulation of the catalytic partial oxidation of methane to synthesis gas in a wire gauze configuration has been previously performed by de Smet et al. [7] but a simple surface reaction model was used.

The understanding of the details of the reactor behavior demands a better agreement between experimental and modeled flow field without neglecting the complex chemistry. Therefore, our present work focuses on setting up an adequate experiment where we use extruded monoliths with straight channels with rectangular cross-section. The model then applies a three-dimensional flow field description coupled with detailed reaction mechanisms and an enthalpy governing equation that includes a heat conducting channel wall.

The inner walls of the single channels of extruded monoliths are frequently coated with a thin layer of washcoat to enlarge the surface. Here, diffusion of reactants and products to and from the catalytic active centers in the washcoat can limit the total heterogeneous reaction rate [8]. Therefore, a set of one-dimensional reaction-diffusion equations is applied to account for washcoat diffusion in the numerical simulation of the monolithic catalyst.

As an example we will discuss experiments and numerical simulations carried out for the partial oxidation of methane to synthesis gas on a rhodium coated extruded monolith.

2. EXPERIMENTAL

The experimental set up was designed in a way that allows the application of detailed models for the physical and chemical processes occurring in the reactor. Experiments were carried out in a tubular quartz reactor, 25 cm long and 2.6 cm in (inner) diameter. The tube contains a 1cm long extruded monolith with a well-defined rectangular cross-section (1mm x 1mm) of its channels. The ceramic monolith made of cordierite is coated with the noble metal rhodium by saturation with an acidic aqueous solution of $Rh_2(SO_4)_3$, 24 hours drying at 100°C, reduction in H_2 at 500°C, and calcination in air at 500°C for 18 hours. Metal loadings are 1 to 3% Rh by weight. Energy dispersive X-ray spectroscopy (EDX) pictures revealed that no sulfur compounds are left on the surface from the impregnation process. For the investigation of washcoat diffusion, cordierite monoliths pre-coated with an alumina washcoat are used.

The entire reactor can be run either auto-thermally or temperature-controlled by a furnace. The gas temperature at the exit of the catalytic monolith is determined by a thermocouple placed inside a thin quartz tube to prevent catalytic reactions. Because of heat losses to the ambience, the measured exit temperature was always significantly lower than the adiabatic reactor temperature would be.

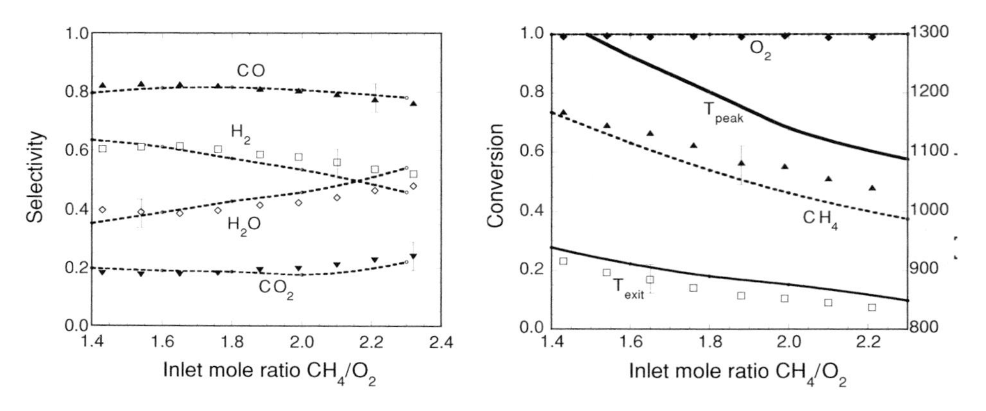

Fig. 1. Effect of methane/oxygen ratio on selectivity, conversion, and peak (simulation) and outlet temperature; symbols = experiment, lines = simulation. No washcoat is used.

The product composition is determined by gas chromatography (TCD, FID) and by quadrupole mass spectroscopy. The latter one can also be used for transient measurements, for instants for studies of ignition and extinction phenomena. The reactor is operated at atmospheric pressure (1.1 bar). The total feed flow rate corresponds to a residence time of few milliseconds.

Methane/oxygen mixtures were fed diluted by argon. In the experiment, we have been studying the effect of composition, flow rate, dilution, and preheat on selectivity and conversion. Exemplary, Fig. 1 exhibits selectivity, conversion, and outlet temperature as function of the methane/oxygen ratio. A good agreement is shown between experimentally determined and numerically predicted data, the latter achieved by the model discussed below.

3. MODELING THE MONOLITH CHANNEL

Even though the experimental measurements reveal that significant heat loss occurs, we simply assume for the model that every channel of the monolith behaves essentially alike. Thus radial profiles over the monolith as a whole are neglected, and only one single channel has to be analyzed. The flow within these small diameter channels is laminar. Because an objective of this study is an appropriate agreement between experimental configuration and flow field model, we solve the three-dimensional Navier-Stokes equations for the simulation of the rectangular shaped channel. These equations are coupled with an energy conservation equation and an additional conservation equation for each chemical species. The energy conservation equation accounts for heat transport by convection and conduction in the gas phase, heat release due to chemical reactions in the gas phase and on the catalytic surface, and heat conduction in the channel walls. Furthermore, we added an external heat loss term at the outer boundary of the channel wall to account for the experimentally occurring heat loss. The temperature-dependent external heat loss was specified so that the predicted outlet temperature agrees with the experimentally measured temperature. Because selectivity and conversion in catalytic partial oxidation of light alkanes strongly depend on the spatial temperature profile, the detailed description of the energy balance is crucial for the understanding of the reaction.

The chemical reactions are modeled by detailed reaction schemes for homogeneous as well as heterogeneous reactions. In the heterogeneous reaction model we apply the mean field approximation. That means that the adsorbates are assumed to be randomly distributed on the surface, which is viewed as being uniform. The state of the catalytic surface is described by the temperature T and a set of surface coverages Θ_i, both depending on the macroscopic position in the reactor, but averaging over microscopic local fluctuations. Balance equations are established to couple the surface processes with the surrounding reactive flow. The production rates $\dot{s}_i$ of surface and gas phase species (due to adsorption and desorption) is then written as

$$\dot{s}_i = \sum_{k=1}^{K_s} \nu_{i_k} k_{f_k} \prod_{i=1}^{N_g+N_s} (c_i)^{\nu'_{i_k}} \qquad (1)$$

with K_s = number of surface reactions including adsorption and desorption, ν_{i_k}, ν'_{i_k} = stoichiometric coefficients, k_{f_k} = forward rate coefficient, N_g (N_s) = number of gaseous (surface) species, c_i = concentration of species i, which is given in mol cm^{-2} for adsorbed species. Because the binding states of adsorption on the surface vary with the surface coverage of all adsorbed species, the expression for the rate coefficient becomes complex:

$$k_{f_k} = A_k T^{\beta_k} \exp\left[\frac{-E_{a_k}}{RT}\right] \prod_{i=1}^{N_s} \Theta_i^{\mu_{i_k}} \exp\left[\frac{\varepsilon_{i_k} \Theta_i}{RT}\right] \tag{2}$$

with A_k = preexponential factor, β_k = temperature exponent, and E_{a_k} = activation energy of reaction k. Coefficients μ_{i_k} and ε_{i_k} describe the dependence of the rate coefficients on the surface coverage of species i. For adsorption reactions, sticking coefficients are commonly used. They are converted to conventional rate coefficients by

$$k_{f_k}^{\text{ads}} = \frac{S_i^0}{\Gamma^\tau} \sqrt{\frac{RT}{2\pi M_i}} \tag{3}$$

with S_i^0 = initial sticking coefficient, Γ = surface site density in mol cm^{-2}, τ = number of sites occupied by the adsorbing species, M_i = molar mass of species i. While the surface site density can be estimated from the catalyst material, the knowledge of the ratio of the active catalytic surface area to geometrical surface area is essential for the model. An exact value for this ratio has to be determined experimentally. In the simulation discussed here we simply use a value of unity, and the surface site density for rhodium is set to be $2.7 \cdot 10^{-9}$ mol cm^{-2}.

In spite of numerous surface science studies on H_2, CO and hydrocarbon oxidation there is still a substantial lack in kinetic data. Nevertheless, several surface reaction mechanisms with associated rate expressions have been published for complete and partial oxidation on noble metal catalysts in the last decade. Even though the mechanisms are often based on few experimental data, which were achieved for a limited range of conditions, they led to a better understanding of the process.

In the present study, the surface chemistry is described by a detailed surface reaction mechanism that is under development for the description of partial as well as complete oxidation of methane on rhodium [9]. The mechanism consists of 38 reactions among 6 gas phase species and further 11 adsorbed species, as shown in Table 1. Because the surface coverage is low for the conditions chosen in the present study, the dependence of the rate coefficients on the surface coverage was neglected. However, it may become important at different conditions. For more details, we refer to a forthcoming paper in which the establishment of the reaction mechanism will be discussed [9]. We would like to note that the present study does not focus on the development of the surface reaction mechanism but rather on its application in multi-dimensional simulations that allow to describe the reactor behavior as adequate as possible.

TABLE 1: SURFACE REACTION MECHANISM

					A	E_a
(1)	H_2	+ 2 Rh(s)	⇒ 2 H(s)		$1.00\cdot10^{-02}$	s.c.
(2)	O_2	+ 2 Rh(s)	⇒ 2 O(s)		$1.00\cdot10^{-02}$	s.c.
(3)	CH_4	+ Rh(s)	⇒ CH_4(s)		$8.00\cdot10^{-03}$	s.c.
(4)	H_2O	+ Rh(s)	⇒ H_2O(s)		$1.00\cdot10^{-01}$	s.c.
(5)	CO_2	+ Rh(s)	⇒ CO_2(s)		$1.00\cdot10^{-05}$	s.c.
(6)	CO	+ Rh(s)	⇒ CO(s)		$5.00\cdot10^{-01}$	s.c.
(7)	2 H(s)		⇒ 2 Rh(s)	+ H_2	$3.00\cdot10^{+21}$	77.8
(8)	2 O(s)		⇒ 2 Rh(s)	+ O_2	$1.30\cdot10^{+22}$	355.2
(9)	H_2O(s)		⇒ H_2O	+ Rh(s)	$3.00\cdot10^{+13}$	45.0
(10)	CO(s)		⇒ CO	+ Rh(s)	$3.50\cdot10^{+13}$	133.4
(11)	CO_2(s)		⇒ CO_2	+ Rh(s)	$1.00\cdot10^{+13}$	21.7
(12)	CH_4(s)		⇒ CH_4	+ Rh(s)	$1.00\cdot10^{+13}$	25.1
(13)	O(s)	+ H(s)	⇒ OH(s)	+ Rh(s)	$5.00\cdot10^{+22}$	83.7
(14)	OH(s)	+ Rh(s)	⇒ O(s)	+ H(s)	$3.00\cdot10^{+20}$	37.7
(15)	H(s)	+ OH(s)	⇒ H_2O(s)	+ Rh(s)	$3.00\cdot10^{+20}$	33.5
(16)	Rh(s)	+ H_2O(s)	⇒ H(s)	+ OH(s)	$5.00\cdot10^{+22}$	104.7
(17)	OH(s)	+ OH(s)	⇒ H_2O(s)	+ O(s)	$3.00\cdot10^{+21}$	100.8
(18)	O(s)	+ H_2O(s)	⇒ OH(s)	+ OH(s)	$3.00\cdot10^{+21}$	171.8
(19)	C(s)	+ O(s)	⇒ CO(s)	+ Rh(s)	$3.00\cdot10^{+22}$	97.9
(20)	CO(s)	+ Rh(s)	⇒ C(s)	+ O(s)	$2.50\cdot10^{+21}$	169.0
(21)	CO(s)	+ O(s)	⇒ CO_2(s)	+ Rh(s)	$1.40\cdot10^{+20}$	121.6
(22)	CO_2(s)	+ Rh(s)	⇒ CO(s)	+ O(s)	$3.00\cdot10^{+21}$	115.3
(23)	CH_4(s)	+ Rh(s)	⇒ CH_3(s)	+ H(s)	$3.70\cdot10^{+21}$	61.0
(24)	CH_3(s)	+ H(s)	⇒ CH_4(s)	+ Rh(s)	$3.70\cdot10^{+21}$	51.0
(25)	CH_3(s)	+ Rh(s)	⇒ CH_2(s)	+ H(s)	$3.70\cdot10^{+24}$	103.0
(26)	CH_2(s)	+ H(s)	⇒ CH_3(s)	+ Rh(s)	$3.70\cdot10^{+21}$	44.0
(27)	CH_2(s)	+ Rh(s)	⇒ CH(s)	+ H(s)	$3.70\cdot10^{+24}$	100.0
(28)	CH(s)	+ H(s)	⇒ CH_2(s)	+ Rh(s)	$3.70\cdot10^{+21}$	68.0
(29)	CH(s)	+ Rh(s)	⇒ C(s)	+ H(s)	$3.70\cdot10^{+21}$	21.0
(30)	C(s)	+ H(s)	⇒ CH(s)	+ Rh(s)	$3.70\cdot10^{+21}$	172.8
(31)	CH_4(s)	+ O(s)	⇒ CH_3(s)	+ OH(s)	$1.70\cdot10^{+24}$	80.3
(32)	CH_3(s)	+ OH(s)	⇒ CH_4(s)	+ O(s)	$3.70\cdot10^{+21}$	24.3
(33)	CH_3(s)	+ O(s)	⇒ CH_2(s)	+ OH(s)	$3.70\cdot10^{+24}$	120.1
(34)	CH_2(s)	+ OH(s)	⇒ CH_3(s)	+ O(s)	$3.70\cdot10^{+21}$	15.1
(35)	CH_2(s)	+ O(s)	⇒ CH(s)	+ OH(s)	$3.70\cdot10^{+24}$	158.4
(36)	CH(s)	+ OH(s)	⇒ CH_2(s)	+ O(s)	$3.70\cdot10^{+21}$	36.8
(37)	CH(s)	+ O(s)	⇒ C(s)	+ OH(s)	$3.70\cdot10^{+21}$	30.1
(38)	C(s)	+ OH(s)	⇒ CH(s)	+ O(s)	$3.70\cdot10^{+21}$	145.5

The units of A are given in terms of [mol, cm, s] and of E_a in [kJ/mol].
s.c. = initial sticking coefficient.

If the inner walls of the single channels are coated with a thin layer of washcoat the catalytic active surface area can easily be increased by a factor of 10^2-10^3 in comparison to the geometrical surface area of the inner wall of the monolith channel. Diffusion and reaction within the washcoat pores leads to local concentration gradients that have to be resolved to calculate the total heterogeneous reaction rate. Therefore, we apply a set of one-dimensional reaction-diffusion equations to calculate the species concentrations and surface coverages within the washcoat as function of the distance (r) from the gas-washcoat interface:

$$\frac{\partial j_i}{\partial r} = \gamma \dot{s}_i = 0 \ . \tag{4}$$

Here, $\dot{s}_i$ = local molar reaction rate of gas phase species i due to adsorption and desorption, γ = active catalytic surface area / washcoat volume, j_i = mass diffusion flux of species i. Depending on the washcoat structure molecular or Knudsen diffusion coefficients have to be used. The number of coupled nonlinear reaction-diffusion equations equals the number of gas phase species. In addition to this equation set, one additional algebraic equation, $\dot{s}_i = 0$, has to be solved for each surface species. The equation simply says that the surface coverage has to be constant when the steady state is reached.

The numerical simulation is based on the computational fluid dynamics code FLUENT [10] which was coupled to the chemistry tool DETCHEM [5, 11] via FLUENT's interface for user defined subroutines. DETCHEM models the chemical processes in the gas phase and on the surface including the surface coverage calculation based on multi-step chemical reaction mechanisms. Additionally, several washcoat models can be included [12]. Due to symmetry, only an eighth of the channel cross section has to be simulated, for numerical reasons we simulated a quarter instead.

4. RESULTS AND DISCUSSION

The single channel of the catalytic monolith is simulated under conditions as chosen in the experiment. The methane/oxygen mixture, diluted by 75 vol.% argon, flows at 300 K and 1.1 bar with a uniform velocity of 0.26 m/s (corresponding to 7 slpm over the whole monolith) in the rectangular shaped monolith channel. The simulated channel is 1.1 cm in length with the first millimeter being non-catalytic. In the simulation and in the experiment, the reaction has to be ignited. In the experiment the reactor is inside a furnace which is heated up to initiate the reaction; after ignition the furnace is switched off, the reactor is operated auto-thermally. In the simulation, the channel wall is given a sufficiently high temperature to ignite the reaction. Then, a heat conducting wall is assumed including external heat loss.

In Figure 1, a good agreement is shown between experimentally determined and computed selectivity and conversion. The syngas selectivity and methane conversion are lower than the data reported by the Schmidt group [1] due to significant heat loss in the reactor. The computed peak temperature is much higher than the exit temperature, not only due to heat loss but also due to endothermic steam reforming.

The three-dimensional simulation including the detailed reaction mechanisms allows us to study this behavior in more detail. In previous studies [5], we used the surface reaction mechanism proposed by Hickman and Schmidt in their pioneering work in 1993 [1], in which steam and CO_2 reforming is not significant. Because of new experimental studies [9], at least steam reforming seems to be an important reaction step. Among other reasons, this fact led to the development of the revised reaction mechanism. In Figure 2, the profiles of the species mass fractions for a CH_4/O_2 vol. ratio of 1.8 reveal fast O_2 consumption at the catalyst leading edge while CH_4 is consumed over almost the whole length of the reactor. Complete oxidation takes place at the catalyst entrance only, where O_2 is still available. Because CH_4 as well as re-adsorbed H_2O decompose into atoms on the surface (Table 1), it cannot be distinguished between direct partial oxidation of CH_4 to H_2 or steam reforming. At least further downstream H_2 is formed via steam reforming only. In contrast to that, CO_2 reforming does not occur. The temperature profiles exhibit the strong axial and radial gradients due to chemical reactions and heat transport.

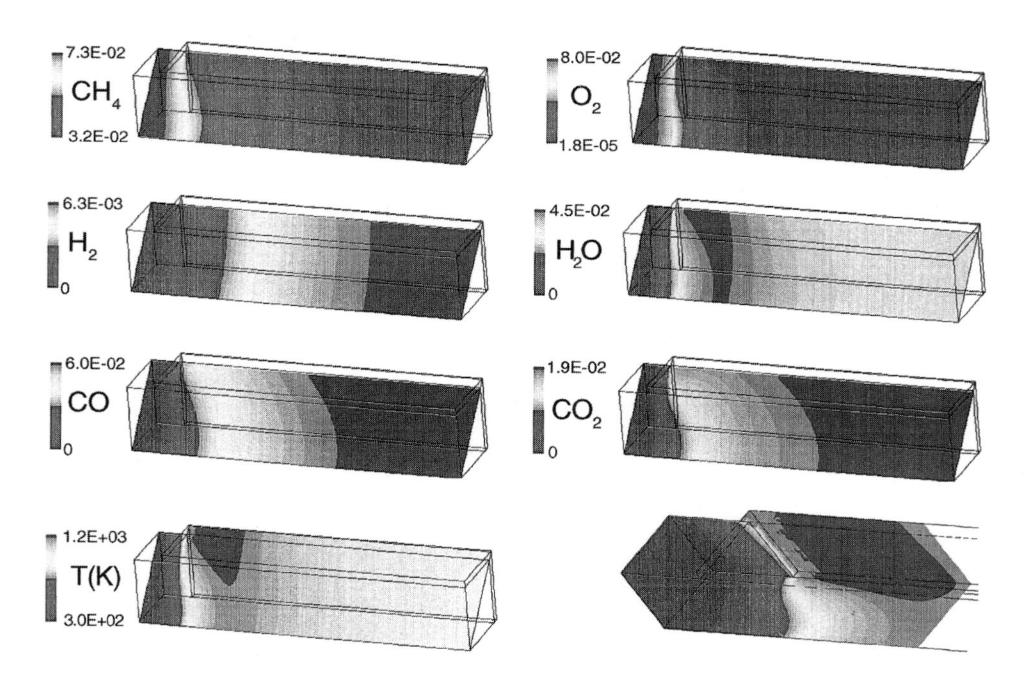

Fig. 2. Species mass fractions and temperature in the monolith channel (1 mm x 1mm). The contour plots represent the diagonal face of the simulated channel section reaching from the inner corner of the catalytic walls to the channel axis; the diagonal coordinate has been enlarged for visual clarity, the total length is 1.1 cm with the first millimeter being non-catalytic (no wall is shown). The lower right figure shows the temperature profiles in the channel wall, the inlet, and the front symmetry face at the catalyst entrance.

The simulation also reveals that chemical reactions in the gas phase are not significant at atmospheric pressure but become important above 10 bar.

If the monolith is coated with an alumina washcoat, the diameters of the pores are on the order of micrometers. Molecular diffusion inside the pores can limit the overall reaction rate. In Figure 3, we exemplary show computed concentration and coverage profiles inside the washcoat at one millimeter behind the catalyst entrance. Here most of the oxygen is already consumed, and also the oxygen concentration inside the washcoat decreases rapidly due to catalytic reaction with methane. Deeper inside the pores, oxygen is vanished and water reacts with methane to form syngas and carbon dioxide. This behavior explains why the *product* water shows its highest concentration at the washcoat inlet.

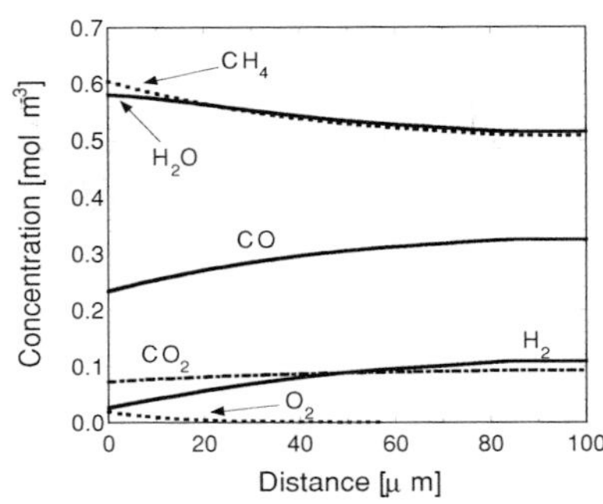

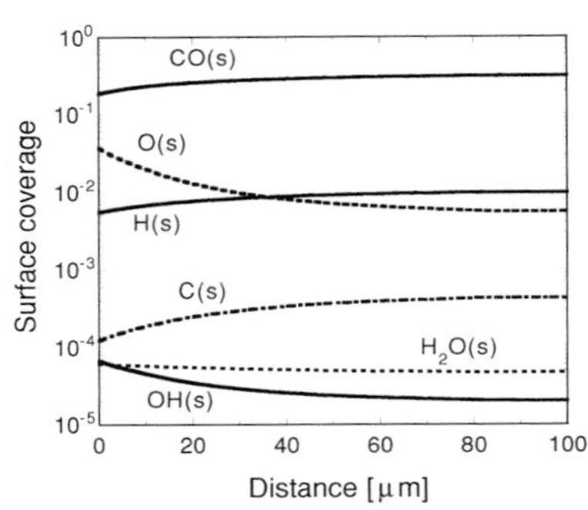

Fig. 3. Species concentrations (left) and surface coverage (right) inside the washcoat. Washcoat parameter: thickness = 100 μm, porosity = 0.51, tortousity = 3, $\gamma = 10^5 m^{-1}$.

REFERENCES

1. D. A. Hickman and L. D. Schmidt, AIChE J. 39 (1993) 1164.
2. M. Huff and L. D. Schmidt, J. Catal. 155 (1995) 82.
3. R. E. Hayes and S. T. Kolaczkowski, Introduction to Catalytic Combustion, Gordon and Breach Science Publ., Amsterdam, 1997.
4. L. L. Raja, R. J. Kee, O. Deutschmann, J. Warnatz, L. D. Schmidt, Catal. Today 59 (2000) 47.
5. O. Deutschmann and L. D. Schmidt, AIChE J. 44 (1998) 2465.
6. D. K. Zerkle, M. D. Allendorf, M. Wolf, O. Deutschmann, J. Catal. 196 (2000) 18.
7. C. R. H. de Smet, M.H.J.M. de Croon, R.J. Berger, G. B. Marin, J. C. Schouten, Appl. Catal. A 187(1) (1999) 33.
8. D. Papadias, L. Edsberg, P. Björnbom, Catalysis Today 60 (2000) 11.
9. S. Tummala, L. D. Schmidt, O. Deutschmann, publication in preparation.
10. Fluent Version. 4.4, Fluent Inc., Lebanon, New Hampshire, 1997.
11. O. Deutschmann, DETCHEM: Computer package for detailed chemistry in CFD codes, http://reaflow.iwr.uni-heidelberg.de/~dmann/DETCHEM.html, 1999.
12. D. Chatterjee, O. Deutschmann, J. Warnatz, Faraday Discussions, accepted for publication.

Studies in Surface Science and Catalysis
J.J. Spivey, E. Iglesia and T.H. Fleisch (Editors)

Direct Synthesis of Acetic Acid from Methane and Carbon Dioxide

E.M. Wilcox[a], M.R. Gogate[b], J.J. Spivey[b*], G.W. Roberts[a]

[a]Dept. Chemical Engineering, NC State University, PO Box 7905, Raleigh, NC 27695 USA
[b]Research Triangle Institute, PO Box 12194, RTP, NC 27709 USA
*Corresponding author, current address: Dept. Chemical Engineering, NC State University, email: jjspivey@ncsu.edu

The direct synthesis of acetic acid from methane and carbon dioxide was investigated. Thermodynamic calculations show that this reaction is unfavorable, with an equilibrium fractional conversion of methane of 1.6×10^{-6} at 1000 K and 150 atm. Diffuse reflectance infrared Fourier transform spectroscopy (DRIFTS) experiments showed the formation of acetate when a 5%Pd/carbon catalyst was exposed to a mixture of methane and carbon dioxide.

1. INTRODUCTION

The conversion of methane to chemicals and fuels is a key component of the long-term strategy for increased methane utilization. Most chemical synthesis routes require the production of syngas from methane. This step is often the most costly part of the overall process. The direct synthesis of chemicals from methane eliminates the need for the intermediate syngas production step, but these types of processes have not been widely reported.

One potential target chemical is acetic acid, which is produced at an annual rate of more than 6 million tons worldwide[1]. Currently, acetic acid is produced industrially by the Monsanto process, which uses a rhodium catalyst for the carbonylation of methanol with carbon monoxide. In principle, acetic acid can be produced by the direct reaction of methane and carbon dioxide:

$$CH_{4(g)} + CO_{2(g)} \leftrightarrow CH_3COOH_{(g)} \qquad (1)$$

$$\Delta H_{298K} = +8.62 \text{ kJ/mol} \qquad \Delta G_{298K} = +16.98 \text{ kJ/mol}$$

Two previous groups have investigated the synthesis of acetic acid from methane and carbon dioxide. Taniguchi et al.[2] studied the reaction of methane with carbon dioxide using an aqueous vanadium catalyst, in conjunction with trifluoroacetic acid (TFA) and peroxydisulfate ($K_2S_2O_8$). Several different catalysts were tested using partial pressures of 5 atm CO_2 and 5 atm CH_4, in a liquid consisting of 5.0 mmol $K_2S_2O_8$, 10 ml TFA and 0.05 mmol of catalyst in a 25 ml autoclave at 80°C for 20 hours. The best catalyst was VO(acetylacetonate)$_2$, which produced a yield of 15.7% based on CH_4. They also demonstrated that both the catalyst and the $K_2S_2O_8$ oxidant were required for the reaction to occur.

A further set of experiments was run to determine the dependence of the yield on the partial pressures of CH_4 and CO_2. The yield increased with an increase in the CH_4 partial pressure. However, the yield was independent of the CO_2 pressure. In fact, the reaction still occurred with no CO_2 present, suggesting that another reaction was occurring to form the acetic acid. The relatively high yield (based on thermodynamics, as discussed later) and the occurrence of the reaction in the absence of CO_2, leaves some doubt that acetic acid was formed by the direct reaction of CH_4 and CO_2.

A South African patent by Freund and Wambach[3] claims a process for the direct synthesis of acetic acid from CO_2 and CH_4 using a solid catalyst. The catalyst is not clearly identified. It is described as containing "one or more metals of groups VIA, VIIA, VIIIA" on a support of an aluminum oxide, aluminum hydroxide, or silicon dioxide. The reaction conditions are between 100 – 600 °C and 0.1 – 20 MPa. The patent claims that under these conditions a selectivity of 70 – 95% and an unspecified "sufficient conversion for commercial scale" can be achieved. No actual data or experimental examples are given in the patent to support the claims.

There is little evidence in the literature that the direct synthesis of acetic acid from CO_2 and CH_4 has been achieved. There is evidence, however, that this reaction has importance and has potential industrial use once successful. Further work is needed to increase the yield, decrease the reaction time, and make an industrially useable system, preferably based on a heterogeneous catalyst.

2. THERMODYNAMICS

Thermodynamic calculations were performed to understand the potential behavior of the reaction between CO_2 and CH_4. The RGIBBS reactor model in AspenPlus™ engineering simulation software was used to perform a Gibbs free energy minimization on the system to give the chemical and phase equilibrium composition at various temperatures and pressures. Although these calculations do not take into account any surface interactions, molecular interactions between molecules are taken into account in the equation of state.

A series of equilibrium calculations using ideal gas, Peng-Robinson, and Redlich-Kwong equations of state shows the thermodynamic limitations for this reaction. An initial composition of 95mol% CO_2 and 5mol% CH_4 was used to help maximize the conversion of CH_4. Above 600 K and below 25 atm, the equilibrium compositions from all three methods agreed to within 3%. As expected, at low temperatures and high pressures (below 500 K and above 50 atm), the ideal gas equation was not in close agreement with the other two. The Peng-Robinson and Redlich-Kwong equations were in close agreement with each other, except at extremely low temperatures and extremely high pressure (below 400 K and above 50 atm).

The equilibrium conversion of methane increased with increasing temperature and increasing pressure. However, even at 1000 K, and 150 atm., the fractional conversion of methane from the Peng-Robinson equation of state was only 1.6×10^{-6}. The fractional conversion of methane at various temperature and pressures can be seen in figure 1.

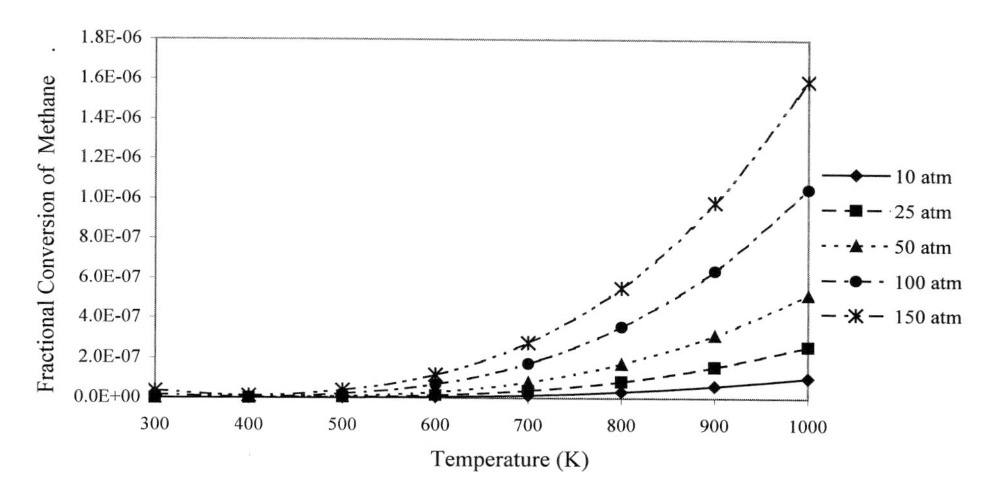

Figure 1: Equilibrium fractional conversion of methane at various temperature and pressures from the reaction of $CO_2 + CH_4 \rightarrow CH_3COOH$ calculated by the RGIBBS reactor model in AspenPlus™

These calculations show that the direct synthesis of acetic acid from carbon dioxide and methane is thermodynamically limited at all conditions of practical interest. In order for this process to be industrially useful, the unfavorable thermodynamics must be overcome. The fractional conversions shown in Figure 1 show that the Freund and Wambach claim of "sufficient conversion for commercial scale" must be based on extremely low conversions.

3. EXPERIMENTAL

DRIFTS studies show the formation of the acetate species from a mixture of CO_2/CH_4 on a 5% Pd/C catalyst. Experiments were carried out by first adsorbing acetic acid on the catalyst to identify the absorption bands corresponding to acetic acid. Then, a CO_2/CH_4 mixture was preadsorbed on a fresh sample of the same catalyst and temperature programmed desorption (TPD)-DRIFTS was carried out. The appearance of spectral bands was matched with those of pure acetic acid.

The results of acetic acid adsorption on the catalyst showed a sharp band at 1,743 cm^{-1} at 25 °C due to a carboxylic group of dimeric acetic acid and a small shoulder to this band at 1,790 cm^{-1} due to the carboxylic group of the monomeric acid[4]. A split broad band at 3,000 cm^{-1} also was observed, corresponding to an overlap of the (CH_3)- and OH stretching vibrations[4]. The pure component acetic acid spectrum[4] also shows these characteristic bands of acetic acid, as well as smaller peaks at 1,420 (OH deformation in the dimer), 1,300 (CO stretch in the dimer), and 1,189 cm^{-1} (CO stretch in the monomer), which we also observed. The TPD at 110, 220, and 320 °C shows a pronounced decrease in intensity in adsorption bands at 3,000, 1,740, 1,420 and 1,300 cm^{-1}. The TPD also shows an increase in intensity in adsorption bands at 3,587 (OH stretch in the monomer)[4], 1,790 and 1,189 cm^{-1} and the appearance of small bands at and 2,362 cm^{-1} due to gas phase carbon dioxide[4]. The change in the intensity of the bands is due to the increase of the monomeric form of the acetate. The appearance of carbon dioxide is due to the decomposition of the acid. Figure 2 shows the spectra of the pure acetic acid taken at different temperatures.

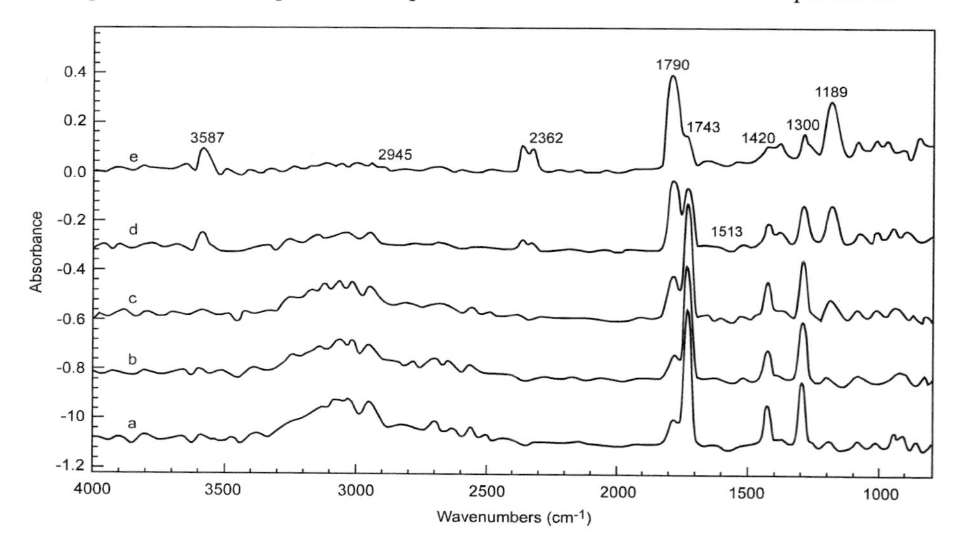

Figure 2: FTIR spectra between 4,000 and 800 cm^{-1} obtained over 5% Pd/C catalyst after exposure to a ca. 2.8% acetic acid/helium mixture 60 min at various temperatures; (a) 25 °C flow-through condition, (b) 25 °C sealed cell, (c) 110 °C sealed cell, (d) 220 °C sealed cell, and (e) 320 °C sealed cell.

The adsorption of an equimolar mixture of CO_2 and CH_4 was examined on an identical catalyst (5% Pd/C)-KBr admixture. In this case, the adsorption was carried out for 60 min and subsequent desorption was performed at 50°/min up to 420 °C. The results (Figure 3) show that the spectra at 25 °C under flow-through (spectrum a) and sealed (spectrum b) conditions are identical.

At 25 °C characteristic bands at 3,729, 3,010, 2,362, and 1,301 cm^{-1} are observed. The bands at 3,010 and 1,301 cm^{-1} can be assigned to gas-phase methane[5] and a pronounced bands at 3,729 and 2,362 cm^{-1} is due to CO_2[4]. A small shoulder band at 2,371 cm^{-1}, visible in the adsorption spectra (spectra a, b) in Figure 2, is due to naturally occurring $^{13}CO_2$[6]. The spectrum at 120 °C (spectrum c) is similar to the methane and CO_2 adsorption spectra, suggesting that no reaction has occurred. However, at higher temperatures of 220 to 420 °C (spectra d to f), small but distinct peaks at 1,790 to 1,740 cm^{-1} region, corresponding to characteristic carboxylic group are observed. Further, the small band at 1,513 can be assigned to the acetate (CH_3COO) species[7].

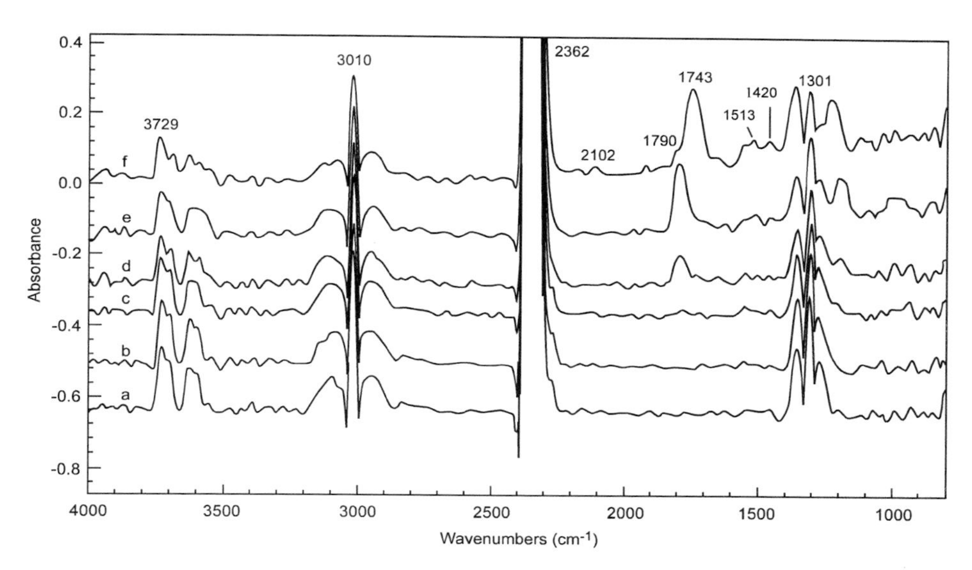

Figure 3: FTIR spectra between 4,000 and 800 cm^{-1} obtained over 5% Pd/C catalyst after exposure to equimolar CH_4/CO_2 mixture for ca. 60 min at various temperatures; (a) 25 °C flow-through condition, (b) 25 °C sealed cell, (c) 110 °C sealed cell, (d) 220 °C sealed cell, (e) 320 °C sealed cell, and (f) 420 °C sealed cell.

These spectra also show that the reaction does not occur through a syngas route: $CO_2 + CH_4 \rightarrow 2CO + 2H_2 \rightarrow CH_3COOH$. For this to occur, the formation of carbon monoxide would be detected on the FTIR spectra. Carbon monoxide displays a distinctive band between 2100 – 2180 cm^{-1} [4], which the spectra do not show.

The formation of methyl formate under these conditions might also seem plausible. However, the thermodynamics for that reaction are even less favorable than that of the acetic acid. The highest fractional conversion of methane calculated was 7.3 x 10^{-12} at 1000K and 150 atm. Additionally, methyl formate would not display the bands at 1,513 and 1,565 cm^{-1}. Additionally, the FTIR spectra of methyl formate has a distinct sharp band around 3000 cm^{-1}

corresponding to the CH_3 stretch. While this peak also corresponds to methane, if methyl formate were indeed present, this peak would increase in intensity as the reaction occurred. This is not observed in the spectra.

4. CONCLUSIONS

Thermodynamic calculations show that the equilibrium for the direct synthesis of acetic acid from methane and carbon dioxide is highly unfavorable. However, the DRIFTS studies show the formation of acetate over a 5%Pd/carbon catalyst exposed to CO_2 and CH_4. The spectra show no evidence of the formation of syngas or methyl formate, which are other possible reactions. We have demonstrated here, the feasibility of the direct synthesis of acetic acid from CO_2 and CH_4 over a solid catalyst. Further work to overcome the thermodynamic limitations of the reaction is underway.

REFERENCES

(1) Chemical Marketing Reporter 37 (1995) 247.
(2) Taniguchi, Y. et al, Adv. in Chem. Conv. For Mitigating CO_2 114 (1998) 439.
(3) Freund, H. J.; Wambach, J. Republic of South Africa Patent No. 95 - 6606 (1995).
(4) Pouchert, C. J. The Aldrich Pure Compound FTIR Spectra-Vapor phase; I ed., 1985.
(5) Zhang, Z. L. et al., J. of Phy. Chem. 100 (1996) 11.
(6) Burkett, H. D. et al., Chem. Phys. Letters 173 (1990) 5.
(7) Viswanathan, B. et al., Indian J. of Chem.: Section A 29 (1990) 3.

Studies in Surface Science and Catalysis
J.J. Spivey, E. Iglesia and T.H. Fleisch (Editors)

Partial Oxidation of Methane to Form Synthesis Gas in a Tubular AC Plasma Reactor

T.A. Caldwell, H. Le, L.L. Lobban, and R.G. Mallinson

Institute for Gas Utilization and Technologies, School of Chemical Engineering and Materials Science, University of Oklahoma, 100 E. Boyd St. T-335, Norman, OK 73019

AC plasma discharges can produce valuable products, namely synthesis gas, from the partial oxidation of methane while maintaining low bulk gas temperatures. The products for this reaction are limited to synthesis gas, ethane, ethylene, acetylene, CO_2, and water. The objective of the study is to maximize the partial oxidation of methane to synthesis gas and/or C_2 species using both pure O_2 and air and to minimize the electrical energy required for conversion.

1. INTRODUCTION

Many of the current and proposed methane conversion technologies involve the initial reforming or partial oxidation of methane to synthesis gas at high temperatures (> 700 °C). The synthesis gas may then be converted to methanol using a Cu/Zn catalyst or to higher hydrocarbons over an iron or cobalt catalyst via Fischer-Tropsch synthesis. These technologies are energy-intensive and are economically feasible only on a limited basis, typically on the largest possible scale. Nearly two-thirds of the cost is assumed in the production of synthesis gas. Lowering the expense of synthesis gas production is paramount to making methane conversion technologies more competitive particularly on smaller scales.

Considerable research has been focused on the use of electrical discharge plasmas to initiate many different chemical reactions. These studies include the conversion of methane to various products (Larkin 1998, Thanyanchotpaiboon 1998). Non-equilibrium electrical discharge plasmas are characterized by a large number of free electrons that are accelerated through a low-temperature gas by a large electric potential applied across a pair of electrodes. Normally, gases act as very good electrical insulators. If, however, an electric potential of sufficient strength is applied across the gas, free electrons within the gas are accelerated to high energies and can collide with the gas molecules resulting in ionization. Ionization produces more free electrons that are also accelerated and repeat the process until an electron "avalanche" occurs between the electrodes. The electron avalanche consists of high-energy electrons that produce a conductive path through the gas allowing electric charge to pass freely between the electrodes. This process is known as electric "breakdown."

The accelerated electrons can gain considerable energy in the electric field and are capable of transferring their kinetic energy to the molecules of the feed gas through inelastic collisions. These collisions can increase the internal energy of the gas molecules and may result in excitation, dissociation, or ionization without significantly increasing the bulk gas temperature. This increase in the internal energy of the gas molecules can overcome the reaction activation energy allowing the species to be converted to various products. Since most of the energetic electrons required to activate the feed gas are located in a relatively narrow conduction channel, special consideration must be taken to design a reactor that will maximize the contact time between the energetic electrons and the neutral feed gas species.

Under the conditions of this study, the high energy electrons initiate chemistry that is predominantly free radical in nature and the energy transfer processes of the inelastic collisions do not involve significant heating. Thus the bulk gas temperature is not in equilibrium with the electrons and remains relatively low.

2. EXPERIMENTAL

This low-temperature plasma reactor consists of two electrodes, spaced axially, in a narrow quartz tube. The top electrode is a rounded point while the bottom electrode is a small flat disk (point-to-plane configuration). The feed gas flows axially from top to bottom through the discharge zone between the electrodes. The feed gas flow rates are controlled using Porter mass flow controllers. A Carle 400 Series gas chromatograph is used to analyze the composition of the exit gases. A soap bubble meter measures the exit gas flow rates.

An Elgar AC power system is supplied with wall current having a voltage of 120 V and a frequency of 60 Hz. Connected to the power supply is a multifunction generator that generates sinusoidal, square, or triangular waveforms and allows the frequency to be varied over the range used for these experiments. The output of the power supply is connected to a high voltage transformer. The high voltage cables from the transformer are connected to the electrodes of the reactor system. An Extech power analyzer that is connected to the low voltage side of the power circuit measures the voltage, frequency, current, power factor, and power consumed by the high voltage transformer and reactor system.

The *visible* discharge within the quartz tube appears as a small cluster of fine bluish filamentous "sparks" between the electrodes. These sparks, which are known as streamers, occupy a relatively small fraction of the total reactor volume. Some fraction of the feed gas bypasses these streamers and may not interact with the energetic electrons needed for activation. In order to decrease the feed gas bypassing the "active" reactor volume, the point electrode is forced to oscillate back and forth in a planar path parallel to the bottom electrode by means of an external magnetic field. The streamer discharges were observed to move through a larger fraction of the reaction zone as the point electrode oscillated. The electrical "efficiency" showed little improvement using this technique, but the formation of coke within the reactor became less problematic, resulting in a more stable discharge.

3. RESULTS AND DISCUSSION

The reaction of pure methane in the tubular AC plasma reactor under the conditions examined in this work is characterized by low conversions (< 5%) and poor electrical efficiencies (~100 eV/molecule CH_4 converted). The gaseous products are limited to C_2 hydrocarbons and hydrogen. It is apparent that the activation of the methane molecule in this environment is not very efficient. The presence of active oxygen species in the feed gas has been shown to considerably enhance the conversion of methane in dielectric barrier discharge plasmas (Larkin 1998). A base "standard" experiment in this study had a feed mixture of methane and oxygen at a 3:1 ratio flowing at 100 SCCM through a quartz tube with an inner diameter of 7 mm. The gas gap between the electrodes was set at 1.0 cm. This resulted in a residence time of 0.231 seconds. The applied voltage was a sine wave with a frequency set at 300 Hz. The power for this experiment was fixed at 10 W.

The total methane conversion for this experiment was 24.4% and the oxygen conversion was 49.2%. These results show that active oxygen species greatly enhance the conversion of methane in this system as well. The methane conversion efficiency was determined to be 6.8 eV per molecule of methane converted. The addition of oxygen to the feed also changes the product distribution. 52.3% of the hydrogen from the reacted methane formed molecular hydrogen. The carbon-based selectivities were as follows: ethane (7.4%), ethylene (15.4%), acetylene (16.7%), carbon monoxide (40.0%), and carbon dioxide (5.2%).

A series of experiments was performed to study the effects of residence time. The reaction parameters were the same for the standard experiment, except the flow rate of the feed gas was varied between 400 and 20 SCCM giving feed gas residence times ranging from 0.057 to 1.128 seconds. The conversion and efficiency data are shown in Figure 1. Although the overall conversion of methane increases as the residence time increases, the reaction becomes less efficient. The overall rate of methane conversion is highest at low residence times. This suggests that the rate of methane conversion is a function of both methane and oxygen concentration. At low residence times, the methane conversion rate is higher because the methane and oxygen concentrations remain high throughout the entire reaction zone. As residence times increase and the overall conversion of methane and oxygen increases, the fractions of methane and oxygen become smaller, thereby reducing the average reaction rate of methane and lowering the overall efficiency of the reaction. The product selectivities also vary with residence time. As the residence increases, the selectivities of ethane, ethylene, CO, and CO_2 increase while acetylene decreases. The fraction of hydrogen produced changes little over the range tested.

3.1 Frequency, Waveform, and System Power

A series of experiments was performed to study the effect of changing the frequency of the applied voltage. All of the reactor parameters were held constant while the frequency was varied between 100 and 800 Hz. Varying the frequency of the voltage waveform caused the power factor and therefore the reaction power to vary as well. The applied voltage was adjusted for each frequency to maintain a constant power of 12 W. The total methane and oxygen conversion varied little for these experiments, suggesting that the total power consumption was the primary determinant for the extent of conversion. The selectivities of the gas products were also nearly the same for each frequency.

The next set of experiments tested how the shape of the voltage waveform might affect the reaction. Experiments used sine, square, and triangular voltage waveforms under the same reaction conditions. The results showed that the methane and oxygen conversions, as well as the product selectivities, were essentially the same for each waveform. The square waveform did, however, produce slightly more stable discharges at voltages just above breakdown.

Other experiments were performed to test the effect of increasing the system power from 10 W to as high as 30 W. At low power levels, both the methane and oxygen conversion increase linearly with power. At higher powers, the oxygen conversion increases to the point where the concentration of active oxygen species becomes limited. Thus, the increase in the methane conversion rate with increasing power becomes less pronounced since active oxygen is needed to enhance the reactivity of methane. The electrical efficiency decreases with increasing system power because more and more energy is required to convert the available methane when active oxygen becomes depleted.

As the power applied to the reactor increases, the bulk gas temperature in the reaction zone increases. At higher temperatures, the thermal dehydrogenation reactions of methane to form olefins and carbon become more prevalent (Mallinson et al. 1992). The acetylene fraction increases considerably with system power. Excessive coking also becomes a problem at these higher power densities. The carbon deposits tend to form along the inner wall of the quartz tube or grow as fine filaments extending from the bottom electrode. Since this coke is electrically conductive, the current tends to flow almost entirely through these carbon deposits. This reduces the number of discharge streamers and limits the number of energetic electrons that can interact with the feed gases in the reaction zone, thereby lowering the conversion of methane in the reactor.

3.2 Pressure

Increasing the pressure within the reactor system is one way to increase the throughput of the feed gas without lowering the residence time. However, increasing the reactor pressure changes the electrical breakdown characteristics within the system. At higher pressure, the mean free path of free electrons accelerated in the applied electric field decreases. This means that the average electron energy is lower in the presence of the same electric field strength. A larger applied voltage is required to achieve electric breakdown and sustain the discharge under these conditions.

An experiment conducted at 2 atm was performed to examine the effect of increased reactor pressure on the conversion of methane and the distribution of products. A back-pressure regulator was used to set the reactor pressure at 15 psig. The methane-oxygen ratio remained at 3:1 while the feed gas flow rate was raised to 200 SCCM in order to maintain a residence time of 0.23 seconds. The minimum power required to obtain a stable discharge was higher at this pressure. The system power was fixed at 14 W. The other reaction conditions were the same as the standard experiment.

Although the high pressure experiment required more power to sustain a stable discharge, its electrical efficiency was better than the atmospheric pressure experiment. The overall methane conversions were similar for each experiment, 22.1% for the high pressure experiment vs. 24.4% for the standard experiment, but the larger throughput for the high pressure system resulted in an electrical efficiency of 5.6 eV/molecule of CH_4 converted vs. 6.8 eV for the standard experiment. The selectivities of several products differed for each system pressure as well. At higher pressure, the fraction of ethane produced was higher while ethylene and acetylene was lower. This was most likely due to the shift in the relative rates of the dehydrogenation-hydrogenation reactions at higher pressure to favor hydrogenated species. This also resulted in lower hydrogen selectivities at higher pressure as well, since more hydrogen is retained on the hydrocarbon products. The selectivities of CO_X species did not change appreciably with changes in system pressure.

3.3 Pure Oxygen and Air

The next series of experiments was designed to examine how the concentration of oxygen in the feed affects the conversion of methane and the distribution of products. The methane - oxygen ratio was varied between 19:1 and 2:1, resulting in oxygen concentration that varied between 5% and 33% (maintained below the flammability limit). The other reactor parameters were the same as the standard experiment. Figure 2 confirms that the addition of oxygen is effective at activating methane. The total methane conversion increases appreciably as the feed concentration of oxygen increases. The methane conversion increases from less than 2% with low oxygen concentration to nearly 45% at higher oxygen concentrations. This means that the methane conversion efficiency improves as well, decreasing from nearly 90 eV per molecule of methane converted at low oxygen fractions to under 6 eV at higher oxygen feed fractions.

Increasing the fraction of oxygen in the feed also changes the distribution of several major products. Figure 3 shows that the hydrogen selectivity does not change significantly for these experiments - remaining between 50% and 60%. But the selectivities of the CO_X compounds increase significantly as the concentration of oxygen in the feed increases. The carbon monoxide selectivity increases from 19% to 55% while the carbon dioxide selectivity increases from less than 1% to 6%. The CO/CO_2 ratio remains high over this range of experiments because the activated oxygen is more likely to react with non-oxygenated species before completely oxidizing carbon monoxide.

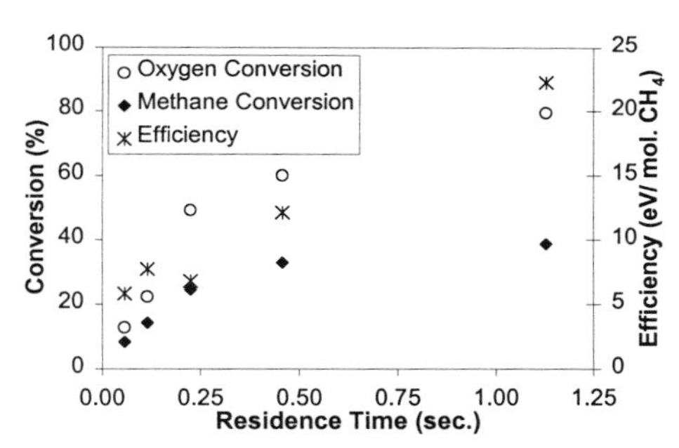

Figure 1 - Conversion and Efficiency vs. Residence Time for a Methane-Oxygen Feed (CH_4:O_2 = 3:1)

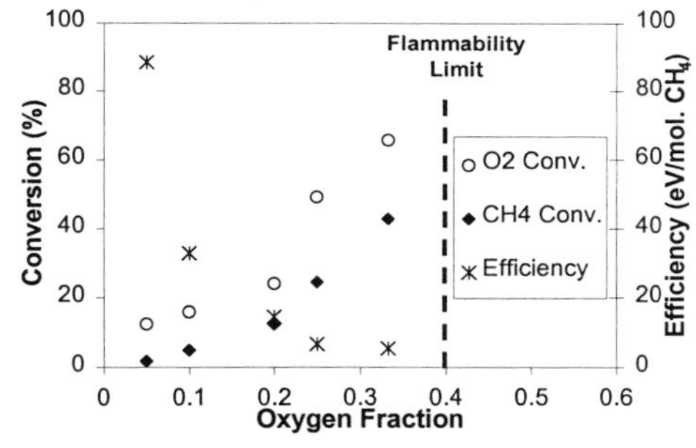

Figure 2 - Conversion and Efficiency vs. Oxygen Fraction in the AC tubular reactor

The selectivity of ethane decreases rapidly as the oxygen concentration increases - from nearly 50% to under 4%. The selectivity of ethylene changes little at oxygen fractions below 20%, remaining between 14% and 18%, but drops to less than 10% as the oxygen fraction in the feed increases to 33%. The acetylene concentration increases rapidly as the oxygen fraction increases from 5% to 10%, but remains relatively constant as the oxygen fraction is raised further. The increase in olefin production at higher oxygen concentrations may be due to an increase in the reactor temperature from exothermic oxidative reactions. The higher temperatures increase the likelihood of dehydrogenation reactions. The slight decrease in the C_2 selectivities at the highest oxygen fractions can be attributed to a significant increase in the relative rates of oxidation reactions.

Using pure oxygen in large quantities can be of prohibitive cost, especially for smaller operations. Since air separation to provide pure O_2 is a significant expense, the use of air for the partial oxidation of methane would appear to be an interesting alternative. Experiments were performed to show how the partial oxidation of methane using air compares to that of using pure O_2. Since the best methane-oxygen results were at a methane-oxygen ratio of 2:1, the feed gas used for the methane-air experiments was a combination of 71% air and 29%

methane, resulting in a methane to oxygen ratio of 2:1. This meant that a majority (~ 56%) of the feed gas was nitrogen. Nitrogen does not react very readily in the discharge environment, so it acts primarily as a diluent and a heat sink for the reaction. A higher system power of 16 W was used for these experiments. The nitrogen absorbed some of the heat generated by the exothermic partial oxidation reactions, making excessive coking less of a problem. Feed gas flow rates were varied between 100 and 800 SCCM. Higher flow rates were used in order to obtain similar methane throughputs as previous residence time experiments. The other reaction parameters were the same as the standard experiment.

Figure 4 shows the methane and oxygen conversion and the conversion efficiency for the methane-air system at the different residence times. Relatively high methane and oxygen conversions were obtained despite the large nitrogen dilution. Since conversions were high, the electrical efficiencies remained reasonably low, being only slightly higher than those obtained for the pure O_2 system. A qualitative analysis to determine the presence of NO_X compounds found that NO_2 was present in the product stream, NO levels were on the order of about 10 ppm, and N_2O was below detectable levels.

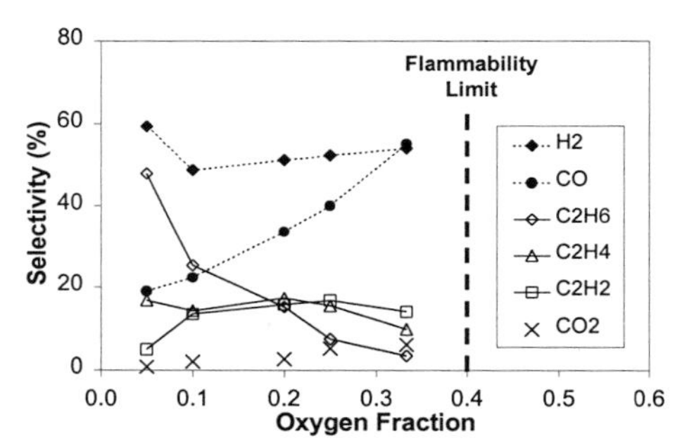

Figure 3 - Product Selectivity vs. Oxygen Fraction in the AC tubular reactor

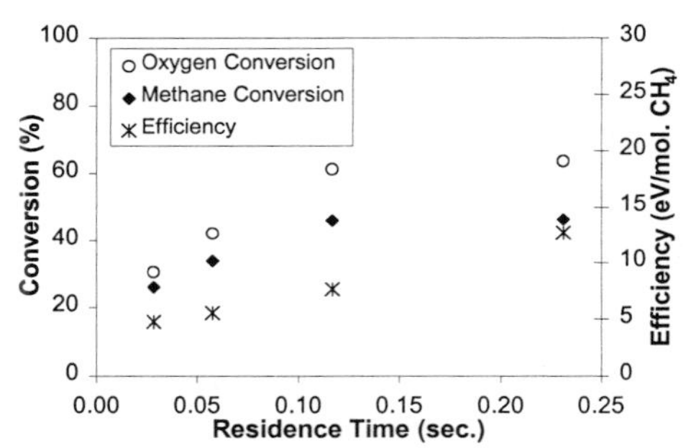

Figure 4 - Conversion and Efficiency vs. Residence Time for a Methane-Air feed (CH_4:O_2 = 2:1)

4. CONCLUSIONS

A range of reaction parameters has been varied to study their effects on methane conversion, product selectivity, and energy consumption in the AC tubular reactor. The results indicate that increasing the oxygen fraction in the feed enhances the activation of methane in the reaction. Also, reducing feed gas bypassing, operating at low power, and increasing the reaction pressure all improve the efficiency of methane conversion.

ACKNOWLEDGEMENTS

Texaco, Inc., the United States Department of Energy under Contract DE-FG21-94MC31170 and the National Science Foundation for a Graduate Traineeship are gratefully acknowledged for their support of this research.

REFERENCES

1. Larkin, D.W., Caldwell, T.A., Lobban, L.L., Mallinson, R.G., "Oxygen Pathways and Carbon Dioxide Utilization in Methane Partial Oxidation in Ambient Temperature Electric Discharges," *Energy and Fuels*, Vol. 12, No. 4, p. 740, 1998.
2. Thanyach, K., Chavadej, S., Caldwell, T.A., Lobban, L.L., Mallinson, R.G., "Conversion of Methane to Higher Hydrocarbons in AC Nonequilibrium Plasmas," *AIChE Journal*, Vol. 44, No. 10, p. 2252, 1998.
3. Mallinson, R.G., Braun, R.L., Westbrook, C.K., Burnham, A.K., "Detailed Chemical Kinetics Study of the Role of Pressure in Butane Pyrolysis," *Ind. Eng. Chem. Res.*, Vol. 31, No.1, p. 37, 1992.

Studies in Surface Science and Catalysis
J.J. Spivey, E. Iglesia and T.H. Fleisch (Editors)

Selective hydrogenation of acetylene to ethylene during the conversion of methane in a catalytic dc plasma reactor

C. L. Gordon, L. L. Lobban, and R. G. Mallinson

Institute for Gas Utilization Technologies, School of Chemical Engineering and Materials Science, University of Oklahoma, 100 East Boyd, Norman, Oklahoma 73019

Previous studies have shown that plasma reactors can achieve high methane conversions while maintaining high selectivities towards acetylene and hydrogen. Palladium catalysts are commonly used in the selective hydrogenation of acetylene to ethylene. This paper discusses the use of Pd-Y zeolite for the in-situ selective hydrogenation of acetylene. Pd loading, presence of oxygen, and temperature play a large role in the selective hydrogenation of the acetylene.

1. INTRODUCTION

It is generally agreed that, for a variety of reasons, natural gas will become increasingly important as a petrochemical feedstock and source of liquid fuels. Natural gas reserves are abundant throughout the world. A limitation to the utilization of the reserves is the transportation of the natural gas to the desired location.

C_2 hydrocarbons consist of acetylene, ethylene, and ethane. Ethane is primarily used for the formation of ethylene by dehydrogenation, and ethylene has largely replaced acetylene as a petrochemical building block. The United States alone produces nearly 60 billion pounds of ethylene per year. Ethylene is highly reactive, due to its double bond, allowing it to be converted to a large assortment of products by addition, oxidative, and polymerization reactions. It is primarily used in the production of plastics, fibers, films, resins, adhesives, and elastomers [1], but is can also be oligimerized to liquid hydrocarbons.

Ethylene can be derived from many different feedstocks. The feedstock, and the resulting process, varies depending on the region. Table 1 lists different feedstocks used in the production of ethylene and their relative amounts in the United States. However, for Western Europe and Japan naphtha is the primary feedstock for the production of ethylene since natural gas is less abundant.

Table 1. Ethylene Feedstocks in the U.S.[2]

Ethane	27-30 billion pounds
Propane	8-10 billion pounds
Naphtha	10 billion pounds
Other	10 billion pounds

The thermal cracking of petroleum based naphtha with steam, known as pyrolysis, is used for over 97% of the worldwide production of ethylene. This process uses a feed stream that is a mixture of hydrocarbons and steam. The stream is preheated to a temperature of 500-650°C, and then raised to 750-875°C in a controlled manner in a radiant tube [3]. In the radiant tube the hydrocarbons crack into the major products: ethylene, olefins, and diolefins. Due to the high temperatures required for the endothermic reaction, an intensive energy input is required to drive the process. The large amount of by-products resulting from the use of the heavy hydrocarbon feedstock also requires oversized equipment for the resulting ethylene quantity.

The production of ethylene from ethane is also done by steam cracking. Not including methane, ethane requires both the highest temperature and the longest residence time to achieve acceptable conversion [3]. A typical ethane process operates near 60 % conversion of ethane and achieves an ethylene selectivity of 85%. This process uses the burning of the unwanted by-products to provide the necessary heat for the endothermic reaction. Over 10% of the ethane is thereby converted into carbon dioxide, and nitrogen oxides are also formed by the combustion. Production of ethylene by steam cracking is a large contributor to greenhouse gases [4]. It is desired, therefore, to introduce technologies that will reduce the emission of greenhouse gases during the production of ethylene.

It is already known that the use of methane as a petrochemical feedstock can lead to environmentally friendly technologies. The production of hydrogen by means of steam reforming is the only significant use of natural gas in the petrochemical industry. Steam reforming of methane is the industry's cleanest option for the production of hydrogen. Steam reforming of methane results in one carbon dioxide released for every four hydrogen molecules produced. On the other hand, the partial oxidation of coal results in a 1:1 production of carbon dioxide and hydrogen.

It has also been shown [5] that plasma reactors can activate methane at low temperatures. Low temperature plasmas offer the potential for efficient processes for direct conversion of methane to higher value products or the widely used intermediate, synthesis gas. The reaction is driven by highly energetic electrons that are created by applying an electric potential across the reaction volume. The low temperature plasma makes use of these excited electrons as initiators in what is predominantly a free radical pathway for conversion. The bulk gas temperature remains relatively low, allowing for unique, non-equilibrium product distributions. This paper will discuss the use of a dc plasma catalytic reactor for the production of C_2 hydrocarbons and/or synthesis gas (H_2 and CO).

2. EXPERIMENTAL

The experimental apparatus is similar to the system that has been described previously [5, 6]. The feed gases consisted of a combination of methane, oxygen, hydrogen, and helium. Helium was only used in initial experiments and for characterization studies of the catalyst. The feed gas flowrates were controlled by Porter mass flow controllers, model 201. The feed gases flowed axially down the reactor tube. The reactor is a quartz tube with a 9.0 mm O.D. and an I.D. of either 4.5 mm or 7.0 mm.

A point to plane electrode configuration was employed, meaning that the top electrode is a wire point electrode and the bottom electrode is a flat plate that also serves as a support for the catalyst. The top electrode is positioned concentrically within the reactor, and the gap between its tip and the plate is 8.0 mm. The dc corona discharge is created using a high voltage power supply (Model 210-50R, Bertan Associates Inc.).

The preparation and characterization of the NaOH treated Y zeolite has been discussed elsewhere [7, 8]. Palladium was added using chemical vapor deposition according to the following procedure. The NaOH Y zeolite was calcined at 400°C, then mixed with the appropriate amount of palladium acetylacetonate to achieve the desired Pd loading (ranging from 0.025 to 1%). The mixture was then heated slowly under vacuum to a temperature of 130°C to disperse the palladium throughout the NaOH Y zeolite. This Pd on NaOH Y zeolite will henceforth be referred to as Pd-Y zeolite. The Pd-Y zeolite was then calcined for the final time to a temperature of 350°C.

The fresh catalyst is pretreated in the reactor before each experiment. The NaOH Y zeolite is heated at 250°C to remove any moisture from the catalyst. The Pd-Y zeolite is reduced in 30 ml of hydrogen for 4 hours. The temperature is slowly ramped to 350°C during the first two hours, then held at 350°C for the final two hours.

As mentioned before, this system operates at low temperatures. A furnace around the reactor is used to heat the system to the desired temperature. However, when the desired operating temperature is below 373 K it is necessary to use cooling air across the tube exterior to control the temperature since the plasma itself does heat the gas to some extent. The temperature measured on the outside has been calibrated against the internal temperature of the reactor, and has been discussed elsewhere [7, 8].

The product gases are passed through a dry ice/acetone bath that allows for any condensable organic liquids to be separated from the product gases. It should be noted that the dc system does not produce any measurable liquids, including water. The effluent gases can be analyzed on-line by either a gas chromatograph or a mass spectrometer.

3. RESULTS and DISCUSSION

The catalytic dc plasma reactor has shown to produce high yields of C_2 hydrocarbons, hydrogen, and carbon monoxide at reasonable methane conversions. It is desired, due to many factors, to shift the acetylene production to ethylene. Pd supported catalysts are commonly used for the selective hydrogenation of acetylene to ethylene. These catalyst usually convert acetylene at low concentrations in ethylene streams to ethylene with a very high selectivity. The Pd-Y zeolite prepared by CVD was used to study the in-situ production of ethylene from acetylene (where the acetylene comes from the conversion of methane). Figure 1 shows the effect of different Pd loadings on methane conversion and acetylene, ethylene, and ethane selectivity.

The methane conversion is fairly independent of Pd loading when compared to the parent Y zeolite. The C_2 selectivities vary greatly depending on the Pd loading. As the Pd loading is increased, the amount of acetylene that is hydrogenated increases until all of the acetylene is hydrogenated at 1.0 wt% Pd. At low Pd loadings, ethylene is the major hydrogenated product, while at the high loadings ethane predominates. This dependence is believed to be due to the amount of Pd available for hydrogenation to ethylene and ethane. It is thought that low amounts of palladium and small Pd ensembles favor the hydrogenation to ethylene, while excess Pd and large Pd clusters allow for the produced ethylene to be further hydrogenated to ethane. Acetylene's adsorption strength is higher than that of ethylene, resulting in the desorption of ethylene before it can be further hydrogenated to ethane when metal site availability is limited. However, the 1.0 wt% Pd loading has larger Pd ensembles that lead to the formation of ethane.

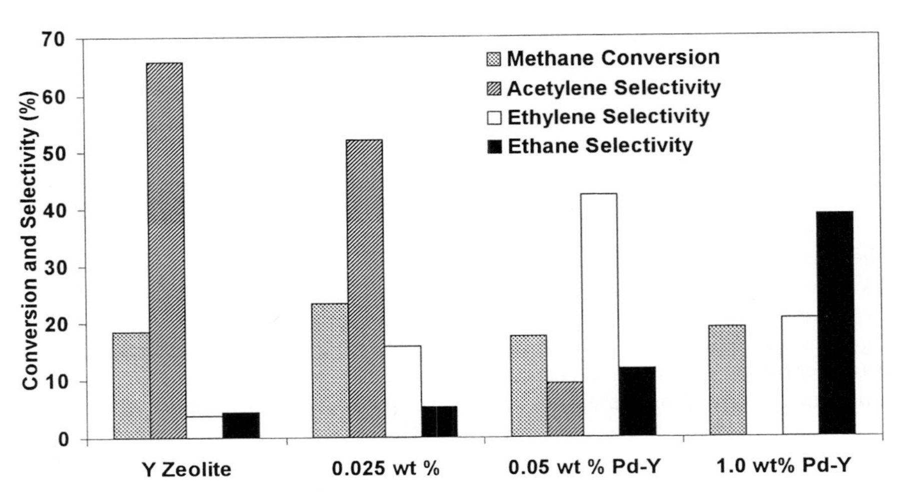

Figure 1. Effect of different Pd loadings on methane conversion and hydrocarbon selectivity. 2/1 H_2/CH_4 with 2% O_2, 7200 hr^{-1}, 3.9 watts, 45°C

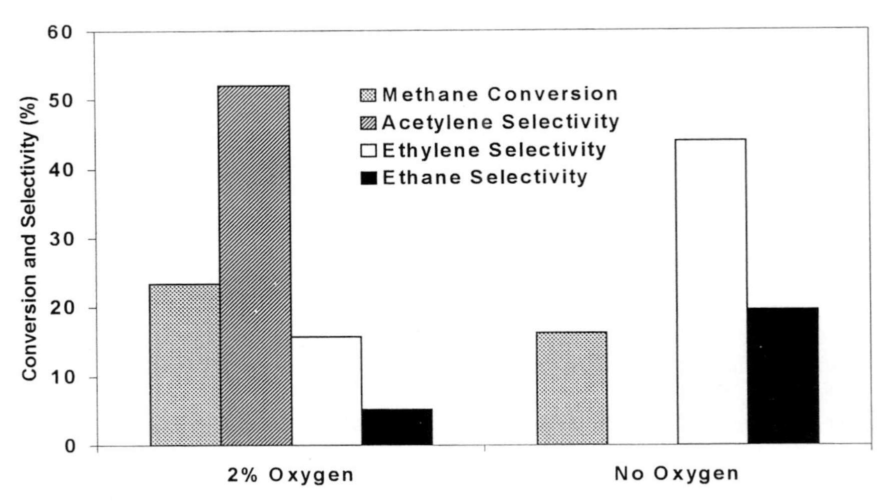

Figure 2. Effect of oxygen on methane conversion and hydrocarbon selectivity. 0.025 wt% Pd-Y zeolite, 2/1 H_2/CH_4 with x% O_2, 7200 hr^{-1}, 3.9 watts, 45°C

It was also found that oxygen plays a major role in the selectivity of the hydrogenation catalyst, probably via production of carbon monoxide. Previous researchers found that carbon monoxide reduced the hydrogenation of ethylene to ethane because of its higher adsorption strength than ethylene.[9]. In our studies we noticed a large effect on selectivity when carbon monoxide is present; however, it was not the usual inhibition of ethylene hydrogenation, but rather acetylene hydrogenation (Figure 2 and 3).

As seen from Figures 2 and 3, carbon monoxide affects the hydrogenation of acetylene to ethylene and ethane. When oxygen is removed from the feed, all of the acetylene is hydrogenated. Carbon monoxide seems to reduce the number of Pd sites that are available for the hydrogenation of acetylene. Similarly, when the system is operating without oxygen and carbon monoxide is injected in the system, acetylene (not present before) becomes the major product and immediately disappears upon removal of carbon monoxide.

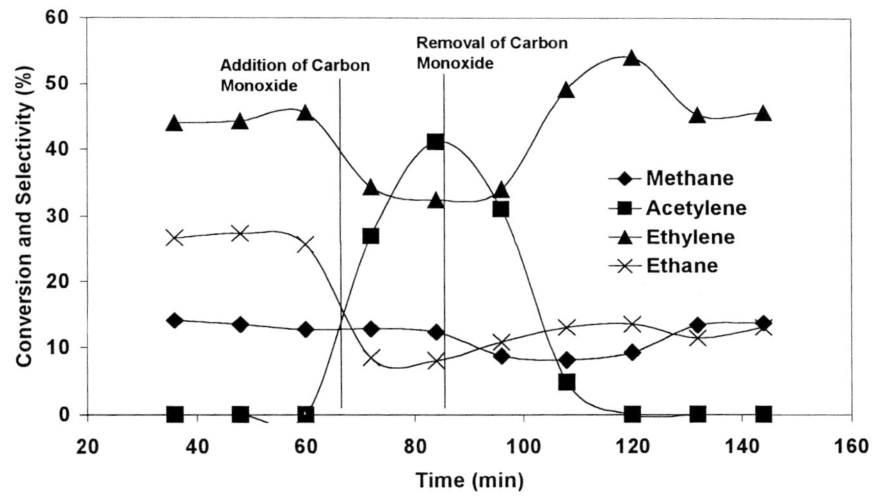

Figure 3. Effect of addition and removal of carbon monoxide on methane conversion and hydrocarbon selectivity. 0.025 wt% Pd-Y zeolite, 2/1 H_2/CH_4 with x% O_2, 7200 hr^{-1}, 3.9 watts, 45°C

The operating temperature plays a large role in the activity of the Pd catalyst, Figure 4. As the temperature is increased from room temperature to 100°C, the acetylene is completely hydrogenated to ethylene and ethane. Ethylene goes through a selectivity maximum at around 50°C. Above 50°C, the selectivity decreases due to the over hydrogenation to ethane. At temperatures above 95°C, ethane is the only C_2 hydrocarbon produced. Therefore, it is desirable to operate at low temperatures at which good conversions can still be achieved in the plasma discharge.

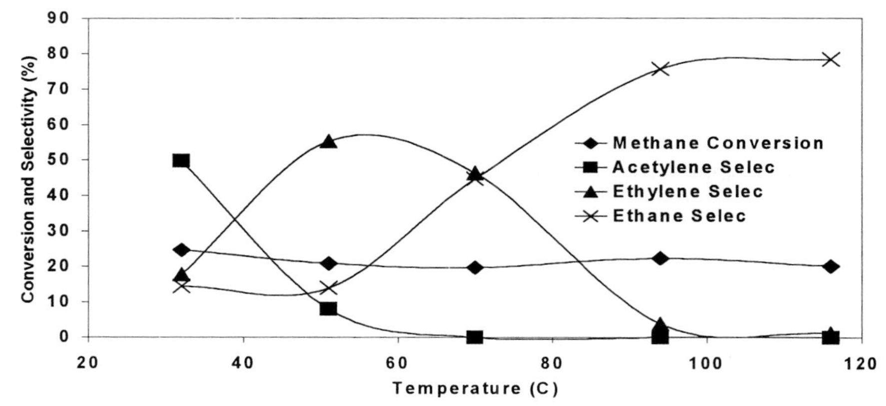

Figure 4. Effect of temperature on methane conversion and hydrocarbon selectivity. 0.025 wt% Pd-Y zeolite, 2/1 H_2/CH_4 with no O_2, 7200 hr^{-1}, 3.9 watts

4. CONCLUSIONS

The addition of palladium to the Y zeolite used in the catalytic dc plasma reactor allows for the hydrogenation of acetylene to ethylene and ethane. Currently, operating around 50°C and without oxygen allows for greater selectivity control to the desired ethylene. Further studies and characterization should allow for further improvements in the selectivity of ethylene in the hydrogenation of acetylene.

ACKNOWLEDGMENTS

The authors would like to express their appreciation to Texaco Inc., the United States Department of Energy (Contract DE-FG21-94MC31170), and the National Science Foundation for sponsoring a Graduate Research Traineeship in Environmentally Friendly Natural Gas Technologies. Duc Le's help on the experimental work is greatly appreciated.

REFERENCES

1. McKetta, J.J., ed. *Encyclopedia of Chemical Processing and Design*. . Vol. 20. 1986, Marcel Dekker: 1984. 88-159.
2. Greenberg, K., *Domestic ethylene market stays strong*, in *Chemical Market Reporter*. 1999. p. 4,12.
3. Gerhartz, W., *Ullman's Encyclopedia of Industrial Chemistry*. Vol. A10. 1987, Weinheim: VCH. 45.
4. Jacoby, M., *New ethylene process is environmentally-friendly*, in *C&EN*. 1999. p. 6.
5. Liu, C.J., R.G. Mallinson, and L.L. Lobban, *Nonoxidative methane conversion to acetylene over zeolite in a low temperature plasma*. Journal of catalysis, 1998. **179**: p. 326-334.
6. Gordon, C.L., L.L. Lobban, and R.G. Mallinson, *The production of hydrogen from methane using tubular plasma reactors*, in *Advances in hydrogen energy*, C. Gregoire-Padro, Editor. 2000, Kulmer.
7. Liu, C.-J., A. Marafee, B.J. Hill, G. Xu, R.G. Mallinson, and L.L. Lobban, *Oxidative coupling of methane with ac and dc corona discharge*. Industrial & Engineering Chemistry Research, 1996. **35**(10): p. 3295-3301.
8. Marafee, A., C.-J. Liu, G. Xu, R.G. Mallinson, and L.L. Lobban, *An experimental study on the oxidative coupling of methane in a direct current corona discharge reactor over Sr/La2O3 catalyst*. Industrial & Engineering Chemiestry Research, 1997. **36**: p. 632-637.
9. Price, G.L. and Y.H. Park, *Deuterium tracer study on the effect of CO on the selective hydrogenation of acetylene over Pd/AL2O3*. Industrial Engineering and Chemical Research, 1991. **30**: p. 1693-1699.

Studies in Surface Science and Catalysis
J.J. Spivey, E. Iglesia and T.H. Fleisch (Editors)

The Nigerian Gas-to-Liquids (GTL) Plant

R. J. Motal[a], S. Shadiya[b], R.C. Burleson[c], Dr. E. Cameron[d]

[a]Chevron Research & Technology Co., 100 Chevron Way, Richmond, California, USA
[b]Chevron Nigeria Ltd, 2 Chevron Drive, Lekki, Lagos, Nigeria
[c]Chevron, Project Resources Group, 11111 South Wilcrest, Houston, Texas, USA
[d]Sasol Chevron Global Joint Venture, 45-50 Portman Square, London, UK

West Africa is geologically rich in oil and gas. Historically the oil was produced for export and local refining, and associated gas in excess of local market needs was flared. Non-associated gas was either processed for condensate removal or left in the ground. Gulf Nigeria, which became part of Chevron in the mid 1980s, began production in the western delta region in Nigeria in 1965. Today Chevron Nigeria operates these concessions producing over 400,000 BPD. Chevron is working to responsibly continue producing hydrocarbons in this region and is committed to putting out the flares. Our concession partner (Nigerian National Petroleum Company), the Nigerian government, and our technology partners, through the Sasol Chevron Global Joint Venture (GJV), are working together to make this happen. This paper will describe the GTL plant, highlight some of the challenges and tell how those challenges were addressed.

Figure 1 Chevron Nigeria-Operated Concessions in Western Niger Delta

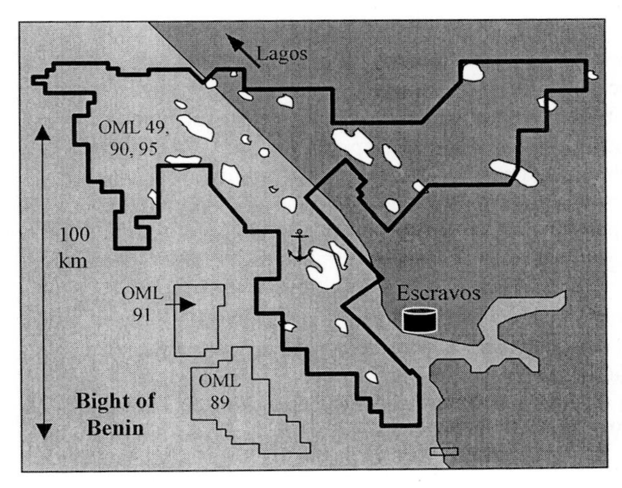

1. GAS SUPPLY

The gas for the GTL project will be gathered from oil mining leases: OML49, OML90 and OML95. These are outlined in bold on Figure 1. It is rich gas associated with current oil production. The feed for the current project and future expansions could be supplemented by gas from OML89, OML91 and future license awards. One of the challenges is to economically gather this gas, which is typically at near-atmospheric pressure, from this geographically large area. Currently a limited amount of this gas is gathered, processed and fed into the Nigerian Gas Company (NGC) pipeline from Escravos to Lagos.

Chevron is expanding the gas gathering and compression system in the swamp area and debottlenecking the existing gas plant. The ultimate plan is to expand the gas gathering to all significant fields in the three concession areas, build a second gas processing plant and flow the residue gas to the GTL plant and the West Africa Gas Pipeline (WAGP). WAGP will transport gas to the neighboring countries to the west. WAGP is a six-company consortium consisting of: NNPC, the state energy companies of Ghana, Benin, and Togo, Chevron, and Shell. Chevron is the Managing Sponsor.

A limited amount of non-associated gas is being tapped to provide a reliable emergency backup to keep the NGC pipeline, WAGP and the GTL plant operating during maintenance or disruptions to the normal associated gas supply. Importantly for plant design purposes the backup non-associated gas tends to be leaner in LPG and richer in condensate.

Together the two gas processing plants will produce about 650 MMSCFD of lean gas along with fractionated LPG and a C_5^+/condensate mix. These gas processing plants will provide a reasonably constant gas composition for the GTL plant containing approximately 90% methane, 7% ethane, with small amounts of nitrogen, carbon dioxide, and heavier hydrocarbons. Without these plants the GTL plant feed would contain 6-7% LPG and higher boiling naphtha. The GTL plant is designed to take this swing in gas composition.

2. GTL TECHNOLOGY SELECTION

The GTL plant will use commercially proven technology provided by the Sasol Chevron Global Joint Venture. Topsoe will provide the gas reforming technology. Sasol will provide the Slurry-Phase Fischer Tropsch technology and Chevron will provide the product workup hydroprocessing technology. Each of these companies is a leader in their respective fields. Topsoe's proprietary catalysts and burner designs have been proven in combined reforming mode at Mossgas in South Africa which is currently the largest GTL facility in the world. Autothermal reforming, which will be used in the Escravos GTL plant, is a modest stepout from combined reforming. The synthesis gas production rates contemplated for Escravos are essentially the same as for the Mossgas reactors.

Sasol has run both iron-based and cobalt-based slurry catalysts in a 100 BPD demonstration reactor. Sasol also has demonstrated its Slurry Phase Fischer Tropsch reactor with iron-based catalyst in their 2,500 BPD reactor at Sasolburg with continuous commercial operations starting in 1993. The Escravos plant will use two 15,000 BPD Fischer-Tropsch reactors. Sasol and their reactor consultants, Washington Group International, are confident that they can predict the performance of the equipment at this 6x scaleup. Sasol already operates four 20,000 BPD high temperature Advanced Synthol Fischer Tropsch reactors at Secunda which are similar size to the slurry reactors proposed for Escravos.

Chevron has designed, built, and operated many Isocrackers® for hydroprocessing conventional gas oils in sizes larger than contemplated for Escravos. Laboratory testing has confirmed that Chevron's catalysts produce high yields of premium quality products from Fischer-Tropsch liquids.

These companies have been working together and with the engineering contractor Foster Wheeler for years to combine these technologies into an integrated cost-effective design.

3. ESCRAVOS GTL SITING ISSUES

The existing Escravos terminal is in a low lying delta swamp on the right bank of the Escravos River where it empties into the Bight of Benin. The terrain slopes very gently seaward both onshore and offshore. The soil is water saturated with little load-carrying capacity. Sand from offshore deposits will be used to raise the ground level and increase the load bearing capacity. The best construction window is during November to April when monthly rainfall drops to 50 to 150 mm/month versus 300 mm/month during the peak of the rainy season in June.

Figure 2 Escravos Terminal/Construction Site

The Escravos terminal is the hub for Chevron Nigeria's production operations. It receives oil production from 24 fields spread across the three concessions. It also currently receives limited gas production from some nearby fields to supply the Nigerian Gas Company pipeline to Lagos. The terminal includes a gas processing plant. Pipelines connect Escravos to 14 offshore platforms, 6 onshore gathering stations, two offshore loading buoys for exporting oil, and a LPG floating storage and offloading vessel.

The location for the GTL plant is adjacent to and upriver from the existing oil terminal and gas plant.

We found cost savings due to process integration with the existing facilities primarily in the LPG area where the GTL plant will rely on the existing LPG handling, storage and export facilities. Spare capacity in the utilities area was small and insufficient to eliminate the need for adding units for all utilities. In other areas the need for stream segregation to preserve the desirable qualities of the GTL liquid products limited integration.

4. PROCESS DESIGN

The GTL plant design has two parallel trains each consisting of a world scale air separation plant, an autothermal reformer, and a Slurry Phase Fischer-Tropsch reactor. The combined output from these two trains feeds a single train Isocracker®.

The GTL plant is designed to convert 300 MMSCFD of gas into 22,300 BPD of ultraclean diesel boiling range fuel, 10,800 BPD of naphtha and 1,000 BPD of LPG. The exact product split and overall plant efficiency is preliminary pending the results of detailed engineering. The ultraclean premium fuel and the naphtha will be stored and shipped separately through a new offshore loading buoy. The mixed LPG from the gas processing plants will be combined with the LPG produced in the GTL plant, fractionated and shipped via the existing LPG floating storage and offloading vessel.

GTL plants have a significant cooling load, which can be provided by air cooling, water cooling or both. The project chose a mixed air/water cooling system to reduce local impacts. A cooling water system is being provided for general process cooling. Air cooled condensers are used to condense exhaust steam from some of the larger steam turbines in the plant.

Figure 3 GTL Onplot/Near Offplot Layout

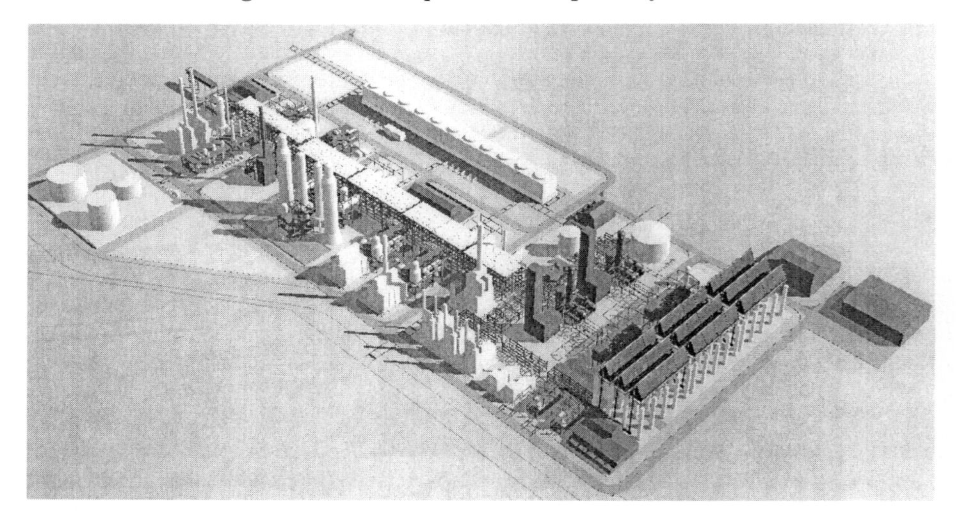

The current design represents a balance between overall plant efficiency and capital cost. Additional work will be done during front end engineering to further optimize the design.

The primary design basis for the Escravos GTL project has always been a large land-based plant although a floating option and small-sized plants were also considered in preliminary studies. Neither looked attractive due to higher overall capital cost. The floating design that was considered preceded the one presented in another paper at this conference. The more recent design reaffirms the previous decision to reject the floating option.

5. ENVIRONMENTAL BENEFITS

The environment is the big winner when the expanded gas gathering/processing plants, the GTL plant and the West Africa Gas Pipeline are all completed. Greenhouse gas emissions will drop by 200-400 million metric tons (expressed as CO_2 equivalents) over a 20 year project timeframe. Approximately one quarter of these emissions reductions is attributable to the West Africa Gas Pipeline. The balance is a result of LPG and condensate recovery in the gas processing plant and the GTL facility.

The environmental benefits are not limited solely to reduced carbon dioxide and methane emissions. All traditionally regulated air pollutants are reduced. The West Africa Gas Pipeline will enable existing and new power plants along the pipeline route to use clean-burning gas instead of burning crude oil. This removes several hundred thousand tons of sulfur oxides and nitrogen oxides from the power plant exhausts. There is also potential that the ultraclean diesel boiling range fuel will help reduce diesel vehicle emissions based on engine testing at Southwest Research Institute and elsewhere.

6. CAPITAL COST

The GTL plant will be installed as part of a $2 billion initiative to put out the flares. This number continues to be refined with ongoing site studies, design optimization, equipment quotations, safety and risk assessment studies, layout modifications, environmental studies, etc. The GTL plant cost is somewhat more than half of the total although the breakdown is somewhat arbitrary since some units provide dual functions. If this plant was a pure greenfield facility, the capital cost for the GTL portion would break down about as follows: 30% synthesis gas production, 15% Fischer-Tropsch synthesis, 20% offsites, 10% product workup, 15% utilities and 10% other process units.

We should say a word about what we include in the total project cost since the literature and conferences like this often do not provide this detail. Total project cost is the sum of the estimated cost of construction (often called the contractor's cost), a project contingency determined by an analysis of the level of engineering that has been completed, and the owner's costs.

- The contractor's cost or EPC cost includes the purchase and delivery to site of the equipment and bulk materials, and the labor directly involved in erecting the equipment and supervising construction. Sales, import and excise taxes as well as duties are included. Site clearing and preparation is included which is significant for this project because of the marshy soil. It includes unallocated and nonmanual labor, temporary housing facilities, construction equipment and consumables. The preparation of contractor bid documents is included. Detailed engineering and procurement costs are included as well as any technology licenses and contractor's insurance.
- Project contingency covers cost increases due to scope omission, uncertainties in design or cost data, construction problems (e.g. unforeseen inclement weather), labor problems, schedule slippage (e.g. late deliveries by vendors) and unforeseen permitting problems.
- Owner's costs are project preoperating expenses, which are normally paid directly by the owner instead of being handled by the contractor. These costs are added to the

contractor's project cost estimate to determine the full cost of the project to be capitalized. Owner's costs cover the first loads of catalyst, chemicals and spare parts, feasibility and front end engineering studies, costs for company personnel involved in project design and construction management, recruitment and training of operators and maintenance staff prior to startup, and plant post construction/prestartup commissioning. Owner's costs also include: environmental reviews, building permits, land and right-of-way acquisitions, third party assistance during startup, project insurance, and local agent representation. Other owner costs include gathering of site-specific engineering data, transport and marketing studies, publicity/inauguration and specialized help on legal, finance and tax issues.

The total project cost includes all that is needed to allow the plant to start up and run. We have included the costs for tankage, modifications to the shipping facilities, wharf modifications required to bring in the equipment and material, relief and blowdown systems and utilities. The GTL plant will essentially have black start capability.

7. PROJECT STATUS

This project received formal project recognition from the Nigerian government in September 2000. Project recognition codifies the fiscal and tax terms for the project. We expect to begin Front End Engineering in the spring of 2001. The current schedule calls for putting the EPC contract out to open bid in 2002.

8. SUMMARY

This project has many winners. Nigeria will see billions in additional revenues generated over the project lifetime through increased taxes and NNPC earnings. Local communities will see high-skill, stable jobs being created in the area. The new infrastructure will improve the economics for future projects thus encouraging further growth. This project and subsequent projects will act as catalysts to generate new service and cottage industries to support these plants.

The environment will win. The plant will produce clean fuel products, which will essentially displace an equivalent amount of crude oil production resulting in less carbon dioxide going into the atmosphere.

Importantly the GTL industry will win. This project will demonstrate equipment, processes and catalysts in the largest GTL plant built to date so we, as an industry, will be able to design even better plants in the future. We will have more confidence in stepping up to the expanded 100,000 BPD GTL facilities. The Escravos plant will provide yet another reliable source of GTL fuels to further assure the market that GTL product supply is both growing and competitive. Users will then make maximum use of the many superior qualities of GTL fuels once they are convinced the supply is there.

And the world's population will be a winner as we utilize some of the world's large gas resources to mitigate the demand on oil resources.

Studies in Surface Science and Catalysis
J.J. Spivey, E. Iglesia and T.H. Fleisch (Editors)

Pt-Promotion of Co/SiO_2 Fischer-Tropsch Synthesis Catalysts

George W. Huber and Calvin H. Bartholomew

BYU Catalysis Lab, Department of Chemical Engineering, Brigham Young University, Provo, UT 84602, E-mail: bartc@byu.edu

A Co/SiO_2 and two Pt-promoted Co/SiO_2 catalysts (containing 11 wt % Co and 0.0, 0.6, and 2.3 wt % Pt) were prepared by controlled-evaporative deposition from acetone solution. In temperature programmed reduction (TPR) experiments Pt-promoted Co/SiO_2 catalysts were reduced at lower temperatures than the unpromoted catalyst. Pt did not significantly affect intrinsic Fischer-Tropsch synthesis (FTS) activity but did decrease methane selectivity. During temperature programmed surface reaction in a 1:1 CO/H_2 mixture, a significant amount of carbon deposited (15 % of the catalyst weight) on the unpromoted catalyst starting at a temperature of 623 K, while a negligible amount of carbon deposited on the promoted catalyst at temperatures up to 773 K. It is proposed that during reaction Pt facilitates removal of carbon from the catalyst surface, thus maintaining the metal in an active state.

1. INTRODUCTION

Among catalysts for FTS, cobalt catalysts are the most active, selective, and stable [1-3]. A number of studies have demonstrated that addition of noble metals to cobalt catalysts can increase reducibility, FTS activity, and selectivity to heavier products [4-14]. Holmen et al. [9,10] reported that Pt enhances the reducibility of Co/Al_2O_3 greatly and Co/SiO_2 significantly, increases metal dispersion of both Co/Al_2O_3 and Co/SiO_2, increases specific activity of both catalyst types for CO hydrogenation at a level of 0.4 wt.% Pt while decreasing the TOF of Co/Al_2O_3 at a 1.0 wt % Pt level (when turnover frequency is based on H_2 chemisorption), but does not affect product selectivity. Using steady state isotopic kinetic analysis (SSITKA) they concluded that the increase in reaction rate of the 0.4 wt. % Pt catalysts was due to an increase in the coverage of reactive intermediates, and that the reaction rate, based on the number of reactive intermediates, is constant. Their catalysts were prepared by conventional impregnation, effects of Pt level on activity and selectivity of Co/SiO_2 were not determined, and resistance to carbon deposition was not addressed. The objectives of this work were to (1) determine effects of Pt promotion at different Pt levels on dispersion and FTS activity and selectivity of Co/SiO_2 FTS catalysts prepared by a non aqueous evaporative deposition method and (2) to address the effects of Pt on resistance to carbon deposition.

2. EXPERIMENTAL

One unpromoted and two Pt-promoted Co/SiO_2 catalysts were prepared by evaporative deposition from acetone solutions of the nitrates onto dehydroxylated SiO_2 (Grace Davison Grade 654) followed by drying at 60°C and reduction in H_2 at 400°C. The Pt salt was added after drying of the cobalt deposit. The cobalt metal loading was 11 wt. % for all three catalysts while Pt loadings were 0.6 and 2.3 wt % for the Pt-promoted catalysts. Details of the preparation method are discussed elsewhere [15,16]. Temperature programmed reduction (TPR), temperature programmed CO hydrogenation and extent of reduction experiments were performed in a Perkin Elmer thermogravimetric analyzer (TGA) Model TGA 7 described elsewhere [15]. H_2 chemisorption measurements were conducted using a flow chemisorption method and apparatus described by Jones and Bartholomew [17]. Activity measurements were conducted in a fixed-bed microreactor described elsewhere [15].

3. RESULTS

3.1 Catalyst Characterization

H_2 chemisorption uptakes, extents of reduction and cobalt dispersions are listed in Table 1. Cobalt dispersion was calculated assuming (a) H_2 adsorbs on both Co and Pt surface atoms with a stoichiometry of $H/M_s = 1$ (M_s refers to a surface metal site) and (b) that Pt has a dispersion of 50% and is present in a separate phase as a result of sequential addition of Pt to Co. These data show that the chemisorption capacity of Co/SiO_2 decreased 35 % after reaction, while the chemisorption capacity of the 2.3 wt % Pt promoted catalyst increased by 17 %. It is evident that Pt-promotion has little or no effect (within experimental error) on cobalt metal dispersion.

Table 1. Effects of Pt on H_2 chemisorption, extent of reduction, and cobalt dispersion. (Values of % dispersion (% fraction exposed) are averages of 3-5 runs for each run.)

Catalyst/Previous Treatment	Extent of Redn. at 400°C (%)	H_2 Uptake (μmole / g_{cat})	D = H/Co[a] (%)
11 wt.% Co/SiO_2, before reaction	89	98.3 ± 3.4	11.2 ± 0.4
11 wt.% Co/SiO_2, after reaction		64.0 ± 6.9	
11 wt.% Co/SiO_2/0.6 wt.% Pt, before reaction	100	95.2 ± 5.3	9.1 ± 0.6
11 wt.% Co/SiO_2/2.3 wt.% Pt, before reaction	100	136.8 ± 6.7	11.1 ± 0.7
11 wt.% Co/SiO_2/2.3 wt.% Pt, after reaction		160.3 ± 25.9	

a. Cobalt dispersion assuming that Pt is 50 % dispersed and in a separate phase from Co.

From the TPR spectra in Figure 1, a substantial growth in the peak at 150°C and a progressive shift in the most intense peak from 240 to 190°C with increasing Pt content is evident, consistent with substantially higher rates of reduction of cobalt nitrate to cobalt oxide and of the oxide to the metal at low temperatures due to Pt addition. Two sometimes overlapping TPR peaks around 300-500°C have been previously reported and assigned to the reduction of Co_3O_4 to CoO and CoO to Co(0) [9,18-22], the position of these peaks shifting significantly to higher temperatures with increasing temperature of calcination prior to reduction. The observation in this study of TPR peaks for 11% Co/SiO_2 at much lower temperatures, i.e., at 150 and 250°C compared to 325 and 400°C for 9% Co/SiO_2 calcined at 400°C [9], can be explained by our unique preparation and pretreatment involving non aqueous impregnation of previously dehydroxylated silica and after drying at 60°C and calcination at 300°C, reduction at 400°C, thereby minimizing interactions of the cobalt precursors with the support.

The ultimate effect of Pt on reducibility of cobalt observed in this study is more pronounced than in previous work [9], i.e., a shift for our 11% Co/SiO_2 of the high temperature TPR peak from 240 to 190°C compared to a shift for 9% Co/SiO_2 [9] from 400 to 355°C. The areas of our TPR peaks for Co and Co-Pt catalysts from 100-700°C (Fig. 1) are the same suggesting that Pt is in the zero valence state prior to reduction of cobalt in H_2, consistent with the data of Zsoldos et al. [21]. Our hypothesis that Pt is in a separate phase is supported by this observation. Pt is thought to enhance reduction of cobalt by providing sites for dissociation of H_2 at relative-ly low reduction temperatures.

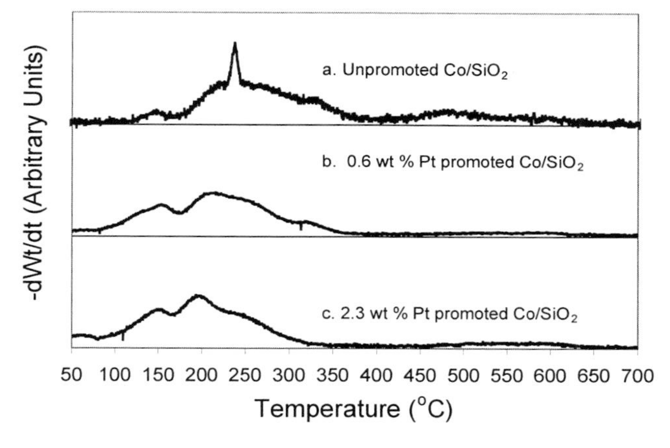

Fig. 1. Effects of Pt promotion on the temperature programmed reduction profile of Co/SiO_2. (Ramp 1°C/min in 10% H_2, GHSV = 500,000. Catalysts previously calcined at a temperature ramp of 1°C/min in 10% O_2 to 300°C and held for 3 h, GHSV = ~ 500,000.)

3.2 Results of Activity and Selectivity Tests

The effects of Pt promotion on the activity and selectivity of cobalt are summarized in Table 2. Turnover frequencies (TOFs or site time yields) are based on H_2 chemisorption uptakes from Table 1 both before and after reaction. The relatively low values of the structural parameter χ [1] provide evidence that the rate data were obtained in the absence of pore diffusional influences. The TOFs for these catalysts are within one standard deviation of those reported previously for cobalt FTS catalysts at conditions similar to ours [23]. Of the TOFs based on the chemisorption uptake before reaction, the TOF for the catalyst containing 0.6 wt % Pt is about 50% less and that for the 2.3 % Pt catalyst is a factor of two higher relative to the TOF of the unpromoted Co/SiO_2. Accordingly, we conclude that the addition of Pt does not significantly change (more than a factor of two) the TOF for FTS. Methane selectivity, however, decreases substantially (from 18.4 to 1.4 mole%) with increasing Pt content.

Table 2. Effects of Pt on activity and molar methane selectivity for FTS in a fixed bed reactor at 20 atm; 200°C; $y_{CO} = 0.2$, $y_{H2} = 0.4$, and $y_{Ar} = 0.4$; after 20 h TOS; catalysts reduced at a ramp of 1°C/min to 400°C and held for 12 h at a GHSV ~ 2000.

Catalyst	GHSV (h^{-1})	X_{CO}[a] (%)	Rate (mol CO/kg_{cat}-h)	CH_4 Sel. (%)	TOF x 10^3 (s^{-1})[b]	TOF x 10^3 (s^{-1})[c]	χ x10^{16} (m^{-1})[d]
Unpromoted	1920	11.3	13.2	18.4	17.4	28.4	6
0.6 wt % Pt[e]	2904	12.7	5.2	8.2	10.4		8
2.3 wt % Pt	2880	12.8	22.4	1.4	39.2	30.1	9

a. CO conversion calculated from: $X_{CO} = \{1- [(y_{CO}/y_{Ar})_{products} / (y_{CO}/y_{Ar})_{feed}]\} \times 100$
b. Based on chemisorption uptake before reaction.
c. Based on chemisorption uptake after reaction for 20-30 h.
d. Structural parameter according to Iglesia et al. [1] which provides a measure of pore diffusional resistance during FTS.
e. Catalyst bulk reduced, passivated and then re-reduced (extent of reduction at 400°C decreased from 100 to 85 % after bulk reduction and passivation). Feed composition: $y_{CO} = 0.3$, $y_{H2} = 0.6$ and $y_{N2} = 0.1$, Reported after 10 h on stream in a micro-fixed bed reactor.

3.3 Temperature Programmed CO Hydrogenation

During temperature programmed CO hydrogenation (TPCOH) in a 1:1 CO/H_2 mixture (see Figure 2) a substantial quantity of carbon forms on the unpromoted Co/SiO_2 catalyst starting at a temperature of 375°C, while a negligible amount of carbon forms on the Pt promoted catalyst at temperatures up to 500°C. Thus addition of Pt to Co/SiO_2 apparently inhibits carbon formation at higher reaction temperatures.

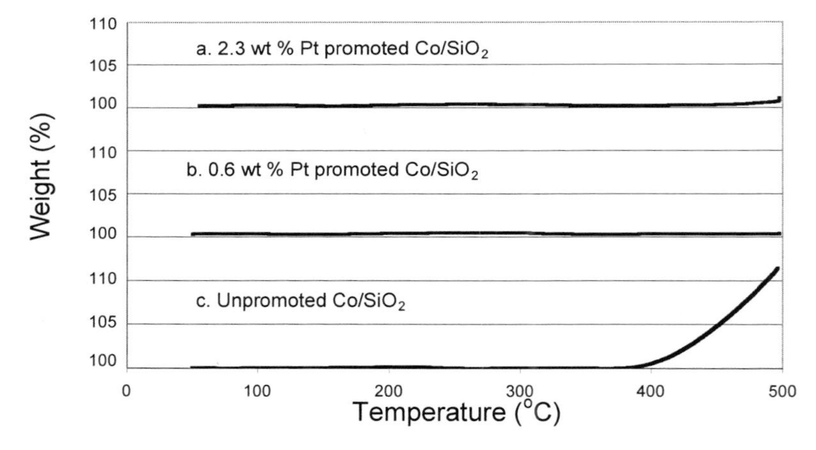

Fig. 2. Effects of Pt and calcination on temperature programmed carbon monoxide hydrogenation (TPCOH). (1:1 mixture of CO and H_2, temperature ramp 6°C/min.)

4. DISCUSSION

The observation in this study that Pt facilitates reduction of Co supported on silica is qualitatively in agreement with the results of Schanke et al. [9] obtained on a similar Co/SiO_2 catalyst and is consistent with the generally observed enhancement in reducibility of supported base metals by noble metal promoters [1,2,8]. The quantitatively larger effect of Pt on the reducibility of Co/SiO_2 observed in this study relative to that that observed by Schanke et al. [9] is probably due in part to the higher Pt loadings used in this study, but is probably also a result of differences in preparation and pretreatment. While Pt is probably in a separate phase as a result of our preparation method, Pt crystallites would have to be well-dispersed and in close proximity of or in contact with the cobalt phase for such an enhancement in cobalt reducibility.

The further observation from this work that addition of Pt to Co/SiO_2 inhibits carbon formation is consistent with very similar results reported by Iglesia et al. [7] for Ru-promoted Co/TiO_2. Iglesia et al. [7] proposed that during reaction Ru facilitates hydrogenative gasification of coke precursors from the metal surface, maintaining Co sites in an active state. Bartholomew et al. [24] showed that addition of Pt decreases the rate of carbon deposition during CO hydrogenation on Ni. The postulate that Pt gasifies coke precursors is consistent with its observed catalytic chemistry [2], i.e., Pt catalyzes hydrogenation, hydrogenolysis and hydro-isomerization (the last reaction to a lesser extent) while neither dissociating CO nor catalyzing chain growth. The ability of Pt to remove coke precursors may best be related to its hydrogenolysis activity. Indeed, the hydrogenolysis activity of $Co\text{-}Ru/SiO_2$ catalysts has been reported to be higher than that of Co/SiO_2 catalysts [7].

The results of this work, showing no significant change in specific activity of Co/SiO_2 with addition of Pt, are in agreement with those of Vada et al. [10]; however, our results showing that Pt addition causes methane selectivity of Co/SiO_2 to decrease significantly contradict those of Holmen et al. [9,10] showing no effect of Pt promotion on methane selectivities of Co/Al_2O_3 and Co/SiO_2. These differences in selectivity behavior may relate to differences in the preparation of catalysts from the two different laboratories as well as the different H_2/CO ratios used in the two studies (Schanke et al. operated at an H_2/CO ratio of 7.3, while our activity tests involved a ratio of 2.0). Addition of other promoters, e.g., Ru, K and Gd, has also been shown to decrease methane selectivity of cobalt [2,7,25].

Three hypotheses may explain the lower CH_4 selectivity observed on Pt-promoted Co/SiO_2 catalysts. The first is that Pt inhibits formation of a graphitic carbon species that leads to methane formation. It has been postulated that FTS reaction occurs on a cobalt carbide surface while methanation may occur on a graphite-covered surface [9]. It is possible that a graphitic carbon was formed over 20-30 hours of reaction on unpromoted Co/SiO_2 in our steady-state activity tests—one having properties similar to the graphitic carbon observed during our TPCOH experiments; the graphitic carbon, formed thusly, may cause higher methane yields. The second hypothesis is that adsorbed ethyl and methyl species formed on the Pt surface by hydrogenolysis enter the polymerization cycle and react with carbon or coke precursors to decrease methane selectivity, while increasing selectivity to heavier products. Finally, the higher extent of reduction of cobalt to the metal due to Pt may increase the propagation rate relative to the rate of termination.

5. CONCLUSIONS

Three different promotional effects of Pt on a Co/SiO_2, prepared by a novel non aqueous evaporative deposition, are observed: (1) increased Co reducibility, (2) decreased CH_4

selectivity, and (3) decreased carbon deposition rate. It is proposed that Pt facilitates reduction of cobalt oxides via hydrogen atom spillover from Pt to cobalt oxide. During reaction Pt appears to facilitate hydrogenative gasification of carbon from the metal surface, maintaining Co sites in an active state. It doesn't, however, affect specific activity.

REFERENCES

1. E. Iglesia, S.L. Soled and R.A. Fiato, J. Catal. 137 (1992) 212.
2. R.J. Farrauto and C.H. Bartholomew, "Fundamentals of Industrial Catalytic Processes", Chapman and Hall (Kluwer Academic), London, 1997.
3. J.J.C. Geerlings, J.H. Wilson, G.J. Kramer, H.P.C.E. Kuipers, A. Hoek and H.M. Huisman, Appl. Catal. A 186 (1999) 27.
4. J. H. E. Glezer, K. P. De Jong, and M. F. M. Post, European Patent EP 0,221,598, (1989).
5. S. Eri, J. G. Goodwin, G. Marcelin, and T. Rilis, U.S. Patent 5,116,879 (1992).
6. H. F. J. van't Blik, D. C. Koningsberger and R. Prins, J. Catal. 97 (1986) 210.
7. E. Iglesia, S.L. Soled, R.A. Fiato and G.H. Via, J. Catal. 143 (1993) 345.
8. L.A. Bruce, M. Hoang, A.E. Hughes and T.W. Turney, Appl. Catal. A 100 (1993) 51.
9. D. Schanke, S. Vada, E.A. Blekkan, A. M. Hilmen, A. Hoff and A. Holmen, J. Catal. 156 (1995) 85.
10. S. Vada, A. Hoff, E. Adnanes, D. Schanke, and A. Holmen, Topics Catalysis 2 (1995) 155.
11. A.L. Lapidus, A.Y. Krylova, V.P. Tonkonogov, D.O.C. Izzuka and M.P. Kapur, Solid Fuel Chemistry 29 (1995) 90.
12. A. Kogelbauer, J.G. Goodwin and R. Oukaci, J. Catal. 160 (1996) 125.
13. A.R. Belambe, R. Oukaci and J.G. Goodwin, J. Catal. 166 (1997) 8.
14. L. Guczi, R. Sundararajan, Z. Kippany, Z. Zsoldos, Z. Schay, F. Mizukami and S. Niwa, J. Catal. 167 (1997) 482.
15. G.W. Huber, Masters, Brigham Young University, 2000.
16. G.W. Huber, C.G. Guymon and C.H. Bartholomew, Paper in preparation.
17. R.D. Jones and C.H. Bartholomew, Appl. Catal. 39 (1988) 77.
18. P. Arnoldy and J.A. Moulijn, J. Catal. 93 (1985) 38.
19. B.A. Sexton, A.E. Hughes and T.W. Turney, J. Catal. 97 (1986) 390.
20. H.-C. Tung, C.-T. Yey, and C.-T. Hong, J. Catal. 122 (1990) 211.
21. Z. Zsoldos, T. Hoffer and L. Guczi, J. Phys. Chem. 95 (1991) 198.
22. A.Y. Khodakov, J. Lynch, D. Bazin, B. Rebours, N. Zanier, B. Moisson and P. Chaumette, J. Catal. 168 (1997) 16.
23. F.H. Ribeiro, A.E.S. von Wittenau, C.H. Bartholomew and G.A. Somorjai, Catal. Review Sci. Technol. 39 (1997) 49.
24. C.H. Bartholomew, G.D. Weatherbee and G.A. Jarvi, Chem. Eng. Commun. 5 (1980) 125.
25. G.W. Huber, S.J. Butala, M. Lee and C.H. Bartholomew, Catalysis Letters, in press.

Studies in Surface Science and Catalysis
J.J. Spivey, E. Iglesia and T.H. Fleisch (Editors)

Catalytic dehydrogenation of propane over a $PtSn/SiO_2$ catalyst with oxygen addition: Selective oxidation of H_2 in the presence of hydrocarbons

L. Låte, J.-I. Rundereim and E.A. Blekkan

Department of Chemical Engineering, Norwegian University of Science and Technology, N-7491 Trondheim, Norway

In this communication we present experimental results from a study of the combustion of hydrogen in the presence of light hydrocarbons over a $PtSn/SiO_2$ catalyst. This reaction is of importance in the development of new dehydrogenation technology based on partial combustion of hydrogen to provide *in situ* heat generation to the endothermic dehydrogenation process. The results show that under conditions with limited amounts of alkene in the gas, the oxygen will selectively react with hydrogen. However, at simulated high dehydrogenation conditions, the oxygen reacts with the hydrocarbons, giving CO and CO_2 as the products.

Introduction

The catalytic dehydrogenation of light alkanes like propane and butane is growing in importance, in the case of propane due to an imbalance between the growth in demand for propene and ethene. The selective production of propene via catalytic dehydrogenation, reaction (1), is difficult to do economically using the existing process technology. The strong

$$C_3H_8 \rightleftarrows C_3H_6 + H_2 \qquad (1)$$

endothermic character of the equilibrium limited reaction (1) ($\Delta H^0_{873K} = 124$ kJ/mole) demands a high reaction temperature, low pressure or strong dilution (often with steam) to achieve high conversions. The propane - propene separation is difficult and energy consuming due to a small boiling-point difference, thus making the recycling of unconverted reactant more expensive. Oxidative dehydrogenation according to eq. (2), which is exothermic and without equilibrium limitations at normal conditions has been studied as an alternative, but so

$$C_3H_8 + \tfrac{1}{2}O_2 \rightarrow C_3H_6 + H_2O \qquad (2)$$

far there has been no breakthrough in the search for catalysts providing the necessary yield of propene [1]. The main problem is the lack of selectivity due to complete combustion, especially of the more reactive alkene. A recently adapted idea is the autothermal dehydrogenation, where a fraction of the H_2 formed in the process is combusted to provide energy to the reaction or re-heating of the process stream [2-5]. This type of process will also suffer under the similar side reactions involving complete combustion of the hydrocarbons, and hence the selectivity is an important issue. In this paper we will show results of experiments performed with the aim of exploring the possibility of feeding oxygen to a conventional platinum-based catalytic dehydrogenation system, studying the formation of the undesired carbon oxides as a function of the reaction conditions. We also report the results of an experimental investigation of the combustion reactions that occur when oxygen is mixed into a mixture of propane, propene and H_2 (simulating high conversions). We have focused on a Pt-Sn catalyst, which is well known to be active both in the dehydrogenation reaction as well as in the combustion reactions.

Experimental

The reactions were studied in a catalytic microreactor (U-shaped) made from quartz, containing 0.15 to 0.45 g of catalyst. The reactor was placed in an electrical furnace, and the temperature was controlled using external and internal thermocouples. Gases were premixed and fed to the reactor using electronic mass flow controllers (Bronkhurst). The product gas was analysed for hydrocarbons (C_1 - C_4), and for permanent gases (O_2, N_2, CO, CO_2) by GC. The total pressure was 1 bar, and the temperature was varied in the range 500 - 550 °C.

The catalyst was prepared by aqueous incipient wetness impregnation of silica (Merck Kieselgel 60, S_{BET} = 483 m^2/g, particle size 0.063 - 0.2 mm), using $SnCl_2{\cdot}2H_2O$ and $H_2PtCl_6{\cdot}6H_2O$. The dissolution of the tin salt was aided by addition of HNO_3. The catalyst was dried in air (120 °C, 24 h) and calcined in flowing air (600 °C, 2 h). Before the catalytic experiments reported here the catalyst was reduced *in situ* (550 °C, 2 h) in a flow of H_2 (GHSV = 3000 h^{-1}). The catalyst used in the experiments reported here is termed $PtSn/SiO_2$ and contained 0.35 wt% Pt and 1.0 wt% Sn.

Results and discussion

In Fig. 1 the catalytic stability of the $PtSn/SiO_2$ is shown for the different gas mixtures (propane, propane/H_2, propane/O_2 and propane/H_2/O_2). The initial catalytic activity is lower when oxygen is present in the feed than the case is for the gas mixtures without oxygen, but the presence of oxygen appears to lead to a more stable catalyst. The reason for this could be that oxygen continuously burns off coke deposited on the catalyst. Co-fed hydrogen also seems to stabilise the catalytic activity to some extent. When only propane is present in the feed, the initial catalytic activity is higher, but the drop in activity over the first few hours on-stream is pronounced.

Table 1 provides a summary of some results obtained when increasing amounts of oxygen were added to this conventional catalytic dehydrogenation system. The experiments were conducted with a small amount (2.0 ml/min) of gaseous H_2 in the reactor feed. Without oxygen in the feed gas the catalyst shows a normal behaviour, with a complete selectivity to propene at these low conversions. The results show that as long as the O_2 is fed to the reactor at a less than stoichiometric amount compared to the molecular hydrogen in the feed, according to the reaction $H_2 + ½O_2 \rightarrow H_2O$, the O_2 reacts with the hydrogen with a high selectivity. But when the O_2 feed-rate exceeds this ratio some of the O_2 reacts with hydrocarbons to form CO and CO_2, with a simultaneous drop in the selectivity to propene. The O_2 conversion is high but not complete in all cases. The propane conversion at first decreases with increasing O_2 feed-rate, followed by an increase when the O_2 feed rate exceeds 1.5 ml/min. This apparently complex relationship is the sum of the two main reactions. The propane conversion to propene passes through a maximum when the O_2 feed-rate is increased.

The rate of formation of CO_2 on the other hand increases steeply above 1.0 ml/min of oxygen, and the sum of these reactions give the observed overall consumption of propene. The selectivity to propene is close to 100% up to 1 ml/min of O_2 in the feed gas, above this level

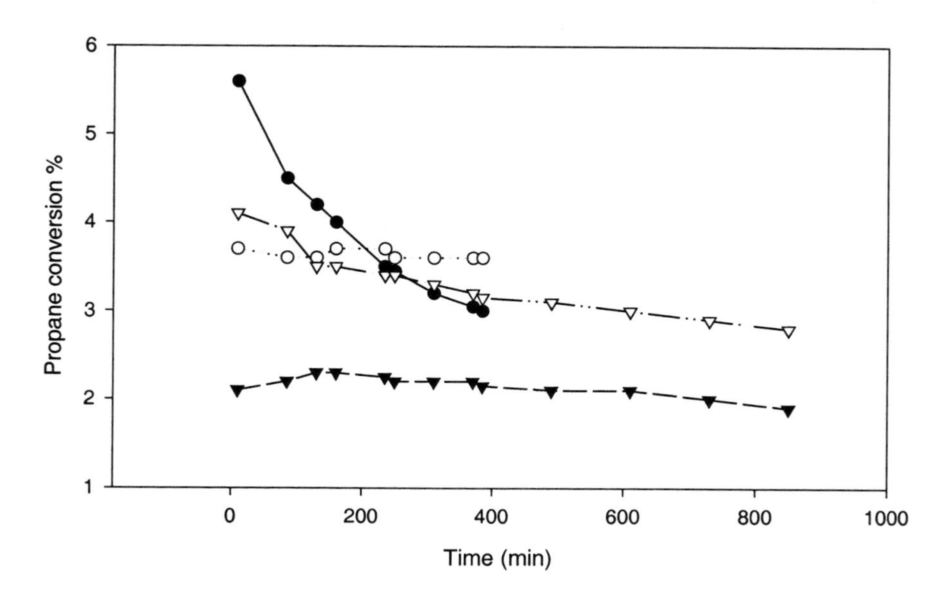

Figure 1. Catalytic stability of $PtSn/SiO_2$ during propane dehydrogenation in different gas mixtures: propane (●), propane/O_2 (10/1) (○), propane/O_2/H_2 (10/2/2) (▼) and propane/H_2 (10/2) (▽). Conditions: 500 °C, 1 atm., 0.30 g catalyst, Feed gas: Always 10 ml/min C_3H_8, H_2 and O_2 as indicated by the ratios, total flow 100 ml/min, balance is He.

Table 1. Summary of catalytic results obtained over the $PtSn/SiO_2$ catalyst with increasing oxygen addition. Conditions: 550 °C, total pressure 1 atm., catalyst mass 0.3 g. Flowrates: 10 ml/min C_3H_8, 2 ml/min H_2, varying flow of O_2 as indicated, balance He up to a total flow of 100 ml/min.

O_2-flow	**Conv. %**		**%-C-selectivities**[1]			**%-O-selectivities**		
ml/min	C_3H_8	O_2	C_3H_6	CO_2	CO	CO_2	CO	H_2O[2]
0	3.4	-	100	-	-	-	-	-
0.5	5.6	94	99.8	-	-	-	-	100
0.75	5.2	98	99.3	-	-	-	-	100
1.0	4.8	98	98.0	-	-	-	-	100
1.25	4.0	98	81.9	10.1	5.9	9.1	2.6	88.3
1.5	3.6	98	69.2	20.3	8.2	13.8	2.8	83.4
2.0	4.4	97	45.2	42.2	11.3	25.9	3.5	70.6

Notes: 1. If the sum is less than 100% the difference is due to hydrocarbons formed in cracking reactions.
2. By difference assuming no oxygenates formed.

the carbon selectivity to propene drops due to the formation of CO and CO_2. In summary, the results indicate that O_2 reacts selectively with hydrogen only under certain conditions, i.e. under a surplus of hydrogen. As soon as this surplus hydrogen is consumed, the excess O_2 reacts with hydrocarbons in a non-selective manner leading to CO/CO_2 which are unwanted products. This consequently leads to a diminished selectivity to propene.

The results reported so far were obtained at differential conditions with respect to propane, and hence with very low partial pressures of propene. For practical purposes, however, a high propene concentration is the critical situation for any process utilising the addition of oxygen or any other oxidant to provide heat. We have therefore studied the system in a situation with a high propene pressure. To avoid the complication of the reverse reaction, we have chosen to use a gas composition close to equilibrium (C_3H_8-C_3H_6, with 2 ml/min H_2) before the O_2 addition. Fig. 2 shows the oxygen selectivities for a series of experiments done at 550 °C.

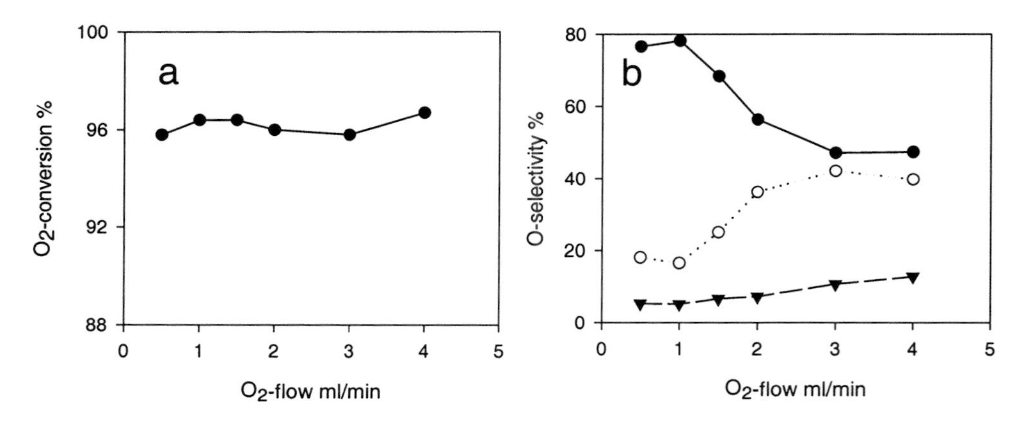

Figure 2. Addition of O_2 to a propane-propene-hydrogen gas mixture. a) O_2 conversion b) O-selectivity to H_2O (•), CO_2 (○) and CO (▼). Conditions: 550 °C, 1 atm., 0.30 g catalyst, Feed gas: 2 ml/min H_2, 5 ml/min C_3H_8, 28.2 ml/min C_3H_6, O_2 as shown in the Figure, total flow 100 ml/min, balance is He.

The flow rate of oxygen was varied from zero via a low flowrate (0.5 ml/min) to high. It can be noted that an oxygen flow of 1 ml/min again corresponds to the stoichiometry of the H_2 - O_2 reaction with the H_2 in the feed-gas. Figure 2a shows that the oxygen conversion is high at all conditions. However, there is some unconverted oxygen, in the range 4-6% of the oxygen fed to the reactor, and this number is constant, independent of the oxygen feed flow. This could indicate a by-pass of some of the reactant gas through the catalyst bed, or as indicated by Beretta et al. [6,7], that the oxygen conversion is limited by diffusion (although their results were obtained in an annular reactor with much higher flow-rates). In a separate experiment an empty reactor gave a very low oxygen conversion at the same conditions (< 10%). A temperature above 625°C was necessary in order to increase the conversion of oxygen in the empty reactor and only at 700°C was the oxygen conversion 95% in the empty reactor with the same flow-rates. This means that the observed selectivities are due to reactions over the catalyst and not gas-phase or wall-catalysed reactions.

Figure 2b shows the oxygen-selectivity to water, CO_2, and CO formation (The selectivity to water is calculated by difference from the elemental balance of oxygen, taking unconverted oxygen, CO and CO_2 into account). At low oxygen flows the selectivity to water is about 80%, and falls significantly when the oxygen flow is increased. Above 2 ml/min O_2 feed rate the selectivity to water is in the range 40 - 50%, which indicates that the hydrocarbons are being combusted. As an example, C_3H_6 combustion would give a H_2O selectivity (based on consumed O_2) of 33 %. There is a clear difference from the results obtained with a pure propane/hydrogen feed, where small amounts of oxygen was found to react with the hydrogen with a high selectivity. At these conditions this is not the case, leading to high CO and CO_2 selectivities.

The hydrocarbon conversion is not presented here. Due to the composition of the gas feed is the main dehydrogenation reaction (1) close to equilibrium, hence there is only limited conversion. However, due to side-reactions, like cracking/hydrogenolysis type reactions, and possibly due to different reactivities towards oxygen is the composition of the hydrocarbons slightly changed at the reactor outlet. There was no clear trend in the cracking product distribution. It could however be observed that the selectivity to methane in most cases was lower than what would be expected if the light products were formed by a simple C-C bond cleavage. The only possible explanation is that the C_1-species is oxidised to carbon oxides, leaving a C_2-unit as a hydrocarbon product. This then must be a catalytic oxidation (the gas-phase reaction will not be selective), and the lower than expected selectivity to CH_4 indicates that the oxygen picks up a C_1-species from the surface, leaving a C_2 unit which subsequently

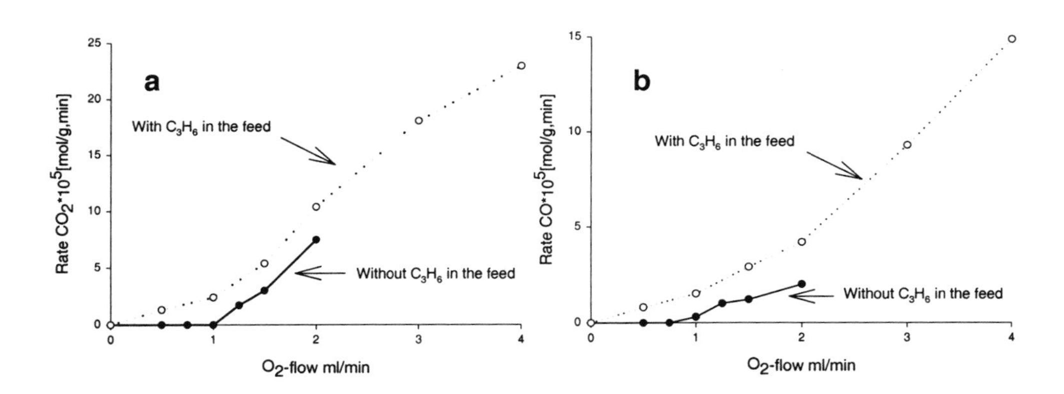

Figure 3. Rates for CO_2 (a) and CO (b) formation. Conditions: 550 °C, 1 atm., 0.30 g $PtSn/SiO_2$ catalyst. Feed gas with C_3H_6: 2 ml/min H_2, 5 ml/min C_3H_8, 28.2 ml/min C_3H_6, O_2 as shown in the Figure, without C_3H_6: 2 ml/min H_2, 10 ml/min C_3H_8, O_2 as shown in the Figure. The total flow is always 100 ml/min, the balance is He.

can desorb as a product.

Fig. 3 shows a comparison of the rates of CO_2 and CO formation using either the mixed feed or a propane-hydrogen mixture. With increasing O_2-feedrate, a clear difference in the behaviour is observed. With only propane as the hydrocarbon in the feed the rate of formation of carbon oxides is negligible up to a point where the hydrogen feed is consumed. (The oxygen conversion is always close to 95%). Only at the situation where excess O_2 is fed is there a measurable CO/CO_2 formation rate. This is contrasted by the experiments with high concentrations of propene in the feed. The CO and CO_2 formation is high and important as soon as O_2 is introduced in the feedstock. This confirms that the problem with the oxidative process over this catalyst is the reactivity of the product, in this case the propene. With low propene concentrations, a selective reaction between H_2 and O_2 is possible. Only when all the H_2 is consumed will O_2 attack the hydrocarbons and form CO and CO_2.

Conclusions

The results show that a selective combustion of H_2 in the presence of propane is possible over the $PtSn/SiO_2$ catalysts used here, but that the presence of high concentrations of propene leads to significant CO_2/CO formation.

Acknowledgements
Financial support from the Norwegian Research Council through the programme "Chemical Conversion of Natural Gas" is gratefully acknowledged.

References

1. See e.g. E.A. Mamedov and V. Cortés Corberán, Appl. Catal. A 127 (1995) 1.
2. C.P. Tagmolila, US Patent No. 5043500 (1991).
3. T. Lægreid, M. Rønnekleiv and Å. Solbakken, Ber.-Deutsch. Wiss.Ges. Erdöl, Erdgas Kohle, Tagungsber. 9305 (1993), 147.
4. P.A. Agaskar, R.K. Grasselli, J.N. Michaels, P.T. Reischman, D.L. Stern and J.G. Tsikoyannis, US Patent No. 5430209 (1995).
5. J.G. Tsikoyannis, D.L. Stern and R.K. Grasselli, J. Catal., 184 (1999) 77.
6. Beretta,A, Gasperini, M.E., Treopiedi,G., Piovesan,L., and Forzatti,P., J. Catal., 184 (1999) 455.
7. A. Beretta, M.E. Gasperini, G. Treopiedi, L. Piovesan, and P. Forzatti, Stud. Surf. Sci. Catal., 119 (Natural Gas Conversion V) (1998) 659.

Studies in Surface Science and Catalysis
J.J. Spivey, E. Iglesia and T.H. Fleisch (Editors)

Selectivity and activity changes upon water addition during Fischer-Tropsch synthesis

Anne-Mette Hilmen [1], Odd Asbjørn Lindvåg [1], Edvard Bergene [1], Dag Schanke[2], Sigrid Eri[2] and Anders Holmen [3*]

[1] SINTEF Applied Chemistry, N-7465 Trondheim, Norway,
[2] Statoil R&D Centre, N-7005 Trondheim, Norway,
[3] Dept. of Chemical Engineering, Norwegian University of Science and Technology (NTNU), N-7491 Trondheim, Norway.

Re-promoted Co/Al_2O_3 catalysts deactivate when water is added to the feed. Model studies indicate that the loss in activity can only partly be recovered by rereduction due to increased Co-alumina interactions. Water decreases the selectivity to CH_4 significantly and the selectivity to C_2-C_4 paraffins is also decreased. As a result the C_{5+} selectivity increases markedly. The effect of water on the CH_4 selectivity is reversible, while the effect on C_2-C_4 selectivity is only partly reversible.

1. INTRODUCTION

The Fischer-Tropsch synthesis is an attractive possibility for converting natural gas into high quality liquid fuels. Cobalt is the preferred catalyst when using natural gas derived synthesis gas. However, cobalt is expensive and it is important that the catalyst has a long life-time and/or can be regenerated. The impact of water on cobalt Fischer-Tropsch catalysts has gained increasing attention in recent years. Water is produced during the Fischer-Tropsch synthesis and will be present in varying quantities during synthesis, depending on the conversion, reactor system and catalyst. The role of water will be especially important in large-scale slurry reactors as a consequence of extensive back-mixing resulting in high water concentrations and low reactant concentrations throughout the reactor.

The effect of water on the activity of Fischer-Tropsch catalysts has been reported for different systems and it has been shown that Al_2O_3-supported Co catalysts deactivate when water is added during the synthesis [1,2,3]. There are also some previous studies on the effect of water on the selectivity of cobalt Fischer-Tropsch catalysts [4,5,6]. It is reported that the C_{5+} selectivity increases when water is added to the feed stream and that the CH_4 selectivity decreases [5,6]. It has also been shown that water inhibits secondary hydrogenation of olefins [4,6]. The inhibition of secondary hydrogenation is completely reversible according to Schultz et al. [6]. It is the purpose of the present work to study the effect of water on the selectivity and activity of Al_2O_3-supported cobalt Fischer-Tropsch catalysts.

---

* Corresponding author: holmen@chembio.ntnu.no

2. EXPERIMENTAL

The catalysts were prepared by incipient wetness (co)impregnation of γ-alumina with aqueous solutions of $Co(NO_3)\cdot H_2O$ and/or $HReO_4$, dried in air overnight at 100-120°C and calcined in air at 300°C for 16h. Further pretreatment was done *in situ*.

The catalysts were tested in an isothermal fixed-bed microreactor. A detailed description of the apparatus has been given elsewhere [1]. The catalyst (1.0-1.6 g diluted with SiC) was loaded to the reactor, flushed with He and reduced in hydrogen at 350°C for 16h (250 Ncm^3/min, 1 atm, 60°C/h heating rate to 350°C). After reduction the reactor was cooled to 170°C, the pressure increased to 20 bar and premixed synthesis gas (31.3 % CO, 65.7 % H_2 and 3 % N_2) was introduced. The reactor temperature was slowly increased to 483 K in order to prevent temperature runaway. Distilled water supplied from a Hi-Tec liquid flow controller was vaporised and mixed with the synthesis gas just before the reactor inlet in order to obtain high partial pressures of water. The feed and products were analysed using an on-line HP 5890 gas chromatograph equipped with thermal conductivity (TCD) and flame ionization (FID) detectors. N_2 (internal standard), CO, CH_4, and CO_2 were separated by a Carbosieve packed column and analyzed on the TCD. Hydrocarbon products were separated by a 0.53 mm i.d. GS-Q capillary column and detected on the FID. CH_4 was used to combine the TCD and FID analyses.

3. RESULTS AND DISCUSSION

3.1. The effect of water on the activity

Previous results [1,2] on the deactivation of $Co\text{-}Re/Al_2O_3$ catalysts by water have indicated that the catalyst deactivates due to reoxidation of small cobalt particles. It has also been observed that the cobalt-alumina interactions increase as a result of water being present [2]. These findings have been further supported by studies of van Berge et al. [3]. When water is added to the feed for an alumina-supported Co catalyst, the deactivation rate increases, as clearly illustrated in Figure 1.

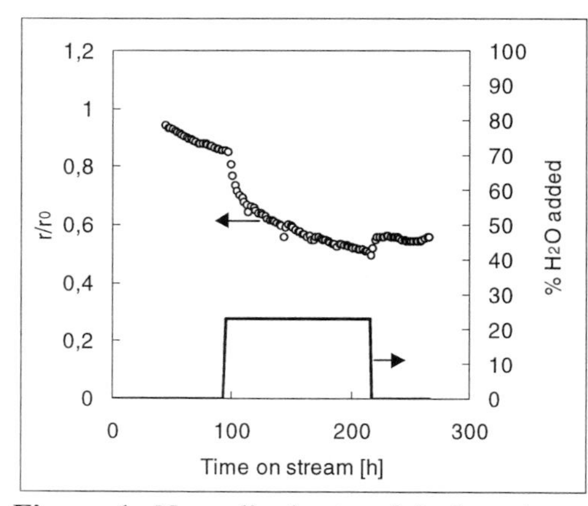

Figure 1. Normalised rate of hydrocarbon formation as a function of time on stream for $20\%Co\text{-}1\%Re/Al_2O_3$ [H_2/CO=2.1, 210°C, 20 bar]

Model studies of catalysts exposed to different $H_2O/H_2/He$ ratios in a microbalance and also using other techniques have been described earlier [2]. Table 1 shows an example of the effect of rereducing the catalyst after such model studies. Parts of the cobalt that were reoxidized during water exposure was not reducible at standard reduction conditions neither in the case of exposure to H_2O/H_2=10 nor after exposure to H_2O/He. This suggests that the deactivation observed due to water may only partly be recovered by rereducing the catalyst and supports the earlier findings (by TPR and XPS) that a phase interacting more strongly with the alumina is formed upon water exposure [2].

This could be a result of parts of the cobalt reacting with the alumina as was found to be thermodynamically feasible by van Berge et al. [3]. As shown in Table 2, the BET surface area of the catalyst decreases from 125 to 97 m^2/g after exposure to H_2O/He which is a strong indication that the changes involves the support. Essentially no change in surface area occurs after exposure to $H_2O/H_2=10$. In this case only a small fraction of Co, of the same order as the dispersion of the catalyst, was reoxidized (Table 1), and large changes in surface area of the catalyst should not be expected.

Table 1. Metallic Co (%) calculated from weight changes recorded in a microbalance during reduction [1°C/min to 350°C, 16 h], exposure to water [10 bar, P_{H2O}=5.5 bar, 250°C,16 h] and rereduction.

Catalyst	Reduction	Metallic Co (%) after : H_2O/H_2 = 10	H_2O/He	Rereduction
20 % Co-1 %Re/Al_2O_3	88.0	74.2	-	83.6
20 % Co-1 %Re/Al_2O_3	89.2	-	29.4	64.6

Table 2. BET surface area of 20%Co-1%Re/Al_2O_3 after exposure to $H_2O/H_2/He$ in a microbalance [10 bar, P_{H2O}=5.5 bar, 250°C, 16h].

Catalyst	Treatment	BET surface area [m^2/g]
20% Co- 1% Re/Al_2O_3	-	125
20% Co- 1% Re/Al_2O_3	H_2O/H_2=10	121
20% Co- 1% Re/Al_2O_3	H_2O/He	97

3.2. Effect of water on the selectivity

Figure 2 shows the effect of CO conversion at dry feed conditions, during water addition and after removal of the water on the selectivities during F-T synthesis. At dry feed conditions, i.e. without adding water to the feed, the C_{5+} selectivity increases as the CO conversion is increased in agreement with previous studies [4]. The olefin selectivity decreases with increasing CO conversion (and bed residence time) and the paraffin selectivity increases. It has been suggested [4] that the increase in C_{5+} selectivity as a function of conversion is to a large extent caused by secondary reactions of primary olefin products at longer residence times (i.e. readsorption and further chain growth). We believe that also the increasing partial pressure of water with increasing CO conversion contributes to the increase in C_{5+} selectivity by inhibiting hydrogenation reactions.

The CH_4 selectivity decreases markedly when water vapour is added to the feed stream (Figure 2) and the increase in C_{5+} selectivity which is observed, is mainly due to the decrease in CH_4 selectivity. The C_2-C_4 paraffin selectivity also decreases when water is added. The C_3 (and C_4) α-olefin selectivity decreases slightly upon water addition, while the C_2 olefin selectivity increases markedly. The observed changes in selectivity have to be interpreted in terms of the reaction network, see Figure 3. The observed C_n olefin selectivity is the net sum of what has grown from C_{n-1} subtracting what has grown further and subtracting what has terminated as C_n paraffins. Even though hydrogenation reactions are inhibited by water, the olefin selectivity decreases (except for C_2) suggesting also increased further chain growth.

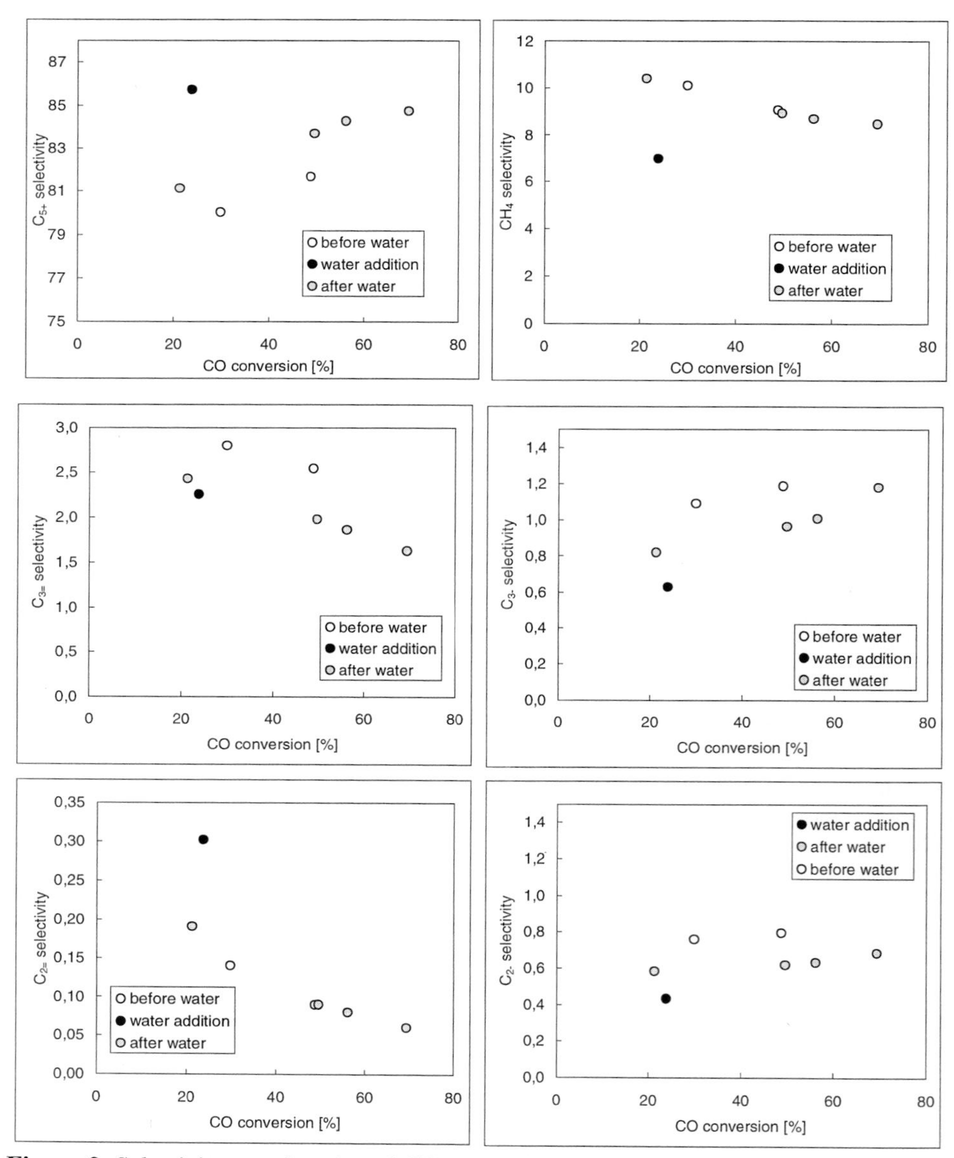

Figure 2. Selectivity as a function of CO conversion before, during and after water addition for a 12 % Co- 0.5 % Re/Al_2O_3 catalyst [H_2/CO=2.1, 210°C, 20 bar].

After water addition is stopped, the C_{5+} selectivity does not return to the same level as before water was added, Figure 2. The effect of water on the methane selectivity is completely reversible, while the C_2-C_4 selectivity remains lower than before water addition. The C_2 olefin selectivity returns to the same level as before water addition, while the C_3 (and C_4) olefin selectivity does not. All the C_2-C_4 paraffin selectivities remain lower than before water was added. Schultz et al. [6] on the other hand found the effect of water to be completely reversible over a SiO_2-supported Co catalyst.

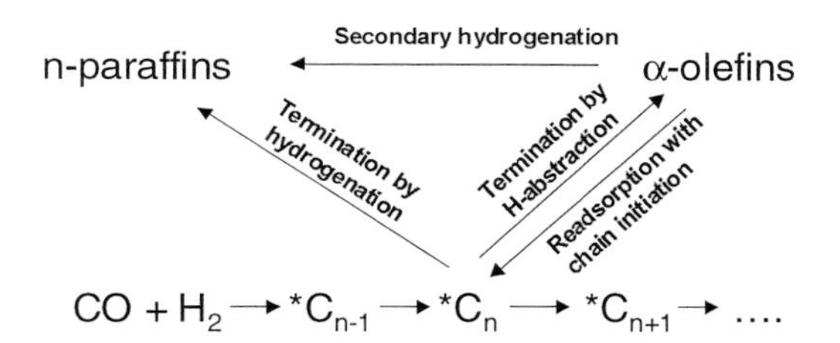

Figure 3. Chain growth, termination and secondary reactions (not all included) in Fischer-Tropsch synthesis.

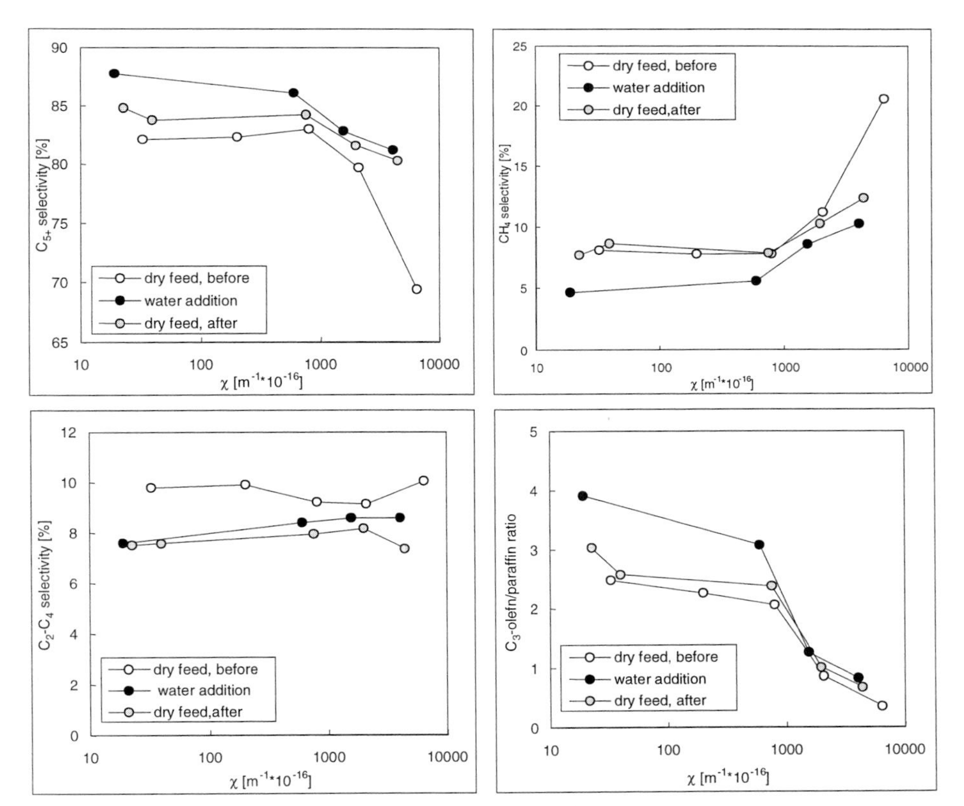

Figure 4. Effect of particle size on the selectivity. The structural parameter χ (see ref. [4]) ($\chi = r_m^2 \theta \phi / r_p$, where r_m=mean particle radius, θ= Co atoms/m^2, ϕ=catalyst porosity, r_p=mean pore radius) has been corrected for the loss of activity [210°C, 20 bar, 20-40 % CO conv.].

3.3. A combination of the effect of water on activity and selectivity - catalysts with different particle size .

Figure 4 shows the effect of increasing catalyst particle size on the product selectivity. At diffusion distances of roughly 0.2-0.3 mm (χ~1000-2000), the C_{5+} selectivity starts to decrease and the CH_4 selectivity increases. There are only small changes in the C_2-C_4 selectivity but a marked decrease in the olefin/paraffin ratio. These results are in agreement with the studies of Iglesia et al. [4], which explained the observations in terms of increased mass transfer restrictions with increasing particle sizes.

Adding water to the feed stream increases the C_{5+} selectivity in all cases, but most markedly for the catalysts with large particles. This is, however, to a large extent explained by the decrease in activity during water addition. The loss of active sites results in a lower structural parameter χ and a shift towards left in the figure. This results in a large effect on the C_{5+} selectivity due to the steepness of the C_{5+} vs. χ curve in this area.

4. CONCLUSION

The addition of water results in catalyst deactivation for alumina-supported cobalt catalysts. Model studies indicate that the catalyst can only partly be regenerated by reduction at 350°C, due to increased Co-alumina interactions.

Water addition has a positive effect on the selectivity of desired products. CH_4 selectivity is significantly decreased and the selectivity to C_2-C_4 paraffins is also decreased upon water addition. As a result the C_{5+} selectivity increases markedly. The effect of water on the CH_4 selectivity is reversible, while the effect on C_2-C_4 selectivity appears to be only partly reversible.

ACKNOWLEDGEMENT

Statoil and the Norwegian Research Council (NFR) are acknowledged for financial support and for the permission to publish these results.

REFERENCES

[1] D. Schanke, A.M. Hilmen, E. Bergene, K. Kinnari, E. Rytter, E. Ådnanes and A. Holmen, Catal. Lett. 34 (1995) 269-284.
[2] A.M. Hilmen, D. Schanke, K.F. Hanssen, A. Holmen, Appl. Catal. 186 (1999) 169-188.
[3] P.J. van Berge, J. van de Loosdrecht, S. Barradas, A.M. van der Kraan, Catal. Today 58 (2000) 321-334.
[4] E. Iglesia, S. Reyes, R. Madon, S. Soled, Adv. in Catalysis 39 (1993) 221-302.
[5] C.J. Kim, US 5,227,407 to Exxon (1993).
[6] H. Schultz, M. Claeys, S. Harms, Stud. Surf. Sci. Catal. 107 (1997) 193-200.

Studies in Surface Science and Catalysis
J.J. Spivey, E. Iglesia and T.H. Fleisch (Editors)

Synthesis gas production by partial oxidation of methane from the cyclic gas-solid reaction using promoted cerium oxide

J.C. Jalibert, M. Fathi, O.A. Rokstad and A. Holmen

Department of Chemical Engineering, Norwegian University of Science and Technology, N-7491 Trondheim, Norway

The partial oxidation of CH_4 has been studied with a $Pt/CeO_2/\gamma\text{-}Al_2O_3$ catalyst at non-steady state conditions by repeated switching from diluted oxygen to diluted methane feed. The production of CO and H_2 highly depends on the degree of reduction of the solid. It has been shown that high selectivities of synthesis gas could be obtained at complete conversion of methane when the degree of reduction of ceria is controlled by the length of the redox cycles.

1. INTRODUCTION

Partial oxidation of CH_4 to H_2 and CO has received much interest during the last years. According to the experimental conditions two mechanisms have been proposed to account for the catalytic conversion of methane and oxygen to synthesis gas [1]. The first route consists of methane combustion followed by subsequent dry and wet reforming [2], while the second proposed route is the direct partial oxidation of methane with oxygen [3]. The main drawback of the direct partial oxidation of methane is the limitation in the synthesis gas yield from the direct oxidation. Even if hydrogen and carbon monoxide can be formed catalytically as primary products, they are more reactive for being further oxidized with oxygen than unconverted reactant methane [4,5] and the product composition will then be determined by equilibrium.

An alternative approach for avoiding over-oxidation of methane is to carry out the partial oxidation anaerobically, i.e., without free oxygen in the gas phase using lattice oxygen in a solid material for the methane oxidation [5]. Better yields could be obtained proceeding with a riser-regenerator or moving bed reactor, as applied in the partial oxidation of butane to maleic anhydride over VPO catalysts [6]. This approach has also recently been proposed [7,8] and demonstrated [9] for the oxidative dehydrogenation of propane over VMgO catalysts. Thus, the capacity to store and release oxygen is a crucial property for the partial oxidation catalysts.

Oxygen storage dynamics of cerium oxide is well described in the literature [10,11]. Cerium oxide is widely used as a solid dynamic oxygen reservoir [11] due to its redox property, its high oxygen storage capacity [12] and the high mobility of lattice oxygen in the solid [13]. Cerium oxide has been studied in context of methane partial oxidation also by other researchers [14,15,16]. The contact between cerium oxide and reactant gases can be improved by impregnating the cerium oxide on top of a high surface area support like $\gamma\text{-}Al_2O_3$. Addition of small amounts of a noble metal such as platinum can provide high activity for hydrocarbon activation [4]. In addition to good oxygen storage dynamics in cerium oxide, it improves the resistance to thermal loss of alumina support surface area and it stabilizes the active precious metals in a finely dispersed state [17,18].

This study deals with the formation of syngas from the reaction between CH_4 and O_2 in a cyclic operation where the oxygen is stored in $Pt/CeO_2/\gamma\text{-}Al_2O_3$. A previous study was conducted with pulses of reactants (CH_4, O_2 or CO_2) over promoted cerium oxide [4] and promising features of this gas-solid reaction system were revealed. For analytical reasons, the materials in the previous study were treated as pulses of reactants rather than as continuous flow. This introduced a large gap to realistic reaction conditions and the results could not be directly used to estimate the magnitude of a suitable GHSV for continuous operation. Further studies with continuous reactant flows were therefore suggested in order to evaluate the potential of such a process under more realistic conditions. This paper deals with syngas production with promoted cerium oxide treated in alternating continuous flows of methane and oxygen.

2. EXPERIMENTAL

2.1 Catalyst preparation

The catalyst sample consisted of 0.5 wt% Pt supported on $CeO_2/\gamma\text{-}Al_2O_3$. The solid was prepared by incipient wetness impregnation of about 16.1% CeO_2 on $\gamma\text{-}Al_2O_3$. The $\gamma\text{-}Al_2O_3$ material was supplied by Alfa Aesar® in pellets and crushed down to particles in the size 300-710 μm before impregnation with $Ce(NO_3)_3 \cdot 6H_2O$ dissolved in distilled water. The material was dried at 90°C. Calcination was performed at 600°C before reimpregnation with $Pt(NH_3)_4(NO_3)_2$ supplied by Aldrich® and recalcination in air for 10 h. Detailed characterization of the $Pt/CeO_2/\gamma\text{-}Al_2O_3$ has been described previously [4].

2.2 Reactor set-up and experimental procedure

The experiments were carried out in a fixed-bed microreactor, which consisted of a 4 mm ID quartz tube placed into a vertical electrically heated tube furnace. A sample of 1.0 g was placed in the center of the quartz tube and supported at the bottom by a layer of quartz wool. A thermocouple was attached to the outside surface of the reactor tube and connected to a temperature controller. The different flows were supplied by an automated 4-way valve allowing to switch abruptly from an oxidizing mixture (O_2/Ar = 2.5/12.5 Nml/min) to a reducing one (CH_4/He = 2.5/12.5 Nml/min) and vice-versa (τ = 4.2 s calculated at 25°C and 1 bar). Relatively high contact time was used in this study in order to analyze in more detail the dynamic behavior of the system. The product gas at the reactor outlet was continuously analyzed by an on-line Balzers QMG 420 quadrupole mass spectrometer (MS). The samples were pretreated under pure O_2. The redox cycles were carried out at atmospheric pressure and at different temperatures (600 to 850°C).

Although calibration and quantitative analysis of H_2O by mass spectrometer are difficult, the analysis enables approximation of the catalyst behavior under O_2 – CH_4 cycling condition.

3. RESULTS AND DISCUSSION

3.1. Reducing conditions

Fig. 1 reports on the product formation rates during the O_2/Ar to CH_4/He switch at 850°C. At time t = 0, the oxidizing atmosphere was replaced by the reducing one. The observed delay between the switch and the drop in the O_2 response (about 25 s) corresponds to the time for the gas to move from the automated 4-way valve to the mass spectrometer. A separate experiment with switches from O_2 to inert to CH_4 as proposed by Pantu et al. [16], was carried out in order to determine the product formation by direct O_2 and CH_4 reactions. No changes in the product formation compared to the case of direct switches from O_2 to CH_4 were observed. It is therefore no effect of purging the system with inert and there is no significant contribution from both CH_4 and O_2 together reacting either directly in the gas phase or with the catalyst without involving bulk oxygen.

A few seconds after CH_4 was introduced to the reactor, CO_2 appeared at the reactor outlet. No CH_4 was observed at the outlet confirming that the conversion was complete. The production of CO_2 was followed by high quantities of H_2O. This observation is in perfect agreement with the previous pulse study [4] where also hydrogen accumulation on the fresh solid was observed. It is important to keep in mind that no reaction was observed over pure $CeO_2/\gamma\text{-}Al_2O_3$ without Pt added as a promoter. It is therefore clear that platinum catalyzes the conversion of CH_4 and it is most likely to believe that this is due to high activity of Pt for activation of methane. Pt-catalyzed methane activation is usually described as a stepwise dehydrogenation to finally yield adsorbed carbon and hydrogen. A previous study [5] revealed that the product formation from methane and oxygen on platinum depended strongly on the concentration and availability of oxygen at the active sites. In this particular system, oxygen is stored in the ceria phase and CH_4 must be activated on the platinum phase. Oxygen migrates from ceria to platinum and reacts with carbon or hydrogen species present on the Pt surface. The initial formation of CO_2 and H_2O is therefore an indication that on a fresh catalyst sample, the oxygen availability is good.

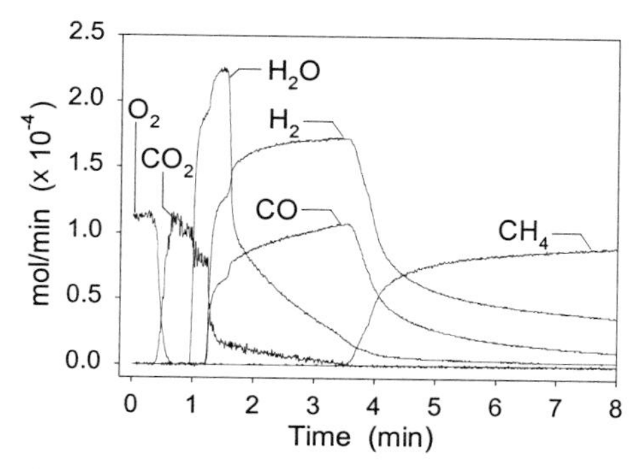

Figure 1. Product formation rates during switching from O_2/Ar to CH_4/He at 850°C.

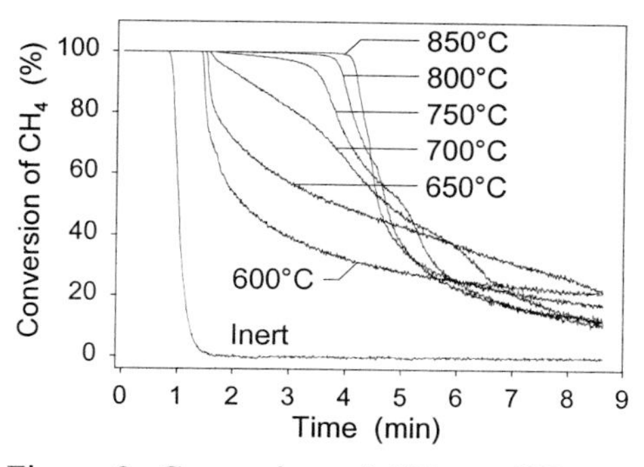

Figure 2. Conversion of CH_4 at different temperature as a function of time.

As the sample was being reduced by CH_4, the product formation switched from CO_2 and H_2O towards more and more CO and H_2. Finally, a syngas with a H_2/CO ratio close to 2.0 was observed before CH_4 slipped through the reactor after about 3.6 min. This is also in agreement with our previous study [4] as well as other studies presented in the literature [14,15,16]. As expected, the product distribution did also under continuous flow of reactants depend strongly on the degree of reduction of the ceria. High selectivity to CO was obtained when the most reactive oxygen had been removed, as was the case after approximately 100 micromoles of methane for 1 g of catalyst. The shoulders located at the top of the H_2O, CO, and H_2 pulses are most probably caused by an effect of partial pressure to which the MS is sensitive. The H_2O shoulder occurs at the same time as the CO_2 production drops at 1.8 min, and in the same way the CO and H_2 shoulders occur when the H_2O production drops at 2 min.

Fig. 2 shows the methane conversion during a switch from O_2 to CH_4 at different temperatures (from 600 to 850°C). The time with complete conversion of CH_4 decreased with decreasing temperatures. At 850°C this time lasted for about 2.7 min., while it was 0.6 min at 600°C. At low temperatures (< 800°C), the conversion of CH_4 decreased rapidly after the drop in formation of CO_2. For these temperatures, a high selectivity to synthesis gas was not obtained simultaneously with complete conversion of CH_4. However, other studies [4,15] with higher weight loading of cerium oxide show that at 700°C there is a larger time gap between the disappearance of CO_2 and the appearance of CH_4.

In order to estimate the progress of the reaction under methane atmosphere, the degree of reduction of CeO_2 by CH_4 was calculated as follows [14]:

$$\text{Degree of reduction} = \frac{\text{amount of CO} + (2\times\text{amount of CO}_2) + \text{amount of H}_2\text{O (mol)}}{\text{amount of oxygen atoms in CeO}_2\text{ (mol)}} \times 100\%$$

Fig. 3 shows how the degree of reduction develops as a function of time under reducing atmosphere for different temperatures. The amounts of CO, CO_2 and H_2O produced correspond to the integration of their curve (Fig. 1). For the first 30 s, the degree of reduction evolved in the same manner independent of the reaction temperature. When the degree of reduction of ceria had been adjusted to ca. 10%, the system produces stoichiometrically CO and H_2.

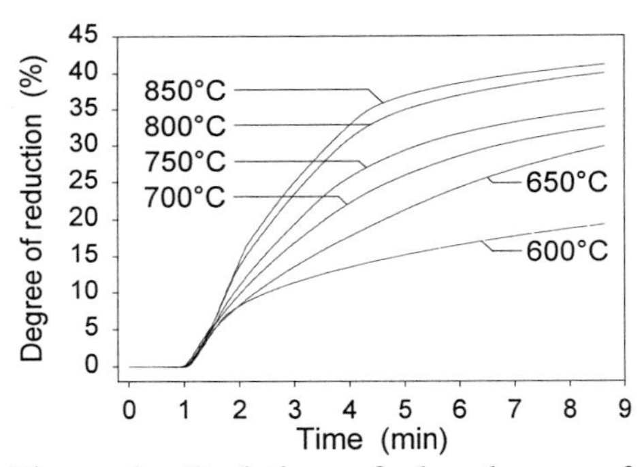

Figure 3. Evolution of the degree of reduction at different temperature as a function of the progress of the reaction.

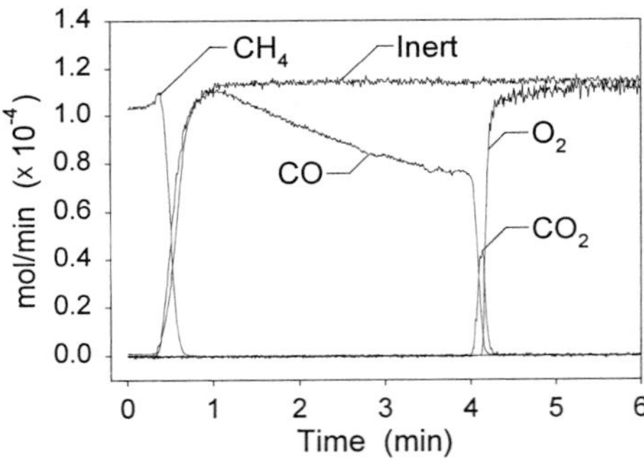

Figure 4. Product formation rates during switching from CH_4/He to O_2/Ar at 800°C.

Carbon deposition on the material is significant during exposure to methane. Steady-state methane partial oxidation over noble metal catalysts under $O_2 + CH_4$ produces synthesis gas with little or no carbon deposition [19,20]. However, under anaerobic conditions the carbon deposition is significant [4]. The H_2/CO ratio became greater than 2.0 after some minutes giving a clear evidence that carbon accumulates in the reactor when the degree of ceria reduction increases. Carbon deposition can occur from different reactants:

$CH_4 \rightleftharpoons C + 2H_2$ $\Delta H_{298K} = 75$ kJ/mol

$2CO \rightleftharpoons C + CO_2$ $\Delta H_{298K} = -172$ kJ/mol (Boudouard reaction)

$CO + H_2 \rightleftharpoons C + H_2O$ $\Delta H_{298K} = -131$ kJ/mol

The reaction enthalpies indicate that at high temperatures carbon is formed from methane. At lower temperatures carbon can also be formed from carbon oxides.

3.2. Oxidizing conditions

Fig. 4 shows the response of CH_4, CO, CO_2 and O_2 for the CH_4/He to O_2/Ar switch after 20 min under reducing conditions at 800°C. Carbon removal with 100% CO selectivity was obtained during the first 3.7 min with complete conversion of oxygen. A very limited quantity of CO_2 was produced just before the O_2-breakthrough. High selectivity to CO is not necessarily a result of direct carbon oxidation to CO, but could also be a result coke oxidation by carbon dioxide or by direct oxidation of reduced ceria by carbon dioxide [21]. During reoxidation of the material in O_2 the ceria reoxidation and carbon removal occur simultaneously:

$$C + CO_2 \rightleftharpoons 2\,CO \qquad \Delta H_{298K} = 172\ kJ.mol^{-1}$$

$$CeO + CO_2 \rightleftharpoons CeO_2 + CO$$

Fig. 1 shows that a complete reactant conversion and a high selectivity to CO and H_2 can be obtained by feeding methane. Fig. 4 shows that successful reoxidation of the ceria and the selective coke gasification to CO can be achieved with O_2. Production of synthesis gas through repeated cycles is therefore experimentally shown to be feasible. By a careful control of the reduction and oxidation time, CO_2 and water formation can be avoided (Fig. 5).

3.3. Repeated cycles of reducing and oxidizing feed

Fig. 5 shows the products formed during repeated cycles of CH_4 and O_2 over $Pt/CeO_2/\gamma\text{-}Al_2O_3$ catalyst at 850°C. Initially, O_2/Ar was switched to CH_4/He which was kept for 1.3 min, then the feed was switched back to oxidizing conditions for 1 min. The following cycles lasted 1 min under CH_4/He and 0.5 min under O_2/Ar.

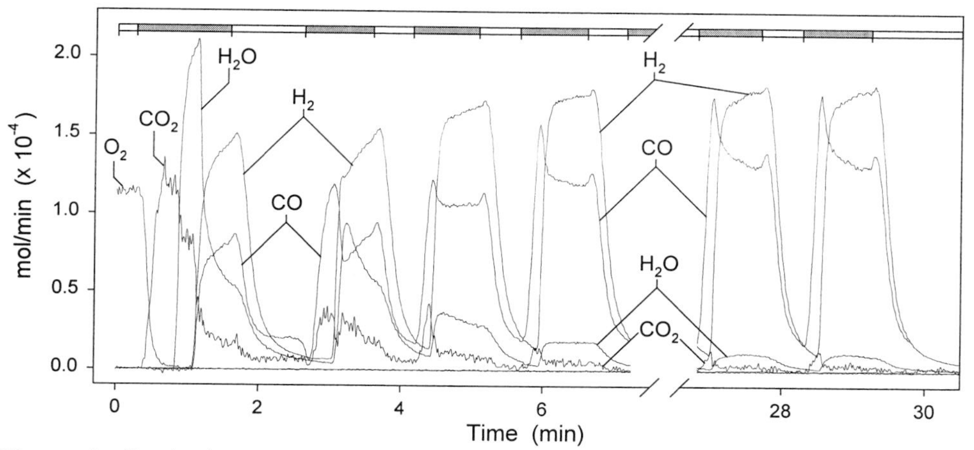

Figure 5. Synthesis gas production during redox cycles between O_2/He (□) and CH_4/Ar (■)atmospheres at 850°C.

The same product formation patterns as those reported in Fig. 1 were observed for the first cycle O_2/Ar - CH_4/He. The conversion of CH_4 was complete during the entire 30 min duration of this experiment. The front peak on the CO top is believed to be caused by reactive oxygen species present on the material surface immediately after the treatment in oxygen. It is also proposed that the level to which the CO production approaches, when stabilizing, corresponds with the average rate of oxygen diffusion from the bulk of the material to the surface. The rate of CO formation from the reaction between CH_4 and a solid oxide is limited by the flux of oxygen through the surface of the material. After 4 redox cycles the product formation cycles were stabilized and continued to oscillate in a reproducible manner as the feed was alternated. The production of CO_2 and H_2O was only minor after the initial stabilization. The time under reducing conditions was twice as long as under oxidizing atmosphere, which is in agreement with the desired reaction stoichiometry.

4. CONCLUSION

The production of synthesis gas highly depends on the degree of reduction of the cerium oxide. Careful control of the degree of reduction could be obtained by manipulating the time in which the material was exposed to the different feeds. The formation of undesired products (CO_2 and H_2O) was successfully avoided, and high selectivities to CO and H_2 with a complete conversion of CH_4 were obtained.

The results show that the partial oxidation of methane on solid $Pt/CeO_2/\gamma\text{-}Al_2O_3$ carried out at non-steady state conditions by repeated redox cycles can be an alternative reaction for synthesis gas production. High selectivities of synthesis gas with complete methane conversion could be reached at relatively low temperatures.

ACKNOWLEDGMENTS

NTNU and SINTEF are gratefully acknowledged for the financial support to J.C.J through the Strong Point Center Kinetics and Catalysis (Kin Cat).

REFERENCES

1. S.C. Tsang, J.B. Claridge and M.L.H. Green, Catal. Today, 23 (1995) 3.
2. M. Prettre, C.H. Eichner and M. Perrin, Trans. Faraday Soc., 43 (1946) 335.
3. D.A. Hickman and L.D. Schmidt, Science, 259 (1993) 343.
4. M. Fathi, E. Bjorgum, T. Viig and O.A. Rokstad, Catal. Today, 63 (2000) 489.
5. M. Fathi, F. Monnet, Y. Schuurman, A. Holmen and C. Mirodatos, J. Catal., 190 (2000) 439.
6. E. Bordes and R.M. Contractor, Topics Catal., 3 (1996) 365.
7. J.C. Jalibert, PhD thesis, Lyon 1 University, 255-99, 1999.
8. H.W. Zanthoff, S.A. Buchholz, A. Pantazidis and C. Mirodatos, Chem. Eng. Sci., 54 (1999) 4397.
9. D. Creaser, B. Andersson, R.R. Hudgins and P.L. Silveston, Chem. Eng. Sci., 54 (1999) 4437.
10. H.C. Yao and Y.F. Yu Yao, J. Catal., 86 (1984) 254.
11. A. Holmgren and B. Andersson, J. Catal., 178 (1998) 14.
12. A. Trovarelli, C. de Leitenburg and G. Dolcetti, Chemtech, (1997) 32.
13. A. Trovarelli, Catal. Rev. Sci. Eng., 38 (1996) 439.
14. K. Otsuka, Y. Wang and M. Nakamura, Appl. Catal. A : General, 183 (1999) 317.
15. K. Otsuka, Y. Wang, E. Sunada and I. Yamanaka, J. Catal., 175 (1998) 152.
16. P. Pantu, K. Kim, G.R. Gavalas, Appl. Catal. A : General, 193 (2000) 203.
17. J.C. Summers, S.A. Ausen, J. Catal., 58 (1979) 131.
18. F.J. Sergey, J.M. Maseller and M.V. Ernest, W.R. Grace Co. U.S. Patent 3 903 020 (1974).
19. P.D.F. Vernon, M.L.H. Green, A.K. Cheetham and A.T. Ashcroft, Catal. Lett., 6 (1990) 181.
20. J.B. Claridge, M.L.H. Green, S.C. Tsang, A.P.E. York, A.T. Ashcroft and P.D. Battle, Catal. Lett., 22 (1993) 299.
21. S. Sharma, S. Hilaire, J.M. Vohs, R.J. Gorte and H.W. Jen, J. Catal., 190 (2000) 1999.

Studies in Surface Science and Catalysis
J.J. Spivey, E. Iglesia and T.H. Fleisch (Editors)

Hydroconversion of a mixture of long chain n-paraffins to middle distillate: effect of the operating parameters and products properties

V.Calemma [(a)], S. Peratello [(a)], S. Pavoni [(b)], G. Clerici [(a)] and C. Perego [(a)]
[(a)]EniTecnologie S.p.A., [(b)]Agip Petroli, Via Maritano 26, 20097 San Donato M.se, Italy

Abstract
Some aspects of the hydrocracking behaviour of a C_{10+} mixture of n-paraffins on a 0.3% platinum/amorphous silica-alumina catalyst have been studied. Particularly, the effect of the operating conditions, that is temperature, pressure and H_2/feed ratio, on the selectivity to middle distillate and its characteristics was investigated. The maximum yield achieved in middle distillate was 82-87%. The isomers content of the kerosene and the gasoil fractions increases considerably during the reaction. As a consequence the Freezing point and Pour point of kerosene and gasoil show a remarkable decrease reaching values of –50 °C and –30 °C respectively. The results indicate that the vapour-liquid equilibrium plays an important role which considerably affects several aspects of the feedstock hydroconversion.

Introduction
During the last years several factors of strategic, economic and environmental nature have lead to a growing interest towards the conversion of natural gas to liquid fuels [1]. It is well known that the conversion of syngas through the Fischer-Tropsch (FT) process leads to the formation of products essentially made up of n-paraffins (>90%), together with smaller percentages of alcohols and olefins, characterised by a wide range of molecular weights whose distribution can be described by the Anderson-Flory-Schulz (AFS) model [2]. An important consequence of the chain growth mechanism is the theoretical impossibility of synthesizing a product with a narrow range of aliphatic chain length. In case of products (C_{5+}) obtained with Co based catalysts of the last generation the weight fraction of middle distillate (C_{10}-C_{22}) is 0.4-0.6 while the remaining is made up of heavier products (0.6-0.4) and naphtha (0.05-0.2). However, due to the nature of FT products the middle distillates so obtained, have very poor cold properties that hamper their use as transportation fuel. Given the impossibility of producing, through FT synthesis, middle distillates with high yields and good cold properties, it is then necessary to subject the FT products to an upgrading step in order to improve the aforementioned facets. At the present time, the achievement of this twofold objective, by the major subjects operating in this field, is pursued by more or less complex hydroconversion processes (FT/HCK) of FT products [3,4]. The kerosene and gasoil produced via FT/HCK show excellent characteristics both as for the specifications needed as transportation fuel and environmental impact [5]. Over the last three years, EniTecnologie has been involved in a research program aimed at developing a suitable catalyst for the hydrocracking of FT products to produce middle distillates. In the present paper we report the

main results of a study concerning the hydrocracking of FT waxes on a Pt/amorphous mesoporous SiO_2-Al_2O_3 catalyst.

Experimental

The catalyst used throughout the experiments consisted of platinum supported on amorphous mesoporous silica-alumina (MSA), with a SiO_2/Al_2O_3 ratio of 100, extruded with a binder

Acid strength							
Weak ($\mu molg^{-1}$) 200°C		Medium ($\mu molg^{-1}$) 300°C		Strong ($\mu molg^{-1}$) 400 °C		V.Strong ($\mu molg^{-1}$) 500 °C	
L	B	L	B	L	B	L	B
210	21	131	9	71	9	41	0

Table 1. Strength, amount and type (B: Brønsted, L: Lewis) of acid sites in the Pt/MSA/E sample.

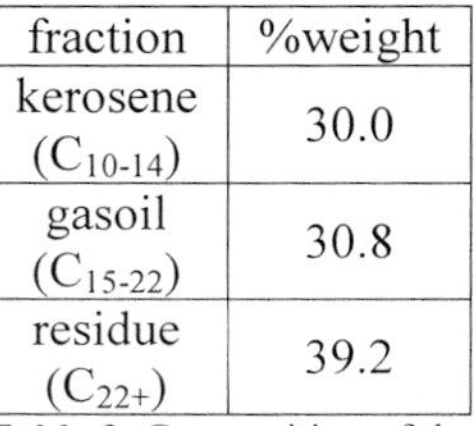

fraction	%weight
kerosene (C_{10-14})	30.0
gasoil (C_{15-22})	30.8
residue (C_{22+})	39.2

Table 2. Composition of the feedstock.

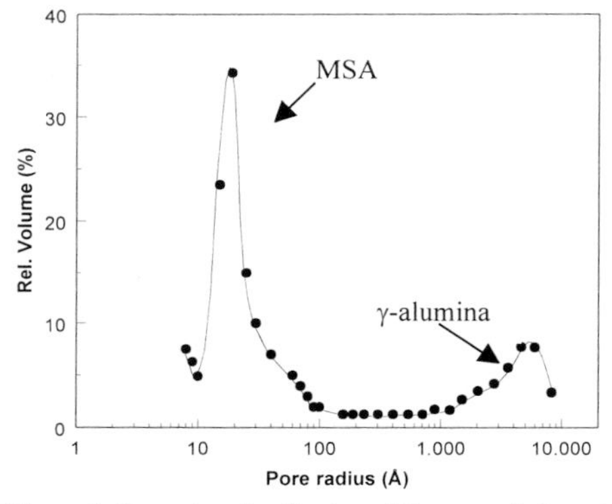

Figure 1. Pore size distribution of the extruded catalyst.

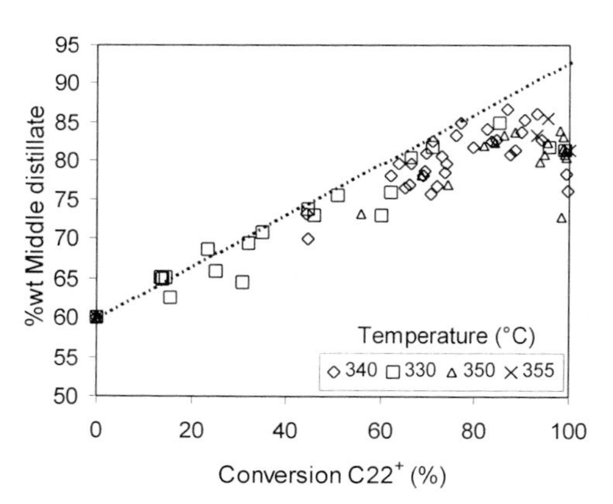

Figure 2. Middle distillate yields vs. conversion of C_{22+} fraction. (Press.:50 atm; H_2/wax: 0.1).

(γ-alumina). The MSA (active phase) was prepared as described in reference [6]. According to reference [7] small MSA particles are dispersed in the binder and then extruded as pellets (Pt/MSA/E). The extrudate is characterized by a B.E.T. surface area of ~550 m^2g^{-1}, a bimodal pore size distribution as shown in Fig. 1 and the predominance of weak acid Brønsted sites (see Table 1). The MSA/γ-alumina weight ratio is 1.5. The noble metal was loaded on the extrudates through impregnation with a solution of H_2PtCl_6, following the procedure described elsewhere [8]. The platinum content was 0.3%. Hydrocracking tests were carried out in a bench scale trickle bed reactor operated in down flow mode. The range of the operating conditions were: temperature= 330-355 °C; pressure= 70-35 atm; H_2/wax= 0.05-0.2 (wt/wt); WHSV= 0.5-3(h^{-1}). The feedstock used in this study was a synthetic mixture of C_{10+} n-paraffins having a weight fraction distribution similar to FT products with α:0.87. The distribution among the different cuts is given in Table 2.

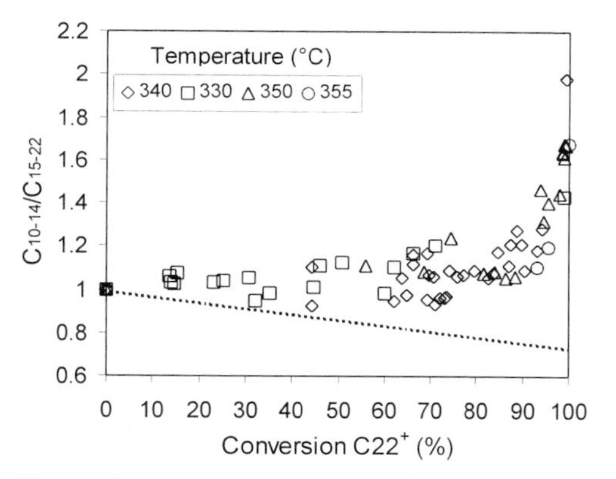

Figure 3. Kerosene/Gasoil ratio vs. conversion of C_{22+} fraction. (Press: 50 atm ; H_2/wax: 0.1)

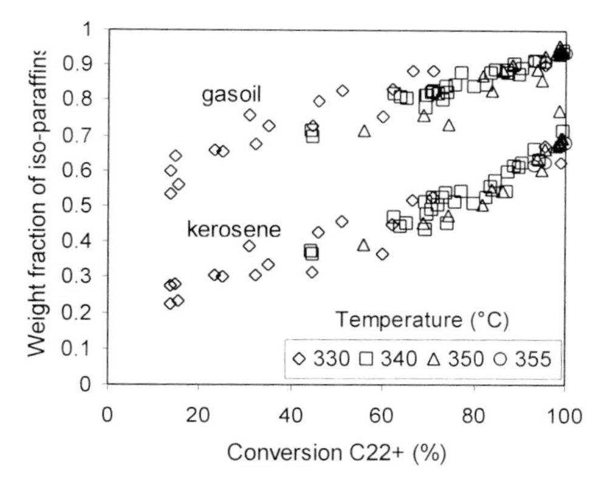

Figure 4. Concentration of iso-paraffins in the kerosene and gasoil fractions vs. conversion of the C_{22+} fraction. (Press.: 50 atm; H_2/wax: 0.1).

Results and discussion

As shown in Fig. 2 the yields of the middle distillate (C_{10-22}) increases up to 85-90% of C_{22+} conversion and thereafter decreases owing to the consecutive hydrocracking reactions which lead to lower molecular weight compounds. The latter are formed by normal and isoparaffins C_{5-9} (~80%), C_{4-3} (~19%) and C_{1-2} (~0.5%). The reaction temperature does not show a significant effect and the conversion into middle distillate seems to be a unique function of the C_{22+} conversion. The maximum yield into middle distillate is in the range 82-87%.

As previously shown by Sie et al.[9], to achieve the highest yields in a given distillation range, two conditions have to be fulfilled:

- The reactivity of components above the desired range should be much higher than those in or below the desired range.
- The mole fraction distribution of products obtained by hydrocracking of each single n-paraffin should be that of pure primary cracking [10].

In our case, assuming that the fraction C_{22-} does not undergo to cracking reaction and an equimolar distribution of products resulting from the cracking of the C_{22+} n-paraffins, except to those resulting from the breakage of α and β bonds, which are supposed zero, the theoretical yields of C_{10-22} fraction can be calculated. The results (dashed line in Fig.2) indicate that the Pt/MSA/E catalyst displays a high selectivity towards the C_{10-22} fraction. At 85-90% of C_{22+} conversion, where the maximum yields are achieved, the selectivity is ~0.93 the theoretical one. A support to the scarce presence of "*overcracking*" reactions up to relatively high values of conversion, is given by the kerosene/gasoil (C_{10-14}/C_{15-22}) weight ratio which, as shown in Fig. 3, exhibits a slight increase up to 85% of C_{22+} conversion. In this case the dashed line in Fig.3 shows the theoretical trend of the C_{10-14}/C_{15-22} weight ratio. The latter was calculated under the same assumptions reported above to determine the theoretical yields of the C_{10-22} fraction. The result point out that under "ideal conditions" the hydrocracking process would lead, with a feedstock having a C_{10-14} /C_{15-22} ratio of 1, to a substantial decrease of the kerosene/gasoil weight ratio in the reaction products. The underlying reason of this trend is that the hydrocracking of the C_{22+} fraction produces kerosene and gasoil with a ratio of ~0.5. During hydroconversion of feedstock essentially made of n-paraffins, a very important role is played

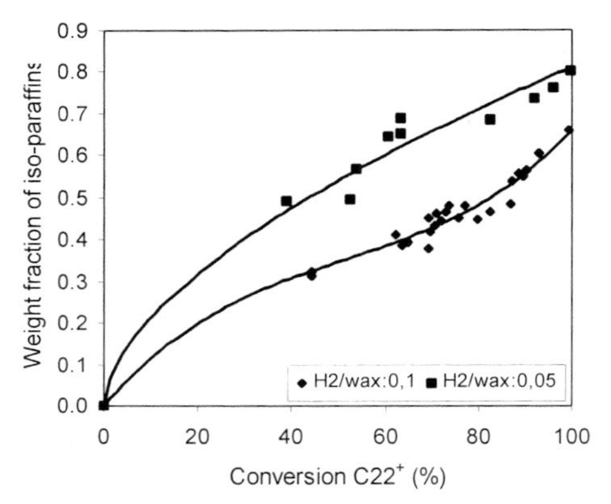

Figure 5. Concentration of isoparaffins in the kerosene vs. conversion of C_{22+} fraction. (Temp: 340 °C; Press.: 50 atm)

by the hydroisomerization reaction because it allows to greatly lower the melting point of paraffinic chains leaving unchanged their molecular weight. As example the melting points of hexadecane and 5-Methyl pentadecane, are 19, –31 °C respectively. For this reason the hydroisomerization capacity of the catalyst has a strong impact on the cold properties of the kerosene and the gasoil fractions which are strictly dependent on their iso-paraffins content. The results (see Fig. 4) show a strong increase of the iso-paraffins content, with the increase of C_{22+} conversion, of both the kerosene and the gasoil fractions. As a consequence, the Freezing point and Pour point of kerosene and gasoil show a remarkable decrease reaching values of –50 °C and –30 °C respectively. Even in this case, the trend does not seem significantly affected by the reaction temperature. The increase of the iso-paraffins concentration of kerosene and gasoil is the result of both the hydrocracking of C_{22+} n-paraffins which gives products partially isomerized and hydroisomerization of middle distillate components. It is reasonable to assume that the C_{22+} hydrocracking products have a constant branching degree as a function of chain length [11]. Therefore the difference of isomerization degree between gasoil and kerosene is mainly determined by the reaction of the middle distillate components. However, its known, that selectivity for hydroisomerization (i.e. formation of feed isomers) of n-paraffins decreases with the increase of chain length [12-14] whereas the experimental data indicate that the isomerization degree of the C_{15-22} fraction is always higher than that of the C_{10-14} fraction. The line of argument presented above suggests that this result is basically caused by the higher reactivity of the C_{15-22} fraction in comparison with the C_{10-14} fraction. Two factors contribute to the different reactivity: the higher reactivity of the longer aliphatic chains [12,14]; the partial evaporation of the feed which leads, as shown in Table 3, to a liquid phase enriched in the heavier components. In this connection, it should be stressed that in a pure trickle flow regime, the catalyst particles are completely covered with flowing liquid and the reaction occurs only at the interface between the surface of the catalyst and the liquid phase which in turn is in equilibrium with the gas phase. The importance of the role played by the vapour-liquid equilibrium in the reaction environment is pointed out by the effect of the H_2/wax ratio on several aspects of the hydroconversion of the feedstock. A decrease of the H_2/wax ratio lead to a decrease of the

H_2/wax	0.05		0.1	
phase	Vapour	Liquid	Vapour	Liquid
%wt	34	66	48	52
C10-14	[69] (60)	[31](14)	[85] (53)	[15] (8)
C15-22	[37] (34)	[63] (29)	[58] (38)	[42] (25)
C22+	[5] (6)	[95] (57)	[11] (9)	[89] (67)

The values between the round brackets refer to the phase composition while those between the square brackets give the percentage of the feed fractions in the two phases.

Table 3. Effect of H_2/wax ratio on vapour-liquid equilibrium at 340 °C and 50 atm.

Conversion (%wt C22+)	*kerosene*	*gasoil*
72(*)	F. P.: -49 °C	P.P.: -22 °C
	S.P.: >50 mm	B.C.N.:76
85 (*)	F.P.: -51 °C	P.P.: -28 °C
	S.P.: >50 mm	B.C.N.:78
78(**)	F.P.: -39 °C	P.P.: -20 °C
	S.P.: >50 mm	B.C.N.:78
96(**)	F.P.: -48 °C	P.P.: -27 °C
	S.P.: >50 mm	B.C.N.:77
(*) H2/wax: 0.05; (**) H2/wax: 0.1		

Table 4. Main properties of middle distillate (S.P.: Smoke Point , F.P.: Freezing Point, P.P.: Pour Point, B.C.N.: Blending Cetane Number.

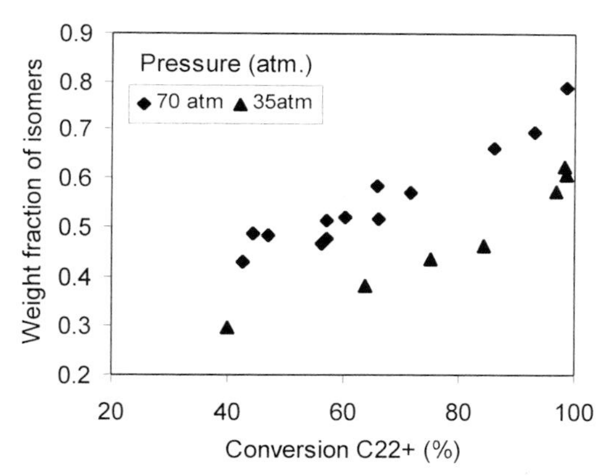

Figure 6. Concentration of isoparaffins in the kerosene vs. conversion of C22+ fraction. (Temp.: 340 °C; H2/wax:0.1 (wt/wt))

conversion rate of the C_{22+} fraction and, at the same conversion levels, an increase of both the iso-paraffin content of the C_{10-22} fraction, and the C_{10-14}/C_{15-22} ratio. An example of the results obtained is reported in Fig. 5. The data in Table 3, calculated by the SRK equation, show that higher values of the H_2/wax ratio leads to a substantial increase of the feed percentage in the vapour phase and a higher concentration of the C_{22+} fraction in the liquid phase. Particularly, the data indicate that the lower isomerization degree of the C_{10-14} fraction at higher H_2/wax ratio is caused by the fact that its percentage in the vapour phase changes from 69 to 85 %. The conversion rate of the C_{22+} is inversely proportional to the hydrogen pressure. This relationship can be explained in terms of bifunctional mechanism [15] where the first step is the formation of the olefin at the metal site. As shown in Fig. 6, the increase of the reaction pressure lead to more isomerized products. Even in this case, the latter effect can be mainly ascribed to the changes of the vapour-liquid equilibrium according to the reaction pressure. The data reported in Table 4 show that kerosene and gasoil produced, in suitable operating conditions, exhibit excellent characteristics, even as regards their cold properties. The kerosene meet the stringent requirement of aviation Jet Fuel while the pour point of gasoil is well beyond the normal refinery specification. The Smoke point of kerosene and the Blending Cetane Number of gasoil are exceptionally high. Moreover the lack of heteroatoms (S, N) and aromatic compounds make these fuels ideal from an environmental point of view.

Conclusions

In this work the hydroconversion behaviour of a mixture of C10-70 n-paraffins over Pt/MSA/E catalyst was investigated. The Pt/MSA/E catalyst displayed a high selectivity for the hydroisomerizzation which lead products characterized by an elevated concentration of isoparaffins. Consequently, under suitable operating conditions both kerosene and gasoil have excellent cold properties. Moreover, the Pt/MSA/E showed a high selectivity for middle distillate. It is known [16,8] that silica-allumina amorphous catalysts show a selectivity higher than zeolite-based catalysts towards the formation of middle distillate. This different behaviour is generally ascribed to the higher content of strong acid sites in zeolite-based catalyst which give a higher activity but at the same time promote the hydrocracking rather than hydroisomerization. In our case, we suggest that the high selectivity for hydroisomerization and middle distillate production are the result of three distinctive characteristics of Pt/MSA/E catalyst, that is the mild Brønsted acidity particularly evident if compared with zeolite catalysts [17,8]; high surface area and pore distribution centred in the mesopore region. Hydrocracking and hydroisomerization of n-paraffins [18] occur in parallel sharing a common intermediate (i.e. iso-alkyl cation) which can desorb or undergo to β-scission. In this case the selectivity for hydroisomerizzation depends on the relative rates of desorption and β-scission [18]which in turn are affected by the strength of acidic sites. In this case it is reasonable to assume that strong acid sites will form reaction intermediates which desorb less easily and so promoting the hydrocracking. As for the textural properties, the high surface area and the pore size distribution centred in the mesopore region (~ 20Å) are factors which favour the diffusion of components and lower the presence of secondary hydrocracking reactions. At last, the results of this study show that the vapour-liquid equilibrium plays an important role which considerably affects several aspects of the feedstock hydroconversion.

References

1. H. Schulz, *Appl. Catal. A: Gen.* 186 (1999) 3
2. R.B. Anderson, *The Fischer-Tropsch Synthesis* Academic Press Inc. N.Y. 1984
3. M.M. Senden, A.D. Punt, A. Hoek in: Natural Gas Conversion V, *Stud. Surf. Sci. Catal.*, vol. 119, p. 961, Eds. A. Parmaliana et al., Elsevier Science BV. 1998
4. B. Eisenberg, R.A. Fiato, C.H. Mauldin, G.R. Say, S.L. Soled in: Natural Gas Conversion V, *Stud. Surf. Sci. Catal.*, vol. 119, p. 943, Eds. A. Parmaliana et al., Elsevier Science BV. 1998
5. T.W. Ryan III Emission Performance of Fischer-Tropsch Diesel Fuel , *Conference on Gas to Liquids Processing '99*, May 17-19, 1999, S. Antonio, Texas
6 G. Bellussi, C. Perego, A. Carati, S. Peratello, S. Previde Massara in Zeolites and Related Microporous Materials: State of the Art 1994, J. Weitkamp, H.G. Karge, H. Pfeifer, W. Hölderich, Eds., *Stud. Surf. Sci. Catal.*, Vol 84, p. 85, Elsevier Science BV. 1994
7. G. Bellussi, C. Perego, S. Peratello, US *Pat.*53 428 14, 1993
8. A. Corma, A. Martinez, S. Pergher, S. Peratello, C. Perego, G. Bellussi *Appl. Catal. A: Gen.* 152 (1997) 107
9. S. T. Sie, M.M. Senden and H.M.H. Van Wechem, *Catal. Today* 8 (1991) 371
10. J. A. Martens, P.A. Jacobs, J. Weitkamp *Appl. Cat.* 20 (1986) 239
11. S. T. Sie, *Ind. Eng. Chem. Res* 32 (1993) 403
12. V. Calemma, S. Peratello, C. Perego, *Appl. Catal. A: Gen* 190 (2000) 207
13. M. Steijns and G. Froment *Ind. Eng. Chem. Prod. Res. Dev.* 20 (1981) 654
14. J. Weitkamp , *Prep. Am. Chem. Soc. Div. Petr. Chem.* 20 (1975) 489
15. H.L. Coonradt, W.E. Garwood, *Ind. Eng. Chem. Proc. Des. Dev.* 3(1) (1964) 3
16. J. Scherzer, A.J. Gruia *Hydrocracking Science and Technology*, Chp. 7, p. 96,Marcel Dekker 1996.
17. C. Perego, S. Amarilli, A. Carati, C. Flego, G. Pazzucconi, C. Rizzo, G. Bellussi *Microporous and Mesoporous Materials* 27 (1999) 345.
18. J. Weitkamp, S. Ernst in :*Guidelines for Mastering the Properties of Molecular Sieves*, p. 343, Eds. D. Barthomeuf et al., Plenum Press, NY, 1990.

Studies in Surface Science and Catalysis
J.J. Spivey, E. Iglesia and T.H. Fleisch (Editors)

Ethylene production via Partial Oxidation and Pyrolysis of Ethane

M. Dente[1], A. Beretta[1], T. Faravelli[1], E. Ranzi[1], A. Abbà[2], M. Notarbartolo[3]

[1]CIIC Department Politecnico di Milano, Italy
[2]Math. Department Politecnico di Milano, Italy
[3]Technip Rome, Italy

This paper presents the theoretical study of a homogeneous process to produce ethylene by ethane dehydrogenation. A hydrogen-oxygen diffusive flame in a conventional burner is used as a hot stream into which the ethane feed is injected. The results indicate that very high selectivities to ethylene can be reached at ethane conversion greater than 70%. The advantages of the process are mainly related to the environmental aspects. The high efficiency of the heat exchange, the use of a clean fuel, like the H_2 produced, and the absence of nitrogen drastically reduce the pollutant emissions. The homogeneous process avoids catalyst poisoning and deactivation. The main limitation in this configuration is the very fast required mixing between the two streams.

INTRODUCTION

The pyrolysis of hydrocarbon feedstocks in the steam cracking process is the most commonly used technology to produce ethylene along with other important chemical commodities. In recent years, the need of increasing the efficiency and reducing the environmental impact of such plants has led to research in this mature technology. In particular the ethylene production via the oxidative dehydrogenation of ethane received a great interest (Bodke et al., 1999; Zerkle et al., 2000). The new process alternative to convert ethane to ethylene is a short contact time reactor (a few milliseconds) in which the reaction takes place in the presence of a foam monolith of alumina supported Pt-Sn catalyst. The process can include a premixed H_2 addition in the feed. The catalyst function is to promote the H_2 oxidation to H_2O avoiding the ethane and ethylene oxidation. CO and CO_2 selectivity significantly decreases and the heat of H_2 oxidation reaction allows the endothermic C_2H_6 dehydrogenation to occur in a very selective way.
The two stages of the process can be then globally summarized as:

$$H_2 + \frac{1}{2} O_2 \rightarrow H_2O$$
$$C_2H_6 \rightarrow C_2H_4 + H_2.$$

$$C_2H_6 + \frac{1}{2} O_2 \rightarrow C_2H_4 + H_2O$$

The overall reaction is exothermic, more hydrogen is produced than the required for autothermal conditions and no additional fuel is needed.
In this scenario, it is then quite evident the advantages offered by such a process. From the environmental point of view, it reduces the greenhouse gas emissions, using hydrogen as a clean fuel, while simultaneously the use of pure oxygen avoids the NOx formation. Moreover, the experimental pilot results seem to indicate that this route is also economically convenient. The reactor is smaller and simpler than the conventional cracking furnaces and because of the higher reactor temperatures, selectivities could also be improved. More than 85% ethylene selectivity was obtained at 70% ethane conversion. On the other hand, higher operation costs are expected due to the required pure oxygen. These considerations together with correct economic evaluations seem to make feasible this new approach (Schmidt et al., 2000).
From a scientific point of view, the reaction mechanism is still under debate. The important role played by gas-phase reactions at this high reactor temperatures has been demonstrated (Beretta et al., 2000). The catalyst mainly drives the hydrogen oxidation, while the pyrolysis reactions occur in

the homogeneous phase. As a matter of fact the same, or even higher, ethylene selectivities at the same ethane conversions can be reached with a very rapid heating of the ethane feed. The high conversion obtained at very short contact times in the catalyzed system can be reasonably explained with active species (radicals) generated by the catalytic surface and released in the gas phase (Schmidt et al., 2000).

On the basis of these results, it is possible to propose a purely homogeneous system which can reach the same performance in terms of both selective ethylene production and effective pollutant reduction. This new process, without catalyst, only involves a primary hydrogen flame followed by ethane injection. Two major problems can be avoided with this homogeneous approach. From one side, the aging and poisoning of the catalyst, that can diminish its activity with a reduction of ethylene selectivity in favor of CO and CO_2 formation. A second and significant improvement of the homogeneous system is related to the increase of plant safety. The diffusive hydrogen flame in a conventional burner allows to avoid the premixing of hydrogen, ethane, and oxygen. The potentially high adiabatic temperatures of the hydrogen oxygen system can be easily managed and can be reduced with steam additions in order to reach the optimum cracking temperature.

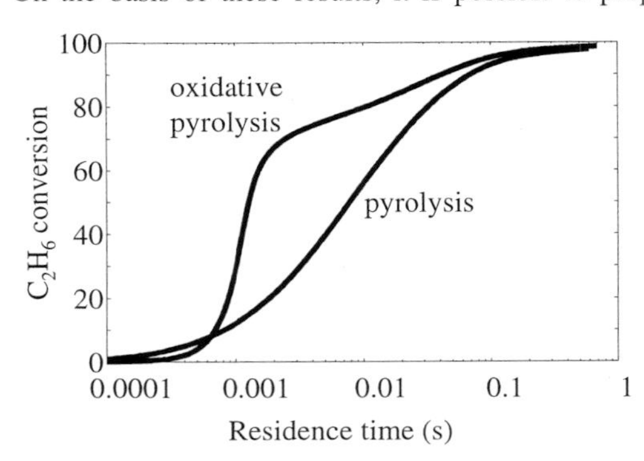

Figure 1. Computed comparison between pyrolysis and oxidative pyrolysis conversion

MODEL RESULTS

The evaluations of the homogeneous oxidative pyrolysis are computed by using detailed kinetic models for the pyrolysis (Dente and Ranzi, 1983) and oxidation of hydrocarbons (Ranzi et al., 2000). The model herein applied is also able to predict the formation of heavier species and especially Poly Aromatic Hydrocarbons (PAH) precursors of coke formation. This model has been already extensively tested and validated by comparison with experimental data carried out in a very wide range of operative conditions. It has been also used to confirm the relevance of the homogeneous gas-phase reactions in the short contact time catalytic reactors (Bodke et al., 1999).

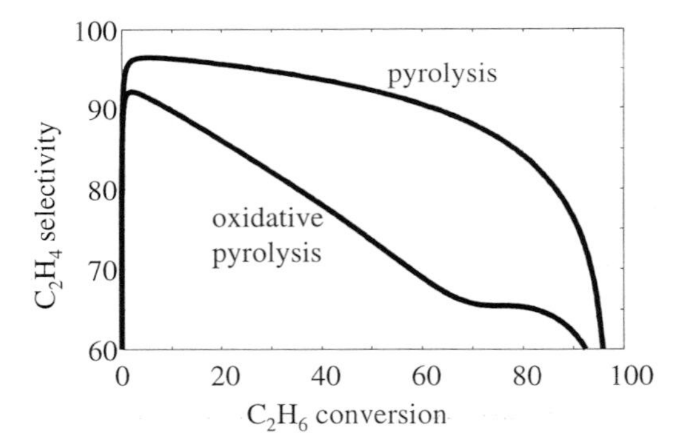

Figure 2. Computed comparison between C_2H_4 selectivity in the pyrolysis and oxidative pyrolysis cases

As already discussed in the past (Chen et al., 1997), the introduction of a small amount of oxygen in the conventional coils of steam cracking processes increases the plant capacity. The heat released from the partial oxidation of hydrocarbons, directly generated inside the reactor together with very active radicals, like OH, enhances the system reactivity.

As an example, fig. 1 shows the conversion of ethane in pyrolysis (C_2H_6:H_2O= 2:1 mol) and partial oxidation systems (C_2H_6:O_2 = 2:1 mol). These data were obtained by simulating an isothermal plug flow reactor at 1200 K. The presence of oxygen increases the conversion and allows to work with shorter contact times. Fig. 2 shows the reduction of ethylene selectivity due to the CO and CO_2 formation. In purely homogeneous conditions, the introduction of an equimolar amount of hydrogen (C_2H_6:O_2:H_2 = 2:1:2) does not significantly modify the performances of the oxidative pyrolysis. This is an implicit confirmation that the higher selectivity of ethylene observed by Bodke et al. (1999) with H_2 addition is due to a catalyzed H_2/O_2 combustion.

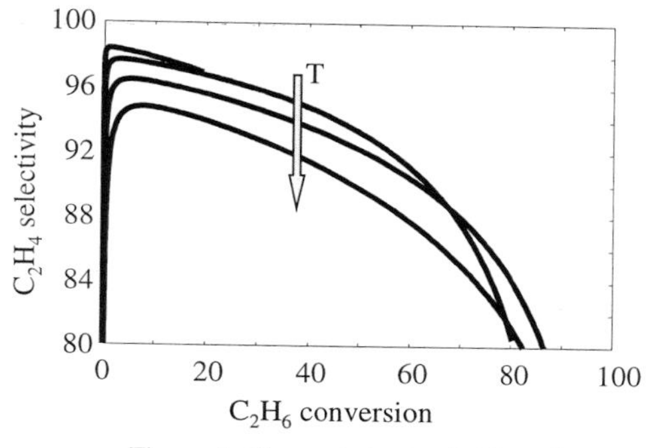

Figure 3. Computed selectivities of isothermal ethane pyrolysis

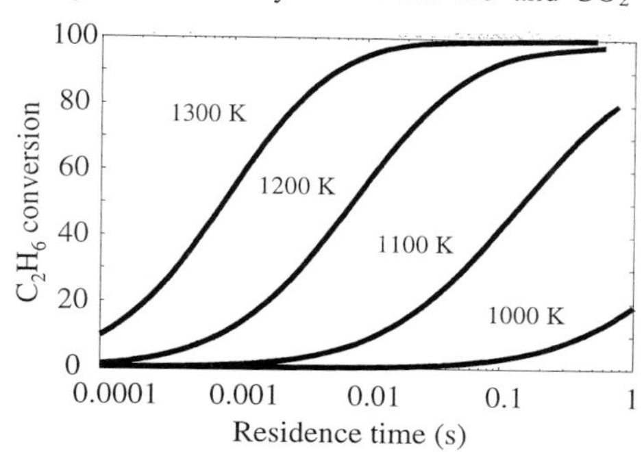

Figure 4. Computed profiles of isothermal ethane pyrolysis

In order to investigate the optimal conditions for the pyrolysis it is convenient to observe the results shown in fig. 3, where the isothermal pure pyrolysis of ethane is investigated at different temperatures. These theoretical results were obtained always with the kinetic scheme here adopted.

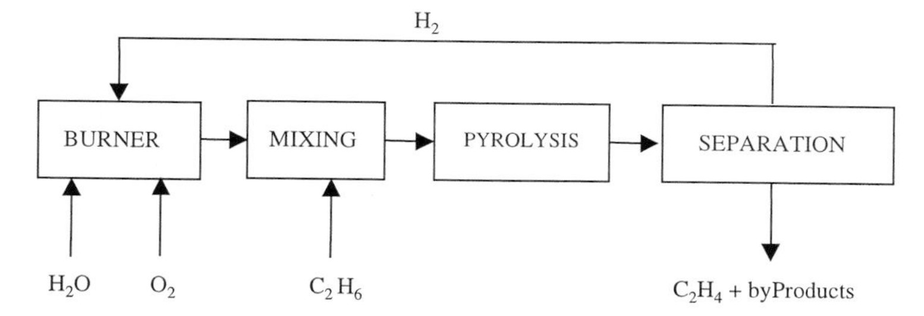

Figure 5. Sketch of the process configuration

The higher temperatures of ethane pyrolysis reduce the ethylene selectivity, in favor of larger amounts of acetylene. The higher temperatures are obviously accompanied by a strong increase in ethane conversion or, in other words, by significant reductions of the reactor contact times (fig. 4). Note that at ethane conversion greater than about 60%, ethylene selectivities show a sharp reduction due to successive condensation reactions. This behavior becomes more evident at lower temperatures. As already mentioned, the reactor temperature can be achieved by a direct injection of ethane in a hot stream, with a very effective mixing. The hot stream can be obtained by recycling and directly burning the same hydrogen produced by the process. Further steam has to be conveniently added to the hydrogen/oxygen flame to reduce the temperatures. The overall exothermicity of the reaction produces energy as byproduct of the process.

Fig 5 sketches the very simple process design in which the ethane feed is directly mixed with the hot flue gases of the H_2-O_2 flame already mixed with a proper amount of dilution steam.

Fig. 6 shows the ethylene selectivities versus ethane conversion in an adiabatic pyrolysis reactor at atmospheric pressure and different steam to reactant weight ratios, assuming the hot gases at 1600 K and the C_2H_6 feed at 1000 K.

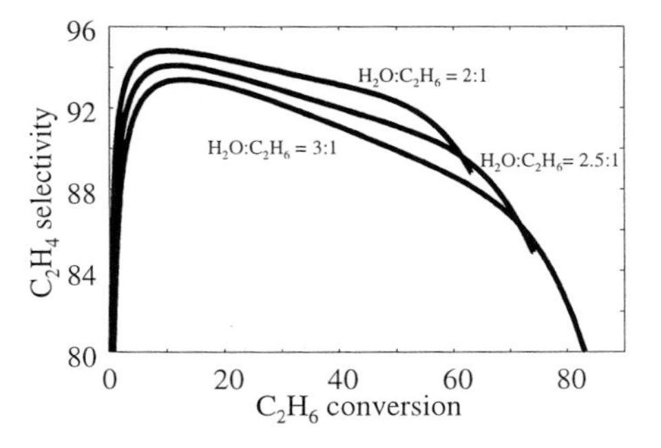

Figure 6. Computed ethylene selectivity at different steam/ feed ratios

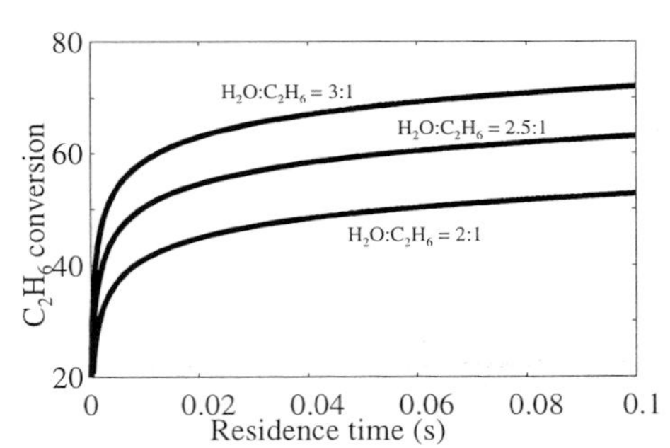

Figure 7. Computed conversion at different steam/ feed ratios

Considering an instantaneous and ideally perfect mixing, the initial pyrolysis temperatures range from about 1380 K in the 3:1 (wt) case to about 1320 K in the 2:1 (wt) case. As already discussed in the isothermal case, also in these conditions, lower temperatures give better initial ethylene selectivities. On the other side, lower temperatures increase the contact time required to obtain the same conversion (fig. 7). The 70% ethane conversion, with about 90% selectivity, is achieved in about 70 ms for the $H_2O:C_2H_6=3:1$ ratio. About 400 ms are required in the $H_2O:C_2H_6=2.5:1$ case. At the lowest steam to ethane ratio, the initial temperature is too low to convert ethane over 65% in reasonable contact times.

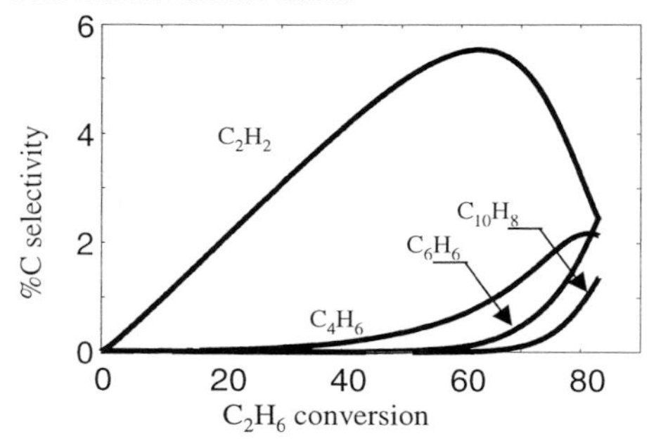

Figure 8. Selectivity of main dehydrogenated species for H2O:C2H6 = 3:1

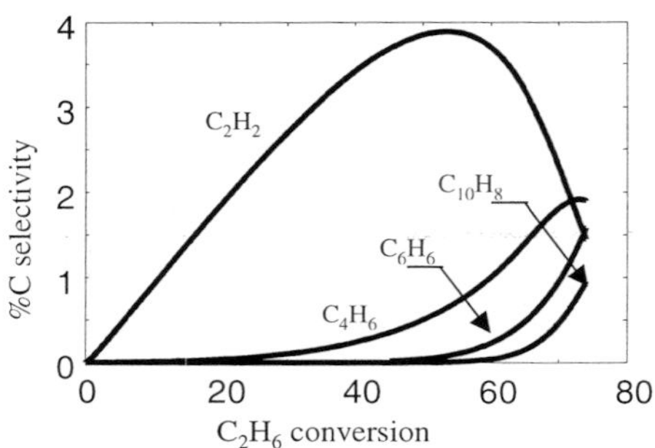

Figure 9. Selectivity of main dehydrogenated species for H2O:C2H6 = 2.5:1

The successive reactions toward more dehydrogenated and heavier species give rise to acetylene, butadiene and benzene. When the first aromatic ring is formed successive interactions with mainly vinyl radical and acetylene as well as cyclopentadiene, produce PAH, precursors of coke. The model predicts the formation of such intermediates as reported in figs. 8 and 9, respectively for the 3:1 and 2.5:1 cases. Benzene and naphthalene selectivities are always lower than 1% even for ethane conversion (i.e. residence times) longer than the optimal conditions.

In the homogeneous conditions herein discussed, the real bottleneck is the fluidynamic effectiveness of the mixing. As a matter of facts, the estimated reaction times are in the order of a few milliseconds, therefore it is necessary to realize an effective mixing in a very short time.

The splitting of the hot gases in two different streams with two successive injections (fig. 10) allows to relax the stringent boundaries of the characteristic mixing times.
In this case the first ideal mixing temperature is lower and it is possible not only to obtain higher selectivity but mainly to operate at relatively higher mixing times reducing in this way the crucial aspects of the injection step. As an example fig. 11 shows the reduction of the maximum reactor temperature of the 2 stage case when compared with the $H_2O:C_2H_6=3:1$ ratio. These results are computed for an initial mixing stage with only 2/3 of the total steam, the remaining steam being injected, always at 1600 K, after about 20 ms. Figs. 12 and 13 show that the conversion of 70% can still be achieved in about 90 ms, with a better ethylene selectivity.
The numerical modeling of the adiabatic mixing of the two reacting streams, hot flame gases and

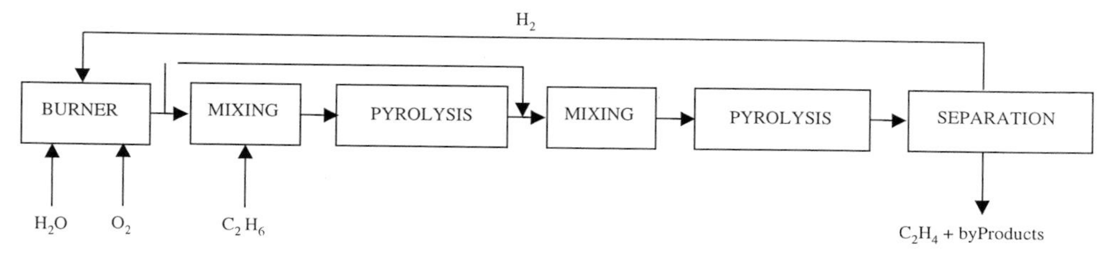

Figure 10. Sketch of the two stage mixing process configuration

C_2H_6 feed, can be conveniently addressed by the proper use of CFD codes. The presence of endothermic reactions, the need of correctly characterizing the influence of mutual interactions between turbulence and chemical reactions, the relative small dimensions of the reactor, and especially its capability to well represent the mixing phenomena make the Large Eddy Simulation (LES) technique an attractive alternative to the traditional methods for the numerical solution of the resulting flows equations (Abbà et al., 1996). The description of the numerical method is beyond the aim of this paper, nevertheless it is worth to highlight the approach we are using to manage the generation term in the species conservation equations.

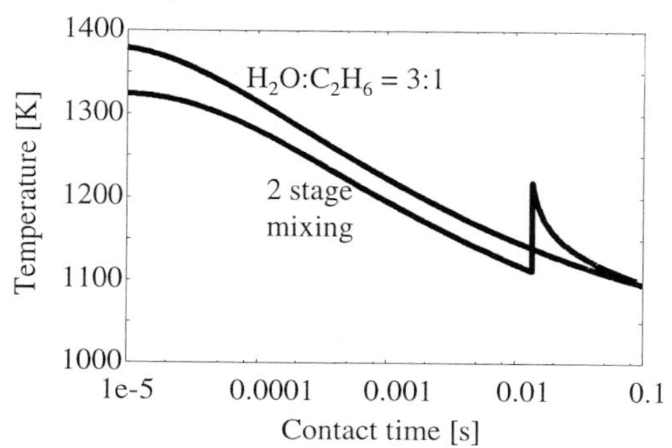

Figure 11. Reactor temperature for the 2 stage mixing

The filtered source term $\overline{\varpi}$ in the species conservation equations is:

$$\overline{\varpi} = \sum_{i=1}^{NR} \tilde{\rho} A_i \overline{e^{-E_i/RT} \prod_{j=1}^{NS} y_j^{\nu_{ij}}}$$

where $\tilde{\rho}$ is the filtered density, NS and NR are respectively the number of species and reactions, A_i, E_i, ν_{ij} the frequency factor, the activation energy and the stoichiometric coefficients of the j-th species in the i-th reaction, and y is the concentration of the chemical species.
Several authors employ a large eddy pdf (probability density function) approach, with a great computational effort (Girimaji et al., 1996).
As a first approximation, it is possible to assume:

$$\overline{e^{-E_i/RT} \prod_{j=1}^{NS} y_j^{\nu_{ij}}} \cong \overline{e^{-E_i/RT}} \prod_{j=1}^{NS-\nu_{ij}} y_j$$

and to approximate the exponential term with the Taylor expansion around $\overline{T}$

$\overline{e^{-E_i/RT}} \cong e^{-E_i/R\overline{T}}\left\{1+\left[\left(\frac{E_i}{R\overline{T}}\right)^2-2\frac{E_i}{R\overline{T}}\right]\frac{\overline{T''^2}}{\overline{T}^2}\right\}$ Since the temperature field is characterized by a spectrum similar to the energy one (Girimaji et al., 1996), it is convenient to apply the scale similarity theory and to take the subgrid-scales temperature fluctuations $\overline{T''}$ similar to turbulent resolved ones $\overline{T''^2} = \overline{\overline{T}^2} - \overline{\overline{T}}^2$

The results of this computations are only at a preliminary stage but already confirm the validity of this approach and the critical aspect of the assumption of an ideal mixing.

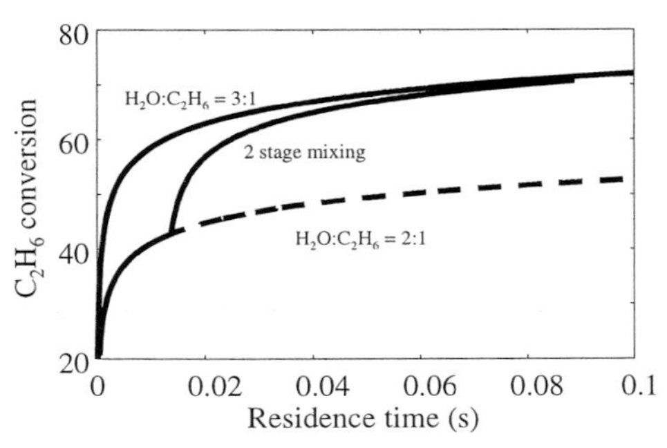

Figure 12. Computed conversion at different steam/ feed ratios and for a 2 stage mixing

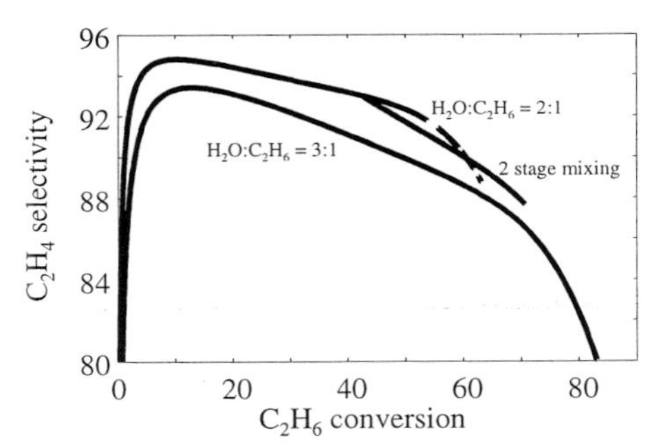

Figure 13. Computed ethylene selectivity at different steam/ feed ratios and for a 2 stage mixing

CONCLUSION

This paper presents an alternative route for ethylene production from ethane. The process, based on a hydrogen/oxygen flame in which a C_2H_6 stream is fed, is autothermal, does not require extra fuel and overall drastically reduces the environmental impact of conventional steam crackers. The emissions of greenhouse gases COx and NOx are drastically cut and also the reactor design is significantly simplified. Moreover the purely homogeneous system makes more safe the process and it avoids the presence of dangerous explosive mixtures.

However, there still remains the need of a careful investigation on the fluidynamic aspects involved in the ideal and very rapid mixing, to make technologically feasible the proposed process.

Finally, this approach could be also extended to different and heavier feedstock, like naphthas and gasoils.

REFERENCES

Abba', A., Cercignani, C., Valdettaro, L., Zanini, P. "LES of turbulent thermal convection", *Proc. of the Second ERCOFTAC* Workshop on Direct and Large Eddy Simulation, (1996)

Bodke, A.S., Olschki, D.A., Schmidt, L.D., Ranzi, E., *Science*, 285: 712 (1999)

Beretta, A., Ranzi, E., Forzatti, P., *Catalysis today*, 2238:1 (2000)

Chen, Q., Schweitzer., E.J.A., Van Den Oosterkamp, P.F., Berger, R.J., De Smet C.R.H., Marin, G.B., *Ind. Eng. Chem. Res.*, 36: 3248 (1997)

Dente, M. and Ranzi, E.. Chapter 7 in '*Pyrolysis Theory and Industrial Practice*' (L. Albright. Bryce and Corcoran Eds.) Academic Press San Diego, (1983).

Girimaji, S.S. , Zhou, Y., *Phys. of Fluids* vol. 8, pp. 1224-1236, (1996)

Ranzi, E., Dente, M., Goldaniga, A., Bozzano, G. and Faravelli, T., Lumping Procedures in Detailed Kinetic Modeling of Gasification, Pyrolysis, Partial Oxidation and Combustion of Hydrocarbon Mixtures, *Prog. Energy Combust. Sci.* (2000)

Schmidt, L.D., Siddall, J., Bearden, M., *AIChE J.*, 46, 8:1492 (2000)

Zerkle, D.K., Allendorf, M.D., Wolf, M., Deutchmann, O., *Proc. Comb. Inst.,* 28 (2000)

Studies in Surface Science and Catalysis
J.J. Spivey, E. Iglesia and T.H. Fleisch (Editors)

The role of gallium oxide in methane partial oxidation catalysts: An experimental and theoretical study

Christopher A. Cooper, Charles R. Hammond, Graham J. Hutchings, Stuart H. Taylor, David J. Willock and Kenji Tabata[†]

Cardiff University, Department of Chemistry, PO Box 912, Cardiff CF10 3TB, UK.
[†]RITE Foundation, 9-2 Kizugawadai, Kizu-Cho, Soraku-Gun, Kyoto, 619-0292, Japan.

Abstract
We present new results on the activation of methane over pure and doped gallium oxide catalysts. This material was selected from our earlier work on methane to methanol conversion over mixed oxide systems. Calculations show that for defect free surfaces of the β-Ga_2O_3 phase only the (010) and (001) facets are stable, suggesting that the real crystal surfaces will be defective. We have attempted to increase the surface concentration of defects further by doping with divalent cations. Zn doping leads to a reduction in the temperature required for methane oxidation over β-Ga_2O_3 while Mg doping increases the intrinsic activity.

1. Introduction

The identification of catalysts showing high activity and selectivity for the direct partial oxidation of methane to methanol and formaldehyde is a major research aim driven by the need to efficiently utilise vast worldwide reserves of natural gas. Many different catalyst systems have been investigated and varying degrees of success have been achieved, however, to date none have demonstrated outstanding performance[1]. Generally, catalysts operate under conditions where gas phase homogeneous reactions dominate and partial oxidation products are unstable. Based on the hydrogen balance current kinetic models predict a limiting selectivity of 67% methanol for the homogeneous process[2]. It is against this background that new approaches for the development of more effective catalysts are required.

Previously, we have reported an approach to develop new methane oxidation catalysts[3]. The approach is to combine components with known methanol combustion, methane activation and oxygen activation properties and so we have investigated activation processes over a wide range of oxides. The activation of methanol, was probed by studying methanol oxidation, as the prime requisite is that a suitable catalyst component must not readily destroy methanol. Methane activation was investigated by probing the rate of CH_4/D_2 exchange, and oxygen by $^{16}O_2/^{18}O_2$ exchange. Thus, simplistically, synergistic combinations may exist in which one component is principally responsible for methane activation and the other for oxygen activation/insertion. This is not without precedent, one of the currently accepted views of the operation of bismuth molybdate catalysts for propene oxidation is that the bismuth component initiates hydrocarbon activation while molybdenum effects oxygen activation and insertion[4].

2. Summary of previous work

Methanol stability studies show that it is most stable over the oxides MoO_3, WO_3, Nb_2O_5, Ta_2O_5 and Sb_2O_3. In particular, the conversion of methanol to carbon oxides over MoO_3 was very low whilst showing high selectivity towards formaldehyde. Additionally the oxygen exchange reaction over MoO_3 takes place with the entire bulk of the oxide, suggesting the diffusion of oxygen through the oxide is facile. This is an important concept for selective oxidation reactions in which labile lattice oxygen is the active oxygen species. Methane activation studies show that Ga_2O_3 is the most active component. Based on these results a Ga_2O_3/MoO_3 catalyst was prepared and tested for methane oxidation (Table 1).

Table 1. Methane oxidation over Ga_2O_3/MoO_3 physical mixture.

Catalyst	Temp (°C)	CH_4 Conv. (%)	Selectivity (%)				CH_3OH Yield[a]
			CH_3OH	CO	CO_2	C_2H_6	
Ga_2O_3/MoO_3	455	3.0	22	50	27	1	0.66
Ga_2O_3	455	1.5	3	27	68	2	0.05
MoO_3	455	0.3	13	69	18	-	0.04
Quartz packing	450	0.1	-	-	100	-	0.00

Notes: $CH_4/O_2/He = 23/3/5$, 15 bar, GHSV = 5,000 h^{-1}, a) Yield per pass.

Ga_2O_3/MoO_3 produced the highest yield of methanol which was greater than the comparative homogeneous gas phase reaction over a quartz chips packed reactor. The methanol yields over MoO_3 and Ga_2O_3 were also low in comparison to Ga_2O_3/MoO_3. Methane oxidation over MoO_3 showed appreciable selectivity to methanol, while Ga_2O_3 was more active, which is consistent with the high activity for methane activation probed by methane/deuterium exchange. These aspects were combined for methane partial oxidation by the physically mixed Ga_2O_3/MoO_3 system. The results obtained indicate that the design approach is valid and the concept has allowed the manufacture of a novel catalyst system based on Ga_2O_3 and MoO_3, which shows promising activity for the partial oxidation of methane to methanol.
The present study has started to investigate further the important steps in the partial oxidation reaction. The high efficacy for methane activation by Ga_2O_3 is the most significant result of these preliminary studies. The present communication details our initial results probing the relationship between the structure/chemistry of Ga_2O_3 and methane activation. Studies have used a combination of experimental and theoretical methods aiming to investigate the effect of doping Ga_2O_3, the role of defects in catalysis and the catalyst surface structure. Gallium oxide has been doped with zinc and magnesium as these have also shown promising activity for methane activation, additionally the ions are of comparable ionic radius to Ga^{3+} and their incorporation in the lattice will increase the defect concentration.

3. Experimental

3.1. Catalyst preparation

A series of gallium oxide based catalysts were prepared. β-Ga_2O_3, the thermodynamically most stable oxide of gallium, was prepared by thermal decomposition of $Ga(NO_3)_3$, by calcination in air, using a temperature programmed regime from ambient to 700°C at 20°C min^{-1}, and then maintained at 700°C for 12 hours. The zinc doped Ga_2O_3 (Zn:Ga=1:99 wt%) was obtained by a coprecipitation method. A mixed solution of gallium and zinc was prepared by dissolving appropriate quantities of the nitrates in deionised water (4.5g $Ga(NO_3)_3$, 0.033g $Zn(NO_3)_2$ in 120 ml). The resulting solution was thoroughly stirred for 30 min and dilute ammonia solution added dropwise until pH 7.5 was attained. The gelatinous precipitate was stirred in the liquor for a further 30 min and then collected by filtration. The precipitant was calcined using the same procedure adopted for β-Ga_2O_3 preparation. The magnesium doped catalyst (Mg:Ga=1:99 wt%) was prepared by wet impregnation of β-Ga_2O_3 with a solution of $Mg(NO_3)_2$, the resulting material was dried at 100°C and calcined by the same procedure described previously.

3.2. Catalyst testing
Catalyst performance was monitored in a stainless steel plug flow microreactor. Studies were performed using a methane/oxygen/helium flow of 23/2/5 ml min^{-1} with a catalyst bed volume of $0.25cm^3$, giving a GHSV=$7200h^{-1}$. Catalysts were pelleted to 0.6-1.0mm and analysis was carried out on-line using gas chromatography with thermal conductivity and flame ionisation detectors.

3.3 Theoretical
The surface structure of gallium oxide was studied via a series of computer simulations using atomistic potentials. The atomic co-ordinates were taken from an experimental determination deposited at the United Kingdom Chemical Database Service[5,6]. Initial calculations used potential parameters from the literature[7], in that work only α-Ga_2O_3 was represented which has exclusively octahedral Ga. β-Ga_2O_3 requires both octahedral (Ga_{oct}) and tetrahedral (Ga_{tet}) gallium and so a new set of potential parameters for the Ga_{tet} ion was fitted using the GULP program[8]. Data for the fitting process was generated from periodic density functional calculations on the bulk structure using the CASTEP code[9]. The bulk and surface structures were only marginally affected by this alteration so the single Ga..O potential may be adequate for both environments.
Conceptually a surface of an ionic material can be created by cleaving the bulk unit cell along any of a large number of Miller planes. The types of surface generated can be broadly classified into three groups: type I in which atomic layers parallel to the surface plane are inherently neutral and without a dipole moment perpendicular to the surface, type II in which a set of layers can be selected which do not have a dipole and form a valid stacking sequence for the bulk and type III surfaces which are inherently dipolar[10]. Simple type I and type II surfaces are electrostatically stable but type III are not. Accordingly we first identify the class of a given Miller surface before optimisation to examine the affect of the termination of the bulk on the local atomic structure.
To calculate surface energies and structures we use METADISE, a code designed for the simulation of interfaces with two dimensional periodicity[11]. The system being simulated is split into four regions, 2,1,1',2'. The interface is between regions 1 and 1' and in these regions all ions are energy minimised. In regions 2 and 2' the atomic structure is held fixed at the structure of the bulk unit cell, as calculated using the same potential parameters. A free surface can be simulated by having regions 1' and 2' empty. The surface energy, E_{surf}, is given by:

$$E_{surf} = \frac{E - \frac{1}{2}E_{bulk}}{S} \quad (3)$$

where E is the energy from the surface calculation, E_{bulk} is the energy calculated with all four regions filled with the oxide structure (no interface) and S is the area of the 2-D unit cell. The surface energies reported here have been converged for the sizes of regions 1 and 2.

4. Results and discussion

4.1. Methane oxidation studies

The activation of methane has been followed by studies of methane oxidation. A summary of results is shown in table 2, these were obtained under highly oxidising conditions and the only products observed were carbon monoxide and carbon dioxide. Data were obtained at low conversion in the kinetic regime and mass and heat transfer limitations were minimised. Studies to probe methane activation using the methane/deuterium isotope exchange reaction were conducted under reducing conditions and the data are not directly comparable.

Table 2. Methane oxidation results

Catalyst	Temperature/°C	Conversion/%	Selectivity/%	
			CO	CO_2
β-Ga_2O_3	300	0	-	-
	400	0.20 ± 0.02	-	100
	490	1.51 ± 0.04	-	100
	580	3.97 ± 0.05	1	99
Zn/β-Ga_2O_3	250	0	-	-
	300	0.17 ± 0.02	-	100
	400	2.70 ± 0.06	1	99
	490	6.12 ± 0.08	3	97
Mg/β-Ga_2O_3	300	0	-	-
	400	0.14 ± 0.02	-	100
	490	3.43 ± 0.07	-	100
blank	300	0	-	-
	400	0.36 ± 0.03	-	100
	490	0.84 ± 0.04	-	100
	580	5.80 ± 0.08	-	100

Notes: $CH_4/O_2/He$ = 23/2/5, GHSV = 7200 h^{-1}. Errors calculated from std. of 5 data points.

Table 3. Methane oxidation rates at 490°C normalised for the effect of surface area

Catalyst	Surface area/m^2g^{-1}	rate/mol s^{-1} m^{-2}
β-Ga_2O_3	2	5.73×10^{16}
Zn/β-Ga_2O_3	16	3.68×10^{16}
Mg/β-Ga_2O_3	1	9.17×10^{17}
blank	-	8.08×10^{16} (mol s^{-1})

The results relating to methane activation in the present system are somewhat inconclusive as there is not an unambiguous increase in the rate of methane oxidation compared with the homogeneous reaction, when any of the gallium based catalysts were employed. However, there are several points to note from the data. It is apparent that the activity of *β*-Ga_2O_3 and the empty tube are broadly similar, both demonstrate initial activity around 400°C with similar product distributions over the active temperature range. The *β*-Ga_2O_3 demonstrates improved ability to activate methane at 490°C whilst ultimately the activity of the empty reactor is greater at 580°C. The lower activity of β-Ga_2O_3 at 580°C relative to the empty tube is not ascribed to deactivation of the catalyst. Rather it is most likely due to a marked reduction in residence time in the heated zone of the reactor for β-Ga_2O_3 compared to the empty tube[12]. The Mg/Ga_2O_3 catalyst also showed initial activity at 400°C with a higher methane conversion of 3.4%. Perhaps the most interesting results were obtained for Zn/β-Ga_2O_3, which showed initial activity *ca.* 100°C lower than either any of the other catalysts or the empty tube, furthermore the conversion at equivalent temperatures is higher than all the other systems evaluated. The lower initial temperature for activity demonstrates the ability to activate methane is enhanced by doping β-Ga_2O_3 with zinc. To account for differences in surface area, the methane oxidation rates at 490°C normalised per unit surface area are shown in table 3. With respect to the normalised rate the Mg/Ga_2O_3 catalyst is the most active by a significant margin. It is therefore evident that the goal of increasing the activity of β-Ga_2O_3 for methane activation by doping to increase the defect capacity of the oxide can be achieved.

4.2. Theoretical studies

A simple analysis of the low index planes of β-Ga_2O_3 based on the experimental structure showed that there is only a single type I surface, the (010), this is composed of a stoichiometric layer of ions and so is charge and dipole neutral. When the bulk cell is relaxed we find other surfaces of type II but only the (001) has stability comparable to the (010). The calculated converged surface energies using the new potentials are given in table 4, the values differ only in the second decimal place from those obtained using the original potentials.

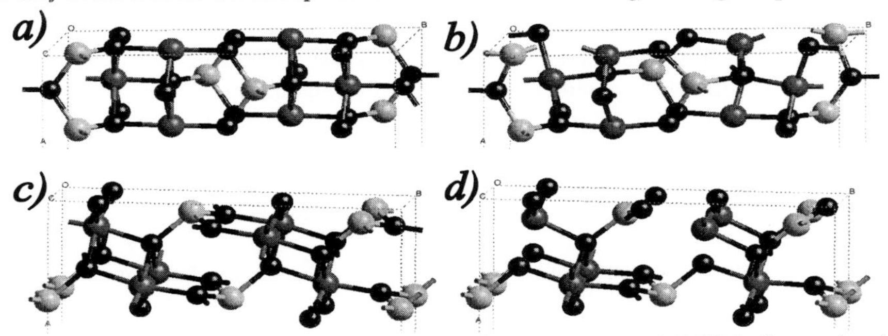

Figure 1 The affect of energy minimisation on surface structures. *a)* (010) bulk termination, *b)* (010) after optimistion, *c)* (001) bulk termination, *d)* (001) after optimistion. Atom shading: O black, Ga_{oct} dark grey Ga_{tet} light grey. Only the top section of region 1 is shown for clarity.

On optimisation the atomic structure of each surface changes considerably compared to the bulk termination starting point. The largest atomic movements in the three cartesian directions (z perpendicular to the surface) are reported in table 4 and the bulk termination and relaxed structures are compared in figure 1. For the (010) surface the bulk termination consists of a stoichiometric layer of co-planer atoms, figure 1a. In this respect, it is similar to the {001} surfaces of rock salt structure oxides such as MgO for which relaxation has very little affect[11].. The (010) β-Ga_2O_3 surface, however, distorts when energy minimised as shown in figure 1b. In general surface oxygen ions have moved out of the surface into the vacuum gap and surface Ga^{3+} ions have moved toward the bulk. The exposed Ga_{oct} ions in this surface have lost 2 neighbouring oxygen anions and this re-arrangement results in bulk Ga_{oct} ions becoming near tetrahedral in the surface. In the (001) surface only Ga_{tet} ions are exposed on the surface (figure 1c) and there is the same general movement as observed for (010). This results in the exposed surface after relaxation being purely oxygen terminated, figure 1d.

Table 4 Comparison of results on bulk and surface structures for potential models.

Bulk results	a (Å)	b (Å)	c (Å)	β (degrees)	Lat. Energy (kJ mol^{-1})
Experiment[6,13]	12.230	3.040	5.800	103.7	60 880
Single Ga pot.	12.213	3.092	5.798	103.4	59 280
New Ga_{oct} Ga_{tet}	12.215	3.094	5.803	103.4	59 250
Surface results	x_{max} (Å)	y_{max} (Å)	z_{max} (Å)		E_{surf} (Jm^{-2})
(010)	0.84	0.83	0.48		0.93
(001)	0.24	0.00	0.67		1.29

5. Conclusions

These results suggest that for defect free surfaces the (010) plane of β-Ga_2O_3 will be dominant in the crystal habit. Although the (010) surface is made up of stoichiometric layers the degree of atomic relaxation on minimisation is considerable leading to a rumpling of the surface. In both the (010) and (001) relaxations anions are exposed and become more accessible to adsorbates. Studies on related catalysts have shown the importance of defects, particularly O^- species for the oxidation of methane[14]. The calculated stability and composition of the clean surface lead us to infer that defect sites will also be important in β-Ga_2O_3 catalysed activation of methane and so catalysts with divalent cation dopants were produced.

Experimental tests show that the temperature required for methane oxidation over β-Ga_2O_3 is lowered in the presence of Zn and that the intrinsic activity is increased by Mg doping.

References

1. T.J.Hall, J.S.J.Hargreaves, G.J.Hutchings, R.W.Joyner and S.H.Taylor, Fuel Proc. Tech., 42 (1995) 151.
2. H.D. Gesser, Hunter, Methane Conversion by Oxidative Processes: Fundamentals and Engineering Aspects, Ed. E.E. Wolf, van Nostrand Reinhold, New York, (1988) 403.
3. G.J. Hutchings, S.H. Taylor, Catal. Today. 49 (1999) 105.
4. J.D. Burrington, C.T. Kartisch, R.K. Grasselli, J. Catal., 87 (1984) 363.
5. D.A.Fletcher,R.F.McMeeking and D.Parkin, J. Chem. Inf. Comp. Sci., 36 (1996) 746.
6. S.Geller, J. Chem.Phys., 33 676 (1960).
7. T.S. Bush, J.D. Gale, C.R.A. Catlow and P.D. Battle, J.Mater.Chem., 4 (1994) 831.
8. J.D.Gale, J.Chem.Soc.Farad. Trans., 93(4) (1997) 629.
9. CASTEP 3.9 academic version, 1999; Rev. Mod.Phys., 64 (1992) 1045.
10. S.C.Parker, E.T.Kelsey, P.M.Oliver and J.O.Titiloye, Farad. Discuss., 95 (1993) 75.
11. G.W. Watson, E.T. Kelsey, N.H.de Leeuw, D.J. Harris and S.C. Parker, J.Chem.Soc., Farad. Trans., 92(3) (1996) 433.
12 J.W.Chun, R.G.Anthony, Int. Eng. Chem. Res., 32 (1993) 259.
13. CRC Handbook of Chemistry and Physics, Ed. D.R.Lide 82nd Edition (1999).
14. R. Orlando, R. Millini, G. Perego and R. Dovesi, J.Mol.Catal.A, Chem., 119 (1997) 253.

Studies in Surface Science and Catalysis
J.J. Spivey, E. Iglesia and T.H. Fleisch (Editors)

LOW TEMPERATURE ROUTES FOR METHANE CONVERSION, AND AN APPROACH BASED ON ORGANOPLATINUM CHEMISTRY

Jay A. Labinger

Beckman Institute and Arnold and Mabel Beckman Laboratories of Chemical Synthesis, California Institute of Technology, Pasadena, California 91125, USA

1. BACKGROUND

The low-temperature, direct conversion of methane presents potential advantages as well as disadvantages in comparison to more traditional high-temperature routes. Foremost among the advantages are possibilities for higher selectivity as a consequence of alternate reaction mechanisms. High-temperature reactions of methane — catalytic and homogeneous oxidation as well as non-oxidative pyrolyses — are dominated by C-H homolysis leading to radical intermediates. Since reaction rates for C-H homolysis are strongly connected to the C-H bond strength, which is higher for methane than for most of the products derived therefrom, the products will be more reactive than methane itself. In other words, the thermodynamically favored products, CO_2 and coke respectively, will almost inevitably be favored by kinetic considerations as well, making it extremely difficult, if not impossible, to achieve good conversion and selectivity simultaneously.

This consideration alone can be used to make semi-quantitative predictions of performance ceilings, which we showed some years ago will be on the order of 30% yield for methane oxidative coupling and 5% yield for oxidation to methanol,[1,2] numbers which have not been (reliably) exceeded experimentally, and which appear to be substantially below economically practical levels. Similar estimations have been extended to a wide range of hydrocarbon oxidations.[3] Of course, this consideration does not always rule out success. For example, one would expect that selective oxidation of propane to acrolein (which has a very weak C-H bond) would not be feasible, whereas reasonable yields have in fact been achieved.[4] But there has been no such breakthrough to date for methane.

Alternatively, one can aim for transformations where the product does *not* have a weaker C-H bond than the reactant, as in the selective oxidation of butane to maleic anhydride. An example involving methane is the dehydrogenative coupling/cyclization to benzene over catalysts based on Mo/ZSM-5,[5] which makes sense as the C-H bond of benzene is *stronger* than that of methane. But even that favorable case can only be operated at "moderate" temperatures (600-700°C) where conversion is thermodynamically limited to around 10%; at higher temperatures severe coking sets in. So the approach of seeking an alternative mechanism, by which desirable methane-derived products are not necessarily more reactive than methane, could be much better suited to optimizing yields.

On the other hand, there are clearly potential disadvantages as well. In particular, most non-oxidative transformations of methane are thermodynamically uphill at low temperatures, while oxidative reactions are incompatible with many of the chemical systems that have been found to activate methane under mild conditions. Process issues are likely to be problematic as well. In the next section, I offer a perspective on recent developments relevant to low-temperature methane conversion, and try to assess their implications for future promise. Following those general remarks I will present some recent findings from our work on platinum complexes, and their implications for low-temperature direct oxidation of methane to methanol.

2. LOW-TEMPERATURE REACTIONS OF METHANE AND RELATED CHEMISTRY

Most of the promising work here involves reactions of transition metal complexes in solution. While there are other cases of low-temperature methane activations, such as reactions in superacidic media[6] and oligomerization on metal surfaces,[7] none of these yet appears to lead to any practical approach. Transition metal chemistry of methane and other alkanes can be conveniently divided into two classes: organometallic, where activation generally involves reaction directly at a metal center; and biological/biomimetic, where oxygen attached to metal seems to be the reacting center. To be sure, there are ambiguous and uncertain cases as well.

2.1 Biological/biomimetic activation

The existence of a class of bacterial enzymes — the methane monooxygenases (MMO) — that catalyze the oxidation of methane selectively to methanol under ambient conditions establishes that it *is* possible. Direct use of bacteria for methane conversion is not too promising: reaction is slow and affords a dilute aqueous solution of methanol which would have to be separated. A prime goal is to design an artificial *biomimetic* catalyst that would effect the same conversion but be more amenable to a practical process, but there are difficulties. First, the stoichiometry of the enzymatic oxidation requires a sacrificial reductant, as only one of the two atoms of O_2 is incorporated into product (Eqn. 1).

$$CH_4 + O_2 + 2\,e^- + 2\,H^+ \xrightarrow{\text{MMO}} CH_3OH + H_2O \qquad (1)$$

Second, the mechanism of methane activation, and the key structural features of the catalyst that are responsible for it, are not at all well understood. In fact, there are (at least) two different classes of MMO, a soluble version that has a non-heme iron active site and a membrane-bound version that appears to involve a copper-based site.[8] The former has been much more extensively studied; the crystal structure is known[9] as well the sequence of steps that leads from the resting form plus O_2 to the probable active form.[10] But the detailed nature of the C-H bond cleaving step has not been definitively characterized.

Another class of enzymes that oxidizes alkanes (but not methane), the cytochromes P-450, has long been thought to react via H-atom abstraction (Eqn. 2).

P-450 often exhibits higher selectivity for more substituted positions, which might be consistent with such a route, whereas MMO prefers terminal positions and especially methane. Hence one might think that a different mechanism for MMO is responsible for the high activity towards methane. More recently, though, it has been shown that the behavior of "radical clock" substrates is *inconsistent* with a radical route for both P-450 and MMO.[11] This and other evidence suggest that the two enzymes, structurally quite different, react by closely related mechanisms, and that the differences in reactivity and selectivity are enforced by the detailed structure of the enzyme surrounding the active site — features that will be very difficult to reproduce artificially.

$$L_nM{=}O + RH \longrightarrow L_nM\text{-}OH + R\cdot \longrightarrow L_nM + ROH \qquad (2)$$

As for non-biological models, there have been many, some structurally analogous to MMO[12] and others not, that catalyze alkane oxidations. But none to date has been found to exhibit a central feature of MMO, preferential reactivity at methane (most won't activate methane *at all*) or even at terminal positions. And in those few cases where detailed mechanistic information is available, it tends to suggest a mechanism involving free radicals,[13] not the one followed by the enzyme (whatever that is). A truly effective biomimetic methane oxidation catalyst is yet to be devised.

2.2 Organometallic activation

Over the last 20 years or so a large number of organometallic systems have been found to activate C-H bonds in alkanes under remarkably mild conditions, leading to new stable metal carbon bonds.[14] The vast majority exhibit one of two stoichiometric patterns, termed *oxidative addition* (Eqn. 3) and *sigma-bond metathesis* (Eqn. 4). We will focus on the former, since the latter appears to offer fewer opportunities for useful conversions, as it generally just exchanges one M-C bond for another.

$$L_nM + RH \longrightarrow L_nM(R)(H) \qquad (3)$$

$$L_nMR + R'H \longrightarrow L_nMR' + RH \qquad (4)$$

Oxidative addition reactions are characteristic of electron-rich, low-valent complexes of the late transition metals. The mechanism clearly does *not* involve C-H homolysis, as there is no correlation with bond strength: aromatic and olefinic compounds are almost always more reactive than saturated ones, and alkanes tend to give terminal alkyl products, often exclusively so. The latter observation would appear to be encouraging for methane activation, which indeed has been demonstrated in a number of cases. But there are complications. Reactions have been shown to proceed in two steps — initial formation of an alkane "sigma" complex followed by C-H bond cleavage — and there is some evidence that the first step is actually *dis*favored for lighter alkanes.[15] When coupled with the low solubility of methane, it is not so clear that methane will necessarily be the most reactive substance in such a system.

A more serious issue is how to proceed to a useful transformation. Having made a species such as $L_nM(R)(H)$, an organometallic chemist would envision inserting something into the metal-carbon bond — such as CO, CO_2, an olefin — leading to overall converion of methane to acetaldehyde, acetic acid, or a higher alkane respectively. But the first two are thermodynamically uphill and can't be run catalytically under the low temperatures for which these systems are suited. The last is thermodynamically allowed, but doesn't look terribly attractive economically, and in any case the olefin is likely to out-compete methane for the C-H bond activation reaction, as noted above. Oxidative transformations seem much more desirable both economically and thermodynamically; but the electron-rich metal centers that effect such chemistry are almost invariably subject to oxidative degradation.

A class of *functionalization* reactions that have been called electrophilic activation (Eqn. 5) appears much more promising.[16] The earliest example was the aqueous platinum system for alkane hydroxylation first reported by Shilov and coworkers (Eqn. 6).[17] Work by our group[18] and others has established the reaction sequence shown in Scheme 1. This system can convert methane to methanol with modest selectivity; estimates for the relative reactivity of methane:methanol range from greater than one[2] to around 0.2.[19] However, as shown it is stoichiometric in Pt; to be useful obviously a cheaper oxidant must be found to recycle the catalyst.

$$L_nMX_2 + RH \longrightarrow L_nM + RX + HX \qquad (5)$$

$$RH + PtCl_6^{2-} + H_2O \xrightarrow[120^\circ]{PtCl_4^{2-}} ROH + PtCl_4^{2-} + 2\,HCl \qquad (6)$$

There have been several reports; by far the most successful is the oxidation of methane by sulfuric acid to methyl bisulfate, catalyzed first by mercuric salts[20] and subsequently by a platinum complex (Scheme 2).[21] The latter affords product in over 70% yield, which implies in turn that methane is *much* more reactive than methyl bisulfate in this system! Nonetheless, at present it does not appear that an overall process — integrating the functionalization with hydrolysis of methyl bisulfate to methanol, oxidation of SO_2 to sulfuric acid and recycling the latter — is economically favorable. Ideally one would like to use O_2 to go directly from methane to a useful product such as methanol, and still maintain the kinds of performance obtained in the sulfuric acid reactions. Whether this is feasible remains to be demonstrated. Our approaches are described in the following section.

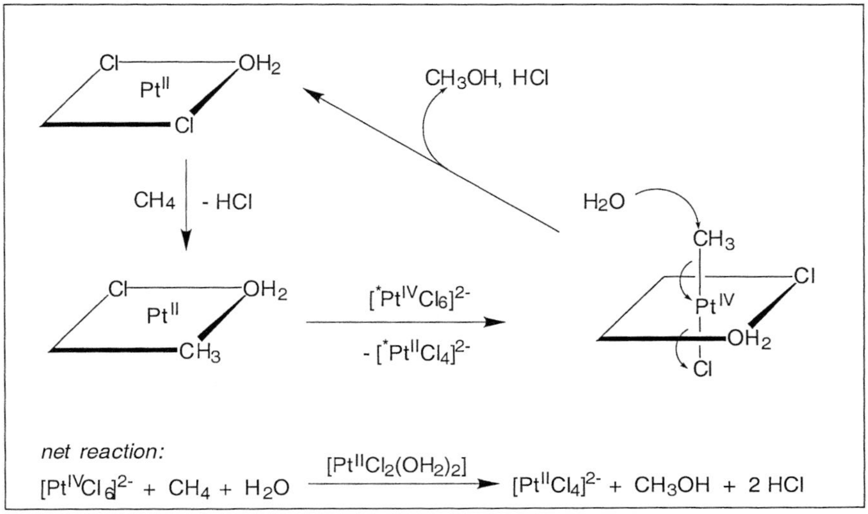

CH_3OH, HCl

CH_4 - HCl

$[^{*}Pt^{IV}Cl_6]^{2-}$

$- [^{*}Pt^{II}Cl_4]^{2-}$

H_2O

net reaction:

$$[Pt^{IV}Cl_6]^{2-} + CH_4 + H_2O \xrightarrow{[Pt^{II}Cl_2(OH_2)_2]} [Pt^{II}Cl_4]^{2-} + CH_3OH + 2\ HCl$$

Scheme 1

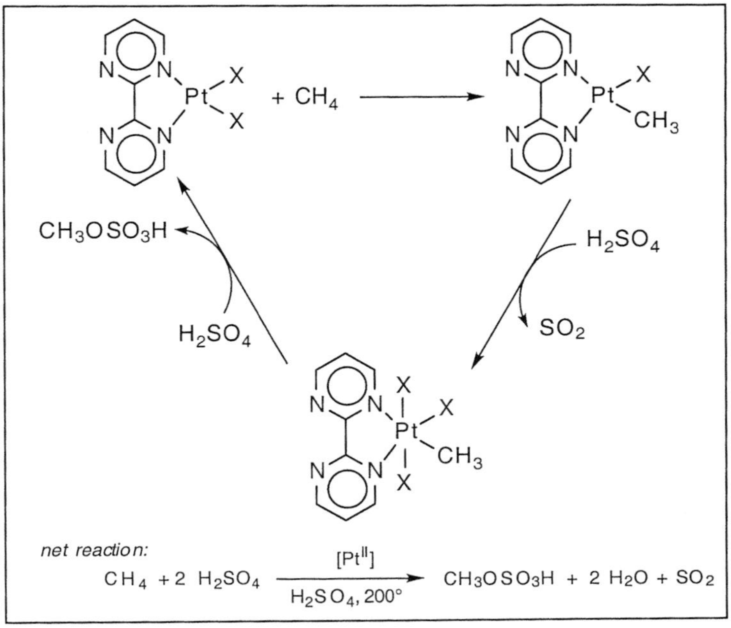

$+ CH_4$

H_2SO_4

SO_2

CH_3OSO_3H

H_2SO_4

net reaction:

$$CH_4 + 2\ H_2SO_4 \xrightarrow[H_2SO_4,\ 200°]{[Pt^{II}]} CH_3OSO_3H + 2\ H_2O + SO_2$$

Scheme 2

3. TOWARDS A PLATINUM-BASED CATALYTIC METHANE OXIDATION

Our starting point is the Shilov system mentioned above. An extensive series of (ongoing) mechanistic studies has provided insight into the key factors required for catalytic methane oxidation.

3.1. C-H activation mechanism

A combination of kinetics and isotopic labeling studies shows that C-H activation proceeds via an oxidative addition-deprotonation sequence (Eqn. 7).[22] A crucial aspect is that the Pt(II)-alkyl can be oxidized to a Pt(IV)-alkyl (Scheme 1) more rapidly than the reverse, protonolytic cleavage of the Pt-C bond.[23] The rate of C-H activation depends strongly on the steric and electronic properties of the ligands on Pt; for example, the rate of reaction of benzene with a series of cationic complexes $[(diimine)Pt(CH_3)(H_2O)]^+$ increases as the diimine ligand becomes more electron-donating.[24]

$$L_2Pt(X)(Y) + RH \longrightarrow L_2Pt(X)(Y)(R)(H) \longrightarrow L_2Pt(Y)(R) + HX \qquad (7)$$

3.2 Oxidation by O_2

The oxidation of Pt(II) to Pt(IV), which is central to the overall methane functionalization (Scheme 1) and requires $PtCl_6^{2-}$ as stoichiometric oxidant in the original Shilov system, can be effected by O_2 for a group of dimethyl-Pt(II) complexes.[25] Mechanistic studies show that this oxidation proceeds via the sequence of steps shown as Eqns. 8 and 9.[26] It is important that these are strictly two-electron transformations, avoiding any one-electron (radical) steps that are common in autoxidations, and that would be highly detrimental to selectivity.

$$L_2Pt(CH_3)_2 + O_2 \xrightarrow{CH_3OH} L_2Pt(CH_3)_2(OCH_3)(OOH) \qquad (8)$$

$$L_2Pt(CH_3)_2(OCH_3)(OOH) + L_2Pt(CH_3)_2 \longrightarrow 2\ L_2Pt(CH_3)_2(OCH_3)(OH) \qquad (9)$$

3.3 Towards catalytic oxidation

Based on the above and related findings we can formulate a hypothetical cycle for catalytic methane oxidation (Scheme 3). Each step (or a close analog) of this cycle has been demonstrated (though not all with the same platinum complex) to take place under mild conditions, 120°C or less in aqueous or alcoholic solutions. Our ability to "tune" the reactivity for at least some of the steps by varying the nature of the ligands on Pt suggests it may well be possible to design a system that can effect *all* of them — for example, both the C-H activation and the oxidation by O_2 are apparently enhanced by increasing electron density. As noted, in this scheme O_2 reacts with Pt by a two-electron mechanism, rather than with organic species by 1-electron steps, so high selectivity may be anticipated. All in all, the results to date offer considerable encouragement for the prospects of this approach to a practical, low-temperature route for selective catalytic oxidation of methane to methanol.

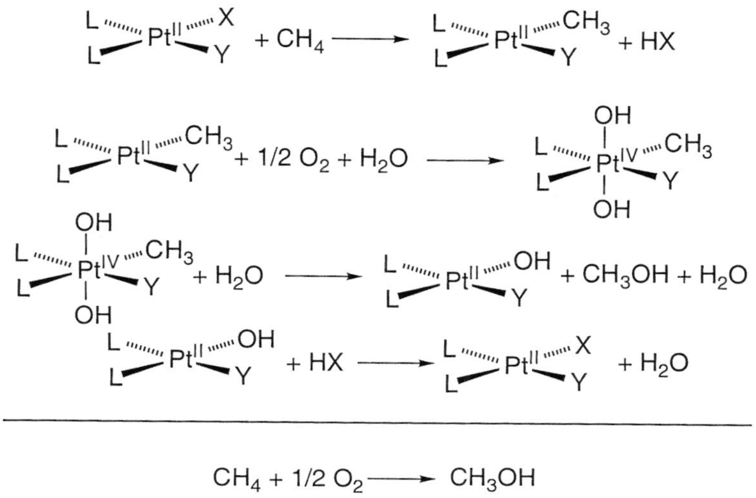

$$L_2Pt^{II}(X)(Y) + CH_4 \longrightarrow L_2Pt^{II}(CH_3)(Y) + HX$$

$$L_2Pt^{II}(CH_3)(Y) + 1/2\ O_2 + H_2O \longrightarrow L_2Pt^{IV}(OH)_2(CH_3)(Y)$$

$$L_2Pt^{IV}(OH)_2(CH_3)(Y) + H_2O \longrightarrow L_2Pt^{II}(OH)(Y) + CH_3OH + H_2O$$

$$L_2Pt^{II}(OH)(Y) + HX \longrightarrow L_2Pt^{II}(X)(Y) + H_2O$$

$$CH_4 + 1/2\ O_2 \longrightarrow CH_3OH$$

Scheme 3

ACKNOWLEDGMENTS

The recent work summarized in part 3 above is a collaborative project with my colleague John Bercaw carried out by students and postdocs Seva Rostovtsev, Antek Wong-Foy, John Scollard, Annita Zhong, Christoph Balzarek, Joseph Sadighi, and Lars Johansson (a student of Mats Tilset at the University of Oslo). Financial support was provided by Akzo Nobel, the US DOE through an OIT grant, the NSF, and BP.

REFERENCES

1. J. A. Labinger, Catal. Lett., 1 (1988) 371.
2. J. A. Labinger, J. E. Bercaw, G. A. Luinstra, D. K. Lyon and A. M. Herring, in Natural Gas Conversion II: Proceedings of the Third International Gas Conversion Symposium, Sydney, July 4-9, 1993 (Eds. R. F. Howe and E. Curry-Hyde), Elsevier, Amsterdam, 1994, p. 515-20.
3. F. E. Cassidy and B. K. Hodnett, CATTECH, 2 (1998) 173.
4. M. Baerns and O. Buyevskaya, Catal. Today, 45 (1998) 13.
5. J. Z. Zhang, M. A.Long and R. F. Howe, Catal. Today, 44 (1998) 293.
6. G. A. Olah, in Activation and Functionalization of Alkanes (Ed. C. L. Hill), Wiley, New York, 1989, chapter 3.
7. T. Koerts and R. A. van Santen, J. Mol. Catal., 74 (1992) 185.
8. H. H. T. Nguyen, A. K. Shiemke, S. J. Jacobs, B. J. Hales, M. E. Lidstrom and S. I. Chan, J. Biol. Chem., 269 (1994) 14995.

9. A. C. Rosenzweig, C. A. Frederick, S. J. Lippard, and P. Nordlund, Nature, 366 (1993) 537.
10. A. M. Valentine, S. S. Stahl and S. J. Lippard, J. Am. Chem. Soc., 121 (1999) 3876.
11. S.-Y. Choi, P. E. Eaton, D. A. Kopp, S. J. Lippard, M. Newcomb and R. Shen, J. Am. Chem. Soc., 121 (1999) 12198.
12. H. Zheng, S. J. Yoo, E. Münck and L. Que, J. Am. Chem. Soc., 122 (2000) 3789.
13. S. Kiani, A. Tapper, R. J. Staples and P. Stavropoulos, J. Am. Chem. Soc., 122 (2000) 7503.
14. B. A. Arndtsen, R. G. Bergman, T. A. Mobley and T. H. Peterson, Acc. Chem. Res., 28 (1995) 154, and references cited therein.
15. B. K.McNamara, J. S. Yeston, R. G. Bergman, and C. B. Moore, J. Am. Chem. Soc., 121 (1999) 6437.
16. S. S. Stahl, J. A. Labinger and J. E. Bercaw, Angew. Chem. Int. Ed., 37 (1998) 2180.
17. A. E. Shilov, Activation of Saturated Hydrocarbons by Transition Metal Complexes, Reidel, Boston, 1984.
18. G. A. Luinstra, L. Wang, S. S. Stahl, J. A. Labinger and J. E. Bercaw, J. Organomet. Chem., 504 (1995) 75, and references cited therein.
19. A. Sen, M. A. Benvenuto, M. R. Lin, A. C. Hutson and N. Basickes, J. Am. Chem. Soc., 116 (1994) 998.
20. R. A. Periana, D. J. Taube, E. R. Evitt, D. G. Loffler, P. R. Wentrcek, G. Voss and T. Masuda, Science, 259 (1993) 340.
21. R. A. Periana, D. J. Taube, S. Gamble, H. Taube, T. Satoh and H. Fujii, Science, 280, (1998) 560.
22. S. S. Stahl, J. A. Labinger and J. E. Bercaw, J. Am. Chem. Soc., 118 (1996) 5961; M. W. Holtcamp, L. M. Henling, M. W. Day, J. A. Labinger and J. E. Bercaw, Inorg. Chim. Acta, 270 (1998) 467.
23. L. Wang, S. S. Stahl, J. A. Labinger and J. E. Bercaw, J. Mol. Catal., 116, (1997) 269.
24. L. Johansson, M. Tilset, J. A. Labinger and J. E. Bercaw, J. Am. Chem. Soc., 122 (2000) 10846; H. A. Zhong, J. A. Labinger and J. E. Bercaw, unpublished results.
25. V. V. Rostovtsev, J. A. Labinger, J. E. Bercaw, T. L. Lasseter and K. I. Goldberg, Organometallics, 17 (1998) 4530.
26. V. V. Rostovtsev, J. A. Labinger and J. E. Bercaw, unpublished results.

Studies in Surface Science and Catalysis
J.J. Spivey, E. Iglesia and T.H. Fleisch (Editors)

NAS (novel aluminosilicates) as catalysts for the selective conversion of propane to fuels and chemicals – effects of crystallinity on catalytic behaviour

C P Nicolaides[a], N P Sincadu[b] and M S Scurrell[b]

[a] Chemical Process Engineering Research Institute/National Centre for Research and Technology-Hellas, P O Box 361, 570 01 Thermi-Thessaloniki, Greece

[b] Molecular Sciences Institute, School of Chemistry, University of the Witwatersrand, P O Wits, Johannesburg 2050, South Africa

The behaviour of a series of ZSM-5-based NAS (novel aluminosilicate) catalysts, with crystallinities (defined by an XRD-based method) ranging between 3 and 100%, for the conversion of propane to aromatic products (mainly benzene, toluene and xylenes (BTX)) is reported for the first time. A pronounced maximum in both conversion and selectivity to BTX is found with the H^+-form of the materials at intermediate crystallinities of about 60 – 80%. The behaviour is modified by the incorporation of zinc(II) by solid-state ion-exchange, in that a maximum conversion is still seen, but at higher conversion levels, as a function of crystallinity, at slightly lower crystallinity values of 55 – 70%. However, the fall in conversion at higher crystallinities is less marked compared with the catalysts in the H^+-form. A similar situation to that found for the conversion is also found for the BTX selectivity of the Zn^{2+}-exchanged and the H^+-exchanged forms. The Zn^{2+}-exchanged solids exhibit activities and selectivities that are among the highest reported for such catalytic systems.

1. INTRODUCTION

The behaviour of aluminosilicate materials having varying degrees of crystallinity as determined by means of XRD methods in catalytic reactions has not been extensively reported. Our recent work [1,2] on NAS materials (novel aluminosilicates) clearly demonstrated that the catalytic properties of partially crystalline, and in some cases substantially amorphous, samples of zeolite-based solids were a strong function of crystallinity. This was found for the catalytic skeletal isomerization of the n-butenes. For preparations of solids of the ZSM-5 type, it was found that maximum yield of isobutene (single-pass basis) was obtained at a crystallinity of 2%. For a series of partially crystalline ferrierite-based materials as catalysts for the same reaction, maximum yield was obtained with samples exhibiting 24% crystallinity. The behaviour of these NAS catalysts in other acid-catalysed systems, such as the aromatization of propane has not been previously reported. The aromatization process has been well studied on catalysts of the ZSM-5 type [3-8] and the beneficial effects of incorporating zinc [3,4,8-14] or gallium [15-20] are well known. The reaction is of direct interest in natural gas conversion technology since the light alkanes associated with methane could be a considerable source of aromatic products. Alkanes could also arise from methane itself by oxidative coupling [21-23] or other reactions such as homologation [24] or alkylation [25,26].

2. EXPERIMENTAL

Highly crystalline batches of ZSM-5 were prepared according to our previously described methods [1,2]. Products with different crystallinity levels, and with the same nominal SiO_2/Al_2O_3 ratio of 70, based on the molar amounts of reagents used in the synthesis, were obtained by varying the hydrothermal synthesis temperature employed in each of these preparations [1,2]. The method used to determine the crystallinity (strictly speaking, the %XRD crystallinity) of individual samples has also been fully described [1]. After hydrothermal synthesis the samples were calcined at 630°C for 3.5h, ion-exchanged with 1M NH_4Cl solution three times at room temperature and subsequently washed with deionised water to remove the chloride and other non-desirable ions. The H^+ - form of the zeolite was obtained by calcination of the NH_4^+ -form at 550°C for 3h. The Zn^{2+}-form of the zeolite was obtained using a solid-state ion-exchange (SSIE) procedure similar to that used by Karge [27]. A homogeneous physical mixture of zinc nitrate and the ZSM-5-based material was heated at 500° C for 3h in a quartz reactor under nitrogen. A zinc loading of 2.8 mass% was used.

The conversion of propane was studied using a stainless steel tubular reactor under the following conditions: propane was passed through the reactor at a rate of $20cm^3min^{-1}$, (with a flow of nitrogen also at $20cm^3$ min^{-1} in the case of the Zn^{2+}-ZSM-5-based samples), using a reaction temperature of 530°C. The reactor bed comprised 0.5g zeolite diluted with 2.0g silica gel (Grace type 432). Analysis of the exit stream was carried out using on-line gas chromatography (Porapak Q column, detection by FID). The selectivity and conversion data

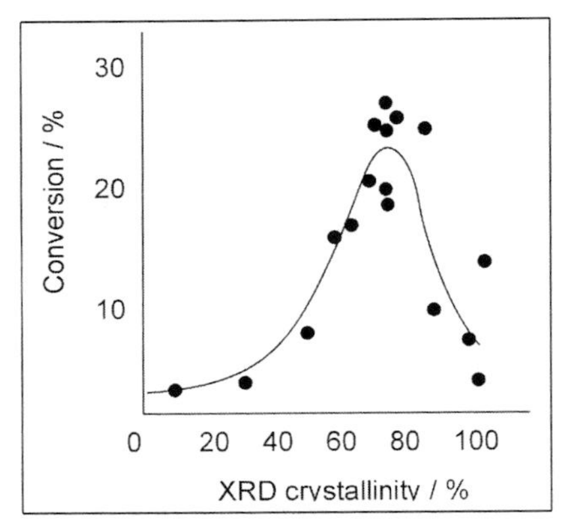

Figure 1 Conversion of propane over H^+-ZSM-5-based samples as a function of crystallinity

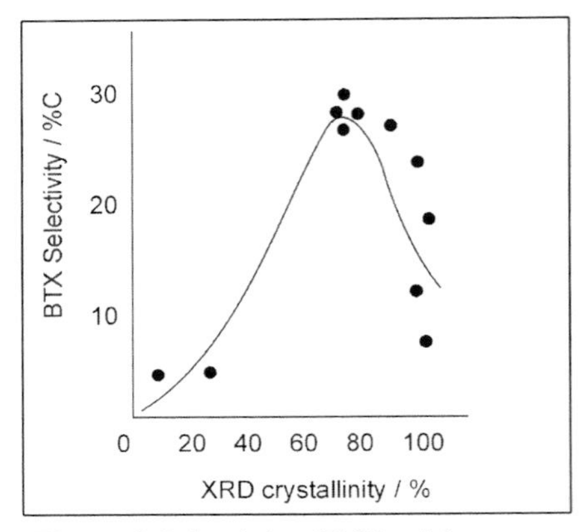

Figure 2 Selectivity (%C basis) to BTX over H^+- ZSM-5-based samples as a a function of crystallinity

reported in this paper were obtained at a time on stream (tos) of 1h. In general, in the present work, for tos values between 1 to 5h, catalyst deactivation was negligible. Further, it is noted that no induction period was seen for any of the catalysts studied.

3. RESULTS

Conversion and BTX selectivities obtained for the H-ZSM-5-based materials as a function of crystallinity are shown in Figures 1 and 2 respectively. It is clear that both conversion and selectivity display a similar behaviour in that maximum values are

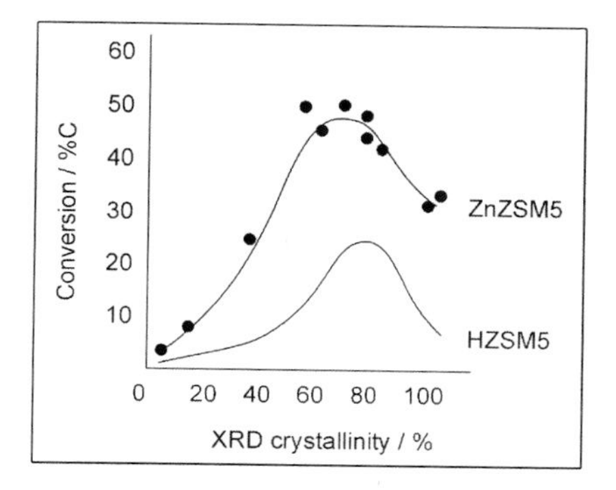

Figure 3 Conversion of propane as a function of crystallinity

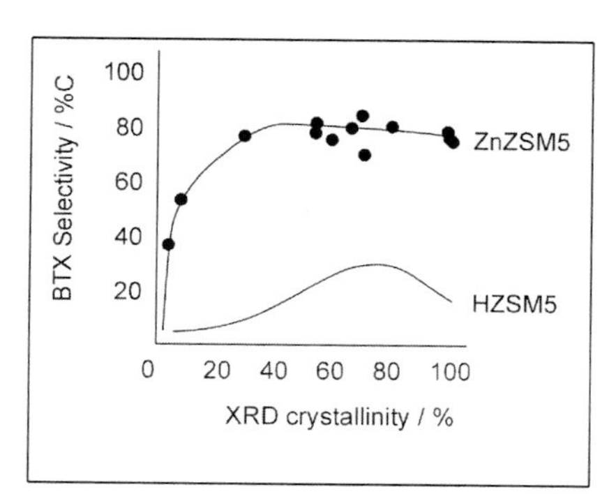

Figure 4 Selectivity (%C basis) to BTX as a function of crystallinity

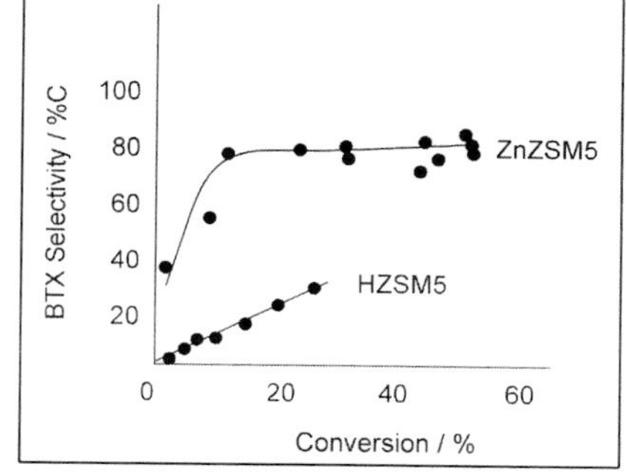

Figure 5 Selectivity (% C basis) to BTX as a function of % conversion

obtained at crystallinity levels in the range 60-80%. For the Zn-ZSM-5 samples there is also a clear maximum in the conversion level as a function of crystallinity (see Figure 3), but the maximum is found at slightly lower crystallinity levels, in the range 55-70%, compared with the situation found for the H-ZSM-5 catalysts (60-80% crystallinity). The conversions found with the Zn-ZSM-5 solids have a considerably higher value than those observed with the H-ZSM-5 materials at all crystallinities. For highly crystalline materials the fall in conversion seen when comparing 100% crystalline catalysts with those exhibiting maximum conversion is somewhat less marked for the Zn containing samples than for the catalysts in the H^+-form. Conversion levels fall by a factor of about 30% for Zn-ZSM-5 but by a factor of about 2 to 3 times for H-ZSM-5. The selectivity of Zn-ZSM-5 samples rises steeply as a function of crystallinity (see Figure 4), reaching a broad maximum value of 70-80% over a wider range of crystallinities and falling slightly to 70% for catalysts having crystallinities close to 100%. A smooth relationship between conversion and selectivity is found for the H-ZSM-5-based materials, as seen in Figure 5. A rise in conversion is associated with a concomitant increase in BTX selectivity. A maximum BTX selectivity of about 27% is found. Highly crystalline samples exhibit markedly lower selectivities in the range 7-17%. In the case of the Zn-ZSM-5 samples, above conversion levels of about 10%, selectivity is almost independent of conversion, in marked contrast to the situation found with H-ZSM-5.

4. DISCUSSION

The dependence of activity (and BTX selectivity) on crystallinity seen for the H^+ -ZSM-5-based samples could be considered "similar" to that reported for the SI of linear butenes [1,2] but for aromatization, maximum activity (and selectivity) occurs at much higher crystallinity levels (60-80% versus 2%). This is directly attributed to the more demanding character of the propane aromatization reaction. For n-hexane cracking it is found [28] that significant activities are associated with samples having crystallinities in excess of 30%. This finding is consistent with n-hexane cracking being more demanding than butene SI but, probably less demanding than propane aromatization in terms of the zeolitic environment requirements and the acid strength requirements of the sites involved or those sites that are associated with the metal. The general concomitant increase of BTX selectivity with conversion level is often observed in this aromatization system [3,4]. The comparatively low activities of the highly crystalline samples are probably associated with high diffusion constraints. In our preparations the highly crystalline samples (90-100%) exhibit crystal sizes that are considerably higher than those associated with samples having about 70-75% crystallinity [29]. The latter have typically a spheroidal shape with diameters close to 0.6 μm. The more highly crystalline samples have dimensions typically some 10-20 times higher with well-defined rectangular block shaped crystallites together with hexagonal and octagonal crystallites. It has been demonstrated by Beschmann et al. [30] that for reactions involving aromatics, activities are markedly reduced as the crystallite size of H-ZSM-5 is increased in the range 0.045 to 4.5 μm for reaction rates in the range 0.5 to 19 x 10^{-6} mol g^{-1} s^{-1} [30]. The absolute rate of conversion found in the present work for the most active H-ZSM-5 catalysts (17% conversion level) is 3.80 x 10^{-6} mol g^{-1} s^{-1} and is, therefore, comparable with the lower end of the range of rates observed by Beschmann et al.

The increase in conversion and selectivity brought about by the incorporation of zinc is typical of the work reported by others using this metal cation. However, new aspects of the effect of crystallinity are revealed in the present work. Clearly the incorporation of Zn is able to render the catalysts less sensitive to crystallinity effects compared with the parent catalysts in the H^+-form as far as selectivity to aromatics is concerned. A similar reduced sensitivity is also apparent in the effect of increased crystallite size on conversion, presumably because in samples prepared by the SSIE process there may well be zinc concentration gradients present with preferential incorporation of Zn towards the outer extremities of the individual crystallites. More characterisation is needed, however, before concrete evidence for this effect can be obtained. The raising of conversion levels by zinc exchange compared with the parent H-ZSM-5 by a factor of about 3 for crystallinities in the range of about 30-60% and by a factor of about 5 for lower crystallinities may be compared with similar factors reported by others. Corresponding factors noted in other work range from 2 – 4 [3,4] to 8 – 10 [8, 11-12]. A comparison [31] of conversion-selectivity data found in the present work with results reported elsewhere for Zn-exchanged catalysts reveals that the present series of catalysts appears to exhibit some of the highest single pass yields of any Zn-ZSM-5 catalysts studied. A similar situation is found for gallium-exchanged samples [31], but the effects of crystallinity on the behaviour of Ga-exchanged catalysts differs somewhat in detail from that found with Zn-exchanged solids.

A further intriguing aspect of the present work is that for catalysts having relatively low crystallinity, the incorporation of zinc by SSIE is able to produce catalysts having substantially higher selectivities and activities than the parent materials (and higher than those reported by others for what appeared to be highly crystalline samples). We believe that by further studying such samples it may be possible to reveal more subtle effects of the role of Zn on the one hand and the zeolitic structural requirements for the aromatization reaction on the other. It would seem that zinc centres can operate in such a way that relatively underdeveloped zeolitic domains and/or weaker non-zeolitic acid sites can be rendered much more effective in bringing about propane conversion. What is not yet clear is whether the zinc centres are predominantly located within or close to the zeolitic domains or whether a more distant juxtaposition of zinc centres and zeolite centres could sustain the enhanced catalytic action observed. More work in this direction may provide further understanding of the bifunctionality [8] of the Zn-ZSM-5 NAS systems in this and other reactions.

ACKNOWLEDGMENTS

Financial support from the Mellon Institute, University of the Witwatersrand, the National Research Foundation (South Africa) and Sasol Technology (Pty) Ltd is gratefully acknowledged.

REFERENCES

1. C. P. Nicolaides, Appl. Catal., 1999.
2. C. P. Nicolaides , US Patent 5 503 818 (Apr 2, 1996).
3. T. Mole, J. R. Anderson,and G. Greer, Appl. Catal., 17 (1985) 141.
4. M. S. Scurrell, Appl Catal 41 (1988) 89.

5. S. B. Abdul Hamid, E. G. Derouane, P. Meriaudeau, C. Naccache and M. A. Yarmo, Stud. Surf. Sci. Catal., 84 (1994) 2335.
6. Y. Ono and K. Kanae, J. C. S. Farad. Trans., 87 (1991) 669.
7. O P Keipert and M. Baerns, Stud. Surf. Sci. Catal., 84 (1994) 1757.
8. J. A. Biscardi and E. Iglesia, J. Catal., 182 (1999) 117.
9. M. Shibata, H. Kitagawa, Y. Sendoda and Y. Ono, Proc. 7th. Int. Zeol. Conf., p. 717, Elsevier, Tokyo 1986.
10. J. A. Biscardi and E. Iglesia, Catal. Today, 31 (1996) 207.
11. H. Berndt, G. Leitz, B. Lucke and J. Volter, Appl. Catal. A, 146 (1996) 351.
12. H. Berndt, G. Lietz and J. Volter, Appl. Catal. A, 146 (1996) 365.
13. N. Viswanatham, A. R. Pradhan, N. Ray, S. C. Vishnoi, U. Shanker and T. S. R. Prasada Rao, Appl. Catal., 137 (1996) 225.
14. E. Iglesia, J. E. Baumgartner and G. L. Price, J. Catal., 134 (1992) 549.
15. K. M. Dooley, C. Chang and G. L. Price, Appl. Catal., 84 (1992) 17.
16. B. S. Kwak and W. M. H. Sachtler, J. Catal., 145 (1994) 456.
17. M. Guisnet and N. S. Gnep, Catal. Today, 31 (1996) 275.
18. I. Nakamura and K. Fujimoto, Catal. Today, 31 (1996) 335.
19. V. R. Choudhary, A. K. Kinage and T. V. Choudhary, Appl. Catal., 162 (1997) 239.
20. V. R. Choudhary and P. Devadas, Appl. Catal., 168 (1998) 187.
21. G. J. Hutchings and M. S. Scurrell, Methane conversion by oxidative processes, ed. E. E. Wolf, van Nostrand Reinhold, New York, NY, 1992, p. 200.
22. C. A. Jones, J. J. Leonard and J. A. Sofranko, Energy and Fuels, 1 (1987) 12.
23. N. D. Parkyns, C. I. Warburton and J. D. Wilson, Catal. Today, 18 (1993) 385.
24. L. Guczi, L. Borko, Zs. Koppany and I. Kiricsi, Stud. Surf. Sci. Catal., 119 (1998) 295.
25. M. S. Scurrell and M. Cooks, Stud. Surf. Sci. Catal., 36 (1988) 433.
26. G. A. Olah, Eur. Pat. Appl., 73 673 (1983).
27. H. G. Karge, Stud. Surf. Sci. Catal., 105 (1997) 1901.
28. C. P. Nicolaides, N. P. Makgoba, M. S. Scurrell and N. P. Sincadu, submitted for publication.
29. C. P. Nicolaides, J. J. Prinsloo, P. B. Ramatsetse and M. E. Lee, submitted for publication; P. B. Ramatsetse, MSc Dissertation, Univ. of the North, Sovenga, 1998.
30. K. Beschmann, L. Riekert and U. Muller, J. Catal., 145 (1994) 243.
31. C. P. Nicolaides, N. P. Sincadu and M. S. Scurrell, Paper presented at CATSA Conference, *Fischer-Tropsch Synthesis on the Eve of the 21st Century,* Kruger National Park, South Africa, Nov 2000, Catal. Today, to be published.

Studies in Surface Science and Catalysis
J.J. Spivey, E. Iglesia and T.H. Fleisch (Editors)

The oxidative dehydrogenation of propane with CO_2 over supported Mo_2C catalyst

F. Solymosi, R. Németh and A. Oszkó

Institute of Solid State and Radiochemistry, The University of Szeged, H-6701 Szeged, P.O. Box 168, Hungary. Fax: + 36 62 420 678; e-mail: fsolym@chem.u-szeged.hu

Following the interaction of propane with Mo_2C/SiO_2 at 373-573 K, the formation of π-bonded and di-σ-bonded propylene, and propylidyne were identified by Fourier transform infrared spectroscopy. The presence of CO_2 exerted very little influence on the formation of these surface species. CO_2 adsorbs weakly and non-dissociatively on Mo_2C catalysts below 773 K. At higher temperature CO_2 oxidizes the Mo_2C; this process was followed by X-ray photoelectron spectroscopy. Mo_2C/SiO_2 is an effective catalyst for the oxidative dehydrogenation of propane. The selectivity to propylene at 773-873 K is 85-90 % The yield of propylene production is ~11% at 943 K. Other hydrocarbon products are ethylene, methane and butane. It is assumed that the Mo oxycarbide formed in the reaction between CO_2 and Mo_2C plays an important role in the activation of propane.

1. INTRODUCTION

Mo_2C deposited on ZSM-5 was found to be an effective catalyst in the high temperature conversion of methane into benzene [1-6]. The selectivity of benzene formation is 80-85% at a conversion level of 8-10% [3,4]. This work was initiated by the results published by Wang et al. [7,8], whose starting catalyst was MoO_3/ZSM-5, and by the observation that MoO_3 has been converted into other compounds during the reaction accompanied by the deposition of a large amount of carbon [9]. As no benzene formation occurred on pure Mo_2C [2,3], it was assumed that the primary role of Mo_2C is to activate the C-H bond of methane and to produce CH_x fragments. Further reactions, particularly the oligomerization of ethylene, proceed on the acidic sites of ZSM-5. The study of the chemistry of the reaction intermediates, CH_2, CH_3 and C_2H_5, on Mo_2C/Mo(100) surface in UHV system by means of several spectroscopic methods disclosed further details of the complex process, namely that CH_3 decomposes to CH_2 on Mo_2C and the recombination of CH_2 can possibly also occur on Mo_2C catalyst [11,12]. Mo_2C/ZSM-5 exhibited a high activity in the aromatization of ethane [13] and propane [14]. Recently, we found the Mo_2C prepared on SiO_2 is an effective catalyst for the oxidative dehydrogenation of ethane using CO_2 as an oxidant [15]. The selectivity to ethylene at 850-923 K was 90-95% at an ethane conversion of 8-30%. In the present paper the catalytic performance of supported Mo_2C is examined in the oxidative dehydrogenation of propane.

2. EXPERIMENTAL

Hexagonal Mo_2C was prepared by the method of Boudart et al. [16]. Supported Mo_2C catalyst was prepared by the carburation of calcined MoO_3/SiO_2 in the catalytic reactor [3].

MoO_3/SiO_2 catalyst was prepared by impregnating silica (Cab-O-Sil, area: 200 m^2/g) with a basic solution of ammonium paramolybdate to yield a nominal 2wt% of MoO_3. The adsorption of reacting gases and that of the products of the reaction was studied by Fourier transform IR spectroscopy using a Biorad spectrometer (FTS 155). For IR measurements the MoO_3/SiO_2 sample was pressed into 10x30 mm self-supporting disc, and the carburation of the disc was performed in the infra red cell. The reaction of propane with supported MoO_3, and the oxidation of Mo_2C with CO_2 were followed in situ by XPS (Kratos XSAM 800).

Catalytic reaction was carried out at 1 atm of pressure in a fixed-bed, continuous flow reactor consisting of a quartz tube connected to a capillary tube [3,14]. The flow rate was in most cases 12 ml/min. The carrier gas was Ar. The propane content was 12.5%, which was kept constant in all experiments. Generally 0.5 g of loosely compressed catalyst sample was applied. Reaction products were analyzed gas chromatographically with a Hewlett-Packard 5890 gas chromatograph and a Porapak QS column [3,14].

3. RESULTS AND DISCUSSION

3.1. Adsorption of CO_2 and C_3H_8

The interaction of CO_2 with supported Mo_2C catalyst has been first examined by FTIR spectroscopy. Adsorption of CO_2 (50 Torr) on Mo_2C/SiO_2 at 300 K produced no absorption bands. Heating the sample in the presence of CO_2 and registering the IR spectrum at 250 K, below the thermal desorption of CO from Mo_2C (Tp = 337 K [17]), we observed weak CO band at 2050 cm^{-1} above 773 K. Mass spectrometric analysis of the gas phase indicated the formation of gaseous CO first at 823 K, suggesting the dissociation of CO_2 and/or its reaction with Mo_2C. The latter process was followed by XPS in the temperature range of C_3H_8 + CO_2 reaction.

In Fig.1A we show the XPS spectra of Mo_2C/SiO_2 treated with CO_2. It shows that CO_2 reacts slowly with Mo_2C at 873 K, but the oxidation of Mo can be clearly seen by the appearance of new peaks at 233.3 and 236.2 eV. In the presence of CO_2 + C_3H_8 gas mixture only very little change is observed in the XPS spectra of the sample, suggesting the occurrence of a fast reaction between Mo-O species and propane.

Before the study of the interaction of propane with Mo_2C, we examined the reaction between MoO_3 and propane. From previous measurements it appeared that MoO_3 is converted into Mo_2C in the methane flow at 973 K [3-5]. This formation of Mo_2C occurred at much lower temperature, at 773 K, with ethane [13]. The carburation was almost complete in 5 hours. Propane proved to be also an effective carburizing compound. In propane flow at 873 K, the characteristic binding energies of Mo_2C already appeared at 10 min. Relevant XPS spectra are presented in Fig. 1B. Characteristic XPS data for different Mo compounds are collected in Table 1.

The absorption bands observed following the adsorption of propane on Mo_2C/SiO_2 at 150-250 K correspond very well to the different vibrations of adsorbed propane. At 573 K, however, new spectral features appeared at 2925, 1429, 1411 and 1374 cm^{-1}, which suggests that a fraction of propane interacted strongly with Mo_2C. Taking into account the spectral features of adsorbed propylene and propylidyne on Mo_2C/SiO_2 [14] and on Pt/SiO_2 [18,19], these absorption bands are attributed to adsorbed propylene formed in the dehydrogenation of propane. At higher temperature two weak bands at 1407 and 1367 cm^{-1} were identified which we order to the vibrations of propylidyne. This suggests that a fraction of adsorbed propylene converted into propylidyne.

Table 1
XPS data for Mo compounds

	Binding energy (eV)		
Sample	$Mo(3d_{5/2})$	$Mo(3d_{3/2})$	C(1s)
Mo foil	227.8-228.0		
MoO_3	233.0	236.2	
MoO_3/SiO_2	232.7	235.7	
MoO_2	229.8	232.9	
Mo_2C(Aldrich)	228.0	231.1	283.85
Mo_2C/SiO_2	227.6	230.7	284.5
$Mo_2C_xO_y$[a]	228.6 (Mo-C) 232.2 (MoO_3)		

[a]ref [20]

This hydrocarbon compound was produced in higher concentration by the adsorption of propylene on Mo_2C/SiO_2 at 300 K, when all its characteristic absorption bands were observed. The above surface species were also detected following the adsorption of propane on Pt/SiO_2 by Sheppard et al. [18,19]. Accordingly, the reactivity of highly dispersed Mo_2C towards propane and propylene is not much less than that of supported Pt catalyst.

The same IR spectroscopic measurements have been carried out in the presence of CO_2. We found practically identical spectral features as in the absence of CO_2 with slight variation in the position and intensities of the absorption bands.

3.2. Catalytic studies

First we tested the catalytic effect of pure Mo_2C. The ratio of the gas mixture, CO_2/C_3H_8, was 1.0. At 873 K we measured only a low conversion 0.5-1.0% of propane. Propylene was produced with a selectivity of ~90%. Similar measurements were performed on silica support, which exhibited even lower activity.

Deposition of Mo_2C on SiO_2 caused a dramatic change in the catalytic performance and produced an effective catalyst for the oxidative dehydrogenation of propane. The main hydrocarbon products were propylene, ethylene and methane. In addition, butane, CO, H_2O, H_2 and very little benzene were found. The H_2O/H_2 ratio varied between 10-15. The products distribution suggests that both the dehydrogenation and oxidative dehydrogenation proceed, but the latter is the dominant process. After the initial decay, the conversion of propane remained constant for the applied measuring time, 4 hours. The reaction began at 773 K, when the steady state conversion was about 2.0%. With the increase of the temperature the conversion gradually increased and at 943 K it attained a value of 20.0%. The selectivity to propylene was very high, 85-90%, and was almost constant in the temperature range 773-873 K. Further increase in the temperature caused a decrease in the selectivity to propylene, but an enhancement in the yield of propylene production. The selectivities of ethylene and methane gradually increased with the rise of the temperature. The effect of reaction temperature on the above features is displayed in Fig. 2A.

Variation of the flow rate showed that the conversion of propane decreased with the rise of space velocity, but the rate of the formation of propylene slightly increased. As a result, the selectivity to propylene remained unchanged (Fig. 2B).

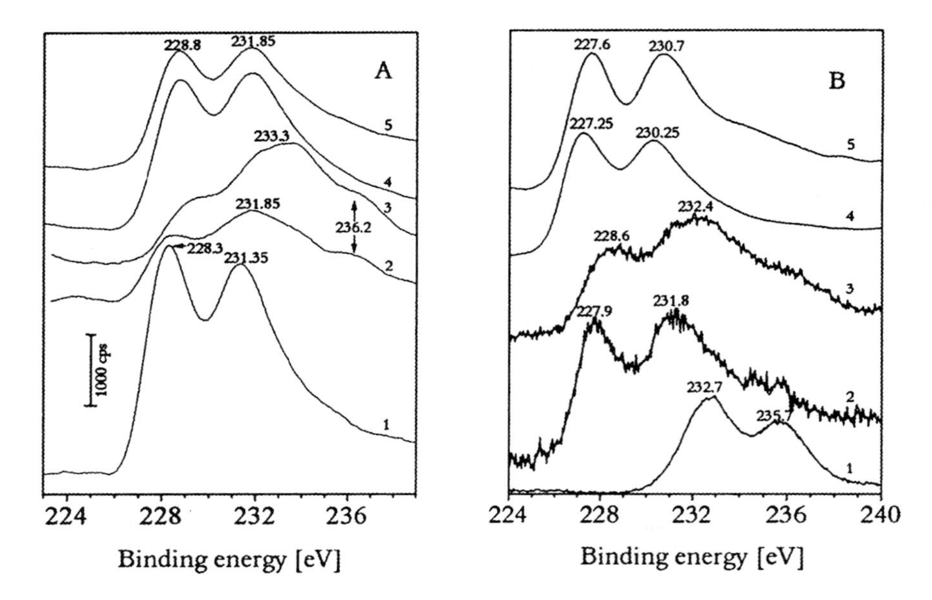

Fig. 1. (A) Effects of CO_2 and $CO_2 + C_3H_8$ (1:2) on the XPS spectra of Mo_2C/SiO_2 (1) 0 min, (2) CO_2, 873 K, 30 min, (3) 180 min, (4) $CO_2 + C_3H_8$, 873 K, 30 min, (5) 180 min. (B) XPS spectra of Mo-containing catalyst. (1) MoO_3/ZSM-5, (2) MoO_3/ZSM-5 treated with CH_4 at 973 K for 2 h. (3) MoO_3/ZSM-5 treated with ethane at 773 K for 5 h (4) MoO_3/SiO_2 treated with propane at 873 K for 10 min (5) Mo_2C/SiO_2.

The effects of the CO_2/C_3H_8 ratio on the rate and selectivities of several products were investigated at 873 K. The amount of propane was kept constant, 12.5%, and the amount of CO_2 was varied. Comparison was made at 60 min of reaction time. Results are presented in Fig. 3. The conversion of propane did not alter with the increase of the CO_2 content, and fell in the range 7-10%. The propylene selectivity remained at a high level, ~85%, and was practically independent of the CO_2/C_3H_8 ratio in the range of 0.5-4.0. In contrast, the rate of the formation of H_2 and the H_2/CO decreased, while the ratio of H_2O/H_2 increased with the increase of CO_2 content suggesting that the oxidative dehydrogenation reaction became more dominant.

In the interpretation of the results of catalytic measurements it appears clearly that CO_2 opened a new route for the formation of propylene. It is very likely that the first step in the oxidative dehydrogenation of propane with CO_2 is the oxidation of Mo_2C with CO_2 and the formation of Mo oxycarbide. Further steps are practically the same as on the oxides. Propane may form a surface complex with the active oxygen of the carbide

$$C_3H_{8(g)} + O_{(a)} \rightleftharpoons C_3H_8O_{(a)} \quad (1)$$

C-H cleavage may proceed on the reduced centers very likely in methylene C-H bonds of propane which are weaker than those of the methyl groups

$$C_3H_8O_{(a)} = C_3H_7O_{(a)} + H_{(a)} \quad (2)$$

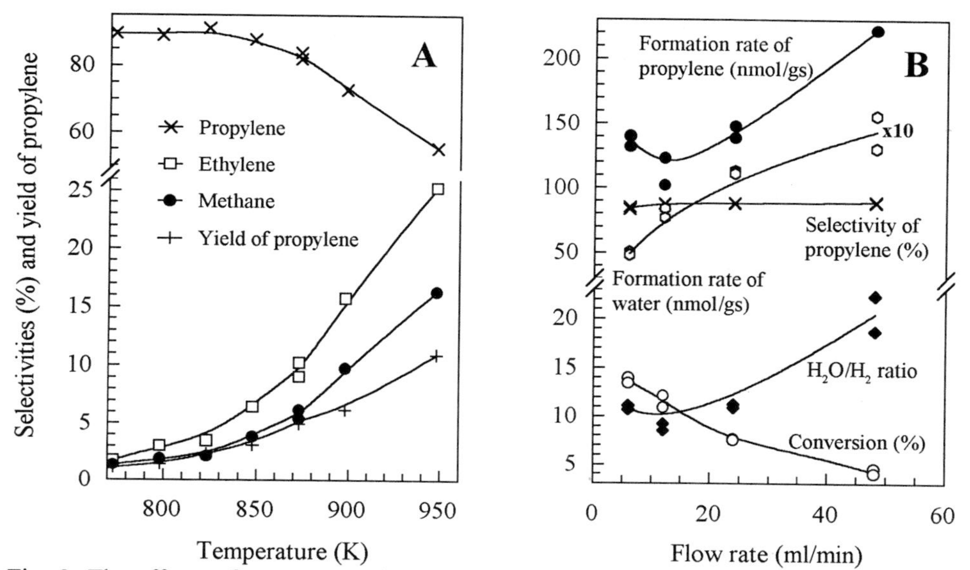

Fig. 2. The effects of temperature (A) and the flow rate (B) on some characteristic data of $C_3H_8 + CO_2$ reaction on Mo_2C/SiO_2.

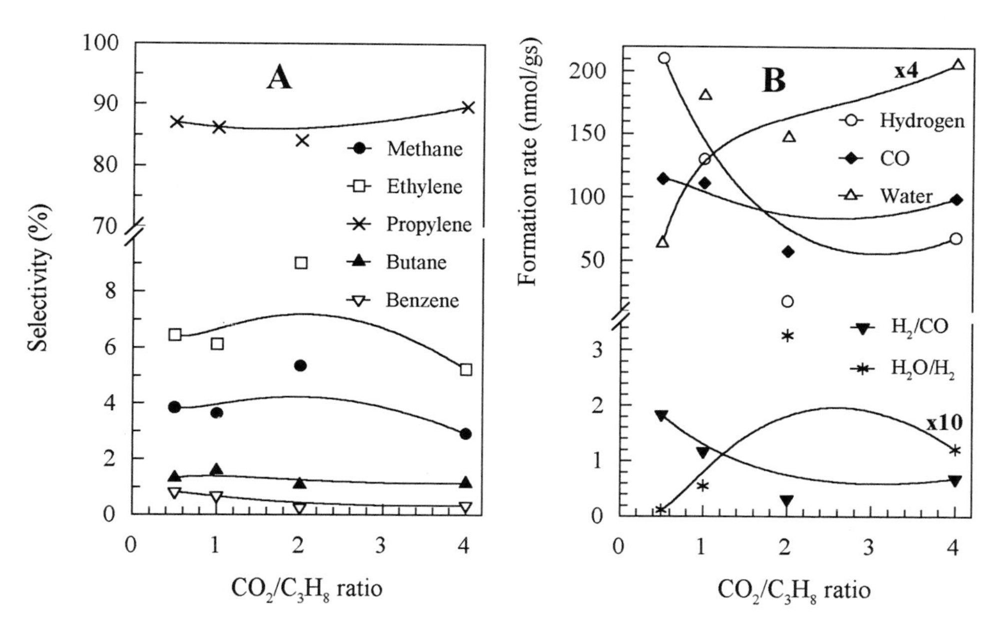

Fig. 3. The effect of the CO_2/C_3H_8 ratio on some characteristic data of $C_3H_8 + CO_2$ reaction on Mo_2C/SiO_2.

or with the participation of neighboring oxygen in Mo oxycarbide

$$C_3H_8O_{(a)} + O_{(a)} = C_3H_7O_{(a)} + OH_{(a)} \quad (3)$$
$$C_3H_7O_{(a)} = C_3H_{6(g)} + OH_{(a)} \quad (4)$$
$$2OH_{(a)} = H_2O_{(g)} + O_{(a)} \quad (5)$$

We may also count with the reactions of hydrogen produced in the dehydrogenation of propane on Mo_2C

$$H_{2(g)} + O_{(a)} = H_2O_{(g)} \quad (6)$$
$$CO_{2(g)} + 4H_{2(g)} = CH_{4(g)} + 2H_2O_{(g)} \quad (7)$$
$$CO_{(g)} + 3H_{2(g)} = CH_{4(g)} + H_2O_{(g)} \quad (8)$$

3.4. Effects of supports

Some experiments were also performed on the effects of supports. Qualitatively, we obtained similar results for Mo_2C/Al_2O_3 and Mo_2C/TiO_2. Basically different behavior was observed for Mo_2C deposited on ZSM-5. On this catalyst, the main reaction pathway was the aromatization of propane. In the absence of CO_2 at 873 K, the aromatic compounds (benzene and toluene) were produced with the highest selectivity, ~45%, at a conversion level of 46%. This was followed by ethylene (12%), propylene (7%) and methane (26%). The presence of CO_2 altered this product distribution, and the oxidative dehydrogenation of propane came into prominence. This is attributed to the partial oxidation of Mo_2C.

REFERENCES

1. D.W. Wang, J.H. Lunsford and M.P. Rosynek, Topics in Catal. 3(4) (1996) 299.
2. F. Solymosi and A. Szőke, Catal. Lett. 39 (1996) 157.
3. F. Solymosi, J. Cserényi, A. Szőke, T. Bánsági and A. Oszkó, J. Catal. 165 (1997) 150.
4. J.H. Lunsford, M.P. Rosynek and D.W. Wang, Stud. Surf. Sci. Catal. 107 (1997) 257.
5. D.W. Wang, J.H. Lunsford and M.P. Rosynek, J. Catal. 169 (1997) 347.
6. S. Liu, Q. Dong, R. Ohnishi and M. Ichikawa, J.C.S. Chem. Commun. 1445 (1997).
7. R.W. Borry III, E.C. Lu, K. Young-ho and E. Iglesia, Stud. Surf. Sci. Catal. 119 (1998) 403.
8. L. Wang, L. Tao, M. Xie, G. Xu, J. Huang and Y. Xu, Catal. Lett. 21 (1993) 35.
9. Y. Xu, L. Wang, M. Xie, and X. Gou, Catal. Lett. 30 (1995) 135.
10. F. Solymosi, A. Erdőhelyi and A. Szőke, Catal. Lett. 32, (1995) 43.
11. F. Solymosi, L. Bugyi and A. Oszkó, Catal. Lett. 57 (1999) 103.
12. F. Solymosi, L. Bugyi, A. Oszkó and I. Horváth, J. Catal. 185 (1999) 160.
13. F. Solymosi and A. Szőke, Appl. Catal. A. 166, (1998) 225.
14. F. Solymosi, R. Németh, L. Óvári and L. Egri, J. Catal in press.
15. F. Solymosi and R. Németh, Catal. Letts., 62 (1999) 197.
16. J.S. Lee, S.T. Oyama and M. Boudart, J. Catal. 106 (1987) 125.
17. F. Solymosi and L. Bugyi, Catal. Lett. 66 (2000) 227.
18. M.A. Chesters, C. De La Cruz, P. Gardner, E.M. McCash, P. Pudney, G., Shahid and N. Shepard, J. Chem. Soc. Faraday Trans. 86(15) (1990) 2757.
19. N. Shepard and C. De La Cruz, Adv. Catal. 42 (1998) 181.
20. T. Shido, K. Asakura, Y . Noguchi, Y. Iwasawa, Appl. Catal. A:Gen. 194-195 (2000) 365.

Studies in Surface Science and Catalysis
J.J. Spivey, E. Iglesia and T.H. Fleisch (Editors)

Decomposition/reformation processes and CH_4 combustion activity of PdO over Al_2O_3 supported catalysts for gas turbine applications.

G. Groppi[*], G. Artioli[**], C. Cristiani[*], L. Lietti[*],and P. Forzatti[*]

[*]Dipartimento di Chimica Industriale - Politecnico di Milano, P.za L. Da Vinci 32, 20133-Milano-Italy, Ph +39-02-23993258; Fax +39-02-70638173; e-mail: gianpiero.groppi@polimi.it;
[**] Dipartimento di Scienza della Terra, Università di Milano e Centro CNR per lo studio della Geodinamica Alpina e Quaternaria - 20133 Milano - Italy

The characteristics of the decomposition/reformation process of PdO supported on La_2O_3/Al_2O_3 and the related variations of CH_4 combustion activity are investigated by means of TG, in-situ XRD synchrotron radiation, Temperature Programmed Oxidation and Temperature Programmed Combustion at high GHSVs. Evidence of the presence of two main PdO species which form according to an in series process during reoxidation of Pd are presented. Hypotheses on the nature of such two species and on their role on the catalytic activity in CH_4 combustion are discussed.

1. INTRODUCTION

Catalytic combustion for gas turbine applications is close to commercialization as the most cost-effective technique to achieve ultra-low emissions of NO_x, CO and UHC [1]. The use of supported PdO catalysts is a key feature issue in this technology in view of the following properties [2]: i) maximum activity in CH_4 combustion achieved with high metal loading ; ii) negligible volatility below 1000°C of all the relevant Pd species; iii) "chemical thermostat", i.e the capability of temperature self control well below the adiabatic reaction level. This latter property, which prevent the catalyst to exceed temperature above 900-950°C, has been associated with the reversible transformation from active PdO to inactive Pd° during methane combustion under adiabatic conditions.
The characteristics of the decomposition/reformation process of PdO, which are still poorly understood, are herein addressed for a catalyst with high palladium content (10% w/w) dispersed on a La_2O_3-stabilised Al_2O_3 support.

2.EXPERIMENTAL

The investigated catalyst have been prepared starting from a commercial γ-Al_2O_3 (Roche) calcined at 700°C with a surface area of 200 m^2/g and a pore volume of 0.7 cm^3/g. 5% w/w of La_2O_3 has been added via dry impregnation with a $La(NO_3)_3$ aqueous solution and the sample has been dried at 110 °C and calcined at 1000°C for 10 h. 10% w/w of Pd has been then added via dry impregnation with a $Pd(NO_3)_2$ solution. The catalyst has been finally re-calcined at 1000°C for 10 h. The resulting sample consisted of crystalline PdO and θ-Al_2O_3 with a surface area of 100 m^2/g and a pore volume of 0.6 cm^3/g. Chemical analysis confirmed the nominal loading of Pd.

The PdO→Pd° reversible transformation during thermal cycles in atmospheres with O_2 partial pressure ranging from 0.01 to 1 bar has been studied by means of TG analyses (SEIKO), high temperature in situ XRD measurements collected at BM08 line at GILDA synchrotron facility in Grenoble and Temperature Programmed Oxidation (TPO) experiments. Details on the XRD experiments and data analysis will be reported in a forthcoming paper [3]. The TPO experiments have been performed using a quartz microreactor loaded with 120 mg of fine catalyst powder (d_p=0.1 mm). Catalyst temperature has been measured with a fixed K-type thermocouple placed in the thin catalyst bed. The reactor has been fed with a gas flow rate ranging from 60 to 120 cm^3/min at STP consisting of 1.0-2.0 % of O_2 in He. The composition of the effluent gases was continuously determined by on line mass spectrometry.

The variations of CH_4 combustion activity related to the PdO decomposition/reformation process have been investigated by means of Temperature Programmed Combustion (TPC) experiments performed over a structured annular reactor [4] specifically designed for investigation of ultra-fast reactions at high GHSV (>10^6 h^{-1}). The reactor consists of one quartz and one ceramic tube coaxially placed to form a narrow annular chamber (0.5 mm) where the gas flows. By a wash-coating technique [5] the surface of the internal ceramic tube is partially coated with a very thin catalyst layer (10-40 μm) which allows to minimise intraparticle diffusional limitations. The catalytic properties of the active layer have been checked to be consistent with those of the catalyst powders in TPO and activity tests. The temperature is measured by a K-type thermocouple sliding in the internal cavity of the ceramic tube. Further experimental and theoretical details on this reactor are reported in [6].

3. RESULTS AND DISCUSSION

3.1 PdO→ Pd° reversible transformation

TPO and TG analyses with repeated thermal cycles have evidenced several complex features of the decomposition/ reformation process of palladium oxide. As it is shown in the TG and DTG profiles reported in Figure 1, decomposition of PdO along the heating ramp occurs through two major steps, observed at different temperatures depending on the O_2 content. One additional minor step at higher temperature has been observed in TPO experiments (see e.g. Figure 3) thanks to the higher sensitivity of the analysis. During the cooling ramp, O_2 uptake associated with reformation of PdO starts markedly below the decomposition temperature, thus originating a characteristic hysteresis that has been first reported by Farrauto and coworkers [7]. Besides PdO reformation is not completed in the cooling ramp and proceeds further in

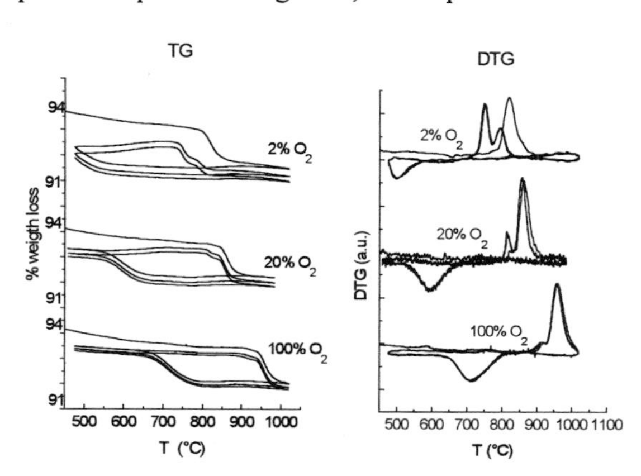

Figure 1. TG and DTG profiles at P_{o2}=0.02-1 bar. Heating/cooling ramp at 20 °C/min.

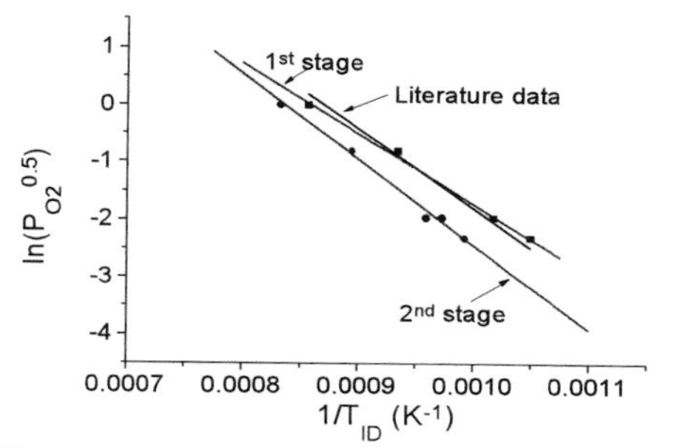

Figure 2. Analysis of decomposition temperatures of supported PdO compared with literature data for bulk PdO.

the following heating ramp.

On increasing P_{O2} the temperature thresholds of the PdO decomposition-reformation stages markedly shift to higher values. In Figure 2 the temperature thresholds of the PdO decomposition (T_{ID}) obtained from TPO and TG experiments are plotted in a $\ln(P_{O2}^{1/2})$ vs $1/T_{ID}$ diagram. For the first step of PdO decomposition a linear trend has been obtained that compares well with literature data for decomposition of bulk PdO. A similar linear correlation has been also obtained for the onset of the second decomposition step, which however is shifted to higher temperatures. This indicates that, under the investigated conditions, PdO decomposition is a thermodynamically driven process which involves two main species (indicated in the following as PdO^{1st} and PdO^{2nd}) having different thermal stability.

On increasing P_{O2} the overall extent of palladium re-oxidation increases as well as the relative amount of the second PdO species. A similar effect was also obtained by decreasing the heating/cooling rate at fixed oxygen partial pressure. Accordingly the relative amount of the different PdO species detected during the oxide decomposition is apparently controlled by the conditions at which re-oxidation is performed. To better clarify this point TPO experiments with holds during the heating ramp at different temperatures and for different time spans have been performed. Results (Figure 3) indicate that the amount of the second PdO species increases with both the duration (Figure 3, left side) and the temperature of the hold (Figure 3, right side), while the opposite is seen for PdO^{1st}. This suggests that oxidation of Pd° occurs via an in series process: Pd° → PdO^{1st} → PdO^{2nd}; the second stage having a high activation energy.

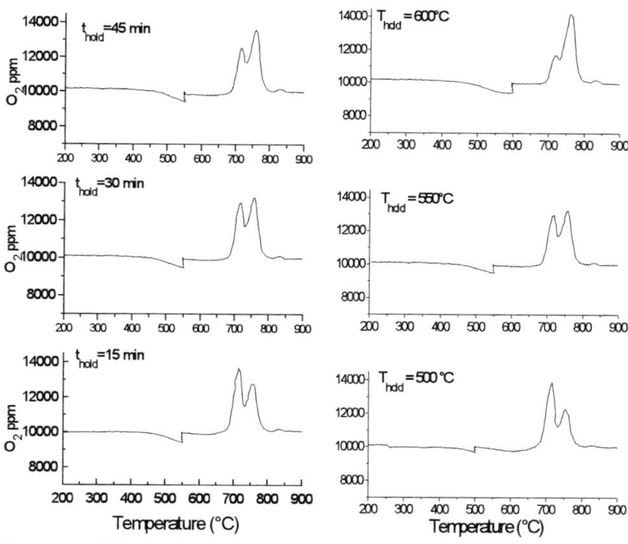

Figure 3. Effect of temperature holds on PdO reformation during heating. Left side: effect of hold time at 550°C; right side: effect of temperature for an hold time of 30 min.

According to the data shown in Figure 1 also P_{O2} has a marked influence on the relative amount of the two PdO species. However in TG experiments the increase of P_{O2} increases the decomposition temperatures of PdO^{1st} and PdO^{2nd} as well, being driven by thermodinamics. Hence, the effect of P_{O2} on the relative amount PdO^{1st}/PdO^{2nd} can not be decoupled from that of the temperature. To address this point, TPO experiments have been performed in which the catalyst has been heated up at 15 °C/min up 550 °C with different O_2 content (1, 20 and 100 % v/v), kept at this temperature for 5 minutes and then heated at 950 °C at 15 °C/min with P_{O2} =0.01 bar. The relative amounts of the different PdO species as a function of P_{O2} are reported in Figure 4. No significant variations have been observed in the range P_{O2}=0.01-1 bar. This suggests that the two PdO species have a similar degree of oxidation, so that the 1st species seems not to be associated with substoichiometric oxides whose presence has been proposed in the literature. Noticeably the third PdO species, corresponding to the medium gray uppermost column section, always represents a very small amount of the overall Pd (below 5% of the total Pd) and, according to literature indication, can be likely associated with Pd-Al_2O_3 surface complexes [8,9].

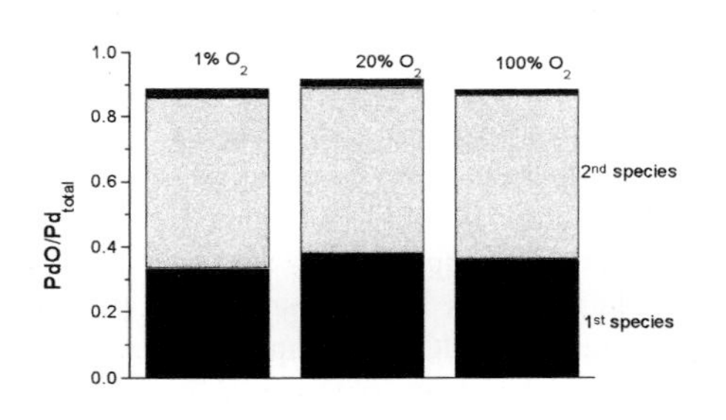

Figure 4 Relative amounts of PdO species

To collect further evidences on the nature of the two main palladium oxide species in-situ high temperature XRD spectra with synchrotron radiation have been performed. The collected spectra plotted in Figure 5 have confirmed the occurrence of the reversible PdO⇔Pd° transformation during thermal cycle in a 2% O_2 atmosphere, with the characteristic hysteresis described above. However only one type of crystalline PdO has been identified, whose ratio to crystalline Pd metal, quantified by Rietveld analysis, well corresponds to the overall extent of reoxidation (PdO^{1st} + PdO^{2nd}) as determined by TG and TPO measurements. Accordingly the 1st oxide species cannot be associated with amorphous PdO_x as proposed by Farrauto and coworkers [7]. Crystallite size of PdO and Pd have been calculated by Rietveld refinement of the XRD spectra [3].

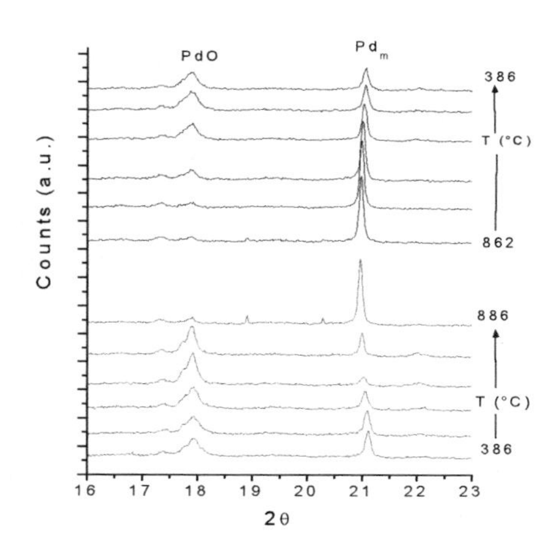

Figure 5. High temperature XRD spectra

Values ranging from 15 to 20 nm have been obtained for PdO whereas for metallic Pd a gradual decrease from 55 to 35 nm and a reverse increase from 35 to 55 nm are associated with the metal re-oxidation and oxide decomposition process respectively. A decrease of Pd crystal size has also been reported to occur during oxidation of metallic Pd supported on transition alumina [8]. This behavior has been associated either with PdO redispersion upon fragmentation of PdO crystallites and spreading onto the alumina surface [9] or with the formation of multiple uncoherent PdO domains from the initial large Pd agglomerates [10]. In view of the high heating/cooling rate adopted in our experiment, reversible spreading and sintering phenomena are unlikely to occur. On the other hand our data are apparently consistent with oxidation of Pd metal occuring via the formation of uncoherent PdO domains or, also, with a fragmenting shrinking core mechanism. In this latter case, due to the large volume changes involved with the transformation from Pd to PdO, oxidation could be reasonably associated with the fragmentation of the PdO shell, which would be responsible for the smaller size PdO crystallites. This process would result in a complex topological picture with PdO crystallites either in boundary contact with metallic Pd, or with other PdO crystallites (or with the support). A similar topological situation could also arise from the growth of multiple uncoherent domains of PdO. These different PdO species can be tentatively associated with the 1st and the 2nd PdO species mentioned above.

3.2 Catalytic activity in CH_4 combustion

Figure 6 shows the CH_4 conversion obtained in a TPC experiment performed using a structured annular reactor [6] with a 0.5/2/1/96.5 $CH_4/O_2/H_2O$/He mixture at extremely high GHSV (10^6 h^{-1}). During the heating ramp the CH_4 consumption starts at 320°C, it increases up to 90% showing a wide maximum at 620°C. Above this temperature, which is slightly lower than the threshold value for the onset of the 1st PdO decomposition stage in TPO experiments with 1% O_2, the CH_4 conversion decreases, passes through a minimum at 730°C and eventually increases again reaching 99% at 860°C. When the temperature ramp is reversed, CH_4 conversion decreases starting from 860°C, passes through a deep minimum at 620°C and then increases, crossing at 500°C the curve obtained during the heating ramp.

As a general picture the activity data herein presented confirm the literature hypothesis [4,7] which assigns PdO as the more active species and Pd metal as the less active one. Along these lines the hysteresis observed in the PdO decomposition/reformation process results in the large conversion hysteresis in Figure 6. On the other hand the data which have been collected at high temperature thanks to the peculiar features of the annular reactor [4,6] provide detailed indications on the catalytic behavior in a temperature range which could not be previously investigated. Although it was not possible to directly quantify the extent of the catalyst reoxidation due to the large variation of the outlet O_2 concentration (associated with the marked increase of CH_4 conversion) which overlaps the O_2 consumption associated with reformation of PdO, comparison with TPO profile with 1% and 2% of O_2 has shown that during the cooling ramp combustion activity increases at the very beginning of the PdO reformation process, and is completely restored at 460°C when only a small fraction of the 1st PdO species has formed. This is in line with literature results [11,12] obtained during low temperature (260-300°C) oxidation of metallic Pd which indicate that activity is completely restored upon formation of 4-7 monolayers of PdO and provides a support to the hypothesis that under the

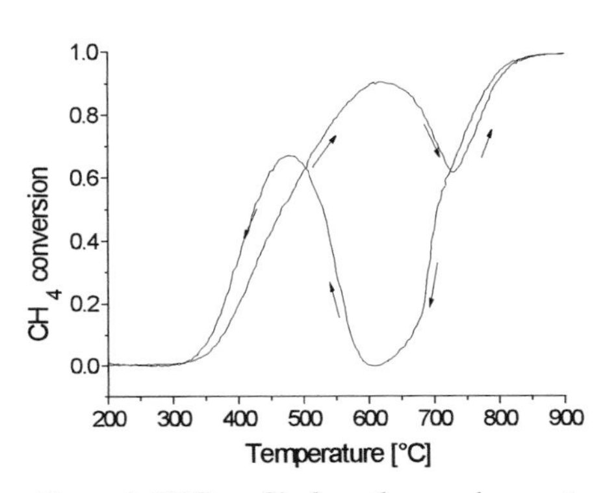

Figure 6. TPC profile from the annular reactor.

condition herein investigated reoxidation occur via shrinking core of the Pd metal. The second major evidence is that during the heating ramp the conversion starts to decrease approximately at the beginning of the PdO decomposition process (note that the conversion minimum in the heating branch is not detectable when performing tests at lower GHSV), but deviations from a classical Arrhenius trend are observed well below the temperature onset of PdO decomposition, i.e. the methane conversion flattens in spite of the fact that the temperature is increasing. Here, the influence of diffusional limitations has been ruled out on the basis of a theoretical analysis of the reactor [6]. To further clarify this point, a TPC experiment with temperature holds of 30 min at 450 and 500°C and of 60 min at 550°C has been performed. A slight decrease of methane conversion has been observed with time during the holds at 450°C and 500°C, whereas a marked decrease (from 60 to 35%) has been observed at 550°C, i.e. well below the temperature onset for PdO decomposition. This clearly indicates that other factors, besides PdO decomposition, play a major role in modifying the catalytic activity. It is noteworthy that the temperature range in which such changes have been observed well correspond to that of the $PdO^{1st} \rightarrow PdO^{2nd}$ transformation, so that a role of this phenomenon could be hypothesized. According to the picture proposed above the 1st PdO species possibly corresponds to PdO in boundary contact with metallic Pd. Such intimate contact has been already invoked in the literature [8,13] to account for increase of catalytic activity upon partial reduction of PdO and was ascribed to the role of metallic sites in the activation of CH_4 through dissociative adsorption.

Alternatively Iglesia and coworkers [14] proposed that methane combustion occurs through a dual site mechanism involving one reduced site (namely an oxygen vacancy in the PdO lattice), that promotes dissociative adsorption of CH_4, and one oxidised site that provides oxygen. Along these lines equilibration of the concentration of the reduced and oxidised active sites with reaction temperature might be invoked to explain the complex experimental activity trend observed during the heating branch below the threshold temperature of PdO decomposition.

Acknowledgement - Financial support for this work has been provided by MURST-Rome

References

[1] P. Evans, Modern Power Systems, March 2000, p. 27
[2] P. Forzatti and G. Groppi, Catal. Today, **54** (1999).49
[3] G. Artioli et al., in preparation.
[4] J. Mc Carty, Catal. Today, **26** (1995) 283
[5] G. Groppi, C. Cristiani, M. Valentini, E. Tronconi, Studies Surf. Sci. Catal., **130** (2000) 2747
[6] G. Groppi, W. Ibashi, M. Valentini and P. Forzatti, Chem. Eng. Sci, in press (2000).
[7] R.J. Farrauto, T. Kennely and E.M. Waterman, Appl. Catal. A:General, **81** (1992) 227
[8] K. Otto, L.P. Haack, and J.E. de Vries, Appl. Catal. B: Environmental, **1** (1992) 1
[9] J.J. Chen and E. Ruckestein, J. Phys. Chem., **85** (1981) 1606
[10] A.K. Datye, J. Bravo, T.R. Nelson, P. Atanasova, M. Lyubovsky, L. Pfefferle, Appl. Catal. A: General **198** (2000) 179
[11] R. Burch, F.J. Urbano, Appl. Catal. A: General, **124** (1995) 121
[12] J.N. Karstens, S.C. Su, A.T. Bell, J. Catal., **176** (1998) 136
[13] M. Lyubovsky , L. Pfefferle, Appl. Catal. A: General, **173** (1998) 107.
[14] K. Fujimoto, F.H. Ribeiro, M. Avalos-Borja and E. Iglesia, J. Catal, **179** (1998) 179

Studies in Surface Science and Catalysis
J.J. Spivey, E. Iglesia and T.H. Fleisch (Editors)

Effect of periodic pulsed operation on product selectivity in Fischer-Tropsch synthesis on Co-ZrO_2/SiO_2

A.A. Nikolopoulos, S.K. Gangwal, and J.J. Spivey*

Research Triangle Institute, P.O. Box 12194, RTP, NC 27709-2194, U.S.A.

The effect of H_2 pulsing on activity and product distribution of a high-α Co/ZrO_2/SiO_2 Fischer-Tropsch (FT) synthesis catalyst was investigated in an attempt to maximize the diesel-range product yield. H_2 pulsing increases CO conversion significantly but only temporarily; catalyst activity decreases gradually towards its steady state until the next pulse. Increasing H_2-pulse frequency enhances the yield of both CH_4 (undesirable) and C_{10}-C_{20} (desirable) products. An optimum H_2-pulse frequency is required in order to maximize the yield of diesel-range FT products without substantially increasing the CH_4 yield.

1. INTRODUCTION

The Fischer-Tropsch synthesis (FTS) can convert solid fuel- or natural gas-derived syngas (CO+H_2) to liquid fuels and high-value products. The extensively reviewed Fischer-Tropsch (FT) reaction [1-3] produces a non-selective distribution of hydrocarbons (C_1-C_{100+}) from syngas. FT catalysts are typically based on Group-VIII metals (Fe, Co, Ni, and Ru), with Fe and Co most frequently used. The product distribution over these catalysts is generally governed by the Schultz-Flory-Anderson (SFA) polymerization kinetics [4].

Currently there is significant commercial interest in producing diesel-fuel range middle distillates (C_{10}-C_{20} paraffins) from natural gas-derived syngas [5]. Increasing the selectivity of FTS to desired products such as diesel (C_{10}-C_{20}) or gasoline (C_5-C_{11}) by altering the SFA distribution is economically attractive. Use of bifunctional catalysts (FT-active metals on zeolite, e.g. ZSM-5) to produce high-octane gasoline-range hydrocarbons (explored in the past 2 decades), has been economically unsuccessful [6-9]. The zeolite cracking activity lowers the chain-growth probability (α), producing gasoline-range products in excess of 48wt% of the total hydrocarbon product; however, it also produces a significant amount of undesirable C_1-C_4 gases (Figure 1).

The present emphasis has shifted towards maximizing the yield of high-cetane C_{10}-C_{20} products from FTS. Increased worldwide demand for low-sulfur diesel has further stressed the importance of development of zero-sulfur FT-diesel products. An alternative approach to the use of bifunctional catalysts to alter selectivity is periodic FT reactor operation (pulsing) [3]. It entails alternatively switching between two predetermined input compositions over the FT catalyst to promote time-average rate, selectivity, and catalyst life [10-12]. Periodic pulsing of H_2 has been examined as a way to limit chain growth by removing the growing hydrocarbon chain from the catalyst surface [13-15]. Experimental studies have shown the

---

* Present address: Chemical Engineering Dept., North Carolina State University, Raleigh, NC 27695

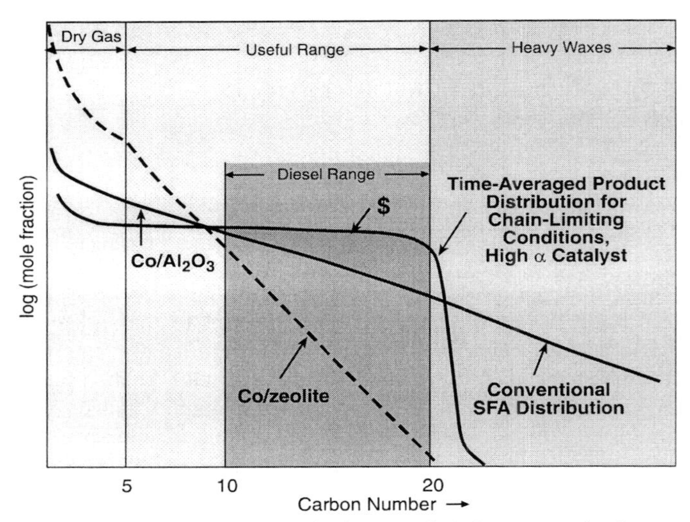

Figure 1. Product distribution (**α**-plot) for FT synthesis

potential to alter the SFA distribution [16,17]; they were performed, however, at conditions of limited industrial interest.

The chain-limiting concept using pulsing to maximize diesel yield is shown in a plot of carbon number vs. mole fraction (Fig. 1). The slope of the curve is determined by the chain-growth probability, **α**. Periodic operation on a high-**α** catalyst may result in removal of the growing chain from the surface at the desired C_{10}-C_{20} length, thereby maximizing diesel yield without increasing the dry gas yield. Thus, the objective of this study is to investigate the effect of H_2 pulsing on the activity and product distribution of a high-**α** (~0.9) $Co/ZrO_2/SiO_2$ FT synthesis catalyst, in an attempt to maximize the C_{10}-C_{20} product yield.

2. EXPERIMENTAL

2.1. Catalyst synthesis and characterization

A 25pbw Co – 18pbw Zr – 100pbw SiO_2 catalyst was synthesized by sequential incipient wetness impregnation of a high-purity, high-surface-area (144-m^2/g) silica support (XS 16080, Norton) [18]. The support (crushed and sieved to a particle size of 100-150 μm) was degassed in vacuum and heated to 80°C. A zirconium tetrapropoxide ($Zr(OCH_2CH_2CH_3)_4$) solution in 1-propanol (Aldrich) was used for the incipient wetness impregnation, performed in two steps. After each impregnation step, the product was dried (120°C, 2 h) and calcined in air (500°C, 1 h). The produced material had a nominal loading of 18pbw Zr – 100pbw silica.

Cobalt was impregnated on the zirconia/silica support using a cobalt nitrate hexahydrate precursor ($Co(NO_3)_2.6H_2O$, Aldrich). The hexahydrate was dissolved in water and the solution was added in a controlled manner to the zirconia/silica support, forming the catalyst with a nominal composition of 25pbw Co – 18pbw Zr – 100pbw SiO_2. Finally, the catalyst was calcined in air at 350°C for 1 hour.

The actual catalyst composition (as measured by ICP-OES) was 14.5%Co-10.5%Zr/SiO_2. Its surface area was measured (by BET method) to be 102±3 m^2/g. Its pore volume was estimated at 0.40±0.01 cc/g (by mercury porosimetry). Its crystalline structure was examined by X-ray diffraction (XRD). The predominant phase was Co_3O_4, with no other Co-O or Zr-O crystalline phases or cobalt silicate present in the diffraction pattern.

2.2. Reaction set-up

The reaction system consisted of the gas-feed, a fixed-bed reactor, and a sampling/ analysis system for the liquid and gaseous products. The feed system blended CO/Ar, H_2, N_2, or other premixed gases in desired concentrations. A time-programmable interface system (Carolina Instrumentation Co.) was used to control a series of actuated valves, so that a (reactant or inert) flow opened / closed automatically and independently of the others. Appropriate periodic switch of these valves offered the capability to perform various pulsing-type experiments with this configuration.

A stainless-steel 3/8-in o.d. (0.305-in i.d.) downflow reactor was enclosed in a three-zone programmable furnace. The liquid products were collected and separated into a wax trap (waxes) maintained at 140°C and a water trap (oil + water) maintained at 25°C. Two sets of these traps, positioned in parallel, enabled continuous operation. A Kammer back-pressure-control valve, located downstream of the traps, controlled the reactor and trap pressure.

An on-line GC-Carle (TCD) analyzed the permanent gases (H_2, CO_2, Ar, N_2, CH_4, CO) periodically (automatic GC injections every 15 min). Argon was used as internal standard, to eliminate any influence of variations in feed composition (due to pulsing) on the data. An on-line GC-FID (100-m Petrocol column, ramped from –25 to 300°C) analyzed the light hydrocarbons (C_1-C_{15}). This analysis was typically performed only once at the end of each run (and for pulse runs after the completion of the last pulse within that run). A third off-line GC-FID (15-m SPB-1 capillary column, 0.1-μm, ramped from 50 to 350°C) analyzed the composite wax and oil collected from the wax and water traps, respectively.

2.3. Reaction procedure

A physical mixture of 2 cc (1.55 g) of the calcined Co-ZrO_2/SiO_2 catalyst and 10 cc (15.91 g) of a low-surface-area (0.2 m^2/g) α-alumina (SA5397, Norton) was loaded into the reactor. The catalyst was reduced *in-situ* under H_2 at 350°C for 14 h, and was cooled and pressurized to ca. 300 psig (19.4 atm). Feeding a 10%Ar/CO gas mix started the FT reaction, establishing the following base reaction conditions:

Syngas (H_2 + CO)=50%, H_2:CO=2:1 (i.e., 33.3% H_2 and 16.7% CO)
Inerts (N_2 + Ar)=50% (1.7% Ar, 48.3% N_2)
P=300 psig, F=200 scc/min, SV=6000 h^{-1}.

The reaction temperature was increased (by 0.5°C/h or less) to 224°C and was stabilized at this value, allowing the reaction to reach a "pseudo-steady state". Pulse runs involved substituting the reactant feed flow (H_2+CO/Ar) with an equal molar flow of a pulse gas. The total molar flow and the reaction pressure were kept constant between base and pulse runs.

3. RESULTS AND DISCUSSION

A "blank" pulse run (i.e., switching between two equal flows of H_2/CO/Ar reactant mix) was performed in order to identify the possible effect of the periodic pressure disturbance (directly related to the applied pulse) due to non-ideal switching of the actuated valves. This run produced no measurable variation on CO conversion, outlet H_2:CO ratio, or product distribution (**α**-value, C_{10}-C_{20} yield). Therefore, pulse runs involving no variations in feed composition have no effect on measurements of the progress of the FT reaction.

A 1-min N_2 (inert) pulse per 1 hour (i.e., substituting the H_2/CO/Ar flow, which is 51.7% of the total, with an equal flow of N_2 for 1 min every hour) was applied so as to examine the effect of inert pulsing on the reaction progress. The N_2 pulse gave only minimal variations in activity (CO conversion) or product selectivity (**α**-value, CH_4 yield, C_{10}-C_{20} yield), implying that short (1-min) disruptions in reactant flow do not substantially affect the FT reaction.

In contrast to the inert pulse, a 1-min H_2 (reactant) pulse caused significant variations in CO conversion and CH_4 selectivity. Effects of varying the H_2 pulse frequency (1-min H_2 per 1, 2, and 4 hours) on CO conversion and C_1 (CH_4 and CO_2) selectivity are shown in the composite plots of Figures 2a and 2b, respectively. These plots are composed of 10-hour segments of a series of sequential runs (typically lasting 48 hours, so as to collect sufficient amounts of oil + wax for the analysis), starting and ending with a base (no pulse) run. The data points correspond to measurements of the reactor effluent gas every 15 minutes.

A 1-min H_2 pulse per 1-hour (10-20-hour segment in Figs. 2a and 2b) caused a significant increase in CO conversion (from 16% to ca. 30%). The measured temperature of the catalyst bed also increased to 226°C, indicating a strong reaction exotherm. The CO conversion decreased *gradually* until the next H_2 pulse. A less-pronounced increase in CO conversion was also observed for the 1-min H_2 pulse per 2-h and 4-h runs.

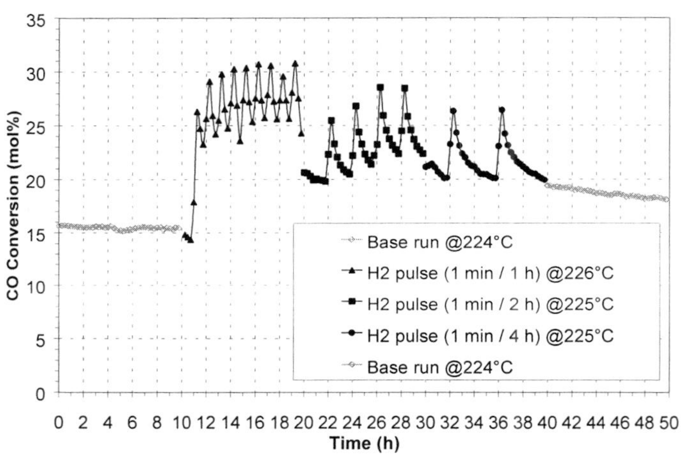

Figure 2a. Effect of H_2 pulse frequency on CO conversion

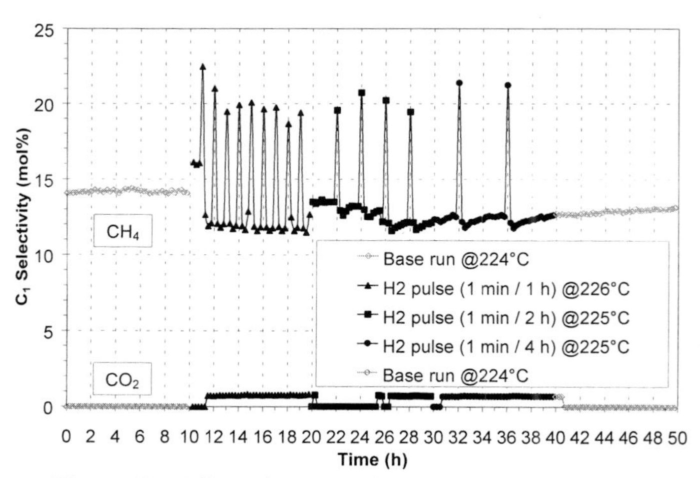

Figure 2b. Effect of H_2 pulse frequency on C_1 selectivity

The observed decrease in CO conversion after the pulse indicates that the activity tends to return to its steady state (comparing also the base runs before and after the 3 pulse runs). The measured changes in CO conversion cannot be attributed to variations in the inlet CO concentration since the conversion was based on comparing the inlet and outlet *ratios* of CO to the inert Ar (fed at a fixed ratio from a single gas cylinder).

The selectivity to CH_4 was observed to increase *instantaneously* after each H_2 pulse (from 13-14% to ca. 20% for all examined pulse runs). It was then quickly restored to its base value (Fig. 2b). Thus, H_2 pulsing increases catalytic

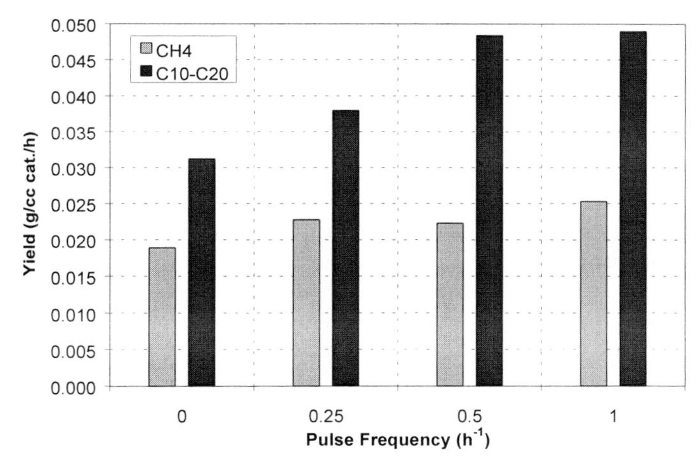

Figure 3. Effect of H_2 pulse frequency on product yield

activity while only briefly increasing the undesirable formation of CH_4.

The effect of varying H_2 pulse frequency on the desired C_{10}-C_{20} yield vs. the undesired CH_4 yield is shown in Figure 3. Pulse frequencies of 1, 0.5, and 0.25 h^{-1} correspond to a 1-min H_2 pulse per 1, 2, and 4 hours, respectively. The zero pulse frequency corresponds to the average of the two no-pulse (base) runs before and after the 3 pulse runs.

Both C_{10}-C_{20} and CH_4 yields increase with H_2-pulse frequency (and so does the yield of C_{21+}), obviously due to the enhancement in catalytic activity caused by the pulsing (Fig. 2a). As seen in Fig. 3, the effect of the 1-min H_2 pulse per 1 hour compared to the (average) base run was to increase the C_{10}-C_{20} yield by ca. 57%, while the CH_4 yield only increased by ca. 34%. Although this comparison entails a temperature change (from 224°C to 226°C), the increase in the C_{10}-C_{20} yield is more than what could be accounted for solely by a 2°C increase in reaction temperature. The CH_4 selectivity in the pulse runs (13-14% on molar basis) is lower than that of the base runs (15.5%), whereas the selectivity to C_{10}-C_{20} and C_{21+} compounds is higher (28-32% vs. 27%, and 23-24% vs. 20%, respectively). The α-values of the pulse runs (based on the molar fractions of C_{10}-C_{65} products) are identical (within experimental error) to that of the base runs (0.890±0.005). Thus, the applied H_2 pulsing apparently does not alter the SFA distribution.

Within the examined pulse frequency range, the greater difference between the yields of the desirable C_{10}-C_{20} and the undesirable CH_4 is obtained at the *intermediate* pulse frequency of 0.5 h^{-1} (1-min H_2 per 2 hours). Also, upon extrapolating to higher H_2-pulse frequencies, we could expect a stronger reaction exotherm and thus an increase in reaction temperature, which is known to cause a shift in FTS product distribution to lower molecular weight compounds and to enhance the methanation reaction [4]. Higher pulse frequencies would thus tend to increase the CH_4 yield much more than the C_{10}-C_{20} yield. An optimum H_2-pulse frequency (depending on catalyst and reaction conditions) would therefore be required for maximizing the C_{10}-C_{20} yield without substantially increasing the CH_4 yield.

Another series of H_2-pulse runs on the Co-ZrO_2/SiO_2 catalyst examined the effect of H_2-pulse duration on activity and product distribution, by varying the pulse duration (1, 2, 4-min of H_2) at a fixed pulse frequency (0.5 h^{-1}). The results of this study (not included here) are qualitatively similar to those of the variable-pulse-frequency study presented here: higher H_2-pulse duration causes an increase in both C_{10}-C_{20} and CH_4 yield, and the greater difference between these yields is obtained at the *intermediate* pulse duration of 2 min. Consequently, optimization of the pulse duration is also important in maximizing the formation of diesel-range FT products.

4. CONCLUSIONS

In contrast to "blank" or inert (N_2) pulsing, pulsing with H_2 has a significant impact on the activity and selectivity of the examined $Co\text{-}ZrO_2/SiO_2$ catalyst. H_2 pulsing causes a significant increase in CO conversion, along with an enhanced reaction exotherm. Then, the CO conversion decreases gradually until the next H_2 pulse, indicating that the catalyst activity tends to return slowly to its steady state, as measured in base (no-pulse) runs. On the other hand, the selectivity to CH_4 increases instantaneously after each H_2 pulse, and gets quickly restored to its steady-state value.

Increasing H_2-pulse frequency has a positive effect on the yield of both CH_4 and $C_{10}\text{-}C_{20}$. The selectivity to $C_{10}\text{-}C_{20}$ and C_{21+} compounds increases with H_2 pulsing compared to the base runs, but the chain-growth probability $\boldsymbol{\alpha}$ is essentially unaffected. An optimum set of H_2-pulse parameters (frequency and duration) appears to be needed to maximize the $C_{10}\text{-}C_{20}$ yield without substantially increasing the CH_4 yield.

5. ACKNOWLEDGEMENTS

Funding for this work (in part) by the US Department of Energy under Contract No. DE-FG26-99FT40680 is gratefully acknowledged.

REFERENCES

1. R.B. Anderson, The Fischer-Tropsch Synthesis, Acad. Press, New York, 1984.
2. M.E. Dry, Appl. Catal. A, 138 (1996) 319.
3. A.A. Adesina, Appl. Catal. A, 138 (1996) 345.
4. M.E. Dry, The Fischer-Tropsch Synthesis, in Catalysis - Science and Technology 1 (J.R. Ander and M. Boudart, eds.), Springer-Verlag, New York, 1981.
5. G. Parkinson, Chem. Eng., 4 (1997) 39.
6. C.D. Chang, W.H. Lang, and A.J. Silvestri, J. Catal., 56 (1979) 268.
7. R.J. Gormley, V.U.S. Rao, R.R. Anderson, R.R. Schehl, and R.D.H. Chi, J. Catal., 113 (1988) 195.
8. S. Bessell, Appl. Catal. A, 126 (1995) 235.
9. K. Jothimurugesan and S.K. Gangwal, Ind. Eng. Chem. Res., 37(4) (1998) 1181.
10. D.L. King, J.A. Cusamano, and R.L. Garten, Catal. Rev. Sci. Eng., 23(1-2) (1981) 233.
11. A.A. Adesina, R.R. Hudgins, and P.L. Silveston, Can. J. Chem. Eng., 25 (1995) 127.
12. J.W. Dun, and E. Gulari, Can. J. Chem. Eng., 64(2) (1986) 260.
13. G. Beer, Gas Conversion Process Using a Chain-Limiting Reactor, WO Patent No. 98/ 19979 (1997).
14. E. Peacock-Lopez and K. Lindenberg, J. Phys. Chem., 88 (1984) 2270.
15. E. Peacock-Lopez and K. Lindenberg, J. Phys. Chem., 90 (1986) 1725.
16. A.A. Khodadadi, R.R. Hudgins, and P.L. Silverston, Canadian J. Chem. Eng., 74 (1996) 695.
17. F.M. Dautzenberg, J.M. Heller, R.A. van Santen, and H. Berbeek, J. Catal., 50 (1977) 8.
18. A. Hoek, M.F.M. Post, J.K. Minderhoud, and P.W. Lednor, Process for the Preparation of a Fischer-Tropsch Catalyst and Preparation of Hydrocarbons from Syngas, US Patent No. 4,499,209 (1985).

Studies in Surface Science and Catalysis
J.J. Spivey, E. Iglesia and T.H. Fleisch (Editors)

Synthesis and Characterization of Proton-Conducting Oxides as Hydrogen Transport Membranes

Lin Li and Enrique Iglesia

Department of Chemical Engineering, University of California at Berkeley,
Division of Materials Sciences, E.O. Lawrence Berkeley National Laboratory,
Berkeley, CA 94720

The hydrogen permeation properties of dense $SrCe_{0.95}Yb_{0.05}O_{3-x}$ membranes were examined under various conditions and described based on the conductivities of available charge carriers using a transport model for mixed conductors. The chemical environment on each side of the membrane influences not only the chemical potential driving force for transport but also the conductivities of the various charge carriers. When one side of the membrane is exposed to oxidizing atmosphere, the H_2 flux is limited by hole conduction. In reducing atmosphere, $SrCe_{0.95}Yb_{0.05}O_{3-x}$ membranes show electron conductivity and the electron transference number increases with increasing H_2 partial pressure. The hydrogen flux-current behavior during electrochemical pumping is caused by an insulator-semiconductor transition that occurs with increasing applied voltage.

1. INTRODUCTION

Cost-effective reactor and separation technologies based on inorganic membranes can lead to significant economic benefits as components of several strategies for the conversion of natural gas. Oxygen-transport membranes based on mixed conductors are being developed for oxygen purification and for coupling with steam reforming reactions in the production of synthesis gas [1]. Oxides with perovskite structures similar to those of oxygen conductors but with immobile oxygen anions can also exhibit protonic or mixed protonic-electronic conductivity, without significant oxygen transport, in hydrogen or water atmospheres [2,3]. Such materials have been less extensively studied than oxygen conductors, but they can be used to extract H_2 from streams in the presence of chemical or electrochemical potential gradients. For example, dense perovskite membranes, especially as thin films with high conductivity, can be used to remove H_2 continuously from reactors used to carry out reactions with unfavorable thermodynamics. H_2 removal during catalytic methane pyrolysis [4] and during steam reforming [5] has been recently suggested in order to increase achievable CH_4 conversions in these endoergic reactions. In this paper, we report the synthesis and transport characteristics of dense disk $SrCe_{0.95}Yb_{0.05}O_{3-x}$ membranes.

2. EXPERIMENTAL METHOD

Combustion methods based on reactions of nitrate groups with metal chelates were used to prepare $SrCe_{0.95}Yb_{0.05}O_{3-x}$ ceramic powders with unimodal size distributions [6]. The powders were treated in air at 5°C/min to 1000°C for 2 h. X-ray diffraction patterns corresponded to the perovskite structure. These powders were then pressed at 142 MPa and sintered in flowing sir at 5°C/min to 1550°C for 2 h in order to obtain dense disks. After sintering, the disk membranes are 20 mm in diameter, 1 mm in thickness, and proved free of cracks by He permeation experiments at room temperature. The density of the sintered membranes were greater than 90% of the skeletal density,

The hydrogen flux through $SrCe_{0.95}Yb_{0.05}O_{3-x}$ membranes was measured using a high-temperature permeation apparatus. Gas flow rates in each side of the membrane disks were metered by electronic mass flow controllers (Porter). One side of the membrane was exposed to H_2 (Airgas, 99.99%) diluted to the desired concentration using He (Airgas, 99.999%). The opposite side was exposed to N_2 (Airgas, 99.999%) or N_2/O_2 (Airgas, 99.6%). The H_2 in the permeate flow was oxidized to water in a hydrogen conversion reactor using a Pt mesh as catalyst, and the hydrogen flux through the membrane was determined by measuring the water concentration using an HMI 38 humidity data processor (Vaisala). When the down side of the membrane was swept by N_2, a make-up O_2 was introduced in the sweep side in order to convert H_2 to water. In order to provide a pathway for electron transport, platinum wires were connected to both sides of membrane using Ag conductor ink (Alfa Aesar). After treating in air at 850°C for 0.5 h, the Ag ink formed a porous electrode layer on the membrane surface. Permeability measurements were performed under both open-circuit and closed-circuit conditions. In addition, electrochemical pumping experiments were carried out by applying a voltage (HP-6334B DC power supply) across the membrane.

3. MASS TRANSFER MODEL

In order to interpret experimental hydrogen transport rates, we adapted a transport model for oxygen separation membranes based on non-equilibrium thermodynamic treatments of diffusion processes [7] to proton-electron mixed conductors. By assuming adsorption-desorption equilibrium at each side of the membrane and protons ($OH^{\cdot}$), electrons (e'), and electron holes ($h^{\cdot}$) as the available charge carriers, the H_2 flux is given by:

$$j_{H_2} = \frac{\sigma_{OH^{\cdot}} I}{2\sigma_T F} - \frac{\sigma_{OH^{\cdot}}(\sigma_{h^{\cdot}} + \sigma_{e'})}{4\sigma_T F^2} \nabla \mu_{H_2} \tag{1}$$

where σ_k is the conductivity of charge carrier k, σ_T is the total conductivity, $\nabla \mu_{H_2}$ is the chemical potential gradient of H_2, and F is the Faraday constant.

For open-circuit conditions, the net current across the membrane is zero (I=0) and Eq.(1) can be simplified to:

$$j_{H_2} = -\frac{\sigma_{OH^{\cdot}} t_{el}}{4F^2} \nabla \mu_{H_2} \tag{2}$$

where $t_{el} = (\sigma_{h^{\cdot}} + \sigma_{e^{\cdot}}) / \sigma_T$ is the electronic transference number. For closed-circuit conditions, the electronic conductivity of the connecting wire is much higher than that of the membrane. Then, the H_2 flux is given by:

$$j_{H_2} = -\frac{\sigma_{OH^{\cdot}}}{4F^2} \nabla \mu_{H_2} \tag{3}$$

In our permeation experiments, the H_2 permeation rate is less than 0.1% of the feed and sweep gas flows; therefore, gas phase compositions on each side of the membrane are similar to those of the respective inlet streams and independent of the nature of the measurement (closed or open circuit). Therefore, $\sigma_{OH^{\cdot}}$ and $\nabla \mu_{H_2}$ in Eq. (2) and (3) are similar for open and closed circuit conditions. By comparing the hydrogen flux under open and closed circuit conditions, we can obtain the electron transference number:

$$t_{el} = \frac{(j_{H_2})_{open-circuit}}{(j_{H_2})_{closed-circuit}} = \frac{\sigma_{h^{\cdot}} + \sigma_{e^{\cdot}}}{\sigma_T} \tag{4}$$

4. RESULTS AND DISCUSSIONS

4.1 H_2 permeation of $SrCe_{0.95}Yb_{0.05}O_{3-x}$ membrane

Figure 1 shows the hydrogen flux through the membrane at 950 K under open and closed circuit conditions when the down side of the membrane is swept by N_2. In this situation, both sides of the membrane are exposed to a reducing atmosphere, although H_2 partial pressure in the sweep side is very low (< 1 kPa). Figure 2 shows the H_2 flux when 10% O_2/N_2 mixture was used as the sweep gas (denoted as asymmetric condition). In both cases, the H_2 flux increased with increasing H_2 concentration, but the measured fluxes were significantly higher with the oxidizing sweep gas (Figs. 1 and 2).

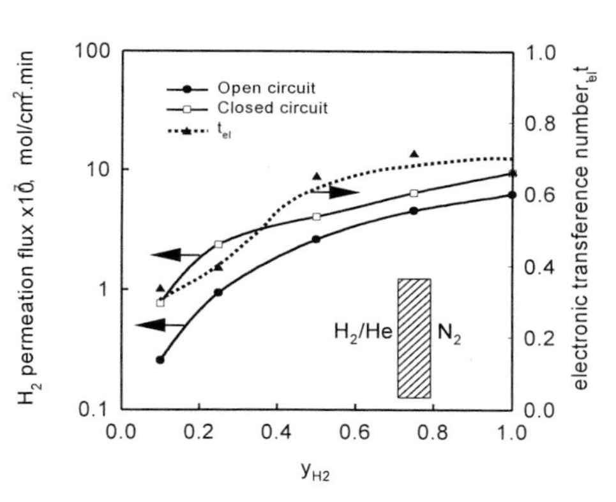

Figure 1. H_2 flux under symmetric conditions T=950 K

These results show that the chemical environment on both sides of $SrCe_{0.95}Yb_{0.05}O_{3-x}$ membranes influences hydrogen permeation rates. These effects can be described by considering the concurrent effects of the chemical environment on the permeation driving force and on the membrane conductivities and charge carrier concentrations. The changes in charge carrier concentrations reflect the thermodynamic equilibrium for the reactions:

$$1/2O_2 + V_O^{\cdot\cdot} = O_O^{\times} + 2h^{\cdot} \quad (5)$$

$$H_2 + 2O_O^{\times} = 2OH^{\cdot} + 2e^{'} \quad (6)$$

$$H_2O + V_O^{\cdot\cdot} + O_O^{\times} = 2OH^{\cdot} \quad (7)$$

In Eq.(2), $\nabla\mu_{H_2}$ is the driving force for mass transfer, and the $(\sigma_{OH} \cdot t_{el})$ can be considered as a resistance to transfer, which is determined by the concentrations and mobilities of charge carriers. When the chemical environment influences only the permeation driving force, Eq. (2) can be integrated with $(\sigma_{OH} \cdot t_{el})$ as a constant. Our symmetric and asymmetric flux measurements, however, show that $(\sigma_{OH} \cdot t_{el})$ deponds on the chemical environment and thus of position within the membrane. It seems, therefore, that a quantitative discussion of the H_2 concentration and of the sweep gas identity will require a detailed description of how $\sigma_{OH^{\cdot}}$ and σ_{el} vary with chemical environment.

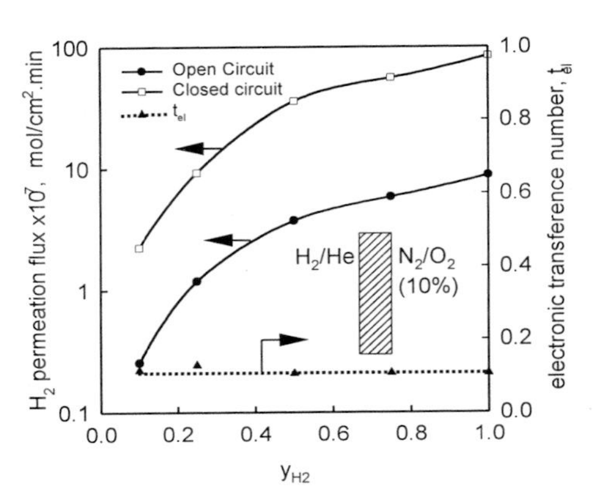

Figure 2. H_2 flux under asymmetrical condition T=950 K

An external circuit led to an obvious increase in H_2 flux, suggesting that electron conductivity limits overall hydrogen transport rates. Therefore, further improvement in hydrogen flux will require an increase in the electronic conductivity of the membrane.

4.2 Electronic transference numbers

The electron transference numbers (t_{el}), calculated from Eq. (4), are shown in Figures 1 and 2. Asymmetric conditions lead to low t_{el} values (~0.1), which are not influenced by H_2 partial pressure. Therefore, hydrogen permeation rates are largely controlled by the ability of the membrane to conduct electrons and to balance the flux of positive charge induced by proton transport.

$SrCe_{0.95}Yb_{0.05}O_{3-x}$ is an acceptor-doped oxide and its electronic conductivity is influenced by the oxygen activity in the environment [8]. From reaction (5) and (6), we can see that under asymmetrical conditions, the major carriers for electronic conduction are electrons at the reducing side and holes at the oxidizing side. The total electronic conductivity of the membrane will be determined by the lower of all the conductivities at each side of the membrane. With increasing H_2 partial pressure, the electronic conductivity in the reducing side of the membrane increases, because of higher electron concentrations, but the electronic conductivity in the oxidizing side remains unchanged because the hole concentration is determined by the O_2 partial pressure in the sweep gas. The constant electronic transference number as H_2 partial pressures varies (Figure 2) indicates that under asymmetric conditions, the electronic conductivity is controlled by the hole conduction in the oxidizing side. This conclusion is consistent with the results of H_2 permeation and conductivity measurements reported by Hamakawa et al [9]. Therefore, under this operation mode hole conduction rate must be increased in order to increase electronic conductivity and hence H_2 fluxes.

From a practical point of view, both sides of the H_2 separation membranes will be exposed to reducing atmospheres. In this situation, we need to maintain electronic conductivity through electron conduction. The electronic transference number results shown in Figure 1 demonstrated that with N_2 as the sweep gas, t_{el} values are significantly higher and they increase with increasing H_2 pressure. This is because in reducing atmospheres, Ce^{4+} can be partially reduced [10,11]; therefore the $SrCe_{0.95}Yb_{0.05}O_{3-x}$ membrane becomes a better electron conductor.

The increase in t_{el} with increasing H_2 partial pressure indicates that with increasing H_2 partial pressure, we increase not only the mass transfer driving force for H_2 transport, but also the electronic conductivity of the membrane, both of which have a positive effect on H_2 flux. For $SrCe_{0.95}Yb_{0.05}O_{3-x}$ membranes Hamakawa et al. reported no detectable H_2 flux when the downside was swept by Ar and concluded that the membrane behaved as a pure proton conductor [9]. The reason for this is that in their experiments, the H_2 partial pressure in the feed was only 1 kPa (vs. 20 - 100kPa in this study); therefore the electron concentration was low. From the above analysis, we conclude that the H_2 flux is limited by the rate of transport of different charge carriers as the chemical environment changes from reducing to oxidizing.

4.3 Electrochemical pumping

The permeation driving force was also varied by applying a direct voltage across the membrane. The H_2 permeation rate increased at low voltages, as predicted by Faraday's law [12] (Fig. 4). At higher applied voltages, H_2 permeation rates are lower than predicted from theory and the deviation becomes larger with increasing applied voltage, indicating that t_{el} concurrently increases. At low voltages, the membrane behaves as an insulator, but it becomes a semiconductor as the applied voltage approaches the band gap of the membrane material. At voltages larger than the band gap, the membrane behaves as an electronic conductor and the contribution of Electronic conduction to the total current becomes larger. Therefore, with increasing current density i, the slope of j_{H_2} -i plot decreases, and ultimately becomes a horizontal line because of the very high electronic conductivity at high applied voltages.

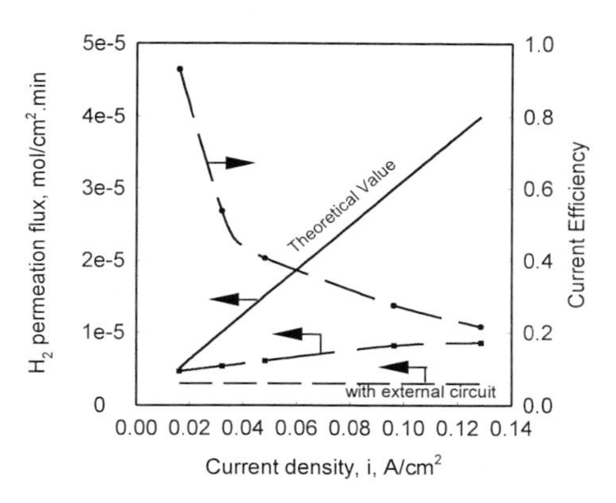

Figure 3. Electrochemical Pumping Results
T=1000 K

5. CONCLUSIONS

The hydrogen flux through $SrCe_{0.95}Yb_{0.05}O_{3-x}$ membrane disks was measured under both open-circuit and closed-circuit conditions, and the hydrogen permeation behavior was analyzed by using a mass transfer model for mixed proton-electron conductor. The change of atmosphere on each sides of the membrane influenced not only the mass transfer driving force for hydrogen transport, but also the membrane transport properties. The increase in hydrogen permeation flux achieved by using an external circuit to provide an additional pathway for electron transfer indicates that the hydrogen permeation rate through the membrane is limited by the electronic conduction within the membrane. By comparing the hydrogen permeation flux under open and closed circuit, the average electron transference number was estimated, and the limiting charge carriers for H_2 transport through the membrane were determined. Under asymmetrical conditions, the electronic transference number is only 0.1 and transport is controlled by hole conduction. In reducing environments the electronic transference number is higher, reflecting a higher electron conduction caused by the partial reduction of Ce^{4+}. These results indicate that in order to maintain high electronic conductivity in H_2 separation membranes, we must introduce reducible cations in order to enhance electron conduction. By applying an external voltage, the membrane can be operated as an electrochemical pump, and the hydrogen flux-current density behavior can be interpreted by considering the insulator to conductor transition of the membrane as the applied voltage increases.

REFERENCES

1. U. Balachandran et al., Catalyst Today, 36 (1997), 265-272
2. U. Iwahara, Solid State Ionics, 86-88 (1996), 9-15
3. X. Qi, Y.-S. Lin, Solid State Ionics, 130 (2000), 149-156
4. R. W. Borry, E. C. Lu, Y. H. Kim, E. Iglesia, Stud. Surf. Sci. Catal., 119 (1998), 403
5. T. Terai, X. H. Li, K. Fujimoto, Chem. Lett., 4 (1999), 323-324
6. E. C. Lu, E. Iglesia, Journal of Materials Science, accepted for publication
7. H. J. M., Bouwmeester, A. J. Burggraaf, Fundamentals of Inorganic Membrane Science and Technology, Ed. By A. J. Burggraaf, pp 435-528, Elsvier Science, Amsterdam, 1996
8. D. M. Smyth, Properties and Applications of Perovskite-type Oxides, Ed. By L. G. Tejuca and J. L. G. Fierro, pp47-72, Marcel Dekker, New York, 1992
9. S. Hamakawa, T. Hibino, H. Iwahara, J. Electrochem. Soc., 141 (1994), 1720-1725
10. H. L. Tuller, Nonstoichiometric Oxides, Ed. By O. Toff Sorensen, pp 271-325, Academic Press, New York, 1981
11. T. Norby, Solid State Ionics, 125(1999), 1-11
12. H. Iwahara, Solid State Ionics, 125(1999), 271-278

Studies in Surface Science and Catalysis
J.J. Spivey, E. Iglesia and T.H. Fleisch (Editors)

Methane to syngas: Development of non-coking catalyst and hydrogen-permselective membrane

J. Galuszka and D. Liu

Natural Resources Canada, CANMET Energy Technology Centre
1 Haanel Drive, Nepean, Ontario, Canada K1A 1M1

Pd/Al_2O_3, Ni/Al pillared clay (Ni/AlPLC) and Ni/large pore zeolite (Ni/LPZ) catalysts were tested for catalytic partial oxidation of methane to syngas. Filamentous carbon was formed in large quantities on Pd/Al_2O_3 . Ni/LPZ showed the least coking propensity. A specific limitation in material transport facilitated by the porous structure of the catalyst support seems to restrict coking propensity of Ni/ALPC and Ni/LPZ. Tubular silica membrane prepared by chemical vapour deposition (CVD) showed commercially attractive $He(H_2)$ permeance about 30 $cm^3/cm^2 \cdot min \cdot atm$ at $He(H_2)/N_2$ separation about 20 at 500°C and outstanding thermal stability. However, hydrothermal stability requires further improvement to reach commercial status.

1. INTRODUCTION

Catalytic partial oxidation of methane using oxygen or CO_2 reforming (dry reforming) may offer alternatives to expensive steam reforming [1] for producing synthesis gas [2,3]. The latter could also have a positive impact on global CO_2 emissions.

During partial oxidation of methane to syngas CH_4 undergoes combustion and CO and H_2 form from secondary reversible reactions of the unreacted CH_4 with H_2O and CO_2 and water gas shift reaction. It was proven [4] that the limit imposed on CH_4 conversion and yields of CO and H_2 by the reversible nature of these reactions could be circumvented in a hydrogen-permselective membrane reactor. CH_4 conversion increased up to 20% and CO and H_2 yield increased to about 18%. However, it was determined that palladium (metallic) membrane used in patented [4] syngas generating technology has no practical application in hydrocarbon or CO containing environment due to rapid "corrosion" caused by filamentous carbon deposition [5]. Therefore, the feasibility of generating syngas in an industrial membrane reactor [6] depends on developing a non-metallic, hydrogen-permselective membrane that is thermally stable and assures high, selective flux of hydrogen. Also, a suitable catalyst with no propensity to coking needs to be developed.

The present study investigated both these aspects in parallel. Since proof of concept of a membrane reactor for syngas generation was already secured [4] there is no real need to study a potentially deficient catalyst in a potentially inferior membrane reactor. Therefore, coking propensity was studied in a conventional, fixed-bed reactor and development of ceramic, hydrogen-permselective membrane was carried out independently.

2. EXPERIMENTAL

A 5.0 wt % Pd/γ-Al_2O_3 catalyst was prepared by incipient wetness impregnation of the γ-Al_2O_3 (BET surface area = 200 m^2/g) support with a solution of $PdCl_2$ salt. The 5 wt % Ni supported on alumina pillared bentonite [7] (AlPILC) was prepared by impregnation. Freshly prepared AlPILC (BET surface area = 215 m^2/g) was suspended in a water solution of $Ni(NO_3)$, stirred and evaporated to dryness over a heating plate. The Ni supported on large pore zeolite (Ni/LPZ) was prepared by a four step impregnation of MCM type material [8] (BET surface area = 685 m^2/g) with $Ni(NO_3)_2$ using methanol as surfactant. Partial oxidation and dry reforming of methane to syngas were carried out in a fixed-bed reactor. The reactor was charged with 0.5-3.0 g of catalyst which was placed centrally and rested on quartz wool.

The reactant streams were fed through mass flow controllers and contained 3:1 CH_4/O_2 (stream I), 3:1:2 (stream II) or 3:2:2 (stream III) $CH_4/O_2/CO_2$ diluted in 70% N_2. Nitrogen contained in the feed streams was used as internal standard for calculating CH_4 conversion and reaction product yields and selectivities. Excess methane was used to simulate the process outside explosive limits for CH_4/O_2 mixtures when N_2 diluent would not be present in the feed stream. In runs with Ni/LPZ, 2:1 CH_4/O_2 ratio was used. Catalyst bed was maintained between 550°C-750°C as measured with a thermocouple located inside the catalyst bed. The streams were analyzed for products and reactants by on-line TCD-gas chromatography.

The starting material for membrane preparation was a 60 cm long α-alumina porous tube with 10 mm o.d. from Coors Ceramics, USA. Reduction of the macropores of the initial tube to about 50 nm was achieved by coating the tube with α-alumina followed by γ-alumina which formed the top layer about 1-2 μm thick. The top silica layer was deposited on γ-alumina layer by chemical vapour deposition (CVD) using tetraethyl-orthosilicate (TEOS). Helium (He was used instead of H_2 for safety reasons) and N_2 permeations of the membranes were monitored in situ during thermal and hydrothermal tests in automated membrane testing facilities at 500°C. During hydrothermal stability tests the membranes were exposed to N_2 flow containing 10 vol % water.

Fresh and spent catalysts were examined by Jeol JSM-5300 scanning electron microscope (SEM). Ceramic substrates and silica membranes were examined by SEM Hitachi S-4700. The amount of carbon on spent catalysts was determined by temperature programmed oxidation (TPO) in a Cahn thermogravimetric analyzer.

3. RESULTS AND DISCUSSION

3.1 Catalytic partial oxidation of methane

3.1.1 Pd/Al_2O_3 catalyst

Figure 1 gives the results for partial oxidation of CH_4 to syngas using O_2 (stream I). The yield of H_2 and CO increased from 32 to 73% and 23 to 60% between 550°C-700°C. Concurrently, H_2O and CO_2 yields decreased to below 2%. H_2/CO ratio was about 2.5 at 700°C. Catalyst activity remained relatively steady during 10-h periods at each temperature. The catalyst showed no signs of deactivation after 40 h on stream.

The addition of CO_2 to the feed stream clearly increased CH_4 conversion by as much as 10% for stream II at 700°C and 40% for stream III at 600°C. Almost complete CH_4 conversion was attained for stream III at 700°C. However, increased concentration of O_2 in stream III had

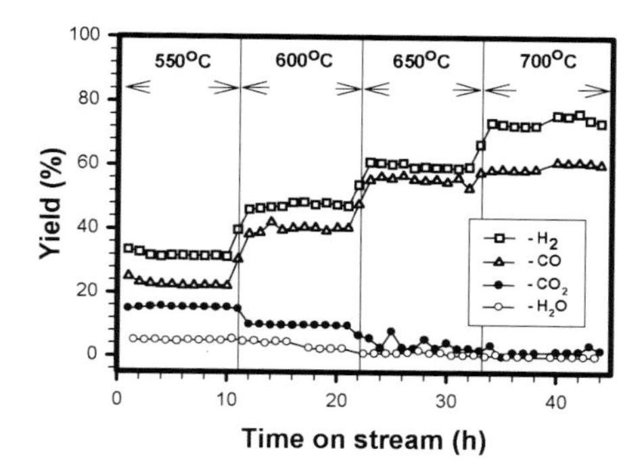

Fig. 1 - H_2, CO, CO_2 and H_2O yields during partial oxidation of CH_4 with O_2 for stream I: total flow – 100 mL/min gave contact time of about 4 s.

a detrimental effect on CO_2 conversion. Net CO_2 conversion for stream III was about 16% at 700°C which was about four times lower than for stream II. Also, at 700°C the H_2/CO ratio decreased from 2.5 for stream I to 1.4 and 1.5 for streams II and III. Although the relative concentrations of reactants in streams II and III were not optimized, the feasibility of increasing CH_4 conversion and adjusting H_2/CO ratios was demonstrated.

It must be emphasized, however, that large amounts of carbon (up to 60% by weight) were identified and measured by TPO on spent Pd/Al_2O_3 catalysts. SEM revealed a filamentous nature of the deposited carbon. Formation of filamentous carbon refreshes the metal surface [9] which explains the relatively insignificant decrease of catalyst activity even after 200 h on stream and encourages revision of literature data that claimed equilibrium gas composition was achieved at high temperatures using supported metal catalysts [2].

3.1.2 Ni/AlPLC catalyst

Figure 2 gives the results of partial oxidation of CH_4 to syngas using oxygen on Ni/AlPLC catalyst. Some fluctuations in the conversion and selectivities were noted with time on stream.

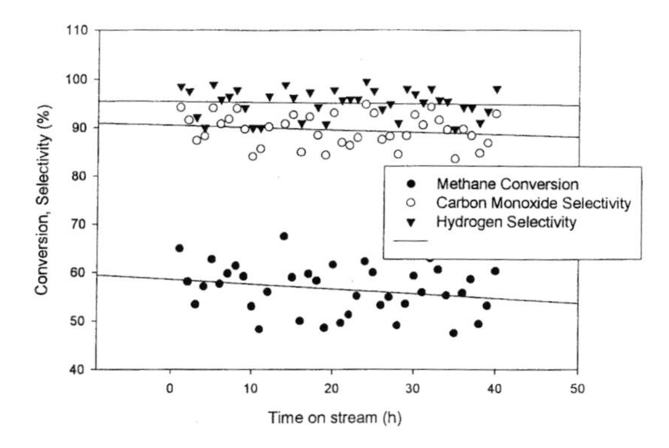

Fig. 2 - Performance of Ni/AlPLC catalyst at 725°C.

In general, CH_4 conversion remained between 55-65%, CO selectivity 90-95% and H_2 selectivity greater than 95%. After about 40 h on stream CH_4 conversion decreased by about 5%. TGA of the spent catalyst revealed only 3.3 wt % carbon. Filamentous carbon was only scarcely observed.

Ni/AlPLC seems to be a promising catalyst showing high activity and selectivity for syngas production. However, its long term stability seems to critically depend on CH_4/O_2 ratio in the feed. High ratio tends to produce carbon deposit on the catalyst, causing its deactivation. Low CH_4/O_2 ratio tends to promote the growth of oxidized phase thus drastically reducing catalyst activity and selectivity. This agrees with the substantial sensitivity of this catalyst to initial treatment that either reduced or oxidized the metal. Porous structure of the catalyst support is thought to play a significant role in limiting carbon deposition. This catalyst must be further optimized to reach more definitive conclusions.

3.1.3 Ni/LPZ catalyst

Time on stream performance of Ni/LPZ catalyst in partial oxidation of methane to syngas with O_2 is shown in Fig. 3. Catalyst activity remained relatively steady although oscillations in

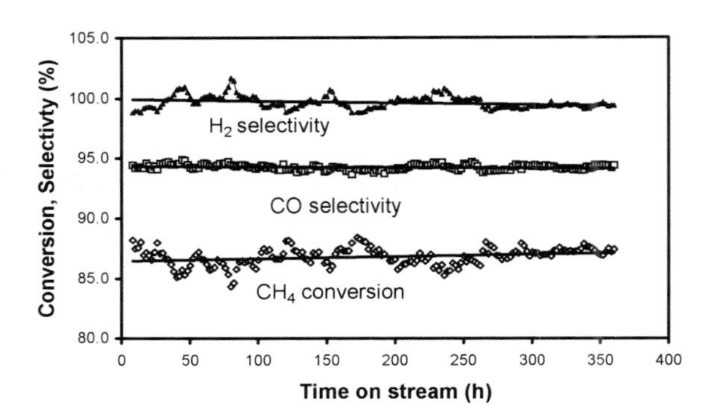

Fig. 3. Performance of Ni/LPZ in partial oxidation of CH_4 to syngas at 750°C

conversion and selectivities were noted with time on stream as shown in Fig. 3. However, Ni/LPZ catalysts showed no deactivation after 360 h on stream.

Carbon accumulated on Ni/LPZ constituted less than 1 wt % after 360 h. Filamentous carbon was hardly observed. Apparently the porous structure of the catalyst support seemed to play a significant role in limiting the carbon deposition like the Ni/AlPLC catalyst. Of course, more data will be needed to fully substantiate this claim. Patent is pending for the Ni/LPZ catalyst.

3.2 Hydrogen permselective membrane

The highest He permeance of our silica membrane measured at 500°C with a single component was 32.0 $cm^3(STP)/cm^2 \cdot min \cdot atm$. He/N_2 selectivity was 17.6 greatly exceeding the Knudsen separation of 2.6. The reported H_2/He permeance ratios (separation) were $\gtrsim$ 1 [10,11]. Therefore, the H_2 permeance of this membrane must be either equal to or slightly higher than the

He permeance. This places this membrane at the leading edge of international development.

Figure 4 shows results of the thermal stability test for the membrane. The relative changes (referenced to performance at time = 0) in He permeance and He/N_2 separation after about 15 d on stream are less than 7%. The initial slight decrease in N_2 permeance pushed the relative selectivity above 100%. However, membrane exposure to stream containing water gradually decreased He permeance and as a result He/N_2 separation. Both decreased irreversibly by about 60% from the initial values during another 11 d on stream before they tended to level off.

Figure 5 shows a cross-section of the hydrothermally treated membrane. Clearly, thermal and hydrothermal treatment cause no mechanical separation between the top γ-alumina layer and the underlying alumina layer as reported by Nijmeijer [12]. This corroborates the observation that hydrothermal decrease in He/N_2 separation was caused mainly by the decrease in He permeance. Densification of the amorphous silica constituting the membrane seems to be the most likely cause [11].

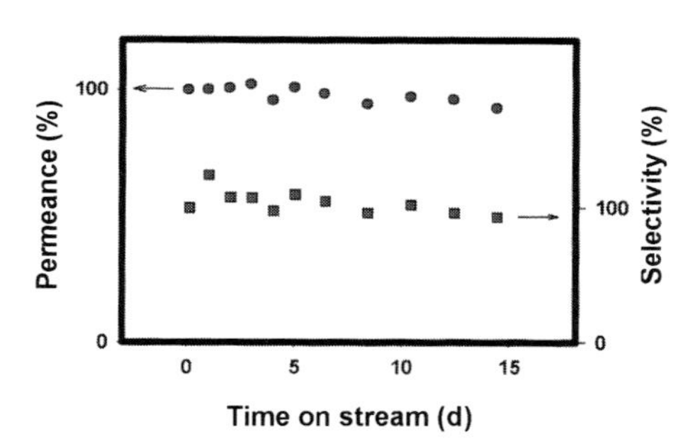

Fig. 4. Results of thermal stability test for hydrogen permselective membrane at 500°C.

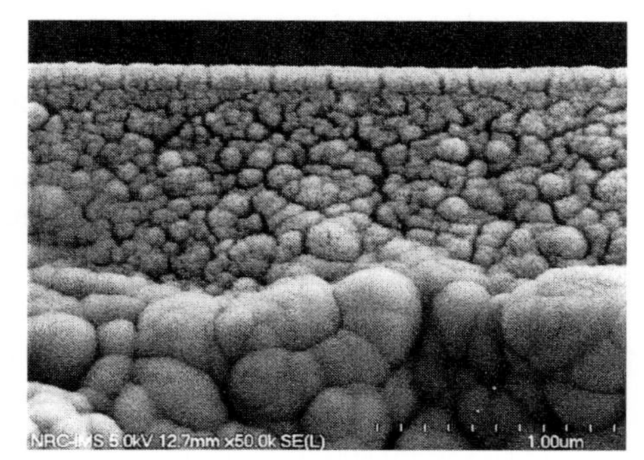

Fig. 5. SEM (X5OK) cross-section of hydrothermally treated membrane.

4.0 CONCLUSIONS

Due to a high probability of filamentous carbon formed on syngas catalysts containing metal on conventional supports such as alumina or silica, the apparent lack of change in CH_4 conversion or CO and H_2 selectivity is an inadequate indicator of catalyst steady performance. Long time on stream studies combined with analysis of the spent catalysts are highly recommended.

The porous structure of AlPLC and LPZ supports seemed to minimize coking propensity of Ni catalysts and filamentous carbon was hardly observed. Although extensive characterization has not yet been done, the most likely mechanism seems to be based on a specific mass tranfer limitation, facilitated by the porous structure of these supports. When the rates of the reagents supply are lower than the rate of methane oxidation a variable CH_4/O_2 ratio is realized that may create either fuel- or oxidant-rich environment on an alternating basis near the catalytically active Ni center. In the oxidant-rich environment surface carbon formed on Ni during fuel-rich cycle may be oxidized. However, oxidation of Ni^o to an inactive NiO may also occur [13]. This may explain the oscillations [13] of activity clearly visible especially for Ni/AlPLC.

Since Ni catalysts supported either on Al_2O_3 or SiO_2 equally underwent extensive coking [2,14,15], support acidity does not seem to be a prevailing factor for the observed suppressed coking propensity. The positive effect of a probable higher dispersion of the metallic phase, especially on suppressing the filamentous carbon formation [9], cannot be ruled out.

Simple stoichiometry allows one to calculate that to support a 1000 t/d methanol plant operation, H_2 contained in syngas must be supplied at $9.7 \cdot 10^8$ cm^3/min. Therefore, the maximum required surface of a hydrogen-permselective membrane having H_2 permeance of 30 $cm^3/cm^2 \cdot min \cdot atm$ in a membrane syngas reactor would be about 3240 m^2. Since not all H_2 needs to permeate through the membrane, lowering the membrane surface by about 30% seems to be quite feasible. Assuming a 5% allowable CH_4 slippage during syngas production sets the required H_2/CH_4 (N_2) separation of the membrane at about 20.

Therefore, a target for hydrogen permeance exceeding 20 $cm^3/cm^2 \cdot min \cdot atm$ at H_2/N_2 separation about 20 at 500°C has already been achieved. However, hydrothermal stability of the membrane needs further improvement.

ACKNOWLEDGEMENTS

Contribution to membrane development by Mr. T. Giddings and Mr. I. Clelland is acknowledged. The authors are indebted to Dr. S. Ahmed of Guelph Chemical Laboratories Ltd. for Ni/AlPLC catalyst testing. SEM analysis was done by Mr. G. Pleizier and Mr. J. Fraser, National Research Council of Canada. Financial support provided by the Consortium on the Conversion of Natural Gas (CCNG) and Federal Program on Energy Research and Development (PERD) is gratefully acknowledged.

REFERENCES

1. D.L. Trimm and M.S. Wainwright, Catal. Today, 18 (1993) 305.
2. S.C. Tsang, J.B.Claridge and M.L.H. Green, Catal. Today, 23 (1995) 3.
3. A.P.E. York, J.B. Claridge, A.J. Brungs, S.C. Tsang and M.L.H. Green, Chem. Commun., 39 (1997).
4. J. Galuszka, S. Fouda, R.N. Pandey and S. Ahmed, US Patent 5,637,259 (1997).
5. J. Galuszka, R.N. Pandey and S. Ahmed, Catal. Today, 46 (1998) 83.
6. G. Saracco, H.W.J.P. Neomagus, G.F. Versteeg and W.P.M. van Swaaij, Chem. Eng. Sci., 54 (1999) 1997.
7. R.T. Yang, J.P. Chen, E.S. Kikkinides, L.S. Cheng and J.E. Cichanowicz, Ind. Eng. Chem. Res., 31(1992)1440.
8. C.J. Guo, C.W. Fairbridge and J-P. Charland, US Patent 5,538,710 (1996).
9. J. Galuszka and M. Back, Carbon, 22(1984)141.
10. J.C.S. Wu, H. Sabol, G.W.Smith, D.L. Flowers and P.K.T. Liu, J. Membr. Sci., 96 (1995) 275.
11. R.M. deVos, Thesis Enschede, ISBN 90 365 11410 (1998).
12. A. Nijmeijer, Thesis Enschede, ISBN 90 365 11410 (1999).
13. Y. H. Hu and E. Ruckenstein, Ind. Eng. Chem. Res., 37 (1998) 2333.
14. J.B. Claridge, M.L.H. Green, S.C. Tsang, A.P.E. York, A.T. Ashcroft and P.D. Battle, Cat. Lett., 22 (1993) 299.
15. P. Grochi, P. Centola and R. Del Rosso, Applied Cat. A: G., 152 (1997) 83.

Studies in Surface Science and Catalysis
J.J. Spivey, E. Iglesia and T.H. Fleisch (Editors)

Site reactivity of Fischer-Tropsch synthesis catalysts studied by $^{12}CO \rightarrow {}^{13}CO$ isotope transients

C. J. Bertole[1], C. A. Mims[1]*, G. Kiss[2], P. Joshi[2]
[1]University of Toronto, [2]ExxonMobil Research and Engineering Company

Abstract

A novel application of $^{12}CO \rightarrow {}^{13}CO$ isotope transients at elevated reactor pressures was used to measure the intrinsic site reactivity of a variety of cobalt Fischer-Tropsch synthesis catalysts (220°C, 5 atm syngas, H_2:CO = 2:1, CO conversion = 12-16%, C_{5+} selectivity > 70%). We found that the *in-situ* measured CO-TOF and methane selectivity values were not affected by either the identity of the support (TiO_2, Al_2O_3, and SiO_2) or the presence of Re promoter. A mild variation is evident in the data, but these variations do not correlate with the metal dispersion, the support or the presence of Re, and thus must be caused by another, as of yet unknown factor.

Introduction

Cobalt Fisher-Tropsch synthesis (FTS) catalysts can contain up to four components: (i) the active cobalt metal, (ii) the support (SiO_2, Al_2O_3 or TiO_2), (iii) a small amount of a second metal as a structural or chemical promoter (Ru or Re), and (iv) an oxide promoter (e.g. ZrO_2, or La_2O_3) [1]. This paper will explore the effect of support and metal promoter on the intrinsic site activity of cobalt FTS catalysts, which will be measured by *in-situ* isotope transient methods.

Steady state isotope transient experiments provide an excellent tool to study the intrinsic performance of heterogeneous catalysts [2, and references therein]. They involve placing isotopic labels on reacting molecules and monitoring labels in a time resolved manner as they become incorporated into reaction products. The transient responses yield *in-situ* information about the number and amounts of catalyst surface intermediates, their surface lifetimes, and the sequence of reaction steps, all without disturbing steady state reaction conditions. We used isotope transient experiments, specifically $^{12}CO \rightarrow {}^{13}CO$ switches, to obtain information about the amounts and reactivity of the active surface carbon intermediates and reversibly adsorbed CO on various cobalt FTS catalyst formulations, and to provide insight about the effect of promoter and support.

There have been many studies documented in the literature that have used $^{12}CO \rightarrow {}^{13}CO$ isotope transients to investigate cobalt FTS; these include works by Goodwin *et al.* [1,3-5], Holmen *et al.*[6-8], Mims *et al.* [9,10] and van Dijk *et al.* [11]. All of these studies have been limited to 1 atm total reaction pressure and most of them to very high H_2:CO feed ratios (from 5:1 to 15:1). Under these conditions cobalt is predominantly a methanation catalyst with 50-80% selectivity to CH_4. The catalyst surface composition under FTS conditions (H_2:CO = 2:1 and elevated pressures) can be quite different (i.e., saturated CO coverage and liquid-filled pores) from that in a high methane selectivity regime. Thus, we have extended the application of CO isotope transient experiments to study cobalt-catalyzed FTS at medium syngas pressures (~5 atm syngas and H_2:CO = 2:1), where selectivity to C_{5+} hydrocarbons is above 70% (Table 1).

Experimental

The reaction system has been described in detail elsewhere [12]. Briefly, reactions were performed in a single pass, differential, down-flow, tubular (4 mm ID) quartz, fixed bed reactor. The catalyst charge (40-200 mg), mixed with diluent (TiO_2, 80-150 mg) to help maintain isothermality, was supported on a quartz wool plug in the reactor tube. Catalysts were sieved below 0.1 mm particle size to avoid mass transfer limitations on the overall CO rates. A K-type thermocouple with a stainless steel sheath, placed in the middle of the catalyst bed, monitored the reaction temperature. Pressure control in the reactor was achieved using a back pressure regulator. The reactor effluent was analyzed using two methods. A differentially pumped quartz capillary leak situated directly under the catalyst bed allowed continuous monitoring of the reactor effluent by MS (UTI 100C), while a GC (HP5880A) equipped with a Porapak QS packed column and FID was used to determine the C_1-C_4 production rates. Another differentially pumped stainless steel capillary leak placed after the GC column provided GC-MS determination of the non-FID sensitive compounds (i.e., CO, Ne and CO_2). A PC-based computer program controlled data collection for the MS. Neon was used as an internal standard in the reactor feed to calculate the overall rate of CO conversion and as a tracer in the isotope transient experiments.

The suite of catalysts studied includes unsupported cobalt, and Co-only and metal-promoted (Re, Ru) Co on TiO_2, Al_2O_3, and SiO_2. The cobalt mass fractions on the supported catalysts ranged from 6 to 30% with cobalt dispersions ranging from 0.001 to 0.055 as deduced from the *in-situ* CO inventories measured in the isotope transient experiments. All catalysts, except for the unsupported cobalt powder, were prepared by the incipient wetness impregnation technique as described in previous studies [13,14]. Before reaction, the catalysts were reduced between 370-450°C for 1-4 h in 1 atm H_2 flowing at 40 ml/min after ramping from room temperature at 0.5°C/s. The catalysts were then cooled below reaction temperature (190-200°C) in H_2, at which point the reactor feed was switched to syngas (H_2:CO:Ne = 2:1:1) and the pressure slowly raised to 6.4 atm. At the same time, the temperature was raised to 220°C. Total gas flow rates ranged from 20 to 46 ml/min (STP). Activity measurements were obtained after ~20 h on-stream, and after steady state operation was established, the isotope transient experiments were performed. Labelled ^{13}CO (mixture of ^{13}CO/Ar at 1:1) was injected using a volume loop on the ^{12}CO/Ne feed line. The Ar provided a second counter-tracer for the transient experiments. Pressures were balanced between the volume loop and the reactor by use of a pressure balancing line that connected the volume loop with the ^{12}CO/Ne feed line upstream of the volume loop. On-line MS monitored the progress of the isotope transient experiment using the following *m/e* ratios: ^{12}CO (28, 30), ^{13}CO (29, 31- the ^{13}CO gas contained ~10% ^{18}O), $^{12}CH_4$ (15 - contributions to *m/e* 15 from C_{2+} hydrocarbons are negligible compared to CH_4 in these experiments), Ne (22) and Ar (40). For an isotope transient experiment, the area between the decay of ^{12}C in CH_4 and the tracer yields the average CH_4 time constant. In the case of a response curve represented by a single exponential, this would represent the time required to washout the old isotope in the active carbon pool for methane formation to 1/e of its initial steady state value. The area between the washout of the old ^{12}CO isotopomer and the tracer determines N.CO/N.Co, which is the ratio of the amount of reversibly adsorbed CO to the total number of cobalt atoms. CO is known to reversibly adsorb onto reduced cobalt metal [9], thus, the isotope transient experiment can provide an *in-situ* measure of metal dispersion, if the surface is saturated with CO (as it is likely to be at high CO partial pressures). Previous studies have mainly used *ex-situ* techniques to measure catalyst dispersion (TEM, or H_2 or CO chemisorption). However, the catalyst is undoubtedly different under synthesis conditions than it is during the *ex-situ* chemisorption measurements (i.e., different gas atmosphere, temperature, and perhaps non-reproducible initial reductions). With the isotope transient experiment, there is no need to halt the reaction, or remove the catalyst from the reactor in order to obtain information about the state of the catalyst surface.

Experimental Results and Discussion

In order to separate the effects of the catalyst and operating conditions on measured performance, it was imperative to test the various catalysts under nearly identical conditions, which includes temperature, pressure, H_2:CO ratio *and* CO conversion. The latter is very important, since CO conversion levels have a profound effect on product selectivity [14,15], i.e., as conversion increases CH_4 and C_{5+} selectivity decrease and increase, respectively. For the data in Table 1, all data were obtained in a narrow CO conversion range of 12.5-16.5%, thus, any observed differences in catalyst performance should relate to differences arising from catalyst properties and not experimental conditions.

Catalyst Sample #	Support	Co wt%	Re wt%	N.CO/N.Co	CO-TOF / s^{-1}	$\langle k_{C*} \rangle$ / s^{-1}	CH_4 selectivity	C_{5+} Selectivity
T-6	TiO_2	5.7	0.0	0.022	0.101	0.191	0.113	0.795
T-14	TiO_2	14	0.0	0.033	0.078	0.226	0.153	0.722
T-11	TiO_2	11	0.9	0.041	0.076	0.183	0.147	0.732
T-18	TiO_2	18	0.6	0.040	0.098	0.222	0.129	0.767
A-12	Al_2O_3	12	1.0	0.056	0.070	0.164	0.133	0.756
A-20	Al_2O_3	20	2.2	0.055	0.066	0.125	0.131	0.754
S-20	SiO_2	20	1.7	0.053	0.108	0.203	0.144	0.759
S-30	SiO_2	30	2.5	0.046	0.085	0.275	0.165	0.721

Table 1: Kinetic parameters obtained on a series of supported cobalt catalysts in a fixed bed micro-reactor at $220^{\circ}C$, $H_2/CO/Ne = 2:1:1$, Ptot = 6.4 atm, CO conversion 12.5-16.5%.

Table 1 contains the results of the isotope transient experiments for several catalyst compositions. Four key kinetic parameters are listed: (1) the carbon-based methane selectivity, (2) the carbon-based C_{5+} selectivity, (3) the carbon monoxide turnover frequency (CO-TOF), obtained by dividing the rate of CO conversion by the steady-state inventory of reversibly adsorbed CO (N.CO/N.Co) and (4) the average reactivity of the surface active carbon, $\langle k_{C*} \rangle$. This last parameter is the inverse of the average time constant for the washout of the old isotope in the methane product (τC^* (CH_4)) during the isotope transient. The carbon transients are fairly well represented as a single exponential with an additional minor component with longer lifetimes. The washout behavior measured in methane is expected to be similar to that in the higher hydrocarbon products, as shown in previous studies [9,11]. Thus, the decay of the old isotope in methane is likely to be representative of the washout of the entire CH_x monomer pool.

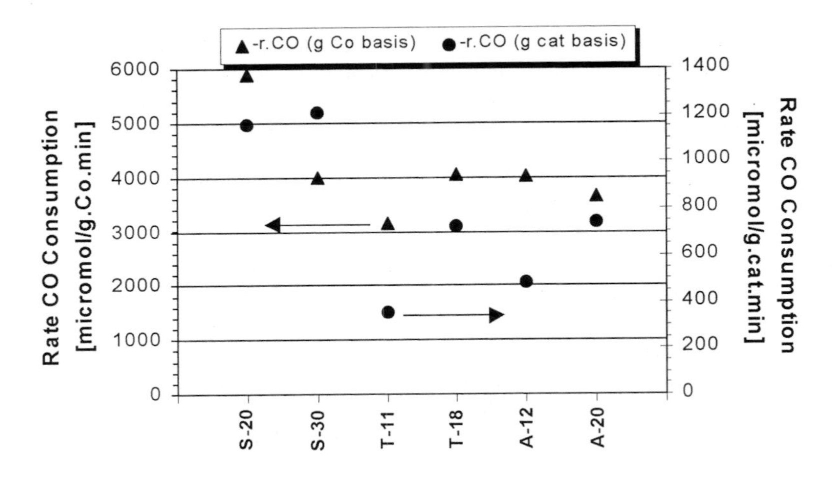

Figure 1: Comparison of the overall activity of various Co-Re catalysts on different supports. S = silica, T = titania, A = alumina; catalyst sample designators are fully explained in Table 1. Experimental conditions were the same as given in Table 1.

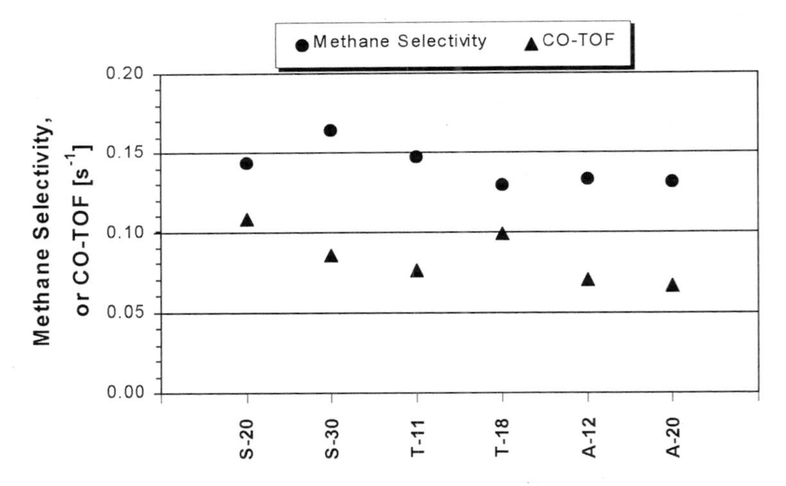

Figure 2: Comparison of Co-Re catalysts on different supports. S = silica, T = titania, A = alumina; catalyst sample designators are fully explained in Table 1. Conditions the same as given in Table 1.

Figure 1 shows the overall activity (mass cobalt and mass catalyst basis) of the various Co-Re catalysts tested in this study; note that the overall rates are comparable on all three supports. Figure 2 clearly shows that the identity of the support has no impact on the site reactivity (CO-TOF and the active carbon reactivity - Table 1) nor does it have a noticeable effect on the methane selectivity, although there is a mild variability noticeable in the CO-TOF values. These observations, including the mild variations in CO-TOF, agree with previous results obtained by Iglesia *et al.* [13-15]. Oukaci *et al.* [16] also demonstrated that site reactivity is independent of the support identity; however, a few of their results do not seem to agree with ours. Their titania-supported catalysts had a much lower overall activity than their silica and alumina counterparts, but this is simply a result of the lower Co dispersion they achieved on titania relative to the other supports. Figure 1 clearly shows that we obtained comparable overall activity for Co-Re catalysts on all three supports, and the dispersion data listed in Table 1 verifies that our titania-supported Co-Re catalysts had dispersion values comparable to the other two supports. Oukaci *et al.* also observed higher methane selectivity for titania-supported catalysts relative to the other supports; as mentioned above, we did not reproduce this behaviour. The reason for this discrepancy is not known.

With respect to the effect of Re on the titania supported cobalt catalysts (Figure 3), Re appears to increase the Co dispersion, which agrees with previous studies [15,16]. However, it had no impact on neither the methane selectivity nor the site reactivity; the former contradicts recent results by Guczi *et al.* [17], who concluded on their results that the presence of Re always leads to enhanced methane production.

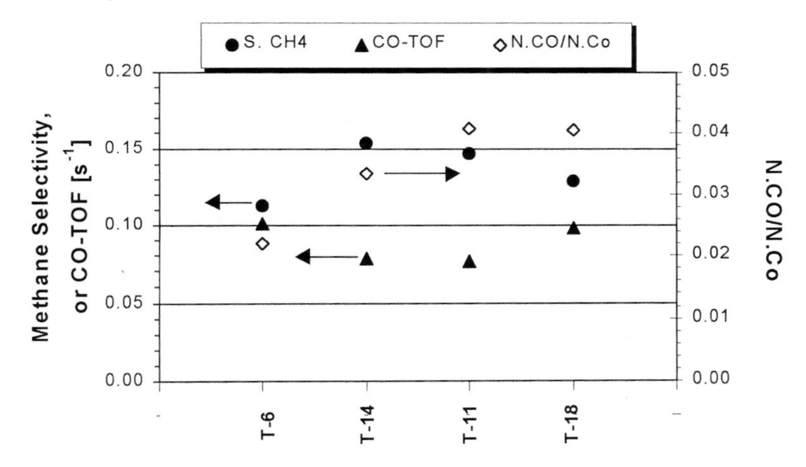

Figure 3: Comparison of the effect of Re in titania supported cobalt catalysts. T = titania; T-6 and T-14 are Co-only catalysts, while T-11 and T-18 contain Co-Re; catalyst sample designators are fully explained in Table 1. Experimental conditions were the same as given in Table 1.

In summary, the various catalysts listed in Table 1, and in Figures 2 and 3, show similar values for all four measured parameters ($S.CH_4$, $S._{C5+}$, CO-TOF and $\langle k_{C*} \rangle$), although a mild variability is evident in the data. The CO-TOF values ranged from 0.066 to 0.11 s^{-1} under these conditions, with the values for most of the catalysts falling into a narrower range of 0.076 to 0.11 s^{-1}. One of the lowest values (0.06 s^{-1}) was obtained on an unsupported cobalt catalyst with very low dispersion (CO(ads)/Co = 0.001, not shown in Table 1). Methane selectivities ranged from 11 to 16%. The average lifetimes of the surface carbon-containing reaction intermediates (measured by $\tau C^*(CH_4)$) was somewhat more variable, ranging from 3.5 to 8 seconds. These modest variabilities, although outside experimental uncertainty, showed no systematic correlation with the identity of the support (Figure 2), the cobalt metal dispersion (Table 1), or the presence of rhenium (Figure 3), mirroring previous studies based on *ex-situ* characterization methods [13-15]. However, the fact remains that there is some variability in the data. As mentioned previously, the CO-TOF values from Iglesia *et al.* [13] also showed a mild variability, which did not correlate with support identity. Their lowest and highest CO-TOF values differed by a factor of 2x for the various catalysts that they tested, but they did not interpret those differences as significant due to the uncertainties associated with their *ex-situ* characterization methods (i.e., different gas atmosphere, temperature, and perhaps non-reproducible initial reductions). Our measurements provide better resolution than previous studies afforded because ours were obtained *in-situ* under "real" FTS conditions; therefore, the observed variations cannot be attributed to problems associated with *ex-situ* characterization techniques. Thus, the mild variation in the fundamental and overall performance of these catalysts, which cannot be assigned to global parameters such as metal dispersion or support, could signify a mild sensitivity to uncharacterized catalyst structural differences or to mild surface contamination. In this regard, the low dispersion of the unsupported cobalt catalyst mentioned makes it more susceptible to surface contamination. A scattered correlation is observed between the surface carbon reactivity and the methane selectivity, suggesting that the variations in performance reflect real differences in the details of the surface reaction kinetics. This is discussed in a companion paper in the conference proceedings.

The authors wish to acknowledge the supply of various catalysts by Stuart L. Soled[c], Claude C. Culross[p], Charles H. Mauldin[p] and Sabato Miseo[c], and the characterization of a selected number of these with TEM by Chris E. Kliewer[c], and with chemisorption by Stuart L. Soled[c] and Joseph E. Baumgartner[c].
[t]Department of Chemical Engineering and Applied Chemistry, University of Toronto, Toronto, Ontario, M5S3E5 Canada; [c]ExxonMobil Corporate Strategic Research Laboratories, Rte 22E, Annandale, NJ, 08801 USA; [p]ExxonMobil Process Research Laboratories, 4045 Scenic Hwy, Baton Rouge, LA, 70821 USA.
*Corresponding author: contact information (01)416-978-4575(Office), (01)416-978-8605(Fax), mims@chem-eng.utoronto.ca

References

[1] Vada, S., Chen, B., Goodwin, J.G.,Jr., *J. Catalysis* 153 (1995) 224-231
[2] Bertole, C.J., Mims, C.A., *J. Catalysis* 184 (1999) 224-235.
[3] Kogelbauer, A., Goodwin, J.G.,Jr., Oukaci, R., *J. Catalysis* 160 (1996) 125-133
[4] Haddad, G.J., Chen, B., Goodwin, J.G.,Jr., *J. Catalysis* 161 (1996) 274-281
[5] Belambe, A.R., Oukaci, R., Goodwin, J.G.,Jr., *J. Catalysis* 166 (1997) 8-15
[6] Schanke,D., Vada,S., Blekkan,E., Hilmen,A., Hoff,A., Holmen,A., *J. Catalysis* 156 (1995) 85-95
[7] Rothmael, M., Hanssen,K.F., Blekkan,E.A., Schanke,D., Holmen,A., *Cat. Today* 38 (1997) 79-84
[8] Hanssen, K.F., Blekkan,E.A., Schanke,D., Holmen,A., *Stud. Surf. Sci. Catal.* 109 (1997) 193-202
[9] Mims, C.A., McCandlish, L.E., *J. Phys. Chem.* 91 (1987) 929-937
[10] Mims, C.A., *Cat. Letters* 1 (1988) 293-298
[11] van Dijk, H.A.J., Hoebink, J.H.B.J., Schouten, J.C., *Stud. Surf. Sci. Catal.* 130 (2000) 383-388
[12] Mauti, R., PhD Thesis, University of Toronto, (1994)
[13] Iglesia, E., Soled, S.L., Fiato, R.A., *J. Catalysis* 137 (1992) 212-224
[14] Iglesia, E., Soled, S.L., Fiato, R.A., Via, G.H., *J. Catalysis* 143 (1993) 345-368
[15] Iglesia, E., *Applied Catalysis A: General* 161 (1997) 59-78
[16] Oukaci,R., Singleton,A.H., Goodwin,J.G.,Jr., *Applied Catalysis A: General* 186 (1999) 129-144
[17] Guczi,L., Stefler,G., Schay,Z., Kirisci,I., Mizukami,F., Toba,M., Niwa,S., *Stud. Surf. Sci. Catal.* 130 (2000) 1097-1102

Studies in Surface Science and Catalysis
J.J. Spivey, E. Iglesia and T.H. Fleisch (Editors)

Surface carbon coverage and selectivity in FT synthesis: a simple model for selectivity correlations

C.A. Mims and C. J. Bertole

Department of Chemical Engineering and Applied Chemistry
University of Toronto
Toronto Ontario M5S3E5, Canada

Abstract

^{12}CO-^{13}CO isotope transients have been performed on a series of cobalt FT catalysts at elevated pressure conditions and at identical conversion levels. A mild variability was observed in the methane selectivity of these catalysts. Lower methane selectivity tended to be observed when the surface was more crowded with active carbon. A simple model based on an extension of simple ASF theory is presented which is consistent with this finding.

Introduction

Selectivity is a central issue in Fischer-Tropsch synthesis (FTS) [1-9]. The selectivity pattern can vary with the composition of the catalyst [4,10,11], and with reaction conditions on any particular catalyst. On a given catalyst, high H_2:CO ratios and temperatures lead to higher methane selectivity and simultaneously to lower chain growth probabilities. Product concentrations, and therefore syngas conversion, also influence the intrinsic selectivities [4]. For this reason, comparisons of selectivity between catalysts must be made at identical reaction conditions (including conversion). We have recently examined a suite of cobalt-based catalysts under such rigorously controlled reaction conditions [12]. We measured reactivity, selectivity and, using ^{12}CO-^{13}CO isotope transients, the amounts of adsorbed CO and active carbon on the surface during the reaction. The CO turnover frequency, given by the ratio of the CO conversion rate to the adsorbed CO inventory, is an intrinsic measure of the site reactivity. We observed a mild variation in both the methane selectivities and the CO turnover frequencies of these catalysts. This variability did not correlate with the identity of the support, the Co metal loading, or the presence of Re. The information about the surface carbon inventory from the isotope transients provides an opportunity to understand the mechanistic origin of these variations.

Product selectivities in FTS are governed by a dynamic balance among the mechanistic steps of the reaction. To a first approximation, the selectivity pattern is described by the Anderson-Shulz-Flory (ASF) distribution that arises from simple polymerization kinetics of a C_1 monomer on the surface. Deviations from the strict ASF distribution (overproduction of methane, underproduction of C_2-C_3 hydrocarbons and higher hydrocarbon distributions consistent with carbon-number-dependent growth probability) are well known [1,2,4-6]. Mass transport effects can clearly influence the selectivity pattern [4] but other factors that change the dynamic balance on the surface can also be responsible. Previous ^{12}CO-^{13}CO isotope transient studies [8,9] revealed that surface residence times for the intermediates leading to methane are similar to those leading to higher hydrocarbons, indicating a common pool of monomer carbon for all products. It might be expected that the amount of active carbon has an effect on the reaction selectivity. This would be the case, for example, if the steps involved in hydrocarbon chain propagation have a higher molecularity in surface carbon intermediates than do the termination steps, including methane formation. In this paper, we develop a simple kinetic model based on this idea, and then compare the model predictions with our observations.

Experimental Procedure and Results

A variety of cobalt catalysts were investigated in a small down-flow tubular reactor containing 0.1 - 0.4 g of the catalyst. The catalysts were made on a variety of supports with Co metal loadings from 10- 30 wt%. The reaction rates, selectivities and ^{12}CO-^{13}CO carbon isotope transients were measured at a common temperature, pressure and feed composition (T = 220°C, p(CO) = 1.7 bar, $p(H_2)$ = 3.4 bar). The residence time was adjusted to produce CO conversions between 10 and 16%. The CO conversions and the individual product formation rates through C_4 were measured directly by GC and GCMS. The rate of C_{5+} hydrocarbon production was obtained by difference. Carbon isotopic transients were performed using on-line mass spectrometry [8,12]. The isotopic response in the methane product provides the amount and the residence time of the active carbon intermediates leading to methane. Assuming that a similar time dependence is exhibited in the higher hydrocarbons, as has been previously determined [8,9], the total amount of active carbon leading to all products can be estimated from the methane transient. The total amount C* is given in this case by C^*_{CH4} / $S(CH_4)$ where C^*_{CH4} is the amount of old isotope reporting to methane during the transient. Trends in C* coverage are not expected to be sensitive to errors in this approximation. More complete details can be obtained in a

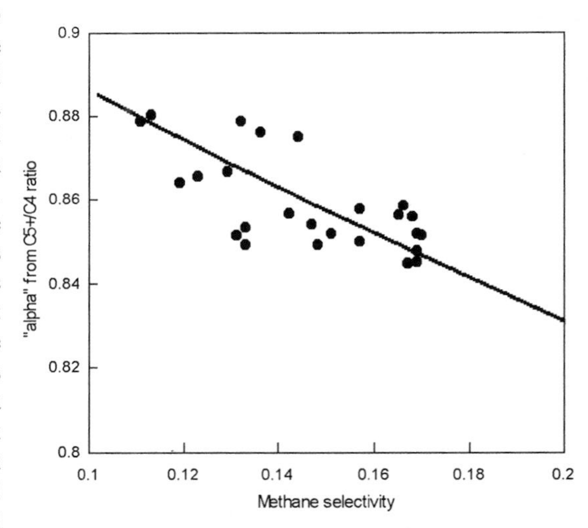

Figure 1: **Chain growth probability (as measured by C_{5+}/C_4 ratio) dependence on methane selectivity. Circles are data from a series of cobalt-based catalysts. Solid line is a correlation predicted by the simple model described in the text.**

companion paper in this volume and elsewhere [12].

As noted above, a mild variation was noted in the methane selectivity, the higher hydrocarbon distributions, the CO turnover frequencies, and the carbon intermediate lifetimes on this group of catalysts [12]. Figure 1 shows the variability in both the methane selectivity and the higher hydrocarbon distribution and that the two are correlated in the usual way. The ordinate is the calculated chain-growth probability, α, required by the Flory distribution to obtain the measured ratio of carbon selectivities, $S(C_5+)/S(C_4)$. This parameter is used as a rough measure of higher hydrocarbon selectivity. A similar trend was observed in the other selectivity ratios (C_4/C_3, etc.) measured in this study. The usual deviations from ASF product distributions, such as low C_2 production, were seen for all catalysts. More complete characterization of the higher hydrocarbon products from a few of the catalysts showed concave deviations on a Flory plot similar to previously published results [4]. The line in Figure 1 is the correlation generated by the simple model developed in the next section.

Model development and comparison with data

The following simple polymerization mechanism considers only the carbon-containing intermediates.

$$\underline{CO} \rightarrow \underline{C^*} \qquad r = r(CO) = k_i\, \theta_{CO} \qquad (1)$$
$$\underline{C^*} \rightarrow CH_4 \qquad r = r(CH_4) = k_1\, \theta_{C^*} \qquad (2)$$
$$2\underline{C^*} \rightarrow \underline{C_2} \qquad r = k_2\, \theta_{C^*}^{\;2} \qquad (3)$$
$$\underline{C_2} \rightarrow C_2\,(g) \qquad r = k_d\, \theta_2 \quad \text{: and by analogy for } \underline{C_n} \rightarrow C_n\,(g) \qquad (4)$$
$$\underline{C_2} + \underline{C^*} \rightarrow \underline{C_3} \qquad r = k_g\, \theta_{C^*}\, \theta_2 \quad \text{: and by analogy for } \underline{C_n} + \underline{C^*} \rightarrow \underline{C_{n+1}}. \qquad (5)$$

Underlined species are surface intermediates; their fractional coverages of the surface are denoted by θ. The rates are expressed as turnover frequencies. We assume simple mass-action rate laws (in C*) for the elementary steps. The rate constants are effective rate constants in which the dependencies on other surface species (H, etc.) are embedded. The surface carbon coverage, which in turn governs the product distribution in the model, is determined by a dynamic balance between CO activation to form active carbon-containing species and the removal of these species by the hydrocarbon synthesis manifold. We assume, as has been indicated in previous isotope transient experiments [8], that the surface carbon exists largely as a C_1 species, $\underline{C^*}$. We also assume that the $\underline{C^*}$ and $\underline{CO}$ compete for the same sites, but that $\underline{CO}$ is in adsorption equilibrium with the gas, while the $\underline{C^*}$ coverage is given by steady state balance. This model is not claimed to be accurate in its details. It is meant to represent a broad class of mechanisms where the elementary surface reaction steps involved in higher hydrocarbon growth have a higher intrinsic molecularity in surface carbon intermediates than do the termination steps.

From the mechanism (equations 1-6) the following expression can be obtained for the selectivity to methane at steady state (by making use of the infinite sum in Flory kinetics for C_{2+}).

$$S(CH_4) = k_1 \ / \ \{k_1 + k_2\, \theta_{C^*}\, (k_g \theta_{C^*}/k_d + 2\,)\}; \tag{7}$$

This is a monotonically decreasing function of θ_{C^*}, provided the rate coefficients (k_1, k_2, k_g, k_d) remain constant. The chain growth probability, α, is given by

$$\alpha = \ k_g \theta_{C^*} / \ (k_g \theta_{C^*} + k_d), \tag{8}$$

which is also a monotonically increasing function of θ_{C^*}.

The steady state carbon coverage, θ_{C^*}, is obtained under our simplifying assumptions by simultaneously applying stationary state constraint on C*, i.e.,

$$d\theta_{C^*} / dt = 0 = k_i \theta_{CO} \ - \ (k_1 \theta_{C^*} + k_2 \theta^2_{C^*} (k_g \theta_{C^*} / k_d + 2\,) \tag{9}$$

and that

$$\theta_{C^*} + \theta_{CO} = 1 \tag{10}$$

This last assumption is not critical – the same trends are developed if C* and CO do not compete for the same sites.

From the equations above and the experimental data, appropriate values of k_i, k_1, k_2 and the ratio k_g/k_d can be determined for each catalyst. The values of k_g and k_d must be much greater than k_2 in order to provide a negligible coverage of the surface by the growing chains. The reaction selectivities are governed only by ratios of the other rate constants. The line in Figure 1 is the model prediction for the following: $k_1/k_2 = 0.43$ and $k_g/k_d = 19.8$. Variation of the ratio k_i/k_1 between 1.5 and 5 provides the prediction in Figure 1.

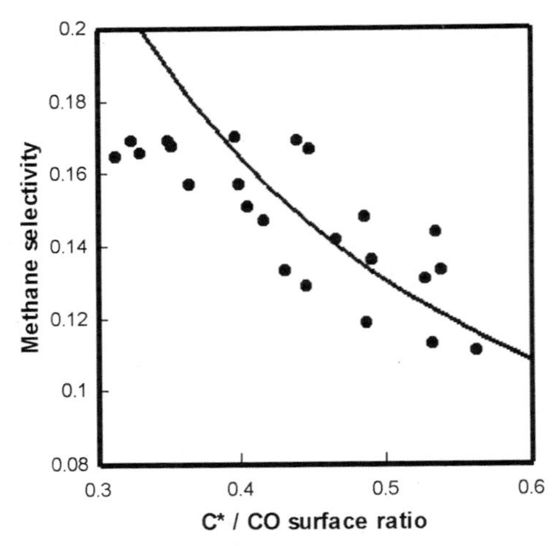

Figure 2: Methane selectivity versus surface abundance ratio of active carbon, C*, to CO. Circles are data from isotope transients; line is predicted by simple model described in the text.

According to the model, the C* coverage is the primary hidden variable that governs the reaction selectivity. Figure 2 shows the methane selectivity plotted against the surface carbon abundance, represented by the C*/CO surface ratios obtained from the carbon isotope transients. A trend is evident in these data. The solid line is the model prediction, using the same rate constant ratios as above. The agreement is intriguing.

Figure 3 plots the chain growth parameter from the C_5+/C_4 ratio against the C*/CO ratio. The solid points are experimental results; the dashed line is a result of the simple model with the rate constants as above. There is a suggestion in the data of the trend predicted by the model, but the result is less convincing than that for methane selectivity. The variation in chain growth probability is predicted to be less dramatic than that for methane selectivity as a consequence of the high-order dependence of higher hydrocarbon formation on θ_{C*}. The imprecision in this selectivity measure also contributes to the scatter in the data.

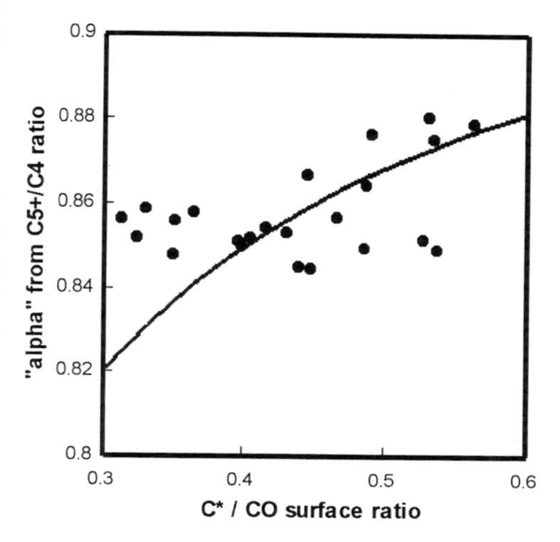

Figure 3: Hydrocarbon growth probability (measured from C_{5+}/C_4 ratio) versus surface abundance ratio of active carbon, C*, to CO, measured in carbon isotope transients (circles): Line is predicted by the simple model described in the text.

Discussion

The preliminary "broad-brush" treatment presented here shows that the simple concept in the model is capable of rationalizing simultaneous changes in several selectivity parameters. By a simple change in the dynamic balance around active carbon during the reaction, the reaction selectivity pattern can be affected without any changes in the rate constants in the hydrocarbon synthesis manifold. We cannot say at this stage that the interpretation is even qualitatively correct, and it should be treated as a working hypothesis. The intuitive nature of the assumptions and the resulting correspondence to the data bodes well, however. The information from isotope transients has been shown here to be an important tool to further our understanding of the mechanistic origins of FT selectivity. The active carbon amounts measured by the isotope transients correlate with the catalyst selectivity variations that, in turn, were not simply related to any catalyst composition attributes.

Given the complexity of FTS, sorting out the origins of selectivity variations is a daunting task. More than one cause can be operating simultaneously, and experimental separation of these is difficult. Global measures of performance, including the lumped rate coefficients in kinetic models, (simple and complex) involve averages over many poorly understood and unmeasured parameters [2,14,15]. Even if all sites are equivalent, the site and neighbouring site co-adsorbate requirements for the individual steps in the mechanism are poorly understood even in an average sense. Dispersion of site properties, such as the role of surface orientation, surface defects, and contamination are all expected to be important at a detailed mechanistic level. The application of isotope transients provides a way of sorting these issues out at a deeper level.

One process that has been shown to have an important influence on selectivity in FT synthesis is the set of secondary reactions involving olefins [2,4,16,17]. Readsorption of alkenes for further chain growth adds complexity to the reaction network. Mass transport effects can influence the significance of these reactions even when reactant supply is not hindered [4]. Alkene readsorption also recycles carbon to the monomer pool [18]. These processes will affect the overall dynamic balance of carbon species on the surface and could have an indirect effect on chain initiation probability. Therefore, these processes can be accommodated in the basic model. Subsequent publications will address variations in the absolute values of these surface rate processes.

Conclusions

A simple model based on ASF kinetics for the FT hydrocarbon product distribution has been presented. The model naturally yields a co-dependence of methane and higher hydrocarbon selectivities on the coverage of the surface by active carbon.. This and related models predict lower methane selectivity and heavier products from a surface more crowded with active carbon – a trend consistent with our results from carbon isotope transient experiments on a series of cobalt FT catalysts.

Acknowledgements

The authors acknowledge support by ExxonMobil Research and Engineering for this study and E. Iglesia, P. Joshi, and G. Kiss for informative discussions

References

[1] R.B. Anderson, The Fischer-Tropsch Synthesis, Academic Press (1984). pp. 186ff.
[2] E.W. Kuipers, C. Scheper, J.H. Wilson, I.H. Vinkenburg, H. Oosterbeek, J. Catal. 158, 288 (1996).
[3] A.A. Adesina, Appl Catal. A: General 138, 345 (1996).
[4] E. Iglesia, Appl. Catal. A: General 161 59 (1997).
[5] M. E. Dry, Appl. Catal. A: General 186, 319 (1999).
[6] H. Schulz, M. Claeys, Appl. Catal. A: General 186 91 (1999).
[7] Herington, E.F.G. Chem. Ind. (London) p. 347 (1946),
[8] C. A. Mims and L.E. McCandlish, J. Phys. Chem. 91, 929 (1987).
[9] H.A.J. van Dijk, J.H.B.J. Hoebink, Surf. Sci. Catal. 130 383 (2000).
[10] E. Iglesia, S. L. Soled, R. A. Fiato, J. Catal. 137, 212 (1992).
[11] C. H. Bartholomew and W.-H. Lee Stud. Surf. Sci. Catal. 130 1151 (2000).
[12] S.L. Shannon, J.G. Goodwin, Chem. Rev. 95, 677 (1995).
[13] C. J. Bertole, C.A. Mims, G. Kiss, P. Joshi, this volume, C. Bertole Ph. D. thesis U. Toronto (2001)..
[14] J.J.C. Geerlings, J.H. Wilson, G.J. Kramer, H.LP.C.E. Kuipers, A. Hoek, H.M. Huisman, Appl. Catal. A: General 186 27 (1999).
[15] H. Schulz, Appl. Catal. A: General 186, 3 (1999).
[16] Ya. T. Eidus, N.D. Zelinskii, N.I. Ershov, Dokl. Akad. Nauk. SSSR 60, 599 (1948).
[17] H. Schulz, M. Claeys, Appl. Catal. A: General 186 71 (1999).
[18] C.A. Mims, J. J. Krajewski, K.D. Rose, M.T. Melchior, Catal. Lett. 7, 119 (1990).

Studies in Surface Science and Catalysis
J.J. Spivey, E. Iglesia and T.H. Fleisch (Editors)

PEROVSKITES AS CATALYSTS PRECURSORS FOR METHANE REFORMING: Ru BASED CATALYSTS.

E. Pietri[a*], A. Barrios[a], O.Gonzalez[a], M.R. Goldwasser[a], M.J. Pérez-Zurita[a], M.L. Cubeiro[a], J. Goldwasser[a], L. Leclercq[b], G. Leclercq[b] and L. Gingembre[b]

a. Centro de Catálisis, Petróleo y Petroquímica, Escuela de Química, Facultad de Ciencias, Universidad Central de Venezuela, Apartado 47102, Los Chaguaramos, Caracas, Venezuela.

b. Université des Sciences et Technologies de Lille, Laboratoire de Catalyse Hétérogene et Homogène, 59655 Villeneuve D'ascq, Cedex, France.

1. ABSTRACT.

Perovskites type oxides $LaMO_3$ (M=Ru, Ni, Mn) were synthesized by the citrate Sol-Gel method and tested as catalysts for the CO_2 reforming of methane. The influence of Ru partial substitution for Ni in the $LaRuO_3$ structure on the activity and selectivity performance was also investigated. The results were compared with those obtained with catalyst samples prepared by wet impregnation. The effects of parameters such as reaction temperature, space velocity, CH_4/CO_2 ratio and time on stream were investigated and optimized to higher yields of syngas. XRD, BET surface area, TEM-EDX, IR, XPS, TPR and H_2 chemisorptions characterized all the solids. Among all the solids investigated, the $LaRu_{0.8}Ni_{0.2}O_3$ precursor was the most active and selective catalyst, reaching values of 89% and 90% in methane conversion and CO selectivity respectively even after 150 hours on stream. A significant decrease in coke deposition for all the catalysts was obtained which constitutes an advantage for future developments of commercial reforming catalysts.

2. INTRODUCTION

The process of carbon dioxide reforming of methane, which converts two of the cheapest carbon containing material into useful products, has received considerable attention in recent years [1-3]. Even though steam reforming of methane is widely industrialized to produce synthesis gas, the dioxide reforming offers advantages such as the production of synthesis gas with a lower H_2/CO ratio more suitable to oxo- and Fischer-Tropsch synthesis, which obviates a water vaporization step to produce steam, an energy consumer process, and the use of CO_2 a major greenhouse gas. Due to the high temperatures involved and the presence of steam, sinterization of the active metal species and coke formation lead to catalysts deactivation and in some cases to plugging of the reactor.

The use of precursors such as perovskites like oxides, which contains the metal distributed in the structure could be the answer to these problems since them, not only present thermal and

---

- To whom correspondence should be addressed
- Acknowledgement: We are grateful to CONICIT-CONIPET for financial support through Agenda Petroleum Project N° 97-003739

hydrothermal stability but also by a careful reduction treatment results in the formation of a finely well dispersed and stable metal particle catalyst [4].
In this work, we have studied the performance of some perovskite like oxides $LaMO_3$ (M= Ru, Mn, Ni) on the CO_2 reforming of methane and compared the results with catalysts prepared by wet impregnation of Ru on $LaMnO_3$. The influence of Ru partial substitution for Ni in the $LaRuO_3$ perovskite on the catalytic activity/selectivity was also investigated.

3. EXPERIMENTAL

The $LaMO_3$ (M=Ru, Ni, Mn) and the $LaRu_{1-x}Ni_xO_3$ (x=0-0.6) perovskite-like oxides were synthesized by a modification of Pechini's citrate technique [5]. Adequate amounts of the precursor of the cation at B position [Mn $(CH_3CO_2)_3.2H_2O$, $RuClxH_2O$ and Ni $(NO_3)_26H_2O$] were dissolved under vigorous stirring in a solution of citric acid (99.5 Riedel-de Haën) with an excess of ethylene glycol (99.5 Riedel-de Haën) as the organic polydentate ligant. The citric acid/ B cation molar ratio was 4, while it was 1.38 for ethylene glycol/citric acid. The mixture was kept at 323-330 K with mild continuous stirring until a clear solution was obtained. At this point, a stoichiometric quantity of the precursor of cation A, La $(NO_3)_3.5H_2O$, was added while keeping the mixture at 330 K. The evaporating process proceeded for 2 days until a viscous resin was formed. The resin was dried at 423K for 24 hours and calcined in air at 973K for 5 hours. Catalyst samples containing 0.3% - 1.5 % by weight of Ru on $LaMnO_3$ were prepared by incipient wetness impregnation with the appropriated amounts of $RuCl_3.xH_2O$ on the perovskite like oxide, followed by drying overnight at 393K.

3.1 Support and catalysts characterization.

The solids were characterized before and after catalytic tests by means of techniques such as chemical analysis, XRD, BET surface area, TEM-EDX, IR, XPS, and H_2 chemisorptions. The IR spectra of the final product were recorded in a Perkin-Elmer 283 spectrometer between 1200 - 400 cm^{-1}. BET surface areas were determined by nitrogen adsorption at 77K with an Ar/N_2 ratio of 70/30 using a Micromeritics model ASAP2010. XRD studies were conducted using a Siemens D-8 advanced diffractometer with a CuKα radiation for crystalline phase detection between 20 and 90° (2θ). Microstructure of samples was studied with a Hitachi H-500 Transmission Electron Microscope coupled to an Electron Dispersive X-Ray device (EDX) for semi quantitative analysis. H_2 chemisorptions isotherms were measured in a glass high-vacuum system by the double isotherm method. For the used catalysts, no pre-treatment was conducted. The XPS analyses were performed with a VG ESCALAB 220XL spectrometer. The monochromatized Al source was operated at 80 Watts. The binding energy scale was initially calibrated using the $Cu3p_{3/2}$ (932.7eV) and $Ag3d_{5/2}$ (368.3eV) peak positions and internal calibration was referenced to C1s line at 285eV. During analysis, the residual vacuum was $\cong 10^{-7}$ Pa. The powder samples were pressed into pellets adapted to the specimen holder. The catalyst reduction in pure H_2 at 673 K for 2 h was performed in the catalyst cell attached to the prelock chamber of the spectrometer.

3.2 Activity Test.

The catalytic tests were carried out using 200 mg of catalyst in a 20-mm ID quartz reactor at atmospheric pressure operated in a fixed-bed continuous flow system (CH_4/CO_2 = 1, 2 and 3; N_2 as diluents, 723-1123 K and various hourly space velocities). Before the reaction, all the catalysts were reduced in situ with a H_2 flow (11 ml/min) at 673K for 12 hours. After reduction, the samples were flushed with N_2 for 15 min and adjusted to the reaction temperature. The water produced during reaction, was condensed before passing the reactants and products to the analyzing system, which consisted of an on-line gas chromatograph (HP 5710A) equipped with a TCD and provided with a Carbosieve S_{II} column (3m). The conversion of CH_4 and CO_2 were defined as the converted CH_4 and CO_2 per total amount of CH_4 and CO_2 feeded respectively. The selectivity for CO was defined as $S_{CO} = n_{CO} / (nCH_4\ (c) + n\ CO_2\ (c))$ and the yield as $Y_{CO} = S_{CO} \times C_{CH4}$, where nCH_4 (c) and nCO_2 (c) are the amount of methane and carbon dioxide converted.

4. RESULTS AND DISCUSSION

4.1.1 Perovskites characterization.

In order to assess the presence of the perovskite-like structures, XRD and IR spectra were recorded. Table 1 list the perovskite compounds synthesized, their measured surface area, EDX analysis, IR bands, cell parameters and their corresponding symmetry.

Catalyst	Area m^2/gr	EDX Metal / La	IR Bands ν_1	ν_2	Symmetry	Cell Parameter (Å) a	b	c
$LaMnO_3$	11	Mn/La = 1.08	600	380	Cubic	3.8851	—	—
$LaRuO_3$	2	Ru/La = 0.92	612	414	Orthorhombic	5.5822	7.673	5.535
$LaRu_{0.8}Ni_{0.2}O_3$	9	Ru/La = 0.72 Ni/La = 0.23	592	417	Orthorhombic	5.5248	7.886	5.540
$LaRu_{0.6}Ni_{0.4}O_3$	15	Ru/La = 0.55 Ni/La = 0.45	608	407	Orthorhombic	5.9384	7.833	5.499
$LaRu_{0.4}Ni_{0.6}O_3$	17	Ru/La = 0.35 Ni/La = 0.59	626	400	Orthorhombic	5.5427	7.881	5.935
$LaNiO_3$	14	Ni/La = 1.02	595	420	Triclinic	10.423	10.428	6.920

It was observed that the citrate method produced solids with high crystallinity and well defined symmetry, in agreement with previously reported values [6]. The perovskite structure was the main phase detected, although some deviations from the perfect stoichiometry resulted in small quantities of other phases. The EDX analysis indicated a close similarity between analytical and targeted values in each case. The BET surface area clearly shown a marked dependence on the degree of substitution in the $LaRu_{1-x}Ni_xO_3$ series, compared to unsubstituted $LaRuO_3$. For substitutions between x = 0.2 - 0.6 the surface area increases constantly from $9 m^2/gr$ to$17\ m^2/gr$. The surface area for $LaNiO_3$ was 14 m^2/gr. The IR spectra for all the solids showed the two broad bands characteristics of the perovskite structure and their positions are in good agreement with those reported in the literature. The XRD spectra after reduction in H_2 flow at 673K revealed that the perovskite structure is still present in most of the solid together with some metallic Ru and/or Ni and La_2O_3. The XRD patterns for the used catalysts showed the complete destruction of the perovskite, leaving metallic Ru and/or Ni, La_2O_3 and a new $La_2O_2CO_3$ phase unambiguously detected in all samples, especially in the $LaRu_{0.8}Ni_{0.2}O_3$. After reoxidation (O_2, 923K, 24 h.), the XRD spectra revealed only the partial regeneration of the perovskite structure. The XPS results of mixed perovskites, showed two well defined peaks for $Ru3d_{5/2}$, one at ≈ 280 eV attributed to Ru° and a second one at ≈ 282.6 eV corresponding to unreduced Ru (Ru III). As the amount of Ni increases on the perovskites, the reducibility of Ru decreases. For the more active perovskite ($LaRu_{0.8}Ni_{0.2}O_3$), the signal attributed to Ru (III) appears at 281.7 eV indicating the existence of a different Ru species. TEM-EDX investigations performed on the fresh, reduced and used perovskite series revealed small and faceted metallic particles, given an average metallic dispersion for the series between 60%-70%, except for the $LaRu_{0.4}Ni_{0.6}O_3$, where the dispersion was close to 9%. All the results are in agreement with the volumetric data and the catalytic activity shown by the solids. These studies also revealed that a continuous and thin layer decorates each metallic particle. EDX analysis performed on this interface revealed a significant amount of La atoms. This layer attributed to dioxomono carbonate $La_2O_2CO_3$, is in fast equilibrium with the carbon dioxide in the gas phase during reaction, reducing the formation of deactivating coke [7-8]. After 30 hours of reaction, the TEM-EDX analysis showed that this layer remains in most of the solids; however, was not detected after longer reaction periods, due essentially to carbon deposition.

4.1.2 Supported Catalysts

For the catalyst samples containing 0.3%-1.5% by weight of Ru on $LaMnO_3$, only small changes were observed in the surface area (11-15 m^2/g). The hydrogen uptake on 1% Ru/$LaMnO_3$, the best-supported catalyst in the series, was relatively low, with dispersion around 40%. Not difference was observed between the as-synthesized and reduced solid. However, a high decrease in dispersion due to sintering of the metal and/or covering by carbon deposits was observed after reaction. XPS analysis show that for 1%Ru/$LaMnO_3$ catalyst, the perovskite structure seems to be homogeneous in depth as assessed by the Mn_{2p}/Mn_{3s}, La_{3d}/La_{4d} and La_{4d}/Mn_{3s} ratios. This homogeneity was maintained even after reduction and reaction. The reduced catalyst shows two Ru signals, one at 280.3 eV due to Ru^0 and a second one at 281.6 eV attributed to partially reduced Ru. After reaction, the presence of two Ru signals at 281.4 eV and 282.7 eV corresponding to partially reduced Ru and to Ru oxide evidenced a reoxidation. The wider shape of the signal corroborates the presence of several Ru species in the catalyst.

4.2 Catalytic Studies.

The effect of parameters such as reaction temperature, space velocity, CH_4/CO_2 ratio, time on stream and metal loading for the supported catalysts were investigated and optimised to higher yields of syngas. The $LaMnO_3$ perovskite-like oxide without Ru showed no catalytic activity for syngas production under the experimental conditions studied. It was observed that the activity on the supported catalysts strongly depended on the Ru loading. The higher activities were obtained on 1%Ru/$LaMnO_3$ catalyst. The steady state conditions for this series were reached rapidly, remaining almost constant for up to 10 hours. CO, H_2 and H_2O were the only products detected in all cases. The activity was very low below 723K, both conversions reaching a maximum of 60% and 65% for CO_2 and methane respectively. The methane conversion was slightly lower than that of the carbon dioxide at all temperatures. This fact indicates that others reactions, notably the reverse water-gas shift (RWGS) reaction were taking place. The CO selectivity for this catalyst was observed to decrease from 78% to 52% as the temperature increased from 723K to 1123K. In this temperature range, the reaction spectra is very wide as reported elsewhere [9,10]

Our results showed that not only the RWGS reaction is taking place, but also that carbon is building-up with temperature due essentially to the Boudouard reaction. In addition, the methane cracking reaction, which becomes thermodynamically favourable at high temperatures is taking place as evidenced by a decrease in the BET surface areas of up to 20% and the lost of activity after 30 hours on stream. However, with a high temperature treatment, the solid recovered its original properties and could be used for several cycles.

When the activity tests were carried out using the perovskite-like oxides under the optimized experimental conditions, the performance exhibited for these solids was superior to that shown by the best Ru supported catalyst. Table 2 compares the CH_4 and CO_2 conversions, CO selectivity, dispersion and carbon balance for the perovskite oxides synthesized and the 1%Ru/$LaMnO_3$ catalyst.

Table 2. CH_4 and CO_2 Conversions, CO Selectivity, Dispersion and Carbon Balance

Catalyst	%C_{CH4}	%C_{CO2}	%S_{CO}	%Dispersion	Carbon Balance
$LaRuO_3$	57.8	67.5	86.2	68.2	98.1
$LaRu_{0.8}Ni_{0.2}O_3$	89.4	71.8	90.4	89.1	98.2
$LaRu_{0.6}Ni_{0.4}O_3$	64.9	60.4	82.3	56.3	94.7
$LaRu_{0.4}Ni_{0.6}O_3$	19.0	9.1	63.2	9.1	96.0
$LaNiO_3$	83.8	65.3	85.0	55.0	94.5
1%Ru/$LaMnO_3$	56.4	60.2	60.8	10.97	—

W= 200 mg, GHSV=24 L/hr.gr, time on stream=5 h, Atmospheric Pressure, T= 923K

Fig 3 and Fig 4 show the dependence of CH_4 and CO_2 conversions and CO selectivity (Sco) and yield (Yco) with the CH_4/CO_2 ratio and reaction temperature using the $LaRu_{0.8}Ni_{0.2}O_3$ oxide as catalyst. As can be seen, the best values were obtained for CH_4/CO_2 = 1 at 923K.

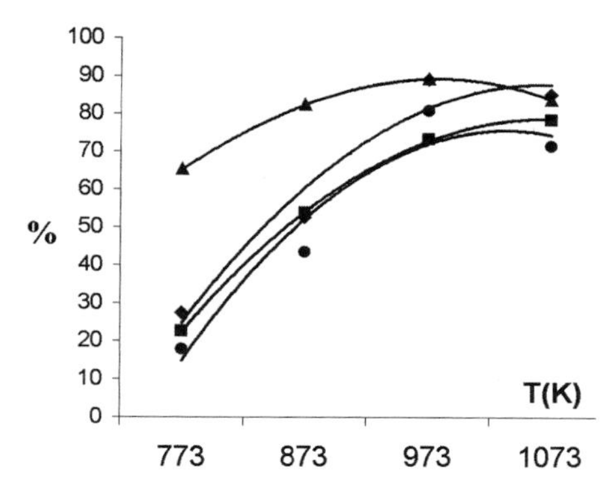

Fig.3.Variation in the CH_4 and CO_2 conversions, CO selectivity and yield over $LaRu_{0.8}Ni_{0.2}O_3$ with temperature.
(♦) CH_4, (■) CO_2, (▲) Sco (•), Yco

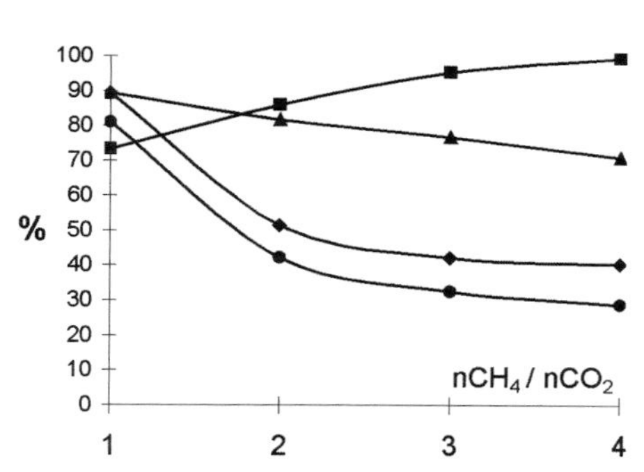

Fig.4. Variation in CH_4 and CO_2 conversions, CO selectivity and yield over $LaRu_{0.8}Ni_{0.2}O_3$ with the CH_4/CO_2 ratio.
(♦) CH_4, (■) CO_2, (▲) Sco (•), Yco

As observed for the supported catalysts, the steady state conditions for the perovskite oxide series were reached rapidly, remaining almost constant for up to 150 hours, except for the

$LaNiO_3$ solid which start loosing some activity after 100 hours due essentially to carbon deposition on the Ni crystallites. With the exception of the $LaRuO_3$ solid, all the perovskites showed a CH_4 conversion superior with respect to that of CO_2 at all temperatures. This could imply that the water reforming of methane and/or methane cracking is taking place. When compared under isoconversion conditions, the $LaRu_{0.8}Ni_{0.2}O_3$ perovskite showed to be the best catalyst. The high activity shown by this solid is attributed to its higher metallic dispersion (89%) compare to $LaRu_{0.4}Ni_{0.6}O_3$ (9%), which showed the lower activity (see table 2). The metal dispersion plays a fundamental role in these reactions as pointed out by G.J. Kim [11] for structure sensitive reactions.

CONCLUSIONS

The synthesis of $LaMO_3$ perovskites by the sol-gel method gave rise to crystalline solids.
TEM-EDX analysis and XRD studies performed on the $LaRu_{1-x}Ni_xO_3$ series revealed the reduction of Ru and/or Ni which are necessary to carry out the reforming reaction to the complete decomposition of the structure during reaction and the formation of small and faceted metallic particles with dispersions between 60%-70% and the existence of a significant amount of a $La_2O_2CO_3$ phase, specially on the $LaRu_{0.8}Ni_{0.2}O_3$. Due to the stability shown by these solids, it is believed that this phase highly contributed to the resistance to deactivation
All perovskites oxides, except $LaMnO_3$ and $LaRu_{0.4}Ni_{0.6}O_3$ were found to be highly active toward the syngas production during the CO_2 reforming of methane. Among the solids tested, the $LaRu_{0.8}Ni_{0.2}O_3$ showed a high potential as a catalyst for this reaction, reaching conversions of 89% and 72% in CH_4 and CO_2 respectively with CO selectivity close to 90% even after 150 hours on stream.

REFERENCES

1. S. Irusta, M. P. Pina, M. Menéndez and J. Santamaría, J. Catal., 179 (1998) 400
2. E. Pietri, A. Barrios; M.R. Goldwasser; M.J. Perez-Zurita; M.L.Cubeiro; J. Goldwasser; L. Leclercq; G. Leclercq and L.Gingembre. Stud. Surf. Sci. Catal, 130 (2000) 3657
3. H. R. Aghabozorg, B. H. Sakakini, A. J. Roberts, J. C. Vickerman and W. R. Flavell Catal. Letters, 39 (1996) 97
4. T. Hayakawa, S. Suzuki, J. Nakamura, T. Uchijima, S. Hamakawa K. Suzuki, T. Shihido and K.Takehira. Applied Catal A: General, 183 (1999) 2:273-285
5. M. P. Pechini, U.S. Patent 3,300,697 (1967)
6. J.B.A. Elermans, B. VanLaar, K. R. Vander Veen and B. O. Loopstra, J. Solid State Chem., 3 (1971) 238
7. N. Matsui, K. Anzai, N.Akamatsu, K. Nakagawa, N. Okikenaga and T. Suzuki Applied catalysis A: General 179(1999): 1-2:247-256
8. Z.L. Zhang, and X.E Verykios, Appl. Catal. A, 138 (1996) 109
9. D. L. Trimm, Catal. Today, 49 (1999) 3
10. Y. Wu, O. Kawaguchi and T. Matsuda, Bull. Chem. Soc. Jpn, 71 (1998) 563
11. G.J. Kim, D.Cho, K. Kim and J. Kim, Catal. Lett. 28 (1994) 41

Studies in Surface Science and Catalysis
J.J. Spivey, E. Iglesia and T.H. Fleisch (Editors)

Fischer-Tropsch synthesis catalysts based on Fe oxide precursors modified by Cu and K: structure and site requirements

Senzi Li[a], George D. Meitzner[b], and Enrique Iglesia[a]

[a]Department of Chemical Engineering, University of California, Berkeley, CA 94720

[b]Edge Analytical, Inc., 2126 Allen Blvd., Middleton, WI 53562

The reduction, carburization, and catalytic properties of Fischer-Tropsch synthesis (FTS) catalysts based on Fe-Cu were examined using kinetic and spectroscopic methods at reaction conditions. Fe_2O_3 precursors reduce to Fe_3O_4 and then carburize to form a mixture of $Fe_{2.5}C$ and Fe_3C in both CO and H_2/CO mixtures at 540-720 K. Oxygen removal initially occurs without FTS reaction as Fe_2O_3 forms inactive O-deficient Fe_2O_3 species during initial contact with synthesis gas at 523 K. FTS reactions start to occur as Fe_3O_4 forms and then rapidly converts to FeC_x. The onset of FTS activity requires only the conversion of surface layers to an active structure, which consists of FeC_x with steady-state surface coverages of oxygen and carbon vacancies formed in CO dissociation and O-removal steps during FTS. The gradual conversion of bulk Fe_3O_4 to FeC_x influences FTS rates and selectivity weakly, suggesting that the catalytic properties of these surface layers are largely independent of the presence of an oxide or carbide core. The presence of Cu and K increases the rate and the extent of Fe_3O_4 carburization during reaction and the Fischer-Tropsch synthesis rates, apparently by decreasing the size of the carbide crystallites formed during reaction.

1. INTRODUCTION

Fe, FeC_x, FeO_y can co-exist when Fe oxides are activated in reactive gases or used in Fischer-Tropsch synthesis (FTS) reactions [1]. The relative abundance of these phases depends on reaction conditions and it can influence catalytic properties. The presence and role of these phases remain controversial [2], because the detection of active sites is usually indirect and the results tend to provide infrequent snapshots of the catalyst structures as a function of time on stream, instead of a continuous record of structural transformations during FTS reactions. The present study addresses the characterization of the structure and stoichiometry of active phases formed from Fe oxide precursors during FTS using a combination of isothermal transient kinetic methods, *in situ* X-ray absorption spectroscopy, and steady-state FTS rate and selectivity measurements.

2. EXPERIMENTAL

Fe oxide precursors were prepared by precipitation of an aqueous solution of $Fe(NO_3)_3$ (Aldrich, 99.99%, 3.0 M) or a mixture of $Fe(NO_3)_3$ and $Zn(NO_3)_2$ (Aldrich, 99.99%, 1.4 M) with $(NH_4)_2CO_3$ (Aldrich, 99.9%, 1 M) at 353 K and at constant pH of 7.0. The precipitates were treated in dry air at 393 K for 12 h and then at 643 K for 4 h. X-ray diffraction measurements confirmed that the crystal structure of resulting Fe oxides was Fe_2O_3. Fe oxide powders were impregnated with an aqueous solution of $Cu(NO_3)_2$ (Aldrich, 99.99%, Cu/Fe=0.01) and/or K_2CO_3 (Aldrich, 99.99%, K/Fe=0.02) using incipient wetness methods. The dried material was treated in air at 673 K for 4 h. Detailed descriptions of these synthesis procedures are reported elsewhere [3]. The Cu-promoted oxide hereafter is designated as Fe_2O_3-Cu, and after K addition, as Fe_2O_3-K-Cu.

A rapid switch transient method was used to determine the extent and rate of reduction and carburization of Fe oxide precursors and the rate of FTS reactions during the initial stages of the catalytic reaction. Samples (0.2g) were pretreated in He (100 cm^3/min) up to 573 K and then cooled to 523 K. A H_2/CO/Ar stream (40/20/40 %, 1 atm, Matheson, 99.99%) was then introduced and the concentrations of gas products in the effluent stream were monitored using on-line mass spectrometry. Here, CH_4 formation rate was used as a surrogate measure of hydrocarbon formation rates because it can be measured accurately and changes with time on stream in parallel with the formation rates of other hydrocarbons. A pre-reduced Fe_3O_4-Cu sample was prepared by treating Fe_2O_3-Cu (0.2g, Cu/Fe=0.01) in 20 % H_2/Ar (100 cm^3/min) while increasing the temperature to 533 K at 0.167 Ks^{-1}. The Fe_3O_4 sample was then cooled to 523 K in He (100 cm^3/min) before exposing it to the H_2/CO/Ar stream. A FeC_x-Cu sample ($Fe_{2.5}C$ and Fe_3C mixture) was also prepared by treating Fe_2O_3-Cu (0.2g, Cu/Fe=0.01) in 20 % CO/Ar (100 cm^3/min, Matheson, 99.99%) while heating to 673 K at 0.083 Ks^{-1}. The presence of Fe_3O_4 and FeC_x phases was confirmed by X-ray diffraction and by detailed oxygen removal and carbon introduction measurements [4].

X-ray absorption spectra were obtained at the Stanford Synchrotron Radiation Laboratory using a wiggler side-station (beamline 4-1). Fe K-edge X-ray absorption spectra were acquired during FTS in synthesis gas using an *in situ* X-ray absorption cell [5]. A precursor oxide sample (8 mg; diluted to 10 wt.% Fe with graphite) was placed within a thin quartz capillary. Synthesis gas (H_2/CO=2) was passed through the sample at 523 K and a space velocity of 30,000 h^{-1}. The spectra were recorded *in situ* as the structure of the catalysts developed with time on stream after exposure to synthesis gas (~14 h). In order to capture the phase evolution of the Fe oxide precursors during initial contact with synthesis gas, XAS spectra were also measured after rapidly cooling samples to room temperature while flowing He through the powder bed. The relative concentrations of Fe carbides and Fe oxides in the samples were obtained using principal component analysis and a linear combination of X-ray absorption near-edge spectra (XANES). The spectra of reference materials, Fe_2O_3, Fe_3O_4, and FeC_x, were fitted to the catalyst XANES in the region between 7.090 and 7.240 keV.

3. RESULTS AND DISCUSSION

Figure 1 shows the temperature-programmed reduction profile of Fe_2O_3-Cu in H_2 and its reduction/carburization in CO. The amount of oxygen removed as a function of temperature indicates that the reduction of Fe_2O_3-Cu in H_2 proceeds in two steps: Fe_2O_3 reduced to Fe_3O_4 (<600 K); then, Fe_3O_4 reduced to Fe at 600-950 K. The oxygen removal and carbon introduction rates as a function of temperature for the Fe_2O_3-Cu sample in CO suggest that reduction/carburization occurred also in two sequential steps, except that reduction started at ~80 K higher temperature in CO than in H_2. It appears that CuO reduced in H_2 at 470 K to form H_2 dissociation sites that increase Fe_2O_3 reduction rates. Carburization did not start until 550 K, as Fe_3O_4 started to reduce. The formation of Fe carbides occurred concurrently with the reduction of Fe_3O_4. X-ray diffraction of samples treated in CO at 550 K and 750 K confirmed that reduction and carburization of Fe oxides in CO proceeds via two sequential steps: Fe_2O_3 reduces to Fe_3O_4, Fe_3O_4 reduces and carburizes to form a mixture of $Fe_{2.5}C$ and Fe_3C.

Mass spectrometric analysis of initial products formed on Fe_2O_3-Cu, Fe_3O_4-Cu and FeC_x-Cu after exposure to synthesis gas at 523 K was used in order to measure the rate of initial reduction and carburization of Fe oxides as well as the FTS rates. Figure 2 shows the oxygen removal by H_2 and CO forming H_2O and CO_2 and the CH_4 formation rates during the initial reduction/carburization in H_2/CO and during subsequent steady-state FTS on Fe_2O_3-Cu at 523 K. During the initial 60 s, oxygen was removed from Fe_2O_3-Cu; reduction occurred without the concurrent formation of CH_4 or other hydrocarbons. The oxygen removed under the first sharp peak arises from the reduction of CuO to Cu. The rest of the oxygen removed during the induction period corresponds to an average stoichiometry of $Fe_2O_{2.8}$. This indicates that O-deficient Fe_2O_3 is not active for FTS reaction. CH_4 formation rates started to increase after this induction period and reached steady-state values after removal of only about 1-2 equivalent layers of O-atoms (assuming that a monolayer consists of 10^{19} O-atom/m^2) in Fe_2O_3, suggesting the facile formation of active sites, which rapidly reached their steady-state site density.

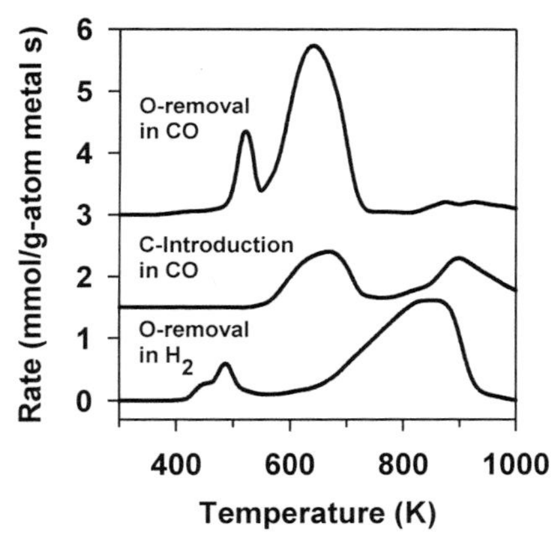

Fig. 1. Temperature-programmed reaction of Fe_2O_3-Cu in H_2 and CO (0.2 g sample; Cu/Fe=0.01, 0.167 Ks^{-1}; 20 % H_2 or CO in Ar; 100 cm^3/min total flow rate).

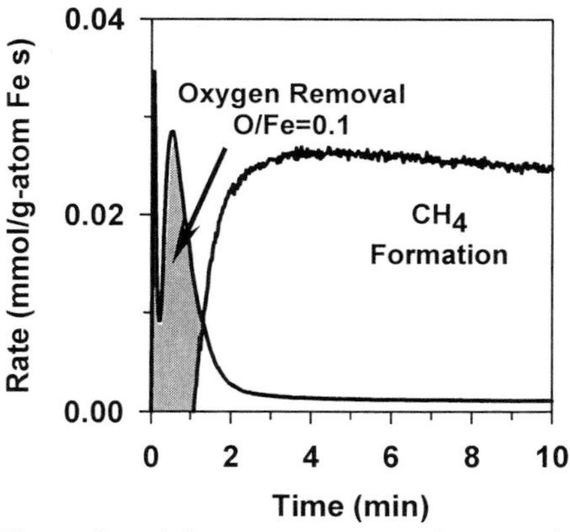

Fig. 2. Mass spectrometric product transients on Fe_2O_3-Cu with time on stream after exposure to synthesis gas at 523 K (0.2 g sample; Cu/Fe=0.01; H_2/CO=2, 100 cm^3/min total flow rate).

Figure 3 shows the results of linear combinations of Fe K-edge XANES for reference compounds (Fe_2O_3, Fe_3O_4 and FeC_x) as a description of the spectra measured for Fe_2O_3-Cu after exposure to synthesis gas at 523 K for various times. Fe_2O_3 rapidly disappeared during the induction period and Fe_3O_4 and Fe carbides formed concurrently; the extent of reduction and carburization increased with time on stream. Fe_3O_4 alone was never detected during FTS reactions, indicating its facile conversion to Fe carbides. Fe metal was not detected at any time during exposure to synthesis gas at 523 K.

The structural changes from Fe_2O_3 to Fe_3O_4, the subsequent facile conversion of Fe_3O_4 to Fe carbides, and the concurrent increase in FTS rates suggest that FTS reactions first occur as Fe_3O_4 is formed and then rapidly converted to FeC_x. The extent of carburization continued to increase with time on stream, without a detectable increase in FTS reaction rates, indicating that only the incipient conversion of Fe_3O_4 to FeC_x was required for FTS reactions to occur at steady-state rates. In effect, only surface layers of Fe carbides appear to be required to form FTS active sites, irrespective of the structure and composition of the bulk phase. The catalytic properties of Fe carbides appear not to be influenced by a remaining Fe oxide core or by its ultimate conversion to FeC_x.

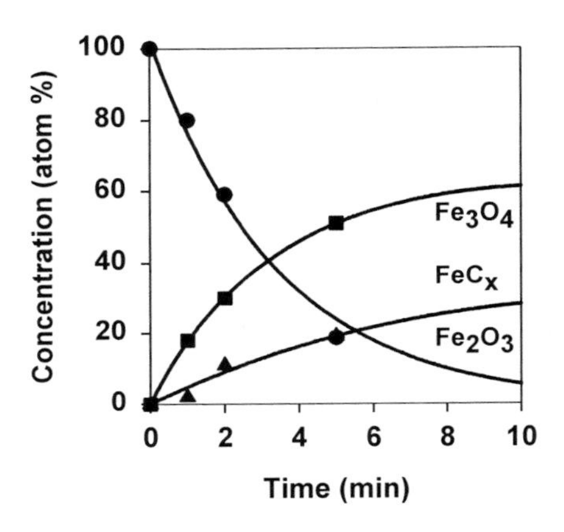

Fig. 3. Fe K-edge X-ray absorption measurements of the phase evolution of Fe_2O_3-Cu oxide with time on stream after exposure to synthesis gas at 523 K (1 mg, precipitated Fe_2O_3, Cu/Fe=0.01, H_2/CO=2, 30000 h^{-1}).

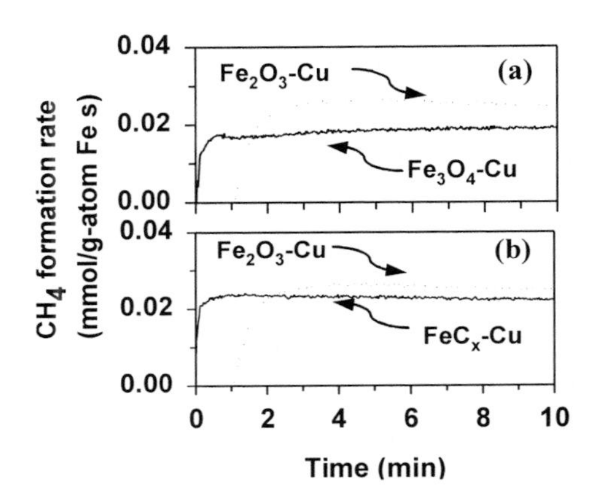

Fig. 4. Mass spectrometric product transients on (a) Fe_3O_4-Cu and (b) Fe_xC-Cu with time on stream after exposure to synthesis gas at 523 K (0.2 g sample; Cu/Fe=0.01; H_2/CO=2, 100 cm^3/min total flow rate).

Figure 4 shows the initial CH_4 transients on pre-carburized Fe carbides (FeC_x-Cu) and pre-reduced Fe oxides (Fe_3O_4-Cu) at 523 K during exposure to synthesis gas. Steady-state reaction rates were reached on both Fe_3O_4-Cu and FeC_x-Cu immediately upon contact with H_2/CO mixtures and without the induction period observed on Fe_2O_3 precursors. FTS rates remained constant even as the gradual removal of oxygen and the introduction of carbon continued to occur on Fe_3O_4-Cu. This indicates that the formation of an active surface occurred only after a few FTS turnovers, irrespective of the initial presence of Fe_3O_4 or FeC_x. This active surface is likely to consist of Fe carbides with a steady-state mixture of surface vacancies and adsorbed C and O atoms formed in the CO dissociation and O-removal steps required to complete a FTS catalytic turnover. The relative concentrations of these species are rapidly established by FTS elementary steps; they depend on the redox properties of the gas phase, on the reaction temperature, and on the presence of any surface promoters that modify the redox properties of the surface.

Figure 5 shows steady-state hydrocarbon formation rates as a function of CO conversion on Fe_2O_3-Zn-Cu and Fe_2O_3-Zn-K-Cu (508 K, 21.4 bar, $H_2/CO=2$). ZnO species in these samples act as a structural promoter to increase FTS rates by inhibiting sintering of the oxide precursors during synthesis. The steady-state reaction data showed that the presence of Cu and K significantly increased FTS rates, suggesting that K and Cu increased the density of active sites formed during activation in H_2/CO mixtures.

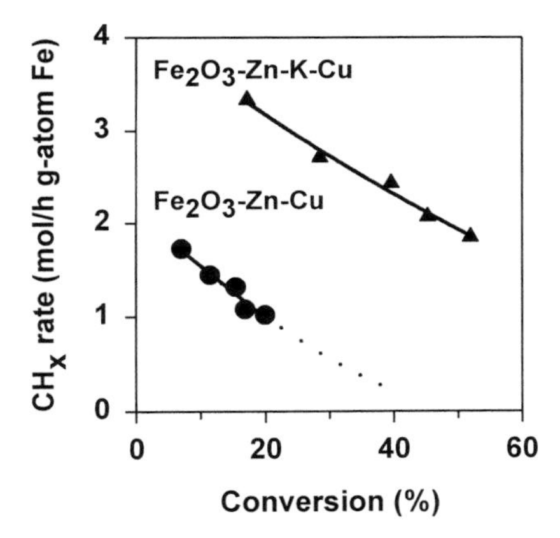

Fig. 5. Hydrocarbon formation rates as a function of CO conversion at steady-state FTS conditions (0.4 g sample; Zn/Fe=0.01, K/Fe=0.02, Cu/Fe=0.01, $H_2/CO=2$; 508 K, 21.4 atm).

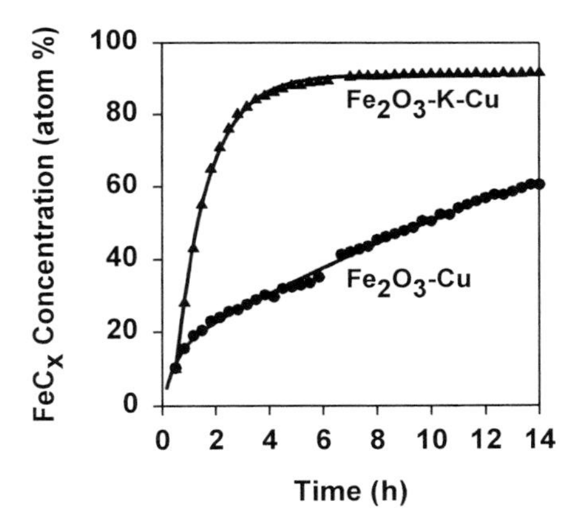

Fig. 6. *In situ* linear combination fits to Fe K-edge XANES of Fe_2O_3-K-Cu and Fe_2O_3-Cu as a function of time in synthesis gas at 523 K (1 mg sample; Cu/Fe=0.01, K/Fe=0.02; $H_2/CO=2$, 30000 h^{-1}).

The presence of Cu and of K increased the rate and the extent of Fe_3O_4 carburization during FTS reactions. The correlation between the rate and extent of carburization of these catalysts and their FTS rates appears to reflect the fact that K and Cu decrease the size of the carbide crystallites formed during reaction and thus increase the number of active sites. Our surface area and CO chemisorption measurements on the samples after FTS reactions confirmed that the surface areas and the amount of CO chemisorbed on the K- and Cu-containing samples are higher than on the samples without K and Cu. This supporting evidence will be described in a separate paper [6]. It appears that the smaller size of the crystallites formed in K- and Cu-containing samples account for their more complete carburization during FTS. It is not, however, their more complete carburization, but their higher surface area that accounts for higher FTS rates obtained on K and Cu promoted catalysts.

4. CONCLUSIONS

Porous Fe_2O_3 precursors sequentially reduce to Fe_3O_4 and then carburize to a mixture of Fe carbides ($Fe_{2.5}C$ and Fe_3C) in CO or in synthesis gas. The incipient conversion from Fe_3O_4 to Fe carbides occurs rapidly, at least in near surface layers, at FTS conditions. Oxygen removal initially occurs without FTS reaction as Fe_2O_3 forms inactive O-deficient Fe_2O_3 species during contact with synthesis gas. FTS reactions occur as Fe_3O_4 is formed and rapidly converted to FeC_x. FTS reactions require only the incipient conversion of surface layers to an active structure, which consists of FeC_x with a steady-state surface coverage of carbon and oxygen vacancies. The catalytic properties of Fe catalysts are not influenced by the gradual conversion of bulk Fe_3O_4 to FeC_x and its ultimate carburization at FTS conditions. The presence of Cu and of K increases the rates and extent of Fe_3O_4 carburization during reaction and the Fischer-Tropsch synthesis rates, apparently by providing multiple nucleation sites that lead to higher surface area FeC_x crystallites during reaction.

5. REFERENCES

1. R. B. Anderson, Catalysis; P. H. Emmett eds.; Van Nostrand-Reinhold: New York, Vol. 4, (1956) 29.
2. M. E. Dry, Catalysis-Science and Technology; J. R. Anderson, and M. Boudart eds.; Springer Verlag: New York, 1, (1981) 196.
3. S. L. Soled, E. Iglesia, S. Miseo, B. A. DeRites, R. A. Fiato, Topics in Catalysis, 2, (1995) 193.
4. S. Li, G. D. Meitzner, E. Iglesia, to be submitted to J. Catal.
5. D. G. Barton, Ph. D dissertation, University of California, Berkeley, 1998.
6. S. Li, G. D. Meitzner, E. Iglesia, to be submitted to J. Phys. Chem.

Studies in Surface Science and Catalysis
J.J. Spivey, E. Iglesia and T.H. Fleisch (Editors)

Catalytic dehydrocondensation of methane towards benzene and naphthalene on zeolite-supported Re and Mo - Templating roles of micropores and novel mechanism-

Ryuichiro Ohnishi, Longya Xu, Kohtaro Issoh, and Masaru Ichikawa

Catalysis Research Center, Hokkaido University, Sapporo 060-0811, Japan

Abstract

The sharp pore-size effect of zeolite support used in methane dehydrocondensation giving the highest yield of benzene was demonstrated, suggesting the templating role of micropores of zeolite support. Further, it was indicated from TPR/TG/Mass studies that aromatics from methane are produced through a different reaction route from MTG process. In the light of previous studies, a novel reaction mechanism is proposed as a direct conversion of surface skeletal carbon ensembles, but not through ethylene oligomerization to benzene and naphthalene on Re and Mo catalysts supported with selected miroporous zeolites. The stable and high methane conversion and the rate of benzene formation were realized at high reaction temperature as 1093K with the help of CO_2 addition into methane feed.

1. Introduction

The direct conversion of methane to fuels and petrochemical feedstocks remains a formidable challenge in natural gas chemical industry. Since 1993, beginning with the work of Wang and others [1], many reports have been dealt with bifunctional catalysis, kinetics and catalytic active phases in the methane dehydrocondensation reaction towards benzene and naphthalene [2] on Mo/HZSM-5 catalysts [3] including our previous works on the effect of CO_2 addition into the methane feed and on metal active phases during the reaction on Re and Mo/HZSM-5 studied by EXAFS and TPR (temperature programmed reaction)/Mass experiments [4]. In this report, we will demonstrate that novel Re catalysts supported with microporous zeolites having an optimum pore size exhibit as high catalytic performances as Mo/HZSM-5 in this reaction and the reaction mechanism is elucidated with TPR/TG/Mass technique.

2. Experimental

The catalytic tests were carried out at 0.3 MPa of methane with or without CO_2 in a continuous flow system with a quartz reactor, which was charged with 0.15-0.30 g of catalyst granules of 20-42 mesh. The feed gas was introduced into the reactor at the flow rate of 7.5-100 ml/min. [space velocity of methane=1440-10000 ml/h/g-cat]. Reaction products including C_1-C_4 hydrocarbons, C_6-C_{11} aromatics, H_2, Ar, CO, and CO_2 were analyzed by use of GCs equipped with FID and TCD similarly as reported elsewhere [4]. Re and Mo catalysts were prepared by the incipient wetness method from various microporous materials such as HZSM-5 and MCM-22 with ammonium perrhenate and ammonium heptamolybdate [4]. TG/Mass/TPR (temperature programmed reaction) study was performed under a methane/He or an ethylene/He stream with a heating rate of 10 K/min at 1 atmospheric pressure. About 20mg of Re catalyst was mounted on a fused alumina holder and evolving products were continuously monitored with a mass spectrometer.

3. Results and Discussion

Table 1 shows the initial stage of catalytic performances of the Re catalysts supported with microporous materials such as HZSM-5, ZRP-1, MCM-22, ZSM-11, H-β, ZSM-12, SAPO-5, ferrierite, and LTL compared with the corresponding Mo catalysts. The trends of catalytic performances on Re and Mo catalysts impregnated on the different microporous materials are similar with each other in all respects such as methane conversion, rate of benzene formation and product selectivities (not shown). Similar result on Mo catalysts has been reported by Zang and others [5], although they did not measure the amount of naphthalene and coke formed. Moreover, it was demonstrated that regardless metal sorts,

Table 1
Catalytic performances of Re and Mo catalysts supported on various porous materials[a)]

support	SiO_2/Al_2O_3	Re[b)] conv.[c)]	Re[b)] R(Bz)[d)]	Mo[b)] conv.[c)]	Mo[b)] R(Bz)[d)]	pore size Å
ZSM-5	39	10.1	1.63	8.1	0.73	5.4×5.6↔5.1×5.5
ZRP-1		7.3	0.13	8.9	0.95	5.4×5.6↔5.1×5.5
MCM-22	36	5.3	0.61	8.0	0.55	5.5×4.0
ZSM-11	38	5.1	0.35	6.3	0.57	5.3×5.4
H-β	37	4.6	0.11	4.1	0.14	5.5×5.5↔6.4×7.6
ZSM-12	40	2.9	0.10	4.3	0.11	5.5×5.9
Ferrierite	64	3.9	0.10	5.3	0.12	4.2×5.4↔3.5×4.8
LTL	6	4.3	0.03	0.4	0.05	7.1
FSM-16	300			16.9[e)]	0.03[e)]	27
Al_2O_3	0	3.1	0.06	8.3	0.05	
SAPO-5		3.2	0.01	4.3	0.05	7.3

a) 993K, CH_4 SV=3000 ml/h/g-cat, 0.3Mpa, 2%CO_2, conv. and R(Bz) are at 2 h of time-on-stream; b) metal loading: 3%Mo and 10%Re; c) conversion of methane in %; d) rate of benzene formation in carbon base in μmole/s/g-cat; e) without CO_2 addition e.g., Mo and Re, supporting porous materials having optimum pore sizes of 5.3-5.6Å (ZSM-5, ZSM-11, ZRP-1 and MCM-22) only provide the effective catalysis of methane dehydrocondensation. The Mo and Re catalysts exhibited higher catalytic performances for the reaction using the porous materials having the optimum SiO_2/Al_2O_3 ratios of 35-50 (not shown), similarly as previously reported for Mo/HZSM-5 catalysts [4]. Accordingly, we propose a templating role of supporting microporous materials for the reaction, which has characteristic pore sizes same as the kinetic diameters of benzene and naphthalene molecules.

Rate of benzene formation on Re/HZSM-5 was greatly decreased to 1/10 of initial value after 6 h of time on stream using pure methane as feed gas at 1023K and methane SV (space velocity) = 5000 ml/h/g-cat as shown in figure 1 [4d], methane only, possibly due to the non-reversible coke deposition. The catalyst performances were effectively stabilized by the addition of a few percent of CO and CO_2 into the methane feed [4d]. Even at high methane space velocity as 5000 ml/h/g-cat, the rate of benzene formation was kept constant by adding 2% CO_2 in the feed for more than 6 h. However, addition of excess CO_2 (more than 4%) results in the marked decrease of rate of benzene formation. It was found that all CO_2 added into methane feed was consumed in forming about twice amount of CO, suggesting the presence of a reforming-like reaction of $CH_x + CO_2 = 2CO + x/2H_2 (x=0-4)$. Used catalysts were subjected to the analysis of deposited coke with TPO/Mass experiments. As expected, the addition of CO_2 in methane feed results in the decrease of coke deposition on the Re catalysts to the level of less than 1/5 of those in pure methane feed. From these results, it is concluded that CO_2 effectively reduces the coke deposition on the catalyst surface to improve the catalytic stability for the benzene production on the Re/HZSM-5 [4d,e].

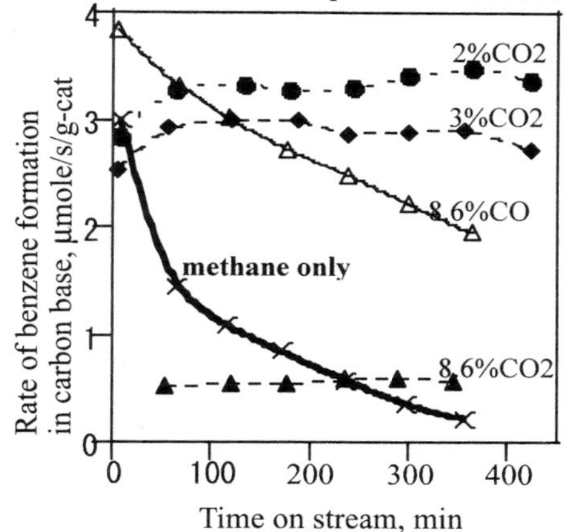

Fig. 1. Effect of CO_2 addition into methane feed on rate of benzene formation. Numbers in fig. refer to mole % of CO_2 in methane feed, at 1023K, 3 atm., methane SV=5000 ml/h/g-cat

Methane conversion at the equilibrium condition was calculated to increase with the reaction temperature. However, it is hard to realize the stable and high methane conversion at higher reaction temperature than 1023 K because of serious coke deposition. Figure 2 shows the variation of benzene formation rate with time on stream on 6%Mo/HZSM-5 at 1098 K and methane SV=2500 ml/h/g-cat. Clearly, as high as 7% of CO_2 added to methane feed was needed to obtain the stable rate of product formation at 1098 K while 1-2% CO_2 added to methane feed was enough to stabilize the rate at 1023 K (see fig.1). Since higher the CO_2% in the methane feed results in lower the rate of benzene formation, there will be a

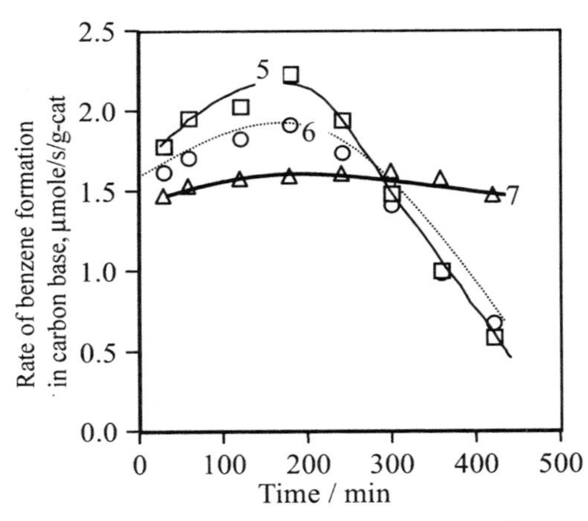

Fig.2. Effect of CO_2 concentration (number in figure) in methane feed on rate of benzene formation at 1098K, CH_4 SV=2500 ml/h/g-cat on 6%Mo/HZSM-5

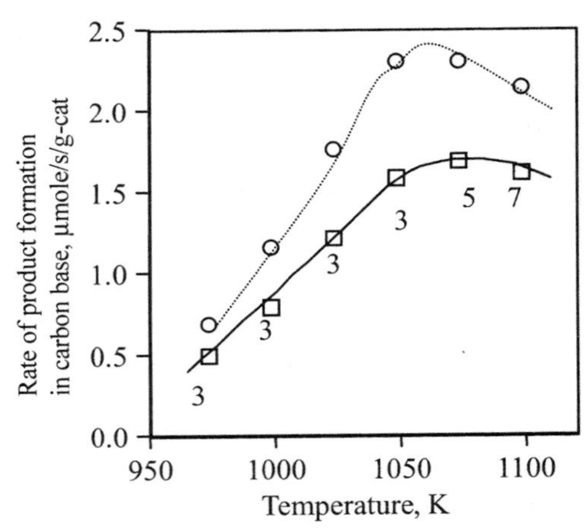

Fig. 3. Temperature dependence of rate of benzene (solid line) and total hydrocarbon (broken line) formations at the steady state conditions, where numbers in figure show CO_2 % giving stable rate

trade-off between the reaction temperature and the rate of benzene formation at steady condition. In fact, a temperature having maximum rate of benzene formation was found to be 1073 K as shown in figure 3.

TPR in methane flow was conducted on the calcinated Re/HZSM-5 using TG/Mass apparatus flowing 10% methane diluted with He while heating from 300 to 973K [4d]. A sharp weight loss in the TG spectrum and the simultaneous evolution of CO, CO_2 and H_2 was observed at 895K, where Re oxide on HZSM-5 was reduced. This initiates the dehydrocondensation of methane above 900 K in forming ethylene, benzene, and naphthalene.

To characterize the active phase of the Re/HZSM-5 responsible for the methane dehydroaromatization to benzene, the Re L_{III}-edge EXAFS/XANES studies were conducted using the 10B beam-line at

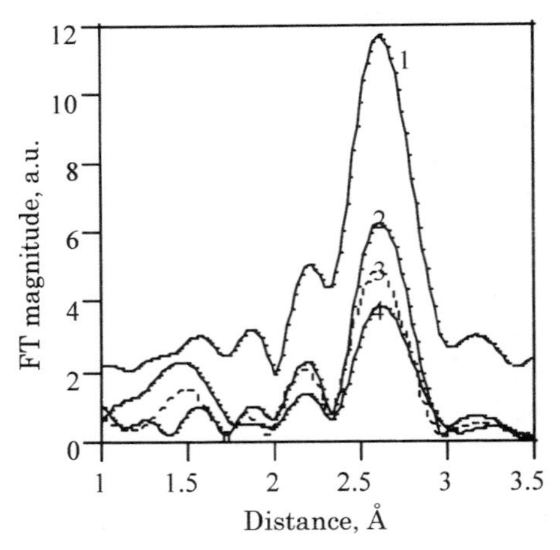

Fig.4. Re L_{III}-edge FT spectra of (1) Re metal (with higher zero level than the others) and (2) Re/HZM-5 reacted at 973K for 24 h and (3) 30min and (4) pre-reduced at 573K for 30min. Signals of (2)-(4) were doubled.

KEK-PF at Tsukuba city [4d]. Figure 4 represents Fourier transform functions of Re L_{III}-edge EXAFS spectra of Re/HZSM-5 samples together with Re metal powder as a reference [4d]. The FT functions of the sample reduced at 573K, reacted with methane at 973 K for 24 h and for 30 min are similar in the shape with those of Re metal powder except for a minor peak at 1.4 Å, which is attributable to Re-O bond. The co-ordination numbers of Re-Re bond for the reduced, the reacted at 973K for 30 min, and the reacted at 973 K for 24 h were calculated to be 4.0, 5.0 and 6.4, respectively, while Re-Re bond lengths were kept unchanged in 2.74-2.75 Å. These results imply that Re-oxide dispersed in HZSM-5 is reduced with hydrogen or methane to highly dispersed Re particles, which is responsible for dehydrocondensation of methane to benzene and naphthalene. By contrast, the relatively stable Mo (oxy)carbide phase characterized by XPS and EXAFS has been previously demonstrated as an active phase on the Mo/HZSM-5 catalyst [3,4].

To elucidate the mechanism of methane dehydrocondensation to benzene, TPR/TG/Mass studies were conducted over pre-reduced Re/HZSM-5 and pre-carburized Mo/HZSM-5 in flowing ethylene or methane. By the pre-reduction and the pre-carburization, formation of active phase was completed prior the TPR experiments. Under the flowing methane, benzene (m/e=78) and naphthalene (m/e=128) were observed at one high temperature region above 800 K (see fig. 5A) while aromatics such as toluene, xylenes (m/e=91), benzene, and naphthalene were formed from ethylene at two temperature regions of 500-650 K and above 800 K (see fig. 5B). At 400-650 K, evolution of hydrogen and weight

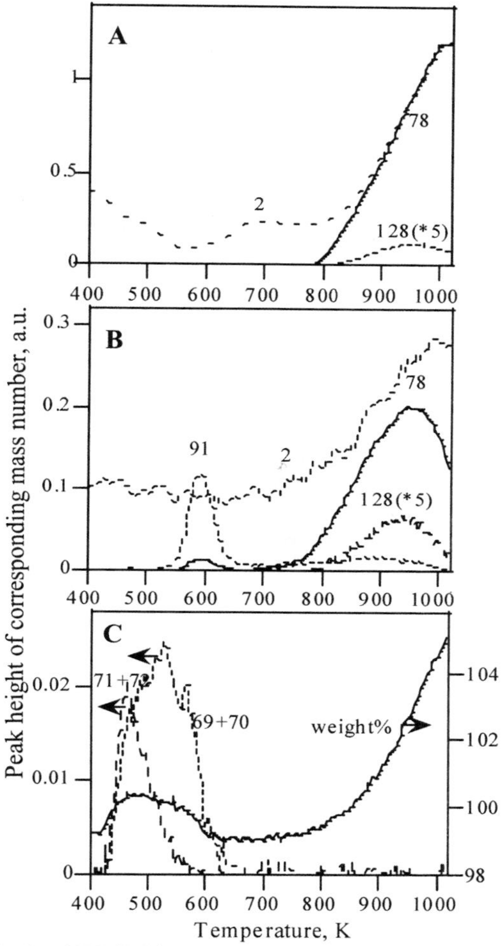

Fig.5. TPR/TG/mass spectra. A) on Re/ HZSM-5 under flowing methane, B) and C) on Mo/HZSM-5 under flowing ethylene, numbers in figure refer to the mass numbers, signals for 128 was multiplied by 5.

change of the catalyst was not observed. At this temperature region, together with aromatics, a large variety of C_3-C_6 aliphatic compounds were formed as shown in fig.5C, where the sum of peak height for C_5 alkanes of m/e=71 and m/e=72 and that for C_5 alkenes of m/e=69 and m/e=70 were presented as the representatives. This product distribution is very similar to that in the methanol-to-gasoline (MTG) process [6], suggesting a common reaction route of oligomerization of C_2 intermediate in the zeolite micropores and dehydroaromatization of these oligomers in forming alkanes and aromatics as shown in route 1 of scheme 1. At the temperature higher than 650 K, neither hydrocarbon product nor change of catalyst weight was observed. However, the production of benzene and naphthalene from ethylene resumed at above 800 K, together with hydrogen but not with higher aliphatic compounds. The characteristics of aromatic formation from methane were almost same as that from ethylene at the high temperature region. The trend of product distribution at the high temperature region and weight change is indicative that aromatics such as benzene and naphthalene are produced through a reaction route different from that at the low temperature region. In the other experiments, a pulse of hydrogen was introduced during methane dehydrocondenzation reaction. As a results, extra formations of benzene together with ethylene were observed. Accordingly, hydrogen deficient intermediates for the formation of aromatics from ethylene at high temperature region and that from methane are proposed as shown in route 2 of scheme 1.

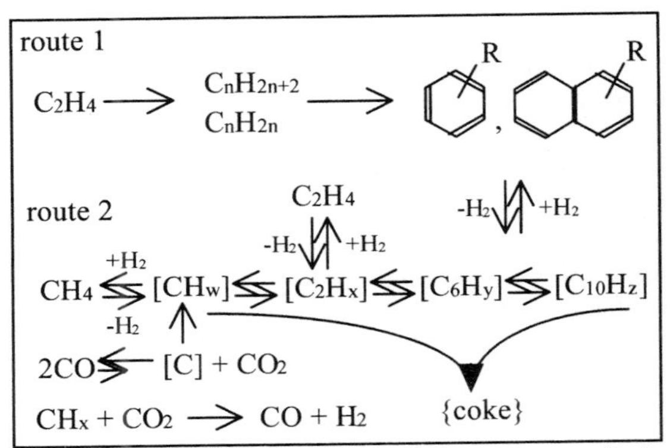

Scheme 1. Proposed reaction routes for producing aromatics

References

1. L.Wang, L.Tao, M.Xie, G.Xu, J.Huang, Y.Xu, Catal. Lett., **30**,135 (1995).
2. F.Solymosi, A.Erdöelyi, J.Cseréyi, J. Catal.,**165**,156 (1997).
3. B.M.Weckhuysen, D.Wang, M.P.Rosynek, J.Lunsford, J. Catal., **175**, 338(1998); , D.Wang, J.Lunsford, M.P.Rosynek, ibid, **169**, 347(1997).
4. a)S.Liu, Q.Dong, R.Ohnishi, M.Ichikawa, J. Chem. Soc., Chem. Commun., 1217(1998); b)S.Liu, L.Wang, R.Ohnishi, M.Ichikawa, J. Catal., **181**, 175 (1999); c)R.Ohnishi, S.Liu, Q.Dong, L.Wang, M.Ichikawa, ibid., **182**, 92 (1999); d)L.Wang, R.Ohnishi, M.Ichikawa, ibid, **190**, 276(2000); e)L.Wang, R.Ohnishi, M.Ichikawa, Catal. Lett., **62**, 29 (1999); f)L.Wang, R.Ohnishi, M.Ichikawa, Catal. Lett.,Kinetics and Catal., **41**, 132 (2000).
5. C.L.Zhang, S.Li, Y.Tuan, W.X.Zhang, T.H.Wu, L.W.Lin, Catal. Lett., **56**, 207(1998).
6. C.D. Chang, in "Handbook of Heterogeniuous Catalysis", Vol 4, p1894 (VCH, Weinheim, 1997), G. Ertl, H. Knözinger, J.Weitkamp eds.

Studies in Surface Science and Catalysis
J.J. Spivey, E. Iglesia and T.H. Fleisch (Editors)

Lurgi's Mega-Methanol technology opens the door for a new era in down-stream applications

Jürgen Haid* and Ulrich Koss, Lurgi Oel•Gas•Chemie GmbH, Germany

ABSTRACT

Lurgi's Mega-Methanol is a new technology for converting natural gas to methanol at low cost in big amounts. This gives the opportunity to replace oil consumption by methanol - so to speak as easy-to-transport liquefied natural gas - in the petrochemical industry as well as in the energy and fuel industry. Two processes are briefly presented and economically assessed, one producing propylene from methanol, the other generating hydrogen.

1. INTRODUCTION

Methanol production is a standard outlet for natural gas. Nevertheless, the methanol price level in the past hindered a possible extension of the usage of methanol. Methanol costs are mainly dictated by feedstock costs and capital-related charges. On one hand the individual methanol synthesis process scheme will influence the efficiency of the usage of the gas, on the other hand the capital-related charges are affected by the economics of plant scale, both being effective parameters for lowering the price of the methanol product. Here, the Mega-Methanol technology contributes its share. It permits the construction of highly efficient single-train plants of at least double the capacity of those implemented until now.

The resulting long-term stable and low methanol prices may pave the way for a wider use of methanol, both in the energy sector and as a feedstock in the petrochemical sector. In the energy sector, fuel cell vehicles with on-board hydrogen generation from methanol are ready to market within the next years. In the chemical sector the route over methanol to olefins is one of the most promising new applications. Another option with high potential is the conversion of methanol to hydrogen.

2. THE MEGA-METHANOL CONCEPT

The presented technology has been developed for world-scale methanol plants with capacities greater than 3000 metric tons per day and combines low environmental impact, cost-optimized energy efficiency and low investment cost [1]. The process concept consists of proven and reliable elements, which reduces the risks normally linked to a new technology nearly to nil.

Synthesis Gas Production: The Synthesis gas production section accounts for about 50% of the investment costs of a methanol plant. This shows that an optimized design of this section is essential.

Conventional steam reforming is economically applied to medium sized methanol plants and the maximum single-train capacity is limited to about 3000 mtpd. Oxygen-blown natural gas

* Author to whom correspondence should be adressed, e-mail: Dr_Juergen_Haid@lurgi.de

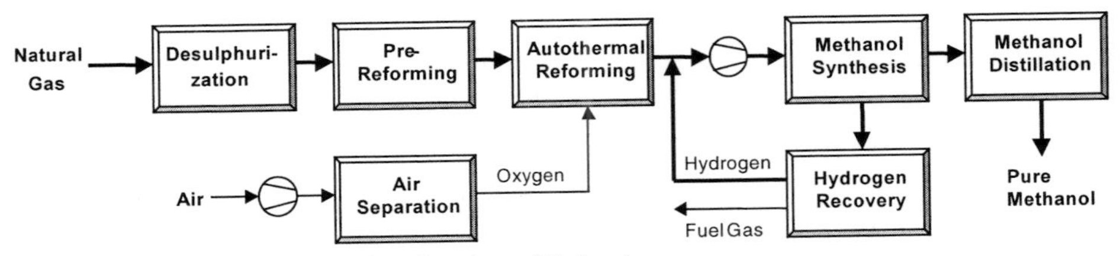

Fig. 1: Mega-Methanol Concept using Autothermal Reforming

reforming, either in combination with steam reforming or as pure autothermal reforming, nowadays is considered to be the best suited technology [2,3]. Whenever light natural gases are to be processed, pure autothermal reforming can be applied.

Autothermal Reforming is a low investment process using a simple reactor design. No tubular steam reformer, till now a sensitive and expensive equipment item of a methanol plant, is required. The capacity limits for single-train plants are shifted due to the considerably lower gas volumes. In addition, the autothermal reforming process achieves CO_2 and NO_x emission reductions of 30 % and 80 % respectively compared to conventional steam reforming.

The process is a combination of steam reforming and partial oxidation. The chemical reactions are defined by Eqs. 1-4, assuming that natural gas consists purely of methane.

Combustion reactions:

$$CH_4 + O_2 = CO + H_2 + H_2O \quad (1)$$

$$CH_4 + 2\,O_2 = CO_2 + 2\,H_2O \quad (2)$$

Thermal and catalytic reactions:

$$CH_4 + H_2O = CO + 3\,H_2 \quad (3)$$

$$CO + H_2O = CO_2 + H_2 \quad (4)$$

The process produces a carbon-free synthesis gas at reformer outlet temperatures typically in the range of 950 - 1000 °C. Commercial experience exists for pressures of up to 40 bar. The synthesis gas is in an equilibrium with respect to the methane reforming reaction (Eq.3) and the shift reaction (Eq.4). The stoichiometric ratio $S_R = (H_2 - CO_2) / (CO + CO_2)$ below 2.0. However the optimum stoichiometric ratio S_R for the methanol synthesis is between 2.05 and 2.1 and will be achieved by recycling hydrogen that can be separated from the purge gas downstream of the methanol synthesis by a membrane unit or pressure swing adsorption unit.

The autothermal reactor is a refractory-lined pressure vessel with a burner in the reactor top section for fast and uniform mixing of the gas/steam mix with oxygen. The combustion and reaction zone is located above the catalyst bed. The catalyst employed is similar to the steam reforming catalysts commonly used in tubular reformers ($NiAl_2O_3$). Because of the operating conditions, establishing a stable flame contour is essential to protect the reactor walls and catalyst from excessive temperatures. Consequent process optimization throughout fifty years of experience as well as the help of a proprietary, three-dimensional Computational Fluid Dynamics (CFD)-model, led to an optimized design of modern Lurgi Autothermal Reformers with regard to efficiency and availability.

Methanol Synthesis Loop: Efficient syngas-to-methanol conversion is essential for low-cost methanol production. Based on the well-known water-cooled tubular methanol reactor and on a new highly active Cu/ZnO catalyst generation [4] a combined converter methanol synthesis was developed mainly for the application in Mega-Methanol plants. (Fig.2).

The first reaction stage is still a water-cooled tubular reactor, nearly identical to the original Lurgi design. However, here it accomplishes partial conversion of the syngas at higher space velocities and higher temperatures. This results in a significant size reduction of the reactor compared to the conventional process while the raising steam is available at a higher pressure (~50 bar).

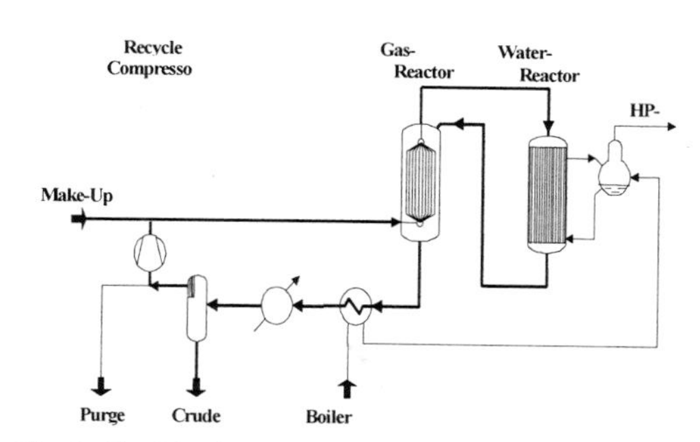

Fig. 2: Combined converter methanol synthesis

The effluent of this water-cooled reactor is directly routed to a second reaction stage, a gas-cooled reactor.
The cooling medium, passed through the empty tubes of this second reactor, is cold feed gas, designated for the first reactor. It is routed in countercurrent flow to the effluent from the first reactor which passes through the catalyst bed outside the tubes. Thus, the temperature of the reacting gas outside the tubes is continuously reduced while passing through the catalyst bed. By this, throughout the whole catalyst bed the equilibrium driving force for the methanol synthesis reaction is maintained (Fig.3).

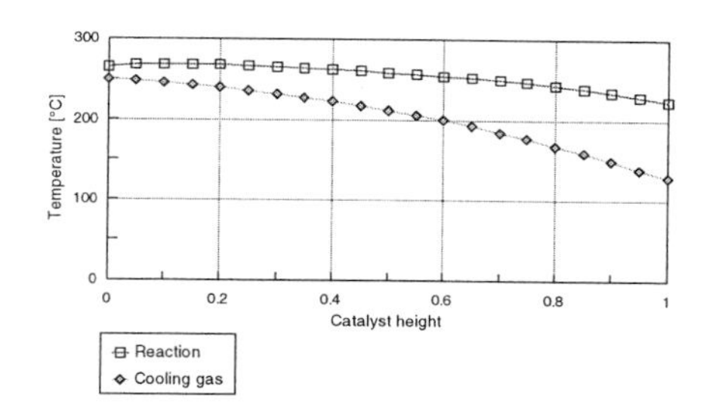

Fig. 3: Temperature Profile Gas-Cooled Reactor

The combined converter methanol synthesis is characterized by single-pass conversion above 80%. Thus the recycle ratio can be reduced by 50% compared to the recycle ratio in a common single-stage, water-cooled reactor.

As fresh synthesis gas is exclusively fed to the first stage, no catalyst poisons enter the second stage which result in a virtually unlimited catalyst service life for the gas-cooled reactor. In addition, two-stage reaction control also prolongs the service life of the catalyst in the water-cooled reactor. If the methanol yield in the water-cooled stage decreases as a result of subsiding catalyst activity, the temperature in the inlet section of the gas-cooled reactor will rise with a resulting improvement in the reaction kinetics and hence, an increased yield in the second reactor stage.
A 50 % reduction of catalyst volume for the water-cooled reactor, the omission of a large feedstock pre-heater and savings for other equipment due to the lower recycle ratio translate into capital cost savings of about 40 % for the synthesis loop.

Economics of Mega-Methanol: As already mentioned, feedstock costs and capital-related charges are the major parameters for the production costs. Table 1 illustrates the feedstock consumption, the capital invest and the resulting production costs for a conventional steam reforming plant and Mega-Methanol plant. The results show that the high efficiency of the process and the low investment costs of a Mega-Methanol plant permit a reduction of the methanol costs to only 80 US$/mt at gate.

		Steam Reforming	Mega-Methanol Concept
Capacity	mtpd	2500	5000
Natural Gas Demand based on LHV	MMBtu/ mt	30	28.5
Total Fixed Cost (EPC)	Million US$	225	300
Production Cost	US$ / mt	113	79
Basis: - Figures include O_2-production, power generation and off-sites - Natural gas price is set to US$ 0.5 / MMBtu - Depreciation is set to 10% of Total Fixed Cost - Return of Investment is set to 20% of Total Fixed Cost - Operating cost for operator staff, plant overhead, maintenance labor and material included			

Table 1: Economics of the Mega-Methanol concept

3. *NEW DOWN-STREAM INDUSTRIES*

A stable methanol price of only 80 US $ gives a new perspective on replacing process routes based on feedstocks derived from crude oil by alternative, methanol based ones. Lurgi faces this challenge especially focussing on two new process developments, which take advantage of their R&D experience and seamlessly integrate into their technology portfolio.

MTP - Methanol To Propylene: The conversion of methanol to olefins is one of the most promising new applications, yielding products with an excellent value-add. Especially methanol-to-propylene seems to have an exciting potential. The global propylene consumption in 1998 was 46 millions tons/year and expected annual growth of consumption is about 6 %. Currently, approximately 98% of the world production is produced as a byproduct from steamcrackers and FCC units. Owners of these plants make big efforts to maximize their propylene output but, however, supply from these sources is limited. Producers of propylene derivates, especially the plastics industry, are looking for new sources of high-quality propylene in order to cover their rising feedstock demand. For them, another attractive aspect of MTP is a possibly shrinking dependency of their industries from the refining industry and thus, the oil price.

The main MTP process feature is an three stage fixed-bed reactor system, where a mixture of methanol and DME generated in a DME pre-reactor is selectively converted to propylene in the presence of steam. Byproducts from the process are a high-RON gasoline fraction, fuel gas, LPG and water. The process operates at a slightly elevated pressure (1.3 to 1.6 bar), moderate steam addition (0.5 – 1.0 kg per kg of methanol) and low reactor inlet temperatures (400 – 450 °C). A very selective tailor-made zeolite catalyst provides maximum propylene selectivity, a very low propane yield and also limited by-product formation.

Because of the extremely low coking tendency of the catalyst, a simple discontinuous in-situ catalyst regeneration by coke burning with nitrogen/air mixture is sufficient, which has to be performed every approximately 400 – 700 hours of operation. The regeneration is carried out at similar temperatures as the reaction itself, hence the catalyst particles do not experience any unusual temperature stress during the in-situ catalyst regeneration procedure.

Recovery of the propylene in the required product quality is relatively cheap and simple in constrast to other propylene producing processes. An expensive and energy consuming ethylene refrigeration system is not required. Problem-bearing compounds like acetylene and di-olefins are not formed in the process. Among other processes in the field of methanol to olefins conversion, the process stands out with its high selectivity, simplicity and low investment required.

The simplified overall mass balance is depicted in Figure 4 based on a combined Mega-Methanol/MTP plant. For a feed rate of 5000 metric tons of methanol per day (1,667 million metric tons annually), approx. 519,000 metric tons of propylene are produced annually.

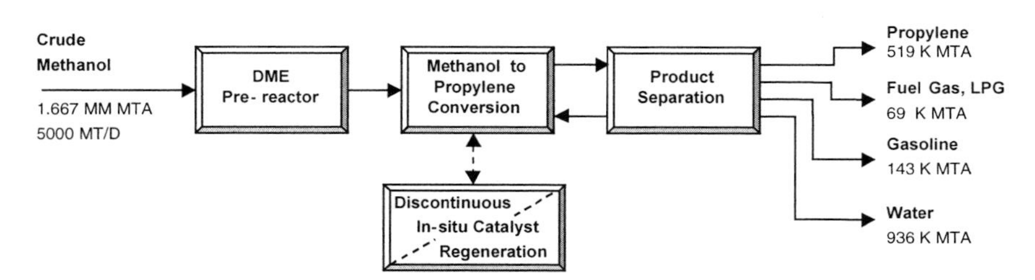

Fig. 4: MTP process route and production figures

The basic process design data were derived from more than 4000 operating hours of a pilot plant at Lurgi's R&D Center. A larger-scale demonstration unit in order to obtain more data on catalyst life and to demonstrate the process to potential customers is expected to go on-line in the summer of 2001.

Based on today's economical conditions, a stand-alone MTP project could yield an internal rate of return of more than 20 % provided that low-cost methanol is available. Integrating this concept with a Mega-Methanol plant is a logical consequence. Fig. 5 gives an impression of the economics of the MTP route, in which methanol is a mere intermediate.

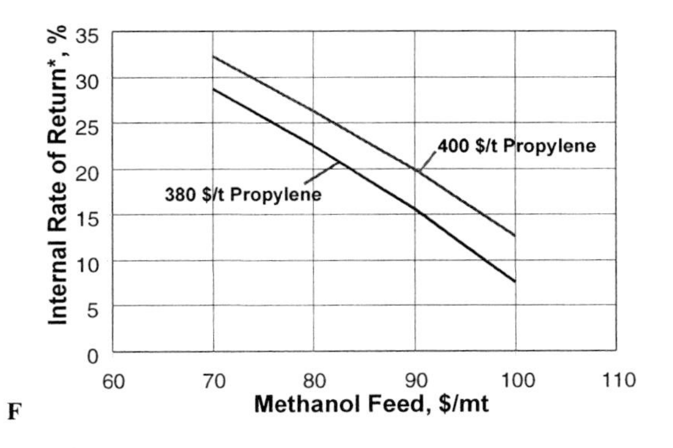

F

The feasibility study is based on the following main assumptions:

- Total investment cost (TIC) budget: 185 million US$
- Equity: 20 %
- Depreciation per year: 10 %

The results concerning internal rates of return (IRR) on total capital employed depend on the methanol feed price and the propylene product price that can be obtained. Different propylene price levels have been considered. Yielding an attractive IRR already at moderate propylene prices, the combination of Lurgi's Mega-Methanol and MTP technology closes the gap from natural gas to propylene.

Methanol to Hydrogen (MTH): Methanol to Hydrogen is a well-known and proven technology that has been applied in more than 50 plants world-wide. All these plants are of capacities smaller than 2.000 m^3_N/h. At the same time more than 100 naphtha based hydrogen plants are operated in India, China and Japan. Their total capacity is equivalent to about 5 million mtpa of methanol. Substituting this naphtha by methanol produced from natural gas can be another challenge.

In the Lurgi MTH plant a methanol / water mixture is converted on a copper based catalyst at low temperatures (250 – 300°C) and medium pressure (25 – 30 bar) into hydrogen, carbon monoxide and carbon dioxide. After cooling the product, the remaining gas impurities are removed by Pressure Swing Adsorption (PSA).

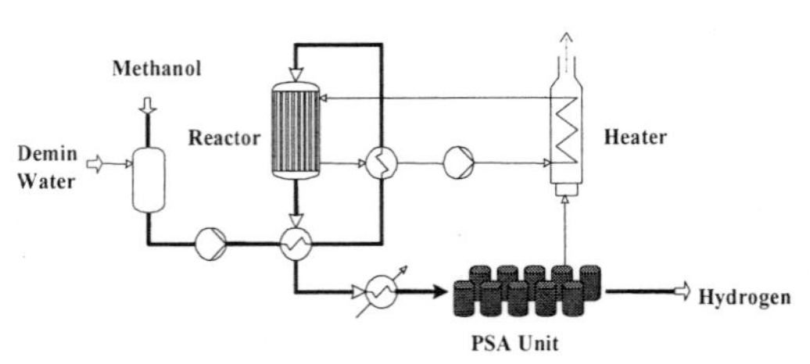

Fig. 6: Flow scheme of a Methanol to Hydrogen plant

Presently, the constraint for the implementation of MTH is the supply with low-cost methanol, but in principle this technology can be scaled up easily to huge capacities, because only standard equipment is used.

	MTH	Steam Reforming of Naphtha
Feedstock price	US$ 80 per mt	US$ 148 per mt
Feedstock demand	60.3 t/h	35.0 t/h
Investment costs	US$ 33 million	US$ 46.5 million
H_2 production costs	US$ 543 per mt	US$ 552 per mt

Table 2: Comparison of Key Operating Data

Comparing MTH with conventional steam reforming for a 100 000 m^3_N/h Hydrogen plant, indicates that methanol can compete against naphtha as feedstock to the steam reforming process. Price stability of the methanol feedstock in comparison to a naphtha feed here also can be considered an additional big "pro" of this route.

4. CONCLUSION

The low methanol price linked to the Mega-Methanol technology gives an option to replace the consumption of oil by natural gas in wide areas of the petrochemical industry. Easy to ship, methanol solves the problem of utilization of natural gas resources located in remote areas. MTP and MTH are only two among other technologies forming the nucleus for future developments. Altogether, these technologies will bring the owners of gas reserves in a position to convert their resources into products creating the maximum value-added.

REFERENCES

1. H. Göhna, "Concepts of modern methanol plants", World Methanol Conference, Tampa (USA) December 8-10, 1997
2. "Low cost routes to higher methanol capacity", Nitrogen, 224, November/December 1996
3. T.T. Christensen and I.I. Primdahl, "Improve Syngas Production using Autothermal Reforming", Hydrocarbon Processing, March 1994
4. N. Ringer, "The importance of reliable catalyst performance for mega-methanol plants ", World Methanol Conference, Copenhagen (Denmark) November 8-10, 2000

Studies in Surface Science and Catalysis
J.J. Spivey, E. Iglesia and T.H. Fleisch (Editors)

Selecting optimum syngas technology and process design for large scale conversion of natural gas into Fischer-Tropsch products (GTL) and Methanol.

by
Roger Hansen, Jostein Sogge, Margrete Hånes Wesenberg, Ola Olsvik

Statoil Research Centre, Trondheim, N-7005 Norway

Abstract
Four different syngas technologies; Conventional Steam Reforming (CSR), Partial Oxidation (POX), Autothermal Reforming (ATR) and Gas Heated Reforming (GHR) have been evaluated for GTL (Fischer-Tropsch) production and compared to a new syngas concept (TGR). The conclusion is that the GHR based syngas route and the new TGR based concept proved to be both more energy efficient (30% lower CO_2 emissions) and have lower investments than the other alternatives.

For methanol production, ATR based syngas production seems to be preferred above 5000 MTPD. However, continued development of steam reformers have increased the maximum capacity range for "Two Step Reforming". A number of new "hybrid" flowsheets have been announced which combine GHR, CSR, and ATR. These flowsheets seem to have a potential for competing with ATR based syngas concepts even up to 10,000 MTPD.

1. Introduction

Gas To Liquids (GTL) technology (Fischer-Tropsch products) and methanol have gained increased interest world-wide as an option to LNG. Non-flaring policies have created a need for associated gas solutions to secure oil production. In addition, Fischer-Tropsch products and methanol represent a potential new business opportunity as clean transportation fuels. The present paper will discuss the suitability of different flowsheets for large scale GTL or methanol production.

2. Optimum syngas technology for GTL (Fischer-Tropsch)

2.1 Syngas Technology

Optimum syngas technology and process design for FT-synthesis has not been studied to the same level as for methanol production, for various reasons.

The present analysis is based on a study which compares four different synthesis gas technologies. These are:

1. CSR = Conventional Steam Reforming (Externally Fired Tubular Reformer)
2. GHR = Gas Heated Steam Reforming (Heat provided from hot syngas from an ATR located downstream)
3. ATR = Autothermal Reforming (Internal oxygen fired steam reforming).
4. POX = Partial Oxidation (Similar concept as ATR with oxygen burner at the top, but with no catalyst for equilibration of gas)

Externally heated steam reforming, (1) generates a syngas rich in hydrogen, while internally fired steam reforming, ((3) and (4) above), generates a syngas with an SN-module lower than

2.0 (SN = $(H_2-CO_2)/(CO+CO_2)$). A combination of the two can provide the theoretical optimum for the overall syngas stoichiometry (SN=2.0) for both the methanol and the F-T-synthesis.

2.2 FT-Synthesis

In contrast to the methanol synthesis, CO is the reactant for the cobalt based low temperature FT-synthesis. Since the water gas shift reaction is not very active, CO_2 will act mainly as an inert. A key issue is therefore to select a syngas technology that will match the syngas composition required by the F-T synthesis.

Figure 1 illustrates the change (delta value) in selectivity to C_5+ components for the FT-synthesis as a function of gas composition (H_2/CO-ratio) and CO- conversion. The curves show that the selectivity to C_5+ products increases with decreasing H_2/CO-ratios. This makes the integration of the syngas section with the FT- synthesis different than for methanol production.

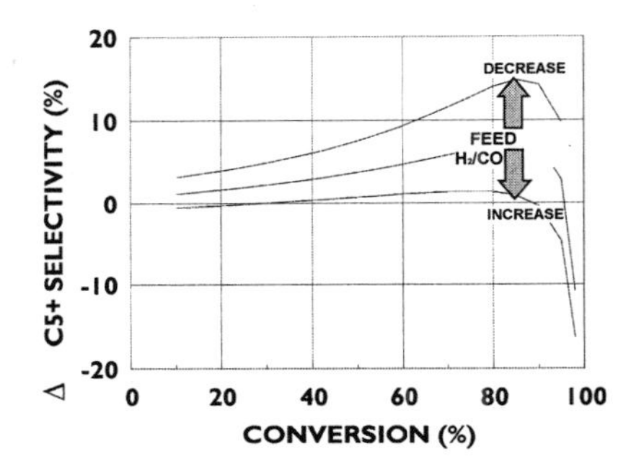

Figure 1 : Typical change in C_{5+} -selectivity (∇C_{5+}) for Co-based Fischer-Tropsch as a function of CO- conversion and H_2/CO-ratio.

The present study is based on an integration of a comprehensive F-T reactor model and process simulation models for different reforming technologies, which enables simulation of the entire GTL-complex. Thus, the impact of using different reforming technologies integrated with the F-T synthesis can be evaluated.

2.3 The general GTL flowsheet

The FT-synthesis is favored by syngas technologies with a high CO selectivity (high CO/CO_2) since CO is the carbon reactant. The CO selectivity is dependent on the steam/carbon-ratio and the outlet temperature. A low steam/carbon-ratio leads to a high CO/CO_2-ratio due to equilibration of the shift reaction.

The oxygen based routes (POX, ATR and GHR+ATR) give the highest outlet temperatures because of the efficiency of internal combustion. POX can operate without steam in the feed whereas operation of an ATR at steam/carbon-ratio as low as 0.60 has been demonstrated (1). Externally heated syngas production (CSR) requires a higher steam/carbon-ratio to avoid coke formation.

The characteristics of the different reforming concepts evaluated here are summarized in Table 1.

Table 1 : Advantages and disadvantages for different syngas technologies for F-T.

Syngas technology	+	-
CSR	no oxygen	very high H_2/CO high s/c-ratio thus high CO_2/CO
ATR	Near optimum H_2/CO low s/c-ratio	high oxygen consumption SN < 2
POX	Near optimum H_2/CO very low s/c-ratio	very high oxygen consumption SN < 2
GHR + ATR	Near optimum H_2/CO moderate oxygen consumption	unproven at s/c-ratios < 2

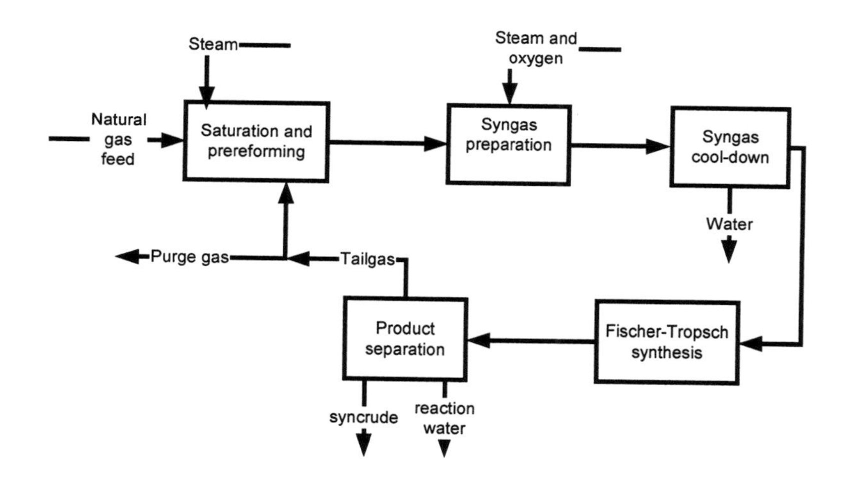

Figure 2 : Basic GTL flowsheet based on Syngas Preparation by; ATR, GHR+ATR, POX or CSR.

Figure 2 shows the integrated GTL flowsheet which has been the basis for the present evaluation. The block diagram illustrates the five basic sections of a GTL plant. For syngas production based on steam reforming (CSR), the second block denoted "Syngas Preparation" will not include addition of oxygen, but of heat.

2.4 A novel GTL flowsheet (Tail Gas Reformer, TGR)

Any flowsheet based on either ATR or POX is overall deficient in hydrogen. Adding steam reforming into the flowsheet can improve the thermal efficiency of the process and reduce the oxygen consumption. A steam reformer could be located upstream or in parallel with the ATR or POX, or on the tail gas from the FT- synthesis (TCR). It will be shown below that the last option, illustrated in Figure 3, provides distinct advantages, ref (2).

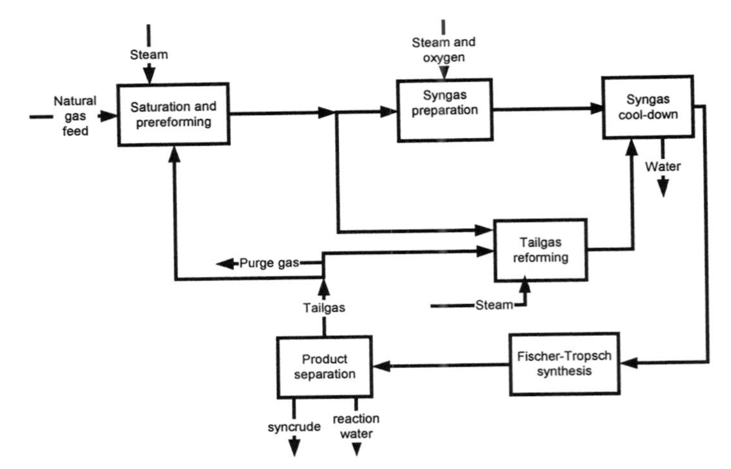

Figure 3 : New GTL concept based on ATR + Tail Gas Reformer

2.5 Process evaluation

The results listed in Table 2 for cases 1, 2, 3 and 4 are all based on a complete process study including all utility systems. Case 2 and 3 are based on GHR+ATR operating at low steam/carbon-ratios. The results for cases 5 and 6 are based on in-house process simulations which included the major utility systems (steam & power, cooling water, fired heaters).

Table 2 : Comparison of syngas technologies for a design capacity of 15,000 BPD C_5+

Syngas technology	ATR	GHR + ATR	GHR + ATR	ATR+ TGR	POX	CSR
Case	1	2	3	4	5	6
NG, feed & fuel ; Sm^3/bbl	235	214	214	217	250	252(*)
steam/c- ratio	0.6	0.9	1.5	0.6 / 2.0	0.2	2.5
Oxygen, TPD	3 626	2 769	2 841	2 694	3 891	0
Dry syngas (Tonne/hr)	378	324	369	376	336	
CO_2 (kg/bbl)	163	113	113	119	200	204
Cost (**) %	100	94	99	99	> 100	> 100

(*) Includes only natural gas (feed and fuel) to the steam reformer.
(**) Costs are relative to case 1 and includes direct costs (ISBL + OSBL).

Our study shows the following :

☒ Steam reforming based GTL (Case 6) will be less efficient than all the other alternatives. Since CSR cannot be scaled up above 10,000 bpd in a single unit, this is not a preferred technology for GTL.

☒ POX (Case 5; s/c-ratio of 0.2) does not compare favorable with ATR (Case 1; s/c = 0.6), due to a lower energy efficiency and a higher oxygen consumption.

☒ Both GHR and TGR based GTL plants (cases 2, 3 and 4) are more energy efficient and have a lower investment than the ATR based concept (case 1). The TGR concept is different from GHR in that the TGR operates with a much higher s/c-ratio.

☒ Case 2, GHR operating at a s/c=0.9 seems to be the best alternative.

3. Optimum syngas for mega methanol production

3.1 Methanol synthesis

The stoichiometric syngas composition for methanol production is

$$SN = (H_2 - CO_2)/(CO + CO_2) = 2.0$$

This number is derived from the chemistry which shows that you need three hydrogen molecules pr. CO_2 and two hydrogen molecules pr. CO. Minimum syngas production pr. tonn methanol produced is obtained when the stoichiometric number equals two. This value can be reached only by the principle of "combined reforming" (two step reforming).

3.2 Process design for mega methanol plants.

Figure 6 shows the relative investments for the three main concepts for methanol production (1).
These are :

1. CSR based "one step reforming"
2. CSR+ATR based "two step reforming"
3. ATR based (with H_2 recovery after or CO_2 removal prior to the methanol synthesis (*))

(*) The methanol synthesis loop cannot operate with an understoichiometric syngas.

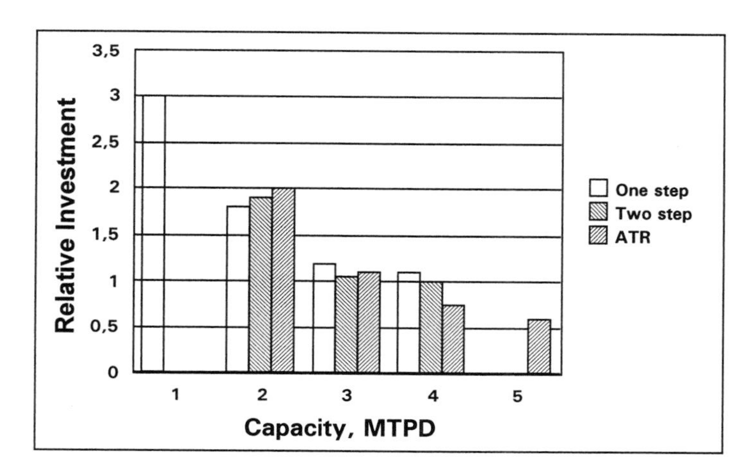

Figure 6 : Capital cost comparison of methanol technologies (1)

The figure implies that for capacities above 5000 MTPD, ATR based syngas production is the preferred technology. This is due to the fact that the cost of oxygen at increasing capacities drops faster than the increased cost of steam reforming. Two step reforming is the best alternative at intermediate capacities ranging from typically 2000 to 3500 MTPD. However, continued development of conventional- and convective- (more compact with improved heat transfer efficiency) steam reformers have increased the maximum capacity range for "Two Step Reforming" (see Figure 7).

Figure 7 lists maximum capacity ranges for some key process units based on information provided by several licensors lately.

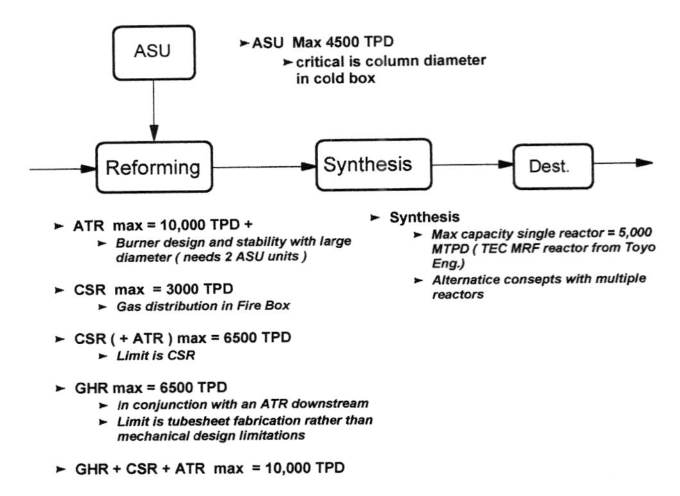

Figure 7 : Maximum capacities for mega methanol technologies

4. Conclusions

The GHR and TGR based syngas routes for GTL show potential for being more economic than all other concepts studied, including ATR. Further, these new alternatives generate in the order of 30% less CO_2 /bbl product, a fact that may gain increased attention in the future with one of the driving forces for GTL being to avoid flaring of gas in remote locations.

For methanol plants, a number of new "hybrid" flowsheets have been announced which combines GHR, CSR, and ATR. These flowsheets may have the potential of competing with ATR based syngas concepts even up to 10,000 MTPD.

REFERENCES

1. W.S. Ernst and S.C. Venables, Sasol Technology, and P.S. Christensen and A.C. Berthelsen, Haldor Topsøe A/S, "Push syngas production limits" Hydrocarbon Prosessing, March 2000

2. NO Patent Application 1999 6091 (1999, Statoil)

3. B.M. Tindall and M.A. Crews, Howe-Baker Engineers, "Alternative technologies to steam-methane reforming" Hydrocarbon Processing, November 1995

Studies in Surface Science and Catalysis
J.J. Spivey, E. Iglesia and T.H. Fleisch (Editors)

CANMET's Integrated Acetic Acid Process: Coproduction of Chemicals and Power from Natural Gas

Andrew McFarlan and Dirkson Liu

Natural Resources Canada, CANMET Energy Technology Centre,
Nepean, Ont. Canada K1A 1M1

ABSTRACT

A direct route for converting natural gas to acetic acid is currently being developed under CANMET's Consortium for the Conversion of Natural Gas program. The novel process integrates methanol production via direct partial oxidation of methane, and acetic acid production via vapour-phase methanol carbonylation over a supported catalyst. An integrated acetic acid plant coproduces bulk chemicals and electrical power, improving carbon utilization and energy efficiency of natural gas conversion. Results of our R&D project are presented which illustrate that the process is technically and economically feasible. The paper also addresses energy efficiency and greenhouse gas reductions which can be achieved via this process scheme.

1. INTRODUCTION

Acetic acid is a major commodity chemical derived largely from natural gas. In 1999, global demand for acetic acid reached 6.4 million tonnes (1). More than 60% of production capacity is based on homogeneously catalyzed methanol carbonylation technology (2). New liquid- and vapour-phase processes which employ heterogeneous catalysts for methanol carbonylation are at various stages of commercialization (3-6).

In 1994, CANMET and several industry partners established a research consortium to develop energy efficient, economically viable technologies for converting natural gas to liquid fuels and petrochemicals. One process currently being developed is a direct route for converting methane to acetic acid. The CANMET integrated acetic acid process differs from conventional acetic acid technology because it avoids synthesis gas production and CO purification, it employs vapour-phase methanol carbonylation over a supported catalyst, and it coproduces electrical power. Results of our R&D project to develop an integrated acetic acid process are presented below.

2. PROCESS DESCRIPTION

In the CANMET acetic acid process, methanol and carbon monoxide feedstocks are produced captively via direct partial oxidation of methane. In the range, 6-12 % conversion of methane per pass, methanol, carbon monoxide, and carbon dioxide are the main products of the gas-phase reaction. Product selectivities vary somewhat, depending on the specific reaction conditions employed (equations 1-3).

	Selectivity range (% C)	
$CH_4 + 1/2\ O_2 \rightarrow CH_3OH$	25-40	(1)
$CH_4 + 3/2\ O_2 \rightarrow CO + 2\ H_2O$	50-65	(2)
$CH_4 + 2\ O_2 \rightarrow CO_2 + 2\ H_2O$	5-20	(3)

The exothermic reactions occur spontaneously when the temperature of the reaction mixture exceeds a threshold of about 400°C. Oxygen is completely consumed, leaving a large fraction of the methane unconverted. However, over a broad range of reaction conditions, the ratio of carbon monoxide to methanol produced in the partial oxidation reaction is about 2:1. A thorough review of direct partial oxidation of methane to methanol has been published by Arutynov et al. (7).

Figure 1 is a schematic of the CANMET acetic acid process (8). Oxygen is mixed with makeup natural gas and recycle gas containing up to 25 vol % inerts (nitrogen and carbon dioxide), preheated, and fed to the partial oxidation (POX) reactor at about 1000 psig .

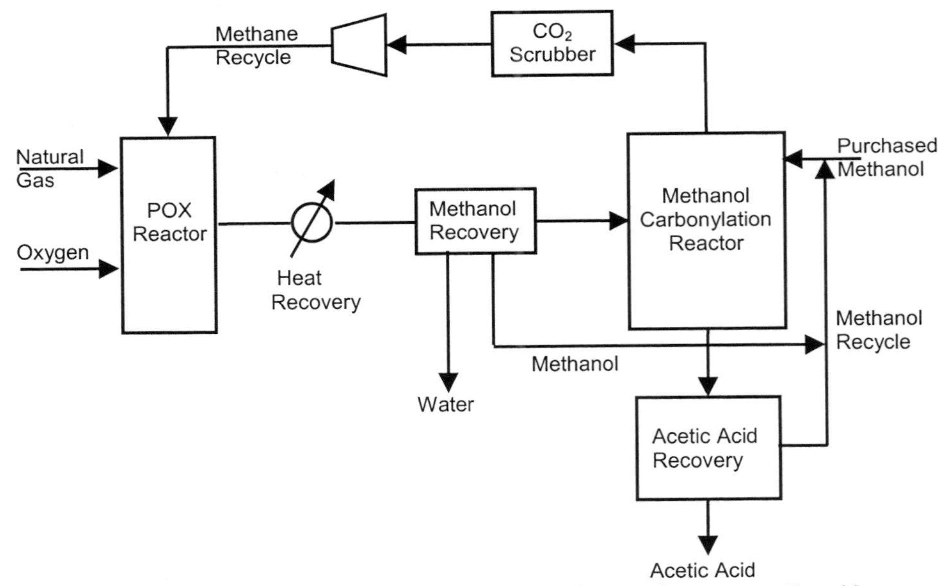

Figure 1. CANMET Integrated process for converting methane to acetic acid

The product stream from the POX reactor is cooled in a waste heat boiler, generating 1000-psig steam that is used to drive compressors and to generate power for export. Methanol and water are recovered in a methanol recovery unit. The gas stream exiting the methanol recovery unit is fed directly to the methanol carbonylation reactor. A liquid recycle stream from the acetic acid recovery unit and purchased methanol, equalling about one half of the total methanol requirement, are vaporized and mixed with the gaseous feed stream at the carbonylation reactor inlet.

In the carbonylation reactor, methanol reacts with carbon monoxide to produce acetic acid and methyl acetate at about 1000 psig and between 200-250°C. The reactions occur in the vapour phase over a supported catalyst such as carbon–supported rhodium and methyl iodide promoter. The gaseous product stream from the carbonylation reactor is cooled, and raw liquid product is collected in a reactor liquid product separator. Acetic acid is purified in a conventional distillation train (9). The cooled methane recycle gas stream, depleted of carbon monoxide, is passed through an absorber column to remove organic vapours and adjust carbon dioxide concentration, and is recycled to the POX reactor.

3. MINIPILOT TESTING OF INTEGRATED ACETIC ACID PROCESS

One of the challenges in a heterogeneous methanol carbonylation system is to prevent catalyst leaching (10). Loss of the precious metal from the catalyst support, by dissolution into the reaction medium, or by volatilization of metal carbonyl complexes into the vapour phase, diminishes catalyst activity necessitating frequent catalyst replacement and recovery of the metals downstream. Process conditions employed in the integrated acetic acid process ensure that the methanol carbonylation reaction operates in the vapour phase to prevent leaching. Catalyst performance and longevity have been demonstrated during hundreds of hours continuous testing in our minipilot plant. Figure 2, traces A, B, and C plots carbon monoxide consumption rate in the vapour-phase methanol carbonylation reaction over a supported rhodium catalyst at different space velocities (A>B>C). Catalyst activity was sustained during 600 h shown in Fig. 2, confirming that vapour-phase carbonylation is technically feasible. Analysis for rhodium by inorganic neutron activation showed no significant change in catalyst rhodium loading after several hundred hours on stream. In the liquid product, rhodium concentration was below the detection limit (<0.1 ppm).

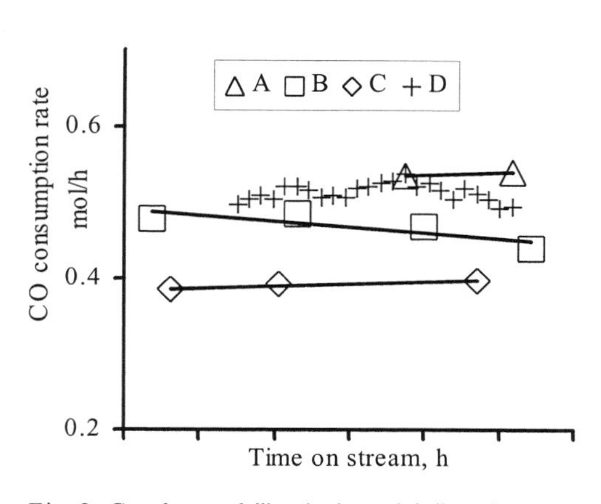

Fig. 2. Catalyst stability during minipilot plant runs. Full scale: 600 h, plots A-C; 300 h, plot D.

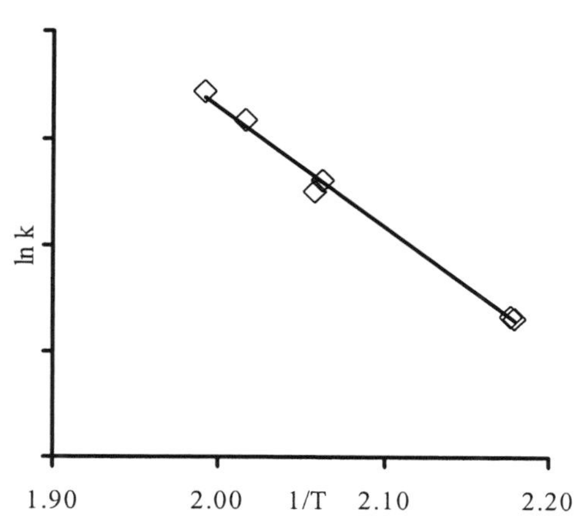

Fig. 3. Arrhenius plot of overall rate constant in vapour-phase methanol carbonylation reaction

Figure 2, trace D, shows catalyst stability during a 300-h minipilot plant run to demonstrate the integrated process shown in Figure 1, excluding acetic acid purification. During 300 h continuous operation, the production rate averaged 32 g/h acetic acid.

Figure 3 shows an Arrhenius plot of overall rate constants measured during a minipilot plant run to test vapour-phase methanol carbonylation over a supported rhodium catalyst. The activation energy as determined from the Arrhenius plot was 46.7 kJ/mol, in reasonable agreement with other published results (11).

4. COMPARISON WITH COMMERCIAL TECHNOLOGIES

4.1 Energy and Carbon Efficiency

Table 1 compares carbon dioxide and NO_x emissions for a world-scale acetic acid plant employing liquid-phase methanol carbonylation with emissions for a world-scale plant employing the CANMET integrated process. Total emissions include emissions attributed to raw materials production (methanol and CO via synthesis gas), and process inefficiencies in acetic acid production.

Table 1
Estimates of emissions for acetic acid production (1000 tonnes/day basis)

	CO_2 1000 t/year	NO_x t/year
Acetic acid via methanol and carbon monoxide:		
methanol feed	64 - 85	26 - 106
CO feed	~ 0	26 - 106
purge gas to fuel	66.0	
total	130 - 150	52 - 212
Acetic acid via CANMET integrated process:		
methanol feed	32 - 42	13 - 53
byproduct CO_2 (captured as pure CO_2)	43	0
total produced GHG	75 - 85	13 - 53
GHG avoided for integrated process:		
byproduct CO_2	43	
export power (18 MWe)	54 - 63	205
total avoided GHG	97 - 106	152 - 192

Thermal efficiencies for methanol production, defined as the ratio of lower heating value (LHV) of 1 tonne methanol divided by the total natural gas requirement, vary depending on the syngas technology employed. Conventional steam methane reforming of natural gas for methanol production is the least energy efficient syngas technology, 62% thermal efficiency. GHG emissions for a 2000 t/d methanol plant using a natural gas-fired steam methane reformer are 320,000 t/y carbon dioxide and 400 t/y NOx . Combined reforming can achieve

67% thermal efficiency for methanol production, lowering emissions of CO_2 by 25%, and NOx by 75% (12). When CO is the only desired syngas product, carbon dioxide and hydrogen can be recycled to the reformer (13). Hydrogen in the reformer chemically shifts CO_2 to CO and provides additional fuel for the reforming reaction. If all of the CO_2 is recycled, the net CO_2 released to the environment approaches zero.

In Table 1, total CO_2 emissions for 1000 t/d acetic acid production capacity, using conventional methanol carbonylation technology, is estimated to be 130,000-150,000 t/y CO_2, of which about one half is due to the production of the raw materials. By comparison, an integrated acetic acid plant is estimated to produce 75,000-85,000 t/y CO_2. NOx emissions attributed to raw materials production are estimated to be 75% lower for the CANMET process.

The 1000 t/d integrated acetic acid plant coproduces 18MWe electrical power for export to the grid. Based on 31% thermal efficiency for conventional steam cycle power generation, and 50% for combined cycle power generation, we estimate 54,000-63,000 t/y CO_2 and 205 t/y NOx emissions are avoided (14). The total quantity of avoided CO_2 could be as much as 100,000 t/y if the byproduct CO_2 captured in the scrubber was reinjected into a natural gas field, used for oil recovery, or sold as a chemical feedstock.

4.2 Process Economics

A comparative economics evaluation showed that CANMET's integrated acetic acid process can compete with conventional commercial technology. The competing process used modern liquid-phase methanol carbonylation technology, and purchased its methanol and carbon monoxide feedstocks. Factored capital cost estimates were obtained for both process, and operating costs were estimated according to current accepted methods (2,15,16).

Table 2 estimates production costs of acetic acid for 100,000 t/y grass roots plants located at Edmonton, Canada. From Table 2, the integrated acetic acid process was estimated to have about a 0.05 $US/lb cost advantage over the liquid-phase methanol carbonylation process, primarily due to lower variable costs. Lower raw materials costs and credit for export power contributed to the favourable process economics.

Table 2
Production cost estimates for acetic acid

Acetic acid process	CANMET	Conventional
Plant capacity, million lb/y	220	220
Production cost, $US/lb		
Variable costs	0.084	0.129
Fixed costs	0.024	0.025
Capital recovery + ROI	0.053	0.055
Total production cost, $US/lb	0.161	0.209

5. CONCLUSIONS

Direct partial oxidation of methane to methanol and carbon monoxide and vapour-phase methanol carbonylation over a supported catalyst have been successfully integrated into a continuous process for acetic acid production. The heterogeneous system for methanol carbonylation to acetic acid demonstrated sustained activity over hundreds of hours continuous operation in a minipilot plant.

CANMET's integrated process for acetic acid production avoids synthesis gas production and purification of carbon monoxide. The process also produces about half of its methanol requirement and is a net energy exporter. For a 100,000 t/y plant, production costs are estimated to be about 0.05 $US/lb lower compared to conventional liquid-phase methanol carbonylation.

Direct routes for converting natural gas to petrochemicals which avoid synthesis gas production offer improved carbon efficiency and lower greenhouse gas emissions. As much as 100,000 t/y carbon dioxide and 200 t/y NOx emissions could be avoided in a world-scale integrated acetic acid process because the process does not use large fired heaters, it captures CO_2 as a byproduct of chemical production, and it coproduces electrical power.

ACKNOWLEDGEMENTS
The authors gratefully acknowledges the financial support provided by Natural Resources Canada and the members of CANMET's Consortium on the Conversion of Natural Gas. The authors acknowledge Alison Janidlo and Bruce Dick who operate the minipilot plant facility.

REFERENCES

1. Chemical Week, Aug 16 2000, 46.
2. Chem Systems, "Acetic Acid/Acetic Anhydride" No. 97/98-1 (1999), 18.
3. N. Yoneda, T. Minami, J. Weizmann, B. Spehlmann, in "Science and Technology in Catalysis 1998", Kodansha Ltd. 1999, p. 93.
4. G.C. Tustin, J.R. Zoeller, H.L. Browning Jr., A.H. Singleton, US Patent No. 5,900,505, Eastman Chemical Company, Jan. 1998.
5. F. Joensen, US Patent No. 5,840,969, Haldor Topsoe A/S, Nov. 1997.
6. A. McFarlan, Proc. AIChE 2000 Spring National Meeting (Gas Conversion to Fuels and Chemicals II) 84A.
7. V.S. Arutynov, V.Ya Basevich, V.I Vedeneev, Russian Chem. Rev. 65(1996), 197.
8. A. McFarlan, U.S. Patent No. 5,659,077, Natural Resources Canada, Aug 1997.
9. M.J. Howard, M.D. Jones, M.S. Roberts, S.A. Taylor, Catal. Today 18 (1993), 325.
10. J. Zoeller, in "Acetic acid and its derivatives" V. Agreda and J. Zoeller eds., Marcel Dekker Inc. 1993.
11. K. Omata, K. Fujimoto, T. Shikada, H. Tominaga, Ind. Eng. Chem. Prod. Res. Dev. 24 (1985), 234.
12. G. L. Farina and E. Supp, Hydrocarbon Proc. Int. ed. 71(3) (1992), 77.
13. B.M. Tindall, M.A. Crews, Hydrocarbon Proc. Int. ed., 74(11) (1995), 75.
14. G. Haupt, D. Jansen, G. Oeljeklaus, R. Pruschek, J.S. Ribberink, G. Zimmerman, Energy Conv. Man. 38 (1997), S153.
15. K.B. Uppal, Hydrocarbon Proc. September (1997), 168-C.
16. D.M. Haseltine, Chem. Eng. June (1996), 26.

Studies in Surface Science and Catalysis
J.J. Spivey, E. Iglesia and T.H. Fleisch (Editors)

Rhenium as a promoter of titania-supported cobalt Fischer-Tropsch catalysts

Charles H. Mauldin and Douglas E. Varnado

ExxonMobil Process Research Laboratories
P.O. Box 2226, Baton Rouge, Louisiana, 70821 USA

A highly active Fischer-Tropsch catalyst is obtained by supporting cobalt on titania and incorporating a small amount of rhenium. Rhenium functions to increase cobalt oxide dispersion during catalyst preparation and to promote the reduction of cobalt oxide to the active metal phase.

1. INTRODUCTION

The opportunity to convert remote natural gas to liquid hydrocarbons has led to renewed interest in cobalt-based Fischer-Tropsch catalysts. Numerous formulations using various supports and promoters have emerged from industrial and academic research over the past twenty years.[1] One of the more active and selective examples consists of cobalt supported on titania and promoted by rhenium.[2] It is the purpose of this paper to describe some of the key features of the role of the rhenium promoter.

2. EXPERIMENTAL PROCEDURES

Most catalysts were prepared by one or more incipient wetness impregnations with aqueous solutions of cobalt nitrate and perrhenic acid, followed by drying and calcining in air, typically at 300°C. The 95% rutile titania supports used in most examples were made by spray-drying Degussa P-25 TiO_2 with 6% of alumina-silica binder[3] and calcining at 1000°C (19 m^2/g BET surface area). Evaporative impregnation of acetone solutions of cobalt nitrate and perrhenic acid was applied on a binder-free rutile support (made by pilling P-25 and calcining at 600°C) for the catalysts in Figure 4. The other supports included in Figure 1 were as follows: silica, 170 m^2/g, made from Degussa Aerosil 380; reforming-grade γ-alumina, 190 m^2/g, made by a commercial vendor.

Catalyst tests were conducted in a 0.25 inch I.D. reactor tube operated at 200°C, 20 atm., with a feed of 64%H_2-32%CO-4%Ne (neon as internal standard). GHSV was adjusted as required to maintain 60-80% CO conversion through 20 hours on stream. Catalysts were tested in powdered form (average sizes of 60-150 microns) to minimize the effects of pore diffusion limitations,[4] diluted with 1-7 parts by volume TiO_2 to minimize temperature gradients, and activated by reduction in hydrogen at 375-450°C, 5,000-20,000 GHSV, 1 atm, unless otherwise noted.

Oxygen chemisorption was determined by a conventional pulse technique. Catalyst was reduced in hydrogen at 450 C, cooled in helium to ambient temperature, and then dosed with small known volumes of oxygen added to the helium carrier gas stream until the catalyst no longer adsorbed oxygen.

High resolution X-ray diffraction was performed using a controlled atmosphere chamber which allowed in-situ reduction and exposure to syngas, installed on beamline X10B at the National Synchrotron Light Source at Brookhaven, Long Island. Catalyst was placed on an electrically heated tantalum ribbon holder, with temperature measured by a thermocouple attached to the underside of the ribbon. For reductions the cell was flushed continuously with pure hydrogen. Diffraction patterns were measured with a beam wavelength of 0.9687 Angstroms.

3. RESULTS

Incorporation of a small amount of rhenium in a Co-TiO_2 catalyst leads to a dramatic improvement in Fischer-Tropsch activity, as illustrated in Figure 1. With a low surface area rutile titania support, Co-Re catalysts show a linear increase in rate as cobalt loading is increased to about 12%. Activity begins to flatten out above 15% loading. Cobalt catalysts prepared on the same titania support but without the rhenium promoter are about half as active. Additional data points in Figure 1 corroborate other reports that rhenium improves relative activity with silica[5] and alumina[6] supports. Rhenium's dominant effect is on activity with selectivity largely unaffected. Methane selectivities (mol % CO converted to CH_4) fall in the 4-6 % range for the titania supported catalysts at the conditions used in this work.

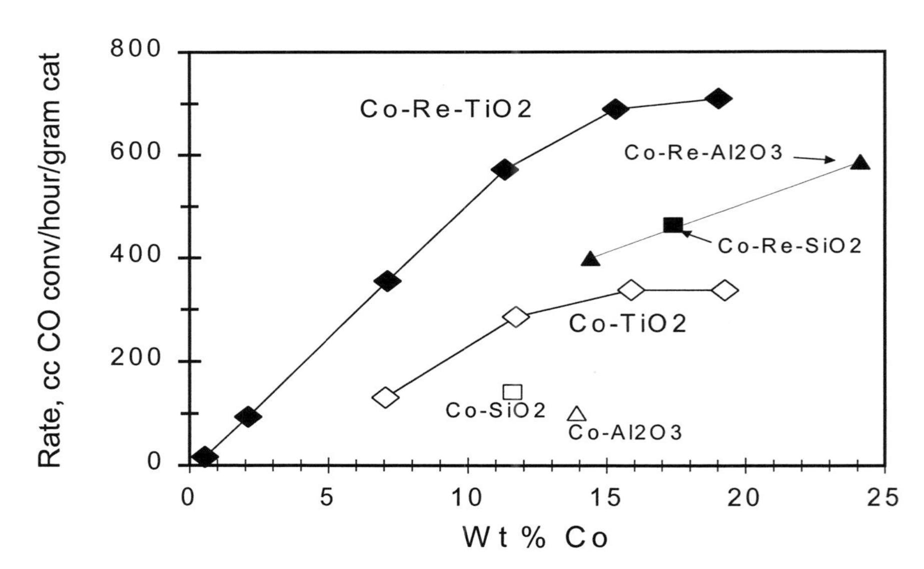

Fig. 1 Rhenium promotes cobalt activity on various supports (Re/Co wt ratio = 0.1; rate determined at 200 C, 280 psig)

This activity improvement arises primarily from a significant increase in cobalt dispersion. The effect of rhenium on cobalt dispersion on titania is shown by oxygen chemisorption data in Figure 2. The O/Co ratio is almost twice as high on Co-Re-TiO_2 compared to Co-TiO_2 at a given cobalt loading. Note that the absolute values of O/Co do not represent a true measure of dispersion since oxygen penetrates beyond the surface and into the bulk. However, the O/Co values are felt to reflect *relative* differences in metal dispersion. True dispersion, based on crystallite size measurement by transmission electron microscopy or by hydrogen chemisorption, is generally below 10 % on these titania supported catalysts.

Cobalt oxide dispersion, and the resulting metal dispersion after reduction, is established during the calcination step of the catalyst preparation. Precisely how rhenium improves cobalt oxide dispersion is not clear. It is speculated that the perrhenate anion is highly dispersed on the titania surface and may affect the wetability of the molten cobalt nitrate as it is dried and decomposed to Co_3O_4 during the calcination. Figure 3 indicates that rhenium's ability to improve activity reaches a plateau at a Re/Co weight ratio of about 0.1, indicating a practical limit to the promotion of cobalt dispersion by rhenium. The optimum ratio corresponds to about 1Re:35Co on an atomic basis.

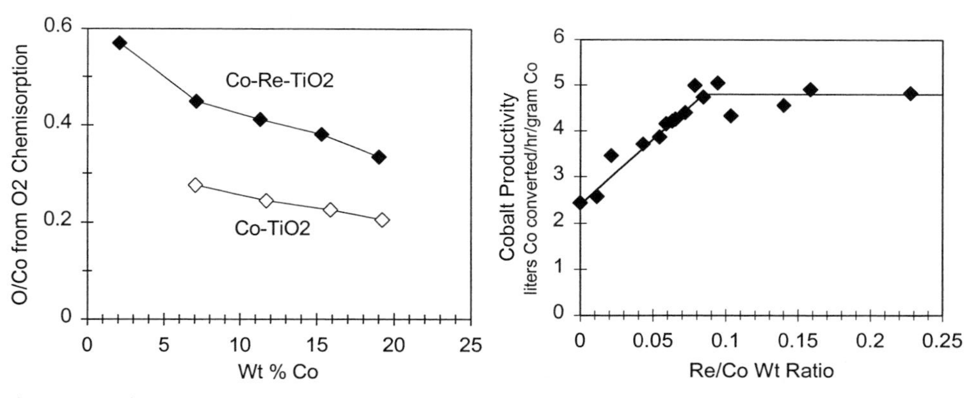

Fig. 2 Rhenium promotes cobalt dispersion (Re/Co wt ratio =0 or 0.1)

Fig. 3 Rhenium effect plateaus (11-12% Co loading)

Rhenium also functions to maintain cobalt oxide dispersion up to 500°C in air, as illustrated in Figure 4 with a plot of rate versus the temperature of a three hour air treatment. Activity of a rhenium-free 11.5% Co-TiO_2 catalyst decreases significantly as calcination temperature is raised. In contrast, a small amount of rhenium stabilizes the catalyst toward oxidizing conditions up to 500°C. Again, as in the *generation* of the oxide dispersion above, it is not clear how rhenium, in the form of either perrhenate or oxide, functions to *preserve* Co_3O_4 dispersion. Some influence on cobalt oxide mobility over titania is implied.

The final role played by rhenium involves the reduction of the cobalt oxide to the active zero valent state. Rhenium lowers the temperature required for essentially complete reduction to below 400°C and has a strong influence on the cobalt crystalline phase

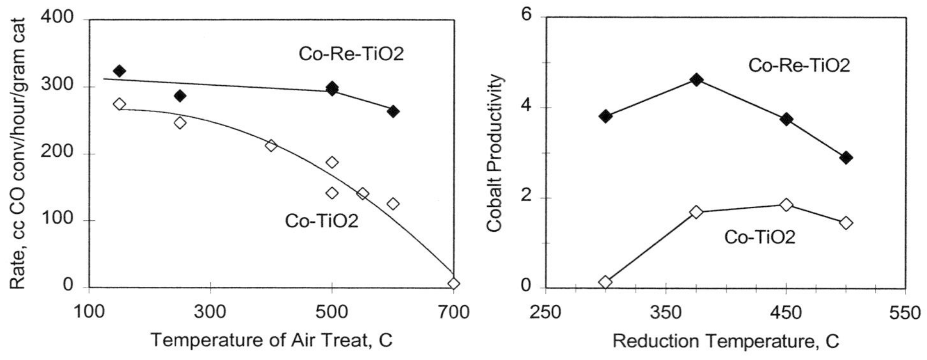

Fig. 4 Rhenium stabilizes catalyst in calcination (11% Co, Re/Co wt ratio = 0 or 0.037)

Fig. 5 Rhenium lowers reduction temp (11% Co, Re/Co wt ratio = 0 or 0.1)

produced. As shown in Figure 5, a catalyst containing rhenium is almost fully activated by reduction at 300°C, whereas a rhenium-free catalyst remains largely unreduced at that temperature. Maximum activities are observed after 375°C H_2 for Co-Re-TiO_2 and 450°C H_2 for Co-TiO_2. By 500°C, both catalysts are beginning to show lower activity, perhaps because of metal sintering. Rhenium's effect on cobalt oxide reducibility is believed to originate from the fact that rhenium on titania is reduced to metal around 250°C, which coincides with the beginning of the reduction of CoO to Co. Early formation of rhenium metal centers provides the system with hydrogen dissociation capability that in turn leads to a catalysis of the reduction of CoO.

Further evidence that rhenium is involved in the formation of Co metal comes from the type of metal phase produced. High resolution X-ray diffraction analyses following in-situ reduction show that the hexagonal close packed (hcp) phase of cobalt is formed in the presence of rhenium, whereas the face centered cubic (fcc) phase is formed in the absence of

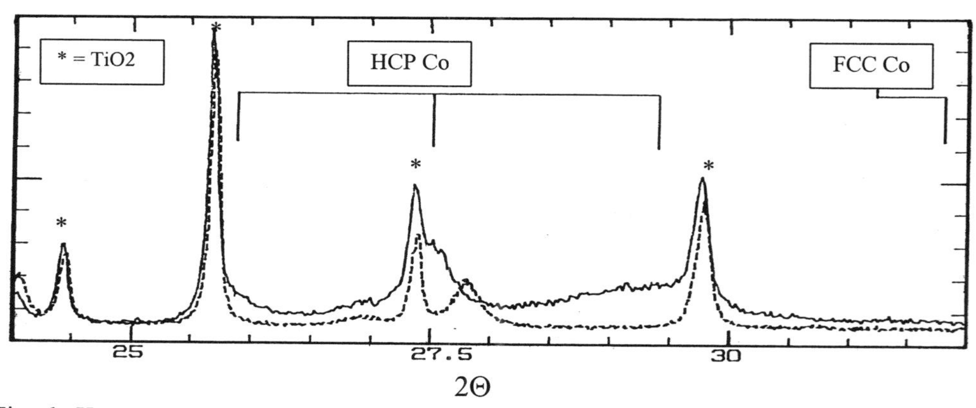

Fig. 6 Hexagonal Co-Re alloy formed on titania (11.6% Co-1.0% Re-TiO_2; fresh oxide, dashed line, compared to reduced at 450°C, solid line)

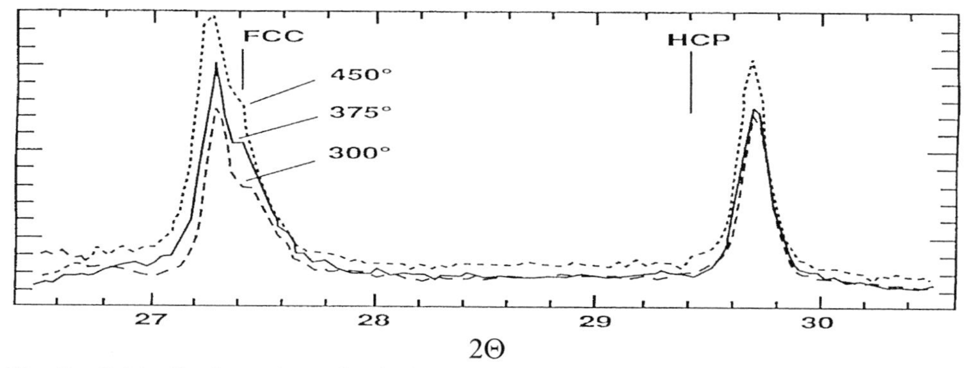

Fig. 7 Cubic Co formed on titania in absence of rhenium (10.8% Co-TiO_2 reduced at temperatures shown)

rhenium. The fact that rhenium metal is hexagonal and readily soluble in Co metal is good evidence that it is alloyed with cobalt on the activated catalyst. Diffraction patterns in Figure 6 show the formation of hcp cobalt from 12% Co-1% Re-TiO_2. The hcp phase did not change after exposure to 2:1 H_2/CO at 200°C, 1 atm. Note that the metal peaks are generally broad and especially so for the 101 peak at 29.4 2Θ, which indicates formation of a hexagonal phase with significant stacking faults.[7] The sharper peaks in the pattern are from the rutile and anatase titania phases present in the support. There is no peak observed at 31.6 2Θ corresponding to the 200 peak of the fcc Co phase. In contrast, as shown in Figure 7, 12% Co-TiO_2 produces only the fcc form of cobalt upon reduction in hydrogen. This is the only metal phase to appear between 300-450°C.

An important question to address is whether there is any intrinsic activity difference between the hexagonal Co-Re alloy on Co-Re-TiO_2 and the cubic Co metal on Co-TiO_2. Evidence in Figure 8 indicates there is no difference. In the figure, a pseudo-turnover

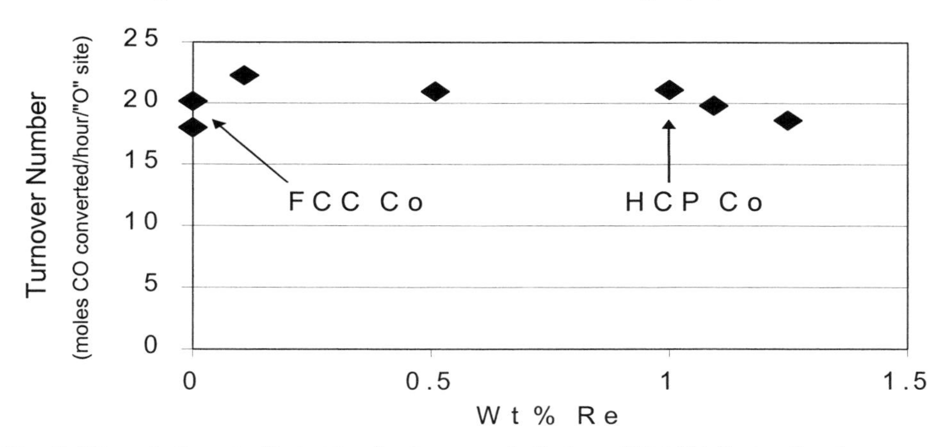

Fig. 8 Site activity not affected by rhenium or cobalt phase (10-12% Co loadings)

number (based on cobalt productivity and oxygen chemisorption data, all obtained after 450°C reductions) is plotted versus wt % Re on a 12% Co-TiO_2 catalyst. Clearly there is little significant change upon going from zero rhenium (i.e. fcc Co) to around 1% Re, where formation of hcp Co is complete. These results agree with a previously reported equivalency of turnover rates (based on H2 chemisorption) between Co-Re-TiO_2 and cobalt on a range of supports, including TiO_2.[8] Similar site activity from the hcp and fcc phases is consistent with (a) an expected enrichment of cobalt on the alloy surface[9] and (b) the reported lack of surface structure sensitivity for cobalt in Fischer-Tropsch.[10] Any possible influence of electronic differences between the Co-Re alloy and Co metal must be negligible as well.

4. CONCLUSIONS

In summary, rhenium has at least three beneficial effects on cobalt on titania: (1) It generates better cobalt oxide dispersion during the decomposition of the nitrate salt in the catalyst preparation; (2) it maintains cobalt oxide dispersion in air at higher temperature; and (3) it promotes the reduction of well-dispersed cobalt oxide at lower temperature, forming a Co-Re alloy as the metallic phase. Interestingly, all of these beneficial effects of rhenium occur *prior* to the catalyst actually being used in the Fischer-Tropsch reaction. Under use, the catalyst behaves with a turnover number that is unaffected by the presence of rhenium or the type of cobalt crystal phase.

5. ACKNOWLEDGMENTS

Besides the authors, a number of people were involved in generating the experimental data reported in this paper. Oxygen chemisorption data was supplied by Jeffrey T. Elks. A team of Kevin L. D'Amico, Hubert E. King, and Ingrid J. Pickering performed the x-ray diffraction experiments at Brookhaven. The support of this research and the collaboration with many others within ExxonMobil is gratefully acknowledged.

REFERENCES

1 R. Oukaci, A. H. Singleton, and J. G. Goodwin, Jr., *Appl. Catal. A:Gen.* 186, 129 (1999)
2 C. H. Mauldin, US Patent 4,568,663 (1986).
3 S. Plecha, C. H. Mauldin, L. E. Pedrick, US Patent 6,087,405 (2000).
4 M. F. M. Post, A. C. van't Hoog, J. K. Minderhoud, and S. T. Sie, *AIChe J.*, 35, 1107 (1989).
5 C. C. Culross and C. H. Mauldin, US Patent 5,856,261 (1999).
6 A. M. Hilmen, D. Schanke, and A. Holmen, *Catal. Lett.* 38, 143 (1996).
7 B. E. Warren, X-ray Diffraction, Dover, N.Y., p. 299 (1990).
8 E. Iglesia, *Appl. Catal. A:Gen.* 161, 59 (1997).
9 M. P. Seah, *J. Catal.* 57, 450 (1979).
10 B. G. Johnson, C. H. Bartholomew, and D. W. Goodman, *J. Catal.* 128, 231 (1991)

Studies in Surface Science and Catalysis
J.J. Spivey, E. Iglesia and T.H. Fleisch (Editors)

Market Led GTL: The Oxygenate Strategy

T. H. Fleisch, R. Puri, R. A. Sills, A. Basu, M. Gradassi and G. R. Jones, Jr.
BP Upstream Technology Group, P.O. Box 3092, Houston, Tx 77253

Abstract Today, natural gas conversion is the basis of a moderately sized, predominantly chemical business whose primary products are ammonia and methanol. A major focus of natural gas conversion R&D over the last few decades has been the manufacture of conventional, yet cleaner transportation fuels. However, Gas To Liquids technology includes more than Fischer Tropsch technology and extends to other liquid fuels, especially in the oxygenate family (methanol, Dimethyl Ether [DME], etc). The focus of this paper will be the promise of oxygenates for power generation, domestic home cooking, transportation, and as economical chemical intermediates.

1. Introduction

Today, natural gas conversion is the basis of a moderately sized, predominantly chemical business whose primary products are ammonia and methanol. Approximately 4 TCF of gas per year are consumed by this industry, representing only about 5% of the global annual gas consumption.

A major focus of natural gas conversion R&D over the last few decades has been the manufacture of conventional, yet cleaner transportation fuels (Fischer-Tropsch [FT] synthesis, Methanol-to-Gasoline). Making products for these large existing fuels markets permits the large-scale monetization of natural gas. However, such fuels are products that have to be price competitive with the fluctuating crude oil derived fuels. Further cost improvements are required before FT synthesis is competitive, since the expectations for the average, long-term price of crude oil is about $16/bbl.

The definition of Gas To Liquids technology includes more than Fischer Tropsch technology and extends to other liquid fuels, especially in the oxygenate family (methanol, DME [1], etc). The focus of this paper will be the promise of oxygenates in the power generation, domestic fuel, transportation fuel and chemical markets.

2. Natural Gas Refinery

Studies of future market demands for clean liquid fuels, fuel additives, and economical chemical feedstocks show the need for oxygenates primarily in power generation (the largest energy sector), diesel engines and hybrids, fuel cell applications and in the olefin industry. These demands can be met by multiple products that are produced from natural gas as the only hydrocarbon feedstock. This concept is also called the "Natural Gas Refinery".

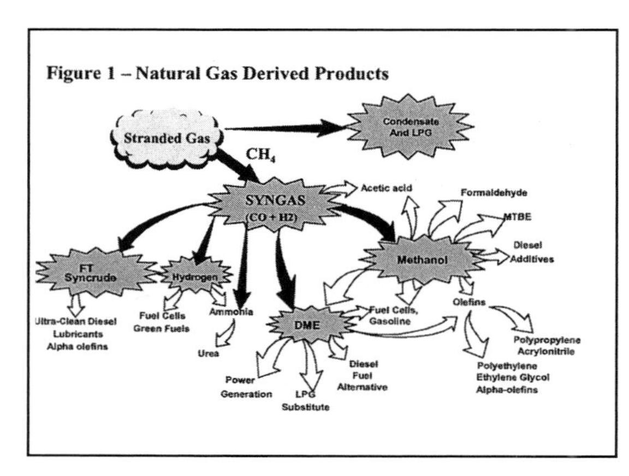

Figure 1 – Natural Gas Derived Products

Oxygenates, particularly methanol and DME, are key intermediate/end-products. As shown in Figure 1, starting from stranded natural gas, liquid hydrocarbons (condensates and LPG) are separated and sold separately. From the remaining main component, methane, all key products, can be produced via intermediate syngas production. Areas in the world that are gas-rich, such as Trinidad, can benefit from an integrated natural gas refinery.

3. Power Market

3.1. DME Demonstrated as a Fuel for Power Generation

BP has successfully completed combustion programs, in cooperation with EPDC of Japan, to demonstrate that DME can be a fuel for modern gas turbines -- offering slightly better performance and emissions than natural gas. Based on these programs, GE will pursue commercial offerings of DME-fired E class and F class heavy-duty gas turbines at standard commercial terms, including guarantees of output and heat rate. Other key findings are: (a) DME can be fired in existing gas turbines, with minor modifications primarily to the fuel delivery system; (b) Critical component life of DME-fired gas turbines is anticipated to be comparable to natural gas-fired machines. Key benefits of DME compared to using liquid fuels include lower heat rate, higher equipment availability and lower maintenance costs.

The performances of a 700 MW combined cycle power plant, using GE PG9171E gas turbines, fueled with DME, natural gas and naphtha are shown below (2):

Fuel	Fuel-grade DME *	Natural Gas *	Naphtha *
Fuel Feed to the Plant Gate	Liquid at 1 atm & -25°C	Gas at +20 °C	Liquid @ +20° C
NOx Emission, ppmvd (15% O2)	25	25	42
NOx Control Method, Burner	Dry Low NOx (DLN-1)	DLN-1	Water Inj.
Fuel Rate, lb./second	110	61	72
Power, Gas Turbines - MW	453.5	444.8	467.4
Power, Steam Turbines - MW	256.9	258.6	261.2
Net Power Produced, MW	692.6	686.6	711.6
Heat Rate, Btu/kwhr (LHV)	6,627	6,731	7,073
Efficiency, % (LHV)	51.5	50.7	48.3
Stack Temperature, °F	177	204	215

* LHV for liquid at 77 F: 6,420 Kcal/kg, LHV for gas: 11,800 Kcal/kg LHV, liquid = 10,670 Kcal/kg

The study showed that the heat rate for DME is about 1.6% and 6.3% lower than those for natural gas and naphtha, respectively. The higher efficiency of DME compared with natural gas is primarily due to the use of low-temperature energy to vaporize DME and lower stack temperature.

According to GE estimates (2), a DME-fueled combined cycle power plant generates about 50% lower NOx than that fueled by distillates and slightly less than that fueled by natural gas. CO emissions are roughly the same for all three fuels. CO2 emissions, which depend on carbon content, are lowest for natural gas and highest for distillate.

3.2. A Case Study – DME Commercialization: India DME Project

BP has been pursuing the fuel market for power generation in India (3) as the first commercial market for DME since only a few power plants would be needed for the entire DME plant production. One large-scale DME plant producing about 5,000 TPD would supply DME for about 900 MW total power plant capacity. This market development can also serve as a supply platform for future use of DME as a transportation fuel (one plant produces 3,400 TPD diesel equivalent), and as a LPG substitute (one plant produces 3,100 TPD LPG equivalent).

In July 1998, BP signed a commercial/technical agreement (Joint Collaboration Agreement) with Indian Oil Company Ltd, the Gas Authority of India Ltd. and the India Institute of Petroleum to collaborate in the development, production and marketing of DME as a multi-purpose fuel for India. In December 1999, Phase 1 was successfully completed with the issuing of the Techno-Economic Feasibility Report (TEFR) for using DME as fuel for power generation. The study showed that DME can be marketed to medium sized power plants in the range of 500-1000 MW capacity that have no access to pipeline natural gas or LNG, such as those along the coast of southern India. Most significantly, the study concluded that the delivered price of DME at the burner tip would be able to compete economically with other conventional fuels.

4. Domestic Heating Market – DME as a Substitute for LPG

The Joint Collaboration Agreement for the India DME Project, signed in 1998, also recognized that DME was an excellent candidate as an LPG substitute in home cooking and commercial applications. India's demand for petroleum products is growing at a fast pace, and even with the addition of large-scale refining capacity in the near future, deficits, particularly in respect of LPG are expected. DME is a candidate alternative for because it has physical characteristics similar to LPG. Comparing DME to LPG, shows DME: high oxygen content allows smokeless combustion with lower CO and total hydrocarbons emissions; has a lower boiling point for very rapid vaporization; and displays a visible blue flame similar to LPG over a wide range of air / fuel ratio which is an important safety characteristics. However, in the same conventional LPG cylinders, DME would weigh 18.7 kg compared to a LPG weight of 14 kg and have 18% less energy. Test conducted in Japan have shown that DME can be used in a city gas stove without modifications (4).

5. Transportation Market

The transportation fuel alternatives depend on the engine/drive train type, as shown in Figure 2. A large variety of transportation fuels can be manufactured from natural gas. Oxygenates can be alternatives for the Otto Engine, Diesel engine and fuel cells. Let's discuss some oxygenate applications for methanol, DME and DMM (Dimethoxymethane).

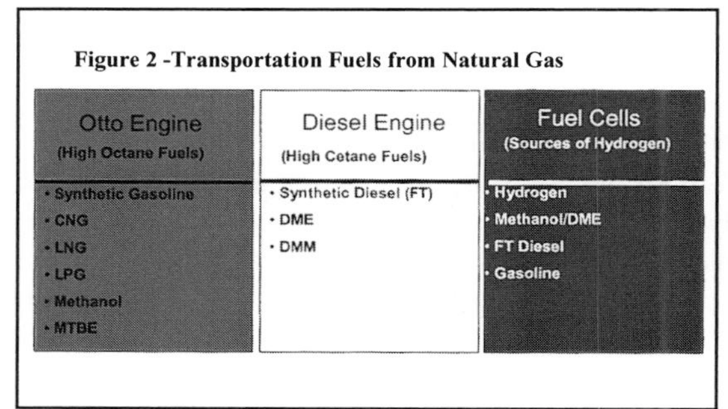

Figure 2 -Transportation Fuels from Natural Gas

A number of development programs are underway using methanol as a potential source of hydrogen for fuel cell vehicles. Methanol can be easily transformed into hydrogen. It can be cost competitive since the higher cost of methanol is compensated by the high efficiency of fuel cells. Methanol must compete with other potential candidates for sources of hydrogen for fuel cells, besides hydrogen itself, such as FT diesel and gasoline. Although, significant developmental efforts for M85 (85% methanol; balance gasoline) as a transportation fuel alternative have been made, methanol cost, new infrastructure requirements for delivery; and safety and health concerns have limited its promise.

BP, along with others such as Haldor Topsoe A/S, and AVL List GMBH, have been advancing the use of DME as an ultra-clean diesel alternative since the first presentation of the concept in 1995 (1). Several programs are underway in Europe, United States and Japan to advance the use of DME as a transportation fuel (5)(6). DME is an excellent fuel for diesel engines operated in urban areas where ultra-low emissions are needed. No basic engine modifications are needed for DME. However, new vehicle fuel storage, handling and injection systems are required and are being developed. Also, because of its high vapor pressure at ambient temperature, large-scale use of DME would necessitate changes to existing fuel delivery and storage systems.

Oxygenates, primarily DME derivatives, such as DMM and APCI's CETANER™, are being proposed as additives for diesel fuel. Tests in light duty diesel engines at Southwest Research Institute (7) have shown that particulate and NOx emissions from a 15% DMM blend are 50% and 96% of the base fuel, respectively. This application has the significant benefit of being able to use existing infrastructure for fuel delivery. Safety risks associated with the lower flash point of DMM need to be addressed.

5.1. Techno-economic Comparisons

When used as a transportation fuel, oxygenates such as DME will need to be cost competitive with market alternatives. For DME these are conventional diesel fuel and gasoline. Figure 3 shows a simplified comparison of end-user costs on a diesel gallon equivalent basis excluding excise taxes for diesel, DME, and gasoline on the U.S. Gulf Coast. End-user costs include fuel price, and fuel transport costs to the filling station, new fuel distribution/infrastructure and vehicle retrofit costs and station operating costs. Diesel and gasoline are produced in a conventional, paid-off, crude oil refinery. Conventional diesel fuel is assigned a relative cost of 100.

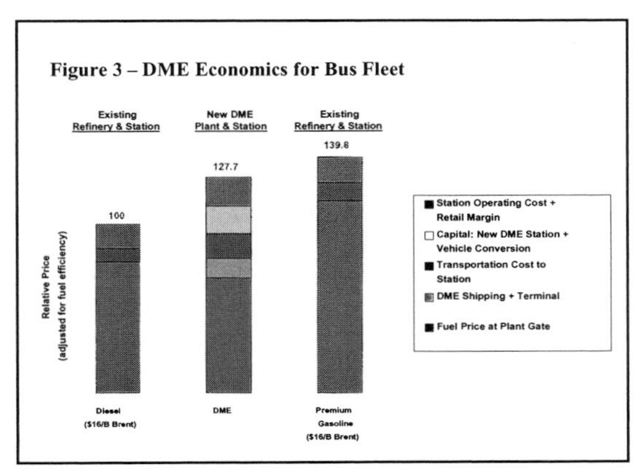

Figure 3 – DME Economics for Bus Fleet

DME is assumed to be manufactured in a large-scale, greenfield plant located in Trinidad. The DME manufacturing capacity is 5,000 MTPD (8), and the natural gas feedstock has an assigned cost of U.S. $1/MMBtu. The relative end-user cost for making DME is 69, or 9 units lower than the relative manufacturing cost of conventional diesel. However, the total cost to the end-user is 28% higher primarily due to new infrastructure and shipping costs. This gap can narrow and even disappear with the added cost of future environmental regulations, higher crude prices, and if new refineries/infrastructure is needed for diesel.

6. Chemical Feedstocks

Today, methanol is a major chemical building block used to manufacture formaldehyde, MTBE, acetic acid and a wide range of other chemical products. Tomorrow's market for methanol could include conversion to ethylene and propylene, which are major chemical building blocks. Since the market for ethylene and propylene are large, the impact of this new use on methanol demand would be dramatic. For example, if methanol were used to meet one year's growth in ethylene demand (3.5 million Tonnes/yr), the production of methanol would have to increase by 14 million Tonnes/yr or almost 50%. Processes for conversion of methanol to olefins (MTO) has been developed both by UOP and Norsk Hydro, and ExxonMobil. MTO was developed as a second step in a two-step process to convert low cost natural gas to ethylene and propylene. DME could also be converted to ethylene and propylene using this technology.

7. Market Potential

The potential new market for oxygenates could be significantly greater than the current market. For example, as shown in the table below, the market for methanol could be about 20 times greater than it's current market. For this example, 600 million Tonnes/yr methanol would be required if 10% of world power were generated from methanol; in the transportation market, 150 million Tonnes/yr would be required if 25% of the world's vehicles are powered by fuel cells using methanol; and 55 million Tonnes/yr would be required if 25% of world's diesel demand is blended @ 15% DMM; and in the chemicals market, if one years annual growth of ethylene demand is met using methanol. The consumption of natural gas would increase significantly; that is, by about 1/3 compared to today's gas consumption of about 80 TCF/yr, as shown in the table.

	Million T/yr methanol	**TCF/yr natural gas feed**
Current Market	35	1.1
Potential New Markets		
• Power	600	19
• Transportation		
Fuel Cell	150	5
Diesel Additive	55	2
• Chemicals	14	0.5

8. Conclusions

GTL is an important option for moving natural gas to market. GTL options include not only the traditional production of FT liquids but also the production of simple, oxygen containing fuels, fuel additives and chemicals, such as methanol, DME, DMM, etc. Several commercialization/development programs around the world have demonstrated that oxygenates can be competitive, environmentally-friendly alternative fuels and feedstocks for the power generation, domestic heating, transportation and chemicals markets. Successful commercialization in these new markets will not only significantly increase the world demand for oxygenates but also have a major impact on monetizing stranded gas.

References

1. Fleisch, T., et.al, SAE Paper No. 950061, Detroit, Michigan, March 1995.
2. Basu, A, Wainwright, J, DME as a Power Generation Fuel: Performance in Gas Turbines, Petrotech-2001, New Delhi, January 2001.
3. Puri, R; Sills, R; A Global First; The India DME Project, International DME Workshop, Tokyo, September 2000.
4. Web site: www.nkk.co.jp/en/environment/dme
5. Sorenson, S.C.; DME in Europe, International DME Workshop, Tokyo, September 2000.
6. Kajitani, S., et. al.; Paper No. 2000-ICE-289, ICE-Vol.34-3, 2000 Spring Technical Conference, ASME 2000.
7. PNGV Test Program, Southwest Research Institute, Diesel Fuel News, Hart Publications, Inc, Vol. 2, No. 10, May 21, 1998.
8. Capital cost based on following paper with contingency adjustments for location. Romani, D; Scozzesi, C.; Holm-Larsen, Helge; Piovesan, L.; Large-Scale Production of Fuel DME from Natural Gas, 2nd International Oil, Gas & Petrochemical Congress, Tehran, May 2000.

Studies in Surface Science and Catalysis
J.J. Spivey, E. Iglesia and T.H. Fleisch (Editors)

Gas-to-Liquids R&D: Setting Cost Reduction Targets

Michael J. Gradassi

Global Gas Technology: Gas to Liquids, Upstream Technology Group, BP Exploration and Production Company, 501 Westlake Park Boulevard, Houston, Texas 77079, USA.

Advances in gas-to-liquids technology reportedly have reduced costs beyond the threshold of commercial attractiveness. Yet, research continues targeting further cost reductions, while few GTL projects are moving forward to construction. To what level do costs need to be reduced to spark solid commercialization of this technology? Is it possible that the drive to reduce costs may actually make a winning technology a loser? When should resource allocation be shifted from primarily R&D to primarily commercialization? Is it possible to avoid destroying monetary value while continuing to make technological improvements?

The results of techno-economic cash flow models of both syncrude and refined product-producing Fischer Tropsch-based GTL projects will be presented to address these key issues. Substantiating cash flow modeling results demonstrating the economic benefit of GTL capital cost reduction versus the R&D expense needed to achieve it will be presented. In addition, the economic benefit brought about by properly timing the commercialization of state-of-the art GTL technology versus one with a lower capital cost in the future will be shown. Example case studies will show how properly and improperly timed commercialization of lower capital cost GTL technology can enhance or destroy the economic value of a gas conversion project.

1. BACKGROUND

Today, there are more than 15 Fischer-Tropsch gas-to-liquids (GTL) projects under commercial consideration with a combined production capacity of 750,000 Bbl/D and a combined equivalent gas input of about 7.5 BCF/D [1]. But, of these two and possibly three may move to commercialization within the next five years. Does this timing represent significant progress or a delay in commercialization? Is GTL technology sufficiently economically viable to warrant its implementation? If the technology is economically ready, should GTL technology R&D continue and, if so, at what level?

The purpose of this paper is to present relative economic information relating to GTL manufacture based solely on a selection of recent literature articles. No separate engineering design or costing was carried out. Parameters measuring the economic value of R&D were calculated using a cash flow analyses. Although the following discussion focuses only on Fischer-Tropsch technology, it should be recognized that it is only one of a family of GTL technologies, including methanol, DME, gas-to-gasoline, and several others that will benefit from the analysis presented here.

2. NPV10: THE R&D VALUE YARDSTICK

The economics methodology used in this paper follows that used earlier [2] for the calculation, shown in Table 1, of after-tax cash flows for a GTL plant described by the parameters listed in Tables 2 and 3.

From the after-tax cash flow calculations, the net present value discounted at 10 percent (NPV10*) was calculated to establish a Base Case economic parameter for later comparison as a measure of value creation. The internal rate of return (IRR†), another common economic parameter, is often used to compare a project's economic return with that of the project developer's cost of capital.

Table 1
After-Tax Cash Flow Calculation

After-Tax Cash Flow Calculation
Revenue
- Natural Gas Feedstock Expense
- Operating Expense
- Freight Expense
- Depreciation
Taxable Income
- Income Tax
Net Income
+ Depreciation
- Plant Capital Cost
After-Tax Cash Flow

Table 2
Base Case Assumptions Summary

Parameter	Value
Plant Capacity	30,000 barrels per stream day
Plant Capital Cost (CAPEX‡)	$750MM ($25,000 per daily barrel) [3]
Natural Gas Consumption	9,500 scf per barrel of liquid product
Natural Gas Price	$0.50 per mscf
Non-Gas Operating Expense (OPEX§)	$4.00 per barrel of liquid product [3,4,5]
Syncrude Value	EIA** Crude Oil Forecast [6]
Naphtha Product Value	At parity with crude oil
Diesel Product Value	EIA Diesel Forecast [6] + $2/Bbl [7,8]
By-product Power	None

3. RESULTS

The Base Case, which assumes a finished product slate of naphtha and diesel fuel, shows a 14% IRR and an NPV10 of $240MM. Given this result, it would appear that today's available GTL technology could very well be found more than adequately attractive by many owners of otherwise stranded gas. It may also help to explain why so many potential projects

* NPV10 is the net present value of the sum of after-tax cash outflows and after-tax cash inflows discounted at a realistic interest rate, chosen here to be 10%.

† IRR is defined as the discount rate that makes the net present value of the after-tax cash outflows and after-tax inflows equal to zero.

‡ CAPEX includes plant capital costs, capitalized owner's costs, capital contingency, interest during construction, first catalyst charges, initial inventory of spare parts, and other capital related costs.

§ OPEX includes fixed and variable operating expenses, such as operating labor, operating maintenance, catalyst expenses, taxes, insurance, and other operating expense items.

** EIA is the Energy Information Administration of the United States of America.

are under discussion. But, are they progressing quickly enough to commercialization? Is it prudent to delay a GTL project while R&D to reduce capital costs is being conducted? Or, is it better to begin project development at the present stage of the technology? What effect is this having on creating monetary value for potential investors?

Table 3
Economic Assumptions

Parameter	Value
Manufacturing Plant Life	25 years
Depreciation Schedule	10 year, Straight Line
Plant Construction Period	3 years
Plant Construction Capital Spending Profile	25%, 50%, 25%
Owner's Equity	100%
General Inflation	None
Escalation above general inflation	None
Federal + State Income Taxes	35%
Plant On-Stream Factor	95%
Plant Stream Day Production Profile	50% year 1, 100% Year 2-25
Working Capital Model	15 Days Liquid Product Inventory
Product Shipping Expense	$2.00 per barrel [9]

3.1. Project Delay: Value creation or value destruction?

Delaying the development of a GTL project, Figure 1, can destroy significant value as measured by NPV10. This arises because a delayed project start postpones revenues and cash flow to the future where they are worth less than if generated today. In this example, a three-year delay may destroy 25 percent of the potential project value, with subsequent delays showing significantly greater value destruction.

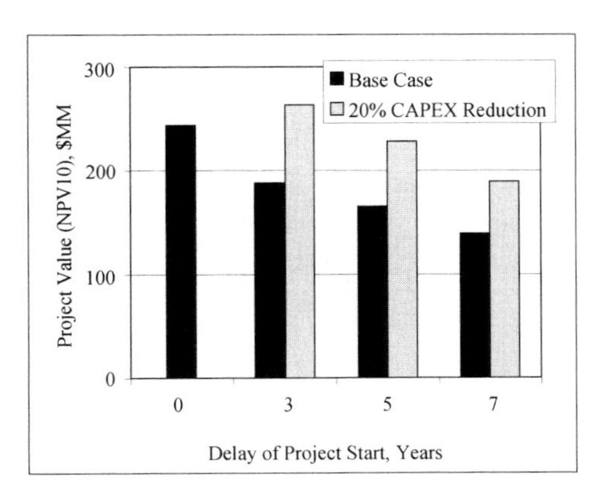

Fig. 1. Delay of project start destroys value unless CAPEX is reduced.

It was previously shown that of all controllable costs, project CAPEX has the single greatest effect on GTL project value [2]. Therefore, a successful R&D program directed towards reducing CAPEX, by say 20 percent, can mitigate value destruction if the R&D can be completed within 3 years as illustrated. Longer completion times would again lead to value destruction. Reductions in OPEX can mitigate value destruction somewhat, but it can be shown that after a 5-year delay in project start substantial reductions would be needed, and with a 7-year delay OPEX would need to be negative. Hence, this example focuses on reducing CAPEX.

But, do project development delays, in fact, destroy value? The answer, of course, depends on the number and type of project options an investing entity has. It is judged that an entity likely to invest in GTL projects has more project options at its disposal than money to invest. So, for such an entity, delaying the development of a GTL project does not necessarily mean that value is destroyed.

3.2. Benefit-to-Cost Ratio

Indeed, R&D continues despite apparent project development delays, for without it, GTL technology would not be at the economic state it is today. But, R&D has a cost. The question is how much to spend to achieve a specified target? The answer, of course, is at the discretion of the entity asking the question. Different entities may set different guidelines. One such R&D value guideline is the Benefit-to-Cost ratio (BCR), which for this discussion is defined as the quotient of the created monetary benefit as measured by NPV10 and the R&D cost to achieve it. Selection of the BCR target value is subjective and is the decision of the entity underwriting the R&D expenditures.

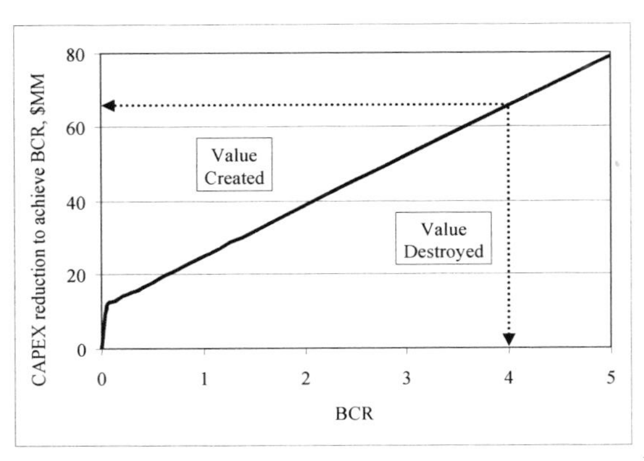

Fig. 2. Example BCR vs. CAPEX reduction for a 3-year, $3MM/a R&D program

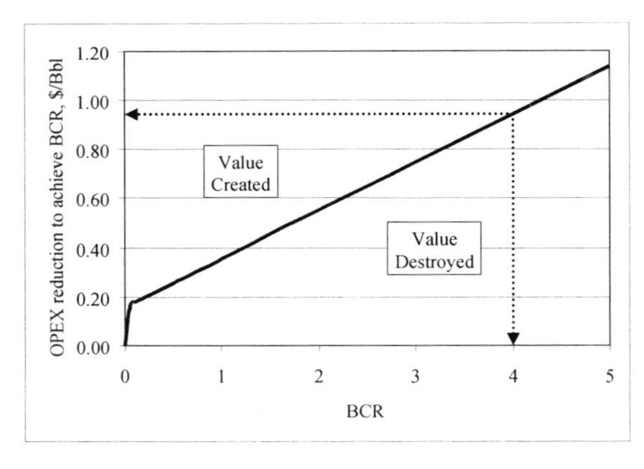

Fig. 3. Example BCR vs. OPEX reduction for a 3-year, $3MM/a R&D program

Assume an entity wishes to conduct R&D to reduce GTL CAPEX. At the same time, it wishes to conduct efficient research and would like to obtain at least a minimum value benefit for its R&D expenditures. Presumably, the minimum BCR should match that of the base project itself, because below that level value would be destroyed. For the example Base Case project of this discussion, the BCR is 0.3. ($240MM NPV10 Benefit / $750MM Project CAPEX = 0.3)

The basis of the illustrated BCR example, Figure 2, was generated using the cash flow analysis discussed above. It assumes a three-year R&D program with annual expenditures of $3MM, for a total *cost* of $9MM. If the entity targets a BCR of 4, successful R&D should create a *benefit* of $36MM in NPV10 over and above that of the Base Case.

Remembering that the GTL Base Case generates an NPV10 of $240MM, the implemented R&D would need to increase this figure by $36MM to $276MM. Figure 2 shows this success can be achieved with R&D that leads to a CAPEX reduction of about $65MM. For any chosen BCR, CAPEX reductions on or above the line create value. Below the line, value is destroyed.

Figure 3 illustrates the BCR for R&D expenditures targeting non-gas operating cost (OPEX) reduction. Using the representative OPEX figure of $4/Bbl listed in Table 2, operating expenditures for a 30,000 B/D GTL project would total $1B over a 25-year project lifetime. Achieving an R&D BCR target of 4 would lead to nearly a $1/Bbl reduction in OPEX, or about $250MM over the of lifetime the project.

Higher or lower cost R&D programs will trace different BCR relationships than those illustrated above and would need to be calculated on a case-by-case basis.

3.3. Finished products or syncrude?

Many descriptions of GTL projects quote a finished product slate of naphtha, jet fuel, and diesel. But, is it obvious that finished products, rather than syncrude should be manufactured? Would it not be less costly to simply manufacture syncrude and ship it to a conventional refinery for finishing? Surprisingly, the answer can be, "No." The literature reports that finishing equipment adds 10 percent in CAPEX to a GTL plant [3]. But, the value that this additional capital creates is significant as shown in Table 4, which compares a 30,000 B/D GTL facility producing syncrude versus one producing finished products. For the additional 10 percent in CAPEX, an additional $190MM in benefit, or value as measured by NPV10, is created, resulting in a BCR of 2.5.

Table 4
BCR: Syncrude vs. Finished Products

Parameter	Syncrude	Finished Products
CAPEX, $MM	675	750
NPV10, $MM	50	240
BCR vs. Syncrude	-	2.5

4. SUMMARY

A discounted cash flow was used to calculate NPV10 for a representative model of a GTL Fischer-Tropsch plant. This economic parameter was used to measure value creation brought about by converting gas to fuel products and to demonstrate value destruction resulting from GTL project start-up delays. It was further demonstrated that value so destroyed is unlikely to be recouped, unless significant capital cost reductions are made and implemented within the span of just a few years. The Benefit-to-Cost ratio, calculated as the quotient of added NPV10 generated and the R&D cost to achieve the GTL process improvement, can be used as a tool to pursue cost effective, value-add research and development. The manufacture of finished GTL products versus syncrude was used as an example to demonstrate it usefulness.

REFERENCES

1. *Gas-to-Liquids News*, December 2000, p 11.
2. Gradassi, M.J., "Economics of Gas to Liquids Manufacture," *Natural Gas Conversion V*, Parmaliana, et al, editors, Elsevier, Amsterdam, 119, 35-44 (1998).
3. "Natural Gas to Liquids Conversion Project," U.S. Trade and Development Agency, PB2000103258, NTIS, Springfield, Virginia, (2000).
4. "Gas to Liquids Technology: Gauging Its Competitive Potential," A.D. Little, February 1998.
5. Anonymous, "Fischer-Tropsch Technology," Howard, Weil, Labouisse, Fredrichs, Inc, 111 Bagby, Suite 2250, Houston, TX, 77002, December 1998.
6. "Annual Energy Outlook 2000," U.S. Energy Information Administration, www.eia.doe.gov/oiaf/aeo/index.html, December 1999.
7. *Gas-to-Liquids News*, November 1999, p 3.
8. Thackeray, F., "Fischer-Tropsch Gas-to-Liquids: Prospects and Implications," SMi Publishing Ltd., London, 2000.
9. BP Marine, Private conversation, September 2000.

Studies in Surface Science and Catalysis
J.J. Spivey, E. Iglesia and T.H. Fleisch (Editors)

Syngas for Large Scale Conversion of Natural Gas to Liquid Fuels

Ib Dybkjær and Thomas S. Christensen
HALDOR TOPSØE A/S

Gas to liquid (GTL) processes require specific syngas properties depending on the specific synthesis process. Autothermal reforming (ATR) at low steam to carbon ratio is an attractive technology for syngas production especially for large plants. New developments in the ATR technology and engineering studies on the benefits of operation at low H_2O/C ratio are reviewed. Furthermore, oxygen- and air-blown ATR for production of syngas for Low Temperature Fischer-Tropsch syngas are compared.

1. INTRODUCTION

Conversion of natural gas to liquid fuels is normally referred to as "Gas-to-Liquids" or GTL-processes. The liquid fuels may be transport fuels (diesel and gasoline) produced via Fischer-Tropsch (FT) synthesis, and also alternative fuels such as methanol (MeOH) and dimethylether (DME). The properties and potential use of these products as transport fuel or fuel for power production are described in (1,2).

For all the products mentioned, the syngas section is the most capital and energy intensive part of the production plant, responsible for most of the energy requirements and for 50-75% of the capital cost (1,3,4). Therefore, major efforts have focused on optimizing the processes for syngas production and on exploring new schemes to replace or complement the existing schemes, which are all based on steam reforming and/or partial oxidation, catalytic or non-catalytic. The various reforming technologies have been described and discussed in several papers (5,6,7,8). The present paper focuses on production of syngas by Autothermal Reforming (ATR) since this technology has been identified as an attractive option for these applications.

2. RECENT DEVELOPMENTS IN ATR TECHNOLOGY

The ATR technology was described in (3,4,5,7,8,9,10,11). The reaction is carried out in a refractory lined vessel containing a burner and a catalyst bed. Especially the design of the burner is critical. Conditions leading to soot formation are not tolerated. Important parameters are inlet and outlet temperatures, feed composition, pressure, and steam to carbon (H_2O/C) ratio.

The feasibility of soot free operation at very low H_2O/C ratio was demonstrated in a Process Demonstration Unit (PDU) in 1997-1999 and reported in (9,10). Operation in industrial scale at an H_2O/C ratio of 0.6 was demonstrated in a plant in the Republic of South Africa in 1999 and reported in (11). In addition to these events further testing has taken place in the PDU. The purpose of this new series of tests was to explore the limits of the ATR technology with respect to feed gas composition and temperatures, and to confirm revised operational procedures:

- The ATR process at low H_2O/C ratio was found to be less sensitive than anticipated to impurities in the feed. However, careful evaluation of specific feedstocks is always necessary.
- The "soot-limit", i.e. the H_2O/C ratio at which soot formation starts, has been investigated by experiments at many combinations of parameters.
- The procedures for start-up and shut-down of ATR reactors in small and large installations have been improved, simplified, and demonstrated.

3. ENGINEERING STUDIES

3.1. Methanol/DME

Syngas for production of MeOH or DME is best characterized by the stoichiometric ratio (or the "Module") $M = (H_2 - CO_2)/(CO + CO_2)$, which should ideally be equal to 2.0. For kinetic reasons and in order to control by-products formation, a value slightly above 2 is normally preferred. The value of M reflects that both CO and CO_2 are reactive in the synthesis, coupled via the shift reaction.

Some years ago, studies were reported on technology selection for large scale production of methanol and DME (1,12). The main conclusions were that:

- for relatively small plants (< about 1500 MTPD MeOH equivalent) single step tubular reforming is advantageous for syngas production.
- for intermediate capacities (corresponding to present world scale capacities, about 1500-3500 MTPD) two step reforming is preferred. This technology was used for a 2400 MTPD MeOH plant in Norway (13).
- for capacities above about 3500 MTPD ATR should be preferred.

Since these conclusions were reached, technology has developed further. Therefore, it was decided to make new studies in order to establish updated break-even points between the various process concepts.

The main technical developments are as follows:

- Single-step tubular reforming has become less expensive (higher heat flux, general optimization), and designs for larger single-stream units have been developed.
- Oxygen plant cost was slightly underestimated in the original study, especially for large capacities. However, proven single-stream capacities for oxygen plants has increased to about 3300 MTPD for oxygen plants corresponding to about 7000 MTPD MeOH equivalent by two step reforming or about 5000 MTPD by ATR.
- ATR has been proven industrially at H_2O/C ratio = 0.6 (11). In the original study the previously proven industrial operation at H_2O/C = 1.3 was used.
- Solutions to avoid metal dusting corrosion in downstream equipment (waste heat boiler and steam superheater) have been identified and demonstrated based on new equipment design and careful selection of construction materials.

- New process concepts for the synthesis loop have been developed especially adapted to syngas production by ATR at low H_2O/C ratio (14).

The findings from several new studies on choice of syngas technology may be summarized as follows (for MeOH):

- Single-step reforming, H_2O/C ratio about 2.5. Capacity range: up to 2500 MTPD. Availability of CO_2 increases capacity where single-step reforming is attractive.
- Two-Step reforming, H_2O/C ratio 1.5-1.8. Capacity range: 1500-7000 MTPD. Heavy natural gas makes two-step reforming less attractive compared to single step reforming. Availability of CO_2 is no advantage with two-step reforming.
- ATR, H_2O/C-ratio 0.6-0.8. Capacity range: 5000-10000 MTPD. Heavy natural gas makes ATR less attractive. Availability of CO_2 is no advantage with ATR.

3.2. Low Temperature Fischer-Tropsch Synthesis

In Low Temperature FT-synthesis with Co-based catalyst, the shift reaction is essentially inactive. Only CO reacts, while CO_2 is inert in the synthesis. The syngas is best characterized by the H_2/CO ratio which should be about 2.0.

The requirements to syngas composition and the advantages of using ATR for production of syngas for Low Temperature FT synthesis are discussed in (8). In (8) it is also shown that production of syngas with $H_2/CO = 2$ requires addition of CO_2 or CO_2-rich gas to the ATR. Without such addition, the ratio is always higher than 2.0 when lean natural gas is used as feed. The ratio increases with increasing H_2O/C ratio, thus increasing the amount of CO_2 required to adjust the syngas composition to the desired H_2/CO ratio. Heavy feed or feed with CO_2 results in lower H_2/CO ratio.

New engineering studies have been performed to quantify the advantages of operation at low H_2O/C ratio. The process concept considered is illustrated in fig. 1. Natural gas is desulphurized, mixed with steam and recycled CO_2 and reacted with oxygen in the ATR. Preheating is done in a fired heater. The exit gas from the ATR is cooled by HP steam production and BFW preheat, and CO_2 is removed as required in an MDEA CO_2-removal unit, compressed, and recycled to the ATR. The syngas after CO_2-removal, with H_2/CO = 2.0, is then available for the FT synthesis.

It has been proposed (15,16) to produce syngas for FT-synthesis by ATR using air or enriched air. In order to evaluate such schemes two cases have been established using air and enriched air (50% oxygen) as oxydant instead of pure (99.5%) oxygen. The steam to carbon ratio was selected at 0.6, and the exit temperature from the ATR was adjusted to give approximately constant CH_4-concentration in the syngas.

In the calculation of energy consumption, feed and fuel is taken at LHV, and the content of CH_4 in the product gas is credited without consideration of possible energy expenses for separation. Export steam is converted to power using normal steam turbine

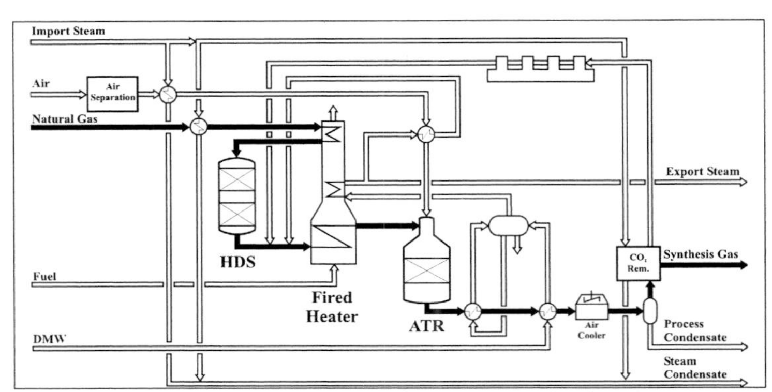

Fig. 1. Process Concept for Syngas Production

efficiencies, and the consumption for CO_2 recovery and recycle is taken as the required power for compression only. Sufficient low level heat for the CO_2 removal unit is available from the process. Oxygen (at pressure) is taken at 0.58 kWh/Nm^3. It was assumed that enriched air is produced by mixing air with pure oxygen and compressing the mixture. Power consumption for production of enriched air is 0.315 kWh/Nm^3 and for compressed air 0.165 kWh/Nm^3. Power is converted at 1 kWh = 2668 kcal.

The main results of the study are shown in Table 1. When comparing the three cases using 99.5% oxygen as oxydant, the advantage of reduced H_2O/C ratio is evident. Both energy consumption and investments decrease with decreasing ratio. It should be noted that application of ATR at $H_2O/C = 0.2$ has not yet been industrially proven.

For the cases with different oxydants it is seen that the energy consumption increases significantly when changing from oxygen to enriched air to air as oxydant. The investments are almost equal for the three schemes. However, the investments are not evaluated in a correct way without taking the quality of the syngas and its effects on the subsequent process steps into consideration, cfr. the discussion below.

The content of NH_3 in the syngas is noteworthy. It is a general experience that NH_3-formation is close to equilibrium at conditions similar to the conditions foreseen in this study. The presence of NH_3 in syngas to FT-synthesis is undesirable, and expensive purification may be required to ensure quantitative removal before the synthesis. Many studies of the FT synthesis assuming use of air for syngas production do not consider this problem.

Little information has been made available concerning design and operating conditions for processes based on the use of enriched air or air. One exception is the study by Hedden et al. (17 a-d,18,19) of a process based on the use of air-blown ATR.

It must be noted that Hedden et al propose the use of Fe-based FT synthesis catalyst, which makes a comparison with modern processes based on Co-catalyst difficult. However, a comparison of consumption figures as shown in Table 2 may still be illustrative. Data for the process using air are calculated from (17 c) whereas data for the process using oxygen are from the study reported in Table 1 (H_2O/C = 0.6) assuming a carbon yield in the process of 70%.

Table 1 - Comparison of Schemes for Production of FT Syngas

Oxydant			99.5% O_2			50% O_2	Air
H_2O-ratio		0.2	0.6	1.0		0.6	0.6
ATR exit Temp., °C		1050	1050	1050		1000	950
Feed + Fuel, Nm^3/h		121497	117096	118558		121026	129711
CH_4 in Product "		10141	3161	1504		4881	4593
Oxydant Flow Nm^3/h		60147	62826	66137		129639	355138
CO_2 rem/recyc, Nm^3/h		1385	12670	23964		14991	17539
Syngas: Flow,	Nm^3/h	318878	312989	312989		381243	604230
Vol %	CO_2	2.28	2.71	3.23		2.55	2.45
"	N_2	0.36	0.35	0.35		17.20	46.59
"	CH_4+Ar	4.18	1.11	0.58		1.57	1.32
ppm	NH_3*	59	-	-		399	454
Net Energy Cons Gcal/h		942.7	970.6	1000.7		998.7	1064.8
Relative Investments		0.9	1.0	1.1		1.0	1.0

Feed and Fuel: Natural Gas (95% CH_4, 3% C_2H_6, 1% CO_2, 1% N_2, 37 kg/cm^2 g)
Syngas: 100.000 Nm^3/h CO, H_2/CO = 2.0, 27 kg / cm^2 g
Preheat Temperatures: Feed: 550°C; Oxydant: 400°C.
Export steam: 400°C, 40 kg/cm^2 g.

* In the wet gas at ATR exit assuming equilibrium for $3H_2 + N_2 \leftrightarrows 2\,NH_3$

Table 2 - Data for O_2 and Air-blown FT Process. Capacity 10250 bbl/d.

		Table 1	Hedden et al. (17c)
Oxydant		99.5% O_2	Air
Nat. Gas Cons.,	Nm^3/h	117096	273300
Oxydant Flow	"	62826	1000500
Power for Oxydant	MW	36.4	167.3

Hedden does not give data on investments. However, it is mentioned that the FT reaction is first order in H_2, which must necessarily mean that required catalyst volume increases due to the dilution of the syngas with N_2. It is further claimed that the dilution with N_2 facilitates the removal of reaction heat in the FT synthesis. Tubular ("ARGE") reactors are assumed. This is not a correct argument since the slurry bed reactors now proposed for all leading Low Temperature FT synthesis processes ensure efficient heat removal even with the concentrated syngas produced with oxygen.

Table 2 clearly shows that the air-blown process described by Hedden is off the mark. The reason is the low conversion of syngas to product in the once-through synthesis section. It does not appear possible to obtain much higher conversion because of the low partial pressure of the reactants. Therefore, an enormous amount of off-gas with a high N_2 content and a very low heating value must be available from the process. Recovery of reactants (H_2, CO, CO_2, CH_4) for recycle from this stream is difficult and very expensive.

The only alternative would be to convert the off gas to power, e.g. in a gas turbine, adding very significant investments and turning the process into co-production of fuel and power. This will in most cases be undesirable since the whole concept of GTL is to convert remote gas fields to transportable energy. In such cases, production of power is not interesting, since no local market exists.

4. CONCLUSION

It has been found that ATR at low H_2O/C ratio is attractive for production of syngas for large GTL facilities. Extensive testing has proven that the technology is ready for application in large industrial scale. Use of pure oxygen as oxydant is most competitive. The disadvantages of using enriched air or air as oxydant (increased energy consumption, increased flow through the units, and dilution of the syngas with N_2) clearly outweigh the advantages (mainly reduction or elimination of the investments related to production of oxygen).

References

1. Dybkjær, I. et al. 4th Int. Nat. Gas Conv. Symp., South Africa, 19-23 November 1995.
2. Dancuart, L. P., Paper AM-00-51, 2000 NPRA Ann. Meeting, USA, 26-28/3-2000.
3. Dybkjær, I. et al. Ind. Conf. on Nat. Gas for Petrochem., Indonesia, 26-27/7-96.
4. Rostrup-Nielsen, J. et al. Symp. Adv. Fischer-Tropsch Chem. Presented before the Div. Petr. Chem. Inc., 219th National Meeting, ACS, USA, March 26-31, 2000.
5. Dybkjær, I., Fuel Processing Technology 42 (1995) 85-107.
6. Rostrup-Nielsen, J.R. et al., J. Jap. Pet. Inst., Vol. 40, No. 5, Sept. 1997.
7. Winter Madsen, S. et al., Int. Symp. Large Chem. Plants (LCP-10), Belgium, Sept. 28-30, 1998.
8. Aasberg-Petersen, K. et al. Special Issue of Appl. Catal. A, Elsevier Science Publ.
9. Christensen, T.S. et al. 5th Nat. Gas Conv. Symp., Italy, 20-25 Sept.1998.
10. Christensen, T.S. et al. 2nd Ann. Conf. "Monetizing Stranded Gas Reserves", USA, Dec.14-16, 1998.
11. Ernst, W.S. et al. Hydrocarbon Processing March 2000, 100-C-100-J.
12. Holm-Larsen, H., 1994 World Meth. Conf., Switzerland, 30/11-1/12-1994.
13. Gedde-Dahl, A. et al. Ammonia Plant Saf. 39 (1999) 14-23.
14. Haugaard, J., Holm-Larsen, H., 1999 World Meth.Conf., USA, 29/11-1/12-1999.
15. www.Syntroleum.com
16. Wilson, G.R., Carr, N.L., Hydrocarbon Engineering April 1998, 43-47.
17. Hedden, K.et al., Erdöl Erdgas Kohle, a: Heft 7/8, Juli/Aug. 1994, 318-321, b: Heft 7/8, Juli/Aug. 1994, 365-370, c: Heft 2, Febr.1995, 67-71, d: Heft 12, Dez.1997, 531-540.
18. Jess, A., Popp, R., Hedden, K., Appl. Catal. A: General 186 (1999) 321-342.
19. Jess, A., Hedden, K., Symp. Adv. Fischer-Tropsch Chem., Presented before the Div. Petr.Chem., Inc., 219th National Meeting, ACS, USA, March 26-31, 2000.

Studies in Surface Science and Catalysis
J.J. Spivey, E. Iglesia and T.H. Fleisch (Editors)

CO_2 reforming for large scale methanol plants – an actual case

H. Holm-Larsen

Haldor Topsøe A/S, 55 Nymøllevej, DK-2800 Lyngby, Denmark

This paper describes the first large-scale industrial application of CO_2 reforming of natural gas, which will occur in a 1 million ton/year methanol plant in Iran. The plant will be located by the Persian Gulf in an existing petrochemical complex at Bandar Imam. At this site excess CO_2 is available from an ammonia plant and an ethylene cracker. The paper also focuses on CO_2 reforming of natural gas in general and the process layout modifications introduced to optimize the CO_2 utilization, the consequences to the plant energy consumption, the efficiency and the plant investment. Variation in the CO_2 feed and CO_2 sequestration is discussed as well.

1. INTRODUCTION

The term "CO_2 reforming" as used in this presentation is short for CO_2 reforming of natural gas. In contrast to steam reforming of natural gas, addition of CO_2 permits optimization of the synthesis gas composition for methanol production. Besides, CO_2 constitutes a less expensive feedstock and CO_2 emission to the environment is reduced. CO_2 is also easier to reform than natural gas, leading to energy and investment savings under the right circumstances.

However, the import of CO_2 may necessitate compression and purification of an extra feedstock, thus adding extra unit operations in the plant. Therefore, CO_2 reforming is only economically feasible where a large and relatively pure amount of CO_2 is available.

The methanol plant described in this paper enjoys these natural preconditions with about 825 ton/day of CO_2 being available from the adjacent petrochemical plants.

2. CO_2 REFORMING FOR METHANOL PRODUCTION

While CO_2 reforming certainly presents a number of advantages, there are certain constraints with respect to the amount of CO_2, which can be added. These constraints include mass balance, kinetic and thermodynamic limitations as discussed below.

2.1 Mass Balance Constraints

Methanol synthesis gas is typically described in terms of its Module, $M=(H_2-CO_2)/(CO+CO_2)$. The syngas is in balance for the methanol reaction when the module is equal to 2. However, industrial applications often operate with a few percent excess of hydrogen, corresponding to a module of around 2.05, in order to suppress formation of by-products.

Theoretically, steam reforming of methane produces an excess of hydrogen, corresponding to a module equal to 3, cf. reaction (1). This excess of hydrogen permits a certain import of CO_2. As visualized by reaction (2), it can be proven that a module equal to 2 is reached with a CO_2 import corresponding to a CO_2/CH_4 ratio in the reformer feed of 1:3.

(1) $CH_4 + H_2O = 3H_2 + CO$ (M=3)

(2) $3CH_4 + CO_2 + 2H_2O = 8H_2 + 4CO$ (M=2)

In the real world, natural gas feedstocks contain higher hydrocarbons as well as CH_4. In this case, a module of 2 is reached with a CO_2 import corresponding to a $CO_2/C_nH_{(2n+2)}$ ratio of 1:3, as visualized in reaction (3). For n=1, reaction (2) and (3) becomes identical.

$$(3) \qquad 3C_nH_{(2n+2)} + CO_2 + (3n-1)H_2O = (6n+2)H_2 + (3n+1)\,CO \qquad (M=2)$$

2.2 Kinetic and Thermodynamic Constraints

On the classic Ni-based reformer catalysts, CH_4 or CO can dissociate into carbon on the surface of the Ni crystals, leading to formation of soot or carbon fibers (whiskers), which will ultimately destroy the catalyst. In order to prevent this, steam is added to a suitably high Steam/Carbon (S/C) ratio. Alternatively, the carbon formation may be suppressed by Sulfur PAssivated ReforminG (SPARG), [1,2], or noble metal reforming, [3]. However, these reforming technologies have proven to be most suitable for production of CO and reducing gas, less so for methanol. Thus, the preferred solution for CO_2 reforming for methanol production is Ni-catalyst based reforming at a suitably high S/C.

The carbon limit of equilibrated syngas is shown in Fig. 1. S/C ratio is plotted along the upper left axis of the graph, CO_2/CH_4 is plotted along the lower left axis. The carbon limit is the S-shaped curve in the middle of the figure. To the right of the carbon limit there is no potential for carbon formation. To the left of the curve, carbon will be formed.

The arrow in the upper right part of the figure illustrates how addition of CO_2 increases the carbon potential because of the lower H/C atomic ratio of the feedstock. However, the figure also shows that the relevant range of operating parameters for the Iranian plant (S/C = 2.5, $CO_2/C_nH_{(2n+2)}$ ratio up to 1:3) are well on the safe side of the carbon limit.

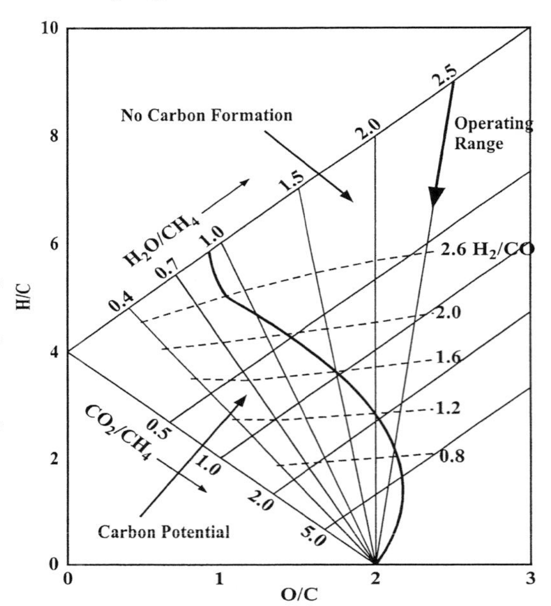

Fig. 1: Carbon limit of equilibrated syngas at 25 bar

The inerts concentration of the syngas is also very important for a methanol plant since the methanol synthesis occurs in a loop with a significant recycle. Therefore, any CH_4 that is not converted in the reforming section (CH_4 leak) will be accumulated in the synthesis loop by a factor up to 10, reducing the partial pressure of the active components.

The CH_4 leak through the reformer is controlled primarily by S/C ratio and reformer outlet temperature as illustrated in Fig. 2. In order to keep an inert concentration in the synthesis loop below 25%, the syngas should typically contain less than 2.5% CH_4, corresponding to a S/C ratio of 2.5 and 920°C outlet the reformer.

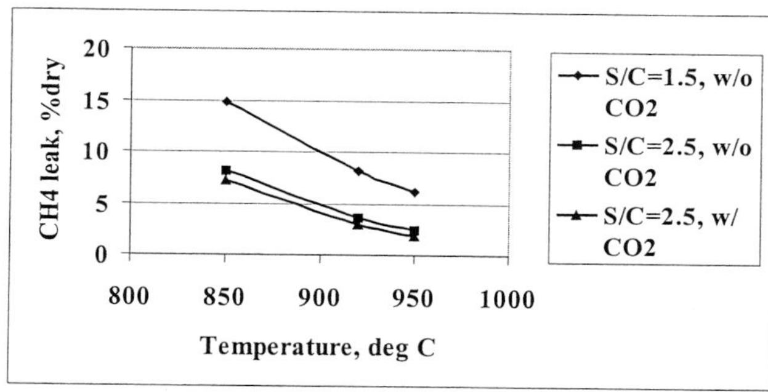

Fig. 2: Reformer methane slip at 25 bar

In summary, the most stringent constraint for a methanol plant proves to be the mass balance constraint. Given the availability of CO_2 and the rich composition of the natural gas for the Iranian plant, the CO_2/hydrocarbon ratio was fixed at about 0.26, resulting in a syngas module of about 2.09.

3. PROCESS CONSIDERATIONS

3.1 Process Layout

The Iranian methanol plant will feature a natural gas desulfurization section (CoMo based hydrogenation and ZnO based sulfur removal) followed by a prereformer and the tubular reformer. The methanol synthesis takes place in boiling water reactors, and the product is purified in a 3 column distillation section.

This process layout, as illustrated in Fig. 3, results in a very energy efficient plant with an energy consumption of only 6.95 Gcal/ton methanol (7.05 Gcal/ ton including CO_2 compression), i.e. 5-10% lower than the energy consumption of a conventional plant.

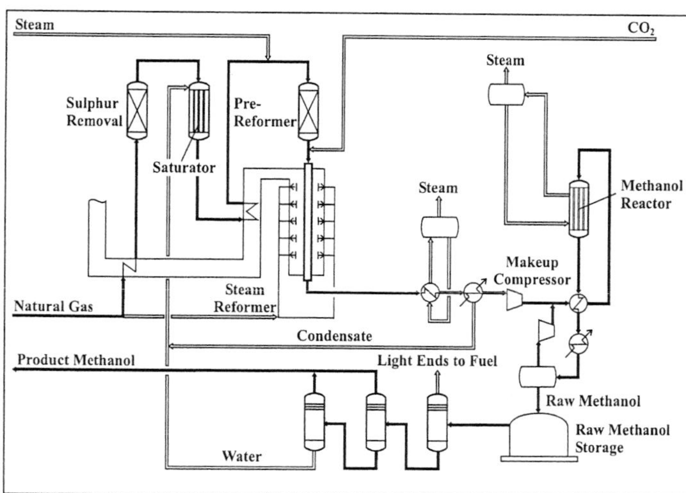

Fig. 3: Process Lay-out

The inclusion of a prereformer serves several purposes [4], of which the most important incentive is that it helps reduce the size of the tubular reformer, since all energy added in connection with the prereformer is subtracted from the tubular reformer. In the present plant, the reduction in the tubular reformer duty is about 13% corresponding to about 75 reformer tubes.

The imported CO_2 may be added to the methanol plant at various locations, typically either upstream the reformer or at the suction point of the syngas compressor. From an overall mass balance point of view, the addition point makes no difference, but from the perspective of the plant subsections such as the tubular reformer or the synthesis loop, the difference is significant. For instance, the reaction rate in the methanol synthesis is highly dependent on the

CO/CO_2 ratio of the syngas, and therefore the synthesis section benefits significantly when the CO_2 is added before the reformer, where most of it is shifted to CO. On the other hand, the addition point upstream the reformer entails a higher reformer load.

In the Iranian plant, the CO_2 is added just after the prereformer. Thereby, the reforming equilibrium in the prereformer is more advantageous, and the maximum reformer feed preheat is achieved, leading to a significantly smaller tubular reformer.

3.2 CO_2 Purification

CO_2 used for feedstock for a catalytic reaction must be free from potential catalyst poisons such as sulfur and halogens. If the CO_2 source is an NH_3 plant this does not pose a problem, but if the CO_2 originates from other sources, purification may be required.

Removal of sulfur from CO_2 is difficult because a significant part of the sulfur may be present as COS. Both H_2S and COS may be removed on traditional ZnO adsorption mass, but unfortunately, the adsorption of COS requires much higher temperature than the adsorption of H_2S (~400°C vs ~200°C respectively). Preheating of the CO_2 stream to 400°C presents several logistic problems, and therefore other alternatives are sought, viz.:

A) Hydrolysis of COS to CO_2 and H_2S

Adding small amounts of water to the CO_2 stream will cause COS to hydrolyze to CO_2 and H_2S. However, water will also increase the sulfur slip from the ZnO adsorbent according to the reaction $ZnO + H_2S = ZnS + H_2O$. Consequently, this solution has certain limitations.

B) Mixing the CO_2 with the natural gas before desulfurization

Mixing the CO_2 and the natural gas introduces another problem. The hydrogen recycle added to the natural gas in order to allow the hydrogenation to take place will also react with the CO_2 (the reverse shift reaction). The water formed by this reaction will cause increased sulfur slip as described under A).

C) Copper based adsorbent

Finally, the CO_2 feed may be treated by a copper based adsorbent mass. Such adsorbent mass has already been applied industrially with success. While the copper based adsorbent is a bit more expensive than the ZnO adsorbent, it has the advantage that it is active at lower temperatures. So far it is considered the most suitable solution for dealing with sulfur in CO_2 feedstock.

4. CONSEQUENCES OF CO_2 REFORMING TO PLANT PERFORMANCE

4.1 Feedstock Variations

When a plant has two different feedstocks, it is necessary to consider the effects, if the balance between the feedstocks changes. Obviously, the plant will have to be designed for a given feed balance, and deviations from this balance will lead to bottlenecks in different places of the plant. The distribution of bottlenecks is illustrated in table 1.

The table shows how the requirements for the reforming section change by about 2% when the available amount of CO_2 changes by 10%. The sensitivity of the recycle compression is due to the increasing difficulty to maintain the carbon efficiency of the synthesis loop as the module approaches 2.

In principle, the bottlenecks have an impact on the production capacity when the feedstock balance changes, i.e. the maximum production will decline when the feedstock balance deviates to either side of the design point. In reality, it is possible to compensate for the worst bottlenecks to a certain extent. The Iranian plant, for instance, is designed with sufficient

CO_2, flow	-10%	Design	+10%
Module	2.16	2.09	2.02
CO_2 comp. power	90	100	110
NG compressor power	102	100	98
Steam reformer duty	102	100	98
MUG comp. power	102	100	98
Recirculator power	87	100	116

Table 1: Relative plant load by plant sub-section

margins in the bottleneck areas to permit the full design capacity over the entire range from −10 to +10% CO_2. In case the import of CO_2 is completely terminated, the plant will still be able to produce about 2,500 MTPD methanol on natural gas alone.

4.2 Comparison to Conventional Layout

The CO_2 addition results in a plant performance quite different from the performance of the conventional layout. The major differences are indicated in table 2.

The lower energy consumption of CO_2 reforming is due to simple substitution of natural gas with CO_2, whereas the lower demineralized water consumption is due to water formed by the reverse shift reaction.

	Steam reforming	CO_2 reforming
Consumption figures:		
Energy, Gcal/ton MeOH	7.4	7.05
Demin. water, m^3/ton MeOH	0.8	0.4
Equipment & line size:		
CO_2 compressor, MW	-	3.7
Reformer size, tubes	720	575
MeOH reactor volume, index	91	100
Dry syngas flow, Nm^3/h	385,000	325,000
Recirculation flow, Nm^3/h	1.2 mill	1.2 mill

Table 2: Typical key figures for a 1 MM ton/yr MeOH plant

With respect to equipment and line size, it is well worth noticing that the reforming section is significantly smaller for the CO_2 reformer plant while the synthesis reactor is slightly larger. With about 60% of the plant investment in the reforming section and only 10-15% in the synthesis section, however, the overall effect is a substantial cost reduction.

5. CONSEQUENCES TO METHANOL PRODUCTION COST

The methanol production cost is affected both by feedstock cost and plant investment. With a natural gas cost of 1 USD/MM BTU and a demineralized water cost of 1 USD/m^3, the operating cost saving is about 2 USD/ton methanol. The plant investment cost saving for a 1 mill ton/year plant is estimated to be about 10 mill USD. With a typical payback time this saving corresponds to a production cost reduction of about 2 USD/ton methanol. The net methanol production cost is therefore reduced by about 4 USD/ton in the present case.

The above saving applies when the CO_2 is available for free. In case the CO_2 carries a fee, the saving will be smaller. Based on the assumptions above, the break-even price for CO_2 is about 15 USD/ton CO_2.

6. CO_2 SEQUESTRATION

The manufacture of methanol partly (or completely) from CO_2 has been discussed extensively as a tool for CO_2 sequestration. It should be realized, however, that the potential for CO_2 sequestration might be seriously limited not only by mass balance limitations, but also by the energy requirement of the manufacturing process.

In a typical methanol plant, a large amount of energy is consumed in the reformer, but the latent heat of the resulting flue gas is recovered to a very large degree. In contrast, most of the energy used in the compression and in the distillation is irreversibly lost.

	Gcal/ton methanol
Theoretical energy consumption	5.5
Irreversible heat loss in reforming section	0.2-0.3
Irreversible heat loss in synthesis section	0.3-0.4
Irreversible heat loss in distillation section	0.7-1.2
Irreversible compression loss in synthesis section	0.2
Net energy requirement	7.0-7.4

Table 3: Typical distribution of energy consumption

A typical breakdown of energy consumption is shown in table 3. The reason for the large range on heat loss in the distillation is that this section often is used as a heat sink for the process, reflecting the efficiency of the reformer section. This table gives a good impression of the CO_2 sequestration obtainable under various circumstances.

If natural gas is the only source of energy for the plant, the CO_2 emission is given by the amount of natural gas required to generate a heating value of 7.0 Gcal/ton methanol. Comparing with the typical energy consumption of 7.4 Gcal/ton and the theoretical energy consumption of 5.5 Gcal/ton, the overall CO_2 emission can thus be reduced by 20-25%.

It is important to realize, however, that even though a methanol plant imports CO_2, this CO_2 is not permanently sequestrated. The CO_2 addition changes the quantity and composition of the purge gas from the methanol synthesis loop, so that the fuel gas (and consequently also the flue gas) for the reformer becomes more carbon rich. As an example, the 825 ton/day of CO_2 imported for the Iranian methanol plant is matched by an increase in flue gas CO_2 content of about 450 ton/day, and therefore the net CO_2 sequestration will be only about 375 ton/day or about 0.12 ton CO_2/ton methanol.

7. CONCLUSION

CO_2 reforming makes good economic sense for large-scale methanol plants when the CO_2 is available free of charge. In case the CO_2 carries a substantial price, the economic justification for CO_2 reforming will be marginal at best.

The CO_2 reforming technology is industrially tested to a certain extent in smaller CO plants and revamped methanol plants. However, the Iranian methanol plant is scheduled to come onstream in year 2002, and by then the technology will be tested in a world scale plant.

With respect to CO_2 sequestration, the CO_2 imported for methanol production tends to be re-emitted in the reformer flue gas to a certain extent. Therefore, the obtainable net sequestration is only about 0.12 ton CO_2/ton methanol.

8. REFERENCES

[1] Rostrup-Nielsen, J.R., J Catal., **85**, 31 (1984)

[2] Udengaard, N.R., Bak-Hansen, J.-H., Hanson, D.C. and Stal, J.A., Oil Gas J., **90**, 10, 62 (1992)

[3] Rostrup-Nielsen, J.R. and Bak-Hansen, J.-H., J. Catal., **144**, 38 (1993)

[4] Christensen, T.S., "Adiabatic Prereforming of Hydrocarbons – An Important Step in Syngas Production", Applied Catalysis, Oct. 30, 1995.

Studies in Surface Science and Catalysis
J.J. Spivey, E. Iglesia and T.H. Fleisch (Editors)

Water-gas shift reaction: reduction kinetics and mechanism of $Cu/ZnO/Al_2O_3$ catalysts

M.J.L. Ginés and C.R. Apesteguía*

Instituto de Investigaciones en Catálisis y Petroquímica (INCAPE), UNL-CONICET, Santiago del Estero 2654, (3000) Santa Fe, Argentina

The reduction of CuO with H_2 in Cu/Zn/Al mixed oxides containing similar Cu concentration than commercial water-gas shift catalysts was studied by thermogravimetry and temperature programmed reduction (TPR) techniques. The reduction kinetics was well interpreted in terms of the unreacted shrinking core model by assuming chemical control at the boundary interface. The model proved to be a useful basis for understanding the effect of various variables on CuO reduction. The reduction of CuO to Cu^0 proceeds in a one-step process without formation of a Cu_2O intermediate. The apparent activation energy, E_a, was determined from TPR profiles by considering a first order reaction with respect to reactant concentrations. The obtained E_a values varied between 67 and 75 kJ/mol, irrespective of catalyst composition, thereby suggesting that the reduction behavior can be related to free CuO.

1. INTRODUCTION

Ternary Cu/Zn/Al mixed oxides are widely employed for catalyzing the water-gas shift (WGS) reaction ($CO + H_2O \rightarrow CO_2 + H_2$) at low temperatures and low concentrations of carbon monoxide [1]. Preparation of $CuO/ZnO/Al_2O_3$ catalysts is carried out in several consecutive steps. The final step consists on the activation of the samples by hydrogen treatment for reducing CuO to metallic copper. The reduction process of bulk metal oxides usually involves consecutive rate-determining steps [2]: i) generation and growth of metallic nuclei; ii) phase boundary-controlled chemical reactions at gas-solid interfaces and/or autocatalytic processes; iii) diffusion of gaseous species through product layers. Although it has been possible to distinguish between rate-determining steps on the reduction of bulk Cu(I, II) oxides [3], few studies have been undertaken to discern between stages of the CuO reduction in $CuO/ZnO/Al_2O_3$ mixed oxides [4]. Depending on both the chemical composition and the preparation method, the CuO phase may be homogeneously distributed across the surface of the support, or exist as islands separated by uncovered support, or form a $CuAl_2O_4$ spinel; also, it may be partially contained in a nonstoichiometric $(CuO)ZnAl_2O_4$ spinel or dissolved in the ZnO phase. A previous detailed characterization of the phase composition of the solid appears therefore as a requisite in achieving a better understanding of the CuO reduction process in copper-supported catalysts.

On the other hand, knowledge regarding the final oxidation state of copper after the reduction step and on reaction conditions is needed. A number of studies have been devoted to the relationship between catalyst components and catalytic activity, but the exact nature of copper active sites (Cu^0 or Cu^{1+} species) for low-temperature CO conversion on Cu/Zn/Al

* Corresponding author. Email: capesteg@fiqus.unl.edu.ar, fax: 54-342-4531068

catalysts is still a matter of debate [5]. Finally activation of fresh catalysts with hydrogen is a critical step at industrial level because catalyst lifetime can be shortened by lack of the adequate precautions.

In this work we investigate in ternary Cu/Zn/Al mixed oxides both the kinetics and mechanism of CuO reduction with H_2 and the effect of the temperature on copper sintering.

2. EXPERIMENTAL

Samples were prepared by coprecipitation from 1.5 molar mixed nitrate solutions with sodium carbonate at 333 K and a constant pH of 7, as stated elsewhere [6]. Three sets of catalysts were prepared. Samples of set B contained the same amount of CuO (42.7 %wt) and different Zn/Al atomic ratios whereas the samples of set C had a Zn/Cu ratio of 1 and different (Zn+Cu)/Al ratios. Samples of set A had the same chemical composition (Cu/Zn = 1; (Cu+Zn)/Al = 4) but were calcined at different temperatures. Main characteristics of all the samples used in this work are shown in Table 1.

Powder X-ray diffraction patterns were collected on a Rich-Seifert diffractometer using nickel filtered Cu$K\alpha$ radiation. Crystallite sizes were calculated from the CuO(111) and ZnO(110) diffraction lines using the Scherrer equation. BET surface areas were measured by N_2 adsorption at 77 K in a Micromeritics Accusorb 2100 sorptometer. The metallic Cu dispersions (D_{Cu}) were measured using the reactive chemisorption of N_2O at 333 K and frontal chromatography [7].

Thermogravimetric studies were carried out in a Cahn 2000 electrobalance using a flow system at 1 atm. Experiments were conducted isothermally at temperatures between 448 and 573 K in a 5% H_2/N_2 gaseous mixture with a flow rate of 20 cm^3/min. Powder samples were placed on a glass container suspended vertically; about 5 mg was loaded to limit diffusion phenomena. Before introducing the reducing mixture, all the samples were pretreated in N_2 at 673 K for 2 h.

The TPR experiments were performed in 5% H_2/Ar gaseous mixture at 28 cm^3/min STP. The sample size was 50-100 mg. Temperature was increased to 1023 K at 5 K/min and the H_2 concentration in the effluent was measured by thermal conductivity.

3. RESULTS AND DISCUSSION

3.1. Catalyst characterization

Hydroxycarbonate precursors were calcined and transformed into mixed oxides by eliminating water and CO_2. The crystalline phases for the obtained mixed oxides are shown in Table 1. All the samples contained CuO and ZnO. Depending on both, the chemical composition and the calcination temperature, the presence of an additional spinel phase of $ZnAl_2O_4$ was observed. Formation of $ZnAl_2O_4$ spinel was favored by increasing the aluminium content on the sample and/or by employing high calcination temperatures. No crystalline diffractogram of $CuAl_2O_4$ was detected.

Qualitatively, similar TPR profiles (not shown here) were obtained for samples of sets A, B, and C. All the samples exhibited a low-temperature peak at ca. 573 K which corresponds to the reduction of CuO to metallic copper. A broad and small band appeared at temperatures higher than 673 K, which is attributed to the partial reduction of ZnO, followed by formation of brass. In brief, characterization by XRD and TPR techniques show that the samples are composed of CuO, ZnO, $ZnAl_2O_4$ and, probably, Al_2O_3, the relative proportion of the phases depending on

Table 1
Main characteristics of the catalysts used in this work

Sample	Cu / Zn / Al (% M_xO_y)			Identified compounds (XRD)	Calcination temperature (K)	Crystallite size L_{CuO} (Å)	Surface areas (m^2/g)
A-673	42.7	43.6	13.7	CuO, ZnO	673	< 60	55
A-773	42.7	43.6	13.7	CuO, ZnO	773	80	64
A-873	42.7	43.6	13.7	CuO, ZnO, $ZnAl_2O_4$	873	130	54
A-973	42.7	43.6	13.7	CuO, ZnO, $ZnAl_2O_4$	973	210	18
B-2	42.7	21.8	35.5	CuO, ZnO, $ZnAl_2O_4$	873	115	50
B-3	42.7	39.0	18.3	CuO, ZnO, $ZnAl_2O_4$	873	80	53
B-4	42.7	43.6	13.7	CuO, ZnO, $ZnAl_2O_4$	873	130	54
B-6	42.7	50.8	6.5	CuO, ZnO	873	170	41
B-∞	42.7	57.3	0	CuO, ZnO	873	210	5
C-1	30.3	31.0	38.7	CuO, ZnO, $ZnAl_2O_4$	873	87	34
C-2	37.6	38.4	24.0	CuO, ZnO, $ZnAl_2O_4$	873	84	38
C-3	40.8	41.7	17.5	CuO, ZnO, $ZnAl_2O_4$	873	79	53
C-4	42.7	43.6	13.7	CuO, ZnO, $ZnAl_2O_4$	873	130	54
C-6	44.7	45.7	9.6	CuO, ZnO	873	153	33
C-∞	49.4	50.6	0	CuO, ZnO	873	250	4

both the chemical composition and the calcination temperature. Treatment of these samples with diluted H_2/N_2 mixtures at temperatures lower than 673 K reduces only the CuO phase.

3.2. Catalyst reduction: kinetics and mechanism

The kinetic studies were performed using thermogravimetry. Experimental results were represented by plotting the reduced CuO fraction, α, as a function of time at constant temperature. Fig. 1 shows the TGA profiles obtained for sample B-4 at different reduction temperatures (T_r), between 448 and 573 K. The reduction curves were analyzed by assuming that only CuO is reduced during the run and that water formed on the catalyst surface is rapidly eliminated. The general patterns of reductograms suggested that the reduction of CuO to Cu^0 occurs in a one-step process. The existence of a Cu^{1+} intermediate has been observed in copper-containing zeolites and in copper/alumina catalysts with low copper loadings [8]. However, and in agreement with our results, several papers have reported that the intermediate Cu_2O phase is not formed during reduction of supported copper catalysts with high copper loadings [9].

On the basis of slope changes, the analysis of TGA curves given in Fig. 1 may be divided in three stages: i) the reaction starts with an induction period where the first Cu^0 nuclei form and the rate is low. This first stage end at a value of $\alpha \cong 0.10$; ii) the second stage is

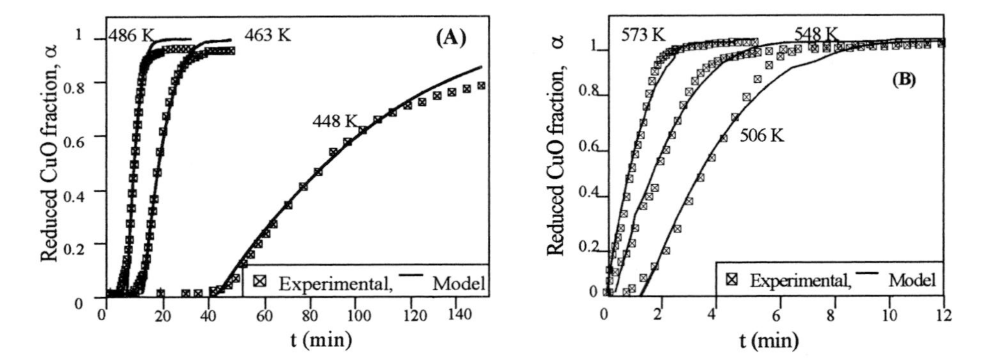

Fig. 1. Reduction kinetics of sample B-4 at different temperatures

characterized by a rapid increase in the reaction rate. The position of the inflection point, which corresponds to the rate maximum, shifts to lower times as T_r is increased; iii) in the third stage ($\alpha > 0.85$), the reduction rate decreases with time monotonically. It is observed that the CuO phase was not completely reduced for T_r values lower than ca. 493 K; i.e. reduction curves stopped at values of α lower than 1 (Fig. 1A).

Several models have been employed for interpreting reduction data of mixed oxides samples [10]. The simplest and more frequently used model is the unreacted shrinking core model. It assumes that reduction of non-porous particles occurs at the metal/metal oxide interface and is characterized by constantly decreasing reaction rates as the substrate particle is consumed in the course of the reaction. Reduction may be controlled by reaction at the phase boundary or by film/ash diffusion. When the rate-limiting step is the reaction interface, Eq. (1) is obtained [11]:

$$\alpha = 1 - [1 - k(t - t_0)]^3 \tag{1}$$

where k is the interface rate constant and t_0 is the induction period length. Equation (1) was used to compare the model predictions with experimental results. The curves fitting was performed by using a least-squares regression which minimizes the objective function $F = \Sigma(\alpha_{exp} - \alpha_{calc})^2$. As it is shown in Fig. 1, the experimental curves were well fitted by the corresponding model curves beyond the induction period. Similar goodness of fit was obtained with the other samples of set B, thereby suggesting chemical control at the boundary interface.

The induction period represents the initial nucleation process, i.e. formation and growth of metallic copper nuclei, and accounts for the sigmoidal shape of the curves obtained at low T_r values. Reaction occurs at the interface between the Cu^0 nuclei and the surrounding CuO until coalition takes place forming metallic shells which encase cores of unreacted oxide. When T_r is increased the nucleation process becomes very fast and causes the almost instantaneous coverage of the oxide grain with a thin metallic layer. Under these conditions, the induction period is not longer observed ($t_0 = 0$) and the reduction isotherm is characterized by a reaction interface that is contracting throughout the reaction. In our case, the induction period was not observed for temperatures higher than ca. 523 K (Fig. 1B).

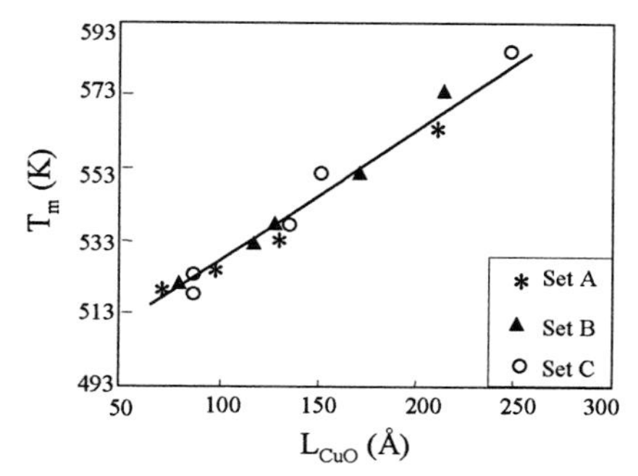

Fig. 2: CuO reduction peak maxima as a function of L_{CuO}

More insight on the CuO reduction mechanism was obtained by TPR characterization. The TPR traces showed that when the CuO crystallite size (L_{CuO}) is increased by raising the calcination temperature the peak width increases and the peak maximum (T_m) shifts to higher temperatures. Results for all the samples are given in Fig. 2; a linear relationship was obtained when T_m was represented as a function of L_{CuO}. This result is consistent with the unreacted shrinking core model predictions. In fact, considering a contracting cube particle, the reaction rate is expressed as the volume diminution rate:

$$dV/dt = d(L_{CuO} - 2\gamma t)^3 / dt \tag{2}$$

where V is the solid volume and γ the linear rate of movement of the reaction interface. Incorporating in Eq. (2) the Arrhenius equation, $\gamma = A \exp(-E_a/RT)$, and the linear heating rate, $T = T_0 + \beta t$, Tonge [12] used computer programs for simulating the effects on dV/dt (proportional to peak height) versus temperature curves of varying L_{CuO} and other parameters. Computer calculations using monodisperse systems predicted that, in agreement with our results, both T_m and the peak width should increase by increasing L_{CuO}. Tonge [12] obtained convincing experimental evidence for the dependence of T_m on particle size by analyzing TPR profiles of different-sized particles of NiO.

The TPR traces were also used for calculating the apparent activation energy, E_a, through a T_m-β analysis. Previous studies have found the CuO reduction to be first order with respect to hydrogen and to unreacted copper oxide [3]. Gentry et al. [13] demonstrated that for a reduction process which is first order with respect to reactant concentrations the variation of T_m with the heating rate, β, is given by:

$$\ln \frac{T_m^2 [H_2]_m}{\beta} = \frac{E_a}{RT_m} + \ln \frac{E_a}{RA} \tag{3}$$

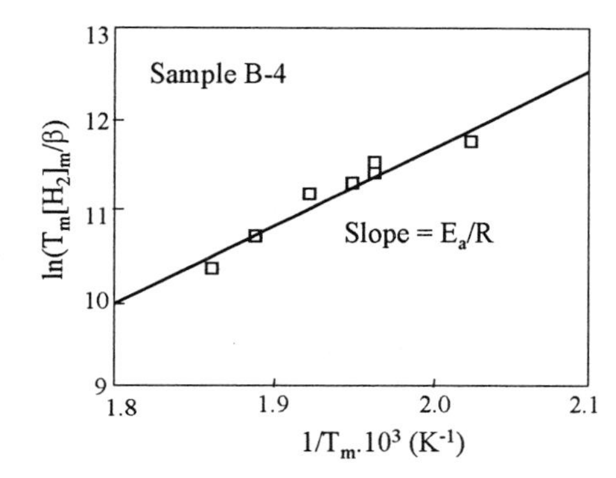

Fig. 3: Determination of E_a from TPR profiles and T_m-β analysis

where $[H_2]_m$ is the hydrogen concentration at T_m. The activation energy is determined from the slope of the straight line obtained when the left-hand side of Eq. (3) is plotted against $1/T_m$. In our experiments we used a 5% H_2/N_2 mixture and β values between 2 and 10 K/min. The plot obtained for sample B-4 is given in Fig. 3; a value of $E_a = 74 \pm 6$ kJ/mol was determined. By using similar T_m-β analysis the apparent activation energy values corresponding to samples of set C were calculated; E_a values between 65 and 75 kJ/mol were obtained. Gusi et al. [4] reported values between 58 and 62 kJ/mol for reduction of Cu/Zn/Al mixed oxides of different composition. Our E_a values are similar to those reported in bibliography for reduction of pure copper oxide. In fact, Gentry et al. [2] found $E_a = 67 \pm 10$ kJ/mol and Boyce et al. [14] $E_a = 77 \pm 10$ kJ/mol for bulk CuO reduction. The reactivity of CuO in the catalysts employed in this study is not significantly influenced by the catalyst composition, thereby suggesting that the reduction behavior can be related to free CuO.

3.3. H_2 activation and copper sintering

Samples B-3, B-4, and C-3 were used for studying the effect of the reduction temperature on copper sintering. All the samples were treated isothermally in consecutive 30 min runs using a 5% H_2/N_2 gaseous mixture at 100 cm^3/min. T_r was varied between 473 and 773 K. In Fig. 4 the values of the relative Cu dispersion, D_{Cu}/D^0_{Cu}, are plotted as a function of T_r; D^0_{Cu} is the metallic dispersion of the samples reduced at 473 K, and D_{Cu} the dispersion of samples reduced at temperature T_r. The three samples exhibited similar qualitative behavior. Although initial sintering of the Cu^0 crystallites was noted at about 548 K, severe loss of the metallic dispersion took place only after 573 K. This is consistent with previous reports which showed that thermal sintering is likely if the temperature exceeds 523 K but it is not severe, however, even at 563 K [15].

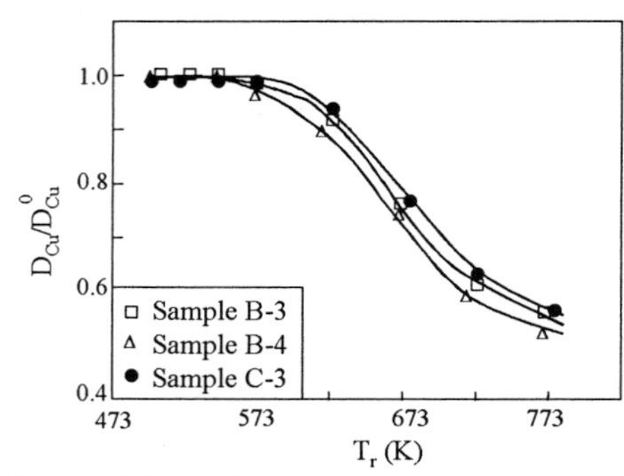

Fig. 4. Effect of the reduction temperature on copper sintering

REFERENCES

1. D.S. Newsome, Catal. Rev. Sci. Eng., 21 (1980) 275.
2. S.J. Gentry, N.W. Hurst and A. Jones, J. Chem. Soc. Faraday Trans. I, 77 (1981) 603.
3. H.H. Voge and L.T. Atkins, J. Catal., 1 (1962) 171.
4. S. Gusi, F. Trifiro and A. Vaccari, React. Solids, 2 (1986) 59.
5. J. Nakamura, J. Campbell and C. Campbell, J. Chem. Soc. Faraday Trans., 86 (1990) 2725.
6. A.J. Marchi, J.I. Di Cosimo and C.R. Apesteguía, Catal. Today, 15 (1992) 383.
7. G.C Chinchen, C.M. Hay, H.D. Vandervell and K.C. Waugh, J. Catal., 103 (1987) 79.
8. J.M. Dumas, C. Geron, A. Kribii and J. Barbier, Appl. Catal., 47 (1989) L9.
9. A.L. Boyce, P.A. Sermon, M.S.W. Vong and M.H. Yates, React. Kinet. Catal. Lett., 44 (1991) 309.
10. N. Pernicone and F. Traina, in Preparation of Catalyst II, Elsevier, Amsterdam, 1979, pp. 321-351.
11. A. Jones and B.D. McNicol, Temperature-Programmed Reduction for Solid Materials Characterization, Chemical Industries/24, Marcel Dekker Inc., N.Y., (1986).
12. K.H. Tonge, Thermochim. Acta, 74 (1984) 151.
13. S.J. Gentry, N.W. Hurst and A. Jones, J. Chem. Soc. Faraday Trans. I, 75 (1979) 1688.
14. A.L. Boyce, S.R. Graville, P. Sermon and M. Vong, React. Kinet. Catal. Lett., 44 (1991) 1.
15. J.S. Campbell, Ind. Eng. Chem. Process Des. Develop., 9 (1970) 588.

Studies in Surface Science and Catalysis
J.J. Spivey, E. Iglesia and T.H. Fleisch (Editors)

Design/Economics of an Associated (or Sub-Quality) Gas Fischer-Tropsch Plant

Gerald N. Choi[a], Anne Helgeson[a], Anthony Lee[a], Sam Tam[a], Venkat Venkataraman[b], and Shelby Rogers[c]
[a] Nexant Inc., 45 Fremont Street, 7th Floor, San Francisco, CA 94105-2210
[b] U.S. DOE/NETL, 3610 Collins Ferry Rd., Morgantown,WV 26507-0880
[c] U.S. DOE/NETL, P.O. Box 10940, Pittsburgh, PA 15236-0490

ABSTRACT
Previous DOE sponsored gas conversion design/economics studies (detailed below) have been updated to incorporate the latest in Fischer-Tropsch (FT) synthesis design technology, including larger slurry-bed reactors and improved catalyst activity. Autothermal reforming is used for synthesis gas (syngas) generation with a gas containing 13% CO_2 as the feedstock. This gas is more representative of a low-cost associated gas and actually helps to produce a stoichiometric syngas. The design for the FT product upgrading section has also been simplified. A once-through FT design with power co-production and a design with tail gas recycle for maximum FT liquid production are compared. The capital servicing and gas feedstock costs remain as the primary factors affecting the overall gas conversion economics. With a reasonable associated gas feed cost, FT products can be produced at a cost that is competitive with crude oil at or below current price levels.

INTRODUCTION
In 1995, Bechtel Technology and Consulting Company (now Nexant Inc.,) developed, on behalf of the U. S. Department of Energy (DOE), a Baseline Design for natural gas conversion using advanced Fischer-Tropsch (FT) technology to produce high-quality, ultra clean, transportation fuels.[1,2] The study has assisted DOE in providing direction for its research and development, focusing on technologies that have the largest impact on the overall gas conversion economics.

The 1995 Baseline study used natural gas (i.e., 95% C_1) as the feed, and the design was intentionally conservative, using a combination of commercially proven technologies of non-catalytic partial oxidation (POX) and steam methane reforming (SMR) for syngas generation. FT plant design was based on the 1991 cobalt catalyst performance data of Satterfield.[3] A total of 24 slurry-bed reactors were used for FT synthesis producing 45,000 barrels per day (BPD) of FT liquids, at a total estimated plant cost of 1.8 billion U. S. dollars. It concluded that in order for the FT gas conversion process to be economical, a low-cost (e.g., $0.50 per million Btu) gas has to be used as the feedstock. Even then, the gas feed cost accounted for over 22% of the total cost of FT production.

A follow-up study showed that power co-production can reduce the FT plant cost at the expense of a small sacrifice in overall thermal efficiency.[4,5] The study also used 95% C_1 natural gas as the feed, but was designed for a smaller (10,000 BPD) plant capacity. Syngas generation was based on enriched-air blown autothermal reforming technology. Despite the loss of economies of scale, gas conversion economics can be improved via power co-production providing there is a demand for the co-produced electricity.

The above studies are now updated using a natural gas containing 13% CO_2 as the feed. This composition is more representative of a low-cost associated (or sub-quality and/or flared) gas feedstock. In addition, the design incorporates the latest FT synthesis performance, slurry-bed reactor design and sizing information. The FT product upgrading section also has been simplified to include only wax hydrocracking, thereby producing an upgraded FT diesel and a raw FT naphtha as the main products. This paper describes the results of this study. It discusses the overall plant design, trade-off analysis between power co-production versus maximizing FT liquid production via tail gas recycling, and the resulting preliminary economics.

DESIGN BASIS

Both the once-through and tail gas recycle designs process slightly over 500 MMSCF/day of associated gas feed to take full advantage of economies of scale. An associated gas containing 13 mole% of CO_2, 81% methane, and 6% higher hydrocarbons is the feed gas. The FT liquefaction plant design is based on the published cobalt-based, high-alpha, FT performance data of Satterfield[3] with an assumed 25% increase in catalyst activity to increase the single pass CO conversion to about 75%, more in line with current advanced catalyst performance. Autothermal reforming using 99.5% oxygen is used for syngas generation. Commercial cryogenic air separation units are used for oxygen production. After satisfying all internal fuel, heat and power requirements, excess tail gas is sent either to a gas turbine to produce export power, or recompressed and recycled for additional FT liquid production.

The plant is assumed to be located at a hypothetical U. S. Gulf Coast site, a common basis for such conceptual studies. For guidance as to how to relate such costs to remote sites, see the earlier natural gas Baseline Study.[1] In developing the present conceptual design, individual plant designs and cost estimates were prorated, where applicable, from the previous studies.

OVERALL PLANT CONFIGURATION

Figure 1 is a simplified block flow diagram showing the process configuration of the FT plant. The ISBL (inside battery limit) plant consists of two main processing areas: Area 100, syngas generation; and Area 200, FT synthesis, product upgrading and fractionation. CO_2 removal and tail gas recompression are included in Area 200. The integrated ISBL and utility plant is simulated using Hysys™, a commercially available process simulation program from Hyprotech Ltd. Preliminary trade-off analyses to determine the optimum process configuration led to the present design of recycling the excess tail gas to the ATR reactors. Tail gas can be recycled to the FT reactors, but a much higher rate is required to recycle all the excess tail gas to extinction.

Figure 1 – Block Flow Diagram of a Conceptual FT Plant Design
Using Associated Gas as Feed

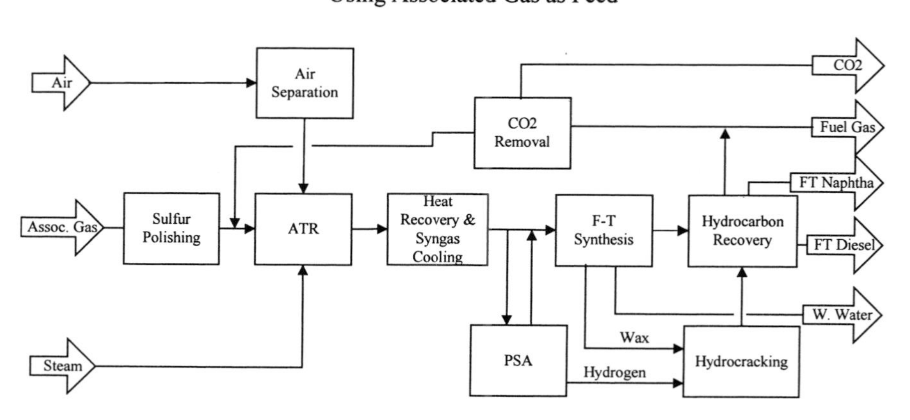

In addition to the ISBL plant, there are eighteen ancillary offsite OSBL (outside battery limit) plants, ranging from interconnecting piping, tankage, and waste water treatment, to power generation. The power generation plant represents a major capital investment, especially for the once-through design. Most of the OSBL plant designs and costs were prorated from previous studies; detailed analysis of the OSBL plants is beyond the current scope of work.

Syngas Generation (Area 100)
This area consists of two major plants: cryogenic air separation, and autothermal reforming (ATR) with waste heat recovery and syngas cooling. Area 100 produces a syngas with a molar H2:CO ratio close to 2.0 to meet the stoichiometric requirement for FT synthesis in Area 200. Sulfur is removed from the associated gas feed before syngas generation by adsorption on ZnO beds. The refractory-lined ATR reactors are designed to operate at a reasonable 1) O_2/carbon ratio of about 0.6 to keep the maximum adiabatic flame temperature below 4000 °F and reactor outlet temperature below 1900 °F, and 2) steam/carbon ratio of about 0.7 to avoid potential soot formation. With 13% CO_2 in the gas feed, the ATR reactor can be operated in a single-pass mode generating a syngas with a H2:CO ratio slightly above 2.0. With tail gas recycling, the excess CO_2 has to be removed. High-purity hydrogen is required for downstream FT wax hydrocracking. It is recovered from a slip stream of cooled syngas by a pressure swing adsorption (PSA) unit. A total of five syngas generation trains is required to process the 507 MMSCFD of associated gas feed.

Fischer-Tropsch Synthesis and Product Upgrading (Area 200)
This area consists of four major sections: FT synthesis, wax hydrocracking, product fractionation, CO_2 removal and tail gas recompression. Five, single-stage, 25' (7.6 meter) diameter by 120' (36 meter) high, slurry-bed reactors are required to process the syngas from Area 100. With tail gas recycling, an additional reactor train is required. Slurry-bed reactor sizing is based on an improved version of the kinetic reactor model originally developed by Viking Systems, International.[6] In this design, the heat generated by the FT synthesis reaction is removed by generation of 165 pound saturated steam in tubes suspended inside the reactor. Each reactor contains about 1,130 tubes of 40 mm in diameter. The reactors operate at 420 °F and 380 psia, with an inlet superficial gas velocity of approximately 0.43 feet (13 cm) per second.

The design of the wax hydrocracking plant, used for upgrading the FT wax, is based on pilot plant data by Mobil and UOP under DOE contract numbers DE-AC22-80PC30022,[7] and DE-AC22-85PC80017.[8] The hydrocarbon product recovery section consists of two conventional fractionation columns. A chilled lean oil absorber, using a slip diesel recycle stream as the lean oil solvent, is used to recover additional C_5+ liquid from the FT effluent vapor stream before sending it to the fuel gas header. In the tail gas recycling case, the excess gas (after satisfying in plant fuel requirement) is sent to a mono-diethanolamine (MDEA) based acid gas removal plant to remove CO_2, before it is recompressed and recycled back to the ATR reactors. A pure CO_2 stream is produced which can be sequestered.

PLANT DESIGN SUMMARY
Effect of Tail Gas Recycle
The effect of tail gas recycling on the overall plant power export and FT liquid production is shown in Figure 2. For a plant to produce just sufficient power to satisfy its in plant use, about 67% of the tail gas has to be recycled. This constitutes to about a 25% increase in gas throughput for Plant 100. Tail gas recycling increases the FT liquid production and the overall FT plant thermal efficiency at the expense of increased plant cost. Figure 3 shows how the plant thermal efficiency and capital cost increase as a function of the percentage of tail gas recycled. The once-through design with power co-production and the 67% tail gas recycling design are compared in more detail in the following sections.

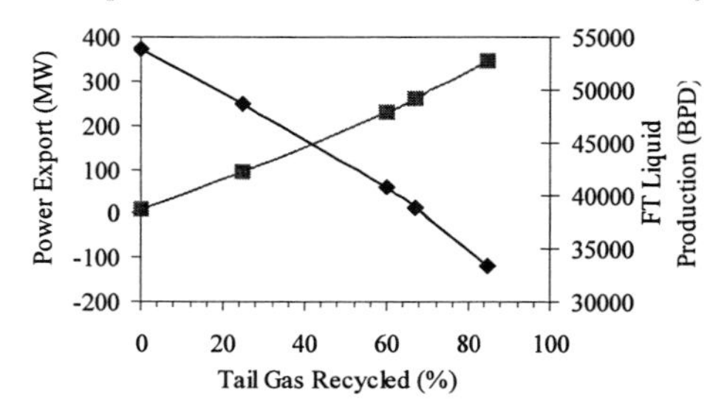

Figure 2 - Trade-Off Between Power Export and Increased FT Liquid Production as a Function of Tail Gas Recycling

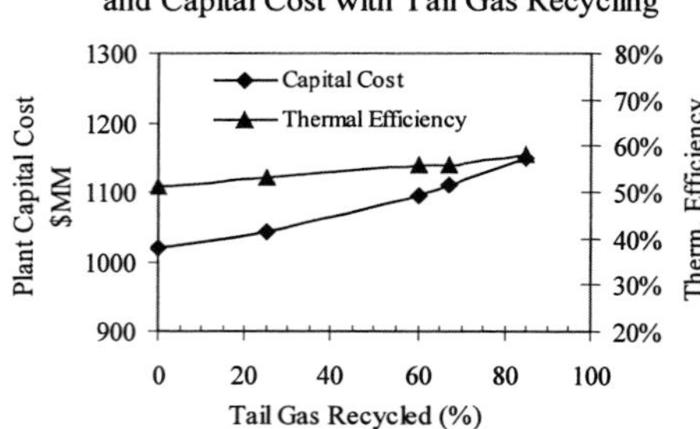

Figure 3 - Increase in Thermal Efficiency and Capital Cost with Tail Gas Recycling

Design Comparison

Both the once-through and the tail gas recycling designs process 507 MMSCF/day of associated gas and produce a C_5-350 °F naphtha and a 350-690 °F distillate product. Both products are essentially free of sulfur and nitrogen containing compounds. The naphtha is a raw product containing a small amount of oxygenates, which can be upgraded to produce a high-quality gasoline blending stock, used as a feedstock for steam cracking to produce light olefins, or used as a liquid fuel for fuel cell automotive application. The distillate is a high-quality diesel meeting all current and projected fuel specifications with no further upgrading required. The only materials delivered to the plant are associated gas, raw water, catalysts and chemicals. Make-up water to the cooling tower represents the major plant water usage. The major feed and product streams entering and leaving the plants are shown in Table 1.

Table 1 – Overall Plant Major Input and Output Flows

	Once-Through Design	Tail Gas Recycle Design
Feed		
Associated Gas (MMSCFD)	507	507
Raw Water (GPM)	3,100	3,900
Primary Products		
FT Naphtha (BPD)	12,100	15,400
FT Distillate (BPD)	26,700	33,800
Electric Power (MW)	372	

Capital Cost Estimate

Total capital costs of the associated gas FT plant are estimated at $1,020 MM and $1,110 MM for the once-through and the 67% tail gas recycle designs respectively. These are order-of-magnitude estimates with an accuracy of +/- 35%, typical of feasibility studies. They include offsite plant costs and allowances for home office costs, services fees and contingency. All ISBL plant cost estimates are based on major equipment costs with the appropriate bulk materials and labor factor applied. Table 2 shows the breakdown of the estimated plant capital costs for both designs.

PRELIMINARY ECONOMIC SENSITIVITY ANALYSIS

While an associated (or sub-quality) gas represents an opportunity feedstock, its cost is very much project and/or site specific. Other factors aside, its upper value may be bounded by its heating content which for the subject associated gas composition is about 13% lower than that of a 95 mole% C_1

Table 2 – Plant Capital Cost Estimate Breakdown

Plant	Description	Once-Through Cost (MM$)	Tail Gas Recycle Cost (MM$)
100	Air Compression & Separation	230.5	248.6
	ATR, Syngas Cooling & PSA	108.4	117.1
200	FT Synthesis	70.7	86.2
	CO2 Removal/Tail Gas Recompression	0	51.9
	Wax Hydrocracking/Product Fractionation	74.8	88.4
Power Plant		126.9	51.4
Other Offsites		210.4	252.1
	Total Field Cost	821.7	895.7
	HO Services/Fee/Contingency	197.2	214.9
	Total Installed Cost	1,018.9	1,110.6

natural gas. On the lower end, a whole spectrum of cost scenarios exists. Lacking an economic outlet, an associated gas commands very little present value. Often, costs are incurred for rejecting it back to the oil-bearing formation (e.g., Alaska Prudhoe Bay). As another example, for a situation where an associated gas is being flared, but will be prohibited to do so in the near future (e.g., Nigeria), environmental constraints or considerations become the main economic driver. The associated gas may very well have a negative value.

A discounted-cash-flow analysis on the FT production cost for a 15% internal rate return on investment was carried out to examine the economics of the associated gas FT designs using the same financial assumptions used in the previous studies[1,3], with the exception that a constant 3 ½% rate of general inflation and gas cost escalation is assumed, and that no premium is assigned to both the FT naphtha and the distillate products. It is well known, however, that FT distillate possesses superior properties and may command a premium over conventional diesel fuels.

Economics are very sensitive to both the purchased price of the associated gas feed and the selling price of the co-produced power. A once-through co-production FT plant has merit only if there is a demand for the produced power. For the present design, a breakeven point for the power-selling price is somewhere between 1 ½ to 3 cents per kWh for purchased gas feed price of 0.5 to 2.0 $/MMBtu. Above this breakeven point, a power co-production design offers a distinct economic advantage. Figure 4 shows the effect of FT production cost with increasing associated gas feed price. With a reasonable associated gas price, FT products can be produced at a cost that is competitive with current crude oil price of $25 to 30/bbl.

SUMMARY

Updating the previous DOE sponsored gas conversion design/economics studies with a combination of the latest Fischer-Tropsch (FT) synthesis performance data, increased slurry-bed reactor size, product upgrading simplification and autothermal reforming for syngas generation, using a gas containing 13% CO_2 which is more representative of a low-cost associated gas, has resulted in a significant reduction in the overall plant cost. A design for a once-through plant, exporting power, has been compared with a design for a plant with 67% tail gas recycle, which is in power balance. The latter produces more FT product but costs more to build. The breakeven point on power selling price is between 1 1/2 to 3 cents per kWh, above which the once-through plant offers a significant advantage.

Both the capital and gas feedstock costs remain as the prevailing factors affecting the overall gas conversion economics. While it is unlikely that much further cost reduction can be achieved beyond

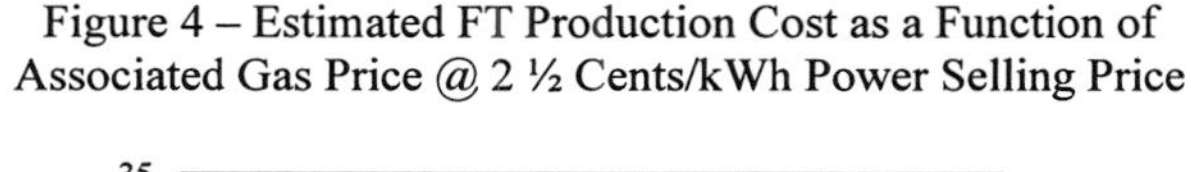

Figure 4 – Estimated FT Production Cost as a Function of Associated Gas Price @ 2 ½ Cents/kWh Power Selling Price

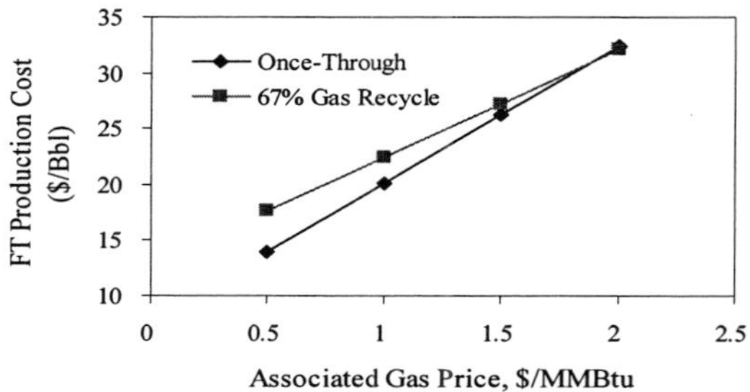

and above the FT synthesis performance and the slurry reactor size used as the basis for the present design study, the same does not apply for syngas generation which still constitutes about 60% of the total capital cost. This is where gas conversion research and development needs to be focused. DOE is currently supporting programs in novel ceramic membrane technology development for both oxygen separation and syngas generation. These types of technology offer opportunities to appreciably reduce the cost of syngas generation which can significantly impact the economics of gas conversion.

REFERENCE

1. G. N. Choi, S. J. Kramer, S. S. Tam, R. Scrivastava and G. Stiegel, *Natural Gas Based Fischer-Tropsch to Liquid Fuels: Economics*, Society of Petroleum Engineers 66th Western Regional Meeting, Anchorage, Alaska, May 22-24, 1996.
2. Report for DOE Contract No. DE-AC22-91PC90027, *Baseline Design/Economics for Advanced Fischer-Tropsch Technology: Topical Report VI: Natural Gas Fischer-Tropsch Case,* March, 1996.
3. Final Report for DOE Contract No. DE-AC-87PC79816, *An Innovative Catalyst System for Slurry-Phase Fischer-Tropsch Synthesis: Cobalt Plus a Water-Gas Shift Catalyst,* July, 1991.
4. G. N. Choi, S. J. Kramer, S. S. Tam, J. M. Fox, N. L. Carr and G. R. Wilson, *Design/Economics of a Once-Through Natural Gas Fischer-Tropsch Plant with Power Co-Production*, Coal Liquefaction & Solid Fuels Contractors Review Conference, Pittsburgh, Pennsylvania, September 3-4, 1997.
5. Report for DOE Contract No. DE-AC22-91PC90027, *Baseline Design/Economics for Advanced Fischer-Tropsch Technology: Topical Report VII: Natural Gas Based Once-Through Fischer-Tropsch Design with Power Co-Production,* March, 1998.
6. Final Report, *Design of Slurry Reactor for Indirect Liquefaction Applications,* DOE Contract No. DE-AC22-89PC89870, December, 1991.
7. J. C. Kuo, et al (Mobil), *Two-Stage Process for Conversion of Synthesis Gas to High Quality Transportation Fuels,* DOE Contract No. DE-AC22-80PC30022, June, 1983.
8. P. P. Shah, et al (UOP), *Fischer-Tropsch Wax Characterization and Upgrading,* DOE Contract No. CE-AC22-85PC80017, December, 1989.

Studies in Surface Science and Catalysis
J.J. Spivey, E. Iglesia and T.H. Fleisch (Editors)

Expanding Markets for GTL Fuels and Specialty Products

Paul F. Schubert[1], Charles A. Bayens[1], Larry Weick[1], Michael O. Haid[2]
[1]Syntroleum Corp., 1350 S. Boulder, Tulsa, OK 74119
[2]TESSAG Edeleanu GmbH, Alzenau, Germany

1. Introduction – Technology Development

In the first part of the 20th century the world was transitioning from a coal-based economy to a petroleum-based economy. This caused significant concerns for developed nations which had abundant coal, but lacked sufficient oil to achieve their economic and political goals. In Germany, a concerted effort was initiated to convert coal and petroleum coke into liquid hydrocarbons. In 1923 Franz Fischer and Hans Tropsch discovered the process which bears their name, and which had potential to meet these objectives. Fischer, and his co-workers in the German petrochemical industry demonstrated the basic concepts and methods still being pursued today such as:

- Tubular reactors for improved temperature control [1];
- Liquid circulation through fixed bed or tubular reactors for improved temperature control and product selectivity [2-4];
- Slurry reactors with micron range particles for improved diffusion and better temperature control [5,6];
- Use of both iron and cobalt catalysts in slurry reactors [1,7];
- Using the special properties of synthetic lubricants for unique applications [8];
- Production of high cetane, sulfur free diesel fuel [8]; and
- Blending synthetic and naturally derived materials to obtain desired properties [1,8];

The Fischer-Tropsch process was first commercialized in Germany in 1936, followed by plants in France, Japan, and China during the World War II period, and then South Africa in the post-war era [9]. Other than in South Africa, interest in the Fischer-Tropsch (FT) process waned as significant oil reserves were discovered, and interest in nuclear power increased.

The energy crisis of the 1970's and concerns about problems with nuclear power caused a brief resurgence in interest in the 1970's. However, sustained interest in the process has largely been associated with the desire to convert stranded natural gas into liquid hydrocarbons, and concerns with the environmental impact of sulfur and aromatics in fuels derived from petroleum. Throughout these periods the technology continued to be developed

The first step in the GTL technology is the production of synthesis gas (H_2 and CO) from natural gas. In a second step synthesis gas (syngas) is converted on a FT catalyst to liquid hydrocarbons (FT liquids). The FT liquids contain no detectable sulfur and can be converted by mild hydrotreating into highly saturated synthetic fuel. Tests with synthetic fuel showed that the emission of environmental pollutants were lowered notably relative to conventional diesel fuel. Another utilization of the FT liquids is the production of lubricants and specialty chemicals. The FT synthesis is essentially the same as for the production of synthetic fuel. The main difference lies in the further processing of the FT liquids. Lubricant production requires hydroisomerization of normal paraffins.

Today, newer GTL technologies are being developed to achieve better economics even at smaller plant sizes. Syntroleum, one of the leading GTL technology suppliers, has now developed their GTL technology to that stage that a commercial plant can be built. The first commercial-scale application of the Syntroleum GTL technology will be a specialty chemicals and lube oil plant, known as the Sweetwater plant. This 10,000 bpd grass roots plant, for which TESSAG started engineering in November 1999, will be located on the Burrup Peninsula in Northwestern Australia.

2. The Syntroleum GTL Technology

Syntroleum's GTL process has been under development since the early 1980's. The process utilizes the same principal steps as other FT processes: (i) production of syngas from natural gas and (ii) conversion of syngas into FT liquids on a catalyst. The goal of Syntroleum's development was to be commercially viable at plant capacities that would allow the use of a significant portion of the world's remote gas reserves. To achieve this goal proprietary cobalt based FT catalysts with high activity and high selectivity were developed and the complexity of each processing step was simplified. This led to several significant differences in the execution of the process steps compared to other FT technologies.

2.1 Syngas Production

Syngas is generally produced from natural gas using either gas phase partial oxidation (POX) or steam methane reforming (SMR). In the POX process natural gas reacts with pure oxygen in an open flame at high temperatures of 1,200 - 1,500 °C. The gas phase POX produces syngas with a H_2 to CO ratio (H_2/CO) of typically below 2:1 on a molar basis. The SMR process converts natural gas with steam on a nickel catalyst at 800 – 1,000 °C to a hydrogen rich syngas. The typical H_2/CO ratio lies above 3:1 (without CO_2 recycle). The natural gas has to be desulfurized to prevent catalyst deactivation. A third commercially applied process - autothermal reforming (ATR) - combines POX and SMR in a single step. The benefits are a lower reaction temperature, lower oxygen consumption and a H_2/CO ratio of 2:1 which is ideally suited for the FT synthesis. However, an expensive cryogenic air separation plant is required for oxygen production in both POX and ATR processes.

The Syntroleum process utilizes a proprietary ATR reactor to catalytically convert natural gas to syngas using air instead of pure oxygen to avoid an air separation plant. Because the system is in heat balance, no heat transfer system is required, and the ATR reactor is much more compact. The H_2/CO ratio of the resulting nitrogen diluted syngas is close to the desired 2:1 for the FT synthesis. The main reactions occurring in the ATR reactor are shown in Eq. 1 – 3. The syngas at the reactor outlet is in equilibrium with respect to the reactions and is soot-free.

$$CH_4 + 0.5\,O_2 \rightarrow CO + 2\,H_2 \qquad \Delta H = -36 \text{ kJ/mole} \qquad (1)$$
$$CH_4 + H_2O \leftrightarrow 3\,H_2 + CO \qquad \Delta H = +206 \text{ kJ/mole} \qquad (2)$$
$$CO + H_2O \leftrightarrow H_2 + CO_2 \qquad \Delta H = -41 \text{ kJ/mole} \qquad (3)$$

The proprietary design allows steam to carbon ratios (S/C) below those typically used in conventional ATR's (1.7:1 to 2.2:1). Lower S/C ratios are favorable since this significantly improves the economics of an ATR-based FT plant [10]. By selecting the right S/C ratio, oxygen to carbon ratio (O_2/C) and operating temperature the syngas composition can be optimized for the downstream FT synthesis.

The viability of the proprietary Syntroleum ATR design was successfully demonstrated in two pilot plants for a broad range of process conditions. In the pilot plant at ARCO's Cherry Point refinery, Washington, USA, more than 6,000 hours of operation was demonstrated. The Cherry Point ATR was sized to generate sufficient syngas for the downstream 70 bpd FT pilot unit. The ATR at Syntroleum's pilot plant in Tulsa, USA, is designed to provide syngas for a 2 bpd FT pilot unit and is in operation since 1991.

2.2 Fischer-Tropsch Synthesis

Generally, the Fischer-Tropsch reaction take places under moderate temperatures (200 – 300 °C) and moderate pressures (10 – 40 bar) utilizing iron or cobalt based catalysts. The chain length of the FT hydrocarbons is dependent on factors such as temperature, type of catalyst and reactor employed. Very similar for iron and cobalt based catalysts [11]. The FT reaction is defined in Eq. 4 in which "-CH_2-" represents a product consisting mainly of linear paraffinic hydrocarbons of variable chain length.

$$CO + 2\,H_2 \rightarrow \text{"-}CH_2\text{-"} + H_2O \qquad \Delta H = -165 \text{ kJ/mole} \qquad (4)$$

Especially with iron catalysts substantial amounts of olefins, alcohols and sometimes aromatics can be produced. Ketones and acids are obtained in minor concentration. With cobalt based catalysts only small amounts of olefins and alcohols and no aromatics are produced. FT liquids are sulfur free since the sulfur is removed from the natural gas feedstock.

Syntroleum developed proprietary catalysts which are highly active and produce a high alpha value. The catalysts are tailored for two different types of reactor systems, for a conventional multi-tubular fixed bed reactor and for a moving-bed reactor. Multi-tubular fixed bed reactors are proven technology in, for example, methanol synthesis. They have low scale-up risks since the scale-up from one tube to several thousand parallel tubes is a straight forward process. Moving bed reactors are simpler in design, less expensive and allow a continuous catalyst regeneration outside the reactor, but have a higher scale-up risk which has now been mitigated by Syntroleum's pilot plant experience.

These pilot plant tests were carried out to demonstrate Syntroleum FT catalyst performance and to gain operating data over a broad range of process conditions. A Syntroleum moving-bed FT reactor with a capacity of 70 bpd of liquid products was successfully operated at ARCO's Cherry Point refinery for more than 4,500 hours. Tests with tubular fixed bed reactors were performed at

Syntroleum's Tulsa pilot plant which has been in operation since 1991. The capacity of the reactors is 2 bpd of liquid products. The pilot plant reactors are equipped with tubes of commercial size length and diameter.

3.0 Synthetic Products

In recent years, the diesel engine development has made significant advances in meeting greater fuel efficiency and lower emission rates of environmental pollutants. To decrease the emission of hydrocarbons (HC) and carbon monoxide (CO), particulate matter (PM) and nitrogen oxides (NO_X) significantly, exhaust gas after treatment with oxidation catalysts and de-NO_X catalysts has to be applied. This however, requires low sulfur levels in the fuel not to poison the catalysts. Low aromatics levels are also environmentally advantageous [11].

Synthetic fuel made from natural gas with the FT process meets those requirements. To maximize fuel production, the high molecular weight hydrocarbons produced can be cracked into fuel range molecules. Several fundamental differences between synthetic fuel (S-2) and conventional fuel (EPA #2) exist. The synthetic fuel contains essentially no sulfur and no aromatics. Since Syntroleum S-2 fuel is almost hydrogen saturated and consists of > 99 % paraffins the cetane index (CI) is exceptionally high. Since paraffins have slightly lower specific gravities versus aromatic molecules with similar carbon number, the API gravity for the synthetic fuel is higher than the values generally obtained for conventional fuels. The higher API gravity results in lower volumetric heating values but higher heating values on a weight basis.

Numerous studies have evaluated emissions from conventional, unmodified diesel engines using a variety of synthetic fuels, including work done at Southwest Research Institute for a comparison of Syntroleum S-2 with EPA #2 fuel using an unmodified heavy duty 5.9 l Cummins engine on a chassis dynamometer and reported elsewhere [12]. The synthetic fuel has significantly lower emission rates than conventional fuel without engine modifications or exhaust gas after treatment. These lower emission rates have been explained [13] by 1) lowering the sulfur content decreasing the sulfate particles in the exhaust gas producing an almost linear decrease of PM with decreasing sulfur content; 2) improved fuel combustion from the high cetane number (CN); and 3) decreasing the aromatics content lowering NO_X and PM emissions.

2.1 Specialty Products

In addition to the synthetic fuels sector the FT liquids can be utilized in the synthetic lubricants and specialty chemicals sector. Depending on the planned utilization of the FT liquids the applied processing steps vary significantly. The processing steps are basically conventional technology. In any case when using cobalt based catalysts the derived final products are of exceptionally high quality due to the lack of sulfur and aromatics in the FT liquids. In Table 3 the main product groups and their typical uses and applications are summarized.

Table 3: Main Product Groups and their Typical Uses and Applications.

Product Group	Typical Uses and Applications
Normal Paraffins	Production of detergent intermediates (LAB, SAS, alcohols) Production of intermediates for plasticizers, auxiliary chemicals, additives, cutting fluids, sealants Manufacturing of film and catalyst carrier Special low polar and odor-free all-purpose solvents and diluents
Mixed Paraffins	Special low polar and odor-free solvents for paints, coatings, dry cleaning, cleaners, insecticide and pesticide formulations, drilling fluids
Synthetic Lubricants	Industrial and automotive lubricant applications including motor oils gear oils, compressor oils, hydraulic fluids, greases
Paraffin Waxes	Manufacturing of candles, crayons, printing inks, potting and cable compounds, cosmetics, pharmaceutics, coatings and packaging

4. Syntroleum's Sweetwater Plant

In November 1999 . Tessag Industrie Anlagen GmbH signed a project development agreement with Syntroleum Sweetwater Operations Ltd. to provide a fixed price for the design and construction of the first commercial-scale application of Syntroleum's GTL technology, known as the Sweetwater plant. This grass roots plant is designed to produce 10,000 bpd of liquid specialty products on a site about four kilometers from the North West Shelf LNG facility on the Burrup Peninsula of Western Australia. Construction is expected to begin in 2001 and the plant to be operational in 2003. Tessag has given a price quote of $506 million for the engineering, procurement and construction (EPC).The quote does not include interest during construction and other owner's costs, which include proprietary catalysts to be supplied by Syntroleum.

The Sweetwater plant has three main areas: syngas production; FT synthesis; and product upgrading. The plant will produce synthetic lubricants, synthetic paraffins and LPG. The syngas production is based on Syntroleum's autothermal reforming technology (ATR). Natural gas is preheated, desulfurized and then fed together with preheated compressed air and steam to the ATR. The hot syngas exiting the ATR is quenched and further cooled down. No air separation and no CO_2 removal is required.

The FT synthesis area consists of a FT reactor section and the downstream FT liquids recovery. A multi-tubular fixed bed reactor design was chosen for several reasons including:

- the effect of the project financing schedule on the design (Syntroleum's moving bed reactor tests with ARCO at the Cherry Point refinery were not completed until July 2000);
- the ease of scale-up from pilot plant tests which employ a limited number of full length commercial diameter tubes to a commercial reactor with thousands of tubes of exactly the same dimensions as the pilot tubes;

- the ability to optimize the fixed bed catalyst to produce the long chain length molecules (high alpha) preferred for lubricant production without the constraints imposed by factors such as attrition resistance, settling rates, etc., present in slurry type reactors; and
- the high value of the refined products resulting from the optimized fixed bed catalyst more than offset the increased cost of the fixed bed reactors relative to slurry reactors at the same scale.

The product upgrading or product refining area in the plant is capable of producing a wide range of specialty products, and flexibility to adapt the product slate to changing market conditions. The FT wax and oil is upgraded in this section of the plant using hydroisomerization technology licensed from Lyondell-Citgo and paraffins separation technology licensed from UOP. The predominant products from this area will be synthetic lubricants. Further products will be LPG, light paraffins, normal paraffins and mixed paraffins.

References

[1] Report on the Petroleum and Synthetic Oil Industry of Germany, Ministry of Fuel and Power, London, His Majesty's Stationery Office, pg. 85, (1947).

[2] Duftschmid, F; Linckh, E.; Winkler, F.; U.S. Patent 2,159,077, (1939).

[3] Duftschmid, F; Linckh, E.; Winkler, F.; U.S. Patent 2,287,092, (1942).

[4] Duftschmid, F; Linckh, E.; Winkler, F.; U.S. Patent 2,318,602, (1943).

[5] Fischer, F.; Kuster, H., Brennstoff-Chemie 14, 3-8, (1933), p 3-8.

[6] Fischer, F.; Roelen, O. Feisst, W.; Pichler, H., Brennstoff-Chemie 13, 461-468 (1932).

[7] Starnburg, W. A.; FIAT Reel K-31, T-459, 1108-1115, (1948).

[8] Report on Investigations by Fuels and Lubricants Teams at the I.G. Farbenindustrie A.G. Works at Leuna, ed. R. Holroyd, CIOS Target No. 30/4.02, Fuels and Lubricants, Item No. 30, File No. XXXII-107Combined Intelligence Objectives Sub-Committee.

[9] Kolbel, H., Chemie-Ing.-Techn., 8, 505, (1957).

[10] T.S. Christensen, P.S. Christensen, I. Dydkjaer, J.H. Bak Hansen and I.I. Primdahl, Stud. Surf. Sci. Catal., 119, 883, (1998).

[11] R.L. Espinoza, A.P. Steynberg, B. Jager and A.C. Vosloo, Applied Catalysis A: General, 186, 13, (1999).

[12] P.V. Snyder, B.J. Russell, P.F. Schubert, Syntroleum Publication at Clean Fuels 2000 Meeting, San Diego, CA, February 7-9, (2000).

[13] Directive 98/70/EC of the European Parliament and of the Council, October, (1998).

[14] Chevron, Diesel Fuels Technical Review (FTR-2), Chevron, San Francisco, CA, USA, (1998).

Studies in Surface Science and Catalysis
J.J. Spivey, E. Iglesia and T.H. Fleisch (Editors)
2001 Elsevier Science B.V.

Development of dense ceramic membranes for hydrogen separation*

U. (Balu) Balachandran,[a] T. H. Lee,[a] G. Zhang,[a] S. E. Dorris,[a] K. S. Rothenberger,[b] B. H. Howard,[b] B. Morreale,[b] A. V. Cugini,[b] R. V. Siriwardane,[c] J. A. Poston Jr.,[c] and E. P. Fisher[c]

[a]Energy Technology Division, Argonne National Laboratory, Argonne, IL 60439

[b]National Energy Technology Laboratory, Pittsburgh, PA 15236

[c]National Energy Technology Laboratory, Morgantown, WV 26507

Novel cermet (i.e., ceramic-metal composite) membranes have been developed for separating hydrogen from gas mixtures at high temperature and pressure. The hydrogen permeation rate in the temperature range of 600-900°C was determined for three classes of cermet membranes (ANL-1, -2, and -3). Among these membranes, ANL-3 showed the highest hydrogen permeation rate, with a maximum flux of 3.2 cm^3/min-cm^2 for a 0.23-mm-thick membrane at 900°C. The effects of membrane thickness and hydrogen partial pressure on permeation rate indicate that bulk diffusion of hydrogen is rate-limiting for ANL-3 membranes with thickness >0.23 mm. At a smaller, presently unknown thickness, interfacial reactions will dominate the permeation kinetics. The lack of degradation in permeation rate during exposure to a simulated syngas mixture suggests that ANL-3 membranes are chemically stable and may be suitable for long-term operation.

1. INTRODUCTION

The DOE Office of Fossil Energy sponsors a wide variety of research, development, and demonstration programs whose goals are to maximize the use of vast domestic fossil resources and to ensure a fuel-diverse energy sector while responding to global environmental concerns. Development of cost-effective membrane-based reactor and separation technologies is of considerable interest for applications in advanced fossil-based power and fuel technologies. Because concerns over global climate change are driving nations to reduce CO_2 emissions, hydrogen is considered the fuel of choice for the electric power and transportation industries. Although it is likely that renewable energy sources will ultimately be used to generate hydrogen, fossil-fuel-based technologies will supply hydrogen in the interim.

Dense, hydrogen permeable membranes are being developed at Argonne National Laboratory (ANL) and National Energy Technology Laboratory (NETL). Dense membranes made from palladium were developed elsewhere and have been commercially available since the early 1960's, but they are expensive and are limited to operating temperatures below

*This work has been supported by the U.S. Department of Energy, National Energy Technology Laboratory, under Contract W-31-109-Eng-38.

≈600°C [1]. The goal at ANL/NETL is to develop a dense, ceramic-based membrane that is highly selective, like palladium-based membranes, but is less expensive and is chemically stable in corrosive environments at operating temperatures up to ≈900°C. In order to be commercially viable, the membrane should provide a hydrogen flux of ≈10 cm^3(STP)/min-cm^2 with a pressure drop of ≈1 atm across the membrane [2].

The efforts at ANL/NETL initially focused on $BaCe_{0.8}Y_{0.2}O_{3-\delta}$ (BCY), because its total electrical conductivity is among the highest of known proton conductors [3,4]. However, its electronic component of conductivity is insufficient for nongalvanic hydrogen permeation [5,6]. In order to produce membranes with increased electronic conductivity, we made ceramic-metal composite (i.e., cermet) membranes by adding various metals to BCY. Nongalvanic hydrogen permeation was observed in two such membranes (ANL-1 and -2) but the hydrogen flux was much higher in ANL-2 because the hydrogen permeability of its metal was much higher than that of the metal in ANL-1 [7,8]. Based on these results, a third class of cermet membrane (ANL-3) was developed in which the metal phase provides the path for hydrogen diffusion while the ceramic phase acts only as a mechanical support.

This paper describes the present status of cermet membrane development at ANL/NETL. The hydrogen permeation properties are compared for three classes of cermet membranes (ANL-1, -2, and -3), and the effects of membrane thickness, temperature, and hydrogen partial pressure on the permeation flux of ANL-3 are presented. In addition, the long-term chemical stability of ANL-3 membranes in simulated "syngas" is demonstrated.

2. EXPERIMENTAL

To prepare ANL-1 and -2 membranes, two different metal powders were first mixed with BCY powder, whose preparation has been described previously [8]. ANL-3 membranes were prepared by first mixing a third metal with a ceramic powder that is reported to be a poor proton conductor [9]. All three membranes were prepared to give 40 vol.% of the metal phase. The powders were mixed with mortar and pestle in isopropyl alcohol, and the alcohol was then evaporated. Pellets were made from the resulting powder mixture by uniaxial pressing at 200 MPa. ANL-1 membranes were sintered at 1400-1420°C for 5 h in 4% H_2/balance N_2. ANL-2 and -3 membranes were sintered in ambient air for 12 h at 1510 and 1350°C, respectively.

For permeation testing, a sintered disk was polished with 600-grit SiC polishing paper and affixed to an alumina (Al_2O_3) tube with spring-loaded alumina rods. A gold ring between the disk and the alumina tube formed a seal when the assembly was heated at 950°C for 5 h. One side of the membrane was purged with 4% H_2/balance He during the sealing procedure, while the other side was purged with 100 ppm H_2/balance N_2 (or Ar). The effective membrane area was ≈1.3 cm^2.

The apparatus for measuring hydrogen permeation has been described elsewhere [8]. Both sweep (≈100 cm^3/min) and feed (≈150 cm^3/min) gases were controlled with MKS 1179A mass flow controllers. A small concentration of hydrogen, 100 ppm, was added to the sweep gas to prevent oxidation of the metal in the membrane and was subtracted in calculating the permeation rate. The gas concentrations in the sweep gas were analyzed with a gas chromatograph (Hewlett-Packard 6890) that was periodically calibrated with standard gases. Hydrogen leakage through the gold seal, or through a small amount of interconnected porosity in the membrane, was corrected by measuring the helium concentration on the sweep side of the membrane and typically represented <10% of the total hydrogen permeation flux. Both dry and wet feed gas were used. For the wet condition, the feed gas was bubbled through a water bath at room temperature, while the dry feed gas was introduced directly into the furnace from the gas cylinder. Simulated "syngas" consisted of 66% H_2, 33% CO, and 1% CO_2 and was purchased from AGA Gas, Inc.

3. RESULTS AND DISCUSSION

Figure 1 compares the hydrogen permeation rates for ANL-1, -2, and -3 membranes, measured with a feed gas of 4% H_2/balance He. In each case, the membrane thickness was nearly the same (≈0.5 mm) and identical measurement conditions were used, so the permeation rates could be directly compared for the three samples. The permeation rate for ANL-3 was ≈3 times higher than that of ANL-1 and ≈30-50% higher than that of ANL-2 over the whole temperature range. Due to its higher hydrogen flux, the remainder of the paper focuses on the ANL-3 membrane.

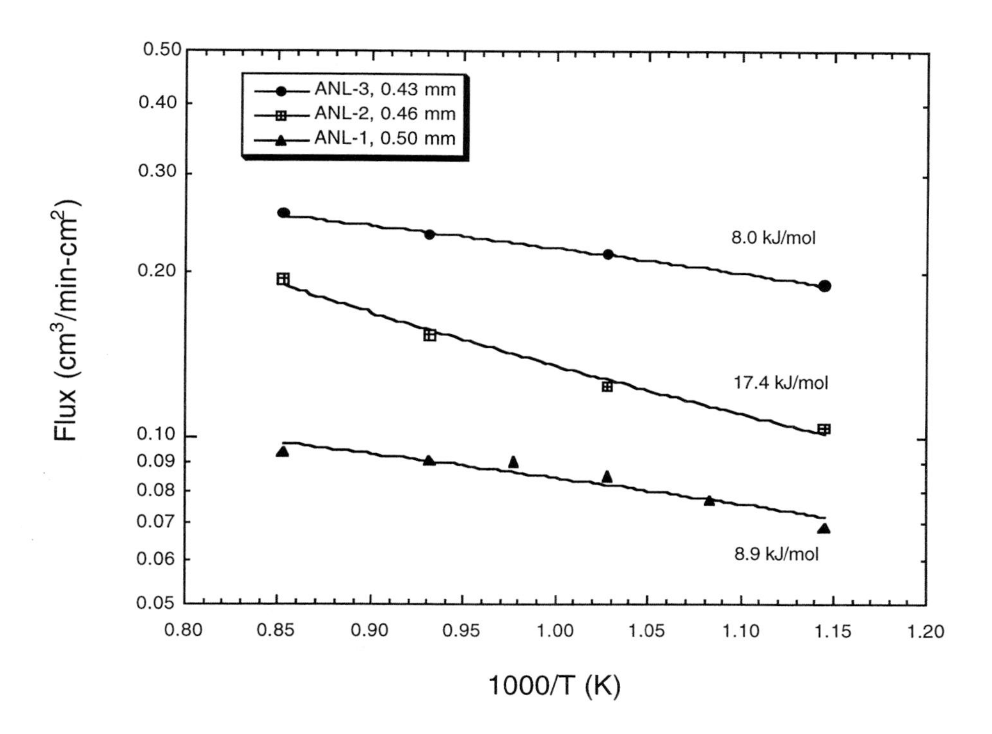

Fig. 1. Hydrogen flux through ANL-1, 2, and -3 membranes using 4% H_2/balance helium as the feed gas. The activation energies for permeation are given in the figure for each membrane.

Figure 2 shows the temperature dependence of hydrogen flux for ANL-3 membranes with thicknesses of 0.23 and 0.43 mm. Measurements were made at temperatures of 600-900°C using wet 100% H_2 as feed gas and 100 ppm H_2/balance N_2 as sweep gas. The hydrogen flux increased with temperature and was approximately proportional to the inverse of membrane thickness over the whole temperature range. This suggests that the bulk diffusion of hydrogen

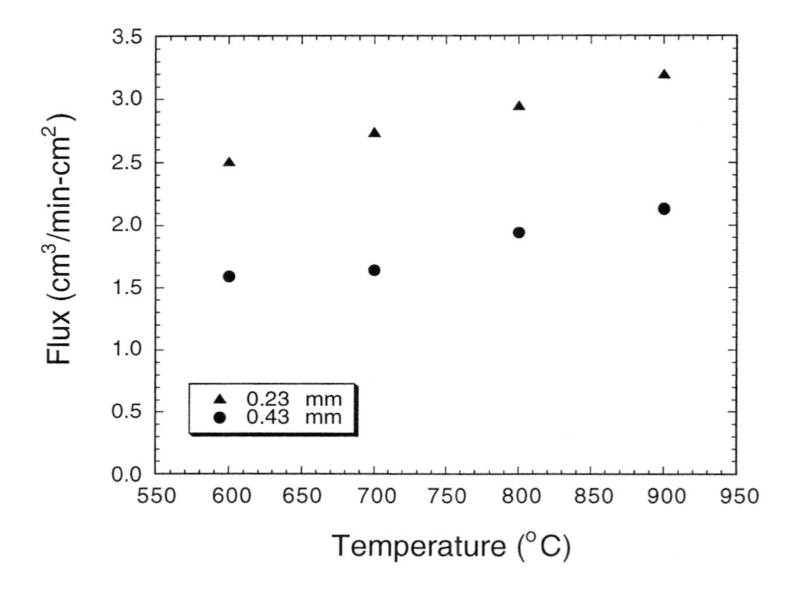

Fig. 2. Hydrogen flux through ANL-3 membranes with wet 100% H_2 as feed gas.

is rate-limiting in ANL-3 membranes, so the flux may be increased further by reducing the membrane thickness. However, below some thickness that is presently unknown, interfacial reactions will dominate the permeation kinetics, and reducing the membrane thickness will not increase the hydrogen flux. A maximum flux of 3.2 cm^3(STP)/min-cm^2 was measured for the 0.23-mm-thick membrane at 900°C.

The effect of hydrogen partial pressure on the permeation flux is shown in Fig. 3 for the 0.23-mm-thick ANL-3 membrane. Measurements were performed at 800 and 900°C. Helium and 100% hydrogen (or syngas) were mixed to obtain desired hydrogen partial pressures in the feed gas. The hydrogen permeation flux is plotted as a function of the difference in square root of hydrogen partial pressure for the feed and sweep sides of the membrane. At both temperatures, the flux is linear with the difference in square root of hydrogen partial pressure. This is characteristic of bulk-limited hydrogen diffusion through metals [10], and suggests that diffusion through the metal phase dominates the permeation process. If, on the other

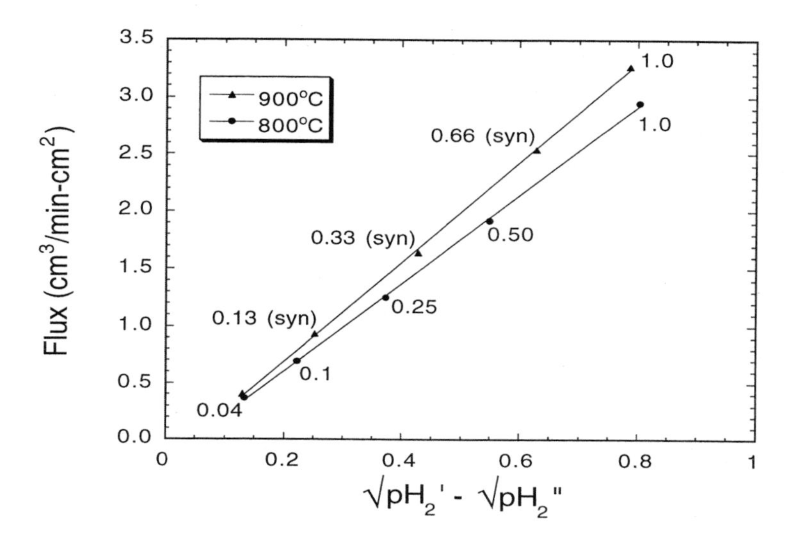

Fig. 3. Hydrogen flux versus difference in partial pressure in feed (pH_2') and sweep (pH_2'') gases for 0.23-mm-thick ANL-3 membrane. Mixtures made with syngas are shown by "syn."

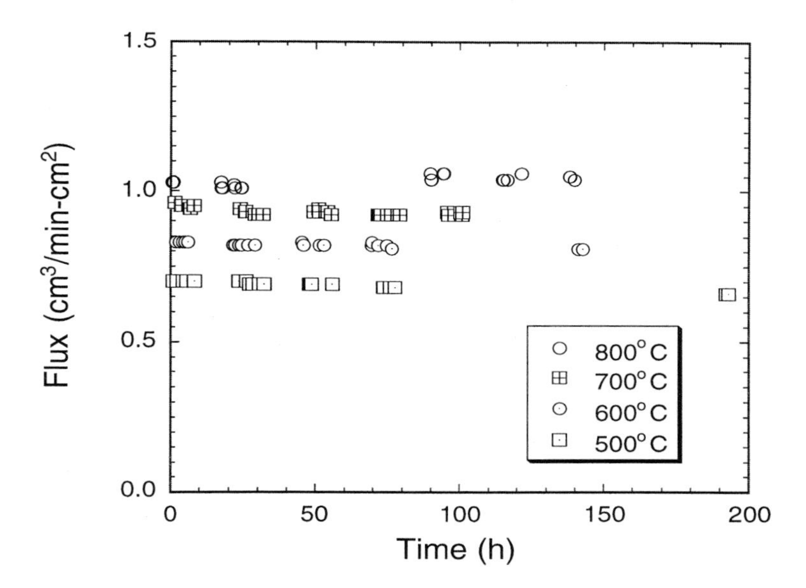

Fig. 4. Hydrogen flux through ANL-3 membrane (0.43-mm-thick) versus time in syngas atmosphere. Feed gas is simulated syngas; sweep gas is 110 ppm H_2/balance Ar.

hand, permeation were dominated by diffusion through the ceramic phase, the hydrogen flux would be expected to have a logarithmic dependence on the ratio of hydrogen partial pressures in the feed and sweep gases [11]. Because our data do not follow such a logarithmic dependence on hydrogen partial pressure, and the ceramic phase in the membrane is reported to be a very poor proton conductor [9], hydrogen permeation through ANL-3 membranes appears to be limited by the bulk diffusion of hydrogen through the metal phase. While bulk diffusion controls the permeation in this thickness range, interfacial reactions will become rate-limiting at some lesser thickness.

The chemical stability of an ANL-3 membrane in syngas was tested by measuring its hydrogen permeation at several temperatures for times up to 190 h, as shown in Fig. 4 for a 0.43-mm-thick membrane. Before and after exposure to syngas at each temperature, a feed gas of 4% H_2/balance He was flowed and the leakage rate of hydrogen was estimated by measuring the helium concentration in the sweep gas. No helium leakage was measured at any of the temperatures. As seen in Fig. 4, no noticeable decrease in flux was observed during exposures to syngas for up to 190 h. Likewise, a 0.23-mm-thick ANL-3 membrane showed no decrease in flux during 120 h of exposure to syngas at 900°C. While the permeation properties are not strongly affected by exposure to syngas, this does not indicate whether the mechanical properties are influenced by syngas; in the future, the effect of syngas and other environments on the mechanical properties of the membrane will be investigated.

4. CONCLUSIONS

We have developed cermet membranes that nongalvanically separate hydrogen from gas mixtures. The highest measured hydrogen flux was 3.2 cm^3(STP)/min-cm^2 for an ANL-3 membrane at 900°C. For ANL-3 membranes with thicknesses of 0.2-0.5 mm, the permeation rate is limited by the bulk diffusion of hydrogen. The effect of hydrogen partial pressure on the permeation rate confirmed this conclusion and indicates that higher permeation rates can be obtained by decreasing the membrane thickness. Permeation rate in a syngas atmosphere for times of up to 190 h showed no degradation in performance, which suggests that ANL-3 membranes may be suitable for long-term, practical hydrogen separations.

REFERENCES

1. S. Uemiya, Separation and Purification Methods, 28(1), 51 (1999).
2. S. Benson, Ceramics for Advanced Power Generation, IEA Coal Research 2000, Pg.43.
3. H. Iwahara, T. Yajima, and H. Uchida, Solid State Ionics, 70/71, 267 (1994).
4. H. Iwahara, Solid State Ionics, 77, 289 (1995).
5. J. Guan, S. E. Dorris, U. Balachandran, and M. Liu, Solid State Ionics, 100, 45 (1997).
6. J. Guan, S. E. Dorris, U. Balachandran, and M. Liu, J. Electrochem. Soc., 145, 1780 (1998).
7. J. Guan, S. E. Dorris, U. Balachandran, and M. Liu, Ceramic Transactions, 92, 1 (1998).
8. U. Balachandran, T. H. Lee and S. E. Dorris, Proc. 6th Annual International Pittsburgh Coal Conf., Pittsburgh, PA, Oct. 11-15, 1999.
9. K.-D. Kreuer, Chem. Mater., 8, 610 (1996).
10. R. E. Buxbaum and T. L. Marker, J. Membrane Science, 85, 29 (1993).
11. M. Liu, Proc. 1st Int'l. Symp. Ionic and Mixed Conducting Ceramics, T. A. Ramanarayana and H. L. Tuller, eds., The Electrochemical Society, Pennington, NJ, Vol. 91-12, 215 (1991).

Studies in Surface Science and Catalysis
J.J. Spivey, E. Iglesia and T.H. Fleisch (Editors)

Methane oxyreforming over the Al_2O_3 supported rhodium catalyst as a promising route of CO and H_2 mixture synthesis

B.Pietruszka[a], M. Najbar[a*], L. Lityńska-Dobrzyńska[b], E. Bielańska[b], M. Zimowska[a], J. Camra[c]

[a]Department of Chemistry, Jagiellonian University, 30 060 Kraków, Ingardena 3, Poland,
[*]Phone: (4812) 6336377 ext.2011, fax: (4812) 6340515, email: mnajbar@chemia.uj.edu.pl

[b]Institute of Metallurgy and Material Chemistry, Polish Academy of Sciences, 30 059 Kraków, ul. Reymonta 9, Poland

[c]Regional Laboratory of Physicochemical Analyses and Structural Research, Jagiellonian University, 30 060 Kraków, Ingardena 3, Poland,

The alumina supported rhodium catalyst deposited on Cr-Al steel foil parings was synthesised. The adhesive alumina layer on the foil was formed by the surface aluminum segregation in air. The stable alumina sol was deposited on the segregated layer and then calcined. The support was impregnated with the rhodium trichloride solution. The physico-chemical characterisation of the catalyst was performed by the use of nitrogen adsorption, SEM, EDS, electron diffraction and XPS. The methane oxidation was investigated using the pulse method. The influence of the oxygen content in the catalyst on its activity and selectivity to hydrogen was determined. The participation of the oxygen coming from support in methane oxidation was shown. Synthesis gas formation is discussed as caused by methane activation on the metallic Rh surface containing oxygen.

1. INTRODUCTION

The partial oxidation of methane to synthesis gas remains one of the most frequently investigated catalytic reactions [1-6]. Lowering the temperature of syngas production is very important for economic and environmental reasons [1]. Selective CH_4 oxidation is often investigated over supported noble metals [2], among which rhodium seems the most promising. This catalyst is very active in methane partial oxidation and its selectivity to synthesis gas seems the highest [3-6]. Knowledge about the surface sites responsible for methane activation and CO and H_2 formation could be a crucial topic in understanding the reaction mechanism.

In the present paper attention is focused on the relation between the oxygen content in the alumina supported rhodium catalyst and its activity in methane oxidation and selectivity toward hydrogen. The investigations were performed on the catalyst deposited on Cr-Al steel foil parings-simulating monolith with a good heat transfer.

2. EXPERIMENTAL

2.1. Catalyst preparation

The Rh/Al_2O_3 catalyst was deposited on twisted parings made from Cr-Al steel foil (ca. 1x15mm) with 2% wt. Rh loading (with respect to the alumina weight) was used for the investigation. The parings were heated to 1110K in air at the rate of 3K/min and then calcinated at this temperature for 20 hours, in order to make the surface of the Cr-Al steel foil adhesive to the alumina support. Heating caused the formation of an Al_2O_3 layer on the surface of the foil due to oxidation-induced aluminum segregation. The stable alumina sol, formed from aluminum tri-sec-butoxide at 350 K [7] was placed on this layer and the foil was then heated similarly as during the segregated layer formation. The alumina layer was impregnated with 1.8×10^{-3} mol/dm^3 $RhCl_3$ solution. The obtained catalyst precursor was heated in air at 623K for 15 hours and subsequently reduced in hydrogen (50 ml/min) at 573K for 2 hours and at 773K for 1 hour. The catalyst loading was equal to ca 2%.

The morphology of the segregated and sol-derived layers was investigated by scanning electron microscopy, while the chemical composition, was examined using X-ray energy dispersive spectrometry. The oxidation state of the Rh was determined by X-ray photoelectron spectroscopy, while the phases present in the catalyst were studied with electron diffraction.

2.2. Scanning Electron Microscopy (SEM) and Energy Dispersive Spectrometry (EDS)

The morphology of the segregated and sol-derived alumina layers, as well as their chemical composition, were investigated by a Philips Scanning Electron Microscope XL with an EDS LINK-ISIS system. To determine the depth of both layers, they were scratched with a quartz knife and the foil was positioned in such a way as to make observation of the layer cross-sections possible. EDS measurements were done using 10 keV accelerating voltage which allowed for the limitation of analysed alumina layers depth to ca. 1.2 μm.

2.3. X-ray photoelectron spectroscopy (XPS)

XPS spectra were measured using an ESCA 150 spectrometer (VSW Scientific Instruments) with a magnesium anode as the source of X-rays, a hemispherical analyser 150 mm in diameter and a Leybold multichannel detector. Samples were prepared for measurement by being pressed into indium disks. The binding energy scale was established by referencing the C (1 s) value of the hydrocarbon portion of the adventitious carbon to 284.6 ± 0.1eV.

2.4. Selected Area Diffraction Patterns (SADP)

The selected area diffraction patterns (SADP) of the alumina support and of the catalyst were taken using a Philips Transmission Electron Microscope CM 20. The alumina support or the alumina supported rhodium catalyst were detached from the steel foil with the quartz knife and placed on carbon film laying on a copper grid. Both alumina layers could be detached by strong scratching, while mild scratching only allowed for separate detachment of the sol-derived layer.

2.5. Catalytic experiments

The methane oxidation was studied at 773K and 823K using the pulse technique. 0.2 g of the catalyst parings was placed on silica wool 2cm above the bottom of a tubular stainless steel reactor, which had a length of 31 cm and a 6mm internal diameter. The internal walls of the reactor were covered with alumina. The reduced catalyst was heated to 773K in He and subjected to interaction with four methane pulses. The lack of distinct interaction of the reduced catalyst with methane was observed. The reduced catalyst was next heated to 773K in He and subjected to interaction with one oxygen pulse (700μl) at this temperature. A few consecutive pulses of CH_4 of the same volume were then introduced into the He stream flowing over the catalyst. The same pulse cycle was completed at 823K. The products of the methane interaction with the oxidised catalyst were separated in the Porapak Q column of a Hewlett-Packard 5890 GC and analysed with a TC detector. H_2, H_2O, CO and CO_2 were found to be the main product of CH_4 oxidation. The value of the selectivity to CO was revealed to be lower than that to H_2 for the methane pulses interacted with catalyst surface enriched in oxygen. Both values were found to become equal to 100% for the catalyst with moderate oxygen content. In separate pulse experiment the products were analysed with quadruple MS [8]. Small amounts of C_3- and C_2- hydrocarbons, increasing with catalyst depletion of oxygen were found in the products. C- balance was revealed to be close to 100% if hydrocarbons were taken into account.

The amount of oxygen taken up from the O_2 pulse was found to be about 50 times greater than that needed for oxidation of the rhodium from its metallic to Rh_2O_3 form.

3. RESULTS

3.1. Physicochemical characterisation of the support and the catalyst

The BET surface area of the segregated alumina layers was found to be equal to about 16 m^2/g and that of the sol- derived alumina to $200 m^2/g$.

In Fig. 1 the secondary electron image (SEI) of the foil with the crack made by the quartz knife is shown. As seen the segregated and sol-derived layers fit very well and the similar thickness of both is equal to ca 2μm.

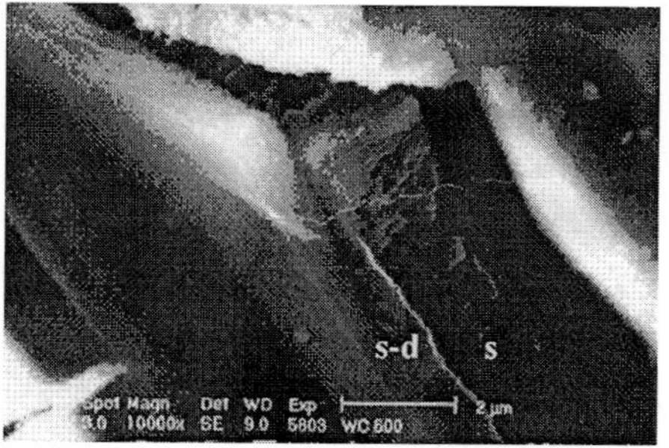

Fig.1 The SEI of the Cr-Al steel foil covered by segregated (s) and sol-derived (s-d) alumina layers.

In Fig. 3: typical SAD patterns of the crystallites of the segregated alumina layer and of polycrystalline alumina of the sol-derived layer are shown. It can be seen that high Fe and Cr concentrations in the alumina of the segregated layer (fig. 2a) allow for the formation of the α–phase at a temperature lower than that of the $\gamma \longrightarrow \alpha$ transition (ca 1300K) [9]. On the contrary, in the sol-derived layer the concentration of iron and chromium is not high enough to cause α alumina formation. β alumina, composed of the layers of γ Al_2O_3 with the spinel structure and the oxygen anion layers containing Cr and Fe cations [9], is thought to occur in intermediate areas between α and γ alumina. This problem will be discussed in details elsewhere [10]. In Fig. 4 typical selected area diffraction patterns of the reduced rhodium catalyst are presented. The presence of rhodium trioxide besides rhodium crystallites is clearly seen.

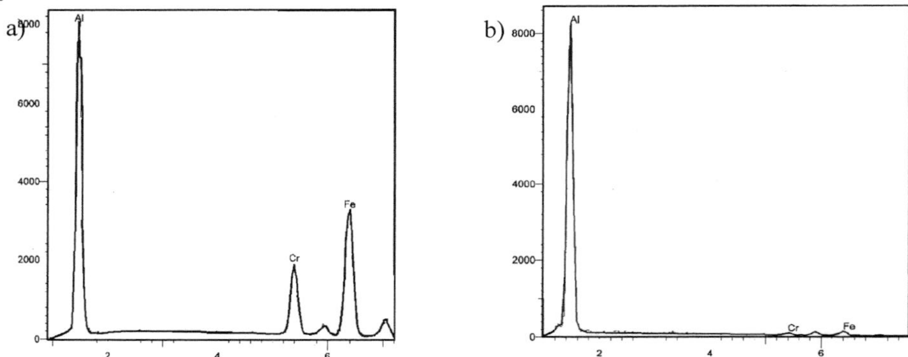

Fig.2. The energy dispersive X-ray spectrum of the foil with: a) the segregated alumina layer and b) the segregated and the sol-derived alumina layers.

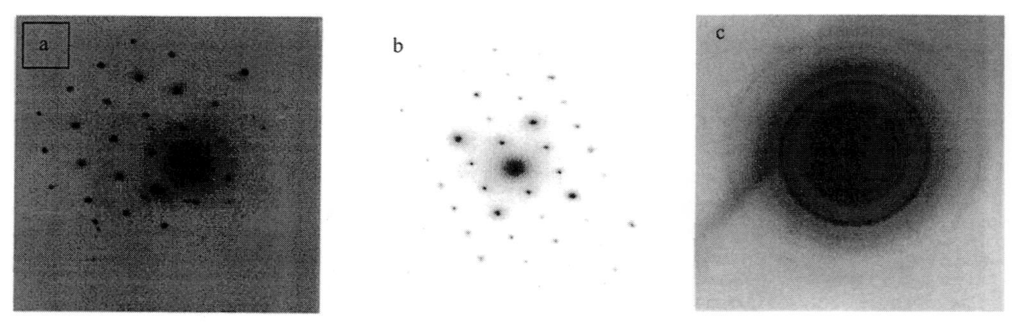

Fig. 3 Typical SAD patterns for crystallites of the segregated alumina layer: a) α Al_2O_3, [44.1] zone, b) β Al_2O_3 [01.0] zone, c) typical diffraction pattern of the polycrystalline γ Al_2O_3 of the sol-derived layer.

Fig. 4 The typical selected area diffraction patterns of the: a) Rh ([011] zone) and b) Rh_2O_3 ([032] zone) in the reduced Rh/Al2O3/Cr-Al foil catalyst.

The experimental (line) Rh 3d $_{3/2}$ and $3d_{5/2}$ peaks (with background subtracted) and simulated spectra (dimmed) of the reduced catalyst are shown in Fig.5.

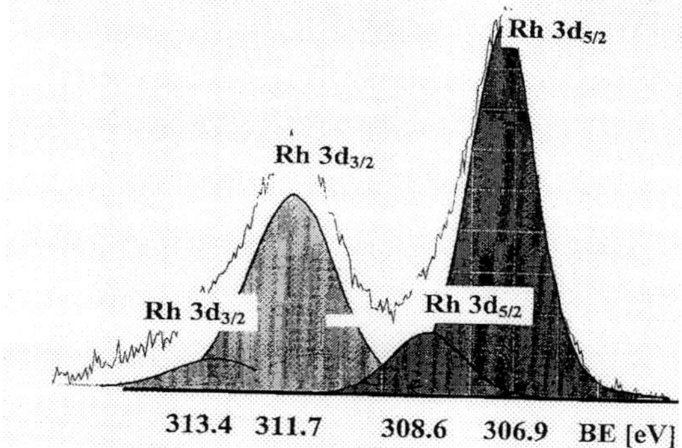

The peaks at 308.6 and 313.4 eV clearly show the presence of the oxide form of rhodium in addition to the metallic one at the catalyst surface. One can think that rhodium trioxide forms intermediate layers between metallic rhodium and alumina support.

Fig.5 The experimental (line) Rh 3d $_{3/2}$ and $3d_{5/2}$ peaks (with background subtracted) and simulated spectra (dimmed) of the reduced catalyst.

In Figures 6 and 7 methane conversion and selectivity to hydrogen in oxidation of the consecutive methane pulses at 773K and 823K, respectively, are shown.

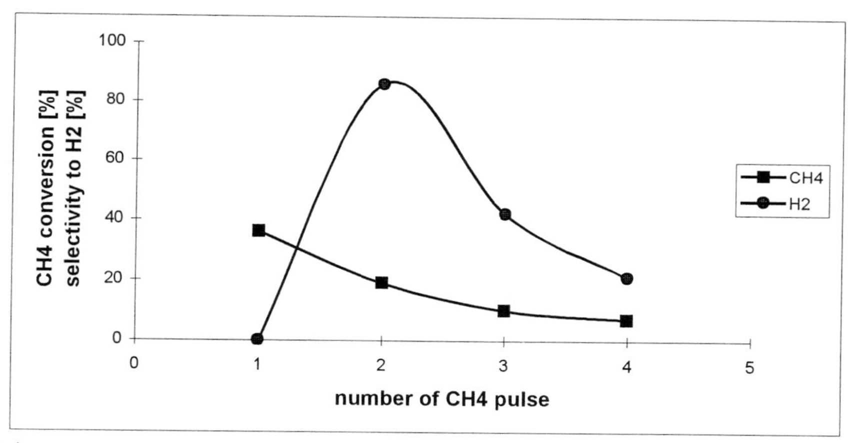

Fig. 6 Methane conversion at 773K and selectivity to hydrogen in consecutive CH_4 pulses.

50 times higher amount of O_2 molecules in oxygen pulse than that needed for total rhodium oxidation indicates that oxygen is uptaken mostly by alumina support. On the other hand, 35 higher amount of oxygen used for conversion of the first methane pulse at 773K than that needed for rhodium oxidation shows that oxygen is supplying from the support to the catalyst surface during catalyst interaction with methane pulse. As methane activation on reduced catalyst does not occur in a distinct way, it can be concluded that oxygen makes possible or accelerates methane activation. It is obvious that the surface oxygen concentration is determined by the temperature and the oxygen content in the support. In the course of the third methane pulse interaction with catalyst at 823K the oxygen concentration is optimal for synthesis gas production. The higher oxygen concentration causes deeper oxidation and lower one leads to the increase of the hydrocarbon content. Methane activation leading to the synthesis gas formation could be ascribed to the active sites composed of the surface oxygen

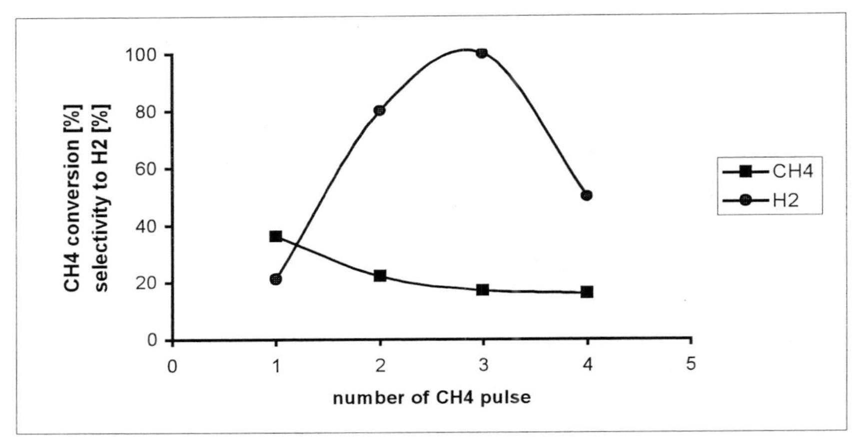

Fig. 7 Methane conversion at 823K and selectivity to hydrogen in consecutive CH_4 pulses.

(probably in interstitial positions) surrounded by rhodium atoms. Further oxidation of CO by the excessive oxygen may be responsible for the higher selectivity to hydrogen than to carbon monoxide observed for the enriched in oxygen catalyst surfaces. T

Decrease of the surface oxygen concentration with temperature explains higher values of the selectivity to hydrogen at 823K than at 773K.

4. CONCLUSIONS.

The interaction of the consecutive methane pulses with the reduced and oxidised surfaces of the alumina supported rhodium catalyst containing the grains composed of Rh and Rh_2O_3 was investigated. It was found that O_2 used for catalyst oxidation was uptaken mainly (> 94%) by the support and it was supplied to the surface in the course of catalyst interaction with methane pulses. The lack of the distinct interaction of the reduced catalyst with methane at 773K was taken as the evidence that the presence of oxygen on the Rh surface makes possible and/or accelerates methane activation. The CH_4 activation on the active sites composed of the interstitial oxygen surrounded by Rh atoms is considered to be responsible for synthesis gas production.

REFERENCES.

1. T. Sundset, J. Sogge, T. Strom, Catal.Today, 21 (1994) 269.
2. P.M. Torniainen, X.Chu, L.D.Schmidt, J. Catal., 146 (1994) 1.
3. O.V. Buyevskaya, K.Walter, D.Wolf, M. Bearns, Catal. Lett., 38 (1996) 81.
4. E.P.J.Mallens, J.H.B. Hoebink, B.Marin, J.Catal. 167 (1997) 43.
5. V.N. Parmon, G.G. Kuvshinov, V.A. Sadykov, V.A. Sobyanin, St. Surf. Sci. & Catal., 119 (1998), 677.
6. C. Elmasides, T. Ioannides, X. E. Verykios, St. Surf. Sci. & Catal., 119 (1998), 801.
7. J. Brinker, G. Scherer, Sol–Gel Science, Academic Press Inc., San Diego, 1990, p. 68.
8. M. Najbar and M. Zimowska, to be published
9. A. F. Wells , Structural Inorganic Chemistry , Oxford University Press, 1990, p. 530
10. M. Najbar, B. Pietruszka, L. Lityńska, E. Bielańska, M. Zimowska, to be published.

Studies in Surface Science and Catalysis
J.J. Spivey, E. Iglesia and T.H. Fleisch (Editors)

Catalytic Partial Oxidation of Ethane to Acetic Acid over $Mo_1V_{0.25}Nb_{0.12}Pd_{0.0005}O_x$ – Catalyst Performance, Reaction Mechanism, Kinetics and Reactor Operation

David Linke[a], Dorit Wolf[a], Manfred Baerns[a]*, Uwe Dingerdissen[b], Sabine Zeyß[b]

[a] Institute for Applied Chemistry Berlin-Adlershof,
Richard-Willstätter-Str. 12, D-12489 Berlin, Germany

[b] Aventis Research and Technologies GmbH & Co KG, D-65926 Frankfurt a.M. , Germany

The oxidation of ethane to acetic acid was studied over the title catalyst at temperatures from 500 K to 580 K and elevated pressures between 1.3 and 1.6 MPa. It was found that a change in reaction mechanism occurs with increasing temperature. While at low temperatures a consecutive reaction scheme with ethylene as intermediate leading to acetic acid dominates, at high temperatures ethylene and acetic acid are mainly formed in parallel. Adding water to the feed increased the selectivity to acetic acid at the expense of ethylene selectivity. This was ascribed to a strong acceleration of ethylene oxidation to acetic acid by the presence of water. The kinetics of the oxidation of ethane to acetic acid were modeled based on experimental data from the integrally operated fixed bed reactor. A kinetic model taking into account surface processes such as catalyst oxidation and reduction and surface hydroxylation was suggested. The model assumes two different catalytic centers, one for the activation of ethane and a second one for the heterogeneous Wacker oxidation of ethylene to acetic acid of which the activity depends on the presence of water. The operation of a fixed-bed and fluidized-bed reactor for the partial oxidation of ethane to acetic acid was modeled and simulated based on the kinetic model.

INTRODUCTION

Acetic acid is one of the most important chemical products with a global total capacity of 8.4 million and an expected supply of 6.4 million tons/year in 2000. Since the discovery of a new iodide-promoted rhodium catalyst with remarkable activity and selectivity for methanol carbonylation by Monsanto researchers in 1968 this process has become the dominating technology for the production of acetic acid (ca. 60 % of total capacity) [1]. Although there is presently no economic alternative to the Monsanto process, a highly selective direct oxidation of ethane to acetic acid might become attractive since ethane is available at low costs as the second major component of natural gas (3 – 20 mole-%) [2]. Moreover, a technology such as the direct partial oxidation of ethane avoids the problem of corrosion found in methanol carbonylation and thus might be attractive.

Former studies on ethane oxidation to acetic acid [3-10] were mainly focussed on catalyst development and to a minor degree on the reaction mechanism, kinetics or reaction

* corresponding author; email: baerns@aca-berlin.de, postal address see above

engineering. The pioneering work goes back to Thorsteinson et al. [3] who studied Mo-V-O catalysts and found a catalyst of the composition $Mo_1V_{0,25}Nb_{0,12}O_x$ to give the highest yield of acetic acid. Kinetic experiments suggested a consecutive reaction scheme for the formation of acetic acid with ethylene as the intermediate product. Surprisingly Ruth et al. [7] reported a parallel formation of acetic acid and ethylene instead of the consecutive formation for a catalyst of the same composition. A breakthrough in catalyst development was achieved by Borchert et al. [11-14] who showed that Pd doping leads to highly selective catalysts for the production of acetic acid. For a catalyst of the composition $Mo_1V_{0.25}Nb_{0.12}Pd_{0.0005}O_x$, which is, apart from Pd, close to the Mo-V-Nb-O catalyst studied by Thorsteinson et al. [3], highly improved selectivities to acetic acid were observed: at 553 K and a total pressure of 1.5 MPa the selectivity to acetic acid was 78 % at 10 % ethane conversion with Pd but only 32 % at 9 % ethane conversion without Pd [11].

In the present study, we have investigated the oxidation of ethane to acetic acid over the $Mo_1V_{0.25}Nb_{0.12}Pd_{0.0005}O_x$ catalyst in a steady-state fixed-bed reactor at elevated pressures of 1.2 to 1.6 MPa. The aim of this study was (i) to derive a kinetic model with a mechanistic basis and (ii) to perform reactor simulations in order to compare the suitability of a fixed bed and a fluidised bed reactor for this reaction. The role of water was particularly taken into consideration since a beneficial influence was reported for the formation of acetic acid [3,8].

EXPERIMENTAL AND MODELLING PROCEDURE

Kinetic Experiments [15]. A stainless steel reactor (ID 12 mm) which was heated by a stirred bath of molten salt was used in the catalytic experiments. The feed gas flows (ethane, oxygen, nitrogen) as well as the flow of water were adjusted by mass flow controllers. The reaction pressure was controlled by a mechanical back pressure controller. After expansion of the gas stream its composition was analysed by on-line GC, which allowed complete separation of all components (H_2O, CO, CO_2, C_2H_4, HOet, Hac, HOac, acOet ; et = C_2H_5, ac = CH_3CO). Measurement of kinetic data was performed in the temperature range from 503 K to 576 K and a total pressure range between 1.3 MPa and 1.6 MPa. The catalyst mass was varied between 1 g and 13.7 g. The catalyst was diluted with particles of quartz of the same size (m_{cat}:m_{quartz} = 1:2) to achieve nearly isothermal operation. Blank experiments showed that no conversion of ethane, ethylene or acetic acid occurs in the reactor at the conditions applied. The catalyst preparation as well as the procedure of evaluating the kinetic data and the kinetic modelling is described elsewhere [15].

Reactor Simulation [15]. The simulation of the fluidized bed reactor and the ideal plug flow reactor was performed applying a Fortran code by Mleczko et al. [16]. For the fluidised bed simulation the bubble assemblage model (BAM) of Kato and Wen [17] is used. The model assumes an isothermal reactor and describes the fluidized bed as a two-phase system consisting of a bubble phase and an emulsion phase. Since the emulsion phase contains a high proportion of solid, the catalytic reaction takes place mainly there whereas in the bubble phase small rates (per volume) are found due to the small concentration of the catalytic material. The model of the fluidized bed reactor describes the mass transfer between the bubbles and the emulsion phase and takes also into account the growth of bubbles with increasing bed height.

RESULTS AND DISCUSSION

The oxidation of ethane over $Mo_1V_{0.25}Nb_{0.12}Pd_{0.0005}O_x$ was investigated at two different temperatures, different water partial pressures and varying space times to analyse the influence of operating conditions on conversion and selectivity and to elucidate the reaction scheme (see Fig. 1). The most important by-product besides the main products acetic acid, ethylene and carbon dioxide was acetaldehyde (selectivity less than 1 %). Further trace products formed were ethyl acetate, acetone, methanol, methane, propylene and ethanol. Carbon monoxide was not detected in the temperature range studied. Due to the low concentration of the trace products they are neglected in the further presentation.

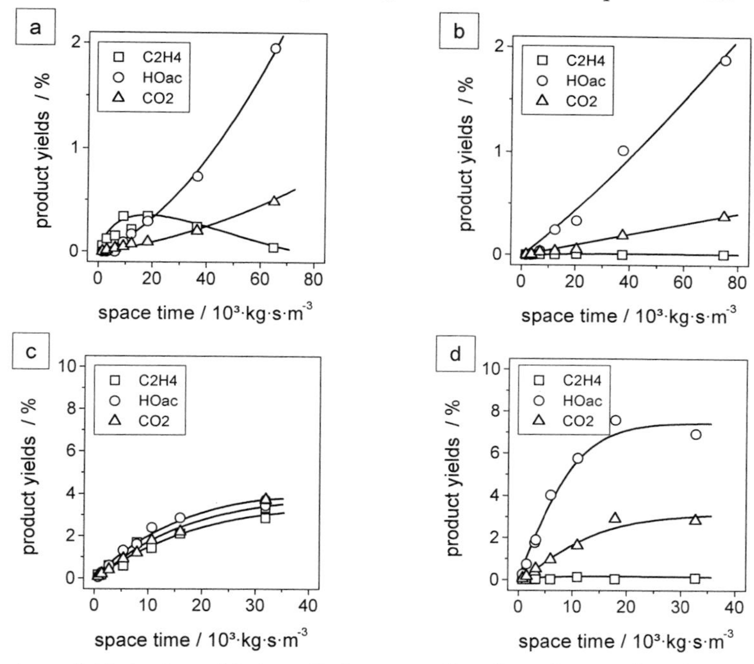

Fig. 1: Product yields in the oxidation of ethane as a function of space time for two different temperatures (top: 503 K, bottom: 576 K) and two different water partial pressures (left - without water added; right - with water added to the feed); P_{tot} = 1.3 MPa / 1.6 MPa for $p_{H2O,in}$ = 0 / 320 kPa, $p_{C2H6,in}$ = 640 / 650 kPa; $p_{O2,in}$ = 128 / 130 kPa

Without water added to the feed at T = 503K ethylene yield shows a maximum with increasing space time (Fig. 1 a). At small contact times ($< 20 \cdot 10^3$ kg s m^{-3}) ethylene is formed as the main product of the ethane oxidation. At higher contact times acetic acid is formed with the highest yield. Thus, ethylene is an intermediate product in the oxidation of ethane to acetic acid. With water added to the feed (Fig. 1 b) lower yields of ethylene were found. Acetic acid is the main product even at small contact times. This indicates that the conversion of ethylene to acetic acid is accelerated by water. This was confirmed in separate experiments with ethylene in the feed [15]. Surprisingly at T = 576 K the formation of ethylene as an intermediate in the oxidation of ethane to acetic acid could not be found (Fig. 1 c). Without water added to the feed both ethylene and acetic acid are formed in parallel. Thus, with increasing temperature an apparent change in the reaction path occurred. While at low temperature the formation of acetic acid is a consecutive reaction with ethylene as intermediate, both products are mainly formed in parallel reactions at high temperature. This change implicates different energies of activation for the direct formation of acetic acid from ethane and the formation via ethylene. The addition of water to the feed at T = 576 K resulted, as before, in a reduction of the ethylene yield and an increase of the acetic acid yield (Fig. 1 d). The selectivity to carbon dioxide was decreased slightly by the addition of water. Further experiments carried out at different water inlet partial pressures showed that an optimum for water partial pressure exist where the rate of ethane conversion is highest.

Based on the catalytic results, TAP experiments and catalyst characterisation [15] a speculative reaction scheme for the oxidation of ethane was developed (Figure 2) which served as a basis for kinetic modelling. This scheme includes a parallel reaction pathway of ethane to ethylene and acetic acid as well as a consecutive reaction pathway to acetic acid with ethylene as intermediate. Carbon dioxide can be formed by unselective oxidation of ethane, ethylene and acetic acid. In the reaction scheme and in the kinetic model two different active sites Z and X are considered. Z represents a redox site where oxidation steps such as oxidative dehydrogenation of ethane, partial and total oxidation of ethylene and acetic acid take place. X is a site of the catalyst which becomes only active in the presence of water. The activated form of X (X-OHOH), which is formed by adsorption of water on oxidised X-sites, leads to the conversion of ethylene to acetic acid via a mechanism similar to the heterogeneous Wacker-oxidation [18-20]. Thus, formation of acetic acid occurs via two different pathways – the partial oxidation of ethane on site Z and a Wacker-like formation on site X. A maximum in the rate of ethane conversion as a function of water partial pressure found in further experiments [15] is expressed in the scheme in Fig. 2 in the following way: Ethylene adsorbs on the same active surface sites Z required for oxidative ethane dehydrogenation and, hence, inhibits the ethane activation. Water, in turn accelerates the conversion of ethylene to acetic acid on the site X. This conversion of ethylene on site X affects the ethane conversion indirectly. The accelerated ethylene conversion leads, in turn, to a lower degree of occupation of sites Z by ethylene. Then, Z is available for oxygen adsorption and ethane activation to a higher extent. Therefore, the rate of ethane oxidation increases slightly with increasing water concentration in the feed. At high water concentration water adsorption on Z sites has to be

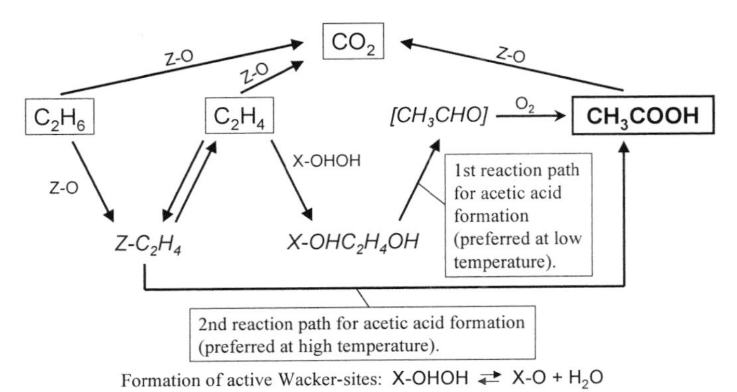

Fig. 2: Reaction scheme and mechanistic basis for the kinetic model involving two different catalytic sites (Z, X)

considered to account for the drop in the rate of ethane conversion. Details of the experimental elucidation of the reaction mechanism, rate equations for the kinetic model, kinetic parameters and the method of analytical derivation of steady-state concentration of surface intermediates as well as the range of model validity are reported in [15].

The simulation of a polytropic fixed-bed reactor showed that heat removal is a critical issue in the oxidation of ethane to acetic acid. Even under conditions which provide high rates of heat transfer a save operation of a fixed bed reactor is only possible if the active catalyst mass per volume is reduced [15]. However, safer reactor-operation modes lead to a decrease of the space time yield for acetic acid. Thus, an alternative might be the application of a fluidised bed reactor due to the opportunities of efficient heat removal.

Simulations of a fluidised bed reactor were performed assuming a reactor diameter of 4 m, a particle size of 60 μm. The simulation results are compared with an ideal plug flow reactor in Fig. 3. In the fluidised bed reactor lower ethylene selectivities are predicted than in the fixed bed reactor. However, the selectivities to acetic acid is higher in the fluidised bed. This is due to the back mixing of water formed in the oxidation reactions, which appears in the fluidised bed only. Since water accelerates the conversion of ethylene to acetic acid, higher acetic acid yields but lower ethylene selectivities are expected for the fluidised bed. However, if the sum of the ethylene and acetic acid selectivities are compared for both reactor types, it can be seen that the fixed bed gives higher selectivities. This is also caused by the back mixing in the fluidised bed which results in a wider residence time distribution and, thus, favours the formation of carbon dioxide being the final product in the reaction sequence. From Fig. 3 it can be seen that the space time yield (moles of acetic acid formed per time and per catalyst mass) is significantly higher in the fixed bed reactor. This is a consequence of the by-passing caused by the bubbles in the fluidised bed.

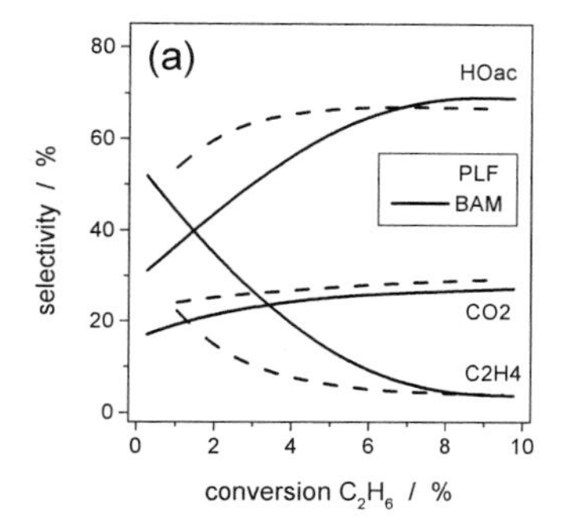

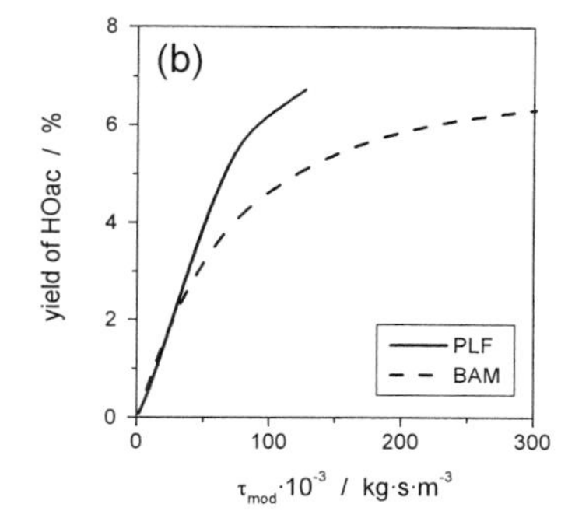

Fig. 3: Comparison of product selectivity depending on ethane conversion (a) and yield of acetic acid depending on modified residence time (b) predicted for the fluidised bed (BAM) and the ideal plug flow reactor (PLF); oxygen conversion reaches up to 99 %; P_{tot} = 1.28 MPa, $\dot{V}_{STP}$ = 4.01 m^3 s^{-1}, $p_{C2H6,0}$ = 640 kPa, $p_{O2,0}$ = 128 kPa, $p_{H2O,0}$ = 0 kPa, T_0 = 533 K, m_{cat} = 1-100 t, u/u_{mf} = 8

CONCLUSIONS

In the oxidation of ethane to acetic acid on $Mo_1V_{0.25}Nb_{0.12}Pd_{0.0005}O_x$ a change in the reaction path way occurs with temperature: While at low temperature the formation of acetic acid is a consecutive reaction with ethylene as intermediate at high temperature ethylene and acetic acid are mainly formed in parallel reactions. Thus, both different reaction pathways discussed in literature [3,7] are consistent with the present results. It was found that adding water to the feed in ethane oxidation to acetic acid increased the acetic acid selectivity at the expense of ethylene selectivity. But increasing selectivity to acetic acid and thereby the acetic acid space time yield by increasing the water concentration is disadvantageous in a chemical process of ethane oxidation to acetic acid since the energy consumption in the separation stages increase strongly with water concentration. Thus, simulation of the whole process must be performed to find out the optimum water concentration.

The operation of a fluidized bed reactor for the partial oxidation of ethane to acetic acid was modeled based on the kinetic model which describes the formation of C_2H_4, CO_2 and acetic acid in the presence as well as in the absence of water and compared with an ideal plug flow reactor. For the fluidized bed reactor higher selectivities to acetic acid are predicted due to back mixing of water. However, the space time yield of acetic acid is significantly lower than in the fixed bed reactor.

REFERENCES

1. Ullmann´s Encyclopedia of Industrial Chemistry, 5th Ed. VCH: Weinheim, 1991.
2. Bañares, M. A., Catal. Today 51 (1999) 319.
3. Thorsteinson, E.M., Wilson, T.P., Young, F.G., Kasai, P.H., J. Catal., 52 (1978) 116.
4. Merzouki, M., Taouk, B., Monceaux, L., Bordes, E., Courtine, P., Stud. Surf. Sci. Catal. 72 (1992) 165.
5. Merzouki, M., Taouk, B., Bordes, E., Courtine, P., Stud. Surf. Sci. Catal., 75 (1993) 753.
6. Burch, R., Kieffer, R., Ruth, K., Topics in Catalysis, 3 (1996) 355.
7. Ruth, K., Burch, R., Kieffer, R., J. Catal., 175 (1998) 27.
8. Tessier, L., Bordes, E., Gubelmann-Bonneau, M., Catal. Today, 24 (1995) 335.
9. Roy, M.; Ponceblanc, H.; Volta, J.C., Topics Catal., 11/12 (2000) 101.
10. Ueda, W.; Oshihara, K., Applied Catal. A, 200 (2000) 135.
11. Borchert, H., Dingerdissen, U. (Hoechst), Ger. Offen. DE 19 630 832 (1998).
12. Borchert, H., Dingerdissen, U. (Hoechst) WO 9805619 (1996), Ger. Offen. DE 197 17 076 (1998), Ger. Offen. DE 197 17 075 (1998), Ger. Offen. DE 197 45 902 (1999).
13. Borchert, H., Dingerdissen, U., Weiguny, J. (Hoechst), WO 9744299 (1996), Ger. Offen. DE 19 620 542 (1997).
14. Borchert, H., Dingerdissen, U., Roesky, R. (Hoechst) WO 9847851 (1998) , WO 9847850 (1998), Ger. Offen. DE 197 17 076 (1998).
15. (a) Linke, D., Wolf, D., Baerns, M., Timpe, O., Schlögl, R., Dingerdissen, U., Zeyß, S., submitted to J. Catal.
    (b) Linke, D., Wolf, D., Baerns, M., Dingerdissen, U., Zeyß, S., Mleczko, L., submitted to Chem. Eng. Sci.
16. Mleczko, L., Ostrowski, T., Wurzel, T., Chem. Eng. Sci. 51 (1996) 3187.
17. Kato, K., Wen, C. H., Chem. Eng. Sci. 24 (1969) 1351.
18. Evnin, A.B., Rabo, J.A., Kasai, P.H., J. Catal., 30 (1973) 109.
19. Seoane, J.L., Boutry, P., Montarnal, R., J. Catal., 63 (1980) 191.
20. Espeel, P.H., De Peuter, G., Tielen, M.C., Jacobs, P.A., J. Phys. Chem. 98 (1994) 11588.

Studies in Surface Science and Catalysis
J.J. Spivey, E. Iglesia and T.H. Fleisch (Editors)

A NON STATIONARY PROCESS FOR H_2 PRODUCTION FROM NATURAL GAS

C. Marquez-Alvarez, E. Odier, L. Pinaeva, Y. Schuurman, C. Millet*, C. Mirodatos.

Institut de Recherches sur la Catalyse, 2 av A. Einstein, 69626 Villeurbanne Cedex, France.
mirodato@catalyse.univ-lyon1.fr
**Air Liquide, Centre de Recherche Claude Delorme, BP 126, Les Loges en Josas, 78353 Jouy en Josas Cedex, France*

INTRODUCTION

The short term industrial application of PEM fuel cells requires new processes for producing hydrogen free of carbon monoxide, either for mobile or stationary application. The conventional reforming of any kind of hydrocarbons (steam reforming, partial oxidation, autothermal reforming) produces CO as a major co-product [1]. The latter has to be converted downstream into CO_2 through complex and costly steps like water gas shift (WGS), selective oxidation of CO in hydrogen rich mixture (SELOX) or pressure swing absorption (PSA) separation before reaching the low CO concentration that can be tolerated by the fuel cell (less than 10 ppm) [2]. An alternative to produce directly CO free hydrogen is the non-stationary catalytic decomposition of methane via a cyclic two-step process as suggested in [3-5] :

$$CH_4 \rightarrow C_{deposited\ on\ catalyst} + 2H_2 \qquad \text{step I}$$

$$C^* + O_2 \rightarrow CO_2 \qquad \text{step II}$$

A high capacity of C storage appears as a prerequisite for such a process which may be ensured by various ways such as : i) C storage by the cracking active phase, such as nickel which easily forms bulk carbide under reforming conditions [6], ii) C storage assisted by the support, which involves a tight interaction and spillover between the metallic phase and the support. For the latter, a low temperature process is expected to be more efficient since the rate of diffusion for the adspecies has to be comparable with the rate of carbon formation during the cracking step. In order to investigate these possibilities, various catalysts were screened under non stationary conditions and a selected formula Pt/CeO_2 was studied more in detail by means of coupled kinetic measurements and *in situ* diffuse reflectance infrared spectroscopy (DRIFT) studies.

EXPERIMENTAL

Materials. The catalysts used in this study were 13.0 wt.% Ni on Al_2O_3, 0.9 wt.% Pt on SiO_2 and 1.1 wt.% Pt on CeO_2 prepared by impregnation with aqueous solutions of $[Ni(NH_3)_6]NO_3$ and $Pt(OH)_2(NH_3)_4$, respectively. They were calcined at 600°C, pelletized, crushed and sieved to 0.2-0.3 mm before use. The catalysts were reduced in situ under hydrogen flow for 2 h at 400 and 600°C for the Pt and Ni based materials, respectively, and then kept under inert (Ar or He) flow before reaction.

Testing procedure. The reaction was carried out under atmospheric pressure either within a quartz tubular micro-reactor (4 mm I.D.) for screening experiments or within the experimental set-up described in Figure 1. A typical experiment consisted in flowing alternatively three gas blends through the catalyst, by turning the appropriate valve (Figure 2). All mixtures had a total flow rate of 50 ml/min (STP): 20% CH_4 in inert gas, pure inert gas (as a flush) and finally 10% O_2 in inert gas. The inert gas was He for screening tests and Ar in the DRIFT experiments. The concentration of H_2, CH_4, CO, CO_2, O_2, H_2O and inert gas in the outlet effluents was continuously monitored by on line mass spectrometry. To characterise the surface of the working catalyst during the forced unsteady-state operation, the catalyst was tested in a high-temperature DRIFT cell. Infrared spectra were recorded with a Nicolet Magna-IR 550 spectrometer, using a MCT detector with a resolution of 4 cm^{-1}.

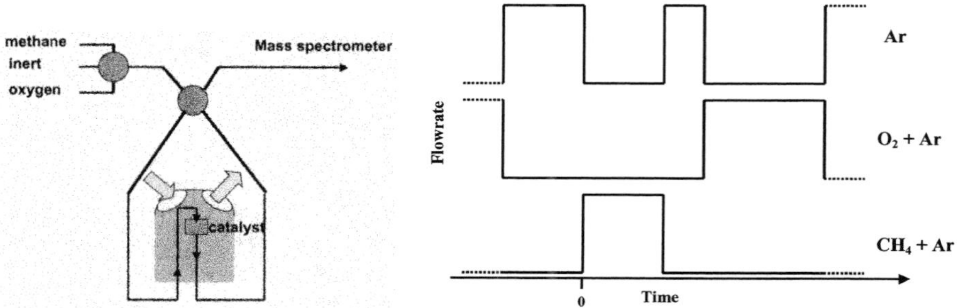

Figure 1: Set-up used for combining forced unsteady-state kinetics and DRIFT spectroscopic analysis.

Figure 2: Sequence of inlet gases sent to the reactor during the forced unsteady-state reaction. The base line for every curve is at zero level.

RESULTS AND DISCUSSION

Catalysts screening

The catalytic behaviour of the three catalysts tested in a tubular microreactor under forced unsteady-state conditions is illustrated in Figures 3-5 and summarized in Table 1.

Table 1 : CH_4 conversion, H_2 and CO selectivity and H_2 yield calculated over the whole cycle

catalyst	T (°C)	X_{CH4} (%)	S_{H2} (%)	Y_{H2} (%)	S_{CO}(%)
Ni/Al_2O_3	700	63	90	57	85
Pt/SiO_2	400	6	78	5	59
Pt/CeO_2	400	44	64	28	<0.1

S_{H2} = formed H_2 / 2(converted CH_4), $Y_{H2} = X_{CH4} \cdot S_{H2}$

For the nickel-based catalyst tested at 700°C (Fig. 3), the formation of bulk nickel carbide and the growth of carbon filaments during the methane cracking step, as demonstrated in reference [6], ensured a high conversion but also lead eventually to reactor plugging, which prevented the regeneration step by O_2. Moreover, CO was produced in non negligible amounts both during the reducing and oxidizing cycles (Fig. 3). This system also revealed the following drawbacks: a high temperature is required to reach complete methane cracking, the storage capacity is limited, hot spots may easily develop within the fixed bed during the regeneration process, and filament formation induces nickel particle fragmentation, leading to weakening and deactivation of the active phase.

Over Pt/SiO_2 tested at 400°C (Fig. 4) the metal surface was quickly saturated by adsorbed species during the methane pulse, rendering the catalyst quite unreactive.

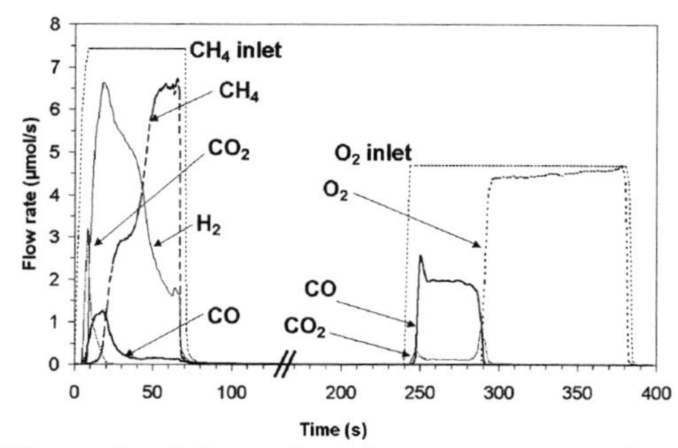

Figure 3 : Inlet and outlet gas concentrations obtained over Ni/Al_2O_3 tested at 700°C under forced unsteady-state conditions.

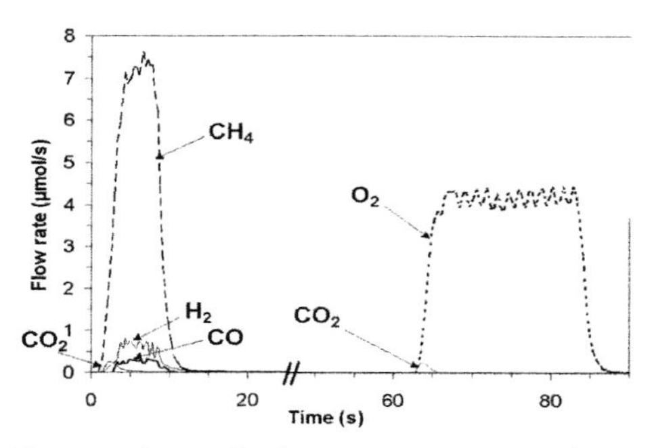

Figure 4 : Outlet gas concentrations obtained over Pt/SiO_2 tested at 400°C under forced unsteady-state conditions.

Over Pt/CeO_2 catalyst tested at 400°C under forced unsteady-state conditions (Fig. 5), a reasonable methane conversion was obtained during the methane pulse, leading to an average hydrogen yield of 28%. No CO was detected during the step of methane cracking into H_2 or during the oxidative regeneration step. Only a weak peak of CO_2 (yield = 1.5%) and a permanent trace release of water were detected during the methane pulse. This catalyst was therefore selected for further investigation and the adsorption/desorption/migration processes of carbonaceous species were followed during the forced unsteady-state step reactions by in situ IR spectroscopy.

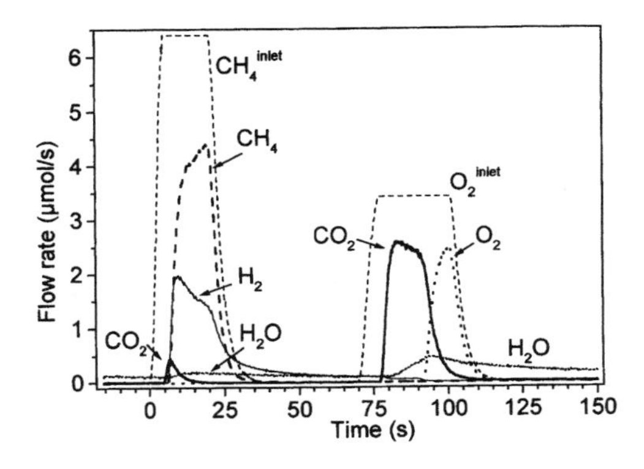

Figure 5 : Outlet gas concentrations obtained over Pt/CeO_2 tested at 400°C under forced unsteady-state conditions

In situ DRIFTS study of Pt/CeO_2 under forced unsteady-state conditions.

Three selected IR spectra corresponding to each of the sequential steps described in Figure 5 are reported in Figure 6 and the main bands are assigned in Table 2 [9-11].

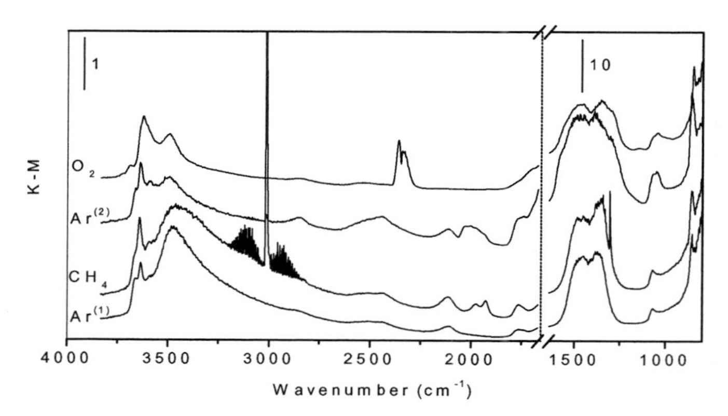

Figure 6: Selected DRIFT spectra obtained at 400°C on Pt/CeO_2 (pre-reduced under H_2 flow at 400°C) during the following reaction cycle : $Ar^{(1)} \rightarrow CH_4 \rightarrow Ar^{(2)} \rightarrow O_2$

Table 2 : Main IR bands observed during the sequential reaction steps over Pt/CeO_2 at 400°C.

3690 cm^{-1}	type I hydroxyl groups	Ce-OH
3665 and 3640 cm^{-1}	type II hydroxyl groups	H O Ce Ce
c.a. 3480 cm^{-1}	H-bonded hydroxyl groups	
3200-2800, 1360-1250 cm^{-1}	gaseous CH_4	
2390-2280 cm^{-1}	gaseous CO_2	
2115 cm^{-1}	Forbidden electronic transition on Ce^{3+} $^2F_{5/2} \rightarrow {}^2F_{7/2}$	
2028-1928 cm^{-1}	carbonyls on Pt (detailed in Table 3)	Pt-CO
1070, 1380, 1460 cm^{-1}	unidentate carbonates on ceria	Ce O C O O
1300, 1560, 2850 cm^{-1}	formate on ceria	Ce O O C H

The main bands in the IR spectrum of the pre-reduced sample correspond to bridging (3665 and 3640 cm^{-1}) and H-bonded hydroxyl groups (broad band at c.a. 3480 cm^{-1}), as well as carbonate species.

When CH_4 is introduced in the cell, the higher intensity of the 3640 cm^{-1} band and broadening of the 3490 cm^{-1} indicated that some part of the H atoms (arising from the activation of methane) react with surface oxygen to give OH and H_2O. Further reaction of CH_4 with CeO_2 produces the decay in intensity for all the OH bands and the sharpening of the 3490 cm^{-1} band. This reveals a progressive dehydration and dehydroxylation of the oxide when exposed to the CH_4 flow for several minutes. The growth of the 2115 cm^{-1} band under reducing atmosphere supports the assignment of this band to Ce^{3+} sites. New bands due to platinum carbonyl species appear under the CH_4 flow in the 2028-1928 cm^{-1} range. Their transient change in intensity is analysed later (Fig. 8 and Table 3). In the 1700-900 cm^{-1} region, the bands assigned to unidentate carbonates progressively increase in intensity and broaden during the CH_4 flow period, suggesting an increase in their concentration and heterogeneity. New bands develop at 1560 and 1300 cm^{-1} (the later one is overlapped in the presence of CH_4 by the strong rotation-vibration band centered at 1303 cm^{-1} and ranging from 1360 to 1250 cm^{-1}, but it is revealed when CH_4 is removed). The 1560 and 1300 cm^{-1} bands could be assigned to both formate and bidentate carbonate on CeO_2. However, the presence of a band at 2850 cm^{-1} supports the assignment to formate species [10].

When CH_4 is replaced by Ar, the strong bands associated to CH_4 in the gas phase are removed but the carbonate and formate bands remain unchanged, as well as the hydroxyl bands. Only bands associated to carbonyl species change with time on Ar stream (see Fig. 8).

When switching to O_2 flow, the bands at 2445, 1770, 1740, 2115 and the carbonyl bands at 2028, 1975 and 1928 are quickly removed. The bands assigned to carbonates quickly decrease, and they stabilise at an intensity lower than that of the original catalyst. The formate bands at 1560, 1300 and 2850 cm^{-1} start to decrease at longer time and ultimately disappear. The transient desorption of CO_2 is also evidenced as its characteristic rotation-vibration band centred at 2350 cm^{-1} evolves. In the presence of O_2, the band at 3665 cm^{-1} disappears and two new bands at 3690 and 3620 cm^{-1} grow within a few seconds after exposure. This can be related to a re-oxidation of the ceria surface and the generation of type I OH groups as a result of the formate decomposition.

Carbon route under forced unsteady-state conditions

Based on the IR characterisation of the working catalyst and the MS analysis of the reaction products, the carbon route can be summarised as follows : gaseous methane is activated on Pt, releasing gaseous H_2 and adsorbed carbon species (CH_x) which are later stored on ceria as formates and carbonates species. Under O_2 this carbon is oxidised and released as CO_2. Obviously, the key step which differentiates this catalyst from the other two catalysts tested is related to the carbon transfer from the platinum phase to ceria.

According to Li et al. [10], the formation of carbonates is due to CO adsorption on fully oxidised ceria while formates species are formed on partly reduced ceria. Since no CO is detected in the gas phase all over the cycle, the C transfer most likely proceeds via the spillover of adsorbed CO. The presence of CO adsorbed on Pt under methane flow is revealed by the three carbonyl bands at 1928, 1975 and 2028 cm^{-1}, which lie, at least for the two highest wavenumbers in the range of bridged to linear platinum carbonyls [10]. The normalised change in intensity for these bands is reported in Figure 8, with the band of free CH_4 at 3015 cm^{-1} as reference. Figure 8 clearly indicates that these species are formed sequentially. When the regenerated catalyst (after oxygen and argon treatment) is contacted with methane, the carbon issued from methane dissociation on platinum is oxidised (by lattice oxygen stored in ceria) first rapidly into multi-bonded (1928 cm^{-1}) and then into bridged carbonyl (1975 cm^{-1}). These species are likely to be located on the most energetic low coordination sites (terraces, edges, corners). Then, CO starts to accumulate on platinum before a slow diffusion to the surrounding ceria proceeds with formation of formates and carbonates (controlled by diffusion). This CO accumulation over Pt particles constrains carbonyl adspecies which become predominantly linear (2028 cm^{-1}) and therefore less strongly adsorbed. This is confirmed by the change in linear carbonyl coverage which follows exactly the change in residual CO concentration (Figure 8) detected in the background level of the MS signal (m/z=28). Under Ar flow, the concentration of all the Pt carbonyls adspecies decreases first rapidly, then slowly. This indicates either a reverse spillover of adsorbed CO from the decomposing formate/carbonate species to the Pt particles, or, more likely, the reaction of carbon species remaining adsorbed on the Pt surface with oxygen provided by the progressively reduced ceria.

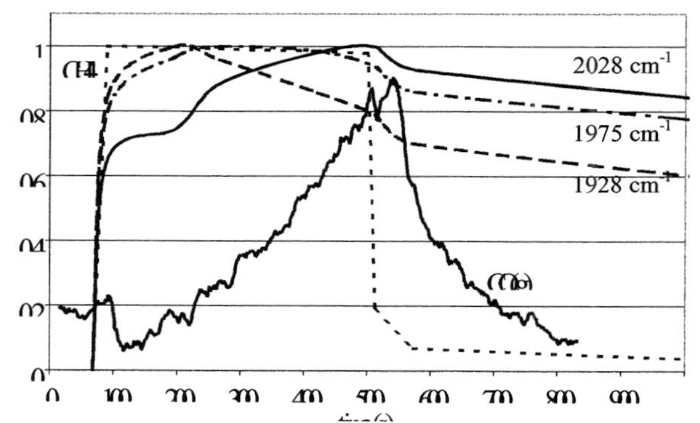

Figure 8 : Normalized intensity of Pt-CO IR bands and MS signal for CO during CH_4 step over Pt/CeO_2 at 400°C.

1928 cm^{-1} : multi-bonded carbonyl (edge and terrace sites)	C=O, Pt Pt
1975 cm^{-1} bridged carbonyl	O, C, Pt Pt
2028 cm^{-1} Linear carbonyl	Pt-C≡O

Table 3 : CO IR bands observed during the sequential reaction steps over Pt/CeO_2 at 400°C.

CONCLUSION

A performing system of CO-free hydrogen production from methane under forced unsteady state conditions was found by considering Pt as a cracking phase active at low temperature and ceria as an active support contributing efficiently to the carbon storage via transfer from the metal phase to the support.

The transient changes in concentration of gaseous and adsorbed carbon containing species during the cycle lead to a preliminary but quantitative description of the main steps which control the various sequences of the process. This study which has to be completed by considering the fate of hydroxyl groups or other types of hydrogen storage has been used for modelling the whole process, which will be shown as a subsequent publication presented elsewhere [13].

REFERENCES

[1] J. N. Armor, *Appl. Cat.* A 176 (1999) 159.

[2] M.A. Pena, J.P. Gomez, J.L.G. Fierro, *Appl. Cat.* A 144 (1996) 7.

[3] T. Zhang, M.D. Amiridis, *Appl. Cat.* A 167 (1998) 161.

[4] N.Z. Muradov, *Energy Fuels* 12 (1998) 161.

[5] T.V. Choudary, D.W. Goodman, *Catal. Lett.* 59 (1999) 93.

[6] V.C.H. Kroll, H.M. Swaan, and C. Mirodatos, *J. Catal.,* 161 (1996) 409.

[7] J.W. Snoeck, G.C. Froment, M. Fowles, *J. Catal.,* 169 (1997) 240.

[8] A. Holmgren, B. Andersson, D. Duprez, *Appl. Catal. B*, 22 (1999) 215.

[9] A. Laachir, V. Perrichon, A. Badri, J. Lamotte, E. Catherine, J.C. Lavalley, J. El Fallah, L. Hilaire, F. le Normand, E. Quéméré, G.N. Sauvion and O. Touret, *J. Chem. Soc. Faraday Trans.*, 87 (1991) 1601.

[10] C. Li, Y. Sakata, K. Domen, K. Maruya, T. Onishi, *J. Chem. Soc. Faraday Trans.*, 85 (1989) 929 & 1451.

[11] J.A. Anderson, *J. Chem. Soc. Faraday Trans.,* 88 (1992) 1197.

[12] T. Jin, T. Okuhara, G.J. Mains, J.M. White, *J. Phys. Chem.*, 91 (1987) 3310.

[13] E. Odier, Y. Schuurman, H. Zanthoff, C. Millet, C. Mirodatos,_to appear in the proceedings of the 3rd International Symposium on Reaction Kinetics and the Development and Operation of Catalytic Processes, Oostende, April 22-25, 2001.

Studies in Surface Science and Catalysis
J.J. Spivey, E. Iglesia and T.H. Fleisch (Editors)

From Natural Gas to Oxygenates for Cleaner Diesel Fuels

M. Marchionna[a], R.Patrini[a], D.Sanfilippo[a], A.Paggini[a], F.Giavazzi[b] and L.Pellegrini[b]

[a] Snamprogetti, Research Laboratories,
Via Maritano 26, 20097 San Donato Milanese, Italy

[b] Agip Petroli, Euron Center Research Laboratories,
Via Maritano 26, 20097 San Donato Milanese, Italy

Different oxygenated compounds were investigated as high quality components for diesel fuels. Among them, di-n-pentyl-ether (DNPE) seemed to represent the best compromise among the availability of a Natural Gas derived feedstock, the existence of efficient production technologies and its excellent properties as a diesel fuel component. An overview on other attractive oxygenates derived from methane will be also given.

1. INTRODUCTION

Technologies that can simultaneously improve emissions performance and fuel economy are receiving special attention for medium term development. It is not easy to satisfy both the requirements but there is no doubt that an increased use of diesel engine represents one of the most viable solutions. This technology is characterized both by low fuel consumption (also considering the whole lifecycle from crude oil) and by excellent performances, recently improved till to the same level of gasoline engines. Emissions have also been strongly reduced although there is a growing environmental concern over the health effect of particulate matter (PM) emissions, especially in the fine particle range. Also nitrogen oxides (NO_x) diesel emissions are relatively high.

According to this general picture, diesel fuel specifications are becoming more and more severe as legislation is adopted to improve air quality. Recent European regulations are particularly stringent: in the near future, diesel fuels would be likely characterized worldwide by higher cetane number, lower density, and lower aromatics, polyaromatics and sulphur content with respect to the current ones. This picture, together with the growing concern for particulate emissions, prompted us to investigate the properties of different oxygenated compounds for diesel fuels.

2. OXYGENATED COMPONENTS FOR CLEANER DIESEL FUELS

Attention to the effect of oxygenates addition to diesel fuels has been recently pointed out by many studies [1]: it is generally reported that it can lead to substantial reductions of exhaust emissions, especially of particulate. This is not surprising as increasing the supply of oxygen during the combustion process does reduce the formation of particulate matter.

Before describing our own results, a few general guidelines have to be given about the ideal characteristics of an oxygenated compound to blend with diesel fuel. The oxygenate must have an adequate (better high) cetane number, it should be high boiling enough to satisfy the flash point specifications, it must not worsen the cold flow properties, it must be miscible with various types of diesel fuels, it should have a suitable density. However, this is not sufficient as the recent MTBE story has pointed out. Other properties are to be strictly taken into consideration: toxicity, biodegradability, ... Finally, economic criteria have also to be satisfied: the raw material for the oxygenate production should be available in huge amounts and the oxygenate production cost should be as low as to be blended in diesel fuels.

These criteria are very severe and it may be difficult to find an ideal product: however, they represent a clear guideline to design a compound that can at least represent the best trade-off among all the requirements.

3. DI-N-PENTYL-ETHER (DNPE) AS BLENDING COMPONENT

Our interest for the addition of oxygenates to fuels is not recent. In fact, the ENI Group, after having pioneered worldwide commercialization of MTBE, in the mid of the 80's turned its attention also to the effect of oxygenates in diesel fuels. A comprehensive study on the blending properties of more than eighty oxygenated compounds pointed out that linear ethers with a relatively long chain (≥ 9 carbon atoms) showed the best compromise among cetane number and cold flow properties. These ethers are high boiling enough to satisfy without any problem the flash point specification [2].

Among the oxygenated derivatives, DNPE seemed to represent the best compromise among the availability of a Natural Gas – derived feedstock (butane and methane) to produce it in large volumes, the existence of selective and efficient production technologies (as reported later) and its excellent properties as a diesel fuel component.

3.1 Product characteristics

As shown in Table 1, DNPE displays an excellent behavior: blending cetane numbers are comprised between 100 and 150 (increasing up to 240 when used together with typical cetane enhancers), cold flow properties are very attractive, density is low and the ether is free from sulphur and aromatics. When it is added to diesel bases (5-20% vol.) all diesel properties are sensibly improved (Table 1); thus, the most advanced specifications could be more easily met by adding this oxygenated compound.

Table 1
DNPE product characteristics

	Diesel	Diesel + DNPE (20%)	DNPE
Density, kg/m^3	848	835	787
Cetane Number	51	62	109*
Pour Point, °C	-9	-12	-25*
Cloud Point, °C	-2	-6	-20*
Cold Filter Plugging Point, °C	-15	-17	-22*
Viscosity$_{15°C}$, cSt	3.6	3.3	1.6
Sulphur, ppm	350	280	-
Aromatics, %	37	29	-

* blending properties

DNPE solubility in water is very low (< 0.3% wt., 12 times lower than MTBE); the ether, although not completely biodegradable, is nearly 15 times more biodegradable than MTBE. Both aspects should limit the risk to run into the problems which have recently affected the MTBE market. Information about the toxicological properties of DNPE are scarce but they do not seem to cause any particular worry.

3.2 Effect on emissions

Extensive engine tests on DNPE-blended diesel fuels, both on Light Duty and on Heavy Duty Engines, have shown that overall exhaust emissions are lowered [3]. The high cetane number generally favors the reduction of CO and HC emissions and improves cold starting performance; there is also a small effect on the reduction of NO_x and PM. The oxygen presence results in reduced emissions of particulate, CO and HC.

However, most of the advantage of the DNPE addition to diesel oil is achieved when engine tuning is optimized to take full gain of the oxygen presence in the molecule: in this case, the trade-off between particulate (but also CO, HC) and NO_x is sensibly improved. This is very important because NO_x and PM represent the major pollutants from diesel engines and generally most measures to reduce NO_x emissions, especially in engine technology, cause an increase in particulate and viceversa. It was also observed that the effect of adding DNPE is even more beneficial when the most advanced engines or the highest quality diesel fuel bases are used.

As regards particulate emissions, several studies recently highlighted that they are strongly influenced by different fuel properties: flame temperature, sulphur presence, cetane and the chemical structure of the fuel; the presence of oxygen, of a high H/C ratio and of a low number of C-C bonds are highly desirable properties to reduce PM emissions. In the case of DNPE, the high cetane plays a major role on emissions although the effect of oxygen is clearly observed for high DNPE concentrations.

3.3 Production process

DNPE can be produced by n-butane via the following process scheme: n-butane is dehydrogenated to n-butenes, which are converted into n-pentanol via reaction with CO/H_2, produced by methane, followed by hydrogenation of the aldehyde intermediate product; finally, the ether is obtained by dehydration of the n-pentanol rich stream. In addition there is a Selective Hydrogenation Unit (SHP), upstream of the Hydroformylation unit, to transform the small amount of butadiene, produced by Dehydrogenation, into n-butenes.

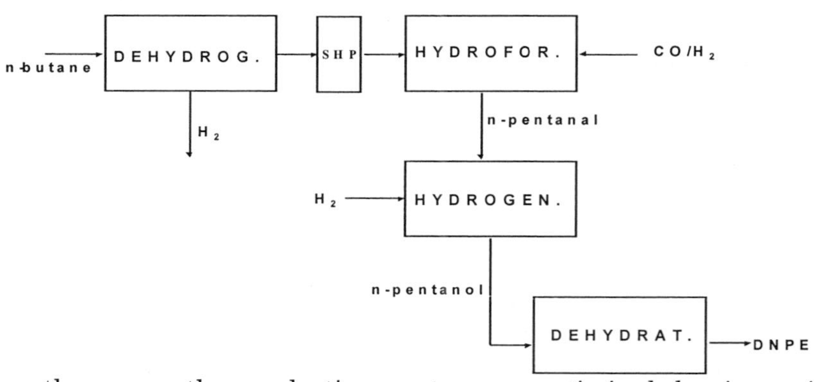

Over the years the production route was optimized by improving selectivity of each step; some more detail will be given in the following.

Dehydrogenation: the catalyst of Snamprogetti FBD-4 commercial technology [4] for the dehydrogenation of isobutane to isobutene was improved, in the n-butane case, to allow to achieve selectivity to n-butenes (comprehensive of the selectively hydrogenated butadiene) of ca. 90%.

Hydroformylation: the choice was addressed towards technologies able to give high selectivity to the linear aldehyde from n-butenes mixtures without producing too much n-butane. Selectivity to n-pentanal is more than 97%.

Dehydration: a new liquid catalyst was found to overcome the environmental problems of the currently used sulphuric acid and to improve selectivity. Nearly total yield (> 95%) to DNPE were achieved.

Economic evaluations indicate that DNPE can be produced on a large scale by merchant plants at a remunerated production cost equivalent to ca. twice the diesel price for a crude oil price of 18 $/bbl (higher crude oil prices definitely improve this figure). This production cost is rather high although a recent preliminary study [3] has pointed out that refineries could be attracted by "merchant" DNPE in the case of very stringent cetane specifications.

4. OXYGENATES FOR DIESEL FUELS FROM METHANE AND FUTURE DIRECTIONS

Taking into account this picture our attention was then addressed to the synthesis of other oxygenated components totally derived from methane, a much cheaper raw material than butane. Actually, several reports recently appeared on the use in diesel engines and on the addition to diesel fuel of oxygen-rich methane-derived components such as di-methyl-ether (DME) [5], di-methoxy-methane (DMM) [6], glycol ethers derivatives [1], or also mixtures of DMM and glycolethers [7]. It is worth noting that all these products have a higher oxygen content with respect to DNPE and a lower number of carbon-carbon bonds; thus their effect on particulate reduction should be even more attractive.

DME allows impressive emissions reductions but, due to its physical properties, it can be used only as an alternative fuel; DMM is liquid at room temperature but is too volatile and its use (as a blending component) does not permit to satisfy the flash point specification for any rate of DMM concentration. As regards glycolethers derivatives, the lighter ones are again too volatile while the heavier ones, such as diglyme, have a very high cetane number but suspects exist about their teratogenic properties.

We have tried to overcome the latter points by studying the conversion of glyme, $CH_3OCH_2CH_2OCH_3$, potentially achievable by DME oxidative coupling, to produce heavier compounds that do not display the problems just described. It was found (scheme below) that glyme can be converted into the corresponding alcohol that, by addition of formaldehyde, can selectively produce the $(CH_3OCH_2CH_2O)_2CH_2$ derivative, known for its good cetane number and cold properties [8].

$$\mathbf{DME} \xrightarrow[+O_2]{} \mathbf{CH_3OCH_2CH_2OCH_3}$$

$$\downarrow$$

$$\mathbf{CH_3OCH_2CH_2OH} \xrightarrow{+CH_2O} \mathbf{(CH_3OCH_2CH_2O)_2CH_2}$$

However, we were not able to achieve significant yields in the oxidative coupling of DME to glyme; the only way we found to produce sufficient yields was the stoichiometric reaction of DME in the presence of di-ter-butyl-peroxide: in this case, selectivities to glyme of 80-90% were achieved with total conversion of the peroxide [9]. These results are promising but the development of such a process seems hard although some similar example exists in the industrial practice (MTBE/propylene oxide co-production).

A more promising approach seems that represented by the production of DMM and higher oligomers (polyoxymethylenes) by reaction of methanol (or DME) with formaldehyde [10,11]. In this case the synthesis process is much easier than the DME oxidative coupling and formaldehyde is a C_1 building block available at a low cost, at least in specific situations.

Similar approaches are under investigation also in our laboratories and results will be reported in the near future. The final goal is to design a cost-effective solution to comply with very stringent specifications concerning both fuel quality and emissions reductions. In the event of success, the clean Oxy-Diesel fuel could prove to be helpful not only for very new and advanced vehicles but also for the whole diesel fleet whose turnover is very slow, particularly in the case of Heavy Duty vehicles.

REFERENCES

1. N. Miyamoto, H. Ogawa, N. Nurun, K. Obata, T. Arima, SAE Paper, (1998), 980506 and references therein contained.
2. G.C. Pecci, M.G. Clerici, F. Giavazzi, F. Ancillotti, M. Marchionna, R. Patrini, IX Int.Symp. Alcohols Fuels, Firenze, Italy, 1 (1991) 321.
3. M. Marchionna, R. Patrini, F. Giavazzi, M. Sposini, P. Garibaldi, 16th World Petrol. Congr., Calgary, Inst. Petr. UK Publ., 3 (2000) and references therein contained.
4. D. Sanfilippo, CatTech, 4(1) (2000) 56.
5. T.H. Fleisch, A. Basu, M.J. Gradassi, J.G. Masin, Stud.Surf. Sci.Catal., 107 (1997) 117.
6. M. Beaujean, "Monetizing Stranded Gas Reserves '98" Conference, San Francisco, December 14-16, 1998; K. Vertin, *ibidem*.
7. Fuel Technol. & Manag., July/August 1998, 26.
8. U. Romano, G. Terzoni, F. Ancillotti, F. Giavazzi, WO Patent No. 86/03511, (1986).
9. H. Naarmann, H.G. Viehe, M. Beaujean, DE Patent No. 2 911 466, (1980); R. Patrini, M. Marchionna, unpublished results.
10. D.S. Moulton, D.W. Naegeli, US Patent No. 5 746 785, (1998).
11. P. Hagen, M.J. Spangler, US Patents No. 6 160 174, 6 160 186, 6 166 266, (2000).

Studies in Surface Science and Catalysis
J.J. Spivey, E. Iglesia and T.H. Fleisch (Editors)

Some Critical Issues in the Analysis of Partial Oxidation Reactions in Monolith Reactors

Sebastián C. Reyes[a,b], John H. Sinfelt[a,c], Ioannis P. Androulakis[a], and Marylin C. Huff[d]

[a]ExxonMobil Research & Engineering Company, 1545 Route 22 East, Annandale, NJ 08801
[b]Corresponding author, screyes@erenj.com.
[c]Senior Scientific Advisor Emeritus
[d]Chemical Engineering Department, Drexel University, Philadelphia, PA 19104

Abstract

In the catalytic oxidation of ethane on platinum-containing monoliths, the high rates of the surface reactions coupled with their high exothermicities lead to a very large axial temperature gradient over a distance comprising only a minor fraction of the reactor length. The temperature reaches a level such that homogeneous gas-phase reactions occur rapidly enough to make a large contribution to the overall conversion occurring in the remainder of the reactor. The gas temperatures generated by the surface reactions are high enough so that ignition delay times for the gas-phase reactions are small compared to total residence times in the reactor. For a quantitative assessment of the extent to which homogeneous gas-phase reactions contribute to the overall conversion, the most important considerations are having a reliable estimate of the temperature gradient generated by the surface catalyzed reactions at the front of the reactor and being careful to utilize a kinetic scheme for the gas-phase reactions that is appropriate for the reaction conditions.

1. INTRODUCTION

In the oxidative dehydrogenation of ethane on a platinum-containing monolith, much of the overall conversion of ethane is due to gas-phase reactions[1]. When ethane and oxygen come into contact with a monolith at a temperature of about 573 K, rapid oxidation of ethane to H_2O, CO, and CO_2 occurs on the surface of the platinum. The heat released by these reactions increases the temperature markedly. In a typical situation, the temperature increases to a level of 1123-1173 K at a distance through the monolith where only about one-fifth of the ethane has been converted. At these temperatures, the rates of homogeneous gas reactions become high enough for them to be responsible for virtually all of the remaining conversion of the ethane along the monolith. The production of ethylene and acetylene is attributed almost exclusively to reactions occurring in the gas phase. The heat required to sustain the endothermic dehydrogenation reactions yielding these products in the tail end of the reaction zone is supplied by exothermic gas-phase oxidation reactions that form additional H_2O and CO.

Existing information on the kinetics of homogeneous gas-phase reactions involved in the high-temperature oxidation of hydrocarbons is very extensive, and complex kinetic networks applicable to a number of different hydrocarbons have been developed[2-8]. For a specific application, the most general network possible on the basis of existing kinetic information may well be far more complex than is necessary for the purpose at hand. Consequently, it has become the practice to utilize simpler networks that are applicable to particular hydrocarbons or hydrocarbon mixtures and to particular ranges of reaction conditions. This is important from the standpoint of utilizing computational resources efficiently. However, one must be very careful to utilize a network that is adequate for the problem under consideration.

In the oxidation of a relatively simple hydrocarbon such as ethane, the most general network available is clearly not needed. Nevertheless, in assessing the degree to which homogeneous gas-phase reactions contribute to the overall conversion of the ethane, one must be confident that the kinetic description is capable of making a reliable estimate of the ignition delay times. This, coupled with a satisfactory calculation of the rate of heat evolution and the corresponding axial temperature gradient in the front section of the monolith, is crucially important for addressing the problem.

2. RESULTS

An important feature of partial oxidation reactions in monolith reactors is that most of the conversion of reactants occurs in a very short time[9]. Despite the high temperatures involved, reactor contact times (which are frequently less than about 10 ms) are generally insufficient to bring the mixture into thermodynamic equilibrium. Figure 1 illustrates this by comparing experimental measurements with equilibrium calculations (C_2H_6/O_2=1.2, 1.2 atm, 1225 K, 20% N_2 dilution). Selectivities of conversion of ethane to the most abundant product species are shown. Oxygen is always totally consumed by the reactions and ethane conversions are high. This figure shows that while equilibrium favors the formation of synthesis gas and some methane, significant amounts of ethylene, acetylene, and water are produced. Irrespective of whether the observed products arise from surface-catalyzed or gas-phase reactions, these results indicate that a quantitative assessment of product composition requires a kinetic description that adequately accounts for intermediate products that would be only minor components at conditions of thermodynamic equilibrium. The partial oxidation experiments clearly reflect this by yielding ethylene, acetylene and water as such products. Given sufficient time, these products could be transformed to H_2 and CO by steam reforming reactions. The presence of water in the products is also an indicator of the high temperatures that develop within these reactors and therefore of the potential involvement of gas-phase reactions.

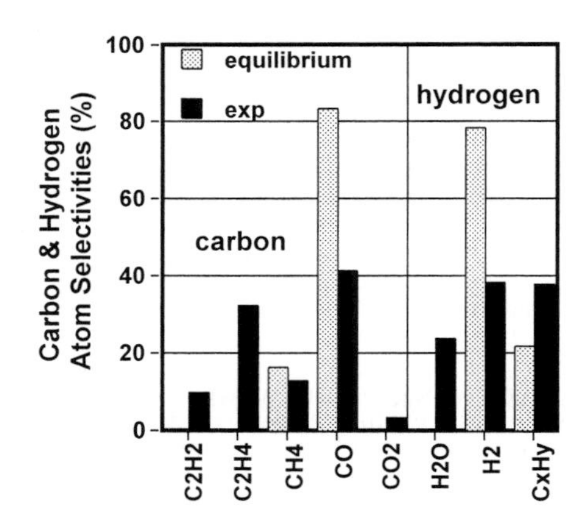

Figure 1. Comparison of experimental and equilibrium carbon and hydrogen atom selectivities (C_xH_y represents sum of C_2H_2, C_2H_4, and CH_4)

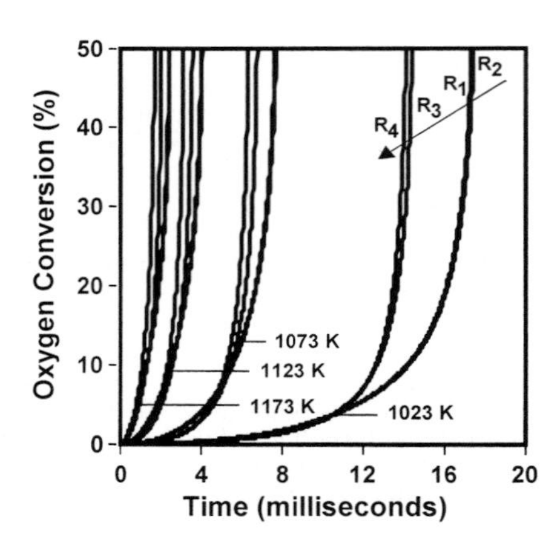

Figure 2. Ignition delay times for various reaction networks and temperatures (Networks R_1, R_2, R_3, and R_4 described in text)

The ignition delay time derived from a reaction network involving free radical intermediates provides a useful diagnostic for assessing the potential involvement of gas-phase reactions. It can be conveniently defined as the time required under adiabatic conditions for the reactions to consume some substantial fraction (say one-half) of the oxygen for given initial conditions of temperature and composition. If this time is shorter than the total residence time in the monolith, then there is a chance that gas-phase reactions may contribute to the conversion process. Figure 2 shows plots of oxygen conversion versus reaction time for various initial temperatures. Results of calculations using networks of varying complexity are included. Times calculated for 50% oxygen conversion are very reasonable measures of ignition delay times since chain reactions rapidly consume the remaining oxygen under the hydrocarbon-rich conditions of interest in partial oxidation. These conversion-time plots are obtained from adiabatic plug-flow calculations at partial pressures that are typical of catalytic experiments employing platinum monoliths (C_2H_6/O_2=1.2, 1.2 atm, 20% N_2). The selected temperatures also represent typical values measured at the outlet of partial oxidation reactors, with 1023 K being a lower limit.

Figure 2 shows that ignition delay times decrease rapidly as temperature increases. They drop from about 15 to 3 ms as the temperature is raised from 1023 K to 1173 K. For temperatures greater than about 1073 K, the ignition delay times are lower than typical residence times in the catalytic monoliths (~10 ms) and the possibility that gas-phase reactions contribute to the conversion is high. Although these ignition delay time calculations provide valuable guidance, they do not represent exactly the delay times in the monoliths. In the monolith, the exothermic surface-catalyzed steps are responsible for the temperature rise that speeds up the ignition process in the gas phase. This catalyst-assisted gas-phase ignition occurs under conditions of partial pressures and heat evolution that differ from those that apply in the absence of a catalyst. As described in further detail later, a proper assessment requires that surface and gas-phase reactions be simultaneously included in the analysis to account correctly for the resulting composition and temperature profiles.

Figure 2 displays ignition delay times for four free-radical reaction networks. R_1 and R_3 correspond to networks taken directly from the literature. R_1 is from Mims et al.[2] (115 species and 447 reactions). R_3 is from Curran et al.[3] (1034 species and 4238 reactions). While the former network includes hydrocarbons up to C_4 species only, and was assembled to study secondary reactions of ethylene in an oxidative environment, the latter includes up to C_8 hydrocarbons and was assembled to study oxidation of fuels in automotive applications. The ignition delay times calculated with these networks differ by only 20%. This level of agreement instills confidence that one can make a very reliable assessment of the contribution of gas-phase reactions. The comparison is included here to highlight the importance of selecting gas-phase networks that are complete enough for the conditions of composition and temperature under study. For practical reasons, gas-phase networks are commonly developed for limited regions of composition and temperature, and their predictive accuracy can be severely compromised if they are used outside their limits of validity. Thus, for example, reaction networks that are developed specifically for combustion or pyrolysis situations will generally not perform well under partial oxidation conditions. This is an issue of great importance because the inability of a network to capture ignition delay times and kinetic trajectories quantitatively can lead to erroneous conclusions about the role of the gas-phase reactions in situations like the one addressed here.

The use of gas-phase networks containing a large number of species can sometimes unduly stress the limits of computational resources. The analysis of experimental data frequently requires estimates of certain parameters. Such an analysis can become prohibitively long when a large number of material balance equations have to be integrated repeatedly in time or space as required by minimization procedures. Since in many situations a relatively small set of reactions is adequate for a description of the system, there is no need to include all the species in the calculations. Network reduction via formal optimization algorithms is of great utility here in providing a means for lowering the number of species and reactions without adversely compromising the integrity of the network[10]. This is illustrated in Figure 2, where ignition delay times are presented for networks R_2 (45 species, 234 reactions) and R_4 (47 species, 237 reactions), which correspond to reduced versions of the parent networks R_1 and R_3, respectively. Corroborating the adequacy of the network reduction procedures for the conditions at hand, both R_2 and R_4 accurately capture the ignition delay times obtained from the extended networks R_1 and R_3, respectively.

Figure 3 illustrates calculated pressure and temperature profiles for a typical situation in ethane partial oxidation in a platinum-containing monolith ($C_2H_6/O_2=1.2$, 1.2 atm, 20% N_2). The calculated outlet partial pressures are in very good agreement with the experimental values[1]. The profiles are obtained by simultaneously integrating heat and material balances along the monolith distance[1]. (In this figure, reaction time is used instead of distance because we wish to highlight how the ignition time relates to the residence time in the reactor). The calculations account for chemical reactions at the catalytic surface and in the surrounding gas phase. Except for the initial dissociative chemisorption of ethane, the kinetics of the elementary surface reactions are precisely those that are applicable in the partial oxidation of methane[11]. Accounting for axial dispersion did not affect the calculated profiles significantly and this is consistent with the observed good agreement between measured and calculated partial pressures and temperature at the reactor outlet when axial dispersion is ignored. The reactor operates nearly adiabatically; heat losses of about 25% were determined from the analysis[1]. Near the front of the reactor, oxygen and ethane are consumed exclusively at the catalyst surface producing H_2O, CO, and CO_2. The exothermicity of these reactions raises the temperature to a level at which the gas-phase reactions occur readily. The change in slope in the profiles of reactant partial pressures and temperature that appear at about 3.5 ms signals the onset of gas-phase ignition. This ignition time is consistent with the value shown in Figure 2 for a temperature of about 1123 K. After ignition, the reactions in the gas phase are largely responsible for the composition and temperature of the evolving mixture. Figure 3 compares calculated profiles for two of the gas-phase networks R_1 (Figure 3a) and R_3 (Figure 3b) discussed earlier. The purpose of this comparison is to emphasize that, in addition to ignition delay times, the kinetic trajectories predicted by the networks are also of importance. This figure shows that R_1 and R_3 lead to profiles that are very similar. However, some differences still remain. Network R_3, for example, predicts a higher degree of ethylene dehydrogenation to acetylene and hydrogen and the endothermicity of this reaction causes a corresponding decrease in temperature. These differences may warrant further investigation if the purpose of the analysis is to make a very accurate quantitative assessment of product distribution. Although not included here, it is noted that the reduced networks R_2 and R_4 lead to profiles that are nearly identical to those in Figures 3a and 3b, respectively.

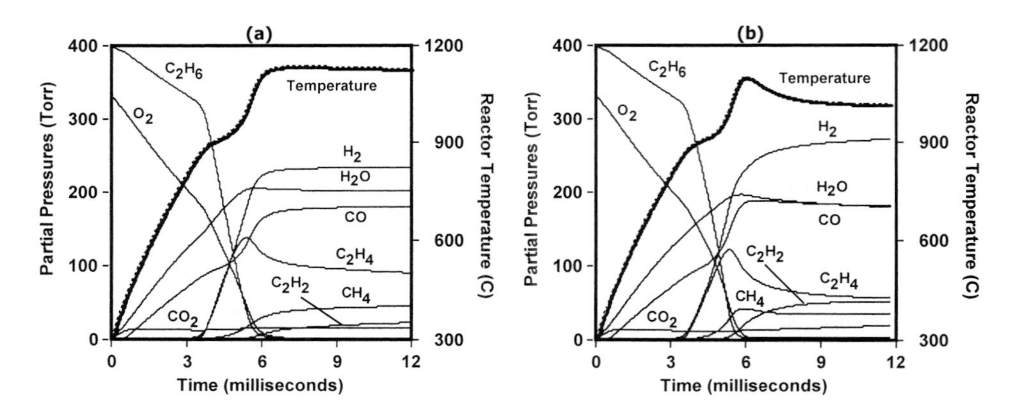

Figure 3. Partial pressures and temperature as a function of reaction time (a: network R_1; b: network R_3) [$C_2H_6/O_2=1.2$, 1.2 atm, 20% N_2 dilution, flow rate of 2 slpm, 5.88wt% Pt/α-Al_2O_3 monolith]

Figure 4 summarizes the salient result of the present analysis: the quantification of the relative contributions of the gas phase and catalytic reactions to the conversion process[1]. For the conditions of Figure 3a, this series of plots shows the region in the reactor where conversion in the gas phase and on the catalyst occurs. It also shows the extent of conversion occurring within these regions. The heat produced by the exothermic catalytic reactions in the front region of the monolith is clearly responsible for initiating gas-phase reactions that dominate the chemical transformations in the remainder of the monolith.

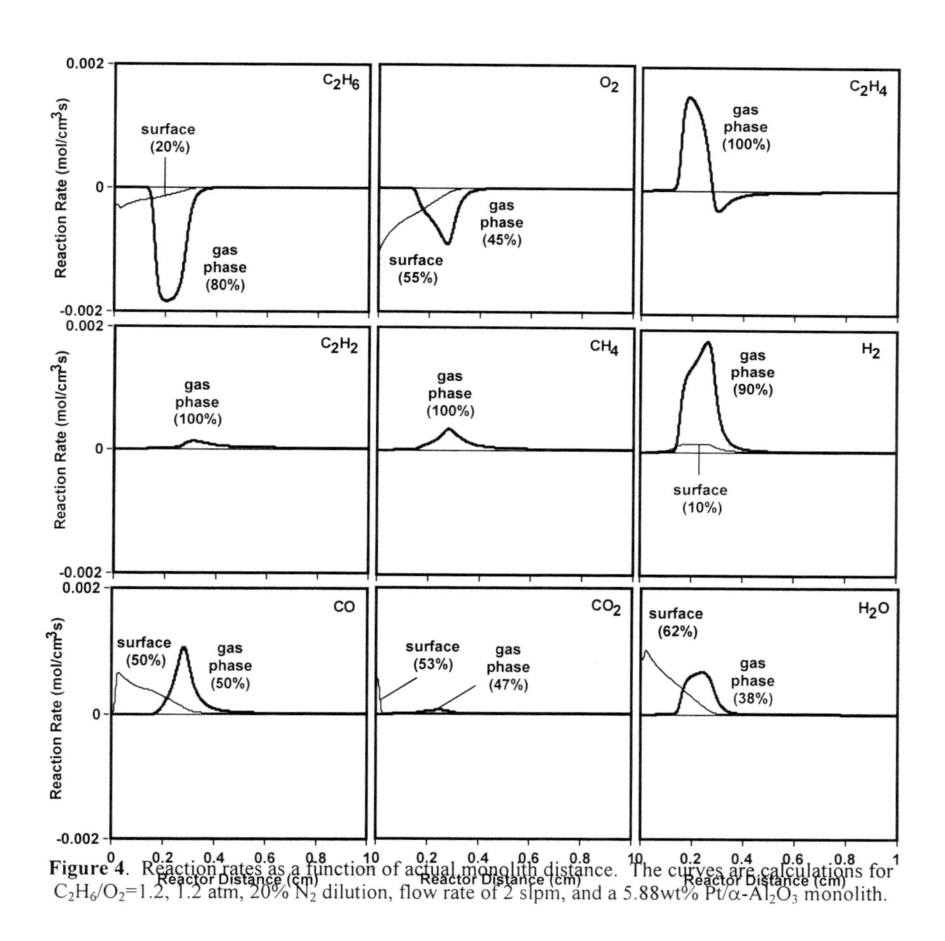

Figure 4. Reaction rates as a function of actual monolith distance. The curves are calculations for $C_2H_6/O_2=1.2$, 1.2 atm, 20% N_2 dilution, flow rate of 2 slpm, and a 5.88wt% Pt/α-Al_2O_3 monolith.

3. CONCLUSIONS

When ethane-oxygen mixtures rich in the hydrocarbon (C_2H_6/O_2 mole ratio = 1.2 to 1.8) are introduced at a temperature of 573 K to a platinum-containing monolith reactor, the conversion of the oxygen is complete and that of the ethane very high in reaction times in the vicinity of 10 ms. The reaction products, which include C_2H_4, C_2H_2, H_2, CO, H_2O, CO_2, and CH_4, do not correspond to thermodynamic equilibrium for such a reaction time, even when the ethane conversion is nearly 100%. The kinetics of the reactions, whether they occur on the surface or in the gas phase, are therefore important in determining the product distribution.

Calculations using extensive kinetic information compiled by Schmidt and others for the surface reactions[11] have made it possible to calculate the rapid rise in temperature in the initial part of the reactor[1]. By the time the reaction gases have traversed roughly one-third of the reactor, the temperature is high enough for gas-phase reactions to occur readily. Ignition delay times calculated with reaction networks developed by various investigators[2,3] are well within the total residence time of the reactor. This finding strongly supports the conclusion that gas-phase reactions contribute significantly to the overall conversion of the ethane.

The determination of the axial temperature gradient and the ignition delay time for the gas-phase reactions are the most important factors in establishing the conclusion. Other issues of importance in the conversion of hydrocarbon-oxygen mixtures in general do not appear to play a significant role when the hydrocarbon is ethane. Such issues include interphase mass and heat transfer limitations, the adsorption and desorption of radicals from the surface, flame speed considerations involving back-diffusion of reactive intermediates, and other mechanistic features of gas-phase reactions that sometimes lead to negative temperature coefficients of reaction rates[3-8].

REFERENCES

1. M.C. Huff, I.P. Androulakis, J.H. Sinfelt, and S.C. Reyes, J. Catal. 46 (2000) 191.
2. C.A. Mims, R. Mauti, A.M. Dean, and K.D. Rose, J. Phys. Chem. 98 (1994) 13357.
3. H.J. Curran, W.J. Pitz, C.K. Westbrook, C.V. Callahan, and F.L. Dryer, Twenty-Seventh Symposium (International) on Combustion, The Combustion Institute (1998) 379.
4. J.F. Griffiths, Prog. Energy Combut. Sci. 21 (1995) 25.
5. M. Nehse, J. Warnatz, and C. Chevalier, Twenty-Sixth Symposium (International) on Combustion, The Combustion Institute (1996) 773.
6. I. Glassman, Combustion, Academic Press, San Diego, 1996.
7. W.C. Gardiner, Jr. (ed), Gas-Phase Combustion Chemistry, Springer-Verlag, NY, 1999.
8. J.W. Bozzelli and W.J. Pitz, Twenty-Fifth Symposium (International) on Combustion, The Combustion Institute (1994) 783.
9. M.C. Huff and L.D. Schmidt, J. Phys. Chem. 97 (1993) 11815.
10. I.P Androulakis, AIChE J. 46 (2000) 361.
11. D.A. Hickman and L.D. Schmidt, AIChE J. 39 (1993) 1164.

Studies in Surface Science and Catalysis
J.J. Spivey, E. Iglesia and T.H. Fleisch (Editors)

Indirect Internal Steam Reforming of Methane in Solid Oxide Fuel Cells

P. Aguiar[a], E. Ramírez-Cabrera[b], N. Lapeña-Rey[a], A. Atkinson[b], L.S. Kershenbaum[a] and D. Chadwick[a]

Department of [a]Chemical Engineering and [b]Materials, Imperial College of Science, Technology and Medicine, London, U.K

A model of indirect internal reforming of methane in SOFCs has been developed. Simulation of SOFC performance demonstrates that mass-transfer controlled catalytic steam reforming can alleviate local temperature minima at the reformer entrance. It is also shown on the basis of measured methane reforming rates, that the use of oxide catalysts such as Ce-Gd-O would lead to smooth temperature profiles.

1. INTRODUCTION

The solid oxide fuel cell (SOFC) operates at high temperatures (700-1000°C) and can use H_2 and CO (and hydrocarbons) as fuel. Improved overall efficiency can be achieved by internal reforming, and it has been shown that there is sufficient heat available for the complete conversion of methane [1]. Direct internal reforming has proved elusive due to deactivation of the electrocatalyst [2]. Indirect internal steam reforming of methane requires efficient thermal coupling of the endothermic reforming reaction to the exothermic oxidation reactions. However, such coupling is not easy to achieve because of the mismatch between the high activity of steam reforming catalysts at typical SOFC temperatures, and the heat available from the fuel cell reactions. Significant local cooling can result leading to thermally-induced fractures of ceramic components. The simple solution of diluting the steam reforming catalyst to reduce its net activity is inadequate because of the inevitable carbon (and sulfur) deposition, which leads to catalyst deactivation [3].

In our approach to this problem, we have developed a model of indirect internal reforming that can be used to define the required catalyst performance and distribution. To meet the desired performance characteristics, we seek to develop oxide-based catalysts, which have a lower activity than conventional steam reforming catalysts while being highly resistant to carbon deposition, or to control the reaction rate by means of mass transfer. The latter can be achieved by the introduction of a diffusive barrier near the outer surface of the catalyst. Under mass transfer control the rate of the reforming reaction is reduced whilst maintaining the overall activity (per unit mass) in the face of possible deactivation [4]. However, it is possible that a combination of approaches may be needed to achieve the desired temperature profile and for ease of fabrication. The paper reports experimental results of steam reforming over an oxide catalyst, $Ce_{0.9}Gd_{0.1}O_{2-x}$, and modelling of indirect internal reforming in an SOFC. Simulations of temperature and methane concentration profiles have been performed using kinetic data of a Ni steam reforming catalyst at various dilutions, measured rate data for the oxide catalyst, and mass transfer limited steam reforming.

2. EXPERIMENTAL

CGO with composition $Ce_{0.9}Gd_{0.1}O_{2-x}$ was supplied by Rhodia and was calcined for 1hr at 1000°C in order to minimize sintering at the maximum reaction temperature. The BET surface area after calcination was $7.1m^2g^{-1}$. TPRx ($25°Cmin^{-1}$) and isothermal reaction were performed in a quartz tube, flow reactor with QMS operating at atmospheric pressure with 1-5%CH_4 in Ar. For steam reforming, gas passed through a water evaporator at 90°C followed by a condenser controlled at a lower temperature to give CH_4:H_2O ratios between 0.6-5.5. Lines downstream were maintained at 70°C to prevent condensation. TPO was carried out in 10%O_2/He at $10°Cmin^{-1}$ after cooling to room temperature under argon.

3. MODELLING

Indirect internal reforming in SOFCs is illustrated schematically in Figure 1. A steam reforming reactor model and a solid oxide fuel cell model are required to be coupled. As our focus of interest is optimisation of the reforming catalyst, we have chosen in this initial work to develop a steady-state model for indirect internal reforming based on a generic configuration: an annular tubular geometry with a single, central reforming reactor channel packed with the reforming catalyst [5].
A conventional steady-state, heterogeneous, 2-d fixed-bed catalytic reactor model [6-8] has been used for the inner reforming reactor. A steady-state, 1-d model is used for the SOFC [9-11].

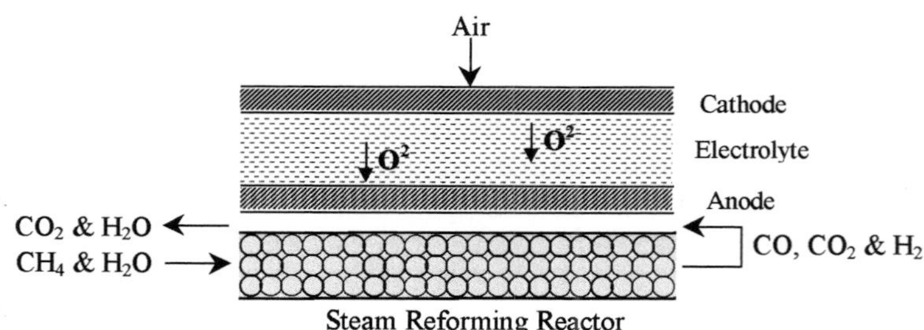

Fig. 1 Schematic diagram of a solid oxide fuel cell with an indirect internal reformer

The SOFC model comprises mass balances of the fuel and air channels, energy balances of the same gas channels and the solid structure (cathode, electrolyte, and anode), and an electrochemical model that relates the gas stream compositions and temperatures to the current density, overpotentials, and cell voltage. The chemical species considered are CH_4, H_2O, CO, H_2, and CO_2 for the fuel, and O_2 and N_2 for the cathode gas. The molar flux in the gas channels is considered to be mainly convective in the flow direction. Simultaneous electrochemical conversion of H_2 and CO to H_2O and CO_2 is accounted for, the electrochemical reactions occurring only at the anode/electrolyte and cathode/electrolyte interfaces. On the anode side, it is assumed that the water gas shift reaction is at equilibrium [10]. The thermal flux in the solid structure is mainly conductive. In the gas channels, it is mainly convective in the gas flow direction and conductive from the channels to the solid parts. An additional convective heat transfer between the anode gas stream and the adjacent inner reformer is also considered. It is assumed that all reaction enthalpies are released at the solid structure [10,12]. Radiation is not considered so far, despite the high temperatures in a SOFC system.

The electrochemical model is based on the approach of Achenbach [10] which assumes the electrochemical reactions operate under kinetic control. The kinetic parameters for reactions at the anode and cathode [10] are used without modification. Full details of the combined model for the SOFC and internal reformer will be presented elsewhere [13]. The resulting system of differential and algebraic equations is solved using gPROMS (Centre for Process Systems Engineering) with the orthogonal collocation on finite elements method [14].

4. RESULTS AND DISCUSSION

4.1. Steam reforming activity of cerias

Interest has centred on doped cerias because of the use of ceria as a key constituent of SOFC anodes. Gd increases the concentration of oxygen vacancies, while Nb increases the concentration of mobile electron carriers. Results are given for Gd doped ceria (CGO) with composition $Ce_{0.9}Gd_{0.1}O_{2-x}$. Nb-doped cerias have been investigated, but as these catalysts proved to have lower steam reforming activity the results are not included here.

Reaction of CGO with dry CH_4 at 900°C gives a H_2/CO ratio of 2.14 demonstrating that the dominant reaction is (1) in agreement with published results on undoped ceria [15].

$$CeO_2 + n\, CH_4 = CeO_{2-n} + n\, CO + 2n\, H_2 \qquad (1)$$

The activation energy was estimated to be 176kJ mol^{-1} and 155kJ mol^{-1} from the H_2 and CO signals respectively in good agreement with the value 160kJ mol^{-1} for ceria [15].

Steam reforming began at 670°C producing H_2, CO with a small amount of CO_2. The rate of CH_4 conversion increased rapidly with temperature passing through a maximum before reaching an approximately constant level at 900°C after about 140 min. The steady-state rate at 900°C was $8.9x10^{-5}$ $molmin^{-1}gm^{-1}$, which is about 10^{-5} of the rate over a Ni steam reforming catalyst. The reforming rate was proportional to the CH_4 concentration, but independent of the steam concentration. The activation energy was determined to be 153kJ mol^{-1}. No carbon deposition could be detected by TPO after any of the experiments involving steam. The rate of CH_4 steam reforming at 900°C and activation energy are approximately equal to the values for the reaction with dry 5%CH_4. This suggests a steam reforming mechanism that is controlled by the reaction between CH_4 and lattice oxygen in the CGO surface.

4.2. Simulation of indirect internal steam reforming

The combined model of indirect internal steam reforming and SOFC described above was used to simulate temperature profiles along the reformer tube using various catalyst options. Interest centres on the occurrence of a local temperature minimum near the reformer entrance. The base case catalyst was a Ni steam reforming catalyst. The kinetics of the steam reforming and shift reactions are based on the work of Xu and Froment [7,8] as were the properties of the Ni catalyst. The length of the module was taken as 0.3m and the diameter of the steam reforming reactor was between 2 - 3mm. The diameter of the steam reforming catalyst was 0.2mm. Steam reforming of CH_4 on the anode was neglected, as the concentration was very low in most cases. In all cases, inlet CH_4/H_2O ratio was 0.5 with small amounts of CO and H_2, and some CO_2, air inlet temperature = 950°C, current density = $4x10^3 A/m^{-2}$, fuel utilization = 0.75, and air ratio =10.

The simulations demonstrate the local cooling associated with indirect internal reforming using the standard Ni catalyst diluted to $2x10^{-3}$, Fig.2a. With a fuel inlet of 900°C, methane

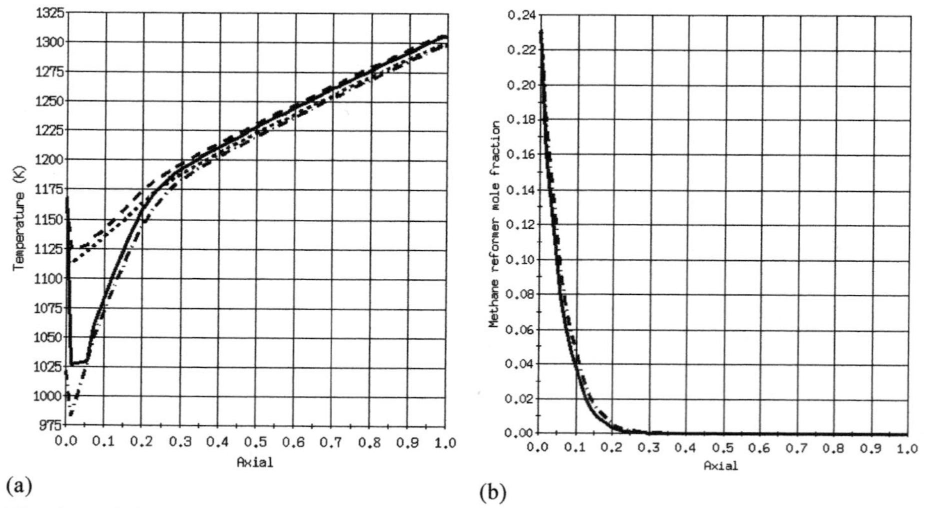

(a) (b)

Fig. 2. Axial profiles using a standard diluted Ni catalyst for fuel inlet temperatures of 900°C and 750°C. (a) Temperature: ——— Reformer - T_0 = 900°C, - - - - - - SOFC fuel channel - T_0 = 900°C, - . - . - . - Reformer - T_0 = 750°C, ... SOFC fuel channel - T_0 = 750°C; (b) Reformer CH_4 mole fraction: ——— T_0 = 900°C, - . - . - . - T_0 = 750°C.

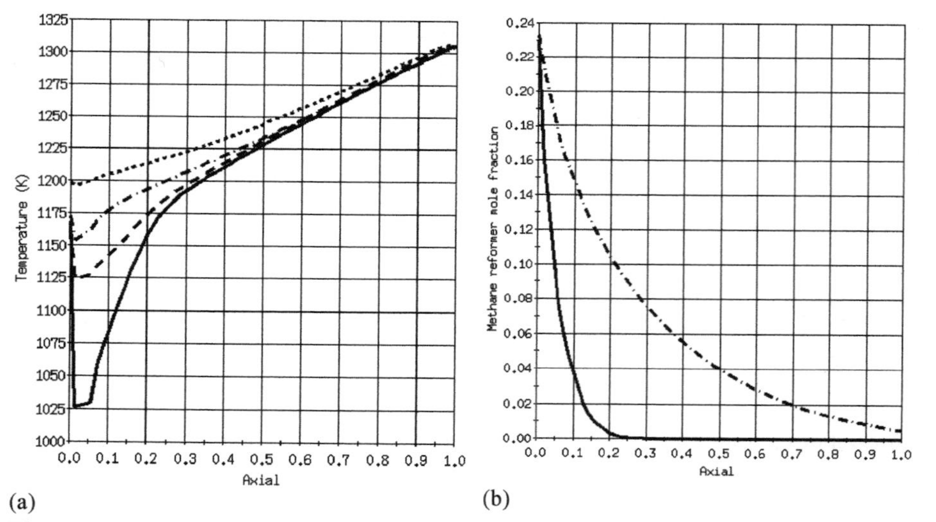

(a) (b)

Fig. 3. Axial profiles using a standard diluted Ni catalyst and 10^{-4} diluted catalyst (Ox) for a fuel inlet temperature of 900°C. (a) Temperature: ——— Reformer - Ni cat., - - - - - - SOFC fuel channel - Ni cat., - . - . - . - Reformer - Ox cat., ... SOFC fuel channel - Ox cat.; (b) Reformer CH_4 mole fraction: ——— Ni cat., - . - . - . - Ox cat..

is rapidly converted in the first 20% of the reformer resulting in a temperature minimum. Greater catalyst concentration results in almost instantaneous CH_4 conversion.

Reducing the fuel inlet temperature while maintaining the air inlet temperature and catalyst content gives a smoother temperature profile, Fig. 2, but leads to an undesirable large temperature rise along the tube.

Simulations using the measured rate data for the $Ce_{0.9}Gd_{0.1}O_{2-x}$ catalyst (but assuming the same form of rate equation as the Ni catalyst), demonstrated that there would be only about 30% methane conversion in the reformer. Increasing the activity to 10^{-4} of the Ni catalyst leads to a smooth temperature profile, Fig. 3, but a few percent of methane is not converted in the reformer. Although the activity is approaching an order of magnitude higher than the particular CGO catalyst studied, it is probably within the achievable range for oxide catalysts. A higher fuel inlet temperature than used here could be required.

The effect of controlling the reaction rate by a mass transfer barrier is shown in Fig 4. The characteristics of the mass transfer barrier and active catalyst distribution within the catalyst particle were taken from Aguiar et al [4], where optimum values were determined to maintain a desired rate of reaction despite a reduction of the intrinsic catalyst activity by 50% because of deactivation. While the predicted temperature profile is much smoother, the characteristics of the barrier require further optimisation. It should be emphasised that an advantage of mass transfer limited steam reforming is the ability to tolerate a degree of catalyst deactivation.

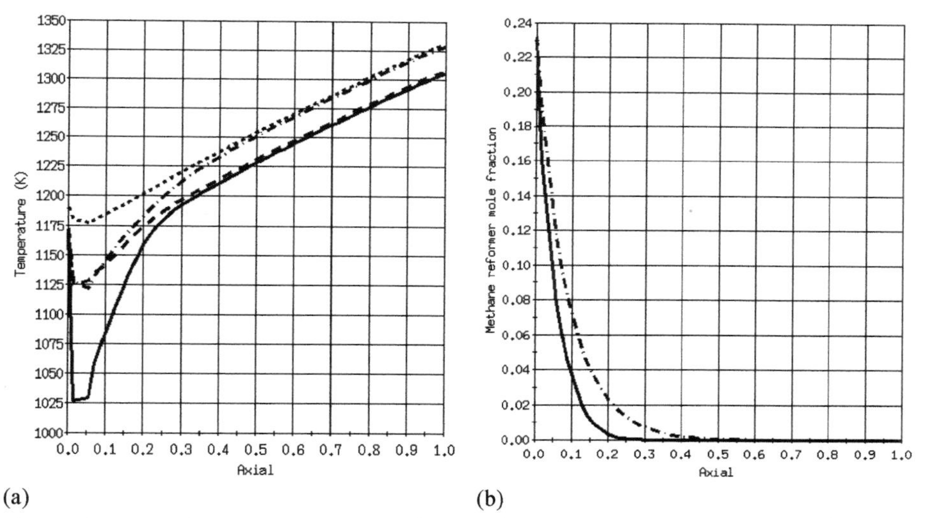

(a) (b)

Fig. 4. Axial profiles using a standard diluted Ni catalyst and with a mass transfer barrier for a fuel inlet temperature of 900°C. (a) Temperature: ——— Reformer - Ni cat., ------ SOFC fuel channel - Ni cat., -.-.-.- Reformer - Ni cat. & mass transfer, ... SOFC fuel channel - Ni cat. & mass transfer; (b) Reformer CH_4 mole fraction: ——— Ni cat., -.-.-.- Ni cat. & mass transfer.

Preliminary work has been carried out to develop the methodology of deposition of a gas-diffusion barrier around an active steam reforming catalyst. The method used for the membrane fabrication is based on Sol-Gel processing techniques, which are widely used in the ceramics field for the production of oxides with a controlled microstructure, shape, density, and porosity. The barrier materials are based on commercially available sols of zirconia and ceria, which satisfy the requirement for high resistance to carbon deposition. The microstructure of the barriers prepared using the sols showed highly dense structures lacking the desired porosity. Porosity has been introduced by mixing the sols with material of larger particle size. Although preliminary results are very promising, the barrier fabrication process is still under optimisation.

5. CONCLUSIONS

Simulation of the performance of an indirect internal reforming SOFC has demonstrated that local temperature minima associated with rapid steam reforming over Ni (and by implication other metal) catalysts can be alleviated by mass transfer control. Oxide catalysts, such as $Ce_{0.9}Gd_{0.1}O_{2-x}$, can also smooth the temperature profile, but do not convert all the methane for the same inlet temperature and reformer length.

6. ACKNOWLEDGEMENTS

The authors are grateful for financial support from the UK Engineering and Physical Sciences Research Council, the Department of Trade and Industry (DTI) and Rolls Royce Plc. The first author would also like to acknowledge financial support from the Portuguese agency FCT through fellowship PRAXIS XXI/BD/15972/98. E. Ramírez-Cabrera thanks CONACyT, Mexico, for the award of a study scholarship.

7. REFERENCES

1. Rostrup-Nielsen, J., L.J. Christiansen, App. Catal. A: General, 126(1995)381
2. Finnerty, C.M., Coe, N.J., Cunningham, R.H., Ormerod, R.M., Catalysis Today, 46(1998)137
3. Rostrup-Nielsen, J.,"Catalytic Steam Reforming", Catalysis: Science and Technology,V5, 1, Springer-Verlag, Berlin, 1984
4. Aguiar, P., Lapeña-Rey, N., Chadwick, D., Kershenbaum, L.S., Int. Symp. Chem. React. Eng. 2000, Poland
5. US Patent 5554454, Sept 1996
6. Froment, G.F., Bischoff, K.B., "Chemical Reactor Analysis and Design", 2nd ed., John Wiley, 1990
7. Xu, J., G.F. Froment, AIChE Journal, 35(1989)88
8. Xu, J., G.F. Froment, AIChE Journal, 35, (1989)97
9. Ferguson, J.R., J.M. Fiard, R. Herbin, Journal of Power Sources, 58(1996)109
10. Achenbach, E., Journal of Power Sources, 49(1994)333
11. Neophytides, S.G., A. Tripakis, Can. Chem. Eng., 74(1996)719
12. Ahmed, S., C. McPheeters, R. Kumar, J. Electrochem. Soc., 138(1991)2712
13. Aguiar, P., Chadwick, D., Kershenbaum, L.S., to be published
14. Oh, M., Pantelides, C.C., Computers Chem. Engng, 20(1996)611
15. Otsuka, K., Ushiyama, T., Yamanaka, I., Chem. Lett., 1517,(1993)

Studies in Surface Science and Catalysis
J.J. Spivey, E. Iglesia and T.H. Fleisch (Editors)

Catalytic Properties of Supported MoO_3 Catalysts for Oxidative Dehydrogenation of Propane

Kaidong Chen, Enrique Iglesia and Alexis T. Bell

Chemical and Materials Sciences Divisions, Lawrence Berkeley National Laboratory, and Department of Chemical Engineering, University of California, Berkeley, CA 94720

The effects of MoO_x structure on propane oxidative dehydrogenation (ODH) rates and selectivity were examined on Al_2O_3-supported MoO_x catalysts with a wide range of surface density (0.4-12 Mo/nm^2), and compared with those obtained on MoO_x/ZrO_2. On MoO_x/Al_2O_3 catalysts, propane turnover rate increased with increasing Mo surface density and reached a maximum value for samples with ~ 4.5 Mo/nm^2. All MoO_x species are exposed at domain surfaces for Mo surface densities below 4.5 Mo/nm^2. Therefore, the observed trends reflect an increase in ODH turnover rates with increasing MoO_x surface density. As Mo surface densities increase above the polymolybdate monolayer value (~ 4.5 Mo/nm^2), ODH turnover rates decreased with increasing Mo surface density, as a result of the formation of MoO_3 crystallites with inaccessible MoO_x species. The ratio of rate constants (k_2/k_1) for propane combustion (k_2) and for propane ODH reactions (k_1) decreased with increasing MoO_x surface density and then remained constant for values above 5 Mo/nm^2. Propene combustion rate constants (k_3) also decreased relative to those for propane ODH (k_1) as two-dimensional structures formed with increasing Mo surface density. These Mo surface density effects on k_2/k_1 and k_3/k_1 ratios were similar on MoO_3/Al_2O_3 and MoO_3/ZrO_2, but the effects of Mo surface density on ODH turnover rates for samples with submonolayer MoO_x contents were opposite on the two catalysts. A comparison of ODH reaction rates and selectivity among MoO_3/Al_2O_3, MoO_3/ZrO_2, bulk MoO_3, $ZrMo_2O_8$, and $Al_2(MoO_4)_3$ suggests that the behavior of supported MoO_x at low surface densities resembles that for the corresponding bulk compounds ($ZrMo_2O_8$, and $Al_2(MoO_4)_3$), while at high surface density the behavior approaches that of bulk MoO_3 on both supports.

1. INTRODUCTION

Many recent studies have explored the oxidative dehydrogenation of light alkanes as a potential route to the corresponding alkenes. Oxidative dehydrogenation (ODH) of alkanes is favored thermodynamically and the presence of O_2 leads to the continuous removal of carbon deposits and to stable reaction rates. Secondary combustion reactions, however, limit alkene yields. For propane oxidative dehydrogenation reactions, the most active and selective catalysts are based on vanadium and molybdenum oxides [1-15]. On both V- and Mo-based catalysts, several studies of the kinetics and reaction mechanisms have shown that propane reactions occur via parallel and sequential oxidation steps (Scheme 1) [1-3, 11-14]. Propene forms via primary ODH reactions limited by the initial activation of the methylene C-H bond in propane (k_1), while CO and CO_2 (CO_x) can form via the combustion of the propene (k_3) formed in step 1 or the primary combustion of propane (k_2). The k_2/k_1 ratio (propane combustion/propane dehydrogenation) is usually low (~0.1) for selective ODH catalysts [13-15]. The alkene yield losses observed with increasing conversion arise, for the most part, from large k_3/k_1 values (propene combustion/propane dehydrogenation ~ 10-50). These large k_3/k_1 values reflect the weaker allylic C-H bond in propene relative to the methylene C-H bond in propane and the higher binding energy of alkenes on oxide surfaces [12-15].

The structure and propane ODH catalytic properties of ZrO_2-supported MoO_3 catalysts were recently described [15]. Al_2O_3-supported MoO_3 catalysts have been widely used in hydrodesulfurization, hydrogenation, and alkene metathesis reactions, and detailed studies of the structure of dispersed MoO_3 on Al_2O_3 have been reported [16]. In contrast, little is known about the reaction pathways and the structural requirements for propane ODH reactions on MoO_x species supported on Al_2O_3 [8]. This work addresses the effect of Mo surface density on the propane ODH properties for MoO_3/Al_2O_3. The catalytic performance results obtained on MoO_3/Al_2O_3 were compared with those reported previously on MoO_3/ZrO_2.

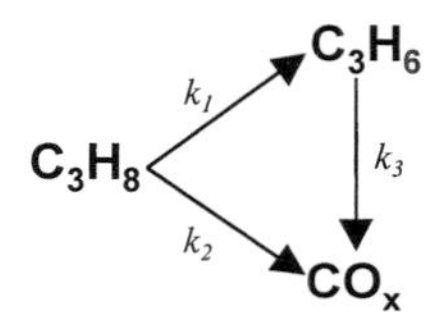

Scheme 1. Reaction network in oxidative dehydrogenation of propane

2. EXPERIMENTAL METHODS

Al_2O_3-supported MoO_x samples were prepared by incipient wetness impregnation of γ-Al_2O_3 (Degussa, AG) with a solution of ammonium heptamolybdate (AHM) (99%, Aldrich, Inc.) at a pH of 5. Impregnated samples were dried overnight in air at 393 K and then treated in dry air (Airgas, zero grade) at 773 K for 3 h. ZrO_2-supported MoO_x samples were also prepared by incipient wetness impregnation method, as described elsewhere [15].

Propane reaction rate and selectivity measurements were carried out at 703 K in a packed-bed tubular quartz reactor using 0.03-0.3 g samples. Propane (14 kPa; Airgas, 99.9%) and oxygen (1.7 kPa; Airgas, 99.999%) with He (Airgas, 99.999%) as a diluent were used as reactants. Reactants and products were analyzed by gas chromatography (Hewlett-Packard 5880 GC) using procedures described previously [13, 14]. C_3H_8 and O_2 conversions were varied by changing reactant space velocity (F/w; w: catalyst mass; F: reactant volumetric flow rate). Typical conversions were < 2% for C_3H_8 and < 20% for O_2. Initial ODH reaction rates and selectivities were obtained by extrapolation of these rate data to zero residence time. The effect of bed residence time on product yields was used in order to calculate rates and rate constants for secondary propene combustion reactions, using procedures reported previously [13, 14].

3. RESULTS AND DISCUSSION

The structures of MoO_3/Al_2O_3 and MoO_3/ZrO_2 catalysts were characterized by BET surface area measurements, X-ray diffraction, and Raman, UV-visible, and X-ray absorption spectroscopy in previous studies [15, 16]. These data showed that the structure and domain size of MoO_x species depend strongly on the Mo surface density and the temperature of treatment in air. For samples treated in dry air below 773 K and with Mo surface densities below "monolayer" values (~4.5 Mo/nm^2), only two-dimensional MoO_x oligomers are detected on Al_2O_3 or ZrO_2 surface. As Mo surface densities exceed this monolayer coverage, crystalline MoO_3 forms. The size of the MoO_x domains increased gradually with increasing Mo surface density.

Propane ODH on Mo-based catalysts occurs via parallel and sequential oxidation pathways (Scheme 1) [15]. The reaction rate constants (k_1, k_2 and k_3) in Scheme 1 can be calculated from the effects of reactant residence time on propene selectivity [14]. Propene yields during propane ODH reactions depend on both k_2/k_1 and k_3/k_1; smaller values of either ratio lead to higher propene selectivity at a given propane conversion. Figure 1 shows the effects of Mo surface density on k_2/k_1 and k_3/k_1 values for MoO_3/Al_2O_3 catalysts. The value of k_2/k_1 reflects the relative rates of initial propane combustion and dehydrogenation. The values of k_2/k_1 decreased with increasing Mo surface density, until it reached a constant value of ~0.05 for surface densities above 5 Mo/nm^2 (Figure 1(a)). This gradual decrease in k_2/k_1 values with increasing MoO_x surface density suggests that Mo-O-Al sites or uncovered Al_2O_3 surfaces near MoO_x species catalyze the unselective conversion of propane to CO_x. This may reflect, in turn, the tendency of such sites to bind alkoxide intermediates more strongly than Mo-O-Mo structures in polymolybdate domains or on the surface of MoO_3 clusters. The complete coverage of Al_2O_3 surfaces by a polymolybdate monolayer leads to a high initial propene selectivity, which resembles that in samples with predominantly MoO_3 species. A similar decrease in k_2/k_1 with increasing surface density of the active oxide was reported previously on MoO_x/ZrO_2 [15] and VO_x/Al_2O_3 [13] catalysts.

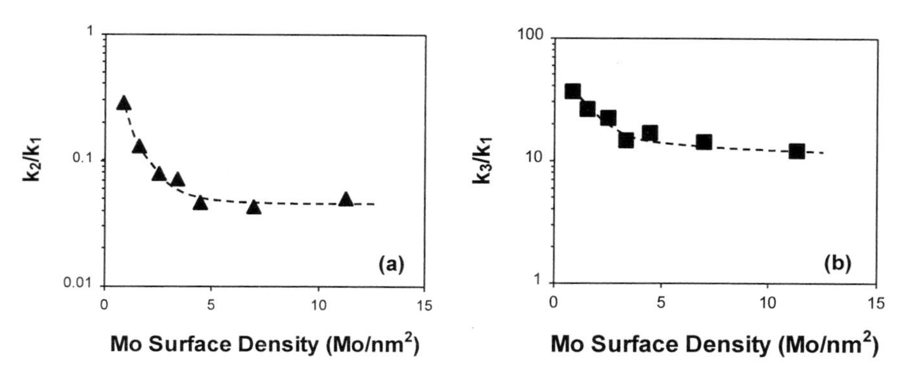

Fig. 1. Dependence of (a) k_2/k_1, and (b) k_3/k_1 on Mo surface density for MoO_x/Al_2O_3 [14 kPa C_3H_8, 1.7 kPa O_2, balance He, 703 K]

The values of k_3/k_1 were much greater than unity on all MoO_x/Al_2O_3 samples (Figure 1(b)), indicating that propene combustion occurs much more rapidly than propane dehydrogenation. It is this large value that causes the significant decrease in propene selectivity with increasing propane conversion. The values of k_3/k_1 (10-40) on these MoO_x/Al_2O_3 catalysts are similar to those measured on MoO_x/ZrO_2 [15]. The k_3/k_1 ratio decreased with increasing Mo surface density and then remained constant for Mo surface densities above 5 Mo/nm^2. The large k_3/k_1 ratio reflects the weaker C-H bonds in propene compared to those in propane, as well as the higher binding energy of propene molecules on Lewis acid sites provided by Mo^{+6} cations present on MoO_3 surfaces [12].

Initial propane reaction rates are reported in Figure 2 as a function of Mo surface density on all MoO_x/Al_2O_3 samples. Propane consumption rates normalized per Mo atom initially increased with increasing Mo surface density and they approached maximum values at surface densities of ~ 4.5 Mo/nm^2 (Figure 2(a)). In this range of surface density, the accessibility of MoO_x at domain surfaces is largely unaffected by Mo surface density, because the Al_2O_3 surface is covered predominantly by two-dimensional MoO_x oligomers. Therefore, the observed increase in reaction rate reflects an increase in the reactivity (turnover rate) of exposed MoO_x active sites with increasing domain size. Propane reaction rates decreased as Mo surface densities exceed ~ 4.5 Mo/nm^2, which corresponds to the approximate surface density in a polymolybdate monolayer. The incipient appearance of three-dimensional MoO_3 structures, with the consequent incorporation of MoO_x into inaccessible positions within such clusters, is likely to account for the observed decrease in apparent turnover rates at higher surface densities (Figure 2(a)).

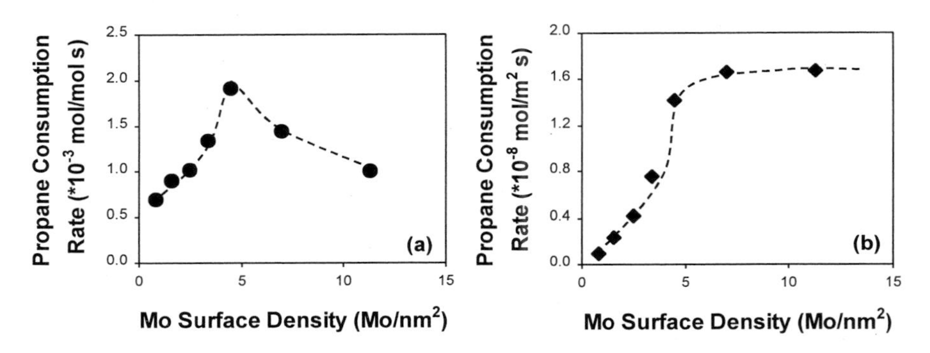

Fig. 2. Effects of Mo surface density on initial propane consumption rate for MoO_x/Al_2O_3 (a) normalized per Mo atoms, and (b) normalized per surface area. [14 kPa C_3H_8, 1.7 kPa O_2, balance He, 703 K]

Figure 2(b)shows the propane consumption rates normalized per BET surface area. These areal rates initially increased sharply with increasing Mo surface density, but then remained almost constant for surface densities above 5 Mo/nm^2. Thus, it appears that the initial increase in propane turnover rates as two-dimensional structures grow reflects the increasing reactivity of MoO_x surface structure on larger oxide domains. Similar domain size effects were observed on Al_2O_3-supported VO_x catalysts [13]. When Mo surface densities exceed monolayer coverages, three-dimensional MoO_3 form and the entire surface of the catalyst becomes covered by either two-dimensional MoO_x domains or MoO_3 clusters with similar surface reactivity. Any additional MoO_x species become inaccessible for propane ODH reactions; therefore, propane reaction rates normalized per Mo atom decreased, but areal rates remain constant with increasing MoO_x surface density.

The observed surface density effects on the catalytic activity of MoO_x/Al_2O_3 and MoO_x/ZrO_2 are different. On MoO_x/ZrO_2, propane turnover rates per Mo decreased with increasing Mo surface density, even below monolayer coverages [15]. Figure 3(a) compares propane consumption rates per Mo atom on MoO_x/Al_2O_3 and MoO_x/ZrO_2. For Mo surface densities above ~ 5 Mo/nm^2, propane turnover rates on MoO_x/Al_2O_3 and MoO_x/ZrO_2 catalysts become similar, because in both cases catalyst surfaces are fully covered by two-dimensional or three-dimensional MoO_x species. Below monolayer coverages (~5 Mo/nm^2), however, propane turnover rates increased on MoO_x/Al_2O_3 but they decreased on MoO_x/ZrO_2 with increasing surface density.

Figure 3(b) shows the corresponding comparison for areal propane reaction rates on these two types of catalysts. Also shown in Figure 3(b) are areal rates on bulk $ZrMo_2O_8$, MoO_3 and $Al_2(MoO_4)_3$. It appears from these data that the catalytic activity of the surface structures in bulk $ZrMo_2O_8$ is considerable higher than on MoO_3 surfaces, which in turn, is higher than that on $Al_2(MoO_4)_3$ surfaces. This suggests that active species with surface structures similar to those on the surface of $ZrMo_2O_8$ may also be more active than those resembling the surfaces of bulk MoO_3, and more active still than those with surfaces resembling $Al_2(MoO_4)_3$. Since Mo surface densities in $ZrMo_2O_8$, MoO_3 and $Al_2(MoO_4)_3$ are similar, propane turnover rates would follow a sequence similar to that of the areal rates shown in Figure 3(b) ($ZrMo_2O_8 > MoO_3 > Al_2(MoO_4)_3$). For low surface density samples, the structures of MoO_x/Al_2O_3 and MoO_x/ZrO_2 surfaces would resemble those in the corresponding $Al_2(MoO_4)_3$ and $ZrMo_2O_8$ bulk phases. On MoO_x/Al_2O_3 catalysts, the surface structure gradually changes from one resembling $Al_2(MoO_4)_3$ to one similar to MoO_3 as the MoO_x surface density increases. Therefore, propane turnover rates increase with increasing Mo surface density for values below monolayer coverages, as surface structures evolve from isolated species with significant Mo-O-Al character to polymolybdate domains resembling in structure and in reactivity of the surface MoO_3 (Figure 3(a)). In contrast, the surface structure of MoO_x/ZrO_2 evolves from one resembling $ZrMo_2O_8$ to one similar to MoO_3 with increasing Mo surface density; as a result, propane turnover rates decrease with increasing Mo surface density (Figure 3(a)).

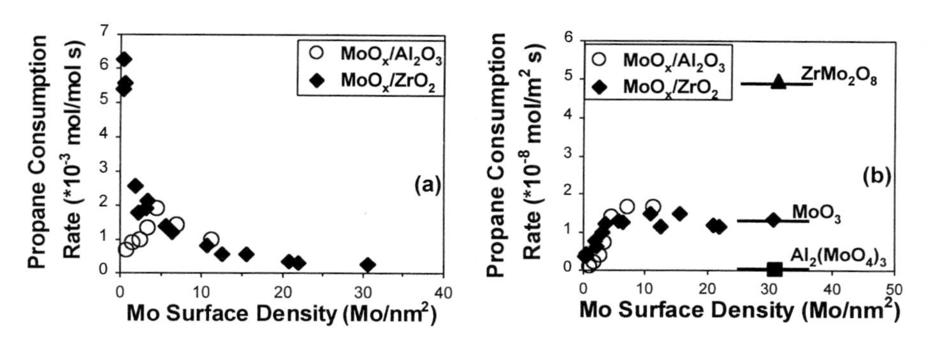

Fig. 3. Effects of Mo surface density on initial propane consumption rate on MoO_x/Al_2O_3, MoO_x/ZrO_2, bulk $ZrMo_2O_8$, MoO_3 and $Al_2(MoO_4)_3$. (a) normalized per Mo atoms, and (b) normalized per surface area. [14 kPa C_3H_8, 1.7 kPa O_2, balance He, 703 K].

Changes of k_2/k_1 and k_3/k_1 ratios as a function of Mo surface density on MoO_x/Al_2O_3 and MoO_x/ZrO_2 catalysts are consistent with the arguments presented above for the evolution of reaction rates with surface density on the two supports. Figure 4 shows k_2/k_1 and k_3/k_1 ratios on the supported MoO_x catalysts and on bulk $ZrMo_2O_8$, MoO_3 and $Al_2(MoO_4)_3$. The

k_2/k_1 and k_3/k_1 ratios on $Al_2(MoO_4)_3$ and on $ZrMo_2O_8$ are higher than on MoO_3. As Mo surface density increases, surface structures evolve from those resembling $Al_2(MoO_4)_3$ or $ZrMo_2O_8$ surfaces to MoO_3-like species; concurrently, k_2/k_1 and k_3/k_1 ratios decrease and approach those measured on MoO_3 (Figures 4(a) and 4(b)). These results suggest that at low surface densities, supported MoO_x species catalyze ODH reactions with turnover rates and selectivities strongly resembling those on the corresponding mixed oxide bulk structure.

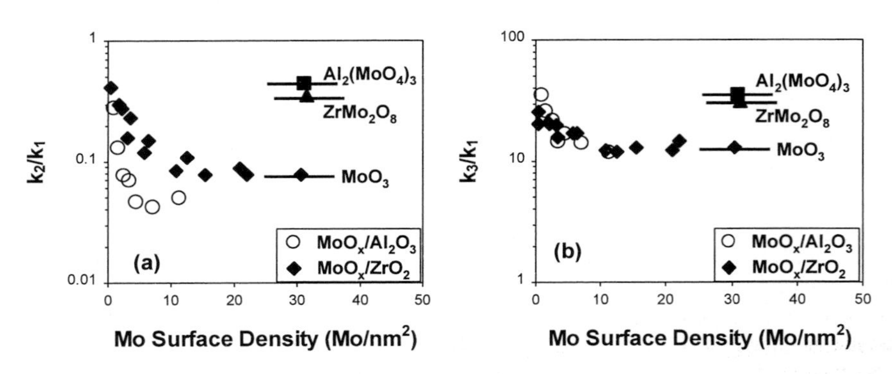

Fig. 4. Effects of Mo surface density on (a) k_2/k_1, and (b) k_3/k_1 ratio for MoO_x/Al_2O_3, MoO_x/ZrO_2, bulk $ZrMo_2O_8$, MoO_3 and $Al_2(MoO_4)_3$. [14 kPa C_3H_8, 1.7 kPa O_2, balance He, 703 K].

ACKNOWLEDGEMENT

This work was supported by the Director, Office of Basic Energy Sciences, Chemical Sciences Division of the U.S. Department of Energy under Contract DE-AC03-76SF00098.

REFERENCES

1. T. Blasko and J. M. López Nieto, Appl. Catal. A 157 (1997) 117.
2. H. H Kung, Adv. Catal. 40 (1994) 1.
3. S. Albonetti, F. Cavani and F. Trifiro, Catal. Rev. -Sci. Eng. 38 (1996) 413.
4. G. Centi and F. Triffiro, Appl. Catal. A 143 (1996) 3.
5. E. A. Mamedov and V. Cortés-Corberan, Appl. Catal. A 127 (1995) 1.
6. F. C. Meunier, A. Yasmeen and J. R. H. Ross, Catal. Today 37 (1997) 33.
7. L. E. Cadus, M. F. Gomez and M. C. Abello, Catal. Lett. 43 (1997) 229.
8. L. Jalowiecki-Duhamel, A. Ponchel and Y. Barbaux, J. Chim. Phys. PCB 94 (1997) 1975.
9. Y. S. Yoon, W. Ueda and Y. Moro-oka, Topics in Catal. 3 (1996) 256.
10. K. H. Lee, Y. S. Yoon, W. Ueda and Y. Moro-oka, Catal. Lett. 46 (1997) 267.
11. K. Chen, A. Khodakov, J. Yang, A. T. Bell and E. Iglesia, J. Catal. 186 (1999) 325.
12. K. Chen, A. T. Bell and E. Iglesia, J. Phys. Chem. B 104 (2000) 1292.
13. A. Khodakov, B. Olthof, A. T. Bell and E. Iglesia, J. Catal. 181 (1999) 205.
14. A. Khodakov, J. Yang, S. Su, E. Iglesia and A. T. Bell, J. Catal. 177 (1998) 343.
15. K. Chen, S. Xie, E. Iglesia and A. T. Bell, J. Catal. 189 (2000) 421.
16. K. Chen, S. Xie, A. T. Bell and E. Iglesia, J. Catal. in press.

Studies in Surface Science and Catalysis
J.J. Spivey, E. Iglesia and T.H. Fleisch (Editors)

Quantitative Comparison of Supported Cobalt and Iron Fischer Tropsch Synthesis Catalysts

Roberto Zennaro[a], Gianni Pederzani[a], Sonia Morselli[a], Shiyong Cheng[b] and Calvin H. Bartholomew[b]

[a]EniTecnologie S.p.A., Via F. Maritano 26, 20097 San Donato Milanese (Milano) Italy
[b]BYU Catalysis Lab, Department of Chemical Engineering, Brigham Young University, Provo, UT 84602, USA

Abstract
A collaborative study between two laboratories was undertaken to determine Fischer-Tropsch activity and selectivity properties of representative cobalt and iron catalysts under conditions relevant to natural-gas to liquids conversion. In FTS activity/selectivity tests of Co/Al_2O_3 and Fe/K/Cu supported on silica, conducted under comparable operating conditions, i.e., $H_2/CO = 2$, $P_{tot} = 20$ bar, GHSV= 2.1 $NL/g_{cat}/h$ and 50-70% CO conversion, 20-25% higher catalyst-mass-based CO conversion activity is observed for Co/Al_2O_3 relative to Fe/K/Cu/Si. Site-based rates (i.e., turnover frequencies based on H_2 chemisorption data) are two times higher for Co/Al_2O_3 relative to Fe/K/Cu/Si. C_{5+} selectivity for Co/Al_2O_3 is 50-60% higher due to its higher propagation probability and its negligible CO_2 selectivity; its C_{2+} productivity is also 50-70% higher because of its higher activity and higher selectivity to C_{2+} hydrocarbons. Activity and selectivity data obtained in the laboratories of Brigham Young University and EniTecnologie using micro and bench scale fixed bed reactors are in the case of Co/Al_2O_3 in excellent agreement and in the case of Fe/K/Cu/Si in fairly good agreement.

1. INTRODUCTION

Wax-crack Fischer-Tropsch synthesis (FTS) is an economically promising, developing technology for conversion of natural gas to middle distillate hydrocarbons. Both cobalt [1-8] and iron [6,7,9] catalysts are active and selective for FTS and have found previous large-scale application: supported cobalt in the production of middle-distillates from natural gas by Shell in Malaysia during the last decade and precipitated iron in the conversion of coal synthesis gas to gasoline and chemicals for the past five decades at Sasol, South Africa. Because of its higher CO conversion activity and hydrocarbon selectivity, high activity maintenance, and potential regenerability, supported cobalt is presently the leading FTS catalyst technology [5]; nevertheless, iron catalysts can reportedly achieve higher productivities at higher space velocities (lower conversions) relative to cobalt catalysts [6,7] and might find application in multi-reactor process designs or as a throw-away catalyst because of their significantly lower cost.

An important factor driving the choice of cobalt or iron catalyst for natural gas-to-liquid (GTL) conversion is the technology of the syngas production process which in turn determines H_2/CO ratio. Commercially available technologies produce syngas from methane with a H_2/CO ratio of 1.7 to 3.8; however, Fe-catalysts, because of their WGS activity, perform best at H_2/CO ratios of 0.6-1.5, while Co-based catalysts have optimum performance at higher H_2/CO ratios (1.5-2.2). Nevertheless it may be possible to operate economically with an iron catalyst or a combination of iron and cobalt catalysts at H_2/CO ratios of 1.7-2.0.

In previous comparisons of cobalt and iron catalysts [6,7], CO conversion and hydrocarbon selectivity data were obtained at a different H_2/CO ratio for each catalyst type; moreover, in none of the previous literature are comparisons of CO conversion rate per gram catalyst or turnover frequency made under comparable high conversion conditions. Nor are comparisons published for well-characterized, supported cobalt and iron catalysts under conditions relevant to natural gas-to-liquids (GTL) conversion.

The aim of this study was to compare activities and selectivities of commercially representative iron and cobalt catalysts under similar operating condition, i.e. at 20 bar, H_2:CO = 2:1, and CO conversions of 50-70%. This study also included comparison of data obtained at Brigham Young University (BYU) and EniTecnologie (ET) laboratories in micro and bench scale fixed-bed reactor units.

2. EXPERIMENTAL

2.1. Catalyst Preparation

A 14% Co catalyst supported on alumina (Co/Al) and a silica supported Fe/K/Cu catalyst (52% wt% Fe, 1.9% K, 2.6% Cu) were prepared at EniTecnologie using *incipient wetness* and *precipitation* techniques respectively.

2.2. Catalyst Characterization

X-ray diffraction of calcined Co/Al showed the crystalline phase to contain 20.3 wt% of spinel phase with an average formula $Co_{2.8}Al_{0.2}O_4$, while the alumina support (Condea) was found to be a mixture of gamma and delta phases. The calcined Fe/K/Cu/Si consisted mainly of well-crystallized hematite (α-Fe_2O_3). BET surface areas and pore volumes of the fresh catalyst are reported in Table 1.

Table 1
Surface area and pore volumes for fresh catalysts

Sample	Surface area (m^2g^{-1})	Pore volume (cm^3g^{-1})
Alumina	174	0.52
Co/Al	152	0.35
Fe/K/Cu/Si	229	0.47

Mössbauer spectra of the Fe/K/Cu/Si, in calcined and activity-tested, passivated forms, were collected as previously described [10]. Spectra of used catalysts were collected at both 295 K and 77 K to better resolve the more complex spectral patterns of the four or five iron phases present. Based on a match of Mössbauer parameters for compounds reported in the literature, peaks from the best fit of each data set were assigned to corresponding iron species.

Selective H_2 chemisorption uptakes were measured by a flow desorption method using a custom flow system with a thermal conductivity detector (TCD) similar to that described by Jones and Bartholomew [11].

2.3 Activity tests

FTS activity tests were conducted in two different fixed-bed reactor systems: (1) a *bench scale* reactor (ET) charged with 20 cm^3 catalyst, shaped into particles of 20-40 mesh, and diluted with quartz chips (1:2 by volume), the latter of which served to minimize the axial temperature gradient in the catalyst bed and control reactor temperature and (2) a *micro scale* reactor (BYU) charged with 2 g of iron catalyst (about 60-100 mesh) diluted with 6 g of quartz chips or with 1 g of cobalt catalyst diluted with 8 g of quartz chips.

Co/Al was activated *in situ* in H_2 at 400°C and 2.0 NL/g_{cat}/h for 16 h. The Fe/K/Cu/Si catalyst was activated *in situ* in CO at 300°C and 1.0 NL/g_{cat}/h for 16 h. The use of CO instead of H_2 for precipitated Fe has been demonstrated [12] to improve the catalyst activity. Catalysts were tested for FTS activity and selectivity at 20 bar, 200-280°C, H_2/CO = 2, and GHSV = 3.75-10 NL/g_{cat}/h (BYU) or GHSV = 2 NL/g_{cat}/h (EniTecnologie).

3. RESULTS AND DISCUSSION

Co supported on alumina is one of the most common FT catalyst formulations cited in the literature [1-5], while impregnation to *incipient wetness* is a common synthesis route for cobalt catalysts [1,3]. The alumina (Condea) used in this study had a particle size and mechanical characteristics suitable for slurry phase application [13].

Fe/K/Cu/Si was prepared according to ref. [9]. Potassium is an important promoter for iron catalysts, increasing both activity and the average molecular weight of products, while decreasing the formation of graphite on the catalyst surface [14,15]. The addition of copper facilitates iron reduction at lower temperatures and improves thermal stability, thereby maintaining higher active surface area. Silica is used as a binder to stabilize iron crystallite size and improve the mechanical strength of the catalyst.

Mössbauer parameters and iron phase distributions of Fe/K/Cu/Si (52 wt% Fe) in its calcined and passivated forms after activity testing are listed in Table 2. The spectrum of the calcined catalyst consists of a doublet of superparamagnetic α–Fe_2O_3 (*sp*-α–Fe_2O_3). After a 16 h pretreatment in CO at 300°C and subsequent FTS reaction conducted at ET, Fe_2O_3 was converted to oxides and carbides of lower oxidation state. The working catalyst (see Table 2) consists of a mixture of superparamagnetic Fe_3O_4 (30.1%), ferromagnetic Fe_3O_4 (13.9%; 4.9% in A-sites and 9.0% in B-sites), and χ-carbide or $Fe_{2.5}C$ (49.2%). The peak of low intensity at about 2.4 mm/s in the spectrum has an isomer shift and quadruple splitting parameters consistent with those of Fe^{2+}[10]; this peak accounts for 6.8% of iron. The estimated percentages of Fe_3O_4, $Fe_{2.5}C$, and Fe^{2+} at 295 K of 44, 49, and 6.8% are in very good agreement with those at 77 K of 44, 47, and 7.8% respectively. Thus, these data provide strong evidence that $Fe_{2.5}C$ and Fe_3O_4 are the predominant phases in the working catalyst.

H_2 chemisorption capacities for the fresh Co/Al and Fe/K/Cu/Si precursors after reduction for 16 h at 400 and 300°C were 52.7±6.6 and 81.4±10.5 μmoles/g respectively; dispersions were 10.1±1.3 and 3.5±0.4. These data are based on extents of reduction of 43.3 and 50% obtained by temperature programmed reduction (TPR) and Mössbauer spectroscopy.

In this study activity/selectivity data were obtained on the same cobalt and iron catalysts under comparable conditions in two different laboratories fixed bed reactors. Activity and selectivity data obtained in a bench-scale, fixed-bed reactor (ET) are compared in Table 3 with those obtained in a fixed-bed microreactor (BYU); excellent agreement between the results obtained in the two laboratories in the fixed bed units of different size (by a factor of 10) is evident. The steady-state TOF from both labs for the Co/Al catalyst at 200°C of $2.2 \times 10^{-2}\ s^{-1}$ (extrapolated using an activation energy of 100 kJ/mol [16]) is the same within experimental error as those reported earlier [5,14,16] for Co catalysts under the same conditions. This validates the comparison of rates between the micro and bench scale reactors for cobalt catalysts. For the Fe catalyst rates/g obtained in the micro reactor were 25-60% higher, while TOF value were only 10-25% higher, i.e., almost the same within experimental error.

While comparison of the TOF values between Fe catalysts of this study may provide a useful relative measure of site activity, a similar comparison between Fe and Co catalysts is questionable, since the iron metal sites were measured by H_2 chemisorption on iron catalysts reduced in H_2, while the active catalytic phase is presumably an iron carbide. On the other hand, comparisons of the TOF values based on H_2 chemisorption among cobalt and ruthenium catalysts reduced in H_2 is warranted, since (1) the reduced metal is known to be the active phase, (2) site density – activity correlations have been shown to be linear [4,5,14,16], and (3) excellent agreement has been observed for TOF values obtained in a number of laboratories for cobalt on different supports at different metal loadings [1,4,5,14,16].

Table 2
Mössbauer parameters of precipitated Fe/Cu/K/Si

Species	Iron Site	IS[a] (mm/s)	ΔE_Q[b] (mm/s)	HFS[c] (KOe)	% Area (295 K)	% Area (77 K)
Fe/K/Cu/Si calcined (Mössbauer 36 h at 295 K)						
Fe_2O_3 (SP)[d]		0.34	0.68	--	100	
Fe/K/Cu/Si after run (Mössbauer 48 h at 295 K)						
Fe_3O_4 (SP)		0.41	1.00	--	30.1	
Fe_3O_4 (FiM)[e]	A	0.34	--	477	4.9	
	B	0.72	--	447	9.0	
χ-$Fe_{2.5}C$	I	0.27	--	176	17.3	
	II	0.36	--	214	17.2	
	III	0.52	--	114	14.7	
Fe^{2+}		1.04	2.03	--	6.8	
Fe/K/Cu/Si after run (Mössbauer 9 h at 77 K)						
Fe_3O_4 (SP)[d]		0.44	1.08	--		33.4
Fe_3O_4 (FiM)[e]	A	0.42	--	493		5.3
	B	0.78	--	466		5.7
χ-$Fe_{2.5}C$	I	0.33	--	196		17.5
	II	0.44	--	240		15.7
	III	0.32	--	109		13.4
Fe^{2+}		1.32	2.30	--		7.8
α-Fe		--	--	342		1.2

[a]Isomer shift relative to α-Fe.
[b] Quadrupole splitting.
[c] Hyperfine (magnetic) field splitting.
[d] Superparamagnetic species.
[e] Ferrimagnetic species.

Table 3
Comparison of activity test results for Co/Al from EniTecnologie and BYU.[a]

Activity Test	T_{av}[b] (°C)	X_{CO} (%)	$-r_{CO}$ (μmol/g·s)	TOF x 10^2 (s^{-1})	S_{CH_4} (%)	S_{CO_2} (%)
EniTecnologie[d]	202	29.2	2.7	1.1	7.3	0.4
	210	54.0	5.3	2.1	7.6	0.4
BYU[e]	202	7.6	2.8	1.2	8.6	n.d.[c]
	212	13.4	5.0	2.1	8.5	n.d.[c]

[a] Reaction conditions: 20 bar; $H_2/CO = 2$.
[b] For bench scale reactor temperature was a geometric average.
[c] CO_2 selectivity was below the detection limit.
[d] GHSV = 2.1 NL/g_{cat}/h; H_2:CO = 2, no He; 20 cm^3 catalyst.
[e] GHSV = 10 NL/g_{cat}/h; H_2:CO:He = 2:1:0.375; 2 cm^3 catalyst.

In a fixed-bed microreactor test at 260°C (BYU), the CO conversion increased sharply to more than 95% during the first 5 h (Fig.1) probably due to formation of active surface sites on the carbide. During the next 37 h, CO conversion decreased steadily from 95 to 62%. The observed steady decrease in catalyst activity at 260°C during the period of 5-42 h may be attributed to: (1) the formation of inactive carbonaceous deposits on the catalyst surface and/or (2) oxidation of the active carbide phase to inactive magnetite.

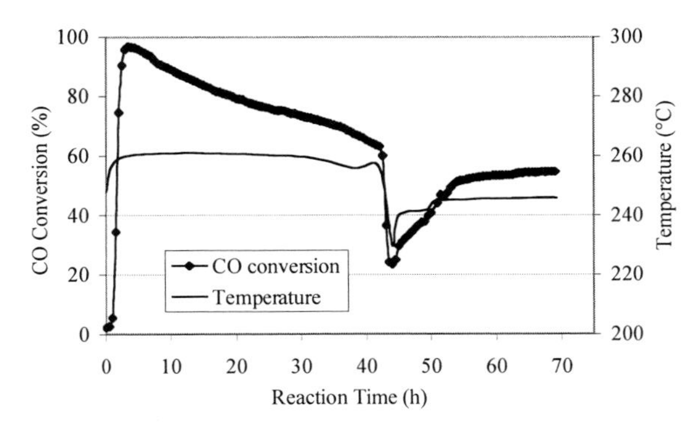

Fig. 1. CO conversion for Fe/K/Cu/Si tested on FB-micro-reactor (BYU).

After 42 h of reaction the temperature was decreased to 230°C and gradually raised to 245°C over the next 10 h and then held at 245°C; CO conversion first decreased and then increased, reaching a steady state value of 55% after a total of 70 h of reaction. By contrast, if the temperature was increased gradually from about 200 to 240°C, the deactivation rate was significantly lower and steady-state conversion could be reached after 60-80 h.

Thus, in comparing the two catalysts (Fe/K/Cu/Si and Co/Al) data were collected at steady state following a gradual increase of the reaction temperature during the initial 20-40 h of reaction and usually after 100 h from the start of the reaction to ensure stable conditions.

Results of selected FTS activity and selectivity tests of Co/Al and Fe/K/Cu/Si catalysts are summarized in Tables 4 and 5. The data in both tables were obtained at high temperatures and conversions typical of commercial operation. From these data it is evident that the cobalt is significantly (20-25%) more active on a catalyst weight basis and significantly (two times) more active on a turnover frequency basis; its C_{2+} productivity is 50-70% higher because of its higher activity, negligible CO_2 selectivity (relative to a high CO_2 selectivity for Fe/K/Cu/Si because of its high water-gas-shift activity), and hence higher selectivity to C_{2+} hydrocarbons. On the other hand the cobalt catalyst has a higher selectivity S (based on at.% C) to methane than the iron catalyst, and its methane selectivity is more temperature dependent (see Table 4). Nevertheless, the selectivity S' of cobalt for C_2-C_4 hydrocarbons (CO_2-free hydrocarbon wt. basis) is lower and its C_{5+} selectivity higher than for Fe, consistent with its higher propagation probability (see Table 5). It should be emphasized that these activity/selectivity data (Tables 4 and 5) were obtained under conditions of low pore diffusional resistance based on common criteria [5,16]; this is especially true of the data obtained at 50% conversion or less.

Table 4
Fixed bed bench scale reactor (EniTecnologie)
Steady-state activity and selectivity data for Fe/K/Cu/Si and Co/Al catalysts.[a]

Catalyst	T_{av} (°C)	X_{CO} (%)	$-r_{CO}\,10^6$ (mol/g/s)	TOF ($10^2 s^{-1}$)	S_{CH_4} [b] (%)	S_{CO_2} [b] (%)	$S_{C_{5+}}$ [b] (%)	α [c] C_{10+}	$r_{C_{2+}}$ [d] ($g_{C_{2+}}/g_{cat}/h$)
Fe/K/Cu/Si	247	68.5	5.3	1.7	5.3	25.5	47.9	0.90	0.15
Fe/K/Cu/Si	221	49.9	3.9	1.3	5.6	18.4	52.2	0.91	0.12
Co/Al	219	64.6	6.6	2.7	10.6	0.6	75.7	0.91	0.23
Co/Al	210	54.0	5.3	2.1	6.9	0.4	84.4	0.92	0.20

[a] Reaction conditions: 20 bar; H_2/CO=2:1; GHSV = 1.82 NL/g_{cat}/h for Fe/K/Cu, 1.82 NL/g_{cat}/h for Co/Al.
[b] Selectivity in atomic C % based on CO conversion.
[c] Propagation probability based on Anderson-Schulz-Flory kinetics.
[d] Productivity in grams C_{2+} hydrocarbon per gram catalyst per hour.

Table 5
Hydrocarbon distribution [a]

Catalyst	T_{av} (°C)	X_{CO} (%)	S'_{CH4} (%)	$S'_{C_2-C_4}$ (%)	$S'_{C_5-C_9}$ (%)	$S'_{C_{10}-C_{21}}$ (%)	$S'_{C_{22+}}$ (%)	S'_{olefin} (%)	$S'_{alcohols}$ [b] (%)
Fe/K/Cu/Si	247	68.5	8.1	21.3	19.4	24.7	20.7	26.7	5.8
Fe/K/Cu/Si	221	49.9	7.8	19.3	14.7	27.4	22.9	25.5	7.9
Co/Al	219	64.6	12.1	9.2	23.1	32.4	19.9	13.7	3.3
Co/Al	210	54.0	8.0	7.2	13.2	35.8	33.1	11.6	2.6

[a] S' denotes selectivity in wt %, based on overall hydrocarbon production (CO_2 not included): $100=\Sigma S'_{hy} - S'_{olefin}$.
[b] Water soluble alcohols not included.

CONCLUSIONS

1. In FTS activity/selectivity tests of Co/Al_2O_3 and Fe/K/Cu supported on silica, conducted under comparable operating conditions, i.e., $H_2/CO = 2$, $P_{tot} = 20$ bar, GHSV= 2.1 $NL/g_{cat}/h$ and 50-70% CO conversion, 20-25% higher catalyst-mass-based CO conversion activity is observed for Co/Al_2O_3 relative to Fe/K/Cu/Si. Site-based rates (i.e., turnover frequencies based on H_2 chemisorption data) are two times higher for Co/Al_2O_3 relative to Fe/K/Cu/Si.

2. C_{5+} selectivity S (based on C at.% from CO conversion) of Co/Al_2O_3 is 50-60% higher due to its higher propagation probability and its negligible CO_2 selectivity; its C_{2+} productivity is also 50-70% higher because of its higher activity and higher selectivity to C_{2+} hydrocarbons. On the other hand the cobalt catalyst has a higher selectivity to methane than the iron catalyst, and its methane selectivity is more temperature dependent. Nevertheless, the selectivity S' of cobalt (on a CO_2-free wt. basis) for C_2-C_4 hydrocarbons is lower and its C_{5+} selectivity higher than for Fe, consistent with its higher propagation probability.

3. Activity and selectivity data obtained in the laboratories of Brigham Young University and EniTecnologie using micro and bench scale fixed bed reactors are in the case of Co/Al_2O_3 in excellent agreement and in the case of Fe/K/Cu/Si in fairly good agreement.

REFERENCES

1. C.H. Bartholomew and R.C. Reuel, Ind. Eng. Chem. Prod. Res. Dev., 24 (1985) 56; J. Catal., 85 (1984) 78.
2. J. H. E. Glezer, K. P. De Jong, and M. F. M. Post, European Patent EP 0,221,598, (1989).
3. E.S. Goodwin , J.G. Marcelin, T. Riis, US Patent 4,801,573 (1989); US Patent 4,857,559. (1989); US Patent 5,102,851 (1992).
4. E. Iglesia, S.L. Soled, R. A. Fiato, J. Catal., 137 (1992) 212-224.
5. E. Iglesia, Applied Catalysis A: General 161 (1997) 59-78.
6. P.J. van Berge and R. C. Everson, Stud. Surf. Sci. and Catal.,107 (1997) 207.
7. A Raje, J. Inga and B. H. Davis, Fuel 76 (1997) 273.
8. J.J.C. Geerlings, J.H. Wilson, G.J. Kramer, H.P.C.E. Kuipers, A. Hoek and H.M. Huisman, Appl. Catal. A 186 (1999) 27.
9. M.E. Dry, Catalysis – Science and Technology, Vol. 1, Springer Verlag, NY (1981) 175.
10. M.W.Stoker, BYU thesis, 1999.
11. R. D. Jones, C.H. Bartholomew, Applied Catalysis 39 (1988) 77.
12. A.K. Datye, M.D. Shroff, M.S. Harrington, A.G. Sault, N.B. Jackson, Stud. Surf. Sci. and Catal.,107 (1997) 169.
13. R.L. Espinoza, EP Patent No.736 326 (1996).
14. C.H. Bartholomew, Stud. Surf. Sci. and Catal.,64 (1991) 158.
15. A.Datye, N.B. Jackson, L. Evans, Stud. Surf. Sci. and Catal.,119 (1998) 137.
16. R. Zennaro, M. Tagliabue, C.H. Bartholomew, Catalysis Today 58 (2000) 309-319.

Studies in Surface Science and Catalysis
J.J. Spivey, E. Iglesia and T.H. Fleisch (Editors)

CO_2 Abatement in Gas-To-Liquid Plants

Harry Audus[a], Gerald Choi[c], Alasdair Heath[b]; and Samuel S. Tam[c] (presenter)

[a]Project Manager, IEA GHG R&D, CRE Group Ltd., Stoke Orchard, Cheltenham, Gloucestershire, GL52 4RZ, United Kingdom
[b]Research & Development, Bechtel Corporation, 45 Fremont Street, 14th Floor, San Francisco, CA 94105, USA
[c]Nexant Inc., A Bechtel Technology and Consulting Company, 45 Fremont Street, 7th Floor, San Francisco, CA 94105, USA

Abstract

The conversion of natural gas at its source to higher-valued liquid fuels is an alternative to the pipeline transportation of treated gas and the dedicated tanker shipping of LNG. This IEA study investigated the effects of reducing carbon emissions from gas-to-liquid (GTL) plants for such remote applications. Three different Fischer-Tropsch GTL plant configurations with characteristics similar to those of Sasol, Shell, and Syntroleum were examined. These three configurations are selected because they cover the major types of GTL technologies currently considered to be either commercially viable or under development.

The design of each of these FT- based GTL plants was based on published information. The purity of the captured CO_2 was better than 95% and is suitable for on-site injection. Plant investment and operating costs were developed from the Bechtel in-house database. The licensors were asked to comment on the results of this study and their comments were included in the Appendix of the study report.

Three steps are generally taken in the task of carbon-emissions abatement: (1) select appropriate carbon-capture technology, (2) identify the location in the process configuration for optimum carbon removal, and (3) modify the process configuration to maximize carbon utilization in liquid fuel products and minimize carbon emission. For each GTL technology, a base case was developed to determine the baseline carbon emissions and the capital and operating cost, prior to CO_2 abatement. The plant was then redesigned to achieve at least 85% reduction in CO_2 emissions, with the capital and operating cost determined as before.

The CO_2 capture technologies investigated here were limited to the commercially-proven processes, such as chemical absorption (amine solutions, etc.), physical adsorption (molecular sieve), and cryogenic separation. Combustion of hydrocarbon fuels in process heating or flaring was minimized to limit CO_2 emissions from the GTL plant.

1. Introduction

This paper summarizes some of the results of a study sponsored by the IEA Greenhouse Gas R&D Programme (IEA GHG)[1]. The aims of the study are to identify technical hurdles and options, to quantify the economical impacts, and to provide data that characterizes the emissions associated with producing road-transport fuels via F-T synthesis. No attempt is made to compare the performance and plant cost of these three GTLplant configurations. The approach of the IEA study is to first develop base cases by selecting three typical GTL plant configurations. Due to

---

[1] IEA Report Number P43/15, November 2000

the difficulties in seeking proprietary information from the technology licensors, these base cases are developed with published data and generally acceptable engineering principles. Then, the CO_2 emission-abatement cases are prepared with similar principles and consistent data. While this approach does not allow readers to compare these three GTL plant configurations, it serves the purposes in illustrating the impacts on the GTL plant performance, cost, and efficiency from CO_2 emission abatement in each GTL plant configuration.

In an F-T process, synthesis gas (CO and H_2) reacts over a catalyst to produce a mixture of straight-chain hydrocarbons which can be used to produce road-transport fuels. The technology has been in existence for many years but has found only limited commercial applications. Historically, the focus has been on the use of F-T as a method of producing transport fuels from coal. The SASOL plants in South Africa are perhaps the best-known example of the technology in commercial use. More recently, F-T technology developed by SASOL was licensed to Mossgas for the conversion of natural gas to liquid fuels in a plant in South Africa. The 12,500 bbl/day Shell MDS (middle distillate synthesis) plant in Bintulu, Malaysia, is a further example of a natural gas processing plant based on F-T technology.

The published plant performance data are selected merely to develop a consistent base case in order to illustrate the impacts of CO_2-abatement on the GTL plant performance, cost, and efficiency. In some cases, the performance data do not represent the latest and/or the highest performance for the select GTL plant configuration. For an example, the efficiency of the Shell-type GTL plant in this study[2,3,4] is lower than the commercially-demonstrated plant efficiency (55% versus 63%). Thus, there are differences in the performance and costs of the GTL processes but they are not likely to have a crucial impact on the overall outcome of a well-to-wheels evaluation of the potential for CO_2 abatement.

2. Scope of Study

In this study, favored combinations of synthesis gas production and F-T synthesis for the production of a liquid road-transport fuel were assessed; techno-economic evaluations were used to establish emissions from, and the costs of, the fuel production process. CO_2 was captured and compressed at appropriate points in the processing scheme.

This study compared three F-T synthesis processes, each including an appropriate synthesis gas production technology. The objective was to establish the range of likely results for the use of such technology. This study also evaluated and compared, at the process plant level, the emissions, efficiencies, and costs of three versions of F-T technology each matched with the most appropriate synthesis gas production technology.

The three technologies evaluated were:

1. The Slurry Phase Distillate (SPD) process, a process based on established SASOL technology, is a low-temperature F-T process using a churn-turbulent bubble column reactor with catalyst contained in process-derived liquid; the synthesis gas is produced using oxygen obtained from an air separation unit (ASU).

---

[2] J. Ansorge and A. Hoek, "Conversion of aNatural gas to clean transportation fuels via the SMDS Process", paper presentedat Spring ACS Natinal Meeting, San Francisco,CA. April 5-10, 1992

[3] M.F.M.Post, et al, "Diffusional Limitations in F-T Catalysts", paper published in AIChE Journal, July 1989, Vol 35, No. 7, p1107

[4] J.W.A. De Swart, et al, "Selection, Design, and Scale-up of the F-T Reactor", published in Natural Gas Conversion, IV, Elsevier

2. The Shell MDS (middle distillate synthesis) process, in which F-T synthesis occurs in fixed-bed tubular reactors; the synthesis gas is produced using oxygen obtained from an ASU.
3. A recently proposed alternative process, the 'Syntroleum' process in which the synthesis gas is produced using air rather than oxygen. The process is claimed to be a lower-cost method for converting natural gas to liquid transport fuels.

3. Study Bases & Assumptions

This study was based on the concept of building a medium-sized GTL facility which was to be a self-contained, grass roots producer of transportation liquids from locally available low-priced natural gas. The flexibility of either exporting or importing small quantities of electric power at low cost either to or from a local grid was assumed. Other site-related and feed stock assumptions are shown in Table 1.

Table 1 Assumptions on Plant Site and Feed Stock

Plant Location:	Saudi Arabian Gulf Coast - product shipped to Northern Europe			
Plant Feedstock:	Natural Gas			
Natural Gas Composition mol%:	CH_4	94.476	$i\text{-}C_5H_{12}$	0.024
	C_2H_6	3.438	$n\text{-}C_5H_{12}$	0.024
	C_3H_8	0.856	CO_2	0.437
	$i\text{-}C_4H_{10}$	0.098	N_2	0.471
	$n\text{-}C_4H_{10}$	0.176	Sulfur (as H_2S)	4 mg/Nm^3
Feed Conditions:	Pressure: 33-45 bar		Temperature: 45°C	
Feed Properties:	Molecular Weight: 17.086		Heating Value, LHV: 834 kJ/mol	
Load Factor:	90% of rated capacity for all operating years			
Ambient Air Temperature:	43°C, 80% relative humidity			

Technical information for each of the three types of F-T technology was obtained from published information, information published under DOE contracts, in-house knowledge, and information developed by Bechtel/Nexant. The yields from, and sizes of, the F-T reactors were based on data of Satterfield et al (DOE contract No. DE-AC22-87PC79816), published F-T cobalt catalyst kinetic data (AIChE J. 35, 7, 1107, 1989), published F-T reactor-design simulation information (Natural Gas Conversion, IV, Elsevier), and engineering analysis. The design of the mild hydrocracking plant, used for upgrading the F-T wax, was based on pilot plant data reported by Mobil, PARC, and UOP under DOE contract Nos. DE-AC22-80PC30022, DE-AC22-89PC88400, and DE-AC22-85PC80017.

4. Sasol-Type GTL Plant Design With and Without CO2 Abatement

A conceptual plant design for a GTL plant based on Sasol-type technology was developed as the base case which was compared with a CO_2 abatement case. Figure 1 summarizes the mass, energy, and carbon balance of the base case. Key parameters and assumptions employed in the process simulation model are summarized in Table 2.

The capital cost estimate was based on a factored estimating technique. The ISBL equipment was sized and materials-of-construction selected based on the particular process configuration, heat and energy balance calculations, and the conditions of the locally available utility streams.

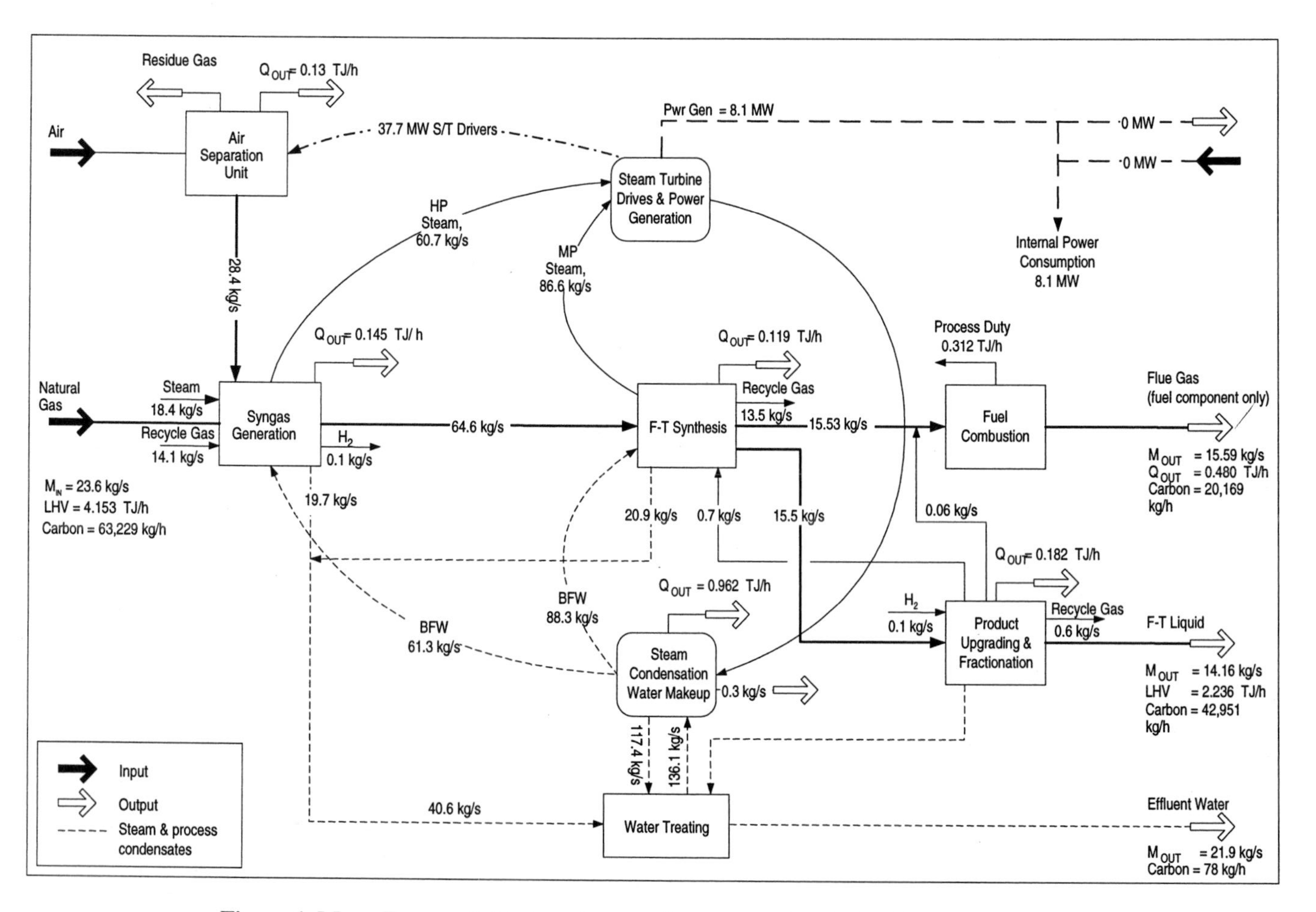

Figure 1 Mass, Energy, & Carbon Balance Summary – Sasol-Type Design - Base Case

Table 2 Key Parameters and Assumptions of the Sasol-Type GTL Plant Base Case

Air Separation Unit	Conventional, single train cryogenic air separation plant	
- Oxygen purity	99.5 mol% O_2	
Syngas Generation	Oxygen-blown, autothermal natural gas reforming	
- Feed ratios:		
- H_2O:C, mole/mole	0.65	
- CO_2:C, mole/mole	0.10	
- O_2:C, mole/mole	0.56	
- Exit conditions:	Pressure: 28 bar	Temperature: 1014°C
- H_2:CO mole ratio	2.04	
Hydrogen Separation	Pressure swing adsorption	
- H_2 purity	> 99.5 mol% H_2	
F-T Synthesis	Single SBCR reactor, cobalt catalyst in F-T synthesis derived liquid, internal heat recovery (steam raising), recycle of part of purge gas to syngas generation, catalyst makeup/activation, catalyst recovery and recycle	
- Operating conditions	Pressure: 26 bar	Temperature: 220°C
- Anderson-Schulz-Flory distribution parameter (α)	Several values used to fit slope of carbon-number distribution for cobalt catalyst	
- CO conversion per pass	76%	
- Steam raising	Saturated – 13 bar, 191°C	
Product Upgrading	Mild hydrocracking of ASTM-D86 350+°C product (wax)	
- Operating conditions	Pressure: 115 bar	Temperature: 370°C
- Reactor LHSV	2 hr^{-1}	

Additional field costs, bulk materials, direct labor, indirect costs, etc., were developed based on cost factors mentioned above. Other field costs, such as sales tax, freight costs, duties, etc., were site specific and developed separately as percentages of the total field cost. The offsite cost estimate was developed from Bechtel in-house data for similar size and type plants in the same site location. The IEA Financial Assessment Criteria was used to develop the costs for home office, fees, and services and plant contingency.

The CO_2 capture and compression area consisted of three plants: (1) feed gas hydrogenation and high-temperature CO shift, (2) MDEA-based CO_2 removal, and (3) CO_2 compression. The intent of this design was capture, prior to emission, the non-product carbon as a single species—CO_2 in this instance—and to deliver it in a pure form and at high pressure to the plant battery limit. This study does not address the collection, transportation, and ultimate disposal/ sequestration of this CO_2 stream.

The purge gas from the F-T synthesis section was hydrogenated to remove the trace amounts of oxygenates and the remaining CO is converted to CO_2 and hydrogen in the shift unit. A 30 wt% aqueous solution of mono-diethanolamine (MDEA) was used to remove CO_2 from the shift reactor effluent gas. The CO_2 from the MDEA regenerator was recovered and compressed to 110 bar for pipeline delivery.

5. Results and Discussion

Due to the page limit for this paper, the details of the Shell and Syntroleum cases are not presented here. Readers who desire to see the details (PFDs, MB, and EB) of this study are asked to contact IEA for the complete report for this study. All three licensors: Shell, Sasol, and Syntroleum declined to comments because of the proprietary nature of their technologies. However, Shell does point out and the authors agree that the performance data of the Shell Bintulu GTL plant are those from the first generation of plant design. The current Bintulu plant efficiency is 63% which corresponds to 12,500 BPD liquid fuels per 100 MM SCFD of gas.

Table 3 includes the final results of all three cases. As shown in Table 3, the reduction in carbon emissions and their costs vary greatly among the three GTL technologies while the same (MDEA) CO_2 capture process is employed. There are two major reasons.

First, the GTL plant designs in the CO_2 abatement cases are not fully optimized. Thus, there are different amounts of CO_2 emitted from these GTL plants as CO_2 in the flue gas streams from direct-fired heaters and furnaces. Second, different amounts of process changes are required in order to maximize the CO_2 capture. For instance, in the Shell-type GTL plant, the steam-methane-reformer (SMR) has to be replaced with an additional POx unit in order to avoid emission of large amounts of CO_2 in the SMR flue gas. Subsequently, the overall thermal efficiency of the GTL plant is increased from 54.8% to 55.6% while the plant capital cost is proportionally higher than those in the other two GTL plants.

The results of this study also indicate that CO_2 capture is costly and difficult to achieve in an air-blown GTL plant. If the DMEA process is replaced with a cryogenic unit, the reduction in carbon emission could be improved from the 57% level. However, this will defeat the intent in avoiding the employment of an air separation plant at the front end of the GTL plant.

Finally, the study has successfully developed a set of consistent emission and cost data in support of IEA GHG Programme in conducting future life cycle analyses for the production of transportation fuels from natural gas. However, additional resources should be allocated to optimize the CO_2 abatement case plant design.

Table 3 GTL Plants CO2 Abatement Costs

	Base Cases			CO_2 Abatement Cases		
GTL Plant Type	Sasol	Shell	Syn-troleum	Sasol	Shell	Syn-troleum
Natural Gas Feed, MM SCFD	100	100	100	100	100	100
Product Rates, BPD: Naphtha	4,136	4,371	4,020	4,063	4,494	4,400
Distillate	6,173	6,093	6,084	6,068	6,257	6,078
Power export/(import), MW	0	3.8	8.9	(10.6)	(7.4)	0
Plant Thermal Efficiency, LHV %	56.1	54.8	53.7	55.0	55.6	54.5
GTL Plant TIC, MM$	346	390	388	389	446	428
Capital Charge, MM$/yr	34.6	38.9	38.8	38.9	44.6	42.8
Annual Operating Cost, MM$/yr	37.5	39.4	49.0	44.2	45.4	55.5
Carbon Emission, MT Carbon per yr x1,000.	160	156	166	48	16	71
Percent Reduction, %				70	90	57
Cost of Reduction, $/MT Carbon				93.9	89.0	124.6

Studies in Surface Science and Catalysis
J.J. Spivey, E. Iglesia and T.H. Fleisch (Editors)

Study on stability of $Co/ZrO_2/SiO_2$ catalyst for F-T synthesis

J.-G. Chen, X.-Z. Wang, H.-W. Xiang and Y.-H. Sun*

State Key Laboratory of Coal Conversion, Institute of Coal Chemistry,
Chinese Academy of Sciences, P.O.Box 165, Taiyuan, PRC

$Co/ZrO_2/SiO_2$ catalyst active for synthesis of heavy hydrocarbons from syngas was subjected to stability test of 500 hours on a laboratory scale. The catalyst showed a considerable decrease in CO conversion while the hydrocarbon distribution hardly changed with time on stream. The regeneration of deactivated catalyst with H_2 at 673K partially restored the activity. XRD, FT-IR and TPR identified a hydrated cobalt silicate species in catalyst near outlet of reactor. Thus, the cause of permanent deactivation was proposed mainly to be formation of inactive hydrated silicate between metallic Co and silica in the presence of high partial pressure of water.

1. INTRODUCTION

The conversion of natural gas to liquid fuel (GTL) via F-T synthesis had been developed by petroleum corporations such as Shell, Exxon and so on. Co-based catalysts were applied and several type of reactors were adopted in these technologies. Among them, the fixed-bed reactor was prominent for simplicity and suitable for the production of waxy hydrocarbons [1]. Recently, we reported a novel Co-based catalyst supported on particular silica, which had merits of high activity and low formation of CH_4. As this catalyst produced a great deal of wax, tubular fixed-bed reactor should be the optimal choice in industrial application. Obviously, the longevity of such a catalyst was of great importance since the productivity of fixed-bed reactor was usually lowered by the laborious and time-consuming replace of catalyst.

Although the deactivation of catalyst was usually considered as loss, poisoning, sintering of active metal as well as coke deposition, the deactivation of Co-based F-T catalyst was related to surface oxidation or compound formation. Schanke [2] studied the possibility of cobalt oxidation during F-T synthesis and found that the deactivation of Co/Al_2O_3 was caused by oxidation of surface cobalt. Van et al [3] showed that the formation of $CoAl_2O_4$ was more thermodynamically favorable than that of cobalt oxide under F-T synthesis conditions. Zhang et al [4] investigated the effect of water vapor on reduction of $Co\text{-}Ru/Al_2O_3$ catalysts, and reported that the water vapor both inhibited the reduction of well-dispersed CoO and facilitated the formation of $CoAl_2O_4$ spinel. Similarly, Co/SiO_2 catalyst was found to form cobalt silicate under hydrothermal conditions [5]. However, these researches took a simulated reaction condition by introduction of water and this might lead to distortion of the reason for deactivation. Thus, a long-term run of $Co/ZrO_2/SiO_2$ catalyst was carried out and the used catalyst was characterized with XRD, FT-IR and TPR to investigate the formation of silicate in real reaction conditions.

* To whom all correspondence should be addressed. Tel: (+86) 351 4049612
E-mail: SKLCC@public. ty. sx. cn.; Fax: (+86) 351 4050320

2. EXPERIMENTAL

2.1 Stability test

$Co/ZrO_2/SiO_2$ catalyst was prepared by impregnation of special silica with aqueous solution of metal nitrate. Details of preparation could be found elsewhere [6]. The said catalyst contained 20wt% Co and 5% Zr. The specific surface area and mean pore diameter of catalyst determined from BET method were $360m^2g^{-1}$ and 6.0nm respectively. TPR provided that the catalyst had a reducibility of 80%.

The stability test of catalyst was carried out in a pressured fixed-bed reactor with a catalyst loading of 5ml at 2MPa, $500h^{-1}$, $H_2/CO=2$ and a temperature range from 459 to 464K. The products were collected with a hot trap and a cold trap in sequence. The tail gas was analyzed by GC. The liquid product was analyzed with GC equipped with an OV-101 capillary column and the solid product by GC with a special injection setup. Both carbon balance and mass balance of reaction were kept at ca 95%.

2.2 Characterization

FT-IR spectra were obtained by scanning KBr pellets of sample in transmission mode with a Nicolet550 spectroscopymeter. TPR was performed with a catalyst loading of 150mg at a heating rate of 10K/min. The carrier gas was diluted hydrogen (5%H_2 in Ar) and the H_2 consumption was detected with TCD. XRD was taken at a Dmax-γ A diffracmeter (Cu target, Ni filter) with a voltage of 40kV and a current of 40mA.

3. RESULTS AND DISCUSSION

Figure 1 shows the performance of the catalyst as a function of time on stream. At the beginning, the catalyst reached a CO conversion of 86.9% at 459K, exhibiting a C_5^+

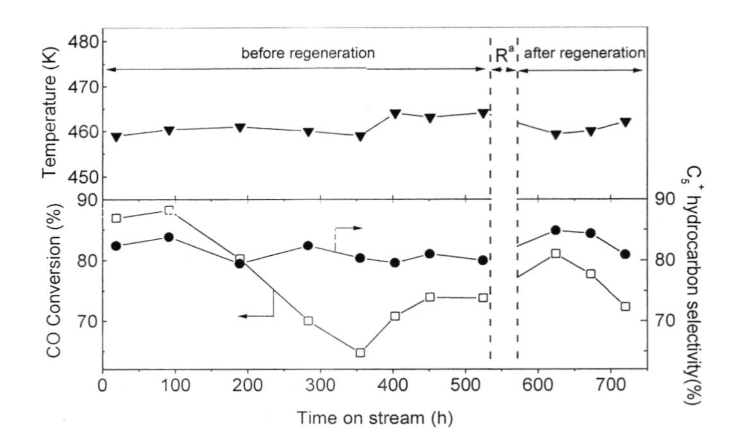

Fig.1 The stability test of $Co/ZrO_2/SiO_2$ catalyst in F-T synthesis. (a: regeneration)

hydrocarbon selectivity in total hydrocarbons up to 82.0 wt %. The hydrocarbon distribution observed Anderson-Flory-Schulz rule well and the chain growth factor approached 0.9 (see Fig.2). The weight percent of CH_4 was about 8%. H_2O was the main byproduct while minor CO_2 as well as oxygenates was detected. As time went on, a considerable drop of CO conversion proceeded though it could be compensated to some degree by a raise in operation temperature. After 500 hours on stream, the catalyst was regenerated by replacing syngas with pure H_2 and increasing temperature to 673K. The activity of catalyst was restored partially after regeneration. In the whole process, the hydrocarbon distribution kept almost unchanged. The catalyst discharged from reactor showed a change in color with bed length. The sample near inlet of reactor was black but that near outlet was pink. Thus, two samples were collected at inlet (named as S2) or outlet (S3) and measured by XRD, FT-IR and TPR. The calcined catalyst (S1) was also measured for the purpose of comparison.

XRD pattern of S1 showed only characteristic peaks of Co_3O_4 as was reported elsewhere [6]. However, the spent catalyst showed rather different profiles (see Fig.3). There was no diffraction peak in S2, suggesting that the cobalt species was highly dispersed. Surprisingly, the XRD pattern of S3 exhibited characteristic peak of cobalt silicate.

FT-IR spectra of S1, S2 and S3 was shown in Fig.4. S1 showed absorption bands at 1096, 800 and 470 cm^{-1}, which was characteristic of SiO_2, and bands at 670 and 581 cm^{-1} assigned to Co_3O_4 [8]. After reaction, FT-IR spectra of used catalyst changed significantly. New bands at 2957,2919,2850 and 1473cm^{-1} evidenced the residue of product hydrocarbons while the disappearance of bands at 670 and 581 cm^{-1} suggested that Co species presented as other form rather than Co_3O_4. Nonetheless, the difference between S2 and S3 was impressive. The latter showed extra peaks at 3624, 1017 and 670 cm^{-1} while the former did not. The peak at 1017 cm^{-1} was assigned to silicate, as was observed in $Na_2SiO_3{\cdot}H_2O$ compound. This assignment was supported by the study on a ferric-substituted silicate in which an adsorption band

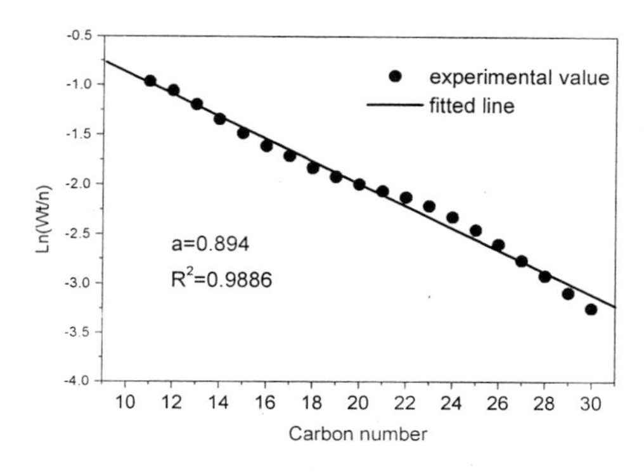

Fig.2 The hydrocarbon distribution of $Co/ZrO_2/SiO_2$ catalyst

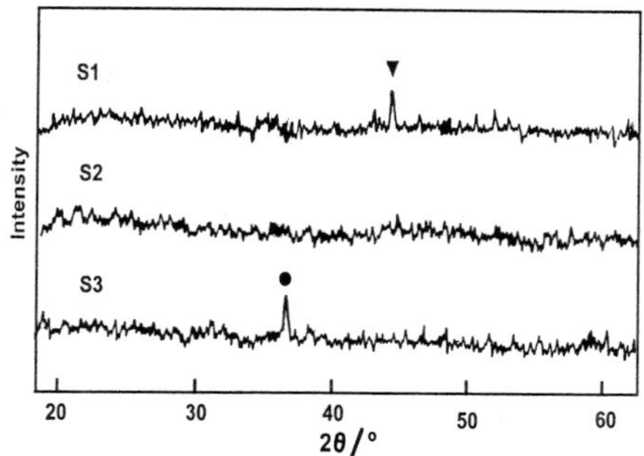

Fig.3 XRD patterns of $Co/ZrO_2/SiO_2$ catalyst after reaction. (▼Co_3O_4, ● Co_2SiO_4)

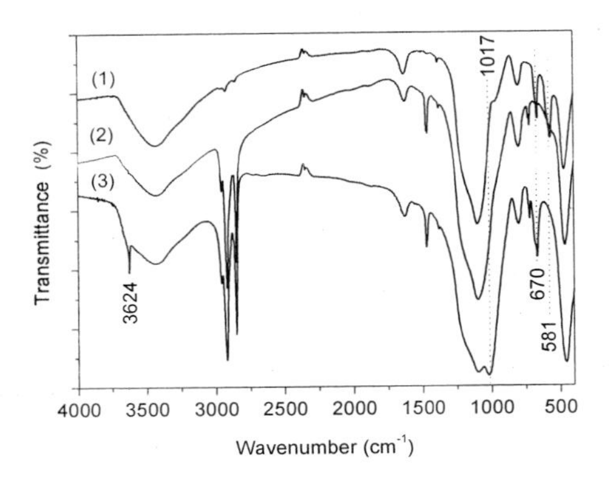

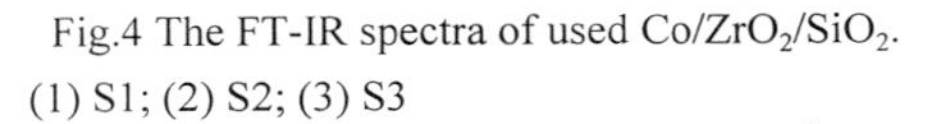
Fig.4 The FT-IR spectra of used $Co/ZrO_2/SiO_2$. (1) S1; (2) S2; (3) S3

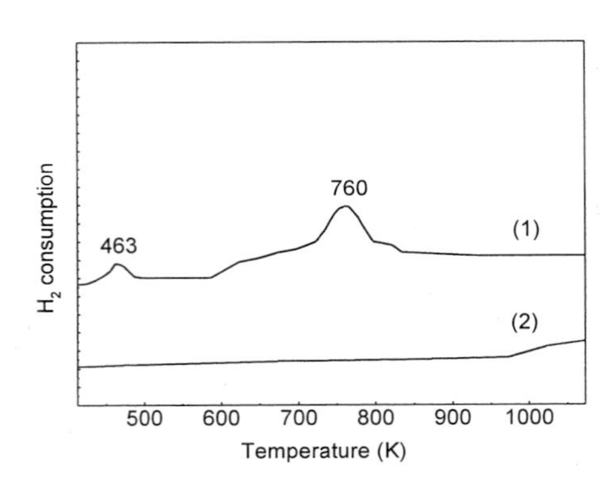

Fig.5 The TPR profiles of deactivated catalysts and reference compound
(1) S3; (2) Co_2SiO_4

at 1034 cm^{-1} was observed [9]. Moreover, Such binary peaks around 1022 cm^{-1} were assigned to vibrations of silicate species interacting with hydroxide species [10]. The 670 cm^{-1} band was related to Co-O bond and 3624cm^{-1} to structural hydroxyl group [11,12], suggesting that the Co in S3 existed as hydrated cobalt silicate. The result of IR was well in line with that of XRD. Both S2 and S3 had identical reaction history except that the latter was near exit where a high concentration of H_2O was expected. Thus, the deactivation was partially supposed to occur by the transformation of cobalt metal into hydrated cobalt silicate in presence of high partial pressure of H_2O. Kondoh et al [13] reported that Co_2SiO_4 could be formed when the SiO_2 surface sustained a significant amount of adsorbed water. The formation of cobalt silicate had also been observed in early industrial practice in which the dark blue Co/diatomaceous earth catalyst turned red and lost activity due to silicate formation when the catalyst was kept in wetness for a long time [14]. Fig.5 showed the reduction behavior of S3 and Co_2SiO_4. The S3 showed a little H_2 consumption at 760K while Co_2SiO_4 did not. Accordingly, the permanent loss of activity, which could not be reversed by hydrogen treatment, was caused by transformation of metallic cobalt into cobalt silicate. The reversible loss of activity might be caused by oxidation of cobalt as was suggested by others[2]. This explained the partial restore of activity of $Co/ZrO_2/SiO_2$ catalyst upon regeneration.

4. CONCLUSION

The stability test of $Co/ZrO_2/SiO_2$ catalyst showed that the activity decreased while the hydrocarbon distribution hardly changed with time on stream. The regeneration of deactivated catalysts with H_2 at 673K partially restored the activity. XRD, FT-IR, and TPR defined a hydrated cobalt silicate, which could be reduced partially. Such a silicate was supposed to be responsible for irreversible deactivation.

REFERENCE

1. J. Eilers, S.A. Posthuma and S.T. Sie, Catal. lett., 7 (1990) 253.
2. D.Schanke, A.M.Hilmen, E.Bergene, K.Kinnari, E.Rytter, E.Adnanes, A.Holmen, Preprints Div. Fuel Chem. Amer. Chem. Soc., 40 (1995) 167.
3. V.P.J. Berge, V. Loosdrecht, S.Barradas, Preprint Div. Petro. Chem. Amer. Chem. Soc., 41 (1999) 84.
4. Y.L.Zhang, D.G. Wei, S. Hammache, J. G. Goodwin, Jr., J. Catal., 188 (1999) 281.
5. A. Kogelbauer, J.C.Weber, J.G. Goodwin Jr., Catal. Lett., 34 (1995) 259.
6. J.G.Chen, H.W. Xiang, Y.H. Sun, Chinese J. Catal., 21 (2000) 169.
7. International Center for Diffraction Data, Powder Diffraction File, 1988. P1230.
8. K. Gaku, W. Hiromu, O. Toshio, M. Makoto, J. Chem. Soc. Faraday Trans., 92 (1996) 3425.
9. V. C.Farmer, J. D.Russell, Spectrochimica Acta, 20 (1964) 1149.
10. K.D. Chuge, Appl. Catal., 103 (1998) 183.
11. Handbook of Spectrascopy edited by H.A.,Szymanski, Plenum Press.
12. Infrared Band Handbook vol(Ⅱ) 99 edited by J. W.Robinson, CRC Press.
13. E.Kondoh, R.A.Donaton, S.Jin, H.Bender, W.Vandervorst, K.Maex, Appl. Surf. Sci., 136 (1998) 87.
14. The Engineering of Synpetroleum, Petroleum Industry Press, Beijing, (1958) P34.

Studies in Surface Science and Catalysis
J.J. Spivey, E. Iglesia and T.H. Fleisch (Editors)

Partial Oxidation of Methane to Formaldehyde on Fe-doped Silica Catalysts

F. Arena [1,2], F. Frusteri [2], J. L. G. Fierro[3] and A. Parmaliana [1,4]

[1] Dipartimento di Chimica Industriale e Ingegneria dei Materiali, Università di Messina, Salita Sperone c.p.29, I-98166, S. Agata (Messina), Italy
[2] Istituto CNR-TAE, Via Salita S. Lucia Sopra Contesse 5, I-98126, S. Lucia (Messina), Italy
[3] Instituto de Catalísis y Petroleoquímica, Campus UAM, Cantoblanco s/n, 28049 Madrid, Spain
[4] Dipartimento di Chimica, Università degli Studi "La Sapienza", P.le Aldo Moro 5, 00185 Roma

Abstract

The effect of the Fe doping on the catalytic pattern of commercial silica samples in the partial oxidation of methane to formaldehyde (MPO) with molecular O_2 at 650°C has been investigated. The influence of loading (Fe_2O_3, 0.02-3.2 wt%) on the activity-selectivity pattern and steady-state of the surface of the Fe/SiO_2 system has been highlighted. EPR measurements of Fe/SiO_2 catalysts indicate the presence of "isolated Fe^{3+} ions", small Fe_2O_3 clusters and Fe_2O_3 particles, their relative concentration depending upon the surface loading. Whatever the loading, Fe promotes the reactivity of the silica surface, while the maximum in HCHO surface productivity, associated with a surface loading of 0.05-0.1 $Fe_{at}\cdot nm^{-2}$, signals that HCHO formation is mostly driven by "isolated Fe species".

Keywords: *Fe/SiO_2 catalysts, Methane Partial Oxidation, Formaldehyde, Iron surface species, Reduced sites*

1. INTRODUCTION

The direct partial oxidation of methane to formaldehyde (MPO) on oxide catalysts is currently one of the most sought goals in the framework of the natural gas conversion to higher added value products. Silica based oxide catalysts denoted a superior functionality in the MPO /1/ and several clues have been provided for relating the surface structure and the coordination of transition supported metal oxide species with their reactivity /2-5/. However, the peculiar reactivity of the silica surface in MPO, ensuring remarkable HCHO productivity values /6-8/, is still an undecided issue /6,8-10/. We contributed to this research subject pointing out that the performance of silica in MPO is controlled by the preparation method /6,8,10/ resulting in the following reactivity scale: *precipitation > sol-gel > pyrolysis* /6/. Notably, the activity of silica catalysts was also correlated with both the concentration of strained siloxane bridges /10/ and density of surface sites under steady state conditions /6/, whereas no relationships between the concentration of alkali and alkaline-earth metal oxides and reactivity have been found /6/. Recently, a comprehensive kinetic study of the MPO on precipitated silica in the range 500-800°C /11-12/ indicated the occurrence of a competitive reaction path of both CH_4 and O_2 molecules on the same active sites of the silica surface according to a "push-pull" redox mechanism. Then, the existence of surface siloxane defect sites generated during dehydroxylation at high temperature has been invoked to account for the peculiar reactivity of the SiO_2 catalyst in MPO /4,8,10/. Kobayashi et al. /13/ reported that the doping of a fumed SiO_2 sample with Fe^{3+} ions greatly enhances the HCHO productivity in the MPO reaction. Fe loading, dispersion and coordination of Fe^{3+} ions on the SiO_2 matrix are crucial for determining the extent of such promoting effect. Moreover, silica surface denotes an intriguing catalytic functionality in other reactions, such as ammoximation of cyclohexanone /14/, n-C_6H_{14} cracking and carbon monoxide oxidation /15/.

Therefore, in order to further contribute to the understanding of the factors controlling the reactivity of the SiO_2 surface in the MPO reaction, this paper is aimed to evaluate the role of Fe^{3+} ions on the surface features and catalytic behaviour of various unpromoted and Fe doped commercial silica samples.

2. EXPERIMENTAL

2.1. Catalyst. Two series of Fe doped silica catalysts have been prepared by impregnation of "fumed" M5 (Cab-O-Sil M5, Cabot product, $S.A._{BET} = 200\ m^2 \cdot g^{-1}$) and "precipitated" Si 4-5P (Si 4-5P, Akzo product, $S.A._{BET}=400\ m^2 \cdot g^{-1}$) silica samples with aqueous solution of $Fe(NO_3)_3$ /16/. The impregnated samples were dried at 90 °C and then calcined at 600°C for 16 h. The list of Fe doped SiO_2 samples along with their code, BET surface area, Fe_2O_3 content and Fe surface loading (S.L.) values are shown in Table 1.

2.2. Catalyst Testing. Catalytic data in the MPO reaction were obtained using a specifically designed batch reactor /7,8/. All the runs were carried out at 650 °C and 1.7 bar using 0.05g of catalyst and a recycle flow rate of 1,000 STP $cm^3 \cdot min^{-1}$. Further details on the experimental procedure and product analysis are reported elsewhere /6/.

2.3. Catalytic Characterization

2.3.1.Reaction Temperature Oxygen Chemisorption (RTOC). In order to probe the state of the catalyst surface under reaction conditions, *in situ* RTOC measurements at 650°C were performed in a pulse mode using He as carrier gas /6,17/. After treatment in the $CH_4/O_2/He$ reaction mixture flow, the sample was purged in the carrier flow and then O_2 pulses (12.9 nmol O_2) were injected until saturation of the sample. The density of reduced sites (ρ, $O_{at} \cdot g^{-1}$) was calculated from the O_2 uptakes by assuming a O_2:"reduced site"=1:2 chemisorption stoichiometry /17/.

2.3.2. Electron Paramagnetic Resonance (EPR). EPR spectra of bare and Fe doped SiO_2 catalysts were recorded at –196°C and r.t. (25°C) with a Bruker ER 200D spectrometer operating in the X-band and calibrated with a DPPH standard (g = 2.0036). A conventional high vacuum line ($<10^{-4}$ torr) was employed for the different treatments. Spectra were recorded after outgassing of the samples at r.t. and 500°C.

Table 1. List of Fe doped SiO_2 catalysts

SiO_2 Support	Code	Fe_2O_3 loading (wt%)	$S.A._{BET}$ ($m^2 \cdot g^{-1}$)	S.L. ($Fe_{at} \cdot nm^{-2}$)
	F1-SI	0.086	400	0.0162
	F2-SI	0.170	400	0.0320
Si 4-5P	F3-SI	0.270	400	0.0508
	F4-SI	1.090	390	0.2105
	F5-SI	3.250	370	0.6613
	F1-M5	0.023	200	0.0087
M5	F2-M5	0.087	200	0.0327
	F3-M5	0.200	203	0.0753
	F4-M5	0.830	200	0.3124

3. RESULTS and DISCUSSION

3.1. Activity of bare and Fe doped SiO_2 catalysts. The catalytic behaviour of Fx-SI and Fx-M5 catalysts is outlined in Figure 1 in terms of reaction rate and selectivity to HCHO and CO_x (CO + CO_2) vs. Fe_2O_3 loading. Addition of Fe^{3+} ions to the precipitated Si4-5P silica yields an enhancement in reaction rate and a concomitant decrease in HCHO selectivity, paralleled by a corresponding increases in CO_x (Fig. 1A). The extent of Fe content plays a critical role in controlling the performance of Fx-SI catalysts. Indeed, when Fe_2O_3 loading rises from 0.045 (Si 4 - 5P) to 0.27 wt % (F3-SI catalyst) a significant promoting effect on the reaction rate from 2.7 to 6.7 $\mu mol_{CH4}\cdot s^{-1}\cdot g_{cat}^{-1}$ along with a modest lowering in HCHO selectivity from 78 to 62% is recorded. Notably, at higher loadings, the F4-SI sample (Fe_2O_3, 1.09 wt %) exhibits a catalytic activity similar to that of the F3-SI system and a considerably lower HCHO selectivity (43%). The highly loaded F5-SI system (Fe_2O_3, 3.25 wt%) features the highest reaction rate value (14.2 $\mu mol_{CH4}\cdot s^{-1}\cdot g_{cat}^{-1}$) associated with the lowest HCHO selectivity (20%).

The doping of fumed M5 SiO_2 with different amount of Fe^{3+} ions implies a progressive promoting effect of the catalytic activity along with a gradual lowering in the selectivity to HCHO (Fig. 1B). Namely, the positive effect of Fe addition to the reactivity of the M5 SiO_2 is analogous to that experienced for the Si 4-5P SiO_2 sample. In fact, we observe a significant promoting action up to a Fe_2O_3 loading of 0.2 wt % while for higher loading no further increase in the activity is observed. Moreover, in spite of the remarkable difference in activity of the undoped M5 (0.2 $\mu mol_{CH4}\cdot s^{-1}\cdot g_{cat}^{-1}$) and Si 4-5P (3.1 $\mu mol_{CH4}\cdot s^{-1}\cdot g_{cat}^{-1}$) silica samples it is evident that the addition of similar amount of Fe^{3+} levels off the activity of the related Fe doped silica catalysts, as documented by the comparable activity of F3-SI (7.0 $\mu mol_{CH4}\cdot s^{-1}\cdot g_{cat}^{-1}$) and F3-M5 (5.3 $\mu mol_{CH4}\cdot s^{-1}\cdot g_{cat}^{-1}$), and F4-SI (7.4 $\mu mol_{CH4}\cdot s^{-1}\cdot g_{cat}^{-1}$) and F4-M5 (6.4 $\mu mol_{CH4}\cdot s^{-1}\cdot g_{cat}^{-1}$) catalyst samples.

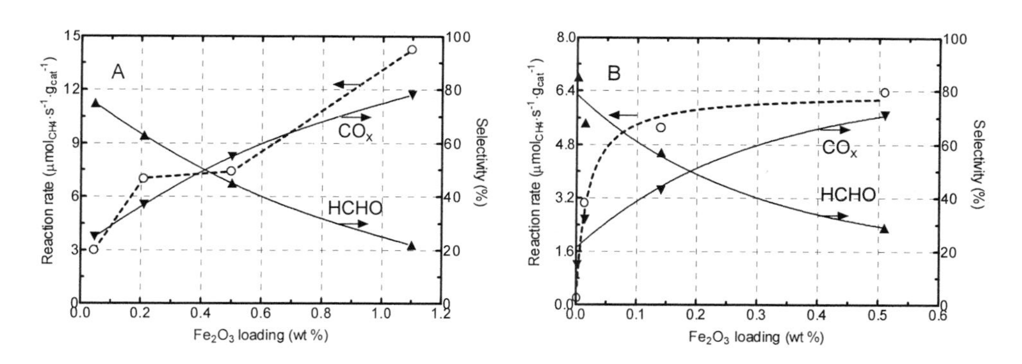

Figure 1. Effect of the Fe_2O_3 loading on reaction rate (O) and product selectivity ((▲) S_{HCHO}; (▼) S_{COx}) of Fe/SiO_2 catalysts in MPO at 650°C: (A) Fx-SI and (B) Fx-M5 catalysts.

On the whole, these findings lead to infer that for Fe_2O_3 loading lower than 0.3 wt % the Fe addition to the silica results in a specific promoting effect, while for higher Fe_2O_3 loadings (>0.3 wt %), where the decrease in HCHO selectivity outweighs the enhancement in reaction rate, we deal with non-selective oxidation catalysts leading to the formation of CO_x with a residual selective oxidation functionality. Then, the catalytic behaviour of low doped Fe silica catalysts (Fe_2O_3< 1.0 wt %) indicates that Fe content is a key factor tuning the activity of the SiO_2 surface in the MPO reaction irrespective of the preparation method and original functionality.

3.2. Surface species (EPR) and redox properties under steady-state (RTOC) of Fe doped SiO_2 catalysts. The total ($\rho_{tot.}$), irreversible ($\rho_{irr.}$) and reversible ($\rho_{rev.}$) density of reduced sites (ρ) with Fe_2O_3 loading for doped Si 4-5P and M5 silica samples are shown in Figures 2. The above data indicate that: i) both the total and irreversible oxygen uptake under steady-state reaction conditions increase monotonically with Fe_2O_3 loading for both Si 4-5P and M5 doped silica samples; ii) the difference between ρ_{tot} and ρ_{rev} for both series of silica samples becomes progressively more pronounced at higher Fe_2O_3 loadings; iii) $\rho_{irr.}$ tends to a maximum value equal to ca. 25 and $15 \cdot 10^{16}$ $s_r \cdot g^{-1}$ on Si 4-5P and M5 silica samples for loadings higher than 2.0 and 0.5 wt %, respectively and iv) at comparable Fe_2O_3 loading levels $\rho_{tot.}$, $\rho_{irr.}$ and $\rho_{rev.}$ attain analogous values. In other words, Fe_2O_3, as a consequence of an easier reduction under reaction conditions /11,12/, involves the stabilization of higher densities of reduced sites along with a promoting effect on the catalytic activity of the silica surface. Moreover, the larger difference between $\rho_{rev.}$ and ρ_{tot} for highly loaded Fe/SiO_2 catalysts, accounting for the capability of such systems to "release" constitutional oxygen /17/, signals a high mobility of lattice oxygen ions likely linked with the formation of "bulk-like" oxide species on the silica surface /17/.

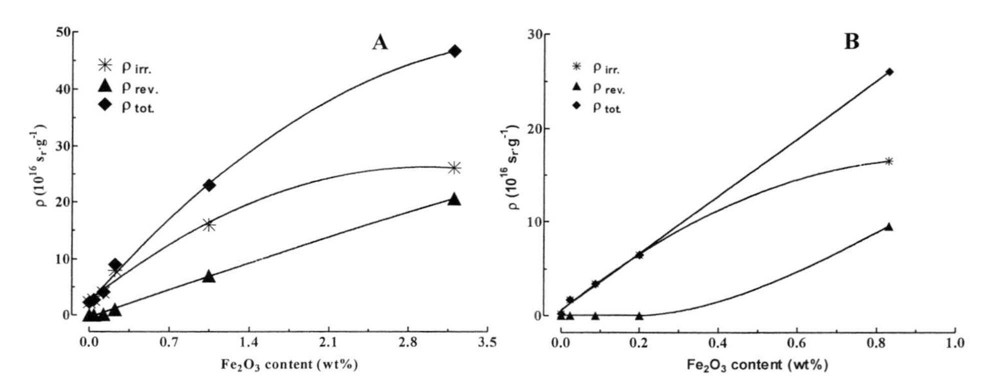

Figure 2. Effect of the Fe_2O_3 loading on density of reduced sites under steady-state of Fe/SiO_2 catalysts at 650°C: (A) Fx-SI and (B) Fx-M5 catalysts.

EPR spectra of SI 4-5P, F3-SI and F5-SI samples, recorded at –196°C are shown in Figure 3(A), while the relative intensity of signals A and B (see *infra*) is compared in Figure 3(B). Two spectral features dominate the spectra; a narrow slightly anisotropic line at $g_{eff} = 4.32$ (signal A) and a broad, almost symmetric signal, whose relative contribution to the spectra is apparently larger, centered at $g_{eff} = 2.24$-2.18 (signal B). Independent experiments (not shown) have evidenced the sensitivity of these signals to an O_2 atmosphere, thus revealing the surface location of the corresponding centers. Apparent decrease of signals A and B, larger for the latter, along with a new weak and narrow isotropic signal at g=2.00 (signal C), arising probably from silica-related structural defects, are produced upon outgassing SI 4-5P and F3-SI samples at 500°C (Figs. 3d-e). For the sample F5-SI, in addition to signal C, a new very large and broad anisotropic signal showing a very large amplitude at low magnetic field, signal D, is produced upon this outgassing treatment (Fig. 3f). Signal A at g_{eff}=4.32 is attributed to isolated Fe^{3+} ions in a rhombic environment /18-22/ since it is difficult to ascertain the symmetry, octahedral or tetrahedral, accounting for such signal. The features observed at g_{eff}=9.0-6.0 are most likely due to the same isolated Fe^{3+} species, since the parallel evolution of these features and signal A; they would correspond to particular energy transitions resulting from the resolution of the spin hamiltonian appropriate for Fe^{3+} high spin $3d^5$ systems like the present ones /21,22/. It cannot be fully discarded, however, that these features belong to other different isolated Fe^{3+} species in an axial symmetry /21/. The large anisotropy and width of signal D (and, in a lower extent, of signal B too) indicate that the species responsible for the signal undergoes a strong anisotropic fields due to magnetic interactions between the spins forming the corresponding oxidic phases. The higher linewidth of signal D with respect to signal B could reflect the difference in the type of oxidic phase, which in the case of signal D might correspond to Fe_3O_4 /23/, formed by reduction of relatively larger Fe_2O_3 particles present in the sample F5-SI.

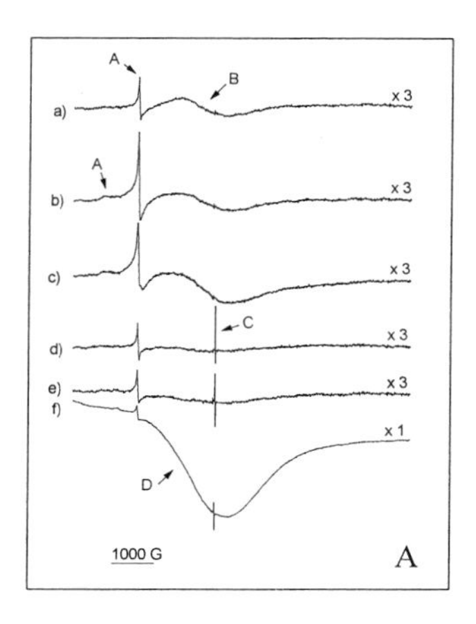

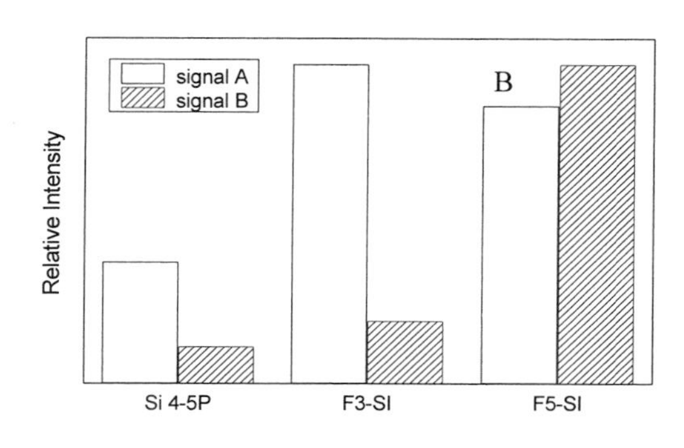

Figure 3. A) EPR spectra at -196°C of samples outgassed at r.t. (a-c) and 500°C (d-f). (a,d) Si 4-5P; (b,e) F3-SI; (c,f) F5-SI and B) relative intensity of signal A and B in the various catalysts.

Thus, EPR data show the presence of different oxidized iron species, whose degree of aggregation growths with iron content. Accordingly, Figure 3B, presenting the relative intensities of EPR signals A and B for the differently loaded Fx-SI samples, signals that low doped F3-SI sample is characterised by the highest concentration of isolated Fe^{3+} species (signal A), while the highest extent of aggregated species is present on the highly loaded F5-SI sample. Thus, considering the activity data shown in Fig. 1, a direct correlation between the amount of Fe^{3+} isolated centres and selective centres for partial oxidation of methane is drawn, while aggregated iron oxide phases promote the total oxidation. These findings are further explicated by the different trends of specific surface activity (SSA, $nmol_{CH4}{\cdot}m^{-2}{\cdot}s^{-1}$) and surface productivity (SY, $g_{HCHO}{\cdot}m^{-2}{\cdot}h^{-1}$) of all the studied catalysts *vs.* the surface Fe loading (Table 1), shown in Figure 4. It is evident that the rising trend of SSA with the SL (Fig. 4A) points to a promoting role of any surface Fe species on the reactivity of the silica surface. However, the fact that the promoting role of Fe on SSA is much more sensitive at very low Fe_2O_3 loading (<0.2 wt%), besides to be in agreement with a more effective dispersion of the promoter on the silica at low SL (<0.1 $Fe_{at}{\cdot}nm^{-2}$), roughly confirms the 2nd-order relationship between reaction rate and concentration of active sites outlined in our mechanistic studies /11,12/. Moreover, it is evident that the specific functionality towards HCHO formation pertains to isolated Fe species, as proved by the maximum in SY value found at SL values ranging between 0.05 and 0.1 $Fe_{at}{\cdot}nm^{-2}$ (Fig. 4B).

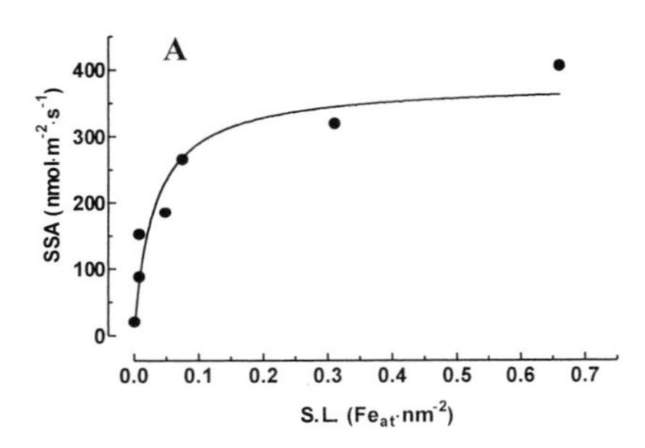

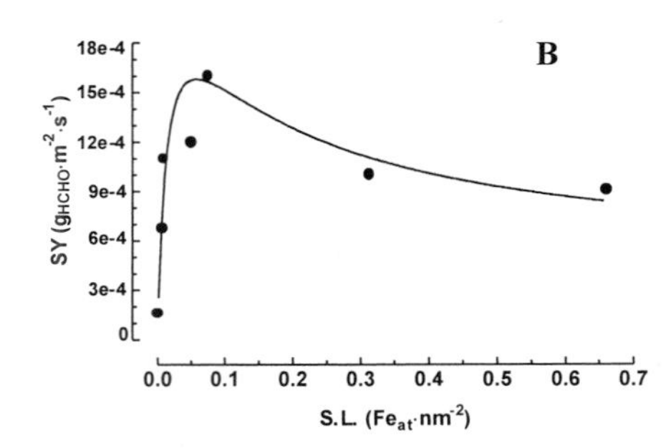

Figure 4. (A) Specific surface activity (SSA) and (B) surface productivity (SY) of both Fx-SI and Fx-M5 catalysts vs. surface Fe loading (SL).

Finally, from the trends depicted in Figure 4, it arises that the surface Fe loading is a key-parameter which allows an unifying rationalisation of the catalytic pattern of the Fe/SiO_2 system in MPO.

4. CONCLUSIONS

a) Addition of Fe^{3+} ions implies a significant promoting effect on the activity of any silica sample, levelling off the difference in the intrinsic activity linked to the preparation method.

b) Isolated Fe^{3+} species, small Fe_2O_3 clusters and large Fe_2O_3 particles coexist on the surface of Fe/SiO_2 catalysts; such species are characterised by different coordination, reducibility and catalytic functionality.

c) A peculiar volcano-shape relationship between surface Fe loading and surface HCHO productivity signals that the best performance of Fe/SiO_2 catalysts is related to the highest density of isolated Fe^{3+} sites on the silica surface.

References

1. R.C. Herman, Q. Sun, C. Shi, K. Klier, C.B. Wang, H. Hu, I.E. Wachs and M.M. Bhasin; *Catal. Today* 37 (1997) 1
2. M. Faraldos, M.A. Bañares, J.A. Anderson, H.Hu, I.E. Wachs and J.L.G. Fierro, *J. Catal.* 160 (1996) 214.
3. Q. Sun, J. M. Jehng, H. Hu, R.G. Herman, I.E. Wachs and K. Klier, *J. Catal.* 165 (1997) 91
4. A. Parmaliana and F. Arena, *J. Catal.* 167(1997) 57
5. F. Arena, N. Giordano and A. Parmaliana, *J. Catal.*167 (1997) 66
6. A. Parmaliana, V. Sokolovskii, D. Miceli, F. Arena and N. Giordano, *J. Catal* 148 (1994) 514
7. A. Parmaliana, F. Arena, F. Frusteri and A. Mezzapica, *Stud. Surf. Sci. Catal.* 119 (1998) 551
8. Q. Sun and R.G. Herman and K. Klier, *Catal. Lett.* 16 (1992)251
9. T. Ono, H. Kudo and J. Maruyama, *Catal. Lett.* 39 (1996)73
10. K. Vikulov, G. Martra, S. Coluccia, D. Miceli, F. Arena, A. Parmaliana, E. Paukshtis, *Catal. Lett.* 37 (1996) 235
11. F. Arena, F. Frusteri and A. Parmaliana, *Appl. Catal. A: General* 197 (2000) 239
12. F. Arena, F. Frusteri and A. Parmaliana, *AIChE J.* 46 (2000) 2285
13. T. Kobayashi, N. Guilhaume, J. Miki, N. Kitamura and M. Haruta, *Catal. Today* 32 (1996) 171
14. D. Pinelli, F. Trifirò, A. Vaccari, E. Giamello and G. Pedulli, *Catal. Lett.* 13 (1992) 21
15. Y. Matsumura and J. B. Moffat, *J. Chem. Soc. Faraday Trans.* 90 (1994) 1177
16. A. Parmaliana, F. Arena, F. Frusteri and A. Mezzapica, German Patent n° 100 544576 assigned to SÜD CHEMIE AG
17. Parmaliana, F. Arena, V. Sokolovskii, F. Frusteri and N. Giordano, *Catal. Today* 28, 363 (1995)
18. S. Bordiga, R. Buzzoni, F. Geobaldo, C. Lamberti, E. Giamello, A. Zecchina, G. Leofanti, G. Petrini, G. Tozzola and G. Vlaic. *J. Catal.*, 158, 486 (1996).
19. E.G. Derouane, M. Mestdagh and L. Vielvoye. *J. Catal.* 33 (1974) 169
20. F. Gazeau, V. Shilov, J.C. Bacri, E. Dubois, F. Gendron, R. Perzynski, Y.L. Raikher and V.I. Stepanov, *J. Magn. Magn. Mater.* 202 (1999) 535
21. R. Aasa. *J. Chem. Phys.* 52 (1970) 3919
22. D.L. Griscom and R.E. Griscom. *J. Chem. Phys.* 47 (1967) 2711
23. M. Hagiwara, K. Nagata and K. Nagata. *J. Phys. Soc. Japan.* 67 (1998) 3590

Studies in Surface Science and Catalysis
J.J. Spivey, E. Iglesia and T.H. Fleisch (Editors)

Production of light olefins from natural gas

B. V. Vora[a], P. R. Pujadó[a], L. W. Miller[a], P. T. Barger[a], H. R. Nilsen[b], S. Kvisle[c], and T. Fuglerud[c]

[a]UOP LLC, P. O. Box 5017, Des Plaines, Illinois 60017-5017, USA
[b]Norsk Hydro ASA, Bygdøy allé 2, N-0240 Oslo, Norway
[c]Norsk Hydro ASA, P. O. Box 2560, N-3901 Porsgrunn, Norway

Introduction

One of the current dreams of the petrochemical industry is the direct conversion of natural gas (mostly methane) to higher-value petrochemical intermediates, in particular light olefins like ethylene and propylene. While such a direct conversion is not yet feasible, technology for the production of methanol from syn gas is readily available, and syn gas can be easily produced from just about any hydrocarbon feedstock, from natural gas to heavy residues or coal. The MTO (methanol-to-olefins) process provides a further means for the downstream conversion of methanol to more valuable petrochemical intermediates.

The MTO process has been jointly developed by UOP LLC and Norsk Hydro ASA for the selective production of ethylene and propylene from either crude or refined methanol. This technology has been extensively demonstrated at a large-scale facility owned by Norsk Hydro in Norway, and is being considered for commercial use in various projects. The catalyst used in the MTO process is based on a silicoaluminophosphate, SAPO-34, that was originally discovered at Union Carbide and later transferred to UOP [1-3].

The MTO process converts methanol to ethylene and propylene at about 75-80% carbon selectivity. The carbon selectivity approaches 90% if butenes are also accounted for as part of the product slate. There is considerable flexibility in the operation of an MTO unit. Typically, the C_2=/C_3= ratio can be modified within a range from about 0.7 to 1.4 by adjusting the operating temperature, with higher temperatures leading to higher C_2=/C_3= ratios [3,4]. The MTO process can operate with low concentration methanol feeds like those obtained from methanol synthesis units before final fractionation and purification. Dimethyl ether (DME) can be used instead of or in combination with methanol, so that DME separation from the methanol feed is not required either.

The economics of MTO have already been extensively discussed elsewhere [5,6]. This paper focuses on the most relevant engineering design features of the MTO process and provides updated economics on a fully integrated gas-to-polymers (GTP) complex for the production of polyethylene and polypropylene based on the use of MTO technology.

Process description

Figure 1 illustrates a typical process flow scheme for a new MTO unit. Methanol feed can be used either totally or in part as scrubbing means to recover small amounts of DME in the reactor effluent before being introduced in the reactor. In the reactor, the conversion of methanol and DME to light olefins proceeds to completion (> 99.5% conversion) in a very short residence time. The unique 3.8 Å pore size of the SAPO-34 catalyst (Figure 2) affords a highly selective conversion to light olefins (C_2= to C_4=, with some C_5=) with the almost total exclusion of heavier and cyclic compounds [4].

Figure 1

MTO Process Flow Scheme

Reactor Regenerator Quench Tower Caustic Wash De-C_2 De-C_1 C_2 Splitter De-C_3 C_3 Splitter De-C_4

Regen Gas, Air, Water, MeOH, DME Removal, Dryer, C_2H_2 Reactor, Tail Gas, Ethylene, Propylene, Mixed C_4, C_5+, Propane, Ethane

UOP 3705-1

The MTO process developed by UOP and Norsk Hydro is significantly different from other approaches, like the pioneering work done by Clarence Chang et al. at Mobil [7], in that those earlier attempts made used of a modified MFI zeolite with a larger pore opening (about 5.5Å with a three-dimensional channel structure). Hence, while still producing significant amounts of olefins, use of MFI zeolites also yields 30% or more of C_5 and heavier gasoline range products. That approach can be best characterized by Mobil's successful commercialization of their methanol-to-gasoline (MTG) process in New Zealand.

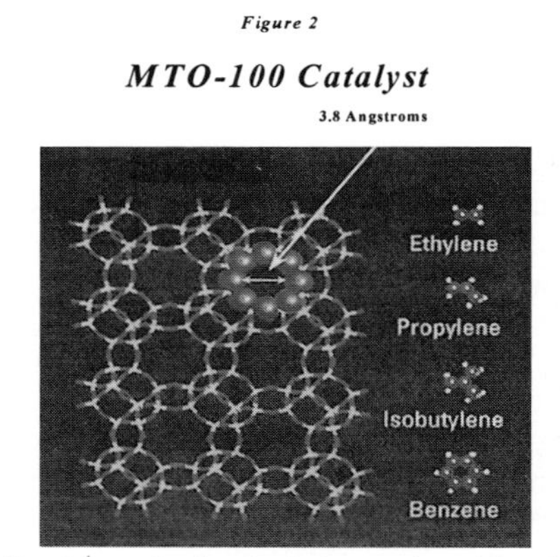

The unique pore size allows the selective conversion to olefins and excludes heavier compounds

UOP 3705-2

In the process of converting methanol to olefins a certain amount of coke is laid on the catalyst. Thus, the method of coke removal and catalyst regeneration dictates the choice of reactor design. A fluidized bed reactor normally provides the best means for catalyst circulation to a separate regenerator in which the coke is combusted and also facilitates the removal of the heat of reaction generated from the methanol conversion.

Different types of fluidized bed reactors can be used for the conversion of methanol. A dense phase reactor with a continuous or periodic removal of catalyst through a slip stream to a regenerator is a first logical choice and offers excellent performance characteristics. By its nature, however, dense phase fluidized bed reactors operate at fairly low superficial gas velocities in the order of 1 m/s, and low space velocities. Thus, it is advantageous to minimize catalyst inventory requirements by employing a higher level of fluidization that in commercial practice can range from a circulating fast fluidized bed reactor to a dilute phase riser type reactor. Broad commercial experience exists in the commercial design and utilization of the latter types of reactor in FCC (fluidized catalytic cracking) applications, either as reactors or as regenerators (see [8], fig. 2). Because of its flexibility, a fast fluidized reactor offers significant operating advantages while reducing the catalyst inventory to levels approaching those in riser reactors.

Reactor operating conditions can be adjusted to the desired product requirements. Pressure is normally dictated by mechanical considerations, with lower methanol partial pressures resulting in higher overall yields. Thus, some yield advantage can be obtained by using a crude methanol feed that typically may contain around 20 wt% water. On the other hand, increased pressures may be favored for the production of higher propylene ratios. Temperature is an important control variable – higher temperatures are required to increase the proportion of ethylene in the product relative to propylene, but higher temperatures also result in a small reduction in the overall light olefin selectivity. Temperature effects are best appreciated in Figure 3. Operating temperatures range from 400 to about 550°C, with more typical values in the 425 to 500 °C range.

Figure 3

Olefins Selectivity vs. Operating Severity

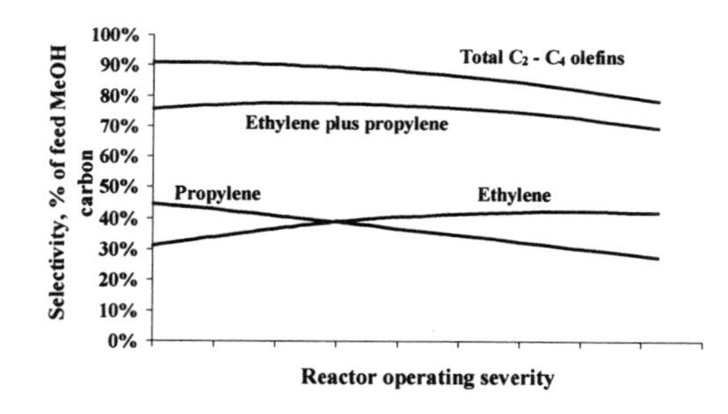

The MTO catalyst operates best at a certain equilibrium level of coke. Fresh or fully regenerated catalyst has very high methanol conversion activity, but cokes up rapidly and has relatively poor selectivity toward the formation of light olefins. Thus, typically, regenerated catalyst is best returned to the reactor at a point above the feed methanol injection point at which the conversion of methanol has already been substantially completed. The slip stream of catalyst to regeneration is calculated to balance off the net yield of coke deposited on the catalyst, so as to maintain a uniform level of coke on the catalyst within the reactor [9].

Because both the conversion of methanol to olefins and the combustion of coke on the catalyst are exothermic, means are provided to remove heat from the reactor and from the regenerator such that the fluidized beds operate at essentially isothermal conditions. Typically, depending on the size of the units and the relative economics, a "back-mix" type catalyst cooler is used in the regenerator and a "flow-through" type catalyst cooler is used in the reactor for the removal of heat and the generation of valuable high-pressure steam that can then be used advantageously elsewhere in the complex.

The product from the MTO reactor contains a substantial percentage of water vapor that is produced as a byproduct from the main methanol condensation and DME dehydration reactions. The water is condensed and separated in a quench tower equipped with suitable heat recovery means; the design of the quench tower is typical of the current art for ethylene crackers. The saturated gas from the quench tower is compressed, processed for the recovery of unconverted DME and methanol, scrubbed with caustic for the removal of CO_2, dried, and fed to the product olefin recovery train. As is usually the case in olefin plants, various recovery schemes are possible; Figure 1 illustrates a typical "deethanizer first" scheme.

Product quality

Light olefins are commonly produced by the thermal cracking of hydrocarbon fractions at very high temperatures (900-1100 °C) in the presence of steam as diluent. Steam cracking reactors produce a broad range of oxygenated impurities, diolefins, and acetylenes that need to be removed in the fractionation train to recover polymer-grade quality ethylene and propylene products. Although based on an oxygenated feedstock like methanol or DME, all the minor oxygenated impurities from the MTO reactions can also be found in most steam cracker effluents. The only difference is that the concentration level of these impurities, though still in the ppm range, is usually higher in an MTO unit than in a conventional cracker. On the other hand, the MTO reaction, because of its milder operating conditions, yields much lower levels of dienes and acetylenic compounds. By way of an example, Figure 4 provides a side-by-side comparison of typical GC analyses of the oxygenates in the condensates from an LPG cracker and from an MTO unit; the parallelism between the two diagrams can be easily appreciated.

Figure 4

Comparison of MTO and LPG Cracking

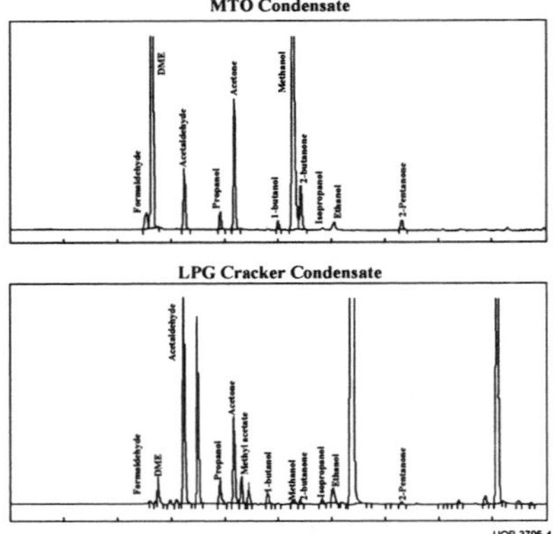

Purification of the MTO products is similar to the methods employed in steam crackers. Selective hydrogenation is used for the removal of trace amounts of acetylene, propadiene (allene), methylacetylene, and butadiene. A proprietary adsorbent is used for the complete removal of oxygenated impurities down to the limit of detection in an Oxygenates Removal Unit (ORU) that can operate either in the vapor phase or in the liquid phase. ORUs have found broad acceptance in highly oxygenate-sensitive catalytic applications, such as isomerization or polymerization, often for the removal of DME or in conjunction with units for the production of MTBE or other light oxygenates.

Table 1 illustrates the typical performance of an ORU in the removal of oxygenated contaminants from a gaseous MTO product stream after scrubbing and drying.

Table 1
Removal of oxygenated contaminants in an ORU

Oxygenated contaminants	ppm before ORU	ppm after ORU
acetaldehyde	700	< 2 (*)
propionaldehyde	190	< 2 (*)
acetone	1260	< 2 (*)
butyraldehyde	100	< 2 (*)
MEK	460	< 2 (*)

(*) typical limit of detection by GC analysis in the gas phase

Economics

MTO projects usually fall in one of three categories:

- Offshore methanol is imported from a location at which plentiful natural gas is available at a nominal cost of recovery. Only MTO and downstream facilities (and offsites) need to be considered.
- A fully integrated gas-to-polymers (GTP) complex is being planned. In this case, the capital cost requirements are augmented by the syn gas and methanol synthesis part of the project.
- MTO is viewed as an add-on to expand the operating capacity of an existing olefins complex. These revamps make use of existing equipment (e.g., the MTO reactor may be a converted FCC riser) and can only be considered on a case-by-case basis.

This paper provides an updated summary of the economics of an integrated GTP complex for the production of polyethylene (PE) and polypropylene (PP) at a 1:1 ratio based on the largest single train size syn gas and methanol synthesis capacity now available (about 1.2 million metric tons per annum [MTA] methanol). This latter restriction is currently being revised upward by a number of companies, with methanol capacities of 5,000 metric tons per day (MTD) or 10,000 MTD being talked about. Lifting the methanol capacity limitations can significantly further improve the future economics of such projects.

Figure 5

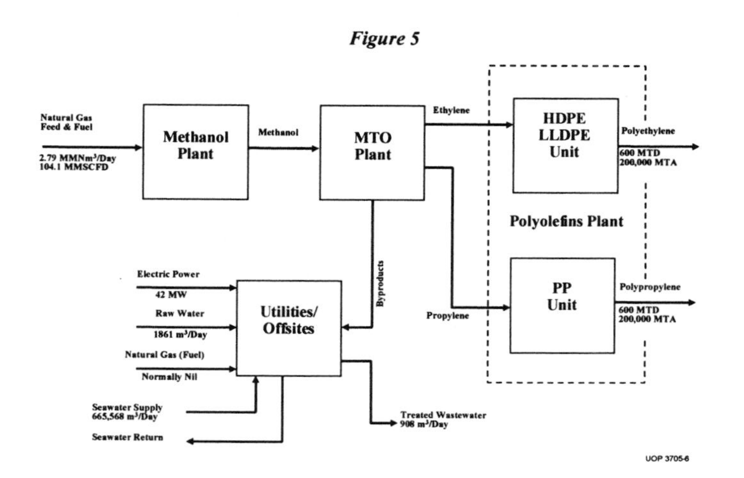

Figure 5 illustrates an integrated flow scheme with flow rates based on a nominal 8000 operating hours per year. On today's basis, with 1.2 million MTA methanol, an MTO unit would yield 204,000 MTA polymer-grade ethylene and 205,000 MTA polymer-grade propylene, sufficient for the production of about 200,000 MTA of PE and 200,000 MTA of PP. The capital investment cost for the entire grass-roots complex on a U.S. Gulf Coast basis is about $835 million, of which about $145 million reflects the ISBL of the MTO unit. These capital cost estimates are based on numbers generated by independent contractors for actual projects and include ISBL, OSBL, license fees, catalyst and adsorbent inventories, and other owner's expenses such as land, permits, etc. Net natural gas feed and fuel requirements are 2.79 million Nm^3/day. Utility imports are limited to electric power (42 MW) and raw water (77.5 m^3/h). Butenes and heavier byproducts were all assumed to be consumed internally as fuel. If natural gas is priced at $0.75/mmBtu and electric power at 3 ¢/kWh, the total cash cost of production for the aggregate of the polyolefins (400,000 MTA) is $235 per MT, or about 10.5 ¢/lb. Clearly, the overall manufacturing cost of polyolefins from natural gas can be very attractive, even with the inclusion of depreciation and return on investment.

Conclusions

The MTO process provides an economically attractive alternative for the production of light olefins, ethylene and propylene, and the corresponding polyolefins from methanol that can be produced at low cost and in large quantities from natural gas or other hydrocarbon sources.

This paper has discussed some of the engineering design features of a typical process unit for the conversion of methanol and/or DME to light olefins. Though based on a relatively novel chemistry, the process configuration and recovery and purification techniques are akin to those currently practiced in industry, either for the production of olefins by fluidized catalytic cracking (FCC) of heavy hydrocarbons or by steam cracking of LPG or naphtha fractions.

References

1. B.M. Lok, C.A. Messina, R.L. Patton, R.T. Gajek, T.R. Cannan, and E.M. Flanigen, Crystalline silicoaluminophosphates, US Patent 4,440,871 (3 April 1984).
2. S.W. Kaiser, Methanol conversion to light olefins over silicoaluminophosphate molecular sieves, Arabian J. Sci. Eng. 10 (1985) 361.
3. J.M.O. Lewis, Methanol to olefins process using aluminophosphate molecular sieve catalysts, Catalysis 1987 (J.W. Ward, ed.), Elsevier, Amsterdam, 1988.
4. P.T. Barger and S.T. Wilson, Converting natural gas to ethylene and propylene by the UOP/Hydro MTO process, Proc. of the 12th Intl. Zeolite Conf., Baltimore, Maryland (5-10 July 1998).
5. B.V. Vora, C.N. Eng, and E.C. Arnold, Natural gas utilization at its best: the UOP/Hydro MTO process, Petrotech 98, Bahrain (13-16 September 1998).
6. E.C. Arnold and J.M. Andersen, The economic case for the methanol route to olefins, 17th Annual World Methanol Conference, San Diego, California (29 November – 1st December 1999).
7. C. D. Chang et al., US Patents 3,894,106 (8 July 1975), 4,025,575 and 4,025,576 (24 May 1977), 4,052,479 (4 October 1977), 4,062,905 (13 December 1977), etc., and numerous additional patents to others within Mobil Oil Corp.
8. B. Miller, J. Warmann, A. Copeland, and T. Stewart, Fast-tracked FCC revamp, Oil & Gas J., 46-53 (5 June 2000).
9. L.W. Miller, Fast fluidized bed reactor for the MTO process, US Patent 6,166,282 (26 December 2000).

Studies in Surface Science and Catalysis
J.J. Spivey, E. Iglesia and T.H. Fleisch (Editors)

Novel Techniques for the Conversion of Methane Hydrates

Dirk D. Link, Charles E. Taylor *, and Edward P. Ladner

U.S. Department of Energy, National Energy Technology Laboratory
P.O. Box 10940, Pittsburgh, PA 15236-0940

* Corresponding author

Abstract

While methane hydrates hold promise as an energy source, methods for the economical recovery of methane from the hydrate must be developed. Effective means of converting the natural gas into a more useful form, such as the photocatalytic oxidation of methane to methanol, may address some of the needs for methane recovery and use. Methanol retains much of the original energy of the methane, and is a liquid at room temperature, which alleviates some of the concerns about fuel transportation and storage. Desired characteristics of the natural gas conversion process include selectivity toward methanol formation, efficiency of conversion, low cost, and ease of use of the conversion method. A method for the conversion of methane to methanol involving a photocatalyst, light, and an electron transfer molecule, is described. Moreover, novel use of the formation of a methane hydrate as a means of maximizing the levels of methane in water, as well as providing the reactants in close proximity, is described. This method demonstrated successful conversion of methane contained in a methane hydrate to methanol.

1. Introduction

The past decade has seen many advances in the study of natural gas hydrates. Initially, hydrates were studied to minimize their formation in natural gas pipelines. However, since the discovery of methane hydrates both on the ocean floor and in permafrost, research has been focused on efficient methods of harvesting these hydrates and using the methane which they contain. Estimates consider the amount of methane contained in these hydrate formations to be greater than all other fossil fuels combined (1). Some issues facing the use of hydrates as fuel sources include the development of harvesting techniques, and ramifications of hydrate harvesting on seafloor stability. Another issue that must be addressed is the development of techniques that make efficient use of these vast methane resources. In many instances, it may be desirable to convert the methane in natural gas hydrates to a more convenient fuel source, such as methanol.

Research conducted at the National Energy Technology Laboratory's methane hydrate laboratory has been directed toward the formation of methane hydrates in the laboratory, and the efficient conversion of methane into a more convenient form for transportation, storage, and use as a fuel source. Conversion of methane to methanol has been suggested to suit these goals, because methanol is a liquid, allowing more convenient storage and transportation while retaining much of the energy properties of the original methane (2-5). Concerns with the conversion of methane to methanol include the selectivity, the efficiency, and the cost of the conversion process. A method for the catalytic conversion of methane to methanol involving mild conditions using light and water is proposed. If successful, this method would provide an inexpensive means for the conversion of methane to methanol.

It has been demonstrated that photochemical oxidation of methane may be a commercially feasible route to methanol (6, 7). It has been shown that methane dissolved in water with a semiconductor catalyst and an electron transfer molecule (methyl viologen dichloride hydrate (MV) (1,1'-dimethyl-4,4'-bipyridinium dichloride)) can be converted to methanol and hydrogen at temperatures greater than 70 °C (8). The conversion proceeds according to the sequence shown in equations 1-6. It was also determined in this previous work that the temperature of the conversion system was one limiting factor for the conversion process. A temperature of at least 70 °C was required in order for the conversion to proceed. Allowing the system to cool below this temperature resulted in a stoppage in the conversion process (3, 4).

$$\mathrm{LaWO_3} \xrightarrow{h\nu(\lambda \geq 410\mathrm{nm})} e^-_{CB} + h^+_{VB} \tag{1}$$

$$e^-_{CB} + \mathrm{MV}^{2+} \rightarrow \mathrm{MV}^{\bullet+} \tag{2}$$

$$h^+_{VB} + \mathrm{H_2O} \rightarrow \mathrm{H}^+ + {}^\bullet\mathrm{OH} \tag{3}$$

$$\mathrm{MV}^{\bullet+} + \mathrm{H}^+ \rightarrow \frac{1}{2}\mathrm{H_2} + \mathrm{MV}^{2+} \tag{4}$$

$$\mathrm{CH_4} + {}^\bullet\mathrm{OH} \rightarrow \mathrm{CH_3^\bullet} + \mathrm{H_2O} \tag{5}$$

$$\mathrm{CH_3^\bullet} + \mathrm{H_2O} \rightarrow \mathrm{CH_3OH} + \frac{1}{2}\mathrm{H_2} \tag{6}$$

where e^-_{CB} represents an electron in the conduction band, h^+_{VB} represents a hole in the valence band, MV represents methyl viologen dichloride hydrate, and $^\bullet$ represents free radical species.

Another limiting factor of the conversion is the amount of methane dissolved in water, because of the limited solubility of methane in water. It is proposed in this study that the amount of methane dissolved in water could be greatly increased by formation of a methane hydrate. Methane hydrates contain large amounts of methane held within a water clathrate structure. It has been inferred that one methane hydrate unit cell may contain up to about 3.9 weight % methane (5, 9). Formation of a methane hydrate would also provide reactants that are immobilized in close proximity, which may facilitate the conversion. By photochemical formation of the hydroxyl radical ($^\bullet$OH) within the methane hydrate, the close proximity and restricted mobility of $^\bullet$OH and CH_4 would favor the formation of CH_3OH. The goals of this work are to form methane hydrate in a pressurized reaction cell, and to subsequently convert the trapped methane into methanol using the photocatalytic process described above. Compared to previous work (3), this approach would demonstrate the use of hydrates to immobilize reactants in a way that favors the desired selectivity, while allowing conversion to occur at much lower temperatures. It would also establish a commercially-feasible process for the efficient conversion of methane from natural gas hydrates to the liquid fuel methanol.

2. Experimental

Methane hydrate formations were performed in a cylindrical, stainless steel, high-pressure reaction cell with an internal volume of approximately 40 mL and a working pressure of 220 MPa (32,000 psia). The cell is equipped with up to five ports to accommodate the fill gas inlet and reaction product outlet, a pressure transducer to monitor the internal pressure of gas inside the cell, and a thermocouple to monitor the temperature of the liquid/hydrate inside the reaction cell. The cell is fitted with two machined end caps, one of which contains a sapphire window for observation of the hydrate using a CCD camera. This sapphire window also allows the contents of the cell to be illuminated by ambient laboratory light for observation purposes, and for illumination by the mercury vapor lamp for initiation of the photocatalytic process. The temperature of the cell is adjusted by the flow of a water-ethylene glycol solution from an external circulating temperature bath through a coil of copper tubing (1/4" OD) which is wrapped around the cell.

For the experiments described in this work, the cell is cleaned, rinsed, dried, and filled with 20 mL of water (plus additives such as salt(s) and sand, if used). To the water is added 0.25 g photocatalyst and 0.01 g methyl viologen. A small Teflon-coated stir bar is also placed inside the cell. The headspace in the cell is then filled with methane gas to a pressure of approximately 1400 psi. A magnetic stirring unit is placed under the cell to continuously agitate the water phase. Using the external circulating temperature bath, the temperature of the water is initially lowered to ~5 °C. Subsequently, the temperature is slowly ramped downward in small increments, typically 0.5 °C. By continuously monitoring the temperature and pressure, the region of maximum methane uptake and hydrate formation can be determined. The solidification of the hydrate is observed with the CCD camera inside the reaction cell, and is typically indicated when the stir bar no longer rotates. Methane uptake continues to be observed even after the solid hydrate has formed. Typically, the hydrate formation experiments are allowed to continue for over 400 hours, so that greater methane uptake can be achieved.

Following formation of the solid methane hydrate, the procedure for methane conversion and analysis is as follows. Using a high pressure mercury vapor lamp emitting both ultraviolet and visible light, the solid hydrate is irradiated for a period of up to 8 hours while keeping the cell at its lowest temperature. Following the period of irradiation, the cell is heated above 70 °C using the recirculating temperature bath. Because any methanol which has been produced will remain solubilized in water, this step is necessary to help drive any methanol out of the liquid phase and into the vapor phase. For analysis of the gaseous products, the outlet of the reaction cell is connected to the inlet of a mass spectrometer using a patented heated inlet capillary. Heating is required so that the gaseous products from the cell will not condense as they travel to the mass spectrometer. The exit valve of the cell is slowly opened and as the cell de-pressurizes, the products are analyzed by mass spectrometry. Opening of the cell for analysis of the gaseous products allows the vapor phase to escape, which promotes further removal of the methanol from the liquid phase.

3. Results and Discussion

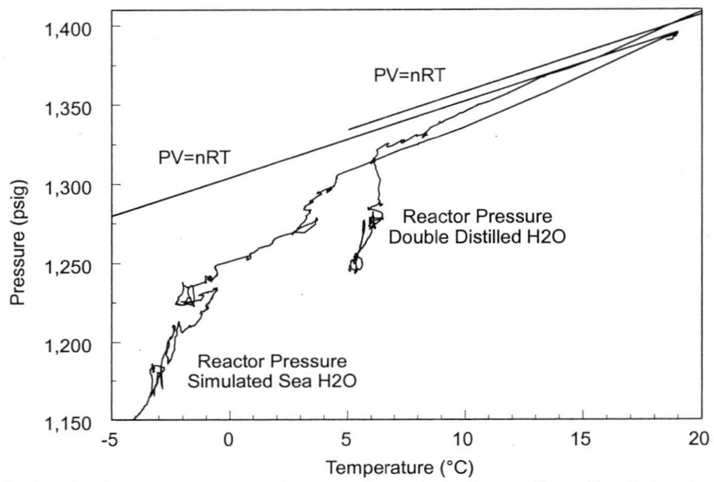

Figure 1. Typical pressure versus temperature plots for methane hydrate formation compared to ideal gas law prediction of pressure versus temperature.

Typical temperature versus pressure plots for the formation of the methane hydrate compared with the temperature and pressure relationship predicted by the ideal gas law are shown in Figure 1. In the initial stage of cooling, the pressure inside the cell follows the ideal gas law, showing a near-linear decrease as the temperature decreases. The point at which the pressure begins to drop lower than predicted indicates increased solubility of methane and initiation of hydrate formation. The drop in cell pressure at a constant temperature indicates the maximum rate of absorption of methane and hydrate formation. After the solid hydrate has formed, the pressure continues to show deviation from the predicted pressure, indicating that methane uptake by the hydrate is continuing, albeit at a slower rate than for initial hydrate formation. The methane in the cell continues to slowly diffuse into the lattice of the methane hydrate.

To determine the degree of conversion of methane by the photocatalytic process, hydrate formation was conducted both in the presence of and in the absence of photocatalyst. Each species was then illuminated with the mercury vapor lamp for several hours, and the gaseous products analyzed by mass spectrometry. The results are compared in Table 1. As can be seen, irradiation of the hydrate which contained the photocatalyst does show significant conversion of the methane to methanol. In the absence of photocatalyst, the amount of conversion of methane to methanol is over five times lower than in the presence of photocatalyst. Other reaction products for methane conversion include hydrogen, oxygen, ethane, formic acid, and carbon dioxide. Hydrogen is produced as a result of the reaction scheme in equations 1 through 6. Side reactions are responsible for the other products observed. Photocatalytic splitting of water can form oxygen, two methyl radicals can combine to form ethane, and formic acid (HCOOH) and carbon dioxide can be formed by further oxidation of methanol. The irradiated photocatalyst-containing hydrate also shows highly elevated levels of these side-reaction products compared to the hydrate which did not contain photocatalyst.

Table 1 also shows the reaction products from the conversion of methane hydrates which contained photocatalyst both with and without mercury lamp irradiation. The irradiated hydrate shows significant conversion of the methane to methanol whereas the non-irradiated hydrate shows almost no conversion. Small amounts of conversion seen in the non-irradiated system, especially the H_2 production, may be due to exposure of the cell contents to ambient laboratory light through the sapphire window. Thus, even the non-irradiated system was still exposed to a small amount of light. These results show that to achieve significant conversion of the methane to methanol, an active photocatalyst must be present in the hydrate and irradiation of the hydrate with the mercury vapor lamp is necessary.

Table 1. Analysis of reaction products for methane hydrate dissociation under different conditions. Concentrations are expressed as percent of total gaseous products.

	Products					
Conditions	CH_3OH	H_2	C_2H_6	HCOOH	O_2	CO_2
with catalyst with illumination	0.44%	4.01%	1.69%	0.74%	0.32%	0.27%
no catalyst with illumination	0.09%	0.94%	0.40%	0.15%	0.06%	0.05%
with catalyst no illumination	0.001%	2.41%	nd*	0.001%	0.07%	0.01%

* Indicates that the presence of species was not detected in mass spectrometric analysis.

4. Summary

Previous studies have shown that the conversion of methane dissolved in water to methanol proceeded only at elevated temperatures (>70 °C), with the solubility of methane in water being a limiting factor. This work has demonstrated that the production of methanol from methane and water in a methane hydrate structure can proceed at low temperatures. This conversion process is enhanced because hydrate formation increases the amount of methane in the water and immobilizes the reactants in close proximity to each other. The possibility exists for further developing the photocatalytic conversion process for use of methane hydrates as a fuel resource. This demonstration also opens the possibility of using photocatalytic processes for converting hydrates that may form in pipelines, thus providing an alternative to the traditional methods of applying heat to the pipeline or pumping large amounts of methanol into the pipeline to minimize formation of hydrates. Hydrates can be solubilized by generation of methanol in situ, which would be a more cost-effective process.

5. Disclaimer

Reference in this work to any specific commercial product, process, or service is to facilitate understanding and does not necessarily imply its endorsement or favoring by the United States Department of Energy.

6. Acknowledgements

This research was supported in part by an appointment to the National Energy Technology Laboratory Postgraduate Research Training Program, sponsored by the U.S. Department of Energy and administered by the Oak Ridge Institute for Science and Education.

References:

(1) Collett, T. S.; Kuuskraa, V. A. *Oil and Gas Journal* **1998**, 90.
(2) Taylor, S. H.; Hargreaves, S. J.; Hutchings, G. J.; Joyner, R. W.; Lembacher, C. W. *Catalysis Today* **1998,** *42*, 217-224.
(3) Noceti, R. P.; Taylor, C. E.; D'Este, J. R. *Catalysis Today* **1997,** *33*, 199-204.
(4) Taylor, C. E.; Noceti, R. P. *Catalysis Today* **2000,** *55*, 259-267.
(5) Wegrzyn, J. E.; Mahajan, D.; Gurevich, M. *Catalysis Today* **1999,** *50*, 97-108.
(6) Ashokkumar, M.; Maruthamuthu, P. *J. Mat. Sci. Lett.* **1988,** *24*, 2135-2139.
(7) Ogura, K.; Kataoka, M. *J. Mol. Cat.* **1988,** *43*, 371-379.
(8) Noceti, R. P.; Taylor, C. E. United States Patent 5,720,858; **1998**.
(9) Sloan, E. D. J. *Clathrate Hydrates of Natural Gases*, 2nd ed.; Marcel Dekker, Inc.: New York, 1997.

List of Authors

STUDIES IN SURFACE SCIENCE AND CATALYSIS

Volume 36 **Methane Conversion.** Proceedings of a Symposium on the Production of Fuels and Chemicals from Natural Gas, Auckland, April 27–30, 1987
edited by **D.M. Bibby, C.D. Chang, R.F. Howe and S. Yurchak**

Volume 37 **Innovation in Zeolite Materials Science.** Proceedings of an International Symposium, Nieuwpoort, September 13–17, 1987
edited by **P.J. Grobet, W.J. Mortier, E.F. Vansant and G. Schulz-Ekloff**

Volume 38 **Catalysis 1987.** Proceedings of the 10th North American Meeting of the Catalysis Society, San Diego, CA, May 17–22, 1987
edited by **J.W. Ward**

Volume 39 **Characterization of Porous Solids.** Proceedings of the IUPAC Symposium (COPS I), Bad Soden a.Ts., April 26–29, 1987
edited by **K.K. Unger, J. Rouquerol, K.S.W. Sing and H. Kral**

Volume 40 **Physics of Solid Surfaces 1987.** Proceedings of the Fourth Symposium on Surface Physics, Bechyne Castle, September 7–11, 1987
edited by **J. Koukal**

Volume 41 **Heterogeneous Catalysis and Fine Chemicals.** Proceedings of an International Symposium, Poitiers, March 15–17, 1988
edited by **M. Guisnet, J. Barrault, C. Bouchoule, D. Duprez, C. Montassier and G. Pérot**

Volume 42 **Laboratory Studies of Heterogeneous Catalytic Processes**
by **E.G. Christoffel,** revised and edited by **Z. Paál**

Volume 43 **Catalytic Processes under Unsteady-State Conditions**
by **Yu. Sh. Matros**

Volume 44 **Successful Design of Catalysts.** Future Requirements and Development. Proceedings of the Worldwide Catalysis Seminars, July, 1988, on the Occasion of the 30th Anniversary of the Catalysis Society of Japan
edited by **T. Inui**

Volume 45 **Transition Metal Oxides.** Surface Chemistry and Catalysis
by **H.H. Kung**

Volume 46 **Zeolites as Catalysts, Sorbents and Detergent Builders.** Applications and Innovations. Proceedings of an International Symposium, Würzburg, September 4–8, 1988
edited by **H.G. Karge and J. Weitkamp**

Volume 47 **Photochemistry on Solid Surfaces**
edited by **M. Anpo and T. Matsuura**

Volume 48 **Structure and Reactivity of Surfaces.** Proceedings of a European Conference, Trieste, September 13–16, 1988
edited by **C. Morterra, A. Zecchina and G. Costa**

Volume 49 **Zeolites: Facts, Figures, Future.** Proceedings of the 8th International Zeolite Conference, Amsterdam, July 10–14, 1989. Parts A and B
edited by **P.A. Jacobs and R.A. van Santen**

Volume 50 **Hydrotreating Catalysts.** Preparation, Characterization and Performance. Proceedings of the Annual International AIChE Meeting, Washington, DC, November 27–December 2, 1988
edited by **M.L. Occelli and R.G. Anthony**

Volume 51 **New Solid Acids and Bases.** Their Catalytic Properties
by **K. Tanabe, M. Misono, Y. Ono and H. Hattori**

Volume 52 **Recent Advances in Zeolite Science.** Proceedings of the 1989 Meeting of the British Zeolite Association, Cambridge, April 17–19, 1989
edited by **J. Klinowsky and P.J. Barrie**

Volume 53 **Catalyst in Petroleum Refining 1989.** Proceedings of the First International Conference on Catalysts in Petroleum Refining, Kuwait, March 5–8, 1989
edited by **D.L. Trimm, S. Akashah, M. Absi-Halabi and A. Bishara**

Volume 54 **Future Opportunities in Catalytic and Separation Technology**
edited by **M. Misono, Y. Moro-oka and S. Kimura**